(continued on inside back cover)

A Survey of Mathematics with Applications

A Survey of Mathematics with Applications

FIFTH EDITION

ALLEN R. ANGEL

Monroe Community College

STUART R. PORTER

Monroe Community College

ADDISON-WESLEY

An imprint of Addison Wesley Longman, Inc.

Reading, Massachusetts • Menlo Park, California • New York • Harlow, England
Don Mills, Ontario • Sydney • Mexico City • Madrid • Amsterdam

Senior Editor • Bill Poole

Editorial Project Manager • Christine O'Brien

Editorial Production Services • Jennifer Bagdigian, Sandra Rigney

Art Coordination • Susan London-Payne

Art Director • Karen Rappaport

Text Designer • Julie Gecha

Cover and Chapter Opener Design • Linda Manly Wade, Wade Design

Chapter Opener Layouts • Sandra Rigney

Senior Marketing Manager • Andrew Fisher

Senior Manufacturing Manager • Roy Logan

Manufacturing Coordinator • Evelyn Beaton

Prepress Services Manager • Sarah McCracken

Composition and Prepress Services • University Graphics, Inc.

Illustrations • Tech Graphics, James A. Bryant

Library of Congress Cataloging-in-Publication Data

Angel, Allen R., 1942–
    A survey of mathematics with applications / Allen Angel, Stuart
Porter. — 5th ed.
      p. cm.
    Includes index.
    ISBN 0-201-84600-4. — ISBN 0-201-85761-8
    1. Mathematics. I. Porter, Stuart R., 1932–   . II. Title.
QA39.2.A54 1996
510—dc20   96-43388
                                            CIP

1 2 3 4 5 6 7 8 9 10—DOW—99989796

To my wife, Kathy, and my sons, Robert and Steven
A.R.A.

To my family: Joyce, Lisa, Tod, Teri, Brian, Adam, Andrew, Matthew, Emily, Molly
S.R.P.

Math: It's All Around Us!

We present *A Survey of Mathematics with Applications*, fifth edition, with that vision in mind. Our primary goal in writing this book was to give students a text that they can read, understand, and enjoy while learning how mathematics affects the world around them. Numerous applied examples motivate topics. A variety of interesting applied exercises demonstrate the real-life nature of mathematics and its importance in the students' lives.

The text is intended for students who require a broad-based general overview of mathematics, especially those majoring in the liberal arts, elementary education, the social sciences, business, nursing, and allied health fields. It is particularly suitable for those courses that satisfy the minimum competency requirement in mathematics for graduation or transfer.

New and Expanded Features

In this edition we made several important improvements in presentation.

- A modified design and increased page size allow for easier reading and referencing of material.
- New Group Projects appear at the end of each chapter to facilitate cooperative/group learning.
- Approximately 40% of the exercises are new.
- The number of writing exercises was increased significantly.
- There is increased focus on data and graphical analysis in both the textual material and exercise sets.
- More problem Solving/Group Activity exercises were added to the section exercises.
- The number of Research Activities was increased.
- The number of examples was increased throughout the text to promote student understanding.

Content Revision

In addition we revised and expanded certain topics to introduce new material and to increase understanding.

- Chapter 1, Critical Thinking Skills, was expanded. Additional examples and explanations are intended to promote student problem solving.
- Chapter 3, Logic, was reorganized for greater clarity.
- Chapter 6, Algebra, Graphs, and Functions, was expanded to include coverage of variation.
- Metric System coverage was expanded to an entire chapter, Chapter 8, and a new section on Dimensional Analysis was added.
- Chapter 9, Geometry, was expanded to include more coverage of Topology.
- Chapter 12, Probability, was expanded to include a new section on the binomial probability formula.
- Chapter 13, Statistics, has been expanded to include a new section 13.8, on Correlation and Regression.

Continuing Features

Several features appear throughout the book, adding interest and provoking thought.

- **Problem Solving** Beginning in Chapter 1, students are introduced to problem solving and critical thinking. The theme of problem solving is then continued throughout the text, and special problem-solving exercises are presented in the exercise sets.

- **Critical Thinking Skills** In addition to a focus on *Problem Solving*, the book also features sections on *Inductive Reasoning* and the important skills of *Estimation* and *Dimensional Analysis*.

- **Profiles in Mathematics** Brief historical sketches and vignettes present the stories of people who have advanced the discipline of mathematics.

- **Chapter Openers** Interesting and motivating photo essays introduce each chapter and illustrate the real-world nature of the chapter topics.

- **Did You Know** ... These colorful, engaging, and lively boxed features highlight the connections of mathematics to history, to the arts and sciences, to technology, and to a broad variety of disciplines and student majors.

Instructor's Supplements

Instructor's Solutions Manual   This manual contains detailed, worked-out solutions to all the exercises in the text, and answers to the Group Projects.

OmniTest³

OmniTest³ is available in DOS-based and Macintosh formats. This new version of this easy-to-use software was developed for Addison-Wesley by ips Publishing, a leader in computerized testing and assessment.

- The Macintosh format makes full use of the Macintosh graphical user interface.

- DOS user interface is easy to learn and operate. Its Windows look-alike structure lets the user easily choose and control the items as well as the format for each test.

- Make-up exams, customized homework assignments, and multiple test forms can be quickly and easily created.

- OmniTest³ is algorithm driven—meaning that the program can automatically handle insertion of new numbers into the same equation—creating hundreds of variations of that equation.
  - **a)** The numbers are constrained to keep answers reasonable, so that a virtually endless supply of parallel versions of the same test can be created.
  - **b)** With this new version of OmniTest the values shown in the model problem may be "locked in."

- OmniTest³ is keyed section by section to the text, allowing selection of questions that test individual objectives from that section.

- Instructor-generated questions may be entered by way of OmniTest³'s sophisticated editor—complete with mathematical notation.

Printed Test Bank   The Test Bank includes three alternative tests per chapter. These items may be used as actual tests or as references for creating tests with or without the computer.

Test Generator/Editor for Mathematics with Quiz Master (CLAST Version)   Is a computerized test generator that lets instructors select test questions by CLAST objective or make use of ready made tests. The software is algorithm driven so that regenerated number values maintain problem types, and provide a large number of test items in both multiple-choice and open-response formats for one or more test forms. The IBM/Windows editor lets instructors modify existing questions or create their own including graphics and accurate mathematics symbols. The Macintosh version allows instructors to add their own questions. Tests created with the Test Generator can be used with Quizmaster, which records student scores as they take tests on a single computer network and prints reports for students, classes, or courses.

Videotapes   Correlated to each important topic are available to departments. Contact your Addison-Wesley Sales Consultant.

Student's Supplements

Student's Solutions Manual   This manual contains detailed worked-out solutions to all the odd-numbered section exercises and to all Review and Chapter Test exercises. Students will find this manual very helpful.

Guide to CLAST Mathematical Competency (State of Florida)   This guide provides all the necessary material to help students prepare for the computational portion of the CLAST test. It includes worked-out examples and practice for CLAST skills, as well as a practice test. Optional topics in trigonometry are provided for those who wish to brush up in this area as well.

Interactive Mathematics Tutorial Software with Management System Is an innovative package that is CLAST objective based, self paced, and algorithm driven to provide unlimited opportunity for review and practice. Tutorial lessons provide examples, progress-check questions, and access to an on-line glossary. Practice problems include hints for first incorrect responses, solutions, and on-line tools to aid in computation and understanding. The optional management system records student scores on disk, and lets instructors print diagnostic reports for individual students or classes. **Student Versions**, which include record keeping and practice tests, may be purchased by students for home use. (IBM DOS/ Windows and Macintosh).

Acknowledgments

We appreciate the suggestions offered by many students and faculty members for improving our text. In particular, we thank those who reviewed this edition of the text: Helen G. Bass, Southern Connecticut State University; Una Bray, Skidmore College; Linda F. Crabtree, Metropolitan Community College; Richard DeCesare, Southern Connecticut State University; Theresa A. Geiger, St. Petersburg Junior College; Leo Andrew Lusk, West Virginia University; Frances O. McDonald, Delgado Community College; Will Miles, Tri-County Technical College; Joanne Peeples, El Paso Community College; and Donna Weir, Union County College. We would also like to thank Larry Clar of Monroe Community College for reading the galleys and checking answers.

Our wives, Kathy and Joyce, and our children, Robert and Steven, and Tod, Teri, Lisa, and Brian deserve our special thanks. Without their support and great sacrifice, this book could not have become a reality.

It is our pleasure to acknowledge the assistance given us by the staff at Addison Wesley Longman, Inc. In particular, we appreciate the advice and encouragement of our acquisitions editor, William Poole and our project manager, Christine O'Brien. We would also like to thank our production editors, Jenny Bagdigian and Sandra Rigney, and our art editor, Susan London-Payne.

We also acknowledge the contribution of our supplement authors: Christine Dunn and Gary Egan of Monroe Community Coilege, for the *Instructor's Solutions Manual* and *Student's Solutions Manual*; Evelyn Woodward for the *CLAST Supplement*; and ips Publishing, Inc., Vancouver, Washington, for OmniTest[3].

We would also like to thank all the reviewers and students who have provided valuable suggestions for past editions of this book. The success of this book is a result of the cumulative constructive suggestions we have received over the years, and we want our reviewers to know we appreciate their help. The following is a list of reviewers for the first through fourth editions, and their affiliation at the time they provided their review.

First Edition — 1981

Rebecca Baum
Lincoln Land Community College, IL

Vivian Baxter
Fort Hays State University, KS

David H. Buckley
Polk Community College, FL

Kent F. Carlson
St. Cloud State University, MN

Donald Catheart
Salisbury State College, MD

Judith L. Gersting
Indiana University—Purdue University at Indianapolis

Lucille Groenke
Mesa Community College, AZ

Daniel Kimborowicz
Massasoit Community College, MA

Don Marsian
Hillsborough Community College, FL

Robert F. Wheeler
Northern Illinois University

Second Edition — 1985

Don Cohen
SUNY Ag & Tech College at Cobleskill

Charles Downey
University of Nebraska

Nancy Johnson
Broward Community College, FL

Peter Lindstrom
North Lake College, TX

James Magliano
Union County College, NJ

Marilyn Mays
North Lake College, TX

Robert McGuigan
Westfield State College, MA

Maurice Monahan
South Dakota State University

Nelson G. Rich
Monroe Community College, NY

Sandra Savage
Orange Coast College, CA

William Trotter, Jr.
University of South Carolina

Sandra Welch
Stephen F. Austin State University, TX

Susan Wirth
Indian River Community College, FL

Michael A. Zwick
Monroe Community College, NY

Third Edition — 1989

Robert C. Bueker
Western Kentucky University

Carl Carlson
Moorhead State University, MN

Ruth Ediden
Morgan State University, MD

Raymond Flagg
McPherson College, KS

Penelope Fowler
Tennessee Wesleyan College

Gilberto Garza
El Paso Community College, TX

John Hornsby
University of New Orleans, LA

Gerald Schultz
Southern Connecticut State University

David Lehmann
Southwest Missouri State University

Minnie Shuler
Chipola Junior College, FL

Julie Monte
Daytona Beach Community College, FL

Steve Sworder
Saddleback College, CA

Wing Park
College of Lake County, IL

Alvin D. Tinsley
Central Missouri State University

Bettye Parnham
Daytona Beach Community College, FL

Shirley Thompson
Morehouse College, GA

Fourth Edition — 1993

Frank Asta
College of DuPage, IL

Joanne Peeples
El Paso Community College, TX

Hughette Bach
California State University—Sacramento

Ronald Ruemmler
Middlesex County College, NJ

Madeline Bates
Bronx Community College, NY

Rosa Rusinek
Queensborough Community College, NY

Una Bray
Skidmore College, NY

John Samoylo
Delaware County Community College, PA

David Dean
Santa Fe Community College, FL

Richard Schwartz
College of Staten Island, NY

Karen Estes
St. Petersburg Junior College, FL

Joyce Wellington
Southeastern Community College, NC

Kurtis Fink
Northwest Missouri State University

Sue Welsch
Sierra Nevada College

Mary Lois King
Tallahassee Community College, FL

James Wooland
Florida State University

Edwin Owens
Pennsylvania College of Technology

Mathematics is an exciting, living study. It has applications that shape the world around you and influence your everyday life. We hope that as you read through this book you will realize just how important mathematics is and gain an appreciation of both its usefulness and its beauty. We also hope to teach you some practical mathematics that you can use in your everyday life and that will prepare you for further courses in mathematics.

Our primary purpose in writing this text was to provide material that you could read, understand, and enjoy. To this end we have used straightforward language and tried to relate mathematical concepts to everyday experiences. We have also provided many detailed examples for you to follow.

The concepts, definitions, and formulas that deserve special attention have been either boxed or set in boldface type. The exercises are graded so that the more difficult problems appear at the end of the exercise set. The problems with exercise numbers set in color are writing exercises. At the end of most exercise sets are Problem Solving/Group Activity exercises that contain challenging or exploratory exercises. At the end of each chapter are Group Projects which reinforce the material learned or provide related material.

Each chapter has a summary, review exercises, and a chapter test. When studying for a test, be sure to read the chapter summary, work the review exercises, and take the chapter test. The answers to the odd-numbered exercises, all review exercises, and all chapter test exercises appear in the Answer section in the back of the text. However, you should use the answers only to check your work.

It is difficult to learn mathematics without becoming involved. To be successful, we suggest you read the text carefully *and work each exercise in each assignment in detail.* Check with your instructor to determine which supplements are available for your use.

We welcome your suggestions and your comments. Our address is Monroe Community College, Rochester, NY 14623. Good luck in your adventure in mathematics!

CONTENTS

Chapter 1 Critical Thinking Skills

Charles Lindbergh, who made the first solo flight across the Atlantic in 1927, had to solve many problems, such as the route over the ocean and the financing of the trip. Most important, the aircraft design had to be modified to carry enough fuel for the 3600-mile trip. To keep the load light, Lindbergh flew without a copilot, parachute, or radio and carried only five sandwiches and a quart of water on his 33-hour, 32-minute trip.

Life constantly presents new problems. The great inventors, scientists, scholars, politicians, and artists make their contributions to civilization by confronting and solving problems. To learn the techniques for solving problems requires practice and patience. But once the basic principles are understood, they can be applied to each new challenge.

The goal of this chapter is to help you master the skills of reasoning, estimating, and problem solving. These skills will aid you in solving the problems in the remainder of this book, as well as problems that you will encounter in everyday life.

Problem solving requires critical thinking: considering the information given, deciding what question must be answered, making and acting on a plan to find the answer, and checking the answer. Mathematics provides the tools needed to think critically and solve problems. Every day, you make decisions that require you to use critical thinking skills. Sometimes, the problem may be one that can be solved by computation. For example, in a drugstore you may be choosing between two bottles of shampoo. One is 6 ounces and costs $2.95, the other is 8 ounces and costs $3.50. Which is the better price? The answer to this problem lies in finding the unit cost: If you compute the cost for 1 ounce of each

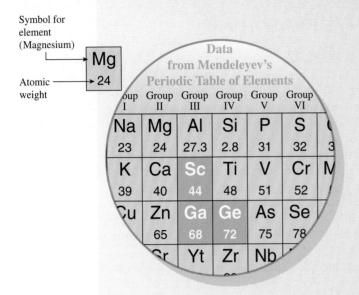

Symbol for element (Magnesium)

Mg

Atomic weight → **24**

Data from Mendeleyev's Periodic Table of Elements

up I	Group II	Group III	Group IV	Group V	Group VI	
Na	Mg	Al	Si	P	S	
23	24	27.3	2.8	31	32	3
K	Ca	Sc	Ti	V	Cr	M
39	40	44	48	51	52	
Cu	Zn	Ga	Ge	As	Se	
	65	68	72	75	78	
Sr	Yt	Zr	Nb			

Seeing patterns often helps in solving problems. In 1869, chemists had isolated 63 of 109 of the chemical elements known today, but had not found any apparent underlying order. In that year, Russian chemist Dmitri Mendeleyev saw a pattern. He proposed a table of the elements organized by increasing atomic weight and grouped according to similar properties. To make this method work, he had to predict the existence of three then-unknown elements: scandium, gallium, and germanium.

shampoo, you will find that the 8-ounce bottle costs the least per ounce.

Sometimes solving a problem may require you to make a reasonable estimate, to look for clues, or to experiment with several possible solutions before choosing the best solution. Often the most important part of solving a problem is just understanding what question must be answered. ■

One way of determining the answer to the question "how many" is to estimate, using a small sampling of a larger group. This technique can be used to guess how many jelly beans are in a jar, or how many people are in attendance at a political rally.

I.I INDUCTIVE REASONING

The goal of this chapter is to help you improve your reasoning and problem-solving skills. This section introduces inductive and deductive reasoning, which are used in problem solving. The next section introduces the concept of estimation. Estimation is a technique that can be used to determine if an answer obtained for a problem or from a calculation is "reasonable." Section 1.3 introduces and applies problem-solving techniques.

Before looking at some examples of inductive reasoning and problem solving, let us first review a few facts about certain numbers. The **natural numbers** or **counting numbers** are the numbers 1, 2, 3, 4, 5, 6, 7, 8, ... The three dots, called an **ellipsis**, mean that 8 is not the last number but that the numbers continue in the same manner. A word that we sometimes use is "divisible." If $a \div b$ has a remainder of zero, then *a is divisible by b*. The counting numbers that are divisible by 2 are 2, 4, 6, 8, ... These are called the *even counting numbers*. The numbers that are not divisible by 2 are 1, 3, 5, 7, 9, ... These are the *odd counting numbers*. When we refer to *odd numbers* or *even numbers*, we mean odd or even counting numbers.

Recognizing patterns is sometimes helpful in solving problems, as Examples 1 and 2 illustrate.

EXAMPLE I

If two odd counting numbers are multiplied together, will the product always be an odd counting number?

Solution: To answer this question, we will examine the products of several pairs of odd numbers to see if there is a pattern.

$$
\begin{array}{lll}
1 \times 3 = 3 & 3 \times 5 = 15 & 5 \times 7 = 35 \\
1 \times 5 = 5 & 3 \times 7 = 21 & 5 \times 9 = 45 \\
1 \times 7 = 7 & 3 \times 9 = 27 & 5 \times 11 = 55 \\
1 \times 9 = 9 & 3 \times 11 = 33 & 5 \times 13 = 65
\end{array}
$$

None of the products is divisible by 2. Thus we might predict from these examples that the product of any two odd numbers is an odd number. ▲

EXAMPLE 2

If an odd and an even counting number are added, will the sum be an odd or an even counting number?

Solution: Let's look at a few examples where one number is odd and the other number is even.

$$
\begin{array}{lll}
1 + 2 = 3 & 9 + 6 = 15 & 23 + 18 = 41 \\
3 + 12 = 15 & 5 + 14 = 19 & 81 + 32 = 113
\end{array}
$$

None of these sums is divisible by 2. Therefore we might predict that the sum of an odd and an even number is an odd number. ▲

In Examples 1 and 2 we cannot conclude that the results are true for all counting numbers. However, from the patterns developed, we can make predictions. This type of reasoning process, arriving at a general conclusion from specific observations or examples, is called **inductive reasoning**, or **induction**.

> **Inductive reasoning** is the process of reasoning to a general conclusion through observations of specific cases.

Induction often involves observing a pattern and from that pattern predicting a conclusion. Imagine an endless row of dominoes. You knock down the first, which knocks down the second, which knocks down the third, and so on. Assuming the pattern will continue uninterrupted, you conclude that eventually all the dominoes will fall, even though you may not witness the event.

Inductive reasoning is often used by mathematicians and scientists to predict answers to complicated problems. For this reason, inductive reasoning is part of the **scientific method**. When a scientist or mathematician makes a prediction based on specific observations, it is called a **hypothesis** or **conjecture**. After looking at the products in Example 1 we might conjecture that the product of two odd counting numbers will be an odd counting number.

Examples 3 and 4 illustrate how we arrive at a conclusion using inductive reasoning.

EXAMPLE 3

What reasoning process has led to the conclusion that no two people have the same fingerprints? This conclusion has resulted in fingerprints being used in courts of law as evidence to convict persons of crimes.

Solution: In millions of tests, no two people have been found to have the same fingerprints. By induction, then, we believe that fingerprints provide a unique identification and can therefore be used in a court of law as evidence. Is it possible that sometime in the future, two people will be found who do have exactly the same fingerprints? ▲

EXAMPLE 4

Consider the conjecture "If the sum of the digits of a number is divisible by 3, then the number is divisible by 3." Test several numbers to determine whether the conjecture appears true or false.

Solution: Let's look at some numbers, the sum of whose digits are divisible by 3.

Number	Sum of the Digits	Sum of the Digits Divided by 3	Number Divided by 3
114	$1 + 1 + 4 = 6$	$6 \div 3 = 2$	$114 \div 3 = 38$
234	$2 + 3 + 4 = 9$	$9 \div 3 = 3$	$234 \div 3 = 78$
7020	$7 + 0 + 2 + 0 = 9$	$9 \div 3 = 3$	$7020 \div 3 = 2340$
2943	$2 + 9 + 4 + 3 = 18$	$18 \div 3 = 6$	$2943 \div 3 = 981$
9873	$9 + 8 + 7 + 3 = 27$	$27 \div 3 = 9$	$9873 \div 3 = 3291$

In each of the examples we find that the sum of the digits is divisible by 3 and the number itself is divisible by 3. From these specific examples we might be tempted to generalize that the conjecture "If the sum of the digits of a number is divisible by 3, then the number is divisible by 3" is true.

EXAMPLE 5

Pick any number, multiply the number by 4, add 6 to the product, divide the sum by 2, and subtract 3 from the quotient. Repeat this procedure for several different numbers and then make a conjecture about the relationship between the original number and the final number.

Solution:　Let's go through this one together.

Pick a number:	say, 5
Multiply the number by 4:	$4 \times 5 = 20$
Add 6 to the product:	$20 + 6 = 26$
Divide the sum by 2:	$26 \div 2 = 13$
Subtract 3 from the quotient:	$13 - 3 = 10$

Note that we started with the number 5 and finished with the number 10. If you start with the number 2, you will end with the number 4. Starting with 3 would result in a final number of 6, 4 would result in 8, and so on. On the basis of these few examples many of you would conjecture that when you follow the given procedure, the number you end with will always be twice the original number.

The result reached by inductive reasoning is often correct for the specific cases studied but not correct for all cases. History has shown that not all conclusions arrived at by inductive reasoning are correct. For example, Aristotle (384–322 B.C.) reasoned inductively that heavy objects fall at a faster rate than light objects. About 2000 years later, Galileo dropped two pieces of metal—one 10 times heavier than the other—from the Leaning Tower of Pisa in Italy. He found that both hit the ground at exactly the same moment, so they must have traveled at the same rate.

When forming a general conclusion using inductive reasoning, you should test it with several special cases to see whether the conclusion appears correct. If a special case is found that satisfies the conditions of the conjecture but produces a different result, such a case is called a **counterexample**. This case proves that the conjecture is false because only one exception is needed to show that a con-

Apollo 15 Astronaut David Scott used the moon as his laboratory to show that a heavy object (a hammer) does indeed fall at the same rate as a light object (a feather). Had Galileo dropped a hammer and feather from the tower of Pisa, the hammer would have fallen more quickly to the ground, and he still would have concluded that a heavy object falls faster than a lighter one. If it is not the object's mass that is affecting the outcome, then what is it? The answer is air resistance or friction: The earth has an atmosphere; the moon does not. ■

clusion is not valid. Galileo's counterexample disproved Aristotle's conjecture. If a counterexample cannot be found, the conjecture is neither proven nor disproven.

A second type of reasoning process is called *deductive reasoning*, or *deduction*. Mathematicians use deductive reasoning to *prove* conjectures true or false.

> **Deductive reasoning** is the process of reasoning to a specific conclusion from a general statement.

EXAMPLE 6

Prove, using deductive reasoning, that the procedure in Example 5 will always result in twice the original number selected.

Solution: To use deductive reasoning we begin with the *general* case rather than specific examples. In Example 5, specific cases were used. Let's select the letter n to represent *any number*.

Pick any number:	n
Multiply the number by 4:	$4n$ (4n means 4 times n)
Add 6 to the product:	$4n + 6$
Divide the sum by 2:	$\dfrac{4n + 6}{2} = \dfrac{\overset{2}{4n}}{\underset{1}{2}} + \dfrac{\overset{3}{6}}{\underset{1}{2}} = 2n + 3$
Subtract 3 from the quotient:	$2n + 3 - 3 = 2n$

Note that, for any number n selected, the result is $2n$, or twice the original number selected. ▲

In Example 5 you may have conjectured, using specific examples and inductive reasoning, that the result would be twice the original number selected. In Example 6 we proved, using deductive reasoning, that the result will always be twice the original number selected.

Section 1.1 Exercises

1. a) List the natural numbers.
  b) What is another name for the natural numbers?

2. a) What does it mean to say, "*a* is divisible by *b*," where *a* and *b* represent natural numbers?
  b) List three natural numbers that are divisible by 5.
  c) List three natural numbers that are divisible by 12.

In Exercises 3–6, explain your answer in a sentence or sentences.

3. What is a conjecture?

4. What is inductive reasoning?

5. What is deductive reasoning?

6. What is a counterexample?

7. In the 1950s doctors noticed that many of their lung cancer patients were also cigarette smokers. Doctors reasoned that cigarette smoking increased a person's chance of getting lung cancer. What type of reasoning did the doctors use? Explain.

8. Bill tells his son that if he continues to drive over the speed limit he will eventually get a traffic ticket. What type of reasoning is Bill using? Explain.

In Exercises 9–12, use inductive reasoning to predict the next line in the pattern.

9. $1 \times 10 + 2 = 12$
$12 \times 10 + 3 = 123$
$123 \times 10 + 4 = 1234$

10. $1 = 1$
$1 + 2 = 3$
$1 + 2 + 3 = 6$
$1 + 2 + 3 + 4 = 10$
$1 + 2 + 3 + 4 + 5 = 15$

11. $11 \times 11 = 121$
$11 \times 12 = 132$
$11 \times 13 = 143$

12.

In Exercises 13–16, draw the next figure in the pattern (or sequence).

13. 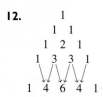 ...

14. ...

15. ...

16. ...

In Exercises 17–26, use inductive reasoning to predict the next three numbers in the pattern (or sequence).

17. $1, 4, 7, 10, \ldots$ **18.** $12, 10, 8, 6, \ldots$

19. $5, 3, 1, -1, -3, \ldots$ **20.** $1, -1, 1, -1, 1, \ldots$

21. $2, -6, 18, -54, \ldots$ **22.** $1, 1/2, 1/4, 1/8, \ldots$

23. $1, 4, 9, 16, 25, \ldots$

24. $1, 1, 2, 3, 5, 8, 13, 21, \ldots$

25. $0, 3, 8, 15, 24, 35, 48, 63, \ldots$

26. $5, -\frac{10}{3}, \frac{20}{9}, -\frac{40}{27}, \ldots$

In Exercises 27 and 28, look for a pattern in the first three products and use it to find the fourth product.

27.

9	99	999	9999
$\times\ 9$	$\times\ 9$	$\times\ \ 9$	$\times\ \ \ 9$
81	891	8991	?

28.

9	909	90909	9090909
$\times\ 9$	$\times\ \ 9$	$\times\ \ \ 9$	$\times\ \ \ \ \ 9$
81	8181	818181	?

29. Study the following entries and use the pattern they exhibit to complete the last two rows.
$1 + 3 = 4$ or 2^2
$1 + 3 + 5 = 9$ or 3^2
$1 + 3 + 5 + 7 = 16$ or 4^2
$1 + 3 + 5 + 7 + 9 = ?$
$1 + 3 + 5 + 7 + 9 + 11 = ?$

30. Consider the number 142,857 and its first four multiples:

142857	142857	142857	142857
$\times\ \ \ \ \ 1$	$\times\ \ \ \ \ 2$	$\times\ \ \ \ \ 3$	$\times\ \ \ \ \ 4$
142857	285714	428571	571428

a) Observe the digits in the product and use inductive reasoning to make a conjecture about the digits that will appear in the product $142{,}857 \times 5$.

b) Multiply 142,857 by 5 to see whether your conjecture appears to be correct.

c) Can you make a more general conjecture about the digits in the product of a multiplication problem where 142,857 is multiplied by a one-digit positive number?

d) Multiply 142,857 by the digits 6 through 8 and see whether your conjecture appears to be correct.

31. The ancient Greeks labeled certain numbers as **triangular numbers**. The numbers 1, 3, 6, 10, 15, 21, and so on are triangular numbers.

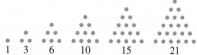

a) Can you determine the next two triangular numbers?

b) Describe a procedure to determine the next five triangular numbers without drawing the figures.

c) Is 72 a triangular number? Explain how you determined your answer.

32. Just as there are triangular numbers, there are also **square numbers**. The numbers 1, 4, 9, 16, 25, and so on are square numbers.

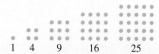

a) Determine the next three square numbers.

b) Describe a procedure to determine the next five square numbers without drawing the figures.

c) Is 72 a square number? Explain how you determined your answer.

33. Four rows of a triangular figure are shown.

a) If you added six additional rows to the bottom of this triangle, using the same pattern displayed, how many triangles would appear in the 10th row?

b) If the triangles in all 10 rows were added, how many triangles would appear in the entire figure?

34. Find the letter that is the 118th entry in the following sequence. Explain how you determined your answer.

Y, R, Y, R, R, Y, R, R, R, Y, R, R, R, R, Y, R, R, . . .

35. The pattern shown here is taken from a quilt design known as a Triple Irish Chain. Complete the color pattern by indicating the color assigned to each square.

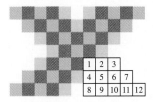

36. Pick a number, multiply the number by 2, add 4 to the product, divide the sum by 2, and subtract 2 from the quotient. See Example 5.

a) What is the relationship between the number you started with and the final number?

b) Arbitrarily select some different numbers and repeat the process, recording the original number and the result.

c) Can you make a conjecture about the relationship between the original number and the final number?

d) Prove, using deductive reasoning, the conjecture you made in part (c). See Example 6.

37. Pick any number and multiply the number by 6. Add 3 to the product. Divide the sum by 3, and subtract 1 from the quotient.

a) What is the relationship between the number you started with and the final answer?

b) Arbitrarily select some different numbers and repeat the process, recording the original number and the results.

c) Can you make a conjecture about the relationship between the original number and the final number?

d) Try to prove, using deductive reasoning, the conjecture you made in part (c).

38. Pick any number and add 1 to it. Find the sum of the new number and the original number. Add 9 to the sum. Divide the sum by 2 and subtract the original number from the quotient.

a) What is the final number?

b) Arbitrarily select some different numbers and repeat the process. Record the results.

c) Can you make a conjecture about the final number?

d) Try to prove, using deductive reasoning, the conjecture you made in part (c).

39. Pick a number, add 5 to the number, divide the sum by 5, subtract 1 from the quotient, and multiply the result by 5.

a) What is the relationship between the number you started with and the final number?

b) Arbitrarily select some different numbers and repeat the process, recording the original number and the result.

c) Can you make a conjecture about the relationship between the original number and the final number?

d) Try to prove, using deductive reasoning, the conjecture you made in part (c).

40. Pick a number and add 10 to the number. Divide the sum by 5. Multiply this quotient by 5. Subtract 10 from the product. Then subtract your original number.

a) What is the result?

b) Arbitrarily select some different numbers and repeat the process, recording the original number and the result.

c) Can you make a conjecture regarding the result when this process is followed?

d) Try to prove, using deductive reasoning, the conjecture you made in part (c).

In Exercises 41–47, find a counterexample to show that each of the statements is incorrect.

41. The product of any two counting numbers is divisible by 2.

42. Every counting number greater than 5 is the sum of either two or three consecutive counting numbers. For example, 9 = 4 + 5 and 17 = 9 + 8.

43. The product of a number multiplied by itself is even.

44. The product of 2 two-digit numbers is a three-digit number.

45. The sum of 3 two-digit numbers is a three-digit number.

46. The sum of any two odd numbers is divisible by 3.

47. When a counting number is added to 3 and the sum is divided by 2, the quotient will be an even number.

48. a) Construct a triangle and measure the three interior angles with a protractor. What is the sum of the measures?
b) Construct three other triangles, measure the angles, and record the sums. Are your answers the same?
c) Make a conjecture about the sum of the measures of the three interior angles of a triangle.

49. a) Construct a quadrilateral (a four-sided figure) and measure the four interior angles with a protractor. What is the sum of their angle measures?
b) Construct three other quadrilaterals, measure the angles, and record the sums. Are your answers the same?
c) Make a conjecture about the sum of the measures of the four interior angles of a quadrilateral.

50. a) Select one- and two-digit numbers and multiply each by 99. Record your results.
b) Find the sum of the digits in each of your products in part (a).
c) Make a conjecture about the sum of the digits when a one- or two-digit number is multiplied by 99.

51. a) Calculate the squares of 15, 25, 35, and 45 and record the results. (Note that the square of 15 is $15 \times 15 = 225$.) Examine the products and see whether you can establish a pattern in relation to the numbers being multiplied.
b) Make a conjecture about how to mentally calculate the square of any number whose unit digit is 5.
c) Calculate the squares of 55, 75, 95, and 105 using your conjecture.

Problem Solving/Group Activities

52. Complete the following square of numbers. Explain how you determined your answer.

1	2	3	4
2	5	10	17
3	10	25	52
4	17	52	?

53. Find the next three numbers in the sequence.
1, 8, 11, 88, 101, 111, 181, 1001, 1111, . . .

54. The following numbers are swogs.

12347, 70523, 56123, 90341, 16325, 34127

The following numbers are not swogs.

1573, 12345, 953, 56213, 56132, 34325

a) Describe the common characteristics of numbers that are swogs.
b) Which two of the following numbers are swogs?

43217, 54323, 52307, 16235, 36521

55. In this section we defined *conjecture*. The following is *Ulm's conjecture*. Pick any positive integer. If the integer is even, divide it by 2. If the integer is odd, multiply it by 3 and add 1. If you continue this process with each answer obtained, the result will eventually be 1. For example, suppose that we begin with the number 5 and carry out the process.

Pick 5	Multiply 5 by 3 and add 1; obtain 16.
16	Divide by 2; obtain 8
8	Divide by 2; obtain 4
4	Divide by 2; obtain 2
2	Divide by 2; obtain 1
1	

a) Select three positive even integers greater than 10 and show that Ulm's conjecture holds for each number.
b) Select three positive odd integers greater than 10 and show that Ulm's conjecture holds for each number.

Research Activities

56. a) Using newspapers, magazines, and other sources, find examples of conclusions arrived at by inductive reasoning.
b) Explain how inductive reasoning was used in arriving at the conclusion.

57. When a jury decides the guilt or innocence of a defendant, do the jurors collectively use primarily inductive reasoning, deductive reasoning, or an equal amount of each? Write a brief report supporting your answer.

1.2 ESTIMATION

An important step in solving mathematical problems—or, in fact, *any* problem—is to make sure that the answer you've arrived at makes sense. One technique for determining whether an answer is reasonable is to estimate. Estimation is the process of arriving at an approximate answer to a question. This section demonstrates several estimation methods.

To estimate, or approximate, an answer we often round off numbers as illustrated in the following examples. The symbol $\approx$ means *is approximately equal to*.

EXAMPLE 1

Estimate the cost of 22 poster boards at $0.89 each.

Solution: We may round off the amounts as follows to obtain an estimate.

$$
\begin{array}{r}
22 \rightarrow \quad 20 \\
\times\ 0.89 \rightarrow \times\ 0.90 \\
\hline
18.00
\end{array}
$$

Thus the 22 poster boards will cost approximately $18.00, written ≈ $18. ▲

In Example 1 we could have rounded the 22 to 20 and the $0.89 to $1.00, which would result in an estimate of $20. The true cost is $0.89 × 22, or $19.58. *Estimates are not meant to give exact values for answers but are a means of determining whether your answer is reasonable.* If you calculated an answer of $19.58 and then did a quick estimate to check it, you would know that the answer is reasonable because it is close to your estimated answer.

EXAMPLE 2

At a local supermarket Amy purchased milk for $2.39, lettuce for $0.89, bread for $0.99, hot dogs for $2.15, ground beef for $4.76, bananas for $0.49, and a green onion for $0.40. The total bill was $16.08. Use estimation to determine whether this amount is reasonable.

Solution: The most expensive item is $4.76 and the least expensive is $0.40. How should we estimate? We will estimate two different ways. First, we will round the cost of each item to the nearest 10 cents. Then we will round the cost to the nearest dollar. Rounding to the nearest 10 cents is more accurate. However, to determine whether the total bill is reasonable we may need to round only to the nearest dollar.

	Rounding to the Nearest 10 Cents		Rounding to the Nearest Dollar	
Milk	$2.39 →	$2.40	$2.39 →	$2.00
Lettuce	0.89 →	0.90	0.89 →	1.00
Bread	0.99 →	1.00	0.99 →	1.00
Hot dogs	2.15 →	2.20	2.15 →	2.00
Ground beef	4.76 →	4.80	4.76 →	5.00
Bananas	0.49 →	0.50	0.49 →	0.00
Onion	0.40 →	0.40	0.40 →	0.00
		$12.20		$11.00

Using either estimate, we find that the bill of $16.08 is quite high. Therefore Amy should check the bill carefully before paying it. Adding the prices of all seven items gives the true cost of $12.07. ▲

EXAMPLE 3

The number of bushels of grapes produced at a vineyard are 62,408 Cabernet Sauvignon, 118,916 French Colombard, 106,490 Chenin Blanc, 5960 Charbono, and 12,104 Chardonnay. Select the best estimates of the total number of bushels produced by the vineyard.

a) 500,000   **b)** 30,000   **c)** 300,000   **d)** 5,000,000

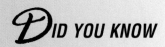

Solution: Following are suggested roundings. On the left, the numbers are rounded to thousands. For a less close estimate, round to ten thousands, as illustrated on the right.

Round to the Nearest Thousand	Round to the Nearest Ten Thousand
$62{,}408 \rightarrow 62{,}000$	$62{,}408 \rightarrow 60{,}000$
$118{,}916 \rightarrow 119{,}000$	$118{,}916 \rightarrow 120{,}000$
$106{,}490 \rightarrow 106{,}000$	$106{,}490 \rightarrow 110{,}000$
$5{,}960 \rightarrow 6{,}000$	$5{,}960 \rightarrow 10{,}000$
$12{,}104 \rightarrow \underline{12{,}000}$	$12{,}104 \rightarrow \underline{10{,}000}$
$305{,}000$	$310{,}000$

Either rounding procedure indicates that the best estimate is (c), or 300,000.

EXAMPLE 4

Which of the answers (a)–(d) is the best estimate of the number of heartbeats in a lifetime? Assume 78 heartbeats per minute and an average life expectancy of 76 years.

    a) 3,000,000,000   **b)** 3,000,000

    c) 300,000,000   **d)** 30,000,000

Solution: Begin by setting up the calculations to determine the number of heartbeats in a lifetime.

78×60	Number of heartbeats per hour
$78 \times 60 \times 24$	Number of heartbeats per day
$78 \times 60 \times 24 \times 365$	Number of heartbeats per year
$78 \times 60 \times 24 \times 365 \times 76$	Number of heartbeats per lifetime

Round the multipliers to obtain an estimate of the number of heartbeats in a lifetime.

$$78 \times 60 \times 24 \times 365 \times 76$$
$$\downarrow \quad \downarrow \quad \downarrow \quad \downarrow \quad \downarrow$$
$$80 \times 60 \times 20 \times 400 \times 80$$

Now move the decimals in each number until each number is a number greater than or equal to 1 and less than 10.

$$8\underset{\smile}{0} \times 6\underset{\smile}{0} \times 2\underset{\smile}{0} \times 4\underset{\smile}{0}\underset{\smile}{0} \times 8\underset{\smile}{0}$$

Multiply the nonzero digits.

$$8 \times 6 \times 2 \times 4 \times 8 = 3072$$

Each number originally multiplied is greater than 10 (the arrows all moved to the left). Therefore we add the total number of zeros in the five numbers multiplied, 6 zeros, to the product 3072 to obtain an estimate of 3,072,000,000. Therefore the best estimate is (a) 3,000,000,000, three billion.

EXAMPLE 5

The odometer of an automobile reads 54,186.2 mi. If the automobile averaged 23.6 miles per gallon (mpg) for that mileage, (a) estimate the number of gallons of gasoline used. (b) If the cost of the gasoline averaged $1.24 per gallon, estimate the total cost of the gasoline.

Solution:

a) To estimate the number of gallons divide the mileage by the number of miles per gallon.

$$\frac{54,186.62}{23.6}$$

Round these numbers to obtain an estimate.

$$\frac{50,000}{20} = 2500$$

Therefore the car used approximately 2500 gallons (gal) of gasoline.

b) Rounding the price of the gasoline to $1.20 per gallon gives the cost of the gasoline as 2500 × $1.20, or $3000.

Now let's look at some different types of estimation problems.

*D*ID YOU KNOW

ESTIMATING TECHNIQUES IN MEDICINE

*E*stimating is one of the diagnostic tools used by the medical profession. Physicians take a small sample of blood, tissue, or body fluids to be representative of the body as a whole. Human blood contains different types of white blood cells that fight infection. When a bacterium or virus gets into the blood, the body responds by producing more of the type of white blood cell whose job it is to destroy that particular invader. Thus an increased level of white blood cells in a sample of blood not only indicates the presence of an infection but also helps identify its type. A trained medical technician estimates the relative number of each kind of white blood cell found in a count of 100 white blood cells. An increase in any one kind indicates the type of infection present. The accuracy of this diagnostic tool is impressive when you consider that there are normally 5 to 9 thousand white blood cells in a dropful of blood (1 cubic millimeter). ■

Monocyte (wbc)

Red blood cells

Eosinophils (wbc)

Neutrophil (wbc)

A high count of esinophils (a particular kind of white blood cell, or wbc) can be an indicator of an allergic reaction.

Most analysis of blood samples done today is performed using computers.

Scale: $\frac{1}{4}$ in. = 70 mi

EXAMPLE 6

The scale on the accompanying map is $\frac{1}{4}$ inch = 70 miles.

a) Estimate the distance from Los Angeles to San Francisco, via Route 101 (the colored route).

b) If your average speed during the trip is likely to be 60 miles per hour (mph), estimate how long the trip will take.

Solution:

a) Begin by measuring the map distance in inches from Los Angeles to San Francisco. You should obtain a length of about $1\frac{1}{2}$ in. (depending on how you estimate the curves). One and one half inches equals 6 one quarter inches. Thus the distance is approximately 6 × 70, or 420 mi.

b) The average speed is found by dividing the distance by the rate. Thus the trip will take about 420 mi ÷ 60 mph, or 7 hr.

EXAMPLE 7

Utility bills sometimes contain graphs illustrating the amount of electricity and gas used. The following graphs show gas and electric use at a specific residence for a period of one year, starting in April 1994, as well as the current month's use by the average residential customer. Using these graphs, can you answer the questions?

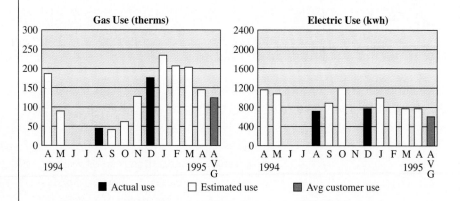

a) How often was an actual gas and electric reading made?

b) Estimate the number of therms of gas used by the average residential customer in April 1995.

c) Estimate the amount of gas used by the resident in April 1995.

d) If the cost of gas is 58.6151 cents per therm, estimate the gas bill in November.

e) In which month was the most electricity used? How many kilowatt hours (kWh) were used in this month?

f) If the cost of electricity is 8.8499 cents per kilowatt hour, estimate the cost of electricity in February.

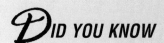

Solution:

a) Actual readings were made in only two months, August and December.

b) Approximately 125 therms were used, as shown by the height of the red bar.

c) Approximately 145 therms were used (slightly less than 150).

d) In November about 125 therms were used. The rate, 58.6151 cents per therm, is the same as $0.586151 per therm. To get a rough approximation round off the rate to $0.60 per therm.

$$0.60 \times 125 = 75*$$

Thus cost of gas used was about $75.

e) The most electricity (approximately 1200 kWh) was used in October.

f) In February about 800 kWh were used. Write 8.8499 cents as $0.08849. Rounding the rate to $0.09 per kilowatt hour and multiplying by 800 yields an estimate of $72.

$$0.09 \times 800 = \$72*$$

▲

EXAMPLE 8

Scientists who are concerned about dwindling animal populations often use aerial photography to make estimates. Estimate the number of zebras in the accompanying photograph.

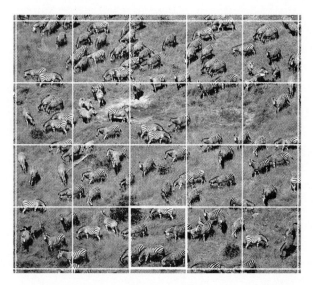

Solution: To estimate the number of zebras, we can divide the photograph into rectangles with equal areas, then select one area that appears to be representative of all the areas. Estimate (or count) the number of zebras in this single area, and then multiply this number by the number of equal areas.

---

*These amounts do not include the basic monthly charge, fuel adjustment, taxes, and other extra charges often included on utility bills.

Let's divide the photo into 20 approximately equal areas. We will select the middle region in the bottom row as the representative region. We enlarge this region and count the zebras in it. If half a zebra is in the region, we count it (see enlargement). There are 6 zebras in this region. Multiplying by 20 gives 20 × 6 = 120. Thus there are about 120 zebras in the photo.

In problems similar to that in Example 8, the number of regions or areas into which you choose to divide the total area is arbitrary. Generally the more regions the better is the approximation, so long as the region selected is representative of the other regions in the map, diagram, or photo.

When you estimate an answer, the amount that your approximation differs from the actual answer will depend on how you round off the numbers. Thus in estimating the product of 196,000 × 0.02520, using the rounded values 195,000 × 0.025 would yield an estimate much closer to the true answer than using the rounded values 200,000 × 0.03. However, without a calculator, the product of 195,000 × 0.025 might be more difficult to find than 200,000 × 0.03. When estimating, you need to determine the accuracy desired in your estimate and round the numbers accordingly.

Section 1.2 Exercises

In Exercises 1–24, estimate the answer. Your answers may vary from the answers given in the back of the text, depending on how you round your answers.

1. 243 + 196.4 + 83.5 + 20.4 + 315.9

2. 1.43 + 100.6 + 156.9 + 179 + 0.23 + 416

3. 192,600 × 4120   4. 1854 × 0.0096

5. $\dfrac{315}{0.062}$ 6. 196.43 − 85.964

7. 0.048 × 1964   8. 9% of 2164

9. 41,640 × 89,264   10. $\dfrac{0.0498}{0.00052}$

11. 592 × 2070 × 992.62   12. 296.3 ÷ 0.0096

13. The cost of eighteen 32-cent postage stamps.

14. The distance traveled when driving 58 mph for 6 hr, 5 minutes (min).

15. One fourth of 82 pounds (lb).

16. The total weight of three steaks if their individual weights are 1.96 lb, 2.21 lb, and 0.82 lb.

17. Your salary if you work for 42.8 hr at $6.85 per hour.

18. The total cost of six grocery items if their prices are $2.49, $3.79, $9.99, $3.21, $2.59, and $13.85.

19. A 6 percent sales tax on a car that sells for $12,894.

20. A 3.84-lb package of ground beef divided into five approximately equal parts.

21. The gas mileage of a car that traveled 235.6 mi on 9.16 gal of gasoline.

22. One third of a profit of $97,250.

23. The total cost of 12 packs of gum that cost 55 cents each.

24. The cost of one bottle of soda if a six-pack costs $3.59.

25. Ed and Dorothy go out for dinner and spend $38.60 for their meal. If they want to leave a 15% tip, estimate the amount that they should leave.

26. Approximately 4.8% of the population has a sexually transmitted disease. Estimate the total number of people with a sexually transmitted disease in a city with 280,000 residents.

27. Blane is a well-trained long-distance runner whose average time per mile is 5 min, 26 sec. Estimate the amount of time Blane will need to finish a marathon (26 mi, 385 yd).

28. Mrs. Sanchez determines that her lawn contains an average of 3.8 grubs per square foot (ft^2). If her rectangular lawn measures 60 ft by 80.7 ft, estimate the total number of grubs in her lawn.

29. If 227 unshelled peanuts can be placed in a quart container, estimate how many unshelled peanuts will fit in a 5-gal container.

30. The Uhligs are planning a vacation in Yosemite National Park. Their round-trip airfare from Memphis, Tennessee, to San Francisco, California, totals $792. Car rental is $39 per day. Lodging is a total of $79 per day, and they estimate a total of $70 per day for food, gas, and other miscellaneous items. If they are planning to stay 6 full days and nights estimate their total expenses.

31. Using the scale on the map of the Orlando, Florida, area, estimate (a) in miles and (b) in kilometers the distance via the route indicated in color, from the Gatorland Zoo to Universal Studios to Sea World.

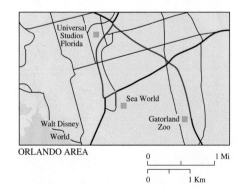

ORLANDO AREA

32. Using the scale on the map of Houston, Texas, estimate (a) in miles and (b) in kilometers the distance of the colored loop, Loop 610, that goes around the city.

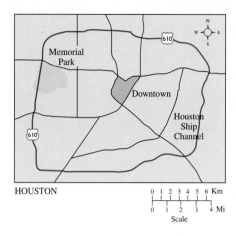

HOUSTON
Scale

33. Using the scale on the map of Washington, D.C., estimate the distance, in miles, tourists walk if they follow the path indicated by the red dashes.

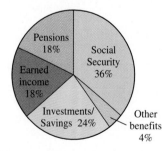

WASHINGTON, D.C.
Scale

34. Consider the circle graph before answering the question. Ms. Weiss, a retiree, has a total yearly retirement income of $36,000 from the various sources shown. Estimate her income from (a) investment/savings and (b) her pension.

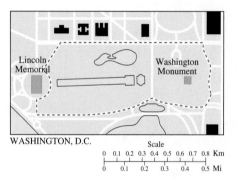

Sources of Retirement Income (for Retirees with at Least $20,000 in Annual Income in 1995.

35. In November 1994, the world's population was estimated to be 5.6 billion. It is projected to almost double by 2050. The explosive growth since 1900 is illustrated in the following graph.

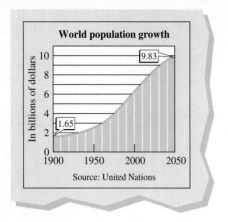

a) Estimate the world's population in 2000.
b) Beginning in 1990, estimate the number of years required for the world population to double (called the *doubling time*).
c) Estimate the doubling time of the world's population beginning in 1960.
d) Estimate the growth of the world's population from the beginning of history until 1950.
e) Estimate the growth of the world's population from 1990 to 2020.
f) By comparing your answers to parts (d) and (e), determine whether the world's population grew more from the beginning of history to 1950 or from 1990 to 2020.

36. The following chart was based on 1994 data provided by the College Savings Bank. Estimated future costs are based on a 7.5% inflation rate. Costs are averages and include tuition, room and board, fees, and other expenses.

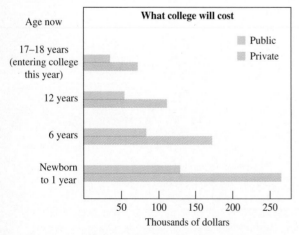

Source: The College Savings Bank

a) Estimate the cost of providing a private college education to a newborn (when she becomes old enough to attend college).
b) Estimate the cost of providing a public college education to a newborn.
c) Estimate the difference in the cost of a college education at a private college for a person entering college in 1994 and for a newborn.
d) Estimate the difference in cost of a college education at a public college for a person entering college in 1994 and for a newborn.

37. The following table is based on information that appeared in an October 1994 issue of *USA Today*.

Over-the-Counter Pain Reliever	Percent of Pain Reliever Market
Tylenol	31.2%
Advil	12.0
Aleve	6.3
Excedrin	6.3
Bayer	4.1
Motrin	3.4

Total sales of over-the-counter pain relievers in 1994 were approximately $2.6 billion.
a) Estimate the dollar sales of Tylenol.
b) Estimate the dollar sales of Bayer.
c) Estimate the dollar sales of all other over-the-counter pain relievers not listed.

38. The following chart is based on data from a University of Michigan study.

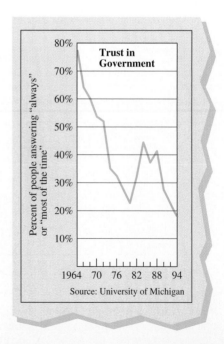

a) In what year between 1975 and 1984 was the people's trust in government the lowest? Estimate the percent of population that answered "always" or "most of the time" in that year.
b) Estimate the highest percent that indicated "always" or "most of the time" between 1982 and 1988.
c) Write a paragraph describing this graph.

In Exercises 39 and 40, estimate the maximum number of smaller figures (at left) that can be placed in the larger figure (at right) without the smaller figures overlapping.

39.
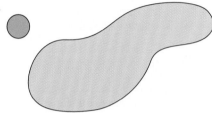

40.

41. Estimate the number of ears of corn shown in the photo.

In Exercises 42–44, estimate, in degrees, the measure of the angles depicted. For comparison purposes a right angle, ⌐, *measures 90°.*

42.

43.

44.

Estimate the percent of area that is shaded in the following figures.

45.
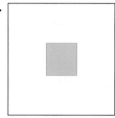

46.
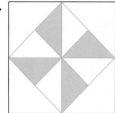

If each square represents one square unit, estimate the area of the shaded figure in square units.

47.
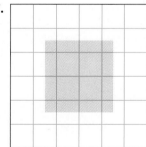

48.
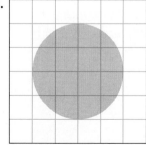

49. Estimate the number of pebbles in the photograph.

50. Estimate the number of people in the photograph.

51. If the hand, which covers the end of the large leaf, is 7 in. long, estimate the length of the leaf.

52. If the diameter of Saturn is 75,100 mi, estimate the diameter of Saturn's largest ring.

53. Estimate, without a ruler, a distance of 12 in. Measure the distance. How good was your estimate?

54. In a bag place objects that you feel have a total weight of 10 lb. Weigh the bag to determine the accuracy of your estimate.

55. Estimate the number of times the phone will ring in 1 min if unanswered. Have a classmate phone you so that you can count the rings and thus test your estimate.

56. Fill a glass with water and estimate the water's temperature. Then use a thermometer to measure the temperature and check your estimate.

57. Estimate the ratio of your height to your neck size. Then have a friend measure you to determine this ratio and check the accuracy of your estimate.

58. Estimate the number of pennies that will fill a 3-ounce (oz) paper cup. Then actually fill a 3-oz paper cup with pennies, counting them to determine the accuracy of your estimate.

59. Estimate how fast you can walk 60 ft. Then mark off a distance of 60 ft and use a watch with a second hand to time yourself walking it. Determine the accuracy of your estimate.

Problem Solving/Group Activities

60. Make a shopping list of 20 items you use regularly that can be purchased at a supermarket. Beside each item write down what you estimate to be its price. Add these price guesses to estimate the total cost of the 20 items. Next, make a trip to your local supermarket and record the actual price of each item. Add these prices to determine the actual total cost. How close was your estimate? (Don't forget to add tax on the taxable items.)

61. Women still earn less than men, but the gap has narrowed during the past decade. Here are the percent of men's weekly wages that women earn at various ages.

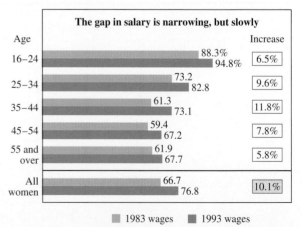

The gap in salary is narrowing, but slowly

Age		Increase
16–24	88.3% / 94.8%	6.5%
25–34	73.2 / 82.8	9.6%
35–44	61.3 / 73.1	11.8%
45–54	59.4 / 67.2	7.8%
55 and over	61.9 / 67.7	5.8%
All women	66.7 / 76.8	10.1%

■ 1983 wages   ■ 1993 wages

Source: Bureau of Labor Statistics

a) If the average men's weekly wage in the 35–44 age group in 1993 was $490, estimate the average women's weekly wage in this age group.

b) If the average women's weekly wage for women 55 and over was $400 in 1983, estimate the average wage for women in this age group in 1993.

c) For women in the 16–24 age group, women's wages are about 95% of the men's wages. However, this percentage drops until the 55-and-over age group where it increases slightly. Explain why you believe this happens.

62. Gary, Sue, Andy, and Megan Gilligan are planning a skiing vacation in the Rockies. Their round-trip airfare from Raleigh, North Carolina, to Denver, Colorado, totals $992. Car rental is $39 per day, ski lift tickets for each person cost $40 per day, lodging is a total of $109 per night, and food and miscellaneous items are estimated to total $90 per day. They plan to fly into Denver on Friday morning, drive to Aspen that same day, begin skiing on Saturday, ski up to and including Wednesday, drive back to Denver on Thursday, and leave from Denver Thursday evening. Estimate the total cost of the vacation.

Research Activities

63. a) About how much water does your household use per day? Use the following data to estimate your household's daily water usage.

HOW MUCH WATER DO YOU USE?	
Activity	**Typical use**
Running clothes washer	40 gal
Bath	35 gal
5-minute shower	25 gal
Doing dishes in sink, water running	20 gal
Running dishwasher	11 gal
Flushing toilet	4 gal
Brushing teeth, water running	2 gal

Sources: U.S. Environmental Protection Agency; CU

b) Determine from your water department (or company) your household's average daily usage by obtaining the total number of gallons used per year and dividing that amount by 365. How close was your estimate in part (a)?

c) Current records indicate that the average household uses about 300 gal of water per day (the average daily usage is 110 gal per person). Based on the number of people in your household, do you feel your household uses more or less than the average amount of water? Explain your answer.

64. Develop a monthly budget by estimating your monthly income and your monthly expenditures. Your monthly income should equal your monthly expenditures.

65. Identify three ways that you use estimation in your daily life. Discuss each of them briefly and give examples.

1.3 PROBLEM SOLVING

Solving mathematical puzzles and real-life mathematical problems can be enjoyable. You should work as many exercises in this section as possible. By doing so, you will sample a variety of problem-solving techniques.

You can approach any problem by using a general procedure developed by George Polya. Before learning Polya's problem-solving procedure, let's consider an example.

EXAMPLE 1

Mike has just returned from a trip to the Grand Canyon and Monument Valley. During the trip he used 20 rolls of slide film, 12 at the Grand Canyon and 8 at Monument Valley. Of the 20 rolls of film, 13 are rolls of 36 exposures and 7 are rolls of 24 exposures. The speed of all the film was 64.

Mike wants to store his slides in slide trays and asks your advice. Slide trays can be purchased in two different sizes. The size that holds 140 slides costs $12.99. The other size that holds 80 slides costs $8.99. How many of each type of tray should Mike purchase to accommodate all his slides yet minimize the cost of the trays?

Solution: The first thing to do is to read the problem carefully. Read it at least twice and be sure you understand the facts given and what you are being asked to find.

Next make a list of the given facts and determine which are relevant to answering the question asked.

Given Information

    20 rolls of film

    12 rolls of Grand Canyon and 8 rolls of Monument Valley

    13 rolls of 36 exposures and 7 rolls of 24 exposures

    64 is the speed of the film.

    Cost of a slide tray that holds 140 slides is $12.99.

    Cost of a slide tray that holds 80 slides is $8.99.

Finding the most economical way to store the slides requires knowing the total number of slides, the number of slides held by each type of tray, and the cost of each type of tray. All other given information is irrelevant to finding the answer.

Relevant Information

    13 rolls of 36 exposures and 7 rolls of 24 exposures

    Cost of a slide tray that holds 140 slides is $12.99.

    Cost of a slide tray that holds 80 slides is $8.99.

The next step is to determine the total number of slides. The total is the sum of the number of slides from the rolls of 36 exposures and from the rolls of 24 exposures.

Number of slides from 36-exposure rolls = 36 $\times$ 13 rolls = 468

Number of slides from 24-exposure rolls = 24 $\times$ 7 rolls $= \dfrac{168}{636}$

At this point we can rephrase the question in simpler form: If slide trays that hold 140 slides cost $12.99, and slide trays that hold 80 slides cost $8.99, what is the least expensive way to store 636 slides in trays?

We now need a plan for solving the problem. One method is to set up a table or chart to compare the costs of different combinations of trays of the two sizes. Mike needs enough trays to hold a total of 636 slides. Start by using all 140-slide trays and work down to all 80-slide trays. The following table illustrates the possible combinations of trays.

Combination of Trays to Hold 636 Slides	Cost of Trays
5 trays of 140 and 0 trays of 80 (700 slides)	$64.95
4 trays of 140 and 1 tray of 80 (640 slides)	$60.95
3 trays of 140 and 3 trays of 80 (660 slides)	$65.94
2 trays of 140 and 5 trays of 80 (680 slides)	$70.93
1 tray of 140 and 7 trays of 80 (700 slides)	$75.92
0 trays of 140 and 8 trays of 80 (640 slides)	$71.92

This table shows that the least expensive way to store the slides is to purchase four trays of 140-slide size and one tray of 80-slide size.   ▲

Following is a general procedure for problem solving as given by George Polya. Note that Example 1 demonstrates many of these guidelines.

Guidelines for Problem Solving

1. *Understand the problem.*
  - Read the problem *carefully* at least twice. In the first reading get a general overview of the problem. In the second reading determine (a) exactly what you are being asked to find, and (b) what information the problem provides.
  - Try to make a sketch to illustrate the problem. Label the information given.
  - Make a list of the given facts. Are they all pertinent to the problem?
  - Determine if the information you are given is sufficient to solve the problem.

2. *Devise a plan to solve the problem.*
  - Have you seen the problem or a similar problem before? Are the procedures you used to solve the similar problem applicable to the new problem?
  - Can you express the problem in terms of an algebraic equation? (We will explain how to write algebraic equations in Chapter 6.)
  - Look for patterns or relationships in the problem that may help in solving it.
  - Can you express the problem more simply?
  - Can you substitute smaller or simpler numbers to make the problem more understandable?
  - Will listing the information in a table help in solving the problem?
  - Can you make an educated guess at the solution? Sometimes if you know an approximate solution you can work backward and eventually determine the correct procedure to solve the problem.

3. *Carry out the plan.*
  Use the plan you devised in step 2 to solve the problem.

4. *Check the results.*
  - Ask yourself, "Does the answer make sense?" and "Is the answer reasonable?" If the answer is not reasonable, recheck your method for solving the problem and your calculations.
  - Can you check the solution using the original statement?
  - Is there an alternative method to arrive at the same conclusion?
  - Can the results of this problem be used to solve other problems?

The following examples show how to apply the guidelines for problem solving.

EXAMPLE 2

A bus operates between John Fitzgerald Kennedy airport and downtown Manhattan, 25 miles away. It makes 10 round trips per day carrying an average of 42 passengers per trip. The fare each way is $13.00. What are the receipts from one day's operation?

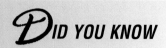

Solution: A careful reading of the problem shows that the task is to find the total receipts from one day's operation. Make a list of all the information given and determine whether it is all pertinent to the problem. The facts given are:

Distance from airport to city = 25 mi
Number of round trips daily = 10
Average number of passengers per trip = 42
Fare each way = $13.00

What information do you need to determine the total receipts for the day? Is all the information given needed in solving the problem? Some thought should reveal that the distance between the airport and the city is not needed to determine the answer. You should realize that the total receipts depend on (a) the number of one-way trips per day, (b) the average number of passengers per trip, and (c) the cost per passenger each way. The product of these three numbers will yield the total daily receipts. For the 10 round trips daily, there are 2 × 10 or 20 one-way trips daily.

$$\begin{pmatrix} \text{Receipts} \\ \text{for one} \\ \text{day} \end{pmatrix} = \begin{pmatrix} \text{Number of} \\ \text{one-way} \\ \text{trips per day} \end{pmatrix} \times \begin{pmatrix} \text{Number of} \\ \text{passengers} \\ \text{per trip} \end{pmatrix} \times \begin{pmatrix} \text{Cost per} \\ \text{passenger} \\ \text{each way} \end{pmatrix}$$

$$= 20 \times 42 \times \$13.00 = \$10,920 \qquad \blacktriangle$$

In Example 2, we could have used 10 round trips at a fare of $26.00 per person to obtain the answer. Why?

Is the answer obtained in Example 2 reasonable for the information given? A quick estimate may be obtained as follows:

Cost of one round trip for 1 person $26
Cost of one round trip for 40 people $26 × 40 = $1040
Cost of 10 round trips for 40 people $1040 × 10 = $10,400

With an estimate of $10,400, the answer $10,920 seems reasonable.

EXAMPLE 3

The cost of Heather's meal before tax is $13.60. A $7\frac{1}{2}\%$ sales tax of $1.02 is then added to bring the check to a total of $14.62.

a) If Heather wants to leave a 10% tip on the *pretax* cost of the meal, how much should she leave?

b) If she wants to leave a 15% tip on the pretax cost of the meal, how much should she leave?

Solution:

a) To find 10% of any number simply move its decimal point one place to the left. Thus a 10% tip would be $1.36, which rounds off to $1.40. (When we tip we rarely give an exact percentage because we generally round the answer.)

b) One method for finding a 15% tip is to find 10% of the cost, as in part (a), then add half that amount. Thus a 15% tip would be about

$$\$1.40 + (\$1.40/2) = \$1.40 + \$0.70 = \$2.10 \qquad \blacktriangle$$

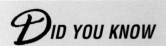

In Example 3(b) can you find another, and perhaps easier, method to deter-mine a 15% tip from the information given?

EXAMPLE 4

The following chart shows the amount of each ingredient recommended to make 2, 4, and 8 servings of Betty Crocker Potato Buds. Determine the amount of each ingredient necessary to make 6 servings of Potato Buds by using the following procedures.

a) Multiply the amount for 2 servings by 3*.

b) Add the amounts for 2 servings to the amounts for 4 servings.

c) Find the average of the amounts for 4 servings and for 8 servings.

d) Subtract the amounts for 2 servings from the amounts for 8 servings.

e) Compare the answers for parts (a)–(d). Are they the same? If not, explain why not.

f) Which is the correct procedure for obtaining 6 servings?

Servings	2	4	8
Water	$\frac{2}{3}$ cup	$1\frac{1}{3}$ cups	$2\frac{2}{3}$ cups
Milk	2 tbsp	$\frac{1}{3}$ cup	$\frac{2}{3}$ cup
Butter or margarine	1 tbsp	2 tbsp	4 tbsp
Salt†	$\frac{1}{4}$ tsp	$\frac{1}{2}$ tsp	1 tsp
Potato Buds®	$\frac{2}{3}$ cup	$1\frac{1}{3}$ cups	$2\frac{2}{3}$ cups

†*Less salt can be used if desired.*

Solution:

a) We multiply the amounts for 2 servings by 3.

Water: $3(\frac{2}{3}) = 2$ cups

Milk: $3(2) = 6$ tablespoons (tbsp)

Butter or margarine: $3(1) = 3$ tbsp

Salt: $3(\frac{1}{4}) = \frac{3}{4}$ teaspoon (tsp)

Potato Buds®: $3(\frac{2}{3}) = 2$ cups

b) We find the amount of each ingredient by adding the amount for 2 and 4 servings.

Water: $\frac{2}{3}$ cup $+ 1\frac{1}{3}$ cup $= 2$ cups

Milk: 2 tbsp $+ \frac{1}{3}$ cup   oh oh!

To add these two amounts, we must convert one of them so that they have the same units. By looking in a cookbook, or a book of conversion factors, we see that 16 tbsp = 1 cup. The milk in part (a) was given in tablespoons, so we convert $\frac{1}{3}$ cup to tablespoons in order to compare an-swers. One third cup equals $\frac{1}{3}(16) = \frac{16}{3}$ or $5\frac{1}{3}$ tbsp. Therefore

Milk: 2 tbsp $+ 5\frac{1}{3}$ tbsp $= 7\frac{1}{3}$ tbsp

---

*Addition, subtraction, multiplication, and division of fractions are discussed in detail in Section 5.3.

Let's continue with the rest of the ingredients:

Butter: 1 tbsp + 2 tbsp = 3 tbsp

Salt: $\frac{1}{4}$ tsp + $\frac{1}{2}$ tsp = $\frac{3}{4}$ tsp

Potato Buds®: $\frac{2}{3}$ cup + $1\frac{1}{3}$ cups = 2 cups

c) We compute the amounts of the ingredients by finding the average of the amounts for 4 and 8 servings. We do so by adding the amounts for each ingredient and dividing the sum by 2.

Water: $\dfrac{1\frac{1}{3} \text{ cups} + 2\frac{2}{3} \text{ cups}}{2} = \dfrac{4 \text{ cups}}{2} = 2$ cups

Milk: $\dfrac{\frac{1}{3} \text{ cup} + \frac{2}{3} \text{ cup}}{2} = \dfrac{1 \text{ cup}}{2} = \frac{1}{2}$ cup (or 8 tbsp)

Butter: $\dfrac{2 \text{ tbsp} + 4 \text{ tbsp}}{2} = \dfrac{6 \text{ tbsp}}{2} = 3$ tbsp

Salt: $\dfrac{\frac{1}{2} \text{ tsp} + 1 \text{ tsp}}{2} = \dfrac{\frac{3}{2} \text{ tsp}}{2} = \frac{3}{4}$ tsp

Potato Buds®: $\dfrac{1\frac{1}{3} \text{ cups} + 2\frac{2}{3} \text{ cups}}{2} = \dfrac{4 \text{ cups}}{2} = 2$ cups

d) We obtain the amounts of ingredients by subtracting the amounts for 2 servings from the amounts for 8 servings.

Water: $2\frac{2}{3}$ cups $- \frac{2}{3}$ cup = 2 cups

Milk: $\frac{2}{3}$ cup $- 2$ tbsp $= \frac{2}{3}(16)$ tbsp $- 2$ tbsp
$= \frac{32}{3}$ tbsp $- \frac{6}{3}$ tbsp
$= \frac{26}{3}$ tbsp, or $5\frac{1}{3}$ tbsp

Butter: 4 tbsp $- 1$ tbsp = 3 tbsp

Salt: 1 tsp $- \frac{1}{4}$ tsp = $\frac{3}{4}$ tsp

Potato Buds®: $2\frac{2}{3}$ cups $- \frac{2}{3}$ cup = 2 cups

e) Comparing the answers in parts (a)–(d), we find that the amounts of all ingredients, except milk, are the same. For milk, we get the following results.

Part (a): Milk = 6 tbsp

Part (b): Milk = $7\frac{1}{3}$ tbsp

Part (c): Milk = 8 tbsp

Part (d): Milk = $5\frac{1}{3}$ tbsp

Why are all these answers different? After rechecking, we find that all our calculations are correct, so we must look deeper. Note that milk is the only ingredient that has different units for 2 servings and 4 servings. Let's check the relationship between 2 tbsp and $\frac{1}{3}$ cup. In going from 2 servings to 4 servings we would expect that $\frac{1}{3}$ cup should be twice 2 tbsp. We know that 1 cup = 16 tbsp, so

$$\frac{1}{3} \text{ cup} = \frac{1}{3}(16) = \frac{16}{3} = 5\frac{1}{3} \text{ tbsp}$$

Therefore, instead of the 4 tbsp of milk we expected for 4 servings, we get $5\frac{1}{3}$ tbsp. This change causes all our calculations for milk to be different.

f) Which is the correct answer? As all our calculations for milk are correct, there is no single correct answer. All our answers are correct. Using 8 tbsp instead of $5\frac{1}{3}$ tbsp might make the Potato Buds® a little thinner. When we cook we generally do not add the *exact* amount recommended. We rely on experience to alter the recommended amounts according to individual taste.

▲

Many real-life problems, such as the one in Example 5, can be solved by using proportions.* A proportion is a statement of equality between two ratios (or fractions).

EXAMPLE 5

The instructions on a bottle of insecticide indicate that 1.5 oz of insecticide should be mixed with 5 gal of water. Mr. Wildas wishes to spray his garden. How much insecticide should he mix with 8 gal of water to get the proper strength solution?

Solution: Use the fact that 1.5 oz of insecticide is to be mixed with 5 gal of water to set up a proportion.

$$\text{Given ratio} \begin{cases} \dfrac{1.5 \text{ oz}}{5 \text{ gal water}} = \dfrac{? \text{ oz}}{8 \text{ gal}} \end{cases}$$

Item to be found

Other information given

Note in the proportion that ounces and gallons are placed in the same relative positions. Often the unknown quantity is replaced with an x. The proportion may be written as follows and solved using cross multiplication.

$$\frac{1.5}{5} = \frac{x}{8}$$

$$1.5(8) = 5x$$

$$12.0 = 5x$$

$$\frac{12.0}{5} = \frac{5x}{5} \qquad \text{Divide both sides of the equation by 5 to solve for } x.$$

$$2.4 = x$$

Thus the gardener must mix 2.4 oz insecticide with 8 gal water to get the proper strength solution.

▲

Most of the problems solved so far have been practical ones. However, many people enjoy solving brainteasers. Two examples of such puzzles follow.

EXAMPLE 6

The odometer of a motor home showed 14,941 miles. The driver said that this number was *palindromic*; that is, it reads the same backward as forward. "Look at this," the driver said to the passengers, "it will be a long time before

---

*Proportions are discussed in greater detail in Section 6.2.

this happens again.'' But after another day's drive, the odometer showed five new palindromic numbers. Can you find the numbers?

Solution: The problem is to find five palindromic numbers larger than 14,941. The numbers must be of the form △ ☐ ◇ ☐ △. The number we use in place of the triangles must be a 1. If the triangles were replaced with a 2, the motor home would have had to travel over 5000 miles in a day—an impossibility. Next we replace the squares and the diamond with numbers. Because the number of miles increased, the next number that could replace the squares is 5. The number now looks like this: 15 ◇ 51. The diamond could be replaced with any number and the number formed would be palindromic. If the diamond is replaced with a 0, the result, 15051, is the smallest palindromic number that is larger than 14941. The remaining four numbers desired are easily found. All that needs to be done is to replace the diamond with 1, 2, 3, and 4. Thus the next five palindromic numbers are 15051, 15151, 15251, 15351, and 15451—all achievable.   ▲

EXAMPLE 7

A magic square is a square array of numbers such that the numbers in all rows, columns, and diagonals have the same sum. Use the digits 1, 2, 3, 4, 5, 6, 7, 8, and 9 to construct a magic square.

Solution: The first step is to create a figure with nine cells as in Fig. 1.1(a). We must place the nine numbers in the cells so that the same sum is obtained in each row, column, and diagonal. Common sense tells us that 7, 8, and 9 cannot be in the same row, column, or diagonal. We need some small and large numbers in the same row, column, and diagonal. To see a relationship, we list the numbers in order:

$$1, 2, 3, 4, 5, 6, 7, 8, 9$$

Note that the middle number is 5 and the smallest and largest numbers are 1 and 9, respectively. The sum of 1, 5, and 9 is 15. If the sum of 2 and 8 is added to 5 the sum is 15. Likewise 3, 5, 7, and 4, 5, 6 have sums of 15. We see that in each group of three numbers the sum is 15 and 5 is a member of the group.

						8		4		8		4	3	8
9	5	1		9	5	1		9	5	1		9	5	1
				2				2		6		2	7	6

(a)   (b)   (c)   (d)

Figure 1.1

Because 5 is the middle number in the list of numbers, place 5 in the center square. Place 9 and 1 to the left and right of 5 as in Fig. 1.1(a). Now we place the 2 and the 8. The 8 cannot be placed next to 9 because $8 + 9 = 17$, which is greater than 15. Place the smaller number 2 next to the larger number 9. We elected to place the 2 in the lower left-hand cell, and the 8 in the upper right-hand cell as in Fig. 1.1(b). The sum of 8 and 1 is 9. To arrive at a sum of 15, we place 6 in the lower right-hand cell as in Fig. 1.1(c). The

> sum of 9 and 2 is 11. To arrive at a sum of 15, we place 4 in the upper left-hand cell as in Fig. 1.1(c). Now the diagonals 2, 5, 8 and 4, 5, 6 have sums of 15. The numbers that remain to be placed in the empty cells are 3 and 7. Using arithmetic, we can see that 3 goes in the top middle cell and 7 goes in the bottom middle cell as in Fig. 1.1(d). A check shows that the sum in all the rows, columns, and diagonals is 15. ▲

The solution to Example 7 is not unique. Other arrangements of the nine numbers in the cells will produce a magic square. Also, other techniques of arriving at a solution for a magic square may be used. In fact, the process described will not work if the number of squares is even—for example, 16 instead of 9. Magic squares are not limited to the operation of addition or to the set of counting numbers.

Section 1.3 Exercises

Solve the following real-life problems.

1. Ships crossing the Panama Canal have to travel 48.2 mi from the Atlantic end of the canal to the Pacific end. The tides have an average height of 3.24 ft per day at the Atlantic end of the canal and 11.63 ft per day at the Pacific end. How many feet higher do the tides average on the Pacific end?

2. A landscape designer is preparing the plan for a town park using a scale of 1 in. = 2.5 yd. He draws a 50.4-in. line to represent the northern boundary of the playground. What actual distance (in yards) does this boundary line represent?

3. An architect is designing a shopping mall. The scale of her plan is 1 in. = 12 ft. If one store in the mall is to have a frontage of 82 ft, how long will the line representing that store's frontage be on the blueprint?

4. At a given time of day the ratio of the height of a tree to its shadow is the same for all trees. If a 3-ft stick in the ground casts a shadow of 1.2 ft, how tall is a tree that casts a shadow of 48.4 ft?

5. A taxicab charges $1.50 for the first 1/8 mi and 10 cents for each additional 1/8 mi. Determine the cost of a 5-mi trip.

6. If 20 lb of fertilizer covers an area of 5000 square feet (ft^2), how much fertilizer is needed to cover an area of 12,000 ft^2?

7. The instructions on a bottle of concentrated liquid cleaner reads "Mix 3 ounces with a gallon of water." If a building custodian wants to mix the solution in a $2\frac{1}{2}$-gal bucket of water, how much concentrate should he use to obtain the proper strength mixture?

8. How long will it take to cut 10 ft of plastic pipe into five 2-ft lengths if each cut takes 3 min?

9. In 1980 the fastest chip available (the 8088) for personal computers could perform about 0.3 million instructions per second (MIPS). In 1994 the fastest chip available for personal computers (the Pentium—100 MHZ) could perform about 200 MIPS. How many times faster is the Pentium chip than the 8088?

10. Jessica makes and sells pillows. On August 4 she bought three large bags of foam rubber. Each bag contained 20 lb of foam rubber. She used 4.25 lb to make each of three large pillows, 10.2 lb to make small pillows, and 9.6 lb to make a mat. How many pounds of foam rubber did she have left?

11. The Main Street Garage charges $1.50 for the first hour of parking and $1.00 for each additional hour. Ron parks his car in the garage from 9 A.M. to 5 P.M., 5 days a week. How much money does he save by paying a weekly parking rate of $30.00?

12. Mr. Greene parked his car in a garage that charged $1.25 for the first hour and $0.80 for each additional hour. When he returned from shopping, he gave the attendant $10.00. The attendant gave him $3.95 change. How many hours was his car in the garage?

13. Bernardo wants to purchase a fax machine that sells for $420. He can either pay the total amount at the time of purchase, or he can agree to pay the store $120 down and $30 a month for 12 months. How much money can he save by paying the total amount at the time of purchase?

14. A person-to-person call from New York City to Denver costs $3.25 for the first 3 min and $0.42 for each additional minute. How much would a 23-min phone call cost?

15. On four exams Cameron's grades were 87, 93, 80, and 76. What grade must he obtain on his fifth exam to have an 80 average?

16. The stage crew built a 4.8 yd by 6.2 yd garden for outdoor scenes in a community theater play. The stage manager bought artificial grass at $3.15 a square yard (yd^2) to cover the garden area. She gave the salesperson a $100 bill. How much change should she receive?

17. In one state lottery game you must select a four-digit number. If your number matches exactly the four-digit number selected by the lottery commission, you win. If you purchase 1 lottery ticket, what is your chance of winning?

18. The following table gives the approximate energy values of some foods, in kilojoules (kJ), and the energy requirements of some activities. How soon would you use up the energy from
a) a fried egg by swimming?
b) a hamburger by walking?
c) a piece of strawberry shortcake by cycling?
d) a hamburger and a chocolate milkshake by walking?

Food	Energy Value (kJ)	Activity	Energy Consumption (kJ/min)
Chocolate milkshake	2200	Walking	25
		Cycling	35
Fried egg	460	Swimming	50
Hamburger	1550	Running	80
Strawberry shortcake	1400		
Glass of skim milk	350		

19. Mary deposited $2000 in an account paying 5% simple interest for 1 year. The interest she received for the year was $100. If Mary had deposited $6000 in an account that was paying 80% of the stated simple interest, how much interest would she have after 1 year?

20. A plane is flying at an altitude of 15,000 ft, where the temperature is −6 degrees Fahrenheit (°F). The nearby airport where the pilot intends to land is at an altitude of 3000 ft, and the control tower there reports precipitation. If the temperature increases 2.5°F for every 1000 ft decrease in altitude, will the precipitation be rain or snow? Assume that rain changes to snow at 32°F.

21. Richard fills his gas tank completely and makes a note that his odometer reads 38,451.4 mi. The next time he stops to put gas in his car, filling his tank takes 12.6 gal, and his odometer reads 38,687.0 mi. Determine the number of miles per gallon that his car gets.

22. a) Quinton works 20 hours per week and makes $5.40 per hour. How much money can he expect to earn in one year (52 weeks)?
b) If he saves all the money that he earns, how long will he have to work to save for a boat that costs $750?

23. Mary purchased 4 tires by mail order. She paid $52.80 per tire plus $5.60 per tire for shipping and handling. There is no sales tax on this purchase because they were purchased out of state. She also had to pay $8.56 per tire for mounting and balancing. At a local tire store, her total for the 4 tires with mounting and balancing would be $324 plus an 8% tax. How much did Mary save by purchasing the tires through the mail?

24. Twelve square posts, 6 in. on a side, are placed in a straight line at 4-ft intervals to construct a fence. How far is the beginning of the first post from the end of the last post?

25. The Federal Income Tax rate schedule for a joint return in 1994 is illustrated in the following table. If the Guardinos paid $53,200 in federal taxes, find the family's adjusted gross income.

Adjusted Gross Income	Taxes
$0–$38,000	15% of income
$30,001–$91,850	$5700 + 28% in excess of $38,000
$91,851–$140,000	$20,778.00 + 31% in excess of $91,800
$140,001–$250,000	$35,704.50 + 36% in excess of $140,000
$250,000 and up	$75,304.50 + 39.6% in excess of $250,000

26. A faucet is leaking at a rate of one drop of water per second. If the volume of one drop of water is 0.1 cubic centimeter (0.1 cm^3), find
a) the volume of water in cubic centimeters lost in a year.
b) how long it would take, in days, to fill a rectangular basin 30 cm by 20 cm by 20 cm.

27. A faucet leaks 1 oz of water per minute.
a) How many gallons of water are wasted in a year? (A gallon contains 128 oz.)
b) If water costs $4.20 per 1000 gal, how much additional money is being spent on the water bill?

28. A gymnasium floor has an area of 2400 yd^2. Each gallon of floor sealant covers an area of 350 ft^2. How many gallons of sealant are needed to cover the gymnasium floor?

29. When a car's tire pressure is 28 pounds per square inch (psi) it averages 15.8 miles per gallon (mpg) of gasoline. If the tire pressure is increased to 30 psi the car averages 16.4 mpg of gasoline.

a) If a commuter drives an average of 15,000 mi per year, how many gallons of gasoline will he save in a year by increasing his tire pressure from 28 to 30 psi?

b) If gasoline costs $1.20 per gallon, how much will he save in a year?

c) If we assume that there are about 140 million cars in the United States and that these changes are typical of each car, how many gallons of gasoline would be saved if all drivers increased their cars' tire pressure?

30. Assume that the rate of inflation is 6% for the next 2 years. What will be the cost of goods 2 years from now, adjusted for inflation, if the goods cost $450.00 today?

31. The cost of a car increases by 20% and then decreases by 20%. Is the resulting price of the car greater than, less than, or equal to the original price of the car?

32. A new high school graduate receives two job offers: Company A offers a starting salary of $15,000 a year with a $600 raise every 6 months. Company B offers $15,000 a year with a $1200 raise every 12 months. Which offer will provide the most income?

33. A video store chain plans to purchase a large number of video tapes. One supplier is selling boxes of 20 tapes for $48, and a second supplier is selling boxes of 12 tapes for $30. Only complete boxes of the tapes are sold.

a) Find the maximum number of tapes that can be purchased for $280. Indicate how many boxes of 20 and how many boxes of 12 will be purchased.

b) How much will the tapes cost?

34. Marty and Marilyn are making marshmallow squares for their class's party. The recipe calls for a total of 160 oz of marshmallows. Marshmallows come in 16-oz, 24-oz, and 36-oz bags. How many different combinations of bags, with no marshmallows left over, could Marty and Marilyn use for the recipe?

35. The following amounts of ingredients are recommended to make various servings of Nabisco Instant Cream of Wheat.

Ingredient	1 Serving	2 Servings	4 Servings
Mix water or milk	1 cup	2 cups	$3\frac{3}{4}$ cups
With salt (optional)	$\frac{1}{8}$ tsp	$\frac{1}{4}$ tsp	$\frac{1}{2}$ tsp
Add Cream of Wheat	3 tbsp	$\frac{1}{2}$ cup	$\frac{3}{4}$ cup

Determine the amount of each ingredient needed to make 3 servings using the following procedure.

a) Multiply the amounts for 1 serving by 3.

b) Find the average of the amounts for 2 and 4 servings.

c) Subtract the amounts for 1 serving from the amounts for 4 servings.

d) Compare the answers obtained in parts (a)–(c) and explain any differences.

36. Following are the amounts of ingredients recommended to make various servings of Uncle Ben's Original Converted Rice.

Ingredient	2 Servings	4 Servings	6 Servings	12 Servings
Rice (cups)	$\frac{1}{2}$	1	$1\frac{1}{2}$	3
Water (cups)	$1\frac{1}{3}$	$2\frac{1}{4}$	$3\frac{1}{3}$	6
Salt (teaspoon)	$\frac{1}{4}$	$\frac{1}{2}$	$\frac{3}{4}$	$1\frac{1}{2}$
Butter or margarine	1 tsp	2 tsp	1 tbsp	2 tbsp

Determine the amount of each ingredient needed to make 8 servings using the following procedures.

a) Multiply the amount for 2 servings by 4.

b) Multiply the amount for 4 servings by 2.

c) Add the amounts for 2 and 6 servings.

d) Compare the answers obtained in parts (a)–(c) and explain any differences.

Solve the following problems.

37. How many square inches, 1 in. by 1 in., fit in an area of 1 square foot, 1 ft by 1 ft?

38. How many cubic inches fit in 1 cubic foot (ft^3)?

39. If the length and width of a rectangle each double, what happens to the area of the rectangle?

40. If the length, width, and height of a cube all double, what happens to the volume of the cube?

41. Fill in the three boxes using the symbols $+$, $-$, $\times$, and $\div$ to make a true statement of equality:

$$7 \ \square \ 7 \ \square \ (7 \ \square \ 7) = 13$$

42. While on a safari in Africa, I saw a small herd of zebras and cranes wandering over the veldt. Counting heads, I got 18. Counting feet, I got 60. How many zebras and how many cranes were on the veldt?

43. A 24-ft-by-24-ft carpet is partitioned into 4-ft-by-4-ft squares. How many squares will there be?

44. The year 1991 was a palindromic year. How many palindromic years will occur between the years 2000 and 3000? List them. See Example 6.

45. If you have a balance scale and only the four weights 1 gram (g), 3 g, 9 g, and 27 g, explain how you could show that an object had the following weights.
a) 5 g **b)** 16 g
(*Hint:* Weights must be added to both sides of the balance scale.)

46. A woman purchased a dress that cost $45 and gave the merchant a $100 bill. After the woman had gone with her dress and her change, the merchant took the $100 bill to the bank. The bank clerk informed him that the $100 bill was counterfeit. What was the financial loss to the merchant?

47. The Sunday morning chef is stuck with a pan that holds only two slices of bread for French toast. Browning one side of a piece of toast takes 30 sec. How can the chef brown both sides of three slices in $1\frac{1}{2}$ min instead of 2 min? (*Hint:* Partially finish two slices and then start the third slice.)

48. Create a magic square by using the numbers 1, 3, 5, 7, 9, 11, 13, 15, and 17. The sum of the numbers in every column, row, and diagonal must be 27.

49. Create a magic square by using the numbers 2, 4, 6, 8, 10, 12, 14, 16, and 18. The sum of the numbers in every column, row, and diagonal must be 30.

In Exercises 50–52, use the three magic squares illustrated to obtain the answers.

6	5	10
11	7	3
4	9	8

3	2	7
8	4	0
1	6	5

10	9	14
15	11	7
8	13	12

50. Examine the 3 by 3 magic squares and find the sum of the four corner entries of each magic square. How can you determine the sum by using a key number in the magic square?

51. For a 3 by 3 magic square, how can you determine the sum of the numbers in any particular row, column, or diagonal by using a key value in the magic square?

52. For a 3 by 3 magic square, how can you determine the sum of all the numbers in the square by using a key value in the magic square?

53. Here is a flat pattern for a cube to be formed by folding. The sides of each square are 6 cm. Find the volume of the cube.

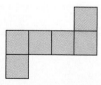

54. Consider a domino with six dots, as shown. Two ways of connecting the three dots on the left with the three dots on the right are illustrated. In how many ways can the three dots on the left be connected with the three dots on the right?

55. Five salespeople gather for a sales meeting. How many handshakes will each person make if each must shake hands with each of the four others?

56. Identical cubes are stacked in the corner of a room, as shown. How many of the cubes are not visible?

57. Digital clocks display numerals with lighted parts of the pattern shown. If each digit 0–9 is displayed once, which part is used least often? Which part is used most often?

58. A TV screen measures 24 in. by 16 in. If each dimension is increased by 20%,
a) by how many square inches is the area increased?
b) by what percent is the area increased?

59. Three open switches are in a row: z can be opened or closed at will; y can be opened or closed only if z is closed; x can be opened or closed only if y is closed and z is open. If all three switches are open, what will be the fewest switch changes needed to get all three in closed positions?

60. How many triangles are in the figure?

61. Place the digits 1–8 in the eight boxes so that each digit is used exactly once and no two consecutive digits touch horizontally, vertically, or diagonally.

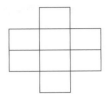

62. Read the three statements and determine which are true and which are false.
   1. There are three numbered statements.
   2. Two of the numbered statements are not true.
   3. This statement is true.

Problem Solving/Group Activity

63. Jessica wants to purchase a twin-size and a full-size futon, 2 matching pillows for each futon, and 3 yd of matching material for the valance. The twin-size futon regularly costs $498, the full-size futon regularly costs $598, each matching pillow regularly costs $15, and each yard of material regularly costs $20. In a package deal Jessica was able to obtain the two futons, four pillows, and 3 yd of material for $972.
   a) How much did she save off the regular price?
   b) What percent of the regular price did she pay?
   c) What percent of the regular price did she save?

64. Mike owns two cars (a Ford Mustang and a Ford Escort), a house, and a rental apartment. He has auto insurance for both cars, a homeowner's policy, and a policy for the rental property. The costs of the policies are

     Mustang: $695 per year

     Escort: $650 per year

     Homeowner's: $412 per year

     Rental property: $597 per year

Mike is considering taking out a $1 million personal umbrella liability policy. The annual cost of the umbrella policy would be $392. However, if he has the umbrella policy he can lower the limits on parts of his auto policies and still have equal or better protection. If Mike purchases the umbrella policy he can reduce his premium on the Mustang by $60 per year and his premium on the Escort by 36%. If he purchases the umbrella policy and reduces the amount he pays for auto insurance, what is the net amount he is actually paying for the umbrella policy?

65. Karita is making a recipe that calls for $\frac{1}{3}$ cup of milk. She just dropped and broke her measuring cup. She is expecting guests in one-half hour and doesn't have the time to go to the store to buy a new one. However, she has a teaspoon measure, and from a cookbook she has she finds that

     3 tsp = 1 tbsp

     8 fluid oz = 1 cup

     $\frac{1}{2}$ fluid oz = 1 tbsp

How many teaspoons of milk will Karita use to get $\frac{1}{3}$ cup?

66. The world population on January 1, 1995, was approximately 5.6 billion. It is projected to increase to 8.5 billion by January 1, 2030. Estimate the average hourly increase in the world population from January 1, 1995, to January 1, 2030. (Don't forget leap year: 1996 and every 4 years thereafter.)

67. An airplane flies in a straight line from location A to location B and then back again. The engine speed is constant and no wind is blowing. Will the travel time be greater, less, or the same if, throughout both flights, at the same engine speed, a constant wind blows from location A toward location B?

68. A videotape can record 2 hr on short play, 4 hr on long play, or 6 hr on extra long play. After recording 32 min on short play and 44 min on long play, how many minutes can it record on extra long play?

69. Rectangle ABCD is made up entirely of squares. The black square has a side of 1 unit. Find the area of ABCD.

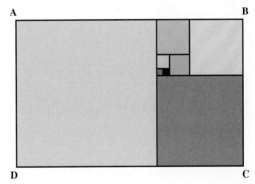

70. Peter, Paul, and Mary are three sports professionals. One is a tennis player, one is a golfer, and one is a skier. They live in three adjacent houses on City View Drive. From the following information determine which is the professional skier. (*Hint:* A table may be helpful.)

     Mary does not play tennis.

     Peter skis and plays tennis, but does not golf.

     The golfer and the skier live beside each other.

     Three years ago, Paul broke his leg skiing and has not tried it since.

     Mary lives in the last house.

     The golfer and the tennis player share a common backyard swimming pool.

71. Consider the following pattern.

$$1 + 2 = 3$$
$$4 + 5 + 6 = 7 + 8$$
$$9 + 10 + 11 + 12 = 13 + 14 + 15$$
$$16 + 17 + 18 + 19 + 20 = 21 + 22 + 23 + 24$$

What will row 50 look like? Can you generalize the result for any row of the triangle? Explain.

Research Activity

72. Select a problem that you need to solve. Use the four-step problem-solving procedure discussed in this section to help solve the problem. Consider those things that are constant (that you cannot change) and those things that are variables (that you can change). Discuss the problem-solving technique used and whether the problem has a single solution, many solutions, or no solution.

CHAPTER I SUMMARY

Key Terms

I.I

conjecture
counterexample
deductive reasoning (deduction)
divisible
inductive reasoning (induction)
scientific method

I.2

estimation

I.3

problem solving

Important Facts

Natural or counting numbers 1, 2, 3, 4, . . .

Guidelines for Problem Solving

1. Understand the problem.
2. Devise a plan to solve the problem.
3. Carry out the plan.
4. Check the results.

Chapter I Review Exercises

I.I*

In Exercises 1–8, use inductive reasoning to predict the next three numbers or figures in the pattern.

1. 3, 5, 7, 9, . . .
2. 1, 4, 9, 16, . . .
3. 4, −8, 16, −32, . . .
4. 5, 7, 10, 14, 19, . . .
5. 25, 24, 22, 19, 15, . . .
6. 1, 1, 2, 3, 5, 8, 13, . . .
7. , . . .
8. , . . .
9. Pick any number and multiply the number by 2. Add 10 to the product. Divide the sum by 2. Subtract 5 from the quotient.

a) What is the relationship between the number you started with and the final number?
b) Arbitrarily select some different numbers and repeat the process, recording the original number and the results.
c) Make a conjecture about the original number and the final number.
d) Prove, using deductive reasoning, the conjecture you made in part (c).

10. Choose a number between 1 and 20. Add 5 to the number. Multiply the sum by 6. Subtract 12 from the product. Divide the difference by 2. Divide the quotient by 3. Subtract the number you started with from the quotient. What is your answer? Try this process with a different number. Make a conjecture as to what your final answer will always be.

11. Find a counterexample to the statement "The difference between two squares is an odd number."

*The number in color indicates the section in which the material is covered.

1.2

In Exercises 12–20, estimate the answer. Your answers may vary from those given in the back of the book, depending on how you round to arrive at the answer.

12. $194{,}600 \times 2024$

13. $\dfrac{18{,}254.5}{624.3}$

14. $146.2 + 96.402 + 1.04 + 897 + 421$

15. 21% of 985

16. Estimate the distance from your wrist to your elbow and estimate the length of your foot. Which do you think is greater? With the help of a friend measure both lengths to determine which is longer.

17. Estimate the cost of eight pairs of socks if each pair cost $1.09.

18. Estimate the amount of a 6% sales tax on an item that costs $192.

19. Estimate your average walking speed in miles per hour if you walked 1.1 mi in 22 min.

20. Estimate the total cost of six grocery items that cost $2.49, $0.79, $1.89, $0.10, $2.19, $6.75.

21. The scale of the map is $\frac{1}{4}$ in. = 0.1 mi. Estimate the distance of the walking path indicated in red.

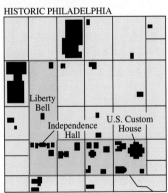

HISTORIC PHILADELPHIA

In Exercises 22 and 23, refer to the following graph.

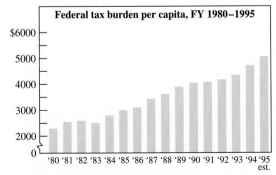

Federal tax burden per capita, FY 1980–1995

Source: Tax Foundation

22. Estimate the federal tax burden per capita in 1980 and in 1994.

23. On the basis of the trend illustrated by the graph make an estimate of the federal tax burden per capita in 1996.

24. If each square represents one square unit, estimate the size of the shaded area.

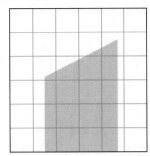

25. The scale of a model railroad is 1 in. = 12.5 ft. Estimate the size of an actual box car if this drawing is the same size as the model box car.

1.3

Solve the following problems.

26. An operator-assisted telephone call to another state costs $10.15 for 15 min. The same call dialed directly costs $0.60 for the first minute and $0.40 for each additional minute. How much money is saved by dialing direct?

27. Mr. Egan parked his car in a lot that charged $2.00 for the first hour and $0.90 for each additional hour. He left the car in the lot for 8 hr. How much change did he receive from a ten-dollar bill?

28. A six-pack of cola costs $3.45. A carton of 4 six-packs costs $12.60. How much will be saved by purchasing the carton rather than 4 individual six-packs?

29. The rental cost of a jet ski from Johnson's Ski Rental is $10 per 15 min, and the cost from Chris' Ski Rental is $25 per half hour. If you plan to rent the jet ski for 2 hr, which is the better deal, and by how much?

30. A taxicab charges $1.10 for the first 1/5 mi and 15 cents for each additional 1/5 mi. Determine the cost of a 10-mi trip.

31. Most insurance companies reduce premiums by 10% until age 25 for people who successfully pass a driver education course. A particular driver education course costs $60. Jason, who just turned 18, has auto insurance that costs $530 per year. By taking the driver education course, how much would he save in auto insurance premiums from the age of 18 until the age of 25?

32. If 1.5 milligrams (mg) of a medicine is to be given for 10 lb of body weight, how many milligrams should be given to a child who weighs 47 lb?

33. Banks will grant an applicant a mortgage if the monthly payments are not greater than 25 percent of the person's take-home pay. What is the maximum monthly mortgage payment you can make if your gross salary is $3800 a month and your payroll deductions are 30% of your gross salary?

34. New York City is on eastern standard time, St. Louis is on central standard time (1 hr earlier than eastern standard time), and Las Vegas is on Pacific standard time (3 hr earlier than eastern standard time). A flight leaves New York City at 9 A.M. eastern standard time, stops for 50 min in St. Louis, and arrives in Las Vegas at 1:35 P.M. Pacific time. How long is the plane actually flying?

35. The international date line is an imaginary line of longitude (from the North Pole to the South Pole) on the earth's surface between Japan and Hawaii in the Pacific Ocean. Crossing the line east to west adds a day to the present date. Crossing the line west to east subtracts a day. At 3:00 P.M. on July 25 in Hawaii, what is the time and date in Tokyo, Japan, which is four time zones to the west?

36. 1 in. = 2.54 cm.
a) How many square centimeters are in a square inch?
b) How many cubic centimeters are in a cubic inch?
c) How long is a centimeter in terms of inches?

37. A large drum filled with water is to be drained by using a small opening at the top. One 1-in. diameter hose, or two $\frac{1}{2}$-in. hoses, to be used in siphoning out the water, can be placed in the small opening. Would it be faster to use the one 1-in. hose or the two $\frac{1}{2}$-in. hoses?

38. If the following pattern is continued, how many dots will be in the hundredth figure?

39. How many different ways can the letters ABC be arranged?

40. How many different ways can the digits of the number 1144 be arranged?

41. a) Find the sum of the first three, four, five, and ten counting numbers.
b) Use the results in part (a) to develop a procedure that can be used to find the sum of the first n counting numbers. (*Hint:* The sum will equal the product of n and a second number expressed in terms of n, all divided by 2.)

c) Use the procedure in part (b) to find the sum of the first 20 counting numbers.
d) Use the same procedure to find the sum of the first 50 counting numbers.

42. Complete the magic square by using the numbers 6–21 exactly once.

21	7		18
10		15	
14	12	11	17
9	19		

43. Create a magic square by using the numbers 13, 15, 17, 19, 21, 23, 25, 27, and 29.

44. Create a magic square by using the numbers 20, 40, 60, 80, 100, 120, 140, 160, and 180.

45. A colony of microbes doubles in number every second. A single microbe is placed in a jar, and in an hour the jar is full. When was the jar half full?

46. Three friends check into a single room in a motel and pay $10 apiece. The room costs $25 instead of $30, so a clerk is sent to the room to give $5 back. The friends each take back $1, and the clerk is given $2 for his trouble. Now each of the friends paid $9, a total of $27, and the clerk received $2. What happened to the missing dollar?

47. Jim has four more brothers than sisters. How many more brothers than sisters does his sister Mary have?

48. Four women in a room have an average weight of 130 lb. A fifth woman who weighs 180 lb enters the room. Find the average weight of all five women.

49. A man who has a garden 10 meters square (10 m by 10 m) wants to know how many posts will be required to enclose his land. If the posts are placed exactly 1 m apart how many are needed? Disregard the thickness of the posts.

50. Can you place the letters A, B, C, D, and E in a 5 by 5 grid so that no letter appears twice in any row, in any column, or in any diagonal?

51. Could a person have $1.15 worth of change in his pocket and still not be able to give someone change for a dollar bill? If so, what coins might he have?

52. You have 13 coins, all of which look alike. Twelve coins weigh exactly the same, but the other one is heavier. You have a pan balance. Tell how to find the heavier coin in just three weighings.

53. Find the sum of the first 500 counting numbers. (*Hint:* Group in pairs.)

54. A red cup contains a pint of water, and a blue cup contains a pint of wine. A teaspoon of water is taken from the red cup and placed in the blue cup, and the contents of the blue cup are mixed thoroughly. A teaspoon of the mixture in the blue cup is returned to the red cup. Is there more wine in the red cup or more water in the blue cup?

55. On a balance scale, three green balls balance six blue balls, two yellow balls balance five blue balls, and six blue balls balance four white balls. How many blue balls are needed to balance four green, two yellow, and two white balls?

56. How many three-digit numbers greater than 100 are palindromes?

57. Describe the fifth figure and the *n*th figure.

58. Place the numbers 1–12 in the 12 circles so that the sum in each of the six rows and at the six points is 26. Use each number from 1 through 12 exactly once.

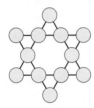

59. Often the name of a puzzle provides a clue to its solution that beginners overlook. The *alphamagic square* shown was invented by Lee Swallows, an electrical engineer from the Netherlands. Every row, column, and diagonal will add up to the same number (162). By taking a clue from the word *alphamagic*, can you discover another property that makes this an interesting magic square? (*Hint:* Count the number of letters in each number, when written out, and construct a magic square using those numbers.)

44	61	57
67	54	41
51	47	64

60. In how many ways can
   a) two people stand in a line?
   b) three people stand in a line?
   c) four people stand in a line?
   d) five people stand in a line?
   e) Using the results from parts (a)–(d), make a conjecture about the number of ways in which *n* people can stand in a line.

61. In each box place a single digit, 0–9, so that no digit repeats and the addition problem formed is correct.

CHAPTER 1 TEST

In Exercises 1 and 2, use inductive reasoning to determine the next three numbers in the pattern.

1. 0, 4, 8, 12, . . . **2.** 1, $\frac{1}{3}$, $\frac{1}{9}$, $\frac{1}{27}$, . . .

3. Pick any number, multiply the number by 5, and add 10 to the number. Divide the sum by 5. Subtract 1 from the quotient.
   a) What is the relationship between the number you started with and the final answer?
   b) Arbitrarily select some different numbers and repeat the process. Record the original number and the results.
   c) Make a conjecture about the relationship between the original number and the final answer.
   d) Prove, using deductive reasoning, the conjecture made in part (c).

In Exercises 4 and 5, estimate the answers.

4. 0.000417 × 930,000 **5.** $\dfrac{89,000}{0.00302}$

6. If each square represents one square unit, estimate the area of the shaded figure.

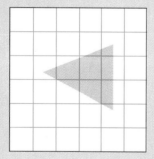

7. The following graph illustrates information obtained at the beginning of 1993.

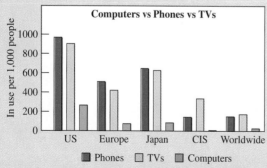

Computers vs Phones vs TVs

In use per 1,000 people

US Europe Japan CIS Worldwide

■ Phones ☐ TVs ■ Computers

Source: 1995 Information Please Almanac

a) Estimate the number of phones, TVs, and computers per 1000 people in the United States.
b) Estimate the difference between the number of TVs and the number of computers per 1000 people in the Commonwealth of Independent States (CIS).
c) Worldwide, the bar for TV is higher than the bar for telephones. Does this difference mean that there are more TVs in use in the world than telephones? Explain your answer.

8. A gas company charges $6.42 for the basic monthly fee, which includes the first 3 therms of gas used. It charges 59 cents for each additional therm used. If the Smiths' gas bill for December was $71.32, how many therms of gas did they use during that month?

9. If three cans of juice cost $1.20, how many cans of juice can be purchased for $10.00? (Only whole cans of juice may be purchased.)

10. How much time does a carpenter need to cut a 4 in. by 4 in. by 8 ft piece of treated lumber into four equal pieces if each cut takes $2\frac{1}{2}$ min?

11. In this photo of the Mona Lisa, 1506, by Leonardo da Vinci, 1 in. equals 12 in. on the actual painting. Find the dimensions of the actual painting.

12. A worker gets $11.25 per hour with time and a half for any time over 44-hr per week. If he works a 49-hr week and gets paid $529.38, by how much was he underpaid?

13. Create a magic square by using the numbers 5, 10, 15, 20, 25, 30, 35, 40, and 45. The sum of the numbers in every row, column, and diagonal must be 75.

14. Ilana drove from her home to the beach that is 30 mi from her house. The first 15 mi she drove at 60 mph, and the next 15 mi she drove at 30 mph. Would the trip take more, less, or the same time if she traveled the entire 30 mi at a steady 45 mph?

15. From the six numbers 2, 6, 8, 9, 11, and 13, pick five that, when multiplied, give 11,232.

16. One guess is off by 9, another guess is off by 17, and yet another guess is off by 31. How many beans are in the jar?

17. Nicole wants to purchase nine herb plants. Countryside Nursery has herbs that are on sale three for $3.99. Nicole has a coupon for 25% off an unlimited number of herb plants at the original price of $1.75 per plant.
  a) Determine the cost of purchasing nine plants at the sale price.
  b) Determine the cost of purchasing nine plants if the coupon is used.
  c) Which is the least expensive way to purchase the nine plants, and by how much?

18. In how many different ways can a panel of four on–off switches be set if no two adjacent switches may be off?

GROUP PROJECTS

Holiday Shopping

1. It is December First, and John needs to begin his holiday shopping. He intends to purchase gifts for his girlfriend Melissa, his mother Ruth, and his father Don. He doesn't want to spend more than a total of $325, including the 7% sales tax.

a) If John were to spend the $325 equally among the three people, approximate the amount that would be spent on each person.

b) If John were to spend the $325 equally among the three people, determine the maximum amount, *before tax*, that he could spend on each person and not exceed the maximum of $325, including tax.

c) John decides to get a new set of wrenches for his father. He sees the specific set he wants on sale at Sears. He calls four Sears stores to see if they have the set of wrenches in stock. They all reply that the set is out of stock. He decides that calling additional Sears stores is useless, for he believes that they will also tell him that the set of wrenches is out of stock. What type of reasoning did John use in arriving at his conclusion? Explain.

d) John finds an equivalent set of wrenches at a True Value hardware store. The set he is considering is a combination set that contains both standard U.S. size and metric size wrenches. Its regular price before tax is $62, but it is selling for 10% off its regular price. He can also purchase the same wrenches by purchasing two separate sets, one for U.S. standard size wrenches and the other for metric sizes. Each of these sets has a regular price, before tax, of $36, but both are on sale for 20% off their regular prices. Can John purchase the combination set or the two individual sets less expensively?

e) How much will John save, *after tax*, by using the less expensive method?

Going on Vacation

2. Bill and Kristen Url and their 4-year-old daughter Betty decide to go on a vacation. They live in San Francisco, California, and plan to drive to New Orleans, Louisiana.

a) Obtain a map that shows routes that they may take from San Francisco to New Orleans. Write directions for them from San Francisco to New Orleans via the shortest distance. Use major highways whenever possible.

b) Use the scale on the map to estimate the one-way distance to New Orleans.

c) If the Urls estimate that they will average 50 mph (including comfort stops), estimate the travel time, in hours, to New Orleans.

d) If the Urls want to travel about 400 miles per day, locate a town in the vicinity of where they will stop each evening.

e) If they begin each segment of the trip each day at 9 A.M., about what time will they look for a hotel each evening?

f) Use the information provided in parts (a)–(e) to estimate the time of day they will arrive in New Orleans.

g) Estimate the mileage of a typical mid-sized car and the cost per gallon of a gallon of regular unleaded gasoline. Then estimate the cost of gasoline for the Urls's trip.

h) Estimate the cost of a typical breakfast, a typical lunch, and a typical dinner for two adults and a 4-year-old child, and the cost of a typical motel room. Then estimate the total cost, including meals, gas, and lodging for the Urls's trip from San Francisco to New Orleans (one-way).

Problem Solving

3. Four acrobats who bill themselves as the "Tumbling Tumbleweeds" finish up their act with the amazing "Human Pillar," in which the acrobats form a tower, each one standing on the shoulders of the one below. Each acrobat (Ernie, Jed, Tex, and Zeke Tumbleweed) wears a different distinctive item of western garb (chaps, holster, Stetson hat, or leather vest) in the act. Can you identify the members of the "Human Pillar," from top to bottom, by name and apparel?

a) Jed Tumbleweed isn't on top, but he's somewhere above the man in the Stetson.

b) Zeke Tumbleweed doesn't wear the holster.

c) The man in the vest isn't on top.

d) The man in the chaps is somewhere above Tex but somewhere below Zeke.

Order	Name	Apparel
_____	_____	_____
_____	_____	_____
_____	_____	_____
_____	_____	_____

A Hot Dog Cart

4. Paula Peterson is a college student who wishes to earn extra income. She is considering purchasing a hot dog cart and selling hot dogs and soda by the beach in the afternoons and on the weekends. Before she purchases the cart she must consider many things including the expenses and the income associated with the business.

a) Make a comprehensive list of items associated with the business that have a fixed monthly expense. *Fixed expenses* are those that must be paid regardless of the number of items sold, such as monthly loan payments for the purchase of the cart and permits that may be required.

b) Make a comprehensive list of items associated with the business that have a variable expense. *Variable expenses* are those that vary with the number of items sold, such as the cost of hot dogs.

c) Paula estimates that she will work 100 hours a month and sell about 2500 hot dogs for $1.50 each and 5500 sodas for $1.00 each. What will Paula's estimated gross monthly income (before expenses) be?

Paula has developed a list of fixed monthly expenses and a list of variable expenses as follows.

Fixed Expenses	**Variable Expenses**
Permit: $50 per month	Hot dogs: 30 cents each
Upkeep and repairs of cart: $40 per month	Rolls: 15 cents each
	Soda: 50 cents per can
	Onions: 3 cents per hot dog (when requested)
Transportation: $40 per month	Sauerkraut: 4 cents per hot dog (when requested)
Loan payment for cart: $100 per month	Relish: 3 cents per hot dog (when requested)
	Mustard: 1 cent per hot dog (when requested)
	Catsup: 1 cent per hot dog (when requested)
	Straw: 1 cent per can of soda (when requested)
	Napkin: 1 cent each

d) Estimate the percent of people who will order the hot dog as indicated in the following chart.

Disregard other possible combinations. The sum of the percents should be 100%.

Hot dog	Percent
Plain	
with onions	
with sauerkraut	
with relish	
with onions and mustard	
with onions and catsup	
with sauerkraut and mustard	
with sauerkraut and catsup	
with relish and mustard	
with relish and catsup	
with onions, relish and mustard	
with onions, relish and catsup	

e) Based upon the information provided earlier estimate the monthly expenses for the soda and straws. Assume 60% of those purchasing soda will take a straw.

f) Estimate the monthly cost of napkins. Assume that a napkin is given with each hot dog and with each soda purchased.

g) Estimate the monthly expense of the hot dogs and the rolls. Assume each hot dog comes in a roll.

h) Using the percents you estimated in part d) estimate the individual monthly expenses of the onions, sauerkraut, relish, mustard, and catsup. Then estimate the total expenses of all items mentioned in this part.

i) Based upon all the information provided in this problem estimate Paula's total monthly variable expenses.

j) Estimate her total monthly fixed expenses.

k) Estimate her total monthly expenses by adding the expenses in part i) and j).

l) Based upon all the information provided, estimate Paula's *net monthly profit* for the month by subtracting her total monthly expenses (part k) from her total monthly income (part c).

m) Estimate Paula's net hourly income.

n) To purchase the cart Paula agreed to pay $300 plus 12 monthly payments of $100. Estimate the number of months when the sum of her net monthly profits will equal the cost of the cart.

Chapter 2 Sets

One of the most basic human impulses is to sort and classify things. People may be categorized by the language they speak, by whether they eat rice or bread, by whether they are brown-haired or blond. Each of these categories allows us to sort a large group of people or objects into smaller groupings.

Consider yourself, for example. How many different sets are you a member of? You might start with some simple categories, such as whether you are male or female, your age group, and the town, city, or state you live in. Then you might think about your family's ethnic group, socioeconomic group, and nationality. Then you could think about your political affiliation, clubs or special-interest groups, and college or university. These are but some of the many ways you could describe yourself to other people.

Of what use is this activity of categorization? As you will see in this chapter, putting elements into sets helps you order and arrange your world. It allows you to deal with large quantities of information. Set building is a learning tool that helps answer the question "What are the characteristics of this group?"

Companies often do market research to determine their customers' profiles. Such customer surveys often ask the age group, the income range, and how the customers heard about the product.

Early civilizations looked at the stars and saw images of things that had meaning for them, like the mythical figure Orion shown above. Astronomers studying the skies today categorize the stars and planets they see by more complex characteristics—for example, a star is categorized by its age, magnitude, and temperature.

Many games depend on forming sets. One of the oldest games in the world is the African game of Mancala, a game where two players place seeds in cells and then use strategic moves to capture their opponent's seeds while protecting their own.

Sets underlie other mathematical topics, such as logic and abstract algebra. In fact, the book *Eléments de Mathématique*, written by a group of French mathematicians under the pseudonym Nicolas Bourbaki, states, "Nowadays it is possible, logically speaking, to derive the whole of known mathematics from a single source, the theory of sets." ■

2.1 SET CONCEPTS

We encounter sets in many different ways every day of our lives. A **set** is a collection of objects, which are called **elements** or **members** of the set. For example, the United States is a collection or set of 50 states. The 50 individual states are the members or elements of the set that is called the United States.

A set is **well defined** if its contents can be clearly determined. The set of justices presently serving on the U.S. Supreme Court is a well-defined set, since its contents, the justices, can be named. The set of the three best cars is not a well-defined set because the word *best* is interpreted differently by different people. In this text we use only well-defined sets.

Three methods are commonly used to indicate a set: (1) description, (2) roster form, and (3) set-builder notation.

The method of indicating a set by **description** is illustrated in Example 1.

EXAMPLE 1

Write a description of the set containing the elements fall, winter, spring, and summer.

Solution: The set is the seasons of the year.   ▲

Listing the elements of a set inside a pair of braces, { }, is called **roster form**. The braces are an essential part of the notation because they identify the contents as a set. For example, {1, 2, 3} is notation for the set whose elements are 1, 2, and 3, but (1, 2, 3) and [1, 2, 3] are not sets because parentheses and brackets do not indicate a set. For a set written in roster form, commas separate the elements of the set. The order in which the elements are listed is not important.

Sets are generally named with capital letters. For example, the name commonly selected for the set of **natural numbers** or **counting numbers** is N.

$$\boxed{\begin{array}{c} \textbf{Natural Numbers} \\ N = \{1, 2, 3, 4, 5, \ldots\} \end{array}}$$

The three dots after the 5, called an *ellipsis*, indicate that the elements in the set continue in the same manner. An ellipsis followed by a last element indicates that the elements continue in the same manner up to and including the last element. This notation is illustrated in Example 2(b).

EXAMPLE 2

Express the following in roster form.

a) Set A is the set of natural numbers less than 4.

b) Set B is the set of natural numbers less than or equal to 80.

c) Set P is the set of planets in the earth's solar system.

Solution:

a) The natural numbers less than 4 are 1, 2, and 3. Thus set A in roster form is $A = \{1, 2, 3\}$.

The planets of the earth's solar system.

b) $B = \{1, 2, 3, 4, \ldots, 80\}$. The 80 after the ellipsis indicates that the elements continue in the same manner through the number 80.

c) $P = \{$Mercury, Venus, Earth, Mars, Jupiter, Saturn, Uranus, Neptune, Pluto$\}$ ▲

EXAMPLE 3

Express the following in roster form.

a) The set of natural numbers between 5 and 8.

b) The set of natural numbers between 5 and 8, inclusive.

Solution:

a) $A = \{6, 7\}$

b) $B = \{5, 6, 7, 8\}$. Note that the word *inclusive* indicates that the values of 5 and 8 are included in the set. ▲

The symbol $\in$ is used to indicate membership in a set. In Example 3, since 6 is an element of set A, we write $6 \in A$. This may also be written $6 \in \{6, 7\}$. We may also write $8 \notin A$, meaning that 8 is not an element of set A.

Set-builder notation (sometimes called *set-generator notation*) may be used to symbolize a set. Set-builder notation is frequently used in algebra. The following example illustrates its form.

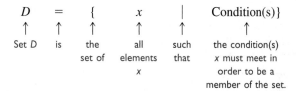

Consider $E = \{x \mid x \in N \text{ and } x > 10\}$. The statement is read: "Set E is the set of all the elements x such that x is a natural number and x is greater than 10." The conditions that x must meet in order to be a member of the set are $x \in N$, which means that x must be a natural number, and $x > 10$, which means that x must be greater than 10. The numbers that meet both conditions are the set of natural numbers greater than 10. The set in roster form is

$$E = \{11, 12, 13, 14, \ldots\}.$$

EXAMPLE 4

a) Write set $B = \{1, 2, 3\}$ in set-builder notation.

b) Write, in words, how you would read set B in set-builder notation.

Solution:

a) Since set B consists of the natural numbers less than 4, we write

$$B = \{x \mid x \in N \text{ and } x < 4\}.$$

Another acceptable answer is $B = \{x \mid x \in N \text{ and } x \leq 3\}$.

b) Set B is the set of all elements x such that x is a natural number and x is less than 4. ▲

EXAMPLE 5

a) Write set D = {Buick, Cadillac, Chevrolet, Oldsmobile, Pontiac, Saturn} in set-builder notation.

b) Write in words how you would read set D in set-builder notation.

Solution:

a) D = {$x | x$ is a car produced by General Motors Corporation}

b) Set D is the set of all elements x such that x is a car produced by General Motors Corporation. ▲

EXAMPLE 6

Write set A = {$x | x \in N$ and $3 \le x < 9$} in roster form.

Solution: A = {3, 4, 5, 6, 7, 8} ▲

EXAMPLE 7

Set C = {$x | x$ is one of the five most populated cities in the United States in 1994}. Use the following information to write set C in roster form.

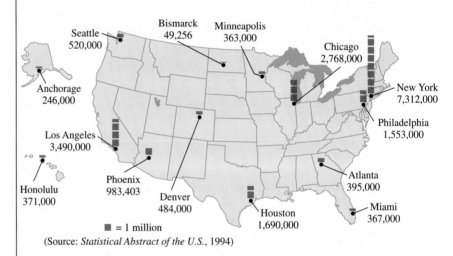

(Source: *Statistical Abstract of the U.S.*, 1994)

Solution: C = {New York, Los Angeles, Chicago, Houston, Philadelphia} ▲

A set is said to be **finite** if it either contains no elements or the number of elements in the set is a natural number. The set B = {2, 4, 6, 8, 10} is a finite set because the number of elements in the set is 5, and 5 is a natural number. A set that is not finite is said to be **infinite**. The set of counting numbers is one example of an infinite set. Infinite sets are discussed in more detail in Section 2.6.

Another important concept is equality of sets.

> Set A is **equal** to set B, symbolized by $A = B$, if and only if they contain exactly the same elements.

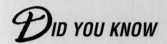
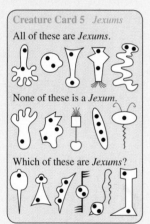
For example, if set $A = \{1, 2, 3\}$ and set $B = \{3, 1, 2\}$, then $A = B$ because they contain exactly the same elements. The order of the elements in the set is not important. If two sets are equal, both must contain the same number of elements. The number of elements in a set is called its *cardinal number*.

> The **cardinal number** of set A, symbolized by $n(A)$, is the number of elements in set A.

Both set $A = \{1, 2, 3\}$ and set $B = \{\text{England, France, Japan}\}$ have a cardinal number of 3; that is, $n(A) = 3$, and $n(B) = 3$. We can say that set A and set B both have a cardinality of 3.

Two sets are said to be *equivalent* if they contain the same number of elements.

> Set A is **equivalent** to set B if and only if $n(A) = n(B)$.

Any sets that are equal must also be equivalent. However, not all sets that are equivalent are equal. The sets $D = \{a, b, c\}$ and $E = \{\text{apple, orange, pear}\}$ are equivalent, since both have the same cardinal number, 3. However, because the elements differ the sets are not equal.

Two sets that are equivalent or have the same cardinality can be placed in one-to-one correspondence. Set A and set B can be placed in one-to-one correspondence if every element of set A can be matched with exactly one element of set B and every element of set B can be matched with exactly one element of set A. For example there is a one-to-one correspondence between the student names on a class list and the student identification numbers because we can match each name with a student identification number.

Consider set C, cowboys, and set H, horses:

$$C = \{\text{Roy Rogers, Gene Autry, Lone Ranger}\},$$
$$H = \{\text{Trigger, Silver, Champion}\}.$$

Two different one-to-one correspondences for sets C and H follow.

$$C = \{\text{Roy Rogers, Gene Autry, Lone Ranger}\},$$
$$H = \{\text{Trigger, Silver, Champion}\}.$$

$$C = \{\text{Roy Rogers, Gene Autry, Lone Ranger}\},$$
$$H = \{\text{Trigger, Silver, Champion}\}.$$

Other one-to-one correspondences between sets C and H are possible. Do you know which cowboy rode which horse?

Null or Empty Set

Some sets do not contain any elements—for example, the set of zebras that are in this room.

The set that contains no elements is called the **empty set** or **null set**, and is symbolized by { } or ∅.

Note that {∅} is not the empty set. This set contains the element ∅ and has a cardinality of 1.

EXAMPLE 8

Indicate the set of natural numbers that satisfies the equation $x + 2 = 0$.

Solution: The values that satisfy the equation are those that make the equation a true statement. Only the number -2 satisfies this equation. Because -2 is not a natural number, the solution set of this equation is { } or ∅.   ▲

Universal Set

Another important set is a *universal set*.

A **universal set**, symbolized by U, is a set that contains all the elements for any specific discussion.

When a universal set is given, only the elements in the universal set may be considered when working the problem. If, for example, the universal set for a particular problem is defined as $U = \{1, 2, 3, 4, \ldots, 10\}$, then only the natural numbers 1 through 10 may be used in that problem.

Section 2.1 Exercises

In Exercises 1–12, answer each question with a complete sentence.

1. What is a set?
2. What is an ellipsis and how is it used?
3. What are the three ways that a set can be written? Give an example of each.
4. What is a finite set?
5. What is an infinite set?
6. What are equal sets?
7. What is the cardinal number of a set?
8. What are equivalent sets?
9. Write the set of counting numbers in roster form.
10. What does a one-to-one correspondence of two sets mean?
11. What is the empty set?
12. What is a universal set?

In Exercises 13–18, determine whether each set is well defined.

13. The set of the best mid-sized automobiles on the market today.
14. The set of people who own large dogs.
15. The set of the four states in the United States having the largest areas.
16. The set of countries that have a common border with Germany.
17. The set of students in this class who own a computer.
18. The set of the most interesting students in this class.

In Exercises 19–24, determine whether each set is finite or infinite.

19. $\{1, 3, 5, 7, \ldots\}$
20. The set of even numbers greater than 12.
21. The set of multiples of 5 between 0 and 50.

22. The set of fractions between 1 and 2.

23. The set of odd numbers greater than 12.

24. The set of crickets chirping in the town park on a warm July 4 evening at 10:00 P.M.

In Exercises 25–34, express each set in roster form. You may need to use a world almanac or some other reference source.

25. The set of continents in the world.

26. The set of countries in South America that share a common border with Chile.

27. The set of natural numbers between 15 and 275.

28. $B = \{x \mid x \in N \text{ and } x \text{ is even}\}$

29. $C = \{x \mid x + 7 = 15\}$

30. $D = \{x \mid x \in N \text{ and } 13 - x = 20\}$

31. The set of states west of the Mississippi that have a common border with the state of Florida.

32. The set of states in the United States that have no common border with any other state.

33. $E = \{x \mid x \in N \text{ and } 5 \le x < 80\}$

34. The set of states in the United States whose names begin with the letter A.

In Exercises 35–42, express each set in set-builder notation.

35. $A = \{1, 2, 3, 4, 5, 6, 7, 8, 9\}$

36. $B = \{4, 5, 6, 7, 8\}$

37. $C = \{2, 4, 6, 8, \ldots\}$

38. $D = \{5, 10, 15, 20, \ldots\}$

39. E is the set of odd natural numbers.

40. A is the set of national holidays in the United States in September.

41. G is the set of the three automobile manufacturers in the United States that manufacture the greatest number of pickup trucks.

42. $F = \{15, 16, 17, \ldots, 100\}$

In Exercises 43–50, write a description of each set.

43. $A = \{1, 2, 3, 4, 5, 6, 7, 8, 9\}$

44. $D = \{4, 8, 12, 16, 20, \ldots\}$

45. $L = \{\text{Superior, Michigan, Huron, Erie, Ontario}\}$

46. $B = \{w, x, y, z\}$

47. $S = \{\text{Bashful, Doc, Dopey, Grumpy, Happy, Sleepy, Sneezy}\}$

48. $C = \{\text{Big Bird, Ernie, Bert, Cookie Monster, Oscar the Grouch, Betty Lou}\}$

49. $E = \{x \mid x \in N \text{ and } x > 18 \text{ and } x < 7\}$

50. $F = \{x \mid x \in N \text{ and } x > 4 \text{ and } x < 17\}$

In Exercises 51–58, state whether each is true or false. If false, give the reason.

51. $\{a\} \in \{a, b, c, d, e, f\}$

52. $g \in \{a, b, c, d, e, f\}$

53. $d \in \{a, b, c, d, e, f\}$

54. Japan $\in \{x \mid x \text{ is a country on the North American continent}\}$

55. $12 \notin \{x \mid x \in N \text{ and } x \text{ is odd}\}$

56. IBM $\in \{x \mid x \text{ is a manufacturer of computers}\}$

57. Peter Pan $\notin \{x \mid x \text{ is a brand of peanut butter}\}$

58. $2 \in \{x \mid x \text{ is an odd natural number}\}$

In Exercises 59–62, for the sets $A = \{2, 4, 6, 8\}$, $B = \{1, 3, 7, 9, 13, 21\}$, $C = \{\ \}$, and $D = \{\#, \&, \%, \square, *\}$, determine

59. $n(A)$

60. $n(B)$

61. $n(C)$

62. $n(D)$

In Exercises 63–68, determine whether the pairs of sets are equal, equivalent, both, or neither.

63. $A = \{5, 6, 7\}$, $B = \{6, 5, 7\}$

64. $A = \{8, 9, 10\}$, $B = \{8, 7, 6\}$

65. $A = \{\text{Donald Duck, Goofy, Mickey Mouse}\}$
    $B = \{\text{Walt Disney, Mickey Mouse, John Wayne, Curly}\}$

66. A is the set of faculty members at the University of Wisconsin.
    B is the set of faculty members at the University of Wisconsin teaching English.

67. A is the set of counties in the state of Oklahoma.
    B is the set of county seats in the state of Oklahoma.

68. A is the set of dog catchers in the United States.
    B is the set of dogs in the United States.

69. Set-builder notation is often more versatile and efficient than listing a set in roster form. This versatility is illustrated with the two sets

$$A = \{x \mid x \in N \text{ and } x > 2\}$$
$$B = \{x \mid x > 2\}$$

a) Write a description of set A and set B.
b) Explain the difference between set A and set B. (*Hint:* Is $4\frac{1}{2} \in A$? Is $4\frac{1}{2} \in B$?)
c) Write set A in roster form.
d) Can set B be written in roster form? Explain your answer.

70. For sets

$$A = \{x \mid 2 < x \leq 5 \text{ and } x \in N\}$$

and

$$B = \{x \mid 2 < x \leq 5\}$$

a) write a description of set A and set B.
b) explain the difference between set A and set B.
c) write set A in roster form.
d) Can set B be written in roster form? Explain your answer.

A cardinal number answers the question "How many?" An **ordinal number** *describes the relative position that an element occupies. For example, Molly's desk is the third desk from the aisle.*

In Exercises 71–74, determine whether the number used is a cardinal number or an ordinal number.

71. There are 34 students in the class.

72. Study the chart on page 25 in the book.

73. Lincoln was the sixteenth president of the United States.

74. Emily paid $35 for her new blouse.

75. Describe three sets of which you are a member.

76. Describe three sets that have no members.

77. Write a short paragraph explaining why the universal set and the empty set are necessary in the study of sets.

Problem Solving/Group Activities

78. a) In a given exercise, a universal set is not specified, but we know that Bill Clinton is a member of the universal set. Describe five different possible universal sets of which Bill Clinton is a member.
b) Write a description of one set that includes all the universal sets in part (a).
c) Are all the universal sets in part (a) also universal sets in part (b)? Explain.

79. a) Have each person in your group describe four universal sets of which they are a member.
b) Write a description of one set that includes all the universal sets of all the members of your group in part (a).
c) Are all the universal sets in part (a) also universal sets in part (b)? Explain.

Research Activity

80. Georg Cantor is recognized as the founder and a leader in the development of set theory. Do research and write a paper on his life and his contributions to set theory and the field of mathematics. References include history of mathematics books and encyclopedias.

2.2 SUBSETS

In our complex world we often break down larger sets into smaller more manageable sets, called subsets. For example, consider the totality of human knowledge as a universal set. Within this universal set are certain broad areas of study: physical science, social science, and liberal arts. Each of these areas may be considered a subset. The subsets can be further separated. For example, the physical sciences can be divided into smaller subsets that include mathematics, biology, physics, chemistry, and geology.

> Set A is a **subset** of set B, symbolized by $A \subseteq B$, if and only if all the elements of set A are also elements of set B.

The symbol $A \subseteq B$ indicates that "Set A is a subset of set B." The symbol $\nsubseteq$ is used to indicate "is not a subset." Thus, $A \nsubseteq B$ indicates that set A is not a subset of set B. *To show that set A is not a subset of set B we must find at least one element of set A that is not an element of set B.*

┌─ **EXAMPLE I**

Determine whether set A is a subset of set B.

a) $A = \{\text{French, German, Russian}\}$
  $B = \{\text{Italian, French, Spanish, German, Russian, Arabic}\}$

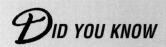

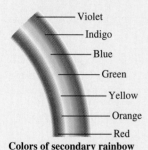
b) $A = \{2, 3, 4, 5\}$ $B = \{2, 3\}$

c) $A = \{x \mid x \text{ is a student at the University of Missouri}\}$
$B = \{x \mid x \text{ is a student at Clark County College}\}$

d) $A = \{\text{Snickers, Baby Ruth, Power House}\}$
$B = \{\text{Power House, Snickers, Baby Ruth}\}$

Solution:

a) All the elements of set A are contained in set B, so $A \subseteq B$.

b) The elements 4 and 5 are in set A but not in set B, so $A \nsubseteq B$ (A is not a subset of B). In this example, however, all the elements of set B are contained in set A; therefore $B \subseteq A$.

c) There are students in set A who are not in set B, so $A \nsubseteq B$.

d) All the elements of set A are contained in set B, so $A \subseteq B$. Note that set A = set B.

Proper Subsets

> Set A is a **proper subset** of set B, symbolized by $A \subset B$, if and only if all the elements of set A are elements of set B and set $A \neq$ set B (that is, set B must contain at least one element not in set A).

Consider the sets $A = \{\text{red, blue, yellow}\}$ and $B = \{\text{red, orange, yellow, green, blue, violet}\}$. Set A is a *subset* of set B, $A \subseteq B$, because every element of set A is also an element of set B. Set A is also a *proper subset* of set B, $A \subset B$, because sets A and B are not equal. Now consider $C = \{\text{car, bus, train}\}$ and $D = \{\text{train, car, bus}\}$. Set C is a subset of set D, $C \subseteq D$, because every element of set C is also an element of set D. However, set C is not a proper subset of set D, $C \not\subset D$, because set C and set D are equal sets.

EXAMPLE 2

Determine whether set A is a proper subset of set B.

a) $A = \{\text{Ford, Dodge, Buick}\}$
$B = \{\text{Ford, Chevrolet, Dodge, Plymouth, Chrysler, Buick}\}$

b) $A = \{a, b, c, d\}$ $B = \{a, c, b, d\}$

Solution:

a) All elements of set A are contained in set B, and sets A and B are not equal; thus $A \subset B$.

b) Set A = set B, so $A \not\subset B$. (However, $A \subseteq B$.)

Every set is a subset of itself, but no set is a proper subset of itself. For all sets A, $A \subseteq A$, but $A \not\subset A$. For example, if $A = \{1, 2, 3\}$ then $A \subseteq A$ because every element of set A is contained in set A. But $A \not\subset A$ because set A = set A.

Let $A = \{\ \}$ and $B = \{1, 2, 3, 4\}$. Is $A \subseteq B$? In order to show $A \nsubseteq B$, you must find at least one element of set A that is not an element of set B. As this cannot be done, $A \subseteq B$ must be true. Using the same reasoning, we can show that the empty set is a subset of every set, including itself.

EXAMPLE 3

Determine whether the following are true or false.

a) $3 \in \{3, 4, 5\}$ b) $\{3\} \in \{3, 4, 5\}$

c) $\{3\} \in \{\{3\}, \{4\}, \{5\}\}$ d) $\{3\} \subseteq \{3, 4, 5\}$

e) $3 \subseteq \{3, 4, 5\}$ f) $\{\ \} \subseteq \{3, 4, 5\}$

Solution:

a) $3 \in \{3, 4, 5\}$ is a true statement because 3 is a member of the set $\{3, 4, 5\}$.

b) $\{3\} \in \{3, 4, 5\}$ is a false statement because $\{3\}$ is a set, and the set $\{3\}$ is not an element of the set $\{3, 4, 5\}$.

c) $\{3\} \in \{\{3\}, \{4\}, \{5\}\}$ is a true statement because $\{3\}$ is an element in the set. The elements of the set $\{\{3\}, \{4\}, \{5\}\}$ are themselves sets.

d) $\{3\} \subseteq \{3, 4, 5\}$ is a true statement because every element of the first set is an element of the second set.

e) $3 \subseteq \{3, 4, 5\}$ is a false statement because the 3 is not in braces, so it is not a set, and thus cannot be a subset. The 3 is an element of the set as indicated in part (a).

f) $\{\ \} \subseteq \{3, 4, 5\}$ is a true statement because the empty set is a subset of every set.

▲

Number of Subsets

How many distinct subsets can be made from a given set? The empty set has no elements and has exactly one subset, the empty set. A set with one element has two subsets. A set with two elements has four subsets. A set with three elements has eight subsets. This information is illustrated in Table 2.1. How many subsets will a set with four elements contain?

Table 2.1

Set	Subsets	Number of Subsets
$\{\ \}$	$\{\ \}$	$1 = 2^0$
$\{a\}$	$\{a\},$ $\{\ \}$	$2 = 2^1$
$\{a, b\}$	$\{a, b\}$ $\{a\}, \{b\},$ $\{\ \}$	$4 = 2 \times 2 = 2^2$
$\{a, b, c\}$	$\{a, b, c\}$ $\{a, b\}, \{a, c\}, \{b, c\}$ $\{a\}, \{b\}, \{c\}$ $\{\ \}$	$8 = 2 \times 2 \times 2 = 2^3$

By continuing this table with larger and larger sets, we can develop a general formula for finding the number of distinct subsets that can be made from any given set.

The **number of distinct subsets** of a finite set A is 2^n, where n is the number of elements in set A.

EXAMPLE 4

a) Determine the number of distinct subsets for the set {H, E, A, T}.

b) List all the distinct subsets for the set {H, E, A, T}.

c) How many of the distinct subsets are proper subsets?

With seven distinct Scrabble tiles, there are 5040 different ways the tiles can be arranged. The four letters H, E, A, T can be arranged in 24 distinct ways.

Solution:

a) Since the number of elements in the set is 4, the number of distinct subsets is $2^4 = 2 \times 2 \times 2 \times 2 = 16$.

b)

{H, E, A, T}	{H, E, A}	{H, E}	{H}	{ }
	{H, E, T}	{H, A}	{E}	
	{H, A, T}	{H, T}	{A}	
	{E, A, T}	{E, A}	{T}	
		{E, T}		
		{A, T}		

c) There are 15 proper subsets. Every subset except {H, E, A, T} is a proper subset. ▲

EXAMPLE 5

Alvaro is purchasing a new computer. The model that he has selected can be purchased with no optional features or with as many as eight optional features. How many different variations of this computer can be made?

Solution: One technique used in problem solving is to consider similar problems that you have solved previously. If you think about this problem you will realize that this problem is the same as "How many distinct subsets can be made from a set with eight elements?" You know how to answer this question. The computer can have zero optional features, or any one of eight optional features, or any two of the eight optional features, and so on, up to all eight optional features. The number of different variations of the computer is the same as the number of possible subsets of a set that has eight elements. There are therefore 2^8 or 256 possible distinct variations of the computer. ▲

DID YOU KNOW

THE LADDER OF LIFE

Scientists use sets to classify and categorize knowledge. In biology, the science of classifying all living things is called *taxonomy* and was probably practiced by the earliest cave-dwellers. Over 2000 years ago Aristotle formalized animal classification with his "ladder of life": higher animals, lower animals, higher plants, lower plants.

Contemporary biologists use a system of classification called the Linnaean system, named after Swedish biologist Carolus Linnaeas (1707–1778). The Linnaean system starts with the smallest unit (member) and assigns it to a specific genus (set) and species (subset). ■

Even more general groupings of living things are made according to shared characteristics. The groupings, from most general to most specific are: kingdom, phylum, class, order, family, genus, and species.

A zebra, *Equus burchelli*, is a member of the genus, *Equus*, as is the horse, *Equus caballus*. Both the zebra and the horse are members of the universal set called the kingdom of animals and the same family, Equidae: however, they are members of different species (*E. burchelli* and *E. caballus*).

Section 2.2 Exercises

In Exercises 1–6, answer each question with a complete sentence.

1. What is a subset?

2. What is a proper subset?

3. Explain the difference between a subset and a proper subset.

4. What is the formula for determining the number of distinct subsets for a set with *n* distinct elements?

5. What is the formula for determining the number of distinct proper subsets for a set with *n* distinct elements?

6. Can any set be a proper subset of itself? Explain.

In Exercises 7–22, answer true or false. If false, give the reason.

7. Bach $\subseteq$ {Bach, Beethoven, Brahms}

8. { } $\in$ {red, green, blue}

9. { } $\subseteq$ {Curly, Larry, Moe}

10. red $\subset$ {red, green, blue}

11. 3 $\notin$ {2, 4, 6}

12. {3, 5, 7} $\subseteq$ {5, 6, 7}

13. { } = {$\varnothing$}

14. {1} $\subseteq$ {1, 5, 9}

15. { } = $\varnothing$

16. $\{1, 5, 9\} \subseteq \{1, 9, 5\}$

17. $\{5\} \in \{5, 6, 7\}$

18. $\{1, 5, 9\} \subset \{1, 9, 5\}$

19. $\{ \ \} \subset \{ \ \}$

20. $\{1\} \in \{\{1\}, \{2\}, \{3\}\}$

21. $\{3, 4, 5\} \not\subset \{5, 4, 3\}$

22. $\{1, 3, 7\} \not\subseteq \{1, 7, 3\}$

In Exercises 23–30, determine whether A = B, A ⊆ B, B ⊆ A, A ⊂ B, B ⊂ A, or none of these. (There may be more than one answer.)

23. $A = \{b, c, e, f\}$
$B = \{c, f\}$

24. Set A is the set of states east of the Mississippi River. Set B is the set of states east of the Pacific Ocean.

25. $A = \{x \,|\, x \in N \text{ and } x < 8\}$
$B = \{x \,|\, x \in N \text{ and } 1 \leq x \leq 7\}$

26. $A = \{x \,|\, x \text{ is a color of the rainbow}\}$
$B = \{\text{blue, green, orange}\}$

27. $A = \{1, 3, 5, 7, 9\}$
$B = \{3, 9, 5, 7, 6\}$

28. $A = \{x \,|\, x \text{ is a sport that uses a ball}\}$
$B = \{\text{basketball, soccer, tennis}\}$

29. Set A is the set of natural numbers between 3 and 8. Set B is the set of natural numbers greater than 3, but less than 8.

30. Set A is the set of white keys on a piano.
Set B is the set of keys on a piano.

In Exercises 31–35, list all the subsets of the sets given.

31. $D = \varnothing$

32. $A = \{\bigcirc\}$

33. $B = \{\text{sheep, goat}\}$

34. $C = \{\text{apple, peach, nectarine}\}$

35. For set $A = \{a, b, c, d\}$
a) list all the subsets of set A.
b) state which of the subsets in part (a) are not proper subsets of set A.

36. A set contains eight elements.
a) How many subsets does it contain?
b) How many proper subsets does it contain?

In Exercises 37–48, if the statement is true for all sets A and B, write true. *If it is not true for all sets A and B, write* false. *Assume that A ≠ ∅, U ≠ ∅, and A ⊂ U.*

37. If $A \subseteq B$, then $A \subset B$.

38. If $A \subset B$, then $A \subseteq B$.

39. $A \subset A$

40. $A \subseteq A$

41. $\varnothing \subseteq A$

42. $\varnothing \subset A$

43. $A \subseteq U$

44. $\varnothing \subset \varnothing$

45. $U \subseteq \varnothing$

46. $\varnothing \subset U$

47. $\varnothing \subseteq \varnothing$

48. $U \subset \varnothing$

49. If $E \subseteq F$ and $F \subseteq E$, what other relationship exists between E and F? Explain.

50. A game warden in South Africa has put together a tourist display of zebra with distinctive markings. He has collected five zebra: an albino (all white), a common zebra with black stripes on a white background, a dappled zebra with spots, an almost all-black zebra, and a zebra with white stripes on a black background. (Yes, they all do exist!) If he decides to display only three at a time in order to keep the tourists coming back, how many groups of three can he create using the five coat patterns?

51. The Schmidt family is purchasing a new clothes washer. The machine that they are selecting can be purchased with no extra features or with as many as five extra features. For the appliance store to display machines with all the possible combinations of features, how many washers must it have on display?

52. Customers ordering hamburgers at Vic and Irv's Hamburger stand are always asked, ''What do you want on it?'' The choices are members of the set {catsup, mustard, relish, hot sauce, onions, lettuce, tomato}. How many different variations are there for ordering a hamburger?

53. A person can order a new car with some, all, or none of the following set of options: {air conditioning, power windows, CD player, leather interior, alarm system, sun roof}. How many different variations of the set of options are possible?

54. At Colombine's Pizzeria a pizza can be ordered with some, all, or none of the following set of toppings: {cheese, pepperoni, peppers, mushrooms, anchovies, olives, sausage}. How many different variations are there for ordering a pizza?

55. For the set $D = \{a, b, c\}$
a) is a an element of set D? Explain.
b) is c a subset of set D? Explain.
c) is $\{a, b\}$ a subset of set D? Explain.

56. How can you determine whether the set of boys is equivalent to the set of girls at a roller skating rink?

57. Describe a set that is equivalent to the set of racing cars in the first lap of the 1995 Indianapolis 500 race.

Problem Solving/Group Activities

58. The executive committee of the student senate consists of Ashton, Bailey, Katz, and Snyder. Each member of the executive committee has exactly one vote (no abstentions), and a simple majority of the committee is needed to pass or reject a motion. If a motion is neither passed nor rejected, then it is considered blocked and will be considered again. Determine the number of specific ways the members can vote so that a motion can be (a) passed, (b) rejected, or (c) blocked.

59. A political science class is trying to determine the outcome of a local election by obtaining a sample of the voting members of the community. Devise a plan to partition the voting members in the community into sets so that each segment of the community can be sampled. The sets should be discrete (nonoverlapping sets)—for example, age or income.

2.3   VENN DIAGRAMS AND SET OPERATIONS

Figure 2.1

A useful technique for picturing set relationships is the Venn diagram, named for the English mathematician John Venn (1834–1923). Venn invented the diagrams and used them to illustrate ideas in his text on symbolic logic, published in 1881.

In a Venn diagram, a rectangle usually represents the universal set, U. The set of points inside the rectangle may be divided into subsets of the universal set. The subsets are usually represented by circles. In Fig. 2.1 the circle labeled A represents set A, which is a subset of the universal set.

Two sets may be represented in a Venn diagram in any of four different ways (see Fig. 2.2). Two sets A and B are **disjoint**, when they have no elements in common. Two disjoint sets A and B are illustrated in Fig. 2.2(a). If set A is a proper subset of set B, $A \subset B$, the two sets may be illustrated as in Fig. 2.2(b). If set A contains exactly the same elements as set B, that is, $A = B$, the two sets may be illustrated as in Fig. 2.2(c). Two sets A and B with some elements in common are shown in Fig. 2.2(d), which is regarded as the most general form of the diagram.

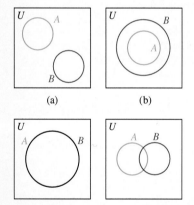

(a)   (b)

(c)   (d)

Figure 2.2

If we label the regions of the diagram in Fig. 2.2(d) using I, II, III, and IV, then we can illustrate the four possible cases with this one diagram, Fig. 2.3.

Case 1: Disjoint Sets When sets A and B are disjoint, they have no elements in common. Therefore region II of Fig. 2.3 is empty.

Case 2: Subsets When $A \subseteq B$, every element of set A is also an element of set B. Thus there can be no elements in region I of Fig. 2.3. However, if $B \subseteq A$, then region III of Fig. 2.3 is empty.

Case 3: Equal Sets When set $A = $ set B, all the elements of set A are elements of set B, and all the elements of set B are elements of set A. Thus regions I and III of Fig. 2.3 are empty.

Case 4: Overlapping Sets When sets A and B have elements in common, those elements are in region II of Fig. 2.3. The elements that belong to set A but not to set B are in region I. The elements that belong to set B but not to set A are in region III.

Figure 2.3

In each of the four cases, any element not belonging to set A or set B is placed in region IV.

Venn diagrams will be helpful in understanding set operations. The operations of arithmetic are $+$, $-$, $\times$, and $\div$. When we see these symbols, we know what procedure to follow to determine the answer. Some of the operations in set theory are $'$, $\cup$, and $\cap$. They represent complement, union, and intersection, respectively.

Figure 2.4

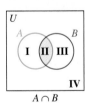

Figure 2.5

Complement

The **complement** of set A, symbolized by A', is the set of all the elements in the universal set that are not in set A.

In Fig. 2.4 the shaded region outside of set A within the universal set represents the complement of set A, or A'.

EXAMPLE 1

Given

$$U = \{1, 2, 3, 4, 5, 6, 7, 8, 9, 10\} \quad \text{and} \quad A = \{1, 2, 4, 6\}$$

Find A' and illustrate the relationship among sets U, A, and A' in a Venn diagram.

Solution: The elements in U that are not in set A are 3, 5, 7, 8, 9, 10. Thus $A' = \{3, 5, 7, 8, 9, 10\}$. The Venn diagram is illustrated in Fig. 2.5.   ▲

Intersection

The word *intersection* brings to mind the area common to two crossing streets. The car in the figure is in the intersection of the two streets. The set operation is defined as follows.

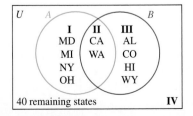

The **intersection** of sets A and B, symbolized by $A \cap B$, is the set containing all the elements that are common to both set A and set B.

The shaded region, region II, in Fig. 2.6 represents the intersection of sets A and B.

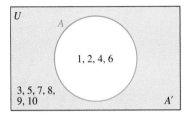

Figure 2.6

EXAMPLE 2

Draw a Venn diagram illustrating the following sets.

Set U is the set of the 50 states in the United States.

Set A is the set of the 6 states in the United States whose governors had the highest salaries in 1994.

Set B is the set of the 6 states in the United States that have the highest points above sea level.

Solution: The 6 states whose governors had the highest salaries in 1994, set A, were California, Maryland, Michigan, New York, Ohio, and Washington. The 6 states in set B are Alaska, California, Colorado, Hawaii, Washington, and Wyoming. First determine the intersection of sets A and B. California and Washington are common to both sets, so

$$A \cap B = \{CA, WA\}$$

Place these elements in region II of Fig. 2.7. Now place the elements in set A that have not been placed in region II, in region I. Therefore MD, MI, NY,

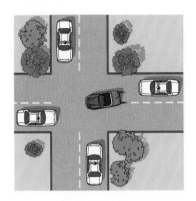

Figure 2.7

and OH go in region I. Complete region III by determining the elements in set B that have not been placed in region II. Thus AK, CO, HI, and WY go in region III. Finally, place those elements in U that are not in either set outside both circles. This includes the remaining 40 states, which go in region IV.

EXAMPLE 3

Given

$$U = \{1, 2, 3, 4, 5, 6, 7, 8, 9, 10\}$$
$$A = \{1, 2, 4, 6\}$$
$$B = \{1, 3, 6, 7, 9\}$$
$$C = \{\ \}$$

Find

a) $A \cap B$ **b)** $A \cap C$ **c)** $A' \cap B$ **d)** $(A \cap B)'$

Solution:

a) $A \cap B = \{1, 2, 4, 6\} \cap \{1, 3, 6, 7, 9\} = \{1, 6\}$. The elements common to both set A and set B are 1 and 6.

b) $A \cap C = \{1, 2, 4, 6\} \cap \{\ \} = \{\ \}$. There are no elements common to both set A and set C.

c) $A' = \{3, 5, 7, 8, 9, 10\}$
$A' \cap B = \{3, 5, 7, 8, 9, 10\} \cap \{1, 3, 6, 7, 9\}$
         $= \{3, 7, 9\}$

d) To find $(A \cap B)'$, first determine $A \cap B$.

$$A \cap B = \{1, 6\} \text{ from part (a)}$$
$$(A \cap B)' = \{1, 6\}' = \{2, 3, 4, 5, 7, 8, 9, 10\}$$

Union

The word *union* means to unite or join together, as in marriage, and that is exactly what is done when we perform the operation of union.

> The **union** of sets A and B, symbolized by $A \cup B$, is the set containing all the elements that are members of set A or of set B (or of both sets).

The three shaded regions of Fig. 2.8, regions I, II, and III, together represent the union of sets A and B. If an element is common to both sets, it is listed only once in the union of the sets.

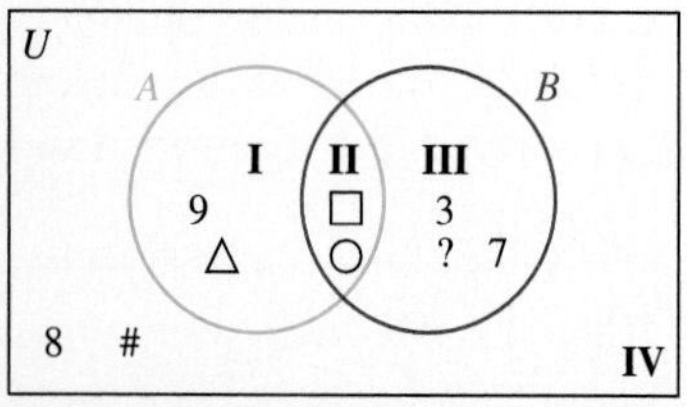

EXAMPLE 4

Use the Venn diagram in Fig. 2.9 to determine the following sets.

a) U **b)** A **c)** B' **d)** $A \cap B$

e) $A \cup B$ **f)** $(A \cup B)'$ **g)** $n(A \cup B)$

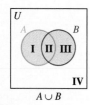

Figure 2.8

Figure 2.9

Solution:

a) The universal set consists of all the elements within the rectangle. Thus $U = \{9, \triangle, \square, \bigcirc, 3, 7, ?, \#, 8\}$.

b) Set A consists of the elements in regions I and II. Thus $A = \{9, \triangle, \square, \bigcirc\}$.

c) B' consists of the elements outside set B, or the elements in regions I and IV. Thus $B' = \{9, \triangle, \#, 8\}$.

d) $A \cap B$ consists of the elements that belong to both set A and set B (region II). Thus $A \cap B = \{\square, \bigcirc\}$.

e) $A \cup B$ consists of the elements that belong to set A or set B (regions I, II, or III). Thus $A \cup B = \{9, \triangle, \square, \bigcirc, 3, 7, ?\}$.

f) $(A \cup B)'$ consists of the elements in U that are not in $A \cup B$. Thus $(A \cup B)' = \{\#, 8\}$.

g) $n(A \cup B)$ represents the *number of elements* in the union of sets A and B. Thus $n(A \cup B) = 7$, as there are seven elements in the union of sets A and B.   ▲

EXAMPLE 5

Given

$$U = \{1, 2, 3, 4, 5, 6, 7, 8, 9, 10\}$$
$$A = \{1, 2, 4, 6\}$$
$$B = \{1, 3, 6, 7, 9\}$$
$$C = \{\ \}$$

Find

a) $A \cup B$ **b)** $A \cup C$ **c)** $A' \cup B$ **d)** $(A \cup B)'$

Solution:

a) $A \cup B = \{1, 2, 4, 6\} \cup \{1, 3, 6, 7, 9\} = \{1, 2, 3, 4, 6, 7, 9\}$

b) $A \cup C = \{1, 2, 4, 6\} \cup \{\ \} = \{1, 2, 4, 6\}$. Note that $A \cup C = A$.

c) In order to determine $A' \cup B$, we must determine A'.

$$A' = \{3, 5, 7, 8, 9, 10\}$$
$$A' \cup B = \{3, 5, 7, 8, 9, 10\} \cup \{1, 3, 6, 7, 9\}$$
$$= \{1, 3, 5, 6, 7, 8, 9, 10\}$$

d) Find $(A \cup B)'$ by first determining $A \cup B$, and then find the complement of $A \cup B$.

$$A \cup B = \{1, 2, 3, 4, 6, 7, 9\} \text{ from part (a)}$$
$$(A \cup B)' = \{1, 2, 3, 4, 6, 7, 9\}' = \{5, 8, 10\}$$ ▲

EXAMPLE 6

Given

$$U = \{a, b, c, d, e, f, g\}$$
$$A = \{a, b, e, g\}$$
$$B = \{a, c, d, e\}$$
$$C = \{b, e, f\}$$

Find

a) $(A \cup B) \cap (A \cup C)$ **b)** $(A \cup B) \cap C'$ **c)** $A' \cap B'$

Solution:

a) $(A \cup B) \cap (A \cup C) = \{a, b, c, d, e, g\} \cap \{a, b, e, f, g\}$
$$= \{a, b, e, g\}$$

b) $(A \cup B) \cap C' = \{a, b, c, d, e, g\} \cap \{a, c, d, g\}$
$$= \{a, c, d, g\}$$

c) $A' \cap B' = \{c, d, f\} \cap \{b, f, g\}$
$$= \{f\}$$

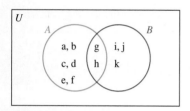

Figure 2.10

Having looked at unions and intersections we can now determine a relationship between $n(A \cup B)$, $n(A)$, $n(B)$, and $n(A \cap B)$. Suppose that set A has eight elements, set B has five elements, and $A \cap B$ has two elements: How many elements are in $A \cup B$? Let's make up some arbitrary sets that meet the criteria specified and draw a Venn diagram. If we let $A = \{a, b, c, d, e, f, g, h\}$, then set B must contain five elements, two of which are also in set A. Let $B = \{g, h, i, j, k\}$. We construct a Venn diagram by filling in the intersection first, as shown in Fig. 2.10. The number of elements in $A \cup B$ is 11. The elements g and h are in both sets, and if we add $n(A) + n(B)$ we are counting these elements twice.

To find the number of elements in the union of sets A and B we can add the number of elements in sets A and B and then subtract the number of elements common to both sets.

For any finite sets A and B

$$n(A \cup B) = n(A) + n(B) - n(A \cap B)$$

EXAMPLE 7

Set A contains 20 elements, set B contains 16 elements, and 4 elements are common to sets A and B. How many elements are in $A \cup B$?

Solution:

$$n(A \cup B) = n(A) + n(B) - n(A \cap B)$$
$$= 20 + 16 - 4$$
$$= 32$$

The Meaning of "*and*" and "*or*"

The words *and* and *or* are very important in many areas of mathematics. We use these words in several chapters in this book, including the probability chapter. The word *or* is generally interpreted to mean *union*, whereas *and* is generally interpreted to mean *intersection*. Suppose $A = \{1, 2, 3, 5, 6, 8\}$ and $B = \{1, 3, 4, 7, 9, 10\}$. Then the elements that belong to set A *or* set B are 1, 2, 3, 4, 5, 6, 7, 8, 9, and 10. These are the elements in the union of the sets. The elements that belong to set A *and* set B are 1 and 3. These are the elements in the intersection of the sets.

EXAMPLE 8

Set A contains six letters and five numbers. Set B contains four letters and nine numbers. Two letters and one number are common to both sets A and B. Find the number of elements in set A or set B.

Solution: You are asked to find the number of elements in set A or set B, which is $n(A \cup B)$. Because $n(A \cup B) = n(A) + n(B) - n(A \cap B)$, if you can determine $n(A)$, $n(B)$, and $n(A \cap B)$ you can solve the problem. Set A contains 6 letters and 5 numbers, so $n(A) = 11$. Set B contains 4 letters and 9 numbers, so $n(B) = 13$. Because 2 letters and 1 number are common to both sets, $n(A \cap B) = 3$.

$$n(A \cup B) = n(A) + n(B) - n(A \cap B)$$
$$= 11 + 13 - 3 = 21$$

Thus the number of elements in set A or set B is 21.

Section 2.3 Exercises

1. For the sets U, A, and B, construct a Venn diagram and place the elements in the proper regions.

   $U = \{a, b, c, d, e, f, g, h, i, j\}$

   $A = \{a, b, c, d, f, h\}$

   $B = \{b, c, e, f, i\}$

2. For the sets U, A, and B, construct a Venn diagram and place the elements in the proper regions.

   $U = \{$rhombus, square, rectangle, parallelogram, trapezoid$\}$

   $A = \{$rhombus, square, rectangle$\}$

   $B = \{$square, trapezoid$\}$

3. Let U represent the set of all the students attending a college or university in Florida during the academic years 1994–1996. Let set A represent the set of all the students who did not attend any social function at the school they were attending during that period. Write a sentence explaining what set A' represents.

4. Let U represent the set of all the rock bands in the United States during the period from 1985 to 1995. Let set A be the set of all the rock bands that have recorded exactly three albums during that period of time. Write a sentence explaining what set A' represents.

In Exercises 5–7, use the following information to describe the set in words.

   U is the set of all the students in your college registered for the same mathematics course as yourself.

   A is the set of students who will pass the course.

5. A'

6. $A \cap A'$

7. $A \cup A'$

In Exercises 8–12, use the following information to describe the set in words.

   U is the set of nurses employed at Community Hospital.

   E is the set of nurses employed at Community Hospital who have experience working in the emergency room.

   L is the set of nurses employed at Community Hospital who have experience working in the delivery room.

8. E' 9. $E \cap L$ 10. $E \cup L$

11. $L \cap E'$ 12. $(E \cap L)'$

In Exercises 13–18, use the following information to describe the set in words.

   U is the set of cities in the State of North Carolina.

   A is the set of cities in the State of North Carolina with a population over 50,000.

   B is the set of cities in the State of North Carolina with a professional sports team.

   C is the set of cities in the State of North Carolina with a philharmonic orchestra.

13. $A \cup C$ 14. $B \cap C$ 15. $B' \cap C$

16. $A \cup B \cup C$ 17. $A' \cup B'$ 18. $A \cap B \cap C$

In Exercises 19–26, use the Venn diagram in Fig. 2.11 to list the set of elements in roster form.

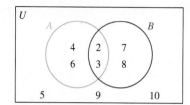

Figure 2.11

19. A

20. B

21. U

22. $A \cup B$

23. $A \cap B$

24. $A \cap B'$

25. $A' \cap B$

26. $(A \cup B)'$

In Exercises 27–34, use the Venn diagram in Fig. 2.12 to list the set of elements in roster form.

U represents the set of the 14 states with the highest average salary for public school teachers for the 1993–1994 academic year.

A represents the set of states (within the universal set) with public school enrollments of more than one million pupils.

B represents the set of states (within the universal set) with an annual public school expenditure of more than \$6300 per student.

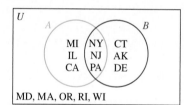

Figure 2.12

27. A

28. B

29. U

30. $A \cup B$

31. $A \cap B$

32. $(A \cap B)'$

33. $(A \cup B)'$

34. $A' \cup B$

In Exercises 35–40, let

$$U = \{1, 2, 3, 4, 5, 6, 7, 8\}$$
$$A = \{1, 2, 4, 5, 8\}$$
$$B = \{2, 3, 4, 6\}$$

Determine the following.

35. $A \cup B$

36. $A \cap B$

37. B'

38. $A \cup B'$

39. $(A \cup B)'$

40. $A' \cap B'$

41. $(A \cup B)' \cap B$

42. $(A \cup B) \cap (A \cup B)'$

43. $(B \cup A)' \cap (B' \cup A')$

44. $A' \cup (A \cap B)$

In Exercises 45–54, let

$$U = \{a, b, c, d, e, f, g, h, i, j, k\}$$
$$A = \{a, c, d, f, g, i\}$$
$$B = \{b, c, d, f, g\}$$
$$C = \{a, b, f, i, j\}$$

Determine the following.

45. A'

46. $B \cup C$

47. $A \cap C$

48. $A' \cup B$

49. $(A \cap C) \cup B$

50. $(A \cap C)'$

51. $A \cup (C \cap B)'$

52. $A' \cap (B \cap C)$

53. $(C \cap B) \cap (A' \cap B)$

54. $A \cup (C' \cup B')$

In Exercises 55–68, let

$$U = \{x \,|\, x \in N \text{ and } x < 10\}$$
$$A = \{x \,|\, x \in N \text{ and } x \text{ is odd and } x < 10\}$$
$$B = \{x \,|\, x \in N \text{ and } x \text{ is even and } x < 10\}$$
$$C = \{x \,|\, x \in N \text{ and } x < 6\}$$

Determine the following.

55. $A \cup B$

56. $A \cap B$

57. $A' \cup B$

58. $(B \cup C)'$

59. $A \cap B'$

60. $A \cap C'$

61. $(B \cap C)'$

62. $(A \cup B) \cap C$

63. $(C \cap B) \cup A$

64. $(C \cup A) \cap B$

65. $(A' \cup C) \cap B$

66. $(A \cap B') \cup C$

67. $(A \cup B)' \cap C$

68. $(A \cap C)' \cap B$

69. When will a set and its complement be disjoint? Explain and give an example.

70. When will $n(A \cap B) = 0$? Explain and give an example.

71. Consider the formula

$$n(A \cup B) = n(A) + n(B) - n(A \cap B)$$

a) Show that this relation holds for $A = \{a, b, c, d\}$ and $B = \{b, d, e, f, g, h\}$.

b) Make up your own sets A and B, each consisting of at least six elements. Using these sets, show that the relation holds.

c) Use a Venn diagram and explain why the relation holds for any two sets A and B.

72. The Venn diagram in Fig. 2.13 shows a technique of labeling the regions to indicate membership of elements in a particular region. Define each of the four regions with a set statement. (*Hint:* $A \cap B'$ defines region I.)

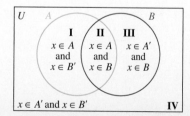

Figure 2.13

*In Exercises 73–82, let U = {0, 1, 2, 3, 4, 5, ...},
A = {1, 2, 3, 4, ...}, B = {4, 8, 12, 16, ...} and
C = {2, 4, 6, 8, ...}. Determine the following.*

73. $A \cup B$ **74.** $A \cap B$

75. $B \cap C$ **76.** $B \cup C$

77. $A \cap C$ **78.** $A' \cap C$

79. $B' \cap C$ **80.** $(B \cup C)' \cup C$

81. $(A \cap C) \cap B'$ **82.** $U' \cap (A \cup B)$

Problem Solving/Group Activities

*In Exercises 83–90, determine whether the answer is $\emptyset$, A,
or U. (Assume that $A \neq \emptyset$, $A \neq U$.)*

83. $A \cup A'$ **84.** $A \cap A'$ **85.** $A \cup \emptyset$

86. $A' \cup U$ **87.** $A \cap \emptyset$ **88.** $A \cup U$

89. $A \cap U$ **90.** $A \cap U'$

*In Exercises 91–95, determine the relationship between set
A and set B if*

91. $A \cap B = B$.   **92.** $A \cup B = B$.   **93.** $A \cap B = \emptyset$.

94. $A \cup B = A$.   **95.** $A \cap B = A$.

*In Exercises 96–99, draw a Venn diagram as in Fig. 2.2
that illustrates the relationship between set A and set B
(assume $A \neq B$) if*

96. $A \subset B$.   **97.** $A \cup B = B$.

98. $A \cap B = B$.   **99.** $A \cap B = \emptyset$.

Another set operation is the **difference of two sets**. The
difference of two sets A and B, symbolized $A - B$, is
defined as

$$A - B = \{x \,|\, x \in A \text{ and } x \notin B\}$$

Thus $A - B$ is the set of elements that belong to set A but
not to set B. For example, if $U = \{1, 2, 3, 4, 5, 6, 7, 8, 9,
10\}$, $A = \{2, 4, 5, 9, 10\}$, and $B = \{1, 3, 4, 5, 6, 7\}$, then
$A - B = \{2, 9, 10\}$ and $B - A = \{1, 3, 6, 7\}$.

*In Exercises 100–103, let U = {a, b, c, d, e, f, g, h, i, j, k},
A = {b, c, e, f, g, h}, and B = {a, b, c, g, i}. Determine the
following.*

100. $A - B$ **101.** $B - A$

102. $A' - B$ **103.** $A - B'$

*In Exercises 104–109, let U = {1, 2, 3, 4, 5, 6, 7, 8, 9, 10,
11, 12, 13, 14, 15}, A = {2, 4, 5, 7, 9, 11, 13}, and
B = {1, 2, 4, 5, 6, 7, 8, 9, 11}. Determine the following.*

104. $A - B$ **105.** $B - A$

106. $(A - B)'$ **107.** $A - B'$

108. $(B - A)'$ **109.** $A \cap (A - B)$

Another set operation is the **Cartesian product**. The
Cartesian product of set A and set B is symbolized $A \times B$
and is read ''A *cross* B.'' The Cartesian product* of set A
and set B is the set of all possible ordered pairs of the form
(a, b) where $a \in A$ and $b \in B$. For example, if $A =$
{George, Paul, John, Ringo} and $B = \{1, 2, 5\}$, then
$A \times B = \{$(George, 1), (George, 2), (George, 5), (Paul, 1),
(Paul, 2), (Paul, 5), (John, 1), (John, 2), (John, 5),
(Ringo, 1), (Ringo, 2), (Ringo, 5)$\}$

*In Exercises 110–115, let set A = {a, b, c} and set
B = {1, 2}.*

110. Determine $A \times B$.

111. Determine $B \times A$.

112. Does $A \times B = B \times A$?

113. Determine $n(A \times B)$.

114. Determine $n(B \times A)$.

115. Does $n(A \times B) = n(B \times A)$?

116. If set $A = \{1, 2, 3, 4\}$, find $A \times A$.

117. If set A has m elements and set B has n elements, how
many elements will be in $A \times B$?

---

*Ordered pairs are discussed in Section 6.7.

2.4   VENN DIAGRAMS WITH THREE SETS AND VERIFICATION OF EQUALITY OF SETS

Venn diagrams can be used to illustrate three or more sets. For three sets, A, B,
and C, the diagram is drawn so the three sets overlap (Fig. 2.14), creating eight
regions. The diagrams in Fig. 2.15 emphasize selected regions of three intersecting sets. *When constructing Venn diagrams with three sets, we generally start with region V and work outward*, as explained in the following procedure.

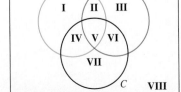

Figure 2.14

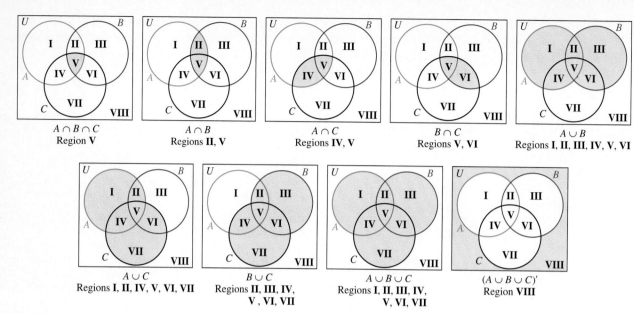

Figure 2.15

General Procedure for Constructing Venn Diagrams with Three Sets, *A*, *B*, and *C*

1. Determine the elements to be placed in region V by finding the elements that are common to all three sets, $A \cap B \cap C$.

2. Determine the elements to be placed in region II. Find the elements in $A \cap B$. The elements in this set belong in regions II and V. Place the elements in the set $A \cap B$ that are not listed in region V in region II. The elements in regions IV and VI are found in a similar manner.

3. Determine the elements to be placed in region I by determining the elements in set *A* that are not in regions II, IV, and V. The elements in regions III and VII are found in a similar manner.

4. Determine the elements to be placed in region VIII by finding the elements in the universal set that are not in regions I through VII.

Example 1 illustrates the general procedure.

EXAMPLE 1

Construct a Venn diagram illustrating the following sets.

$$U = \{1, 2, 3, 4, 5, 6, 7, 8, 9, 10, 11, 12, 13, 14, 15\}$$
$$A = \{1, 2, 3, 4, 7, 9, 11\}$$
$$B = \{2, 3, 4, 5, 10, 12, 14\}$$
$$C = \{1, 2, 4, 8, 9\}$$

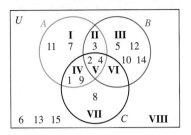

Figure 2.16

Solution: First find the intersection of all three sets. Because the elements 2 and 4 are in all three sets, $A \cap B \cap C = \{2, 4\}$. The elements 2 and 4 are placed in region V (Fig. 2.16). Next complete region II by determining the intersection of sets A and B.

$$A \cap B = \{2, 3, 4\}$$

$A \cap B$ consists of regions II and V. The elements 2 and 4 have already been placed in region V, so 3 must be placed in region II. In a similar manner, complete regions IV and VI. Now complete set A. The only elements of set A that have not previously been placed in regions II, IV, and V are 7 and 11. Therefore place the elements 7 and 11 in region I. The elements in region I are only in set A. Complete regions III and VII in a similar manner. To determine the elements in region VIII, find the elements in U that have not been placed in regions I–VII. The elements 6, 13, and 15 have not been placed in regions I–VII, so place them in region VIII. ▲

Venn diagrams can be used to illustrate and analyze many everyday problems. One example follows.

EXAMPLE 2

Human blood is classified (typed) according to the presence or absence of the specific antigens A, B, and Rh in the red blood cells. Antigens are highly specified proteins and carbohydrates that will trigger the production of antibodies in the blood to fight infection. Blood lacking the Rh antigen is labeled negative and blood lacking both A and B antigens is type O. Sketch a Venn diagram with three sets A, B, and Rh and place each type of blood listed in the proper region. A person has only one type of blood.

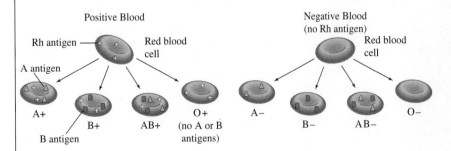

Solution: As illustrated in Chapter 1, the first thing to do is to read the question carefully and make sure you understand what is given and what you are asked to find. There are three antigens A, B, and Rh. Therefore begin by naming the three circles in a Venn diagram with the three antigens, Fig. 2.17.

Any blood containing the Rh antigen is positive. Therefore all blood in the Rh circle is positive, and all blood outside the Rh circle is negative. The intersection of all three sets, region V, is AB+. Region II contains only antigens A and B and is therefore AB−. Region I is A− because it contains only antigen A. Region III is B−, region IV is A+, and region VI is B+. Region VII is O+, containing only the Rh antigen. Region VIII, which lacks all three antigens, is O−. ▲

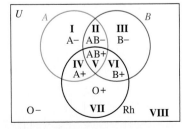

Figure 2.17

Verification of Equality of Sets

Consider the question, Is $A' \cup B = A' \cap B$ *for all sets A and B*? For the specific sets $U = \{1, 2, 3, 4, 5\}$, $A = \{1, 3\}$, and $B = \{2, 4, 5\}$ is $A' \cup B = A' \cap B$? To answer the question, we do the following.

Find $A' \cup B$	Find $A' \cap B$
$A' = \{2, 4, 5\}$	$A' = \{2, 4, 5\}$
$A' \cup B = \{2, 4, 5\}$	$A' \cap B = \{2, 4, 5\}$

For these sets $A' \cup B = A' \cap B$, as both sets are equal to $\{2, 4, 5\}$. At this point you may believe that $A' \cup B = A' \cap B$ for all sets A and B.

If we select the sets $U = \{1, 2, 3, 4, 5\}$, $A = \{1, 3, 5\}$, and $B = \{2, 3\}$, we see that $A' \cup B = \{2, 3, 4\}$ and $A' \cap B = \{2\}$. For this case $A' \cup B \neq A' \cap B$. Thus we have proved that $A' \cup B \neq A' \cap B$ for all sets A and B by using a *counterexample*. A counterexample, as explained in Chapter 1, is an example that shows a statement is not true.

In Chapter 1 we explained that proving a statement involves the use of deductive reasoning. Recall that deductive reasoning begins with a general statement and works to a specific conclusion. To verify, or determine whether set statements are equal for any two sets selected, we will use deductive reasoning with Venn diagrams. Venn diagrams are used because they can illustrate general cases. To determine if statements that contain sets, such as $(A \cup B)'$ and $A' \cap B'$ are equal for all sets A and B we use the regions of Venn diagrams. If both statements represent the same regions of the Venn diagram, then the statements are equal for all sets A and B. See Example 3.

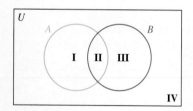

Figure 2.18

EXAMPLE 3

Determine whether $(A \cup B)' = A' \cap B'$ for all sets A and B.

Solution: Draw a Venn diagram with two sets A and B, as in Fig. 2.18. Label the regions as indicated.

Find $(A \cup B)'$

Set	Corresponding Regions
A	I, II
B	II, III
$A \cup B$	I, II, III
$(A \cup B)'$	IV

Find $A' \cap B'$

Set	Corresponding Regions
A'	III, IV
B'	I, IV
$A' \cap B'$	IV

Both statements are represented by the same region, IV, of the Venn diagram. Thus, $(A \cup B)' = A' \cap B'$ for all sets A and B.   ▲

In proving that $(A \cup B)' = A' \cap B'$ we started with two general sets and worked to the specific conclusion that both statements represented the same regions of the Venn diagram. We can also use Venn diagrams to prove statements involving three sets.

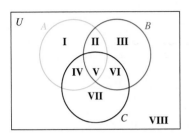

Figure 2.19

EXAMPLE 4

Determine whether $A \cap (B \cup C) = (A \cap B) \cup (A \cap C)$ for all sets, A, B, and C.

Solution: Because the statements include three sets, A, B, and C, three circles must be used. The Venn diagram illustrating the eight regions is shown in Fig. 2.19.

First we will find the regions that correspond to $A \cap (B \cup C)$; then we will find the regions that correspond to $(A \cap B) \cup (A \cap C)$. If both answers are the same, the statements are equal.

Find $A \cap (B \cup C)$

Set	Corresponding Regions
A	I, II, IV, V
$B \cup C$	II, III, IV, V, VI, VII
$A \cap (B \cup C)$	II, IV, V

Find $(A \cap B) \cup (A \cap C)$

Set	Corresponding Regions
$A \cap B$	II, V
$A \cap C$	IV, V
$(A \cap B) \cup (A \cap C)$	II, IV, V

The regions that correspond to $A \cap (B \cup C)$ are II, IV, and V, and the regions that correspond to $(A \cap B) \cup (A \cap C)$ are also II, IV, and V.

The results show that both statements are represented by the same regions, namely, II, IV, and V, and therefore $A \cap (B \cup C) = (A \cap B) \cup (A \cap C)$ for all sets A, B, and C. ▲

De Morgan's Laws

In set theory, logic, and other branches of mathematics a pair of related theorems known as De Morgan's laws make it possible to transform statements and formulas into alternative and often more convenient forms. In set theory, De Morgan's laws are symbolized as

1. $(A \cup B)' = A' \cap B'$

2. $(A \cap B)' = A' \cup B'$

Law 1 was verified in Example 3. We suggest that you verify law 2 at this time. The laws were expressed verbally by William of Ockham in the fourteenth century. In the nineteenth century, Augustus De Morgan expressed them mathematically. De Morgan's laws will be discussed more thoroughly in the chapter on logic.

Section 2.4 Exercises

1. Construct a Venn diagram illustrating the following sets.

$$U = \{a, b, c, d, e, f, g, h, i, j, k\}$$
$$A = \{b, c, e, g, h, k\}$$
$$B = \{a, b, c, d, g, i\}$$
$$C = \{a, b, c, d, g, k\}$$

2. Construct a Venn diagram illustrating the following sets. The elements of the sets are the days Amy, Peter, and Carlos work at Al's Sandwich Shop.

$U = \{$Sunday, Monday, Tuesday, Wednesday, Thursday, Friday, Saturday$\}$

Amy = $\{$Monday, Tuesday, Wednesday, Thursday$\}$

Peter = $\{$Sunday, Monday, Tuesday, Wednesday, Thursday, Friday$\}$

Carlos = $\{$Tuesday, Sunday$\}$

3. Construct a Venn diagram illustrating the following sets.

$U = \{$Jan, Feb, March, April, May, June, July, Aug, Sept, Oct, Nov, Dec$\}$

$A = \{$Jan, April, Aug, Sept, Oct, Dec$\}$

$B = \{$Feb, Aug, Oct, Nov, Dec$\}$

$C = \{$Feb, March, June, Dec$\}$

4. Construct a Venn diagram illustrating the following sets.

$U = \{$Whitney Houston, Alan Menken, Tim Rice, Eric Clapton, Natalie Cole, Irving Gordon, Phil Collins, Quincy Jones, Julie Gold, Bill Poole$\}$

$A = \{$Whitney Houston, Eric Clapton, Natalie Cole, Phil Collins$\}$

$B = \{$Whitney Houston, Eric Clapton, Natalie Cole, Quincy Jones$\}$

$C = \{$Alan Menken, Tim Rice, Eric Clapton, Irving Gordon, Julie Gold$\}$

5. Construct a Venn diagram illustrating the following sets.

$U = \{$football, basketball, baseball, gymnastics, lacrosse, soccer, tennis, volleyball, swimming, wrestling, cross-country, track, golf, fencing$\}$

$A = \{$football, basketball, soccer, lacrosse, volleyball$\}$

$B = \{$baseball, lacrosse, tennis, golf, volleyball$\}$

$C = \{$swimming, gymnastics, fencing, basketball, volleyball$\}$

6. Construct a Venn diagram illustrating the following sets.

$U = \{$hickory, black walnut, hazel, English walnut, pecan, beech, maple, ash, birch, red oak, white oak, cedar, white pine, yellow pine, blue spruce, aspen, red wood$\}$

$A = \{$hickory, black walnut, hazel, English walnut, pecan, beech$\}$

$B = \{$maple, beech, black walnut, ash, birch, white oak, cedar, white pine, yellow pine$\}$

$C = \{$blue spruce, aspen, beech, white pine, yellow pine, hazel$\}$

In Exercises 7–16, refer to Fig. 2.20. Listed below it are the names of the five top female (LPGA) golfers for 1988–1993 and the five top male (PGA) golfers for 1989–1993. In parentheses after each name is the year the player was the top LPGA or PGA golfer. These rankings are based on the top money winners for that year, referred to as Tour Winnings, as reported by the Professional Golf Associations. In addition to the tour winnings for the year they were top golfers are their career winnings through 1993. Because Betsy King was the top LPGA golfer in 1989 and 1993, we added the 1988 winner. In Exercises 7–16, indicate the region in Fig. 2.20 in which each golfer would be placed.

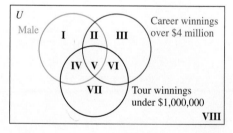

Figure 2.20

Female	Tour Winnings	Career Winnings
7. Betsy King (1993)	$595,992	$4,642,165
8. Dottie Mochrie (1992)	693,335	2,390,888
9. Pat Bradley (1991)	763,118	4,654,277
10. Beth Daniel (1990)	863,578	4,230,557
11. Sherri Turner (1988)	350,851	956,828

Male	Tour Winnings	Career Winnings
12. Nick Price (1993)	$1,478,557	$5,797,418
13. Fred Couples (1992)	1,344,188	6,544,089
14. Corey Pavin (1991)	979,430	5,835,444
15. Greg Norman (1990)	1,165,477	7,726,060
16. Tom Kite (1989)	1,395,278	8,973,382

Country	Gold	Silver	Bronze	Total
29. Norway	10	11	5	26
30. Germany	9	7	8	24
31. Russia	11	8	4	23
32. Italy	7	5	8	20
33. United States	6	5	2	13
34. Canada	3	6	4	13

In Exercises 17–28, indicate in Fig. 2.21 the region in which each of the figures would be placed.

In Exercises 35–48, use the Venn diagram in Fig. 2.23 to list the sets in roster form.

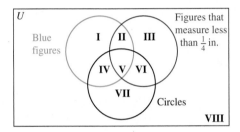

Figure 2.21

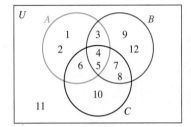

Figure 2.23

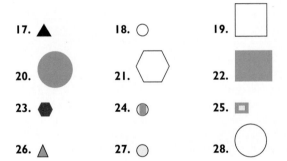

17. ▲ **18.** ○ **19.** ☐

20. ⬤ **21.** ⬡ **22.** ⬛

23. ⬡ **24.** ◖ **25.** ▣

26. △ **27.** ○ **28.** ◯

35. *A*   **36.** *B*   **37.** *C*

38. *U*   **39.** *A* ∩ *B*   **40.** *A* ∩ *C*

41. (*B* ∩ *C*)'   **42.** *A* ∩ *B* ∩ *C*   **43.** *A* ∪ *B*

44. *B* ∪ *C*   **45.** (*A* ∪ *C*)'   **46.** *A* ∪ *B* ∪ *C*

47. *A*'   **48.** (*A* ∪ *B* ∪ *C*)'

49. A hematology text gives the following information on percentages of the different types of blood worldwide.

Type	Positive Blood, %	Negative Blood, %
A	37	6
O	32	6.5
B	11	2
AB	5	0.5

Construct a Venn diagram similar to the one in Example 2 and place the correct percent in each of the eight regions.

In Exercises 29–34, refer to Fig. 2.22. Listed above and to the right are the six countries whose Olympic teams won the greatest number of medals in the 1994 Winter Olympics. The number of gold, silver, and bronze medals are listed for each country. In Exercises 29–34, indicate the region in Fig. 2.22 in which the country would be placed.

In Exercises 50–57, use Venn diagrams to determine whether the following statements are equal for all sets A and B.

50. (*A* ∪ *B*)',   *A*' ∩ *B*'

51. (*A* ∪ *B*)',   *A*' ∩ *B*

52. *A*' ∪ *B*',   *A* ∩ *B*

53. (*A* ∪ *B*)',   (*A* ∩ *B*)'

54. *A*' ∪ *B*',   (*A* ∪ *B*)'

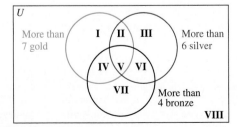

Figure 2.22

55. $A \cap B'$, $A' \cup B$

56. $(A \cap B')'$, $A' \cup B$

57. $A' \cap B'$, $(A' \cap B')'$

In Exercises 58–67, use Venn diagrams to determine whether the following statements are equal for all sets A, B, and C.

58. $A \cup (B \cap C)$, $(A \cup B) \cap C$

59. $A \cup (B \cap C)$, $(B \cap C) \cup A$

60. $A \cap (B \cup C)$, $(B \cup C) \cap A$

61. $A' \cup (B \cap C)$, $A \cap (B \cup C)'$

62. $A \cap (B \cup C)$, $(A \cap B) \cup (A \cap C)$

63. $A \cup (B \cap C)$, $(A \cup B) \cap (A \cup C)$

64. $A \cap (B \cup C)'$, $A \cap (B' \cap C')$

65. $(A \cup B) \cap (B \cup C)$, $B \cup (A \cap C)$

66. $(C \cap B)' \cup (A \cap B)'$, $A \cap (B \cap C)$

67. $(A \cup B)' \cap C$, $(A' \cup C) \cap (B' \cup C)$

68. Let

$$U = \{1, 2, 3, 4, 5, 6, 7, 8, 9, 10\},$$
$$A = \{1, 2, 3, 4\},$$
$$B = \{3, 6, 7\},$$
$$C = \{6, 7, 9\}.$$

a) Show that $(A \cup B) \cap C = (A \cap C) \cup (B \cap C)$ for these sets.

b) Make up your own sets A, B, and C. Verify that $(A \cup B) \cap C = (A \cap C) \cup (B \cap C)$ for your sets A, B, and C.

c) Use Venn diagrams to verify that $(A \cup B) \cap C = (A \cap C) \cup (B \cap C)$ for all sets A, B, and C.

69. Let

$$U = \{a, b, c, d, e, f, g, h, i\}$$
$$A = \{a, c, d, e, f\},$$
$$B = \{c, d\},$$
$$C = \{a, b, c, d, e\}.$$

a) Determine whether $(A \cup C)' \cap B = (A \cap C)' \cap B$ for these sets.

b) Make up your own sets, A, B, and C. Determine whether $(A \cup C)' \cap B = (A \cap C)' \cap B$ for your sets.

c) Determine whether $(A \cup C)' \cap B = (A \cap C)' \cap B$ for all sets A, B, and C.

70. For sets U, A, B, and C, construct a Venn diagram and place the elements in the proper regions.

$$U = \{1, 2, 3, 4, 5, 6, 7, 8, 9, 10, 11, 12\}$$

Set A is the set of even natural numbers less than or equal to 12.

Set B is the set of odd natural numbers less than or equal to 12.

Set C is the set of natural numbers less than or equal to 12 that are multiples of 3. (A multiple of 3 is any number that is divisible by 3.)

71. For sets U, A, B, and C, construct a Venn diagram and place the elements in the proper regions.

$$U = \{1, 2, 3, 4, 5, 6, 7, 8, 9, 10, 11, 12\}$$

Set A is the set of even natural numbers less than or equal to 12.

Set B is the set of natural numbers less than or equal to 12 that are multiples of 3.

Set C is the set of natural numbers that are factors of 12. (A factor of 12 is a number that divides 12.)

72. Define each of the eight regions in Fig. 2.24 using sets A, B, and C and a set operation. (*Hint:* $A \cap B' \cap C'$ defines region I.)

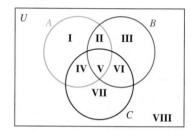

Figure 2.24

Problem Solving/Group Activities

73. The personnel department at a plant manufacturing plastic containers wants to classify employees according to income, education, and gender. The results are intended to show whether the person is male or female, whether the person has a college degree and whether the person is earning less than $30,000 per year.

a) Construct a Venn diagram that will enable the personnel officer at the plant to sort the workers by gender, by whether they have a college degree, and whether they earn less than $30,000.

b) Shade the region of the diagram that identifies the male employees who earn less than $30,000 and have a college degree. Describe the region using sets A, B and C with the operations union, intersection, and complement.

c) Shade the region of the diagram that identifies the female employees who earn $30,000 or more and have a college degree. Describe the region using sets A, B and C with the operations union, intersection, and complement.

d) Shade the region of the diagram that identifies the male employees who do not have a college degree and who earn less than $30,000. Describe the region using sets A, B and C with the operations union, intersection, and complement.

e) Shade the region of the diagram that identifies the male employees who do not have a college degree and who earn $30,000 or more. Describe the region using sets *A*, *B* and *C* with the operations union, intersection, and complement.

f) Shade the region of the diagram that identifies the female employees who earn less than $30,000 and do not have a college degree. Describe the region using sets *A*, *B* and *C* with the operations union, intersection, and complement.

74. a) Construct a Venn diagram illustrating four sets, *A*, *B*, *C*, and *D*. (*Hint:* Four circles cannot be used, and you should end up with 16 *distinct* regions.) Have fun!

b) Label each region with a set statement (see Exercise 72). Check all 16 regions to make sure that *each is distinct*.

75. You were able to determine the number of elements in the union of two sets with the formula

$$n(A \cup B) = n(A) + n(B) - n(A \cap B)$$

Can you determine a formula for finding the number of elements in the union of three sets? In other words, write a formula to determine $n(A \cup B \cup C)$. [*Hint:* The formula will contain each of the following: $n(A)$, $n(B)$, $n(C)$, $n(A \cap B \cap C')$, $n(A \cap B' \cap C)$, $n(A' \cap B \cap C)$, and $2n(A \cap B \cap C)$.]

Research Activity

76. The two Venn diagrams illustrate what happens when colors are added or subtracted. Do research in an art text, an encyclopedia, or another source and write a report explaining the creation of the colors in the Venn diagrams, using such terms as union of colors and subtraction (or difference) of colors.

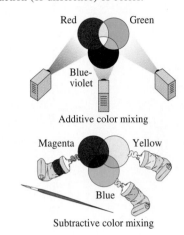

Additive color mixing

Subtractive color mixing

2.5 APPLICATIONS OF SETS

We can solve practical problems involving sets by using the problem-solving process discussed in Chapter 1: Understand the problem, devise a plan, carry out the plan, and then examine and check the results. First determine: What is the problem? or What am I looking for? To devise the plan list all the facts that are given and how they are related. *Look for key words or phrases* like: "only set *A*," "set *A* and set *B*," "set *A* or set *B*," "set *A* and set *B* and not set *C*." Remember that *and* means intersection, *or* means union, and *not* means complement. The problems we will solve in this section contain two or three sets of elements, which can be represented in a Venn diagram. Our plan will generally include drawing a Venn diagram, labeling the diagram, and filling in the regions of the diagram.

Whenever possible follow the procedure in Section 2.4 for completing the Venn diagram and then answer the questions. Remember, when drawing the Venn diagrams, we generally start with the intersection of the sets and work outward.

EXAMPLE 1

At a shopping mall 190 adult drivers were selected at random and asked whether they enjoy driving a minivan or a car. The results of the survey showed that

110 enjoy driving a car,

90 enjoy driving a minivan, and

30 enjoy driving both a car and a minivan.

a) Of those surveyed how many people do not enjoy driving a car or a minivan?

b) How many people enjoy driving a car but do not enjoy driving a minivan?

c) How many people enjoy driving a minivan but do not enjoy driving a car?

d) How many people enjoy driving a car or a minivan?

Solution: The problem statement provides the following information.

The number of people surveyed is 190: $n(U) = 190$.

The number of people who enjoy driving a car is 110: $n(C) = 110$.

The number of people who enjoy driving a minivan is 90: $n(M) = 90$.

The number of people who enjoy driving both a car and a minivan is 30: $n(C \cap M) = 30$.

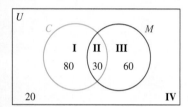

Figure 2.25

We illustrate this information on the Venn diagram shown in Fig. 2.25. We already know that $C \cap M$ corresponds to region II. As $n(C \cap M) = 30$, we write 30 in region II. Set C consists of regions I and II. We know that set C, the people who enjoy driving a car, contains 110 people. Therefore region I contains $110 - 30$, or 80 people, and we write the number 80 in region I. Set M consists of regions II and III. As $n(M) = 90$, the total in these two regions must be 90. Region II contains 30, leaving 60 for region III.

The total number of people who enjoy driving a car or a minivan is $n(C \cup M) = 80 + 30 + 60 = 170$. The number of people in region IV is the difference between $n(U)$ and $n(C \cup M)$. There are $190 - 170$, or 20 people in region IV.

a) The people who do not enjoy driving a car or a minivan are those members of the universal set who are not contained in set C or set M. The 20 people in region IV do not enjoy driving a car or a minivan.

b) The 80 people in region I are those who enjoy driving a car but not a minivan.

c) The 60 people in region III are those who enjoy driving a minivan but not a car.

d) The people in regions I, II, or III enjoy driving a car or a minivan. Thus $80 + 30 + 60$, or 170 people enjoy driving one or both types of vehicles.

▲

Similar problems involving three sets can be solved, as illustrated in Example 2.

EXAMPLE 2

We know the following facts about the first 42 presidents of the United States.

8 held a cabinet post.

14 served as vice-president.

15 served in the U.S. Senate.

2 served in a cabinet post and as vice-president.

4 served in a cabinet post and in the U.S. Senate.

6 served in the U.S. Senate and as vice-president.

1 served in all three positions.

How many presidents served in

a) none of these positions?

b) only the U.S. Senate?

c) at least one of the three positions?

d) exactly two positions?

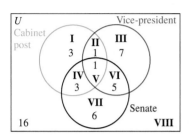

Figure 2.26

Solution: Construct a Venn diagram with three circles: cabinet post, vice-president, and Senate (Fig. 2.26). Label the eight regions.

Whenever possible, work from the center outward. First, find the number of presidents that appear in all three sets. Since one president served in all three positions, place a 1 in region V. Next, determine the number to be placed in region II. Two presidents served in a cabinet post and as vice-president. Regions II and V represent the intersection of presidents who served both in a cabinet post and as vice-president. Therefore the sum of the elements in these two regions must be 2. A 1 was previously placed in region V, leaving a 1 for region II. Four presidents served in a cabinet post and in the Senate. The sum of regions IV and V must therefore be 4. There is already a 1 in region V, so place a 3 in region IV. Using similar reasoning, place a 5 in region VI.

Now find the numbers in regions I, III, and VII. A total of 8 presidents served in a cabinet post. The sum of regions II, IV, and V is 5, so there are 3 in region I. The total number of presidents who served as vice-president was 14. The sum of the numbers in regions II, V, and VI is 7, so there must be 7 in region III. Similarly, the number in region VII must be $15 - 9$, or 6. So far we have accounted for a total of 26 presidents in regions I through VII. There must be $42 - 26$, or 16 presidents who did not serve in any of these capacities. Place 16 in region VIII. Now use the completed Venn diagram to determine the answers to parts (a)–(d).

a) Sixteen presidents did not serve in any of the three positions. This number is shown in region VIII.

b) Region VII represents the presidents who served only in the Senate. There were six.

c) "At least one" means one or more. Thus the number of presidents who served in at least one of the three positions is found by summing the numbers in regions I through VII. Twenty-six presidents served in at least one of these positions.

d) The presidents represented in regions II, IV, and VI served in exactly two positions. Summing the numbers in the three regions shows that nine presidents served in exactly two positions. ▲

When you are solving problems of this type, be sure to check your Venn diagram carefully. *The most common error made by students is forgetting to subtract the number in region V from the respective values in determining the numbers to place in regions II, IV, and VI.*

┌─ **EXAMPLE 3**

Eric Clapton, Natalie Cole, and Whitney Houston have all won Grammy Awards. One hundred twenty students were asked to select which of the three vocalists they enjoyed.

42 selected Eric Clapton.

47 selected Natalie Cole.

41 selected Whitney Houston.

11 selected Eric Clapton and Natalie Cole.

12 selected Natalie Cole and Whitney Houston.

10 selected Whitney Houston and Eric Clapton.

4 selected Clapton, Cole, and Houston.

How many of those surveyed selected

a) none of these vocalists?

b) Clapton, but neither of the other two vocalists?

c) exactly one of the vocalists?

d) Cole and Houston, but not Clapton?

e) Cole or Houston, but not Clapton?

f) exactly two of the vocalists?

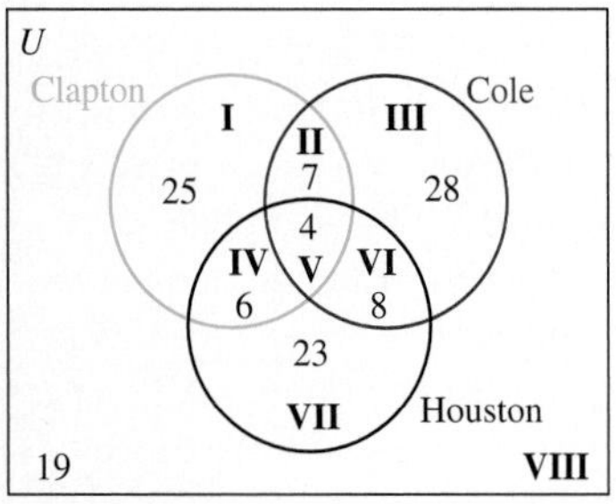

Solution: Construct a Venn diagram with three sets, as in Example 2. You should go through the process of developing the Venn diagram before consulting Fig. 2.27.

a) None of the three vocalists was selected by 19 students.

b) The 25 students in region I selected only Clapton.

c) Those in regions I, III, and VII selected only one vocalist. The sum of the numbers in these regions, 25 + 28 + 23 or 76, is the number of students who selected exactly one vocalist.

d) The 8 students in region VI selected Cole and Houston, but not Clapton.

e) The word *or* in this type of problem means one or the other or both. All the students in regions II, III, IV, V, VI, and VII selected Cole or Houston or both. Those in regions II, IV, and V also selected Clapton. The 59 students who selected Cole or Houston but not Clapton are found by adding the numbers in regions III, VI, and VII.

f) The students in regions II, IV, and VI—a total of 21—selected exactly two of the three vocalists.

Figure 2.27

Section 2.5 Exercises

1. A travel agent interviewed 180 people to determine whether they preferred traveling long distances by airplane or by automobile and learned that

120 preferred to travel by airplane,

90 preferred to travel by automobile, and

50 preferred to travel by both airplane and automobile.

Of those interviewed,

a) how many preferred to travel only by airplane?

b) how many preferred to travel only by automobile?

c) how many preferred not to travel by airplane or automobile?

2. A survey of students at Franklin High School showed that

65 students studied air science or earth science,

50 studied earth science, and

10 studied both air science and earth science.

How many studied air science?

3. A congressional representative polled her constituents regarding two bills. The results showed that of 200 persons in the sample

55 favored bill I,

100 favored bill II, and

40 favored both bills.

a) How many favored only bill I?
b) How many were not in favor of either bill?
c) Did the majority of those surveyed favor bill II?

4. The results of a survey of 50 students showed the following preferences in writing instruments.

16 liked pencils.

18 liked felt-tipped pens.

24 liked ballpoint pens.

4 liked pencils and felt-tipped pens.

5 liked felt-tipped pens and ballpoint pens.

8 liked pencils and ballpoint pens.

3 liked all three.

Determine the number of students who liked
a) none of the three writing instruments.
b) pencils and felt-tipped pens but did not like ballpoint pens.
c) only pencils.
d) only ballpoint pens.
e) felt-tipped pens or ballpoint pens.

5. The results of a survey of 400 former professional football players were as follows.

94 had knee surgery.

65 had ankle surgery.

57 had shoulder surgery.

26 had knee and ankle surgery.

31 had knee and shoulder surgery.

18 had ankle and shoulder surgery.

12 had all three types of surgery.

How many had
a) none of these types of surgery?
b) knee surgery only?
c) at least one of these types of surgery?
d) knee surgery or ankle surgery, but not shoulder surgery?
e) knee surgery and shoulder surgery, but not ankle surgery?

6. The results of a survey of 875 students at Youngstown State University were as follows.

213 were wearing blue jeans.

217 were freshmen.

316 were taking an economics course.

92 were freshmen and wearing blue jeans.

110 were freshmen and taking an economics course.

100 were wearing blue jeans and taking an economics course.

70 were freshmen, wearing jeans, and taking an economics course.

How many were
a) doing none of these things?
b) taking only an economics course?
c) taking an economics course and wearing blue jeans but were not freshmen?
d) taking an economics course or wearing blue jeans but were not freshmen?
e) doing at least one of the three?

7. In a recent blood drive the following data on donors were recorded (see Example 2 in Section 2.4).

243 had the A antigen.

93 had the B antigen.

28 had both the A and B antigens.

80 had both the B and Rh antigens.

325 had the Rh antigen.

33 had none of the antigens.

210 had both the A and Rh antigens.

25 had all three antigens.

How many donors had
a) A+ blood?
b) B− blood?
c) AB+ blood?
d) How many donors are represented in the blood drive?

8. The three major grain crops raised in the world are wheat, maize, and rice. A survey of 40 countries that raise grain yielded the following results.

18 countries raised wheat.

16 countries raised maize.

12 countries raised rice.

9 raised wheat and maize.

3 raised maize and rice.

3 raised wheat and rice.

2 raised all three crops.

How many countries raised
a) none of the three crops?
b) exactly one of the three crops?
c) exactly two of the three crops?
d) wheat and maize, but not rice?
e) maize or rice, but not wheat?

9. A survey of 200 recent traffic accidents reported to an insurance company gave the following results.

107 involved more than $2000 in damages.

52 required an ambulance at the scene.

64 involved alcohol as a contributing factor.

33 involved more than $2000 in damages and required an ambulance at the scene.

40 involved more than $2000 in damages and involved alcohol as a contributing factor.

37 required an ambulance at the scene and involved alcohol as a contributing factor.

29 involved more than $2000 in damages, required an ambulance at the scene, and involved alcohol as a contributing factor.

How many of the accidents

a) involved $2000 or less in damages?

b) involved none of these things?

c) involved more than $2000 in damages and required an ambulance at the scene, but did not involve alcohol?

d) involved more than $2000 in damages or required an ambulance at the scene?

e) involved at least one of these things?

10. A fast-food restaurant compiled the following information regarding its employees.

9 cooked.

6 worked the drive-through window.

23 operated the cash register.

2 cooked and worked the drive-through window at different times.

3 cooked and operated the cash register at different times.

6 worked the drive-through window and operated the cash register at different times.

2 did all three jobs.

4 did none of the three jobs.

How many of the employees

a) did at least two of these jobs?

b) worked at the restaurant?

c) did only one of the jobs?

d) worked the drive-through window or operated the cash register?

e) worked the drive-through window and operated the cash register, but did not cook?

11. Dreamy Cream Ice Cream hired Molly to find out what kind of ice cream people liked. She surveyed 100 people, with the following results: 78 liked hard ice cream, 61 liked soft ice cream, and 40 liked both hard ice cream and soft ice cream. Every person interviewed liked one or the other or both kinds of ice cream. Does this result seem right? Explain your answer.

12. An immigration agent sampled cars going from the United States into Canada. In his report he indicated that of the 85 cars sampled,

35 cars were driven by women.

53 cars were driven by U.S. citizens.

43 cars had two or more passengers.

27 cars were driven by women who are U.S. citizens.

25 cars were driven by women and had two or more passengers.

20 cars were driven by U.S. citizens and had two or more passengers.

15 cars were driven by women who are U.S. citizens and had two or more passengers.

After his supervisor reads the report, she explains to the agent that he made a mistake. Explain how his supervisor knew that the agent's report contained an error.

Problem Solving/Group Activities

13. At a rodeo, 24 of the contestants entered three events—calf roping, steer wrestling, bronco riding.

8 who roped calves did not wrestle steers.

8 who wrestled steers did not rope calves.

3 of the 8 who entered the bronco-riding event entered only that event.

6 entered only the calf-roping contest.

1 entered all three events.

How many contestants entered

a) only the steer-wrestling event?

b) only one event?

c) the steer-wrestling or the calf-roping or the bronco-riding events?

d) the calf-roping and the bronco-riding events, or the steer-wrestling event?

14. At an airplane service center 60 planes were inspected. Assume that 33 planes needed new tires and 44 planes needed new brakes.

a) What is the least number of planes that could have needed both tires and brakes?

b) What is the greatest number of planes that could have needed both?

c) What is the greatest number of planes that could have needed neither?

15. A survey of 500 farmers in a midwestern state showed the following.

125 grew only wheat.

110 grew only corn.

90 grew only oats.

200 grew wheat.

60 grew wheat and corn.

50 grew wheat and oats.

180 grew corn.

Find the number who
a) grew at least one of the three.
b) grew all three.
c) did not grow any of the three.
d) grew exactly two of the three.

16. Brothers Tom and Rob Reardon and their respective families vacation together annually. When the two families discussed where to go this year they found that, of all their children,

> 8 refused to go to the family cottage,
>
> 7 refused to go to the ocean,
>
> 9 refused to go camping,
>
> 3 will go neither to the family cottage nor to the ocean,
>
> 4 will go neither to the ocean nor camping,
>
> 6 will go neither to the family cottage nor camping,
>
> 2 will not go to the family cottage or the ocean or camping, and

zero children will agree to go to all three places.

What is the total number of children in the two families?

Research Activity

17. Using the most recent available statistics, determine the total number of deaths in the United States related to heart disease.

a) Determine the number of males, the number of people over 45, and the number of males over 45 who were victims of heart disease. Let all the people in the United States be the universal set. Draw a Venn diagram that illustrates this information, including the number of people in each of the four regions.

b) Describe the meaning of the numbers in each region of the Venn diagram.

(Sources of information include the *Universal Almanac* and the *Statistical Abstract of the United States*.)

2.6 INFINITE SETS

On page 46 we state that a finite set is a set in which the number of elements is zero or can be expressed as a natural number. On page 47 we define a one-to-one correspondence. To determine the number of elements in a finite set we can place it in a one-to-one correspondence with a subset of the set of counting numbers. For example, the set $A = \{\#, ?, \$\}$ can be placed in one-to-one correspondence with set $B = \{1, 2, 3\}$, a subset of the set of counting numbers.

$$A = \{ \#, ?, \$ \}$$
$$\downarrow \quad \downarrow \quad \downarrow$$
$$B = \{ 1, 2, 3 \}$$

Because the cardinal number of set B is 3, the cardinal number of set A is also 3. Any two sets such as A and B that can be placed in a one-to-one correspondence must have the same number of elements (therefore the same cardinality) and must be equivalent sets. Note that $n(A)$ and $n(B)$ both equal 3.

The German mathematician Georg Cantor (1845–1918), known as the father of set theory, thought about sets that were not bounded. An unbounded set he called an *infinite set* and provided the following definition.

> An **infinite set** is a set that can be placed in a one-to-one correspondence with a proper subset of itself.

In Example 1 we use Cantor's definition of infinite sets to show that the set of counting numbers is infinite.

EXAMPLE 1

Show that $N = \{1, 2, 3, 4, 5, \ldots, n, \ldots\}$ is an infinite set.

Solution: To show that the set N is infinite we establish a one-to-one correspondence between the counting numbers and a proper subset of itself. By removing the first element from the set of counting numbers, we get the set $\{2, 3, 4, 5, \ldots\}$, which is a proper subset of the set of counting numbers. Now we establish the one-to-one correspondence.

$$
\begin{array}{l}
\text{Counting numbers} = \{\ 1,\ \ 2,\ \ 3,\ \ 4,\ \ 5,\ldots,\quad n\quad,\ldots\} \\
\qquad\qquad\qquad\quad\ \downarrow\ \ \downarrow\ \ \downarrow\ \ \downarrow\ \ \downarrow\qquad\quad\ \downarrow \\
\text{Proper subset}\quad\ = \{\ 2,\ \ 3,\ \ 4,\ \ 5,\ \ 6,\ldots, n+1,\ldots\}
\end{array}
$$

Note that for any number, n, in the set of counting numbers, its corresponding number in the proper subset is one greater, or $n + 1$. We have now shown the desired one-to-one correspondence and thus the set of counting numbers is infinite.

Note in Example 1 that we showed the pairing of the general terms $n \rightarrow (n + 1)$. Showing a one-to-one correspondence of infinite sets requires showing the pairing of the general terms in the two infinite sets.

In the set of counting numbers n represents the general term. For any other set of numbers the general term will be different. A general term should be written in terms of n such that when 1 is substituted for n in the general term we get the first number in the set; when 2 is substituted for n in the general term we get the second number in the set; when 6 is substituted for n in the general term we get the sixth number in the set; and so on.

Consider the set $\{4, 9, 14, 19, \ldots\}$. Suppose that we want to write a general term for this set (or sequence) of numbers. What would the general term be? The numbers differ by 5, so the general term will be of the form $5n$ plus or minus some number. Substituting 1 for n yields $5(1)$, or 5. Because the first number in the set is 4, we need to subtract 1 from the 5. Thus the general term is $5n - 1$. Note that when $n = 1$, the value is $5(1) - 1$ or 4; when $n = 2$, the value is $5(2) - 1$ or 9; when $n = 3$, the value is $5(3) - 1$ or 14; and so on. Therefore we write the set of numbers with a general term as

$$\{4, 9, 14, 19, \ldots, 5n - 1, \ldots\}$$

Now that you are aware of how to determine the general term of a set of numbers, we can do some more problems involving sets.

EXAMPLE 2

Show that the set of even counting numbers $\{2, 4, 6, \ldots, 2n, \ldots\}$ is an infinite set.

Solution: First, create a proper subset of the set of even counting numbers by removing the first number from the set. Then establish a one-to-one correspondence.

$$
\begin{array}{l}
\text{Even counting numbers:}\quad \{\ 2,\ \ 4,\ \ 6,\ \ 8,\ldots,\quad 2n\quad,\ldots\} \\
\qquad\qquad\qquad\qquad\qquad\quad\ \downarrow\ \ \downarrow\ \ \downarrow\ \ \downarrow\qquad\quad\ \downarrow \\
\text{Proper subset:}\qquad\qquad\ \{\ 4,\ \ 6,\ \ 8,\ 10,\ldots, 2n+2,\ldots\}
\end{array}
$$

A one-to-one correspondence exists between the two sets, so the set of even counting numbers is infinite.

EXAMPLE 3

Show that the set $\{5, 10, 15, 20, \ldots, 5n, \ldots\}$ is an infinite set.

Solution:

Given set: $\{\ 5\ ,\ 10,\ 15,\ 20,\ 25, \ldots,\quad 5n\quad, \ldots\}$

Proper subset: $\{10,\ 15,\ 20,\ 25,\ 30, \ldots, 5n + 5, \ldots\}$

Therefore the given set is an infinite set. ▲

Countable Sets

In his work with infinite sets, Cantor developed ideas on how to determine the cardinal number of an infinite set. He called the cardinal number of infinite sets "transfinite cardinal numbers" or "transfinite powers." He defined a set as countable if it is finite or if it can be placed in a one-to-one correspondence with the set of counting numbers. All infinite sets that can be placed in a one-to-one correspondence with the set of counting numbers have cardinal number, aleph-null, symbolized $\aleph_0$ (the first Hebrew letter, aleph, with a zero subscript, read "null").

EXAMPLE 4

Show that the set of even counting numbers has cardinal number $\aleph_0$.

Solution: In Example 2 we showed that a set of even counting numbers is infinite by setting up a one-to-one correspondence between the set and a proper subset of itself.

Now we will show that it is countable and has cardinality $\aleph_0$ by setting up a one-to-one correspondence between the set of counting numbers and the set of even counting numbers.

Counting numbers: $N = \{\ 1,\ 2,\ 3,\ 4, \ldots,\ n, \ldots\}$

Even counting numbers: $E = \{\ 2,\ 4,\ 6,\ 8, \ldots, 2n, \ldots\}$

For each number n in the set of counting numbers, its corresponding number is $2n$. Since we found a one-to-one correspondence, the set of even counting numbers is not only infinite, it is also countable. Thus the cardinal number of the set of even counting numbers is $\aleph_0$; that is, $n(E) = \aleph_0$. ▲

> Any set that can be placed in a one-to-one correspondence with the set of counting numbers has cardinality $\aleph_0$ and is countable.

EXAMPLE 5

Show that the set of odd counting numbers has cardinality $\aleph_0$.

Solution: To show that the set of odd counting numbers has cardinality $\aleph_0$, we need to show a one-to-one correspondence between the counting numbers and the odd counting numbers.

Counting numbers: $N = \{\ 1,\ 2,\ 3,\ 4,\ 5,\dots,\quad n\quad,\dots\}$

Odd counting numbers: $O = \{\ 1,\ 3,\ 5,\ 7,\ 9,\dots,\ 2n-1,\dots\}$

Since there is a one-to-one correspondence, the odd counting numbers have cardinality $\aleph_0$; that is, $n(O) = \aleph_0$.

We have shown that both the odd and even counting numbers have cardinality $\aleph_0$. Merging the odd counting numbers with the even counting numbers gives the set of counting numbers, and we may reason that

$$\aleph_0 + \aleph_0 = \aleph_0.$$

... where there's always room for one more ...

This result may seem strange, but it is true. However, what could such a statement mean? Well, consider a hotel with infinitely many rooms. If all the rooms are occupied then the hotel is, of course, full. If more guests appear wanting accommodations, will they be turned away? The answer is *no*, for if the room clerk were to reassign each guest to a new room with a room number twice that of the present room, then all the odd-numbered rooms would become unoccupied and there would be space for more guests!

In Cantor's work he showed that there are different orders of infinity. Sets that are countable and have cardinal number $\aleph_0$ are the lowest order of infinity. Cantor showed that the set of integers and the set of rational numbers (fractions of the form p/q, where $q \neq 0$) are infinite sets with cardinality $\aleph_0$. He also showed that the set of real numbers (to be discussed in Chapter 5) could not be placed in a one-to-one correspondence with the set of counting numbers, and that they have a higher order of infinity, aleph-one, $\aleph_1$.

Section 2.6 Exercises

In Exercises 1–10, show that the set is infinite by placing it in a one-to-one correspondence with a proper subset of itself. Be sure to show the pairing of the general terms in the sets.

1. $\{3, 4, 5, 6, 7, \dots\}$ **2.** $\{4, 8, 12, 16, 20, \dots\}$

3. $\{2, 3, 4, 5, 6, \dots\}$ **4.** $\{3, 5, 7, 9, 11, \dots\}$

5. $\{4, 7, 10, 13, 16, \dots\}$ **6.** $\{6, 11, 16, 21, 26, \dots\}$

7. $\{8, 10, 12, 14, 16, \dots\}$ **8.** $\{1, \frac{1}{2}, \frac{1}{3}, \frac{1}{4}, \frac{1}{5}, \dots\}$

9. $\{1, \frac{1}{3}, \frac{1}{5}, \frac{1}{7}, \frac{1}{9}, \dots\}$ **10.** $\{\frac{5}{8}, \frac{6}{8}, \frac{7}{8}, \frac{8}{8}, \frac{9}{8}, \dots\}$

In Exercises 11–20, show that the set has cardinal number $\aleph_0$ by establishing a one-to-one correspondence between the set of counting numbers and the given set. Be sure to show the pairing of the general terms in the sets.

11. $\{3, 6, 9, 12, 15, \dots\}$

12. $\{100, 101, 102, 103, 104, \dots\}$

13. $\{4, 6, 8, 10, 12, \dots\}$ **14.** $\{0, 2, 4, 6, 8, \dots\}$

15. $\{2, 5, 8, 11, 14, \dots\}$ **16.** $\{4, 9, 14, 19, 24, \dots\}$

17. $\{5, 8, 11, 14, 17, \dots\}$ **18.** $\{\frac{1}{2}, \frac{1}{4}, \frac{1}{6}, \frac{1}{8}, \dots\}$

19. $\{\frac{1}{3}, \frac{1}{4}, \frac{1}{5}, \frac{1}{6}, \frac{1}{7}, \dots\}$ **20.** $\{\frac{1}{2}, \frac{2}{3}, \frac{3}{4}, \frac{4}{5}, \frac{5}{6}, \dots\}$

Problem Solving/Group Activities

In Exercises 21–24, show that the set has cardinality $\aleph_0$ by establishing a one-to-one correspondence between the set of counting numbers and the given set.

21. $\{1, 4, 9, 16, 25, 36, \dots\}$

22. $\{2, 4, 8, 16, 32, \dots\}$

23. $\{3, 9, 27, 81, 243, \dots\}$

24. $\{\frac{1}{3}, \frac{1}{6}, \frac{1}{12}, \frac{1}{24}, \frac{1}{48}, \dots\}$

Research Activities

25. Do research to explain how Cantor proved that the set of rational numbers has cardinal number $\aleph_0$.

26. Do research to explain how it can be shown that the real numbers do not have cardinal number $\aleph_0$.

CHAPTER 2 SUMMARY

Key Terms

2.1

cardinal number
counting number
description of a set
element of a set
empty set
equal sets
equivalent sets
finite set
infinite set
member of a set
natural number
null set
one-to-one correspondence
ordinal number
roster form
set
set-builder notation
universal set
well-defined set

2.2

proper subset
subset

2.3

Cartesian product
complement of a set
difference of sets
disjoint sets
intersection of sets
union of sets

2.4

De Morgan's laws

2.6

aleph-null
countable set
infinite set

Important Facts

For any sets A and B:
$$n(A \cup B) = n(A) + n(B) - n(A \cap B).$$
Number of distinct subsets of a finite set $= 2^n$.

Chapter 2 Review Exercises

2.1, 2.2, 2.6

In Exercises 1–14, answer "true" or "false." If false, give a reason.

1. The set of Kings of Egypt who served in the United States Senate is a well-defined set.

2. The set of bears at Yellowstone National Park who eat too much "people food" is a well-defined set.

3. crow $\in$ {robin, sparrow, cardinal, wren, owl}

4. { } $\subset \varnothing$

5. {3, 6, 9, 12, . . .} and {2, 4, 6, 8, . . .} are disjoint sets.

6. {23, 33, 43, 53, . . .} is an example of a set in roster form.

7. {computer, calculator, pencil} = {calculator, computer, diskette}

8. {1, 3, 5, 9} is equivalent to {3, 9, 15, 27}.

9. If A = {a, e, i, o, u} then $n(A)$ = 5.

10. A = {1, 4, 9, 16, . . .} is a countable set.

11. A = {1, 4, 7, 10, . . . , 31} is a finite set.

12. {3, 6, 7} $\subseteq$ {7, 6, 3, 5}.

13. $\{x \mid x \in N$ and $3 < x \le 5\}$ is a set in set-builder form.

14. $\{x \mid x \in N$ and $5 < x \le 15\} \subseteq \{1, 2, 3, 4, 5, . . . , 20\}$.

In Exercises 15–18, express each set in roster form.

15. Set A is the set of odd natural numbers between 3 and 18.

16. Set B is the set of states that border the Gulf of Mexico.

17. $C = \{x \mid x \in N$ and $x < 454\}$

18. $D = \{x \mid x \in N$ and $3 < x \le 35\}$

In Exercises 19–22, express each set in set-builder notation.

19. Set A is the set of natural numbers between 25 and 150.

20. Set B is the set of natural numbers greater than 79.

21. Set C is the set of natural numbers less than 3.

22. Set *D* is the set of natural numbers between 23 and 41, inclusive.

In Exercises 23–26, express each set with a written description.

23. *A* = {*x* | *x* is a letter of the English alphabet from E through M inclusive}

24. *B* = {penny, nickel, dime, quarter, half-dollar}

25. *C* = {ABC, CBS, NBC}

26. *D* = {*x* | *x* is an English teacher who owns a red car and does not live in Nebraska}

In Exercises 27–32, let

$$U = \{1, 3, 5, 7, 9, 11, 13, 15\}$$
$$A = \{1, 3, 5, 7\}$$
$$B = \{5, 7, 9, 13\}$$
$$C = \{1, 7, 13\}$$

Determine the following.

27. $A \cap B$ **28.** $A \cup B'$

29. $A' \cap B$ **30.** $(A \cup B)' \cup C$

31. The number of subsets of set *B*.

32. The number of proper subsets of set *A*.

33. For the following sets, construct a Venn diagram and place the elements in the proper regions.

   U = {penny, nickel, dime, quarter, half-dollar, dollar}

   A = {dime, half-dollar, dollar}

   B = {quarter, half-dollar, dollar}

In Exercises 34–39, use Fig. 2.28 to determine the sets.

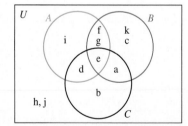

Figure 2.28

34. $A \cup B$ **35.** $A \cap B'$ **36.** $A \cup B \cup C$

37. $A \cap B \cap C$ **38.** $(A \cup B) \cap C$ **39.** $(A \cap B) \cup C$

2.4

Construct a Venn diagram to determine whether the following statements are true.

40. $(A' \cup B')' = A \cap B$

41. $(A \cup B') \cup (A \cup C') = A \cup (B \cap C)'$

2.5

42. A pizza chain was willing to pay $1 to each person interviewed about his or her likes and dislikes of types of pizza crust. Of the people interviewed, 200 liked thin crust, 270 liked thick crust, 70 liked both, and 50 did not like pizza at all. What was the total cost of the survey?

43. The Cookie Shoppe conducted a survey to determine its customers' preferences.

   200 people liked chocolate chip cookies.

   190 people liked peanut butter cookies.

   210 people liked sugar cookies.

   100 people liked chocolate chip and peanut butter cookies.

   150 people liked peanut butter and sugar cookies.

   110 people liked chocolate chip and sugar cookies.

   70 people liked all three.

   5 people liked none of these cookies.

How many people

   a) completed the survey?

   b) liked only peanut butter cookies?

   c) liked peanut butter and chocolate chip, but not sugar cookies?

   d) liked peanut butter cookies or sugar cookies, but not chocolate chip cookies?

44. A survey was taken of 62 people who live around a lake. Its aim was to determine their daily involvement in water activities in the summer months. The survey found that

   15 only swam,

   19 only sailed,

   10 water skied,

   7 swam and water skied,

   5 sailed and water skied,

   8 swam and sailed, and

   3 swam, sailed, and water skied.

Determine the number of people who

   a) participated in none of the water sports.

   b) participated in exactly two of the water sports.

   c) participated in exactly one of the water sports.

   d) swam and water skied, but did not sail.

   e) water skied, but did not swim or sail.

2.6

In Exercises 45 and 46, show that the sets are infinite by placing each set in a one-to-one correspondence with a proper subset of itself.

45. {2, 4, 6, 8, 10, . . .}   **46.** {3, 5, 7, 9, 11, . . .}

In Exercises 47 and 48, show that each set has cardinal number $\aleph_0$ by setting up a one-to-one correspondence between the set of counting numbers and the given set.

47. {5, 8, 11, 14, 17, . . .}   **48.** {4, 9, 14, 19, 24, . . .}

CHAPTER TEST

In Exercises 1–12, state whether each is "true" or "false." If the statement is false explain why.

1. $\{3, a, 7, g\}$ is equivalent to $\{a, 7, 3, f\}$.

2. $\{1, 3, 7, p\} = \{p, 3, 7, 2\}$

3. $\{g, h, i\} \subset \{a, r, g, h, i, p\}$

4. $\{3\} \in \{2, \{3\}, 4, 8, 9\}$

5. $\{7\} \subseteq \{x \mid x \in N \text{ and } x < 6\}$

6. $\{\ \} \not\subset \{\ \}$

7. $\{x, y, z\}$ has seven subsets.

8. If $A \cup B = \{\ \}$, then A and B are disjoint sets.

9. For any set A, $A \cup A' = \{\ \}$.

10. For any set A, $A \cap U = A$.

In Exercises 11 and 12, use set $A = \{x \mid x \in N \text{ and } x < 6\}$.

11. Write set A in roster form.

12. Write a description of set A.

In Exercises 13–17, use the following information.

$U = \{3, 5, 7, 9, 11, 13, 15\}$
$A = \{3, 5, 7, 9\}$
$B = \{7, 9, 11, 13\}$
$C = \{3, 11, 15\}$

Determine the following.

13. $A \cup B$

14. $A \cap C'$

15. $A \cap (B \cap C)'$

16. $n(A \cap B')$

17. Draw a Venn diagram illustrating the relationship among sets U, A, B, and C.

18. Use a Venn diagram to determine whether

$$A \cap (B \cup C') = (A \cap B) \cup (A \cap C')$$

for all sets A, B, and C.

19. In a survey, 500 people at a shopping mall were asked to indicate which sporting events they enjoyed watching on television. The three choices were football, baseball, and hockey. The results showed that

224 enjoyed watching football,
196 enjoyed watching baseball,
73 enjoyed watching hockey,
155 enjoyed watching football and baseball,
59 enjoyed watching football and hockey,
27 enjoyed watching baseball and hockey, and
23 enjoyed watching all three sports.

a) Construct a Venn diagram and record the appropriate number in each region.
b) How many enjoyed watching only one of these sports?
c) How many enjoyed watching none of the sports?
d) How many enjoyed watching exactly two of these sports?
e) How many enjoyed watching at least two of these sports?
f) How many enjoyed watching football or hockey, but not baseball?
g) How many enjoyed watching football and hockey, but not baseball?

20. Show that the following set is infinite by setting up a one-to-one correspondence between the set and a proper subset of itself.

$$\{7, 8, 9, 10, \ldots\}$$

21. Show that the following set has cardinal number $\aleph_0$ by setting up a one-to-one correspondence between the set of counting numbers and the set.

$$\{1, 3, 5, 7, \ldots\}$$

GROUP PROJECTS

Endangered Species

1. On June 3, 1994, the U.S. Wildlife Service provided a worldwide list of 1331 endangered and threatened species. The lists of endangered and threatened species are disjoint. A research group would like to have the lists separated into the following groups: endangered species, birds, and those species that live in the United States. Use the following information to construct a Venn diagram and answer the questions.

    1131 species are on the endangered list.

    200 species are on the threatened list.

    217 species are birds.

    637 species that live in the United States are endangered.

    210 species are birds on the endangered list.

    0 birds that are on the threatened species list live in foreign countries.

    56 species are birds on the United States endangered list.

    532 species are on the foreign lists.

    a) How many of the endangered species in the United States are not birds?

    b) How many birds are on the threatened species list?

    c) How many threatened species that are not birds live in foreign countries?

    d) How many threatened species that are not birds live in the United States?

Selecting a Family Pet

2. The Geater family is considering buying a dog. They have established several criteria for the family dog: It must be pedigreed, must not shed, and must be less than 16 in. tall, and its level of sociability with children must be good.

    a) Using the information in the following table*, construct a Venn diagram in which the universal set is the pedigreed dogs listed. Indicate the set of dogs to be placed in each region of the Venn diagram.

    b) From the Venn diagram constructed in part (a) determine which dogs will meet the criteria set by the Geater family. Explain.

    c) Suppose that the universal set is the set of dogs listed in the table that are less than 16 in. tall. Also, suppose that in addition to the criteria already stated the Geater family requires the dog to be rated good in obedience. Construct a Venn diagram and indicate the set of dogs to be placed in each region.

Pedigreed Breed	Shed		Height		Sociability with Children			Obedience		
	Yes	No	Less than 16"	16" or more	Good	Fair	Poor	Good	Fair	Poor
Airedale		×		×	×				×	
Basset hound	×		×		×				×	
Beagle	×		×		×					×
Border terrier		×	×		×			×		
Cairn terrier		×	×			×			×	
Cocker spaniel	×		×		×			×		
Collie	×			×	×			×		
Dachshund	×		×				×			×
Poodle, miniature		×	×			×		×		
Schnauzer, miniature		×	×				×		×	
Scottish terrier		×	×			×				×
Wirehaired fox terrier		×	×			×				×

*The information is a collection of the opinions of an animal psychologist, Dr. Daniel Tortora, and a group of veterinarians.

d) From the Venn diagram constructed in part (c) determine which dogs will meet the criteria set by the Geater family. Explain.

Classification of the Domestic Cat

3. Read the Did You Know feature on page 54. Do research and indicate the name of the following groupings to which the domestic cat belongs.

- **a)** Kingdom
- **b)** Phylum
- **c)** Class
- **d)** Order
- **e)** Family
- **f)** Genus
- **g)** Species

Who Lives Where

4. On Diplomat Row, a suburb of Washington, D.C., there are five houses. Each owner is a different nationality, each has a different pet, each has a favorite food and a different favorite drink, and each house is painted a different color.

> The green house is directly to the right of the ivory house.

The Senegalese has the red house.

The dog belongs to the Spaniard.

The Afghanistani drinks tea.

The person who eats cheese lives next door to the fox.

The Japanese eats fish.

Milk is drunk in the middle house.

Apples are eaten in the house next to the horse.

Ale is drunk in the green house.

The Norwegian lives in the first house.

The peach eater drinks whiskey.

Apples are eaten in the yellow house.

The banana eater owns a snail.

The Norwegian lives next door to the blue house.

For each house find

- **a)** the color.
- **b)** the nationality of the occupant.
- **c)** the type of food eaten.
- **d)** the owner's favorite drink.
- **e)** the owner's pet.
- **f)** Finally, the crucial question is: Does the zebra's owner drink vodka or ale?

Chapter 3 Logic

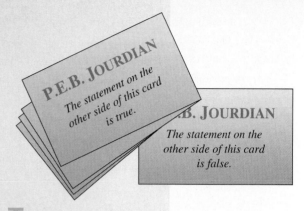

That's highly illogical, Captain," says the Vulcan, Mr. Spock, on the TV reruns of "Star Trek." His statement is full of scorn for the often illogical and emotional workings of human interaction.

However, by using deductive reasoning we too can analyze complicated situations and come to a reasonable conclusion from a given set of information.

The ancient Greeks were the first people to analyze systematically the way people think and arrive at a conclusion. Aristotle, whose study of logic is presented in a work called *Organon*, is called the "father of logic." Since Aristotle's time, the study of logic has been continued by other great mathematicians, including Wilhelm Leibniz and George Boole. Perhaps the most popular logician of all time was Charles Dodgson, better known by his pen name, Lewis Carroll, author of *Alice's Adventures in Wonderland* and *Through the Looking Glass*. Though written for children, Carroll's books contain many puzzles and references to concepts of logic.

Though most people believe that logic deals with the way people think, it is important to realize that logic does not, in fact, describe the actual way in which the human thought process works. Even the most rational people often think irrationally or use false information. Would we want to live in a world where everyone's thoughts followed the strict rules of logical thinking? It is unlikely, since some of the world's greatest inventions, discoveries, and works of art have resulted from creative, nonrational inspiration. For example, consider the German chemist

Logical reasoning can tell us whether a conclusion follows from a set of premises, but not whether those premises are true. For example, Greek astronomers, using the assumption that the planets revolved around the earth, correctly predicted the positions of the planets even though their premise was false.

"and how do you know that you're mad?"

"To begin with," said the Cat, "a dog's not mad. You grant that?"

"I suppose so," said Alice.

"Well then," the Cat went on, "you see a dog growls when it's angry, and wags its tail when it's pleased. Now I growl when I'm pleased, and wag my tail when I'm angry. Therefore I'm mad."

Friedrich Kekulé, who reported that his discovery of the ring structure of benzene came to him in a dream in which dancing snakes formed a ring by taking one another's tails in their mouths.

If human thought does not always follow the rules of logic, then why do we study it? Logic enables us to communicate effectively, to make more convincing arguments, and to develop patterns of reasoning for decision making. The study of logic also prepares an individual to better understand other areas of mathematics, computer programming and design, and in general the thought process involved in learning any subject. ■

Every day, trillions of microscopic switches turn off and on as computers everywhere process and store information. Here is a microscopic view of only a few of the more than 4 million switches found in a computer chip. The encoding and manipulation of data by the simple on/off of a switch had its beginning in the work of English mathematician George Boole.

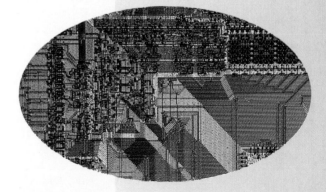

3.1 STATEMENTS AND LOGICAL CONNECTIVES

History

The ancient Greeks were the first people to analyze systematically the way humans think and arrive at conclusions. Aristotle (384–322 B.C.) organized the study of logic for the first time in a work called *Organon*. As a result of his work, Aristotle is called the Father of Logic. The logic from this period, called Aristotelian logic, has been taught and studied for more than 2000 years.

Since Aristotle's time the study of logic has been continued by other great philosophers and mathematicians. Gottfried Wilhelm Leibniz (1646–1716) had a deep conviction that all mathematical and scientific concepts could be derived from logic. As a result he became the first serious student of symbolic logic. One difference between symbolic logic and Aristotelian logic is that in symbolic logic, as its name implies, symbols (usually letters) represent written statements. The forms of the statements in the two types of logic are different. The self-educated English mathematician George Boole (1815–1864) is considered to be the founder of symbolic logic because of his impressive work in this area. Among Boole's publications are *The Mathematical Analysis of Logic* (1847) and *An Investigation of the Law of Thought* (1854).

British philosophers Alfred North Whitehead (1861–1947) and Bertrand Russell (1872–1970) wrote a detailed development of arithmetic starting with only undefined concepts and assumptions of logic. Their work appeared in three large volumes called *Principia Mathematica*, published between 1910 and 1913. Charles Dodgson, better known as Lewis Carroll, incorporated many interesting ideas from logic into his books *Alice's Adventures in Wonderland* and *Through the Looking Glass* and his other children's stories.

Logic has been studied through the ages to exercise the mind's ability to reason. Understanding logic will enable you to think clearly, communicate effectively, make more convincing arguments, and develop patterns of reasoning that will help you in making decisions. It will also help you to detect the fallacies in the reasoning or arguments of others. Studying logic has practical applications as well, such as helping you to understand wills, contracts, and other legal documents.

The study of logic is also good preparation for other areas of mathematics. If you preview Chapter 11, on probability, you will see formulas for the probability of *a* or *b* and the probability of *a* and *b*, symbolized as $P(A \text{ or } B)$ and $P(A \text{ and } B)$, respectively. Special meanings of common words such as *or* and *and* apply to all areas of mathematics. The meaning of these and other special words will be discussed in this chapter.

Logic and the English Language

In reading, writing, and speaking we use many words such as *and*, *or*, and *if . . . then . . .* to connect thoughts. In logic we call these words connectives. How are these words interpreted in daily communication? A judge announces to a convicted offender: ''I hereby sentence you to five months of community service *and* a fine of \$100.'' In this case we normally interpret the word *and* to indicate that *both* events will take place. That is, the person must do community service and must also pay a fine.

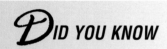

DID YOU KNOW

PLAYING ON WORDS

*B*oole, De Morgan, and other mathematicians of the nineteenth century were anxious to make logic an abstract science that would operate like algebra but be applicable to all fields. One of the problems logicians faced was that verbal language could be ambiguous and easily lead to confusion and contradiction. Abbott and Costello had fun with the ambiguity of language in their skit about the baseball players: "Who's on first, What's on second, I Don't Know is on third—Yeah, but who's on first?" (in the film *The Naughty Nineties*, 1945). ■

Now suppose that a judge states: "I sentence you to six months in prison *or* 10 months of community service." In this case we interpret the connective *or* as meaning the convicted person must either spend time in jail or do community service, but not both. The word *or* in this case is the exclusive or. It excludes one of the possibilities named; one or the other event takes place, but *not both*.

In a restaurant a waiter asks: "May I interest you in a cup of soup or a sandwich?" This question offers three possibilities: You may order soup, you may order a sandwich, or you may order both soup and a sandwich. The *or* in this case is the inclusive or. It includes the possibilities named individually or in combination; one or the other *or both* events can take place. In this chapter, when we use the word *or* in a logic statement it will mean the *inclusive or* unless stated otherwise.

If–then statements are often used to relate two ideas, as in the bank policy statement "If the average daily balance is greater than $500 then there will be no service charge." If–then statements are also used to emphasize a point or add humor, as in the statement "If the Cubs win then I will be a monkey's uncle."

Now let's look at logic from a mathematical point of view.

Statements and Logical Connectives

A sentence that can be judged either true or false is called a statement. Labeling a statement true or false is called *assigning a truth value*. Here are some examples of statements.

1. The famous Gateway Arch in St. Louis is exactly as wide as it is tall.

2. The Apple Power Macintosh is a computer.

3. New Orleans is on the Snake River.

In each case we can say that the sentence is either true or false. Statement 1 is true because the height and width of the arch are both 630 feet. Statement 2 is true. By looking at a map we can see that statement 3 is false; New Orleans is on the Mississippi River, not the Snake River.

The three sentences are examples of simple statements because they convey one idea. Sentences combining two or more ideas that can be assigned a truth value are called compound statements. Compound statements will be discussed shortly.

Quantifiers

Sometimes it is necessary to change a statement to its opposite meaning. To do so we use the negation of a statement. For example the negation of the statement "Emily is at home" is "Emily is not at home." The negation of a true statement is always a false statement and the negation of a false statement is always a true statement. We must use special caution when negating statements containing the words all, none (or no), and some. These words are referred to as quantifiers.

Consider the statement "All lakes contain fresh water." We know that this statement is false because the Great Salt Lake in Utah contains salt water. Its negation must therefore be true. We may be tempted to write its negation as "No lake contains fresh water," but this statement is also false because Lake Superior contains fresh water. Therefore "No lakes contain fresh water" is not the negation of "All lakes contain fresh water." The correct negation of "All lakes contain fresh water" is "Not all lakes contain fresh water" or "At least one lake does not contain fresh water'' or ''Some lakes do not contain fresh water.'' These statements all imply that at least one lake does not contain fresh water, which is a true statement.

Now consider the statement ''No birds can swim.'' This statement is false, since at least one bird, the penguin, can swim. Therefore the negation of this statement must be true. We may be tempted to write the negation as ''All birds can swim,'' but because it is also false it cannot be the negation. The correct negation of the statement is ''Some birds can swim'' or ''At least one bird can swim,'' which are true statements.

Now let's consider statements involving the quantifier *some*, as in ''Some students have a driver's license.'' This is a true statement, meaning that at least one student has a driver's license.'' The negation of this statement must therefore be false. The negation is ''No student has a driver's license,'' which is a false statement.

Consider the statement ''Some students do not ride motorcycles.'' This statement is true because it means ''At least one student does not ride a motorcycle.'' The negation of this statement must therefore be false. The negation is ''All students ride motorcycles,'' which is a false statement.

The negation of quantified statements is summarized as follows:

Form of Statement	Form of Negation
All are.	Some are not.
None are.	Some are.
Some are.	None are.
Some are not.	All are.

The following diagram might help you to remember the statements and their negations:

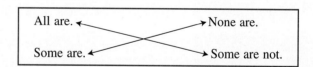

The quantifiers diagonally opposite each other are the negations of each other.

EXAMPLE 1

Write the negation of each statement.

a) Some dogs have tails.

b) All books have pictures.

Solution:

a) Since *some* means ''at least one,'' the statement ''Some dogs have tails'' is the same as ''At least one dog has a tail.'' Because it is a true statement, its negation must be false. The negation is ''No dogs have tails,'' which is a false statement.

b) The statement ''All books have pictures'' is false. Its negation must therefore be true. The negation may be written as ''Some books do not have pictures,'' or ''Not all books have pictures,'' or ''At least one book does not have pictures.'' Each of these statements is true.

Compound Statements

Statements consisting of two or more simple statements are compound statements. The connectives often used to join two simple statements are *and*, *or*, *if . . . then . . .*, and *if and only if*. In addition, we consider a simple statement that has been negated to be a compound statement.

To reduce the amount of writing in logic it is common to represent each simple statement with a lowercase letter and each connective with a symbol. It is customary to use the letters *p*, *q*, *r*, and *s* to represent simple statements. However, other letters may be used instead. Let's now look at the connectives used to make compound statements.

Not Statements

The negation is symbolized by $\sim$ and read "not." For example, the negation of the statement "Steve is a college student" is "Steve is not a college student." If *p* represents the simple statement "Steve is a college student," then $\sim p$ represents the compound statement "Steve is not a college student." For any statement *p*, $\sim(\sim p) = p$. For example, the negation of the statement "Steve is not a college student" is "Steve is a college student."

Consider the statement "Maria is not at home." This statement contains the word *not*, which indicates that it is a negation. To write this statement symbolically, we let *p* represent "Maria *is* at home." Then $\sim p$ would be "Maria is not at home." *We will use this convention of letting letters such as p, q, or r represent statements that are not negated. We will represent negated statements with the negation symbol, $\sim$.*

And Statements

The conjunction is symbolized by $\wedge$ and read "and." Let *p* and *q* represent the simple statements

    p: You will perform five months of community service.
    q: You will pay a $100 fine.

Then the following is the conjunction written in symbolic form.

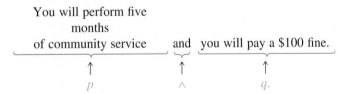

The conjunction is generally expressed as *and*. Other words sometimes used to express a conjunction are *but*, *however*, or *nevertheless*.

EXAMPLE 2

Write the following conjunction in symbolic form. The zebra is striped, but its mane is not white.

Solution: Let *z* and *m* represent the simple statements.

    z: The zebra is striped.
    m: Its mane is white.

In symbolic form the compound statement is $z \wedge \sim m$.

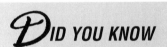

Or Statements

The **disjunction** is symbolized by ∨ and read ''or.'' The *or* we will use in this book (except where indicated in the exercise sets) is the *inclusive or* described on page 89.

EXAMPLE 3

Let   *p*:   Joyce will go to the movie.
                                         q: Joyce will eat popcorn.

Write the following statements in symbolic form.

a) Joyce will go to the movie or Joyce will eat popcorn.

b) Joyce will eat popcorn or Joyce will go to the movie.

c) Joyce will not eat popcorn or Joyce will go to the movie.

Solution:

a) $p \vee q$ **b)** $q \vee p$ **c)** $\sim q \vee p$ ▲

Because *or* represents the inclusive or, the statement ''Joyce will go to the movie or Joyce will eat popcorn'' in Example 3 may mean that Joyce will go to the movie or that Joyce will eat popcorn or that Joyce will go to the movie and also eat popcorn.

When a compound statement contains more than one connective, a comma can be used to indicate which simple statements are to be grouped together. When we write the compound statement symbolically, *the simple statements on the same side of the comma are to be grouped together within parentheses.*

For example, ''Long Shot is a horse (*h*) or Patti is a pilot (*p*), and Carl owns a car (*c*)'' is written $(h \vee p) \wedge c$. Note that *h* and *p* are both on the same side of the comma and are grouped together within parentheses. The statement ''Long Shot is a horse, or Patti is a pilot and Carl owns a car'' is written $h \vee (p \wedge c)$. In this case *p* and *c* are on the same side of the comma and are grouped together within parentheses.

EXAMPLE 4

Let   *p*:   Dinner includes soup.
                                         q: Dinner includes salad.
                                         r: Dinner includes the vegetable of the day.

Write the following statements in symbolic form.

a) Dinner includes soup, and salad or the vegetable of the day.

b) Dinner includes soup and salad, or the vegetable of the day.

Solution:

a) The comma tells us to group the statement ''Dinner includes salad'' with the statement ''Dinner includes the vegetable of the day.'' Note that both statements are on the same side of the comma. The statement in symbolic form is $p \wedge (q \vee r)$. In mathematics we always evaluate the information within the parentheses first. Since the conjunction, ∧, is outside the parentheses and is evaluated last, this statement is considered a conjunction.

b) The comma tells us to group the statement "Dinner includes soup" with the statement "Dinner includes salad." Note that both statements are on the same side of the comma. The statement in symbolic form is $(p \wedge q) \vee r$. Since the disjunction, $\vee$, is outside the parentheses and is evaluated last, this statement is considered a disjunction.   ▲

An important point to remember is that a negation has the effect of negating only the statement that directly follows it. To negate a compound statement, we must use parentheses. When a negation symbol is placed in front of a statement in parentheses, it negates the entire statement in parentheses. The negation symbol in this case is read, "It is not true that . . ." or "It is false that. . . ."

EXAMPLE 5

Let p: Tanya is the star in the home movies.
        q: Matthew is operating the camcorder.

Write the following symbolic statements in words.

a) $\sim p \wedge \sim q$ **b)** $\sim (p \vee q)$

Solution:

a) Tanya is not the star in the home movies and Matthew is not operating the camcorder.

b) It is false that Tanya is the star in the home movies or Matthew is operating the camcorder.   ▲

Part (a) of Example 5 is a conjunction, since it can be written $(\sim p) \wedge (\sim q)$. Part (b) is a negation, since it negates the entire statement. The similarity of these two statements is discussed in Section 3.4.

Occasionally, we come across a **neither–nor** statement. For example, "John is neither handsome nor rich." This statement means that John is not handsome *and* John is not rich. If p represents "John is handsome" and q represents "John is rich," this statement is symbolized by $\sim p \wedge \sim q$.

If–Then Statements

The **conditional** is symbolized by $\rightarrow$ and is read "if–then." The statement $p \rightarrow q$ is read "If p, then q"* The conditional statement consists of two parts; the part that precedes the arrow is the **antecedent**, and the part that follows the arrow is the **consequent**.† In the conditional statement $p \rightarrow q$, p is the antecedent and q is the consequent.

In the conditional statement $\sim (p \vee q) \rightarrow (p \wedge q)$, the antecedent is $\sim (p \vee q)$ and the consequent is $(p \wedge q)$. An example of a conditional statement is "If you drink your milk, then you will grow up to be healthy." A conditional symbol may be placed between any two statements even if the statements are not related.

Sometimes the word *then* in a conditional statement is not explicitly stated. For example, the statement "If you pass this course, I will buy you a car" is a conditional statement for it actually means "If you pass this course, then I will buy you a car."

---

*Some books indicate that $p \rightarrow q$ may also be read "p implies q." However, many higher level mathematics books indicate that $p \rightarrow q$ may be read "p implies q" only under certain conditions. Implications are discussed in Section 3.3.

†Some books refer to the antecedent as the hypothesis or premise, and the consequent as the conclusion.

EXAMPLE 6

Let

p: I go into the bookstore.
q: I will spend money.

Write the following statements symbolically.

a) If I go into the bookstore, then I will spend money.

b) If I do not go into the bookstore, then I will not spend money.

c) It is false that if I go into the bookstore then I will spend money.

Solution:

a) $p \rightarrow q$ **b)** $\sim p \rightarrow \sim q$ **c)** $\sim(p \rightarrow q)$ ▲

EXAMPLE 7

Let

p: Sally is enrolled in calculus.
q: Sally's major is nursing.
r: Sally's major is engineering.

Write the following symbolic statements in words and indicate whether the statement is a negation, conjunction, disjunction, or conditional.

a) $(q \rightarrow \sim p) \vee r$ **b)** $q \rightarrow (\sim p \vee r)$

Solution: The parentheses indicate where to place the commas in the sentences.

a) "If Sally's major is nursing then Sally is not enrolled in calculus, or Sally's major is engineering." This statement is a disjunction because $\vee$ is outside the parentheses.

b) "If Sally's major is nursing, then Sally is not enrolled in calculus or Sally's major is engineering." This is a conditional statement because $\rightarrow$ is outside the parentheses.   ▲

If and Only If Statements

The biconditional is symbolized by $\leftrightarrow$ and is read "if and only if." The phrase *if and only if* is sometimes abbreviated as "iff." The statement $p \leftrightarrow q$ is read "p if and only if q."

EXAMPLE 8

Let

p: The cow is brown.
q: The milk is chocolate.

Write the following symbolic statements in words.

a) $q \leftrightarrow p$ **b)** $\sim(p \leftrightarrow \sim q)$

Solution:

a) The milk is chocolate if and only if the cow is brown.

b) It is false that the cow is brown if and only if the milk is not chocolate.   ▲

You will learn later that $p \leftrightarrow q$ means the same as $(p \to q) \wedge (q \to p)$. Therefore the statement "I will go to college if and only if I can pay the tuition" has the same logical meaning as "If I go to college then I can pay the tuition and If I can pay the tuition then I will go to college."

The following is a summary of the connectives discussed in this section.

Formal Name	Symbol	Read	Symbolic Form
Negation	$\sim$	"Not"	$\sim p$
Conjunction	$\wedge$	"And"	$p \wedge q$
Disjunction	$\vee$	"Or"	$p \vee q$
Conditional	$\to$	"If–then"	$p \to q$
Biconditional	$\leftrightarrow$	"If and only if"	$p \leftrightarrow q$

Dominance of Connectives

What is the answer to the problem $2 + 3 \times 4$? Some of you might say 20, but others might say 14. If you evaluate $2 + 3 \times 4$ on a calculator by pressing

$$2 \boxed{+} 3 \boxed{\times} 4$$

some may give you the answer 14, whereas others may give you the answer 20. Which is the correct answer? In mathematics, unless otherwise changed by parentheses or some other grouping symbol, multiplication is *always* performed before addition. Thus

$$2 + 3 \times 4 = 2 + (3 \times 4) = 14$$

The calculators that gave the incorrect answer of 20 are basic calculators that are not programmed according to the order of operations used in mathematics.

Just as an order of operations exists in the evaluation of arithmetic expressions, a dominance of connectives is used in the evaluation of logic statements. How do we evaluate a symbolic logic statement when no parentheses are used? For example, does $p \vee q \to r$ mean $(p \vee q) \to r$, or does it mean $p \vee (q \to r)$? If we are given a symbolic logic statement for which grouping has not been indicated by parentheses, or a written logic statement for which grouping has not been indicated by a comma, then we use the dominance of connectives shown in Table 3.1. Note that *the least dominant connective is the negation and that the most dominant is the biconditional.*

Table 3.1 Dominance of Connectives

Least dominant →	1. Negation, $\sim$	← Evaluate first
	2. Conjunction, $\wedge$; disjunction, $\vee$	
	3. Conditional, $\to$	
Most dominant →	4. Biconditional, $\leftrightarrow$	← Evaluate last

As is indicated in Table 3.1, the conjunction and disjunction have the same level of dominance. Thus to determine whether the symbolic statement $p \wedge q \vee r$ is a conjunction or a disjunction, we have to use grouping symbols (parentheses). When evaluating a symbolic statement that does not contain parentheses, we *evaluate the least dominant connective first and the most dominant connective last.* For example,

Statement	Statement Means	Type of Statement
$\sim p \vee q$	$(\sim p) \vee q$	Disjunction
$p \rightarrow q \vee r$	$p \rightarrow (q \vee r)$	Conditional
$p \wedge q \rightarrow r$	$(p \wedge q) \rightarrow r$	Conditional
$p \rightarrow q \leftrightarrow r$	$(p \rightarrow q) \leftrightarrow r$	Biconditional
$p \vee r \leftrightarrow r \rightarrow \sim p$	$(p \vee r) \leftrightarrow (r \rightarrow \sim p)$	Biconditional
$p \rightarrow r \leftrightarrow s \wedge p$	$(p \rightarrow r) \leftrightarrow (s \wedge p)$	Biconditional

EXAMPLE 9

Use the dominance of connectives to add parentheses to each statement. Then indicate whether each statement is a negation, conjunction, disjunction, conditional, or biconditional.

a) $p \rightarrow q \vee r$ **b)** $\sim p \wedge q \leftrightarrow r \vee p$

Solution:

a) The conditional has greater dominance than the disjunction, so we place parentheses around $q \vee r$, as follows:

$$p \rightarrow (q \vee r)$$

It is a conditional statement because the conditional symbol is outside the parentheses.

b) The biconditional has greater dominance, so we place parentheses as follows:

$$(\sim p \wedge q) \leftrightarrow (r \vee p)$$

It is a biconditional statement because the biconditional symbol is outside the parentheses.   ▲

EXAMPLE 10

Use the dominance of connectives and parentheses to write each statement symbolically. Then indicate whether each statement is a negation, conjunction, disjunction, conditional, or biconditional.

a) If you are late in paying your rent or you have damaged the apartment then you may be evicted.

b) You are late in paying your rent, or if you have damaged the apartment then you may be evicted.

Solution:

a) Let
        p: You are late in paying your rent.
        q: You have damaged the apartment.
        r: You may be evicted.

No commas appear in the sentence, so we will evaluate it by using the dominance of connectives. Because the conditional has higher dominance than the disjunction, the conditional statement will be evaluated last. There-fore this statement written symbolically with parentheses is

$$(p \lor q) \to r$$

This statement is a conditional.

b) A comma is used in this statement to indicate grouping, just as parentheses do in arithmetic. The placement of the comma indicates that the statements ''you have damaged the apartment'' and ''you may be evicted'' are to be grouped together. Therefore this statement written symbolically is

$$p \lor (q \to r)$$

This statement is a disjunction. Note that the comma overrides the domi-nance of connectives and tells us to evaluate the conditional statement before the disjunction. ▲

Section 3.1 Exercises

1. What is a simple statement?

2. List the words identified as quantifiers.

3. Write the general form of the negation for statements of the form
   a) none are.
   b) some are not.
   c) all are.
   d) some are.

4. What are compound statements?

5. Represent the statement ''The ink is not purple'' symbolically. Explain your answer.

6. Draw the symbol used to represent the
   a) conditional.
   b) disjunction.
   c) conjunction.
   d) negation.
   e) biconditional.

7. Explain how a comma is used to indicate the grouping of simple statements.

8. List the dominance of connectives from the most dominant to the least dominant.

In Exercises 9–22, indicate whether the statement is a simple or a compound statement. If it is a compound statement, indicate whether it is a negation, conjunction, disjunction, conditional, or biconditional by using both the word and its appropriate symbol (for example, ''a negation,'' ~).

9. The lettuce is in the refrigerator and the bread is on the counter.

10. The stapler is not out of staples.

11. The figure is a triangle if and only if it has three sides.

12. If the phone rings, then you will answer it.

13. The store is on Texas Street or Nebraska Avenue.

14. The leather jacket is neither tan nor red.

15. The stop light stays red for 40 seconds.

16. Angie will pay the bill on the due date or she will be charged interest.

17. It is false that the grub killer will be applied to the lawn or we will get grubs.

18. If the dog barks, then the neighbors will get angry.

19. We decided not to go to the party, but we called to let the host know.

20. Nobody studied in the library, nevertheless everybody passed the test.

21. If the door is open, then the class has not started or the teacher is not there.

22. It is false that if there is a loss of power then the security system will not work.

In Exercises 23–36, write the negation of the statement.

23. Some toys cost more than $15.

24. No penguins fly.

25. Some dogs do not have fleas.

26. No item in the store costs more than $1.

27. All plants are living things.

28. Some doctors make house calls.

29. Some orchestras are having financial difficulty.

30. All tires are properly inflated.

31. No one likes a bully.

32. Some students maintain a straight A average.

33. Some rain forests are not being destroyed.

34. All people who earn money pay taxes.

35. All nurses have a degree.

36. Some people who earn money do not pay taxes.

In Exercises 37–42, write the statement in symbolic form. Let

  p: The motor vehicle department office is open.
  q: The line of people extends out the door.

37. The line of people does not extend out the door.

38. The motor vehicle department office is open and the line of people extends out the door.

39. The line of people does not extend out the door or the motor vehicle department office is not open.

40. The line of people does not extend out the door if and only if the motor vehicle department office is not open.

41. If the motor vehicle department office is not open, then the line of people does not extend out the door.

42. The line of people does not extend out the door, however the motor vehicle department office is open.

In Exercises 43–46, write the statement in symbolic form. Let

  p: Kate is a member of the marching band.
  q: Luke is a member of the jazz band.

43. If Kate is a member of the marching band, then Luke is a member of the jazz band.

44. Luke is a member of the jazz band but Kate is not a member of the marching band.

45. Neither is Kate a member of the marching band nor is Luke a member of the jazz band.

46. It is false that Kate is a member of the marching band and Luke is a member of the jazz band.

In Exercises 47–56, write the compound statement in words. Let

  p: The horse is five years old.
  q: The horse won the race.

47. ~*p*   48. ~*q*
49. *p* ∧ *q*   50. *q* ∨ *p*
51. ~*p* → *q*   52. ~*p* ↔ ~*q*
53. ~(*q* ∨ *p*)   54. ~*p* ∧ *q*
55. ~*p* ∨ ~*q*   56. ~(*p* ∧ *q*)

In Exercises 57–66, write the statement in symbolic form. Let

  p: The temperature is 90°.
  q: The air conditioner is working.
  r: The apartment is hot.

57. The temperature is 90° and the air conditioner is working, or the apartment is hot.

58. If the temperature is 90° or the air conditioner is not working, then the apartment is hot.

59. If the temperature is 90°, then the air conditioner is working or the apartment is not hot.

60. The apartment is hot if and only if the temperature is not 90°, or the apartment is hot.

61. The temperature is not 90° if and only if the air conditioner is not working, or the apartment is not hot.

62. If the apartment is hot and the air conditioner is working, then the temperature is 90°.

63. The apartment is hot if and only if the air conditioner is working, and the temperature is 90°.

64. It is false that if the apartment is hot then the air conditioner is not working.

65. The apartment is hot or the air conditioner is not working, if and only if the temperature is 90°.

66. If the air conditioner is working, then the temperature is 90° if and only if the apartment is hot.

In Exercises 67–76, write each symbolic statement in words. Let

  p: The water is 70°.
  q: The sun is shining.
  r: We go swimming.

67. (*p* ∨ *q*) ∧ *r*   68. (~*p* ∨ *q*) ∧ ~*r*
69. (*q* → *p*) ∨ *r*   70. ~*p* ∧ (*q* ∨ *r*)
71. ~*r* → (*q* ∧ *p*)   72. (*q* ∧ *r*) → *p*
73. (*r* → *q*) ∧ *p*   74. ~*p* → (*q* ∨ *r*)
75. (*q* ↔ *p*) ∧ *r*   76. *q* → (*p* ↔ *r*)

In Exercises 77–80, use the following information to arrive at your answers. Many restaurant dinner menus include statements such as the following. All dinners are served with a choice of: Soup or Salad, and Potatoes or Pasta, and Carrots or Peas. Which of the following selections are permissible? If a selection is not permissible, explain why. See the discussion of the exclusive or *on page 89.*

77. Soup, salad, and peas

78. Salad, pasta, and carrots

79. Soup, potatoes, pasta, and peas

80. Soup, pasta, and potatoes

In Exercises 81–94, (a) add parentheses by using the dominance of connectives, and (b) indicate whether the statement is a negation, conjunction, disjunction, conditional, or biconditional (see Example 9).

81. ~*r* → *p*   82. ~*p* ∧ *r* ↔ ~*q*
83. *q* ∧ ~*r*   84. ~*p* ∨ *q*
85. *p* ∨ *q* → *r*   86. *q* → *p* ∧ ~*r*
87. *r* → *p* ∨ *q*   88. *q* → *p* ↔ *p* → *q*

89. $\sim p \leftrightarrow \sim q \rightarrow r$

90. $\sim q \rightarrow r \wedge p$

91. $r \wedge \sim q \rightarrow q \wedge \sim p$

92. $\sim[p \rightarrow q \vee r]$

93. $\sim[p \wedge q \leftrightarrow p \vee r]$

94. $\sim[r \wedge \sim q \rightarrow q \wedge r]$

In Exercises 95–104, (a) select letters to represent the simple statements and write each statement symbolically by using parentheses, and (b) indicate whether the statement is a negation, conjunction, disjunction, conditional, or biconditional.

95. The frog did not jump or it was sleeping.

96. If the moon is out then it is evening or the sun is not shining.

97. If the sun is not shining or it is evening then the moon is out.

98. If the moon is out then it is evening, or the sun is not shining.

99. It is false that if the door is closed then the residents are not at home.

100. If today is Tuesday then tomorrow is Wednesday if and only if today is not Tuesday.

101. The store is closed if and only if today is Sunday or it is after 5 P.M.

102. If a number is divisible by 2 and the number is divisible by 3 then the number is divisible by 6.

103. The store is closed if and only if today is Sunday, or it is after 5 P.M.

104. It is false that if a number is divisible by 6 then the number is divisible by 2 and the number is divisible by 3.

In Exercises 105–111, translate the following statements into symbolic form. Indicate the letters you use to represent each simple statement.

105. ''Each language has a beauty of its own and forms of expression which are duplicated nowhere else.'' (Margaret Mead)

106. ''Our constitution is in actual operation, everything appears to promise that it will last, but in this world nothing is certain but death and taxes.'' (Benjamin Franklin)

107. The Constitution requires that the president be at least 35 years old, a natural-born citizen of the United States, and a resident of the country for 14 years.

108. ''You can fool all of the people some of the time and you can fool some of the people all of the time, but you can't fool all of the people all of the time.'' (Abraham Lincoln)

109. ''If you want to kill an idea in the world today, then get a committee working on it.'' (C. F. Kettering)

110. ''Happiness is beneficial for the body but it is grief that develops the powers of the mind.'' (Marcel Proust)

111. ''I did not know the dignity of their birth, but I do know the glory of their death.'' (Douglas MacArthur)

Problem Solving/Group Activities

112. If Zeus could do anything, could he build a wall that he could not jump over? Explain your answer.

In Exercises 113 and 114, place parentheses in the statement according to the dominance of connectives. Indicate whether the statement is a negation, conjunction, disjunction, conditional, or biconditional.

113. $\sim q \rightarrow r \vee p \leftrightarrow \sim r \wedge q$

114. $\sim[\sim r \rightarrow p \wedge q \leftrightarrow \sim p \vee r]$

115. a) We cannot place parentheses in the statement $p \vee q \wedge r$. Explain why.

b) Make up three simple statements and label them p, q, and r. Then write compound statements to represent $(p \vee q) \wedge r$ and $p \vee (q \wedge r)$.

c) Do you think that the statements for $(p \vee q) \wedge r$ and $p \vee (q \wedge r)$ mean the same thing? Explain.

116. Just as there is a dominance of connectives for logic statements there is an order of operations to be followed when evaluating arithmetic expressions. Using this book or other sources

a) list the order of operations to be followed when evaluating arithmetic expressions.

Then evaluate the following expressions by using the order you gave in part (a).

b) $4 + 12 \div 3 - 9 \times 4$

c) $36 - 4^2 \div 8 \times 10 + 50$

d) $-\{50 - [4 - (3^3 - 25) \times 3] + 9\}$

Research Activities

117. Obtain a legal document such as a will or rental agreement and copy one page of the document. Circle every connective used. Then list the number of times each connective appeared. Be sure to include conditional statements from which the word *then* was omitted from the sentence. Give the page and your listing to your instructor.

118. Write a report on the life and accomplishments of George Boole, who was an important contributor to the development of logic. In your report indicate how his work eventually led to development of the computer. References include encyclopedias and history of mathematics books.

3.2 TRUTH TABLES FOR NEGATION, CONJUNCTION, AND DISJUNCTION

A truth table is a device used to determine when a compound statement is true or false. Five basic truth tables are used in constructing other truth tables. Three will be discussed in this section (Tables 3.2, 3.4, and 3.7), and two will be discussed in the next section. Section 3.5 uses truth tables in determining whether a logical argument is valid or invalid.

Negation

Table 3.2 Negation

	p	$\sim p$
Case 1	T	F
Case 2	F	T

The first truth table is for *negation*. If p is a true statement, then the negation of p, "not p," is a false statement. If p is a false statement, then "not p" is a true statement. For example, if the statement "The shirt is blue" is true, then the statement "The shirt is not blue" is false. These relationships are summarized in Table 3.2. For a simple statement there are exactly two true–false cases, as shown.

If a compound statement consists of two simple statements p and q there are four possible cases, as illustrated in Table 3.3. Consider the statement "The test is today and the test covers Chapter 5." The simple statement "The test is today" has two possible truth values, true or false. The simple statement "The test covers Chapter 5" also has two truth values, true or false. Thus for these two simple statements there are four distinct possible true–false arrangements. Whenever we construct a truth table for a compound statement that consists of two simple statements we begin by listing the four true–false cases.

Table 3.3

	p	q
Case 1	T	T
Case 2	T	F
Case 3	F	T
Case 4	F	F

Conjunction

To illustrate the conjunction, consider the following situation. You have recently purchased a new house. To decorate it you ordered a new carpet and new furniture from the same store. You explain to the salesperson that the carpet must be delivered before the furniture. He promises that the carpet will be delivered on Thursday and that the furniture will be delivered on Friday.

To help determine whether the salesperson kept his promise we assign letters to each simple statement. Let p be "The carpet will be delivered on Thursday" and q be "The furniture will be delivered on Friday." The salesperson's statement written in symbolic form is $p \wedge q$. There are four possible true–false situations to be considered (Table 3.4).

Table 3.4 Conjunction

	p	q	$p \wedge q$
Case 1	T	T	T
Case 2	T	F	F
Case 3	F	T	F
Case 4	F	F	F

Case 1: p is true and q is true. The carpet is delivered on Thursday and the furniture is delivered on Friday. The salesperson has kept his promise and the compound statement is true.

Case 2: p is true and q is false. The carpet is delivered on Thursday but the furniture is not delivered on Friday. Since the furniture was not delivered as promised, the compound statement is false.

Case 3: p is false and q is true. The carpet is not delivered on Thursday but the furniture is delivered on Friday. Since the carpet was not delivered on Thursday as promised, the compound statement is false.

Case 4: p is false and q is false. The carpet is not delivered on Thursday and the furniture is not delivered on Friday. Since the carpet and furniture were not delivered as promised, the compound statement is false.

Examining the four cases, we see that in only one case did the salesperson keep his promise: in case 1. Therefore case 1 (T, T) is true. In cases 2, 3, and 4 the salesperson did not keep his promise and the compound statement is false. The results are summarized in Table 3.4, the truth table for the conjunction.

> The **conjunction** $p \wedge q$ is true only when both p and q are true.

EXAMPLE 1

Construct a truth table for $p \wedge \sim q$.

Solution: Because there are two statements, p and q, construct a truth table with four cases (Table 3.5a). Then write the truth values under the p in the

Table 3.5

(a)

	p	q	$p \wedge \sim q$
Case 1	T	T	
Case 2	T	F	
Case 3	F	T	
Case 4	F	F	

(b)

p	q	$p \wedge \sim q$
T	T	T
T	F	T
F	T	F
F	F	F
		1

(c)

p	q	$p \wedge$	$\sim q$
T	T	T	T
T	F	T	F
F	T	F	T
F	F	F	F
		1	2

(d)

p	q	p	$\wedge$	$\sim$	q
T	T	T		F	T
T	F	T		T	F
F	T	F		F	T
F	F	F		T	F
		1		3	2

(e)

p	q	p	$\wedge$	$\sim$	q
T	T	T	F	F	T
T	F	T	T	T	F
F	T	F	F	F	T
F	F	F	F	T	F
		1	4	3	2

compound statement and label this column 1 (Table 3.5b). Copy these truth values directly from the p column on the left. Write the corresponding truth values under the q in the compound statement and call this column 2 (Table 3.5c). Copy the truth values for column 2 directly from the q column on the left. Now find the truth values of $\sim q$ by negating the truth values in column 2 and call this column 3 (Table 3.5d). Use the conjunction table, Table 3.4, and the entries in columns 1 and 3 to complete column 4 (Table 3.5e). The results in column 4 are obtained as follows:

Row 1: $T \wedge F$ is F.

Row 2: $T \wedge T$ is T.

Row 3: $F \wedge F$ is F.

Row 4: $F \wedge T$ is F.

The answer is always the last column completed. Columns 1, 2, and 3 are only aids in arriving at the answer in column 4.   ▲

The statement $p \land \sim q$ in Example 1 actually means $p \land (\sim q)$. In the future, instead of listing a column for q and a separate column for its negation, we will make one column for $\sim q$, which will have the opposite values of those in the q column on the left. Similarly, when we evaluate $\sim p$ we will use the opposite values of those in the p column on the left. This procedure is illustrated in Example 2.

EXAMPLE 2

a) Construct a truth table for the following statement.

> The furnace is not on and we are not wasting energy.

b) Under which conditions will the compound statement be true?

c) Suppose that "The furnace is on" is a false statement and that "We are wasting energy" is a true statement. Is the compound statement given in part (a) true or false?

Solution:

a) First write the simple statements in symbolic form by using simple non-negated statements.

Let p: The furnace is on.
            q: We are wasting energy.

Therefore the compound statement may be written $\sim p \land \sim q$. Now construct a truth table with four cases, as shown in Table 3.6.

Fill in the table column labeled 1 by negating the truth values under p on the far left. Fill in the column labeled 2 by negating the values under q in the second column from the left. Fill in the column labeled 3 by using the columns labeled 1 and 2 and the definition of conjunction.

b) The compound statement in part (a) will be true only in case 4 (circled in blue) when both simple statements, p and q, are false. That is, when the furnace is not on and we are not wasting energy.

c) We are told that "The furnace is on," p, is a false statement and that "We are wasting energy," q, is a true statement. From the truth table (Table 3.6) we can determine that when p is false and q is true, case 3, the compound statement, is false (circled in red). ▲

Table 3.6

p	q	$\sim p$	$\land$	$\sim q$
T	T	F	F	F
T	F	F	F	T
F	T	T	(F)	F
F	F	T	(T)	T
		1	3	2

Disjunction

Consider the job description that contains the following requirements.

> **Civil Technician**
>
> Municipal program for redevelopment seeks on-site technician. **The applicant must have a two-year college degree in civil technology or five years of experience in the field**. Interested candidates please call 555-1234.

Who qualifies for the job? To help analyze the statement translate it into symbolic form. Let p be "A requirement for the job is a two-year college degree in civil

Table 3.7 Disjunction

p	q	$p \vee q$
T	T	T
T	F	T
F	T	T
F	F	F

technology'' and q be ''A requirement for the job is five years of experience in the field.'' The statement in symbolic form is $p \vee q$. For the two simple statements there are four distinct cases (see Table 3.7).

Case 1: p is true and q is true. A candidate has a two-year college degree in civil technology and five years of experience in the field. The candidate has both requirements, and qualifies for the job. Consider qualifying for the job as a true statement and not qualifying as a false statement.

Case 2: p is true and q is false. A candidate has a two-year college degree in civil technology but does not have five years of experience in the field. The candidate still qualifies for the job with the two-year college degree.

Case 3: p is false and q is true. The candidate does not have a two-year college degree in civil technology but does have five years experience in the field. The candidate qualifies for the job with the five years of experience in the field.

Case 4: p is false and q is false. The candidate does not have a two-year college degree in civil technology and does not have five years of experience in the field. The candidate does not meet either of the two requirements and therefore does not qualify for the job.

In examining the four cases we see that there is only one case in which the candidate does not qualify for the job: case 4. As this example indicates, an *or* statement will be true in every case, except when both simple statements are false. The results are summarized in Table 3.7, the truth table for the disjunction.

> The **disjunction**, $p \vee q$, is true when either p is true, q is true, or both p and q are true.

The disjunction $p \vee q$ is false only when p and q are both false.

In Example 2 we wrote ''in the column labeled 2.'' In Example 3 and other examples that follow, we may shorten the description and write ''in column 2.'' Therefore if we write ''in column 3'' it means ''in the column labeled 3.''

EXAMPLE 3

Construct a truth table for $\sim(q \vee \sim p)$.

Table 3.8

p	q	$\sim$	(q	$\vee$	$\sim p$)
T	T	F	T	T	F
T	F	T	F	F	F
F	T	F	T	T	T
F	F	F	F	T	T
		4	1	3	2

Solution: First construct the standard truth table listing the four cases. Work within parentheses first, determining the truth values for that statement. The order to be followed is indicated by the numbers below the columns (see Table 3.8). In column 1, copy the values from the q column on the left. Under $\sim p$, column 2, write the negation of the p column on the left. Next complete the *or* column, column 3, using columns 1 and 2 and the truth table for the disjunction. The *or* column is false only when both statements are false, as in case 2. Finally negate the values in the *or* column, and place these negated values in column 4. By examining the truth table you can see that the compound statement $\sim(q \vee \sim p)$ is true only in case 2, that is, when p is true and q is false. ▲

A General Procedure for Constructing Truth Tables

1. Study the compound statement and determine whether it is a negation, conjunction, disjunction, conditional, or biconditional statement, as was done in Section 3.1. The answer to the truth table will appear under $\sim$ if the statement is a negation, under $\wedge$ if the statement is a conjunction, under $\vee$ if the statement is a disjunction, under $\rightarrow$ if the statement is a conditional, and under $\leftrightarrow$ if the statement is a biconditional.

2. Complete the columns under the simple statements, p, q, r, and their negations, $\sim p$, $\sim q$, $\sim r$, within parentheses. If there are nested parentheses (one pair of parentheses within another pair) work with the innermost pair first.

3. Complete the column under the connective within the parentheses. You will use these truth values in determining the final answer in step 5.

4. Complete the columns under any remaining statements and their negations.

5. Complete the column under any remaining connectives. Recall that the answer will appear under the column determined in step 1. If the statement is a conjunction, disjunction, conditional, or biconditional, you will obtain the truth values for the connective by using the last column completed on the left side and on the right side of the connective. If the statement is a negation, you will obtain the truth values by negating the truth values of the last column completed within the grouping symbols on the right side of the negation. Be sure to circle or highlight your answer column, or number the columns in the order they were completed.

Table 3.9

p	q	$(\sim p$	$\vee$	$q)$	$\wedge$	$\sim p$
T	T	F	T	T	F	F
T	F	F	F	F	F	F
F	T	T	T	T	T	T
F	F	T	T	F	T	T
		1	3	2	5	4

Table 3.10

	p	q	r
Case 1	T	T	T
Case 2	T	T	F
Case 3	T	F	T
Case 4	T	F	F
Case 5	F	T	T
Case 6	F	T	F
Case 7	F	F	T
Case 8	F	F	F

EXAMPLE 4

Construct a truth table for the statement $(\sim p \vee q) \wedge \sim p$.

Solution: We will follow the general procedure outlined in the box. This statement is a conjunction and so the answer will be under the conjunction symbol. Complete columns under $\sim p$ and q within the parentheses and call these columns 1 and 2, respectively (see Table 3.9). Complete the column under the disjunction, $\vee$, using the values in columns 1 and 2, and call this column 3. Next, complete the column under $\sim p$, and call this column 4. The answer column, 5, is determined from the definition of the conjunction and the truth values in column 3, the last column completed on the left side of the conjunction, and column 4.   ▲

So far all the truth tables that we have constructed have contained at most two simple statements. Now we will explain how to construct a truth table that consists of three simple statements, such as $(p \wedge q) \wedge r$. When a compound statement consists of three simple statements there are eight different true–false possibilities, as illustrated in Table 3.10. To begin such a truth table write four Ts and four Fs in the column under p. Under the second statement, q, pairs of Ts alternate with pairs of Fs. Under the third statement, r, T alternates with F. This technique is not the only way of listing the cases, but it ensures that each case is unique and that no cases are omitted.

EXAMPLE 5

a) Construct a truth table for the statement "Matthew is cutting the grass and he will not water the lawn, or he will weed the flower beds."

b) Suppose that "Matthew is cutting the grass" is a false statement, that "Matthew will water the lawn" is a true statement, and that "Matthew will weed the flower beds" is a true statement. Is the compound statement in part (a) true or false?

Solution:

a) First we will translate the statement into symbolic form. Let

p: Matthew is cutting the grass.
q: Matthew will water the lawn.
r: Matthew will weed the flower beds.

Thus in symbolic form the statement is $(p \wedge {\sim}q) \vee r$.

Since this statement is composed of three simple statements, there are eight cases (Table 3.11). By examining the statement you can see that it is a disjunction. Therefore the answer will be under the $\vee$ column. Fill out the truth table by working in parentheses first. Place values under p, column 1, and ${\sim}q$, column 2. Then find the conjunctions of columns 1 and 2 to obtain column 3. Place the values of r in column 4. To obtain the answer, column 5, use columns 3 and 4 and your knowledge of the disjunction.

Table 3.11

p	q	r	$(p$	$\wedge$	${\sim}q)$	$\vee$	r
T	T	T	T	F	F	T	T
T	T	F	T	F	F	F	F
T	F	T	T	T	T	T	T
T	F	F	T	T	T	T	F
F	T	T	F	F	F	Ⓣ	T
F	T	F	F	F	F	F	F
F	F	T	F	F	T	T	T
F	F	F	F	F	T	F	F
			1	3	2	5	4

b) In case 5 of the truth table p, q, and r are F, T, and T, respectively. Therefore under these conditions this statement is true (as circled in the table). ▲

The number of distinct cases in a truth table with n distinct simple statements is 2^n. The compound statement $(p \vee q) \vee (r \wedge {\sim}s)$ has four simple statements p, q, r, s. Thus a truth table for this compound statement would have 2^4, or 16, distinct cases.

When we construct a truth table we determine the truth values of a compound statement for every possible case. If we want to find the truth value of the compound statement for any specific case when we know the truth values of the simple statements, we do not have to develop the entire table. For example, to determine the truth value for the statement

$$2 + 3 = 5 \quad \text{and} \quad 1 + 1 = 3$$

we let p be $2 + 3 = 5$ and q be $1 + 1 = 3$. Now we can write the compound statement as $p \wedge q$. We know that p is a true statement and q is a false statement. Thus we can substitute T for p and F for q and evaluate the statement:

$$p \wedge q$$
$$T \wedge F$$
$$F$$

Therefore this compound statement is false.

EXAMPLE 6

Determine the truth value for each simple statement. Then, using these truth values, determine the truth value of the compound statement.

a) 3 is greater than or equal to 2 ($3 \geq 2$).

b) Canada is south of Mexico and New York City is east of the Mississippi River, or Dallas is not a city in Texas.

Solution:

a) Let
p: 3 is greater than 2.
q: 3 is equal to 2.

The statement "3 is greater than or equal to 2" means that 3 is greater than 2 or 3 is equal to 2. The compound statement can be expressed as $p \vee q$. We know that p is a true statement and that q is a false statement. Substitute T for p and F for q and evaluate the statement:

$$p \vee q$$
$$T \vee F$$
$$T$$

Therefore the compound statement "3 is greater than or equal to 2" is a true statement.

b) Let
p: Canada is south of Mexico.
q: New York City is east of the Mississippi River.
r: Dallas is a city in Texas.

The compound statement can be written in symbolic form as $(p \wedge q) \vee \sim r$. We know that p is a false statement because Canada is north of Mexico and that q is a true statement because New York City is east of the Mississippi River. Dallas is a city in Texas, so r is a true statement. Its negation, $\sim r$, is therefore a false statement. Substitute F for p, T for q, and F for $\sim r$ and then evaluate the statement:

$$(p \wedge q) \vee \sim r$$
$$(F \wedge T) \vee F$$
$$F \quad \vee F$$
$$F$$

Therefore the compound statement is a false statement.

ᗞID YOU KNOW

APPLICATIONS OF LOGIC

Ꮐeorge Boole provided the key that would unlock the door to modern computing. However, it wasn't until 1938 that Claude Shannon, in his master's thesis for MIT, proposed uniting the on–off capability of electrical switches with Boole's two-value system of 0's and 1's. The operations AND, OR, NOT, and the rules of logic laid the foundation for computer gates. Such gates determine whether the current will pass. The closed switch (current flow) is represented as 1, and the open switch (no current flow) is represented as 0.

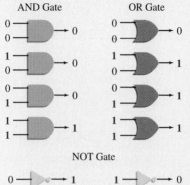

Rear Admiral Grace Hopper (1906–1992), a pioneer of computer science and the inventor of the computer language COBOL, worked on one of the earliest computer projects, Harvard's Mach II.

Early computers were large, noisy, and fairly slow, and used actual relay switches that mechanically clicked open or shut, often causing breakdowns. During Grace Hopper's work on the Mach II, one such breakdown was found to be caused by a moth caught in a relay. She used the word *bug* in describing the problem. The term *bug* is now generally used to describe many computer problems. ■

As you can see from the accompanying diagrams, the gates function in essentially the same way as a truth table. The gates shown here are the simplest ones, representing simple statements. There are other gates, such as the NAND and NOR gates, that are combinations of NOT, AND, and OR gates. The microprocessing unit of a computer uses thousands of these switches.

Section 3.2 Exercises

1. **a)** How many distinct cases must be listed in a truth table that contains two simple statements?
   b) List all the cases.

2. **a)** How many distinct cases must be listed in a truth table that contains three simple statements?
   b) List all the cases.

3. **a)** Construct the truth table for the conjunction, $p \wedge q$.
   b) Under what circumstances is the *and* table true?

4. **a)** Construct the truth table for the disjunction, $p \vee q$.
   b) Under what circumstances is the *or* table false?

In Exercises 5–20, construct a truth table for the statement.

5. $p \wedge \sim p$

6. $p \vee \sim p$

7. $q \vee \sim p$

8. $p \wedge \sim q$

9. $\sim(p \vee \sim q)$

10. $\sim p \vee \sim q$

11. $\sim(p \wedge \sim q)$

12. $\sim(\sim p \wedge \sim q)$

13. $(q \wedge p) \vee \sim q$

14. $(p \vee \sim q) \wedge r$

15. $r \vee (p \wedge \sim q)$

16. $(r \wedge q) \wedge \sim p$

17. $\sim q \wedge (r \vee \sim p)$

18. $\sim p \wedge (q \vee r)$

19. $r \vee (p \wedge \sim q)$

20. $\sim r \vee (\sim p \wedge q)$

In Exercises 21–30, write the statement in symbolic form and construct a truth table.

21. The book is 400 pages but it is the required book.

22. The cake is round and the cookie is not square.

23. Ken is a member of the glee club and Chris is not a member of the chess club.

24. It is false that at least 200 tickets must be sold or the concert will be canceled.

25. Ricardo will take mathematics or history, but he will not take psychology.

26. The dress is blue and the shoes are not exactly what I wanted, however I need them for a meeting on Monday.

27. The soda is coke, but I wanted root beer and I wanted a diet soda.

28. Jorge uses America On Line or Compuserve, but he does not use the internet.

29. The password is pistachio, and the gate is closed, or the gate is not closed.

30. The words are not in black print, or my glasses need new lenses or my glasses are not clean.

In Exercises 31–36, if p is true, q is false, and r is true, determine the truth value of the statement.

31. $(\sim p \wedge r) \wedge q$ **32.** $\sim p \vee (q \wedge r)$

33. $(\sim q \wedge \sim p) \vee \sim r$ **34.** $(\sim p \vee \sim q) \vee \sim r$

35. $(p \wedge \sim q) \vee r$ **36.** $(p \vee \sim q) \wedge \sim(p \wedge \sim r)$

In Exercises 37–42, if p is false, q is true, and r is false, determine the truth value of the statement.

37. $(\sim r \wedge p) \vee q$ **38.** $\sim q \vee (r \wedge p)$

39. $(\sim q \vee \sim p) \wedge r$ **40.** $(\sim r \vee \sim p) \vee \sim q$

41. $(\sim p \vee \sim q) \vee (\sim r \vee q)$ **42.** $(\sim r \wedge \sim q) \wedge (\sim r \vee \sim p)$

In Exercises 43–50, determine the truth value for each simple statement. Then use these truth values to determine the truth value of the compound statement. (You may have to use a reference source such as an almanac.)

43. $5 + 3 = 7$ or $6 + 4 = 15$

44. $9 - 6 = 15$ and $7 - 4 = 3$

45. There are five Great Lakes in the United States or the country of Brazil is in Africa.

46. The capital of Texas is Austin or Ohio is west of the Mississippi River, and Los Angeles is not the capital of California.

47. Eleven and 13 are prime numbers, but 2 is the only even prime number. (A prime number is a natural number that has exactly two distinct divisors: 1 and itself.)

48. Rome is in France or Paris is not in Germany, and London is in England.

49. The cheetah is the fastest land animal or the giraffe is the tallest land animal, and the elephant is the largest land animal.

50. Mars is a planet and the sun is a star, or the moon is not a star.

In Exercises 51–54, let

> *p*: Tanisha owns a convertible.
> *q*: Joan owns a Volvo.

Translate each statement into symbols. Then construct a truth table for each, and indicate under what conditions the compound statement is true.

51. Tanisha does not own a convertible, but Joan owns a Volvo.

52. Tanisha owns a convertible and Joan does not own a Volvo.

53. Tanisha owns a convertible or Joan does not own a Volvo.

54. Tanisha does not own a convertible or Joan does not own a Volvo.

In Exercises 55–58, let

> *p*: The house is owned by an engineer.
> *q*: The heat is solar generated.
> *r*: The car is run by electric power.

Translate each statement into symbols. Then construct a truth table for each and indicate under what conditions the compound statement is true.

55. The house is owned by an engineer and the heat is solar generated, or the car is run by electric power.

56. The car is run by electric power or the heat is solar generated, but the house is owned by an engineer.

57. The heat is solar generated, or the house is owned by an engineer and the car is not run by electric power.

58. The house is not owned by an engineer, and the car is not run by electric power and the heat is solar generated.

In Exercises 59 and 60, read the requirements and each applicant's qualifications for obtaining a loan.

> **a)** *Identify which of the applicants would qualify for the loan.*
> **b)** *For the applicants who do not qualify for the loan, explain why.*

59. To qualify for a loan of $30,000 an applicant must have a gross income of $25,000 if single, $40,000 combined income if married, and assets of at least $5,000.

Mr. Vapmek, married with three children, makes $37,000 on his job. Mrs. Vapmek does not have an income. The Vapmeks have assets of $50,000.

Ms. Bell is not married, works in sales, and earns $29,000. She has assets of $8,000.

Mrs. Spatz and her husband have total assets of $35,000. One earns $25,000 and the other earns $23,500.

60. To qualify for a loan of $45,000 an applicant must have a gross income of $30,000 if single, $50,000 combined income if married, and assets of at least $10,000.

> Mr. Argento, married with two children, makes $37,000 on his job. Mrs. Argento earns $15,000 at a part-time job. The Argentos have assets of $25,000.
>
> Tina McVey, single, has assets of $19,000. She works in a store and earns $25,000.
>
> Mr. Henke earns $24,000 and Mrs. Henke earns $28,000. Their assets total $8,000.

61. An airline advertisement states, "To get the special fare you must purchase your tickets between January 1 and February 15 and fly round trip between March 1 and April 1. You must depart on a Monday, Tuesday, or Wednesday, and return on a Tuesday, Wednesday, or Thursday, and stay over at least one Saturday.

a) Determine which of the following individuals will qualify for the special fare.

b) If the person does not qualify for the special fare, explain why.

> Mr. James plans to purchase his ticket on January 15, depart on Monday, March 3, and return on Tuesday, March 18.
>
> Ms. Vela plans to purchase her ticket on February 1, depart on Wednesday, March 10, and return on Thursday, April 2.
>
> Mr. Davis plans to purchase his ticket on February 14, depart on Tuesday, March 5, and return on Monday, March 18.
>
> Ms. Chang plans to purchase her ticket on January 4, depart on Monday, March 8, and return on Thursday, March 11.
>
> Ms. Ghandi plans to purchase her ticket on January 1, depart on Monday, March 3, and return on Monday, March 10.

Problem Solving/Group Activity

In Exercises 62 and 63, construct a truth table for the symbolic statement.

62. $\sim[(\sim(p \vee q)) \vee (q \wedge r)]$

63. $[(q \wedge \sim r) \wedge (\sim p \vee \sim q)] \vee (p \vee \sim r)$

64. On page 105 we indicated that a compound statement consisting of n simple statements has 2^n distinct true–false cases.

a) How many distinct true–false cases does a truth table containing simple statements p, q, r, and s have?

b) List all possible true–false cases for a truth table containing the simple statements p, q, r, and s.

c) Use the list in part (b) to construct a truth table for $(q \wedge p) \vee (\sim r \wedge s)$.

d) Construct a truth table for $(\sim r \wedge \sim s) \wedge (\sim p \vee q)$.

65. Must $(p \wedge \sim q) \vee r$ and $(q \wedge \sim r) \vee p$ have the same number of trues in their answer columns? Explain.

Research Activities

66. How does the computer make use of logic? Visit the computer center at your school, and ask the person in charge for specific references that can assist you in answering this question.

67. Digital computers use gates that work like switches to perform calculations. Information is fed into the gates and information leaves the gates, according to the type of gate. The three basic gates used in computers are the NOT gate, the AND gate, and the OR gate. Do research on the three types of gates.

a) Explain how each gate works.

b) Explain the relationship between each gate and the corresponding logic connectives *not*, *and*, and *or*.

c) Illustrate how two or more gates can be combined to form a more complex gate.

68. Write a report on the life and accomplishments of Rear Admiral Grace Hopper.

3.3 TRUTH TABLES FOR THE CONDITIONAL AND BICONDITIONAL

Conditional

In Section 3.1 we mentioned that the statement preceding the conditional symbol is called the *antecedent** and that the statement following the conditional symbol is called the *consequent*†. For example, consider $(p \vee q) \rightarrow [\sim(q \wedge r)]$. In this statement, $(p \vee q)$ is the antecedent and $[\sim(q \wedge r)]$ is the consequent.

---

*The terms *premise* or *hypothesis* are sometimes used in place of the term *antecedent*.

†The term *conclusion* is sometimes used in place of the word *consequent*.

Now we will look at the truth table for the conditional. Consider the statement "If you get an A in class, then I will buy you a car." Assume that this statement is true except when I have actually broken my promise to you. Let

p: You get an A.
q: I buy you a car.

Table 3.12 Conditional

p	q	$p \rightarrow q$
T	T	T
T	F	F
F	T	T
F	F	T

Translated into symbolic form, the statement becomes $p \rightarrow q$. Let's examine the four cases shown in Table 3.12.

Case 1: (T, T) You get an A, and I buy a car for you. I have met my commitment, and the statement is true.

Case 2: (T, F) You get an A, and I do not buy a car for you. I have broken my promise, and the statement is false.

What happens if you don't get an A? If you don't get an A, I no longer have a commitment to you, and therefore I cannot break my promise.

Case 3: (F, T) You do not get an A, and I buy you a car. I have not broken my promise, and therefore the statement is true.

Case 4: (F, F) You do not get an A, and I don't buy you a car. I have not broken my promise, and therefore the statement is true.

The conditional statement is false when the antecedent is true and the consequent is false. In every other case the conditional statement is true.

> The **conditional statement** $p \rightarrow q$ is true in every case except when p is a true statement and q is a false statement.

Table 3.13

p	q	p	$\rightarrow$	$\sim q$
T	T	T	F	F
T	F	T	T	T
F	T	F	T	F
F	F	F	T	T
		1	3	2

EXAMPLE 1

Construct a truth table for the statement $p \rightarrow \sim q$.

Solution: Since this is a conditional statement the answer will lie under the $\rightarrow$. Fill out the truth table by placing the appropriate values under p, column 1, and under $\sim q$, column 2 (Table 3.13). Then, using the information given in the truth table for the conditional and the truth values in columns 1 and 2, determine the solution, column 3. In row 1 the antecedent, p, is true and the consequent, $\sim q$, is false. Row 1 is T $\rightarrow$ F, which is F. Row 2 is T $\rightarrow$ T, which is T. Row 3 is F $\rightarrow$ F, which is T. Row 4 is F $\rightarrow$ T, which is T.   ▲

EXAMPLE 2

Construct a truth table for the statement $p \rightarrow (q \wedge \sim r)$.

Solution: Since this is a conditional statement the answer will lie under the $\rightarrow$. Work within the parentheses first. Place truth values under q, column 1, and $\sim r$, column 2 (Table 3.14 on page 111). Then take the conjunction of columns 1 and 2 to obtain column 3. Place the values of p in column 4. To obtain the answer, column 5, use columns 3 and 4 and your knowledge of the conditional statement. Column 4 represents the truth values of the antecedent, and column 3 represents the truth values of the consequent. Remember that the conditional is false only when the antecedent is true and the consequent is false, as in cases 1, 3, and 4 of column 5.

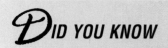

Table 3.14

p	q	r	p	→	(q	∧	~r)
T	T	T	T	F	T	F	F
T	T	F	T	T	T	T	T
T	F	T	T	F	F	F	F
T	F	F	T	F	F	F	T
F	T	T	F	T	T	F	F
F	T	F	F	T	T	T	T
F	F	T	F	T	F	F	F
F	F	F	F	T	F	F	T

           4   5   1   3   2   ▲

EXAMPLE 3

An advertisement in a magazine makes the following claim: "If you use Sunshine Face Lotion, then your skin will be radiant and have no wrinkles." Translate the statement into symbolic form and construct a truth table.

Solution: Let   *p*:   You use Sunshine Face Lotion.
                     q: Your skin will be radiant.
                     r: Your skin will have wrinkles.

In symbolic form the statement is

$$p \to (q \land \sim r)$$

This symbolic statement is identical to the statement in Table 3.14, and the truth tables are the same. Column 3 represents the truth values of $(q \land \sim r)$, which corresponds to the statement "Your skin will be radiant and have no wrinkles." Note that column 3 is true in cases (rows) 2 and 6. In case 2, since *p* is true, Sunshine Face Lotion is used. But in case 6, since *p* is false, Sunshine Face Lotion is not used. From this information we can conclude that it is possible to get skin that is radiant and has no wrinkles without Sunshine Face Lotion (case 6). By examining column 5 we see that the truth table is false in cases 1, 3, and 4. From these cases we can see that it is possible to use Sunshine Face Lotion and not have skin that is radiant and has no wrinkles.   ▲

A truth table alone cannot tell us whether a statement is true or false. However, it can be used to examine the various possibilities.

Biconditional

The *biconditional statement*, $p \leftrightarrow q$, means that $p \to q$ and $q \to p$, or, symbolically, $(p \to q) \land (q \to p)$. To determine the truth table for $p \leftrightarrow q$, we will construct the truth table for $(p \to q) \land (q \to p)$ (Table 3.15). Table 3.16 shows the truth values for the biconditional statement.

Table 3.15

p	q	(p	→	q)	∧	(q	→	p)
T	T	T	T	T	T	T	T	T
T	F	T	F	F	F	F	T	T
F	T	F	T	T	F	T	F	F
F	F	F	T	F	T	F	T	F

        1   3   2   7   4   6   5

Table 3.16   Biconditional

p	q	p ↔ q
T	T	T
T	F	F
F	T	F
F	F	T

The **biconditional statement**, $p \leftrightarrow q$, is true only when both p and q have the same truth value—that is, when both are true or both are false.

EXAMPLE 4

Construct a truth table for the statement $p \leftrightarrow (q \rightarrow \sim r)$.

Solution: Since there are three letters, there must be eight cases. The parentheses indicate that the answer must be under the biconditional (Table 3.17). Use columns 3 and 4 to obtain the answer in column 5. When columns 3 and 4 have the same truth values, place a T in column 5. When columns 3 and 4 have different truth values, place an F in column 5.

Table 3.17

p	q	r	p	$\leftrightarrow$	$(q$	$\rightarrow$	$\sim r)$
T	T	T	T	F	T	F	F
T	T	F	T	T	T	T	T
T	F	T	T	T	F	T	F
T	F	F	T	T	F	T	T
F	T	T	F	T	T	F	F
F	T	F	F	F	T	T	T
F	F	T	F	F	F	T	F
F	F	F	F	F	F	T	T
			4	5	1	3	2

In the preceding section we showed that finding the truth value of a compound statement for a specific case does not require constructing an entire truth table. Example 5 illustrates this technique for the conditional and the biconditional.

EXAMPLE 5

Determine the truth value of the statement $(q \leftrightarrow r) \rightarrow (\sim p \wedge r)$ when p is true, q is false, and r is true.

Solution: Substitute the truth value for each simple statement:

$$(q \leftrightarrow r) \rightarrow (\sim p \wedge r)$$
$$(F \leftrightarrow T) \rightarrow (F \wedge T)$$
$$F \quad \rightarrow \quad F$$
$$T$$

For this specific case the statement is true.

EXAMPLE 6

Determine the truth value for each simple statement. Then use the truth values to determine the truth value of the compound statement.

a) If $3 + 3 = 8$, then $5 + 2 = 11$

b) George Washington was the first president of the United States and George Washington was born in Texas, if and only if Thomas Jefferson is credited with designing our monetary system.

Solution:

a) Let

$$p: \quad 3 + 3 = 8$$
$$q: \quad 5 + 2 = 11$$

Then the statement "If $3 + 3 = 8$, then $5 + 2 = 11$" can be written $p \rightarrow q$. We know that p is a false statement and q is a false statement. Substitute F for p and F for q and evaluate the statement:

$$p \rightarrow q$$
$$F \rightarrow F$$
$$T$$

Therefore "If $3 + 3 = 8$, then $5 + 2 = 11$" is a true statement.

b) Let p be "George Washington was the first president of the United States," q be "George Washington was born in Texas," and r be "Thomas Jefferson is credited with designing our monetary system." The compound statement can be written $(p \wedge q) \leftrightarrow r$. We know that p is a true statement. Statement q is false since Washington was born in Virginia. Statement r is true. Substitute T for p, F for q, and T for r and evaluate the statement:

$$(p \wedge q) \leftrightarrow r$$
$$(T \wedge F) \leftrightarrow T$$
$$F \quad \leftrightarrow T$$
$$F$$

Therefore the given compound statement is false. ▲

Self-Contradictions, Tautologies, and Implications

Two special situations can occur in the truth table of a compound statement: The statement may always be true or the statement may always be false. We give such statements special names.

> A **self-contradiction** is a compound statement that is always false.

When every truth value in the answer column of the truth table is false, then the statement is a self-contradiction.

EXAMPLE 7

Construct a truth table for the statement $(p \leftrightarrow q) \wedge (p \leftrightarrow \sim q)$.

Solution: See Table 3.18. In this example, the truth values are false in each case of column 5. This is an example of a self-contradiction or a *logically false statement.* ▲

Table 3.18

p	q	$(p \leftrightarrow q)$	$\wedge$	$(p$	$\leftrightarrow$	$\sim q)$
T	T	T	F	T	F	F
T	F	F	F	T	T	T
F	T	F	F	F	T	F
F	F	T	F	F	F	T
		1	5	2	4	3

> A **tautology** is a compound statement that is always true.

"Heads I win, tails you lose." Do you think that this statement is a tautology, self-contradiction, or neither? See Problem-Solving Exercise 75.

When every truth value in the answer column of the truth table is true, the statement is a tautology.

EXAMPLE 8

Construct a truth table for the statement $(p \wedge q) \rightarrow (p \vee r)$.

Solution: The answer is given in column 3 of Table 3.19. The truth values are true in every case. Thus the statement is an example of a tautology or a *logically true statement*.

Table 3.19

p	q	r	$(p \wedge q)$	$\rightarrow$	$(p \vee r)$
T	T	T	T	T	T
T	T	F	T	T	T
T	F	T	F	T	T
T	F	F	F	T	T
F	T	T	F	T	T
F	T	F	F	T	F
F	F	T	F	T	T
F	F	F	F	T	F
			1	3	2

The conditional statement $(p \wedge q) \rightarrow (p \vee q)$ is a tautology. Conditional statements that are tautologies are called *implications*. In Example 8 we can say that $p \wedge q$ implies $p \vee q$.

> An **implication** is a conditional statement that is a tautology.

In any implication the antecedent of the conditional statement implies the consequent. In other words, if the antecedent is true, then the consequent must also be true. That is, the consequent will be true whenever the antecedent is true.

EXAMPLE 9

Determine whether the conditional statement $[(p \wedge q) \wedge p] \rightarrow q$ is an implication.

Solution: If the truth table of the conditional statement is a tautology, the conditional statement is an implication. Because the truth table is a tautology (see Table 3.20), the conditional statement is an implication. The antecedent $[(p \wedge q) \wedge p]$ implies the consequent q. Note that the antecedent is true only in case 1 and that the consequent is also true in case 1.

Table 3.20

p	q	$[(p \wedge q)$	$\wedge$	$p]$	$\rightarrow$	q
T	T	T	T	T	T	T
T	F	F	F	T	T	F
F	T	F	F	F	T	T
F	F	F	F	F	T	F
		1	3	2	5	4

Section 3.3 Exercises

1. a) Construct the truth table for the conditional statement, $p \rightarrow q$.

b) Explain when the conditional statement is true and when it is false.

2. a) Construct the truth table for the biconditional statement, $p \leftrightarrow q$.

b) Explain when the biconditional statement is true and when it is false.

3. a) Explain the procedure to determine the truth value of a compound statement when specific truth values are provided for the simple statements.

b) Follow the procedure in part (a) and determine the truth value of the symbolic statement

$$[(p \rightarrow q) \vee (\sim q \rightarrow r)] \rightarrow \sim r$$

when p = True, q = False, and r = True.

4. What is a self-contradiction?

5. What is a tautology?

6. What is an implication?

In Exercises 7–16, construct a truth table for the statement.

7. $p \rightarrow \sim q$ **8.** $\sim q \rightarrow \sim p$

9. $\sim(q \leftrightarrow p)$ **10.** $\sim(p \rightarrow q)$

11. $\sim q \leftrightarrow p$ **12.** $(p \leftrightarrow q) \rightarrow p$

13. $q \leftrightarrow (q \vee p)$ **14.** $(\sim q \wedge p) \rightarrow \sim q$

15. $q \rightarrow (p \rightarrow \sim q)$ **16.** $(p \vee q) \leftrightarrow (p \wedge q)$

In Exercises 17–26, construct a truth table for the statement.

17. $p \rightarrow (r \vee q)$ **18.** $r \wedge (\sim p \rightarrow q)$

19. $r \leftrightarrow (q \wedge p)$ **20.** $(q \leftrightarrow p) \wedge \sim r$

21. $(q \vee \sim r) \leftrightarrow \sim p$ **22.** $(p \vee r) \rightarrow (q \wedge r)$

23. $(\sim p \vee \sim q) \rightarrow r$ **24.** $[r \wedge (q \vee \sim p)] \leftrightarrow \sim p$

25. $(p \rightarrow q) \leftrightarrow (\sim q \rightarrow \sim r)$

26. $(\sim p \leftrightarrow \sim q) \rightarrow (\sim q \leftrightarrow r)$

In Exercises 27–32, write the statement in symbolic form. Then construct a truth table for it.

27. If the test is tomorrow, then I will not go out tonight and I will study for the test.

28. If the cable is out, then we cannot watch TV or we will read a novel.

29. The pizza will be delivered if and only if the driver finds the house, or the pizza will not be hot.

30. If it rains then the roof will leak, and if the sun shines then the roof will not leak.

31. If the pond is not polluted then we will go fishing, or we will not eat the fish.

32. It is false that if I read the assignment, then it is less than 10 pages and there is no movie on television.

In Exercises 33–38, determine whether the statement is a tautology, a self-contradiction, or neither.

33. $\sim p \rightarrow q$

34. $(p \vee q) \vee \sim p$

35. $(p \wedge q) \wedge \sim p$

36. $(\sim p \vee \sim q) \rightarrow p$

37. $(\sim p \rightarrow q) \vee \sim p$

38. $[(p \wedge q) \wedge \sim r] \leftrightarrow [(p \vee q) \wedge r]$

In Exercises 39–44, determine whether the statement is an implication.

39. $(p \vee q) \rightarrow q$

40. $q \rightarrow (p \vee q)$

41. $(p \wedge q) \rightarrow (q \wedge p)$

42. $(p \vee q) \rightarrow (p \vee \sim q)$

43. $[(p \rightarrow q) \wedge (q \rightarrow p)] \rightarrow (p \leftrightarrow q)$

44. $[(p \vee q) \wedge r] \rightarrow (p \vee q)$

In Exercises 45–56, if p is true, q is false, and r is true, find the truth value of the statement.

45. $\sim p \rightarrow (q \vee \sim r)$ **46.** $\sim p \rightarrow (q \wedge \sim r)$

47. $p \leftrightarrow (\sim q \wedge r)$ **48.** $(q \wedge \sim p) \rightarrow \sim r$

49. $(p \wedge \sim q) \wedge \sim r$ **50.** $\sim[p \rightarrow (q \wedge r)]$

51. $(p \vee r) \leftrightarrow (p \wedge \sim q)$ **52.** $(\sim p \vee q) \rightarrow \sim r$

53. $(\sim p \leftrightarrow r) \vee (\sim q \leftrightarrow r)$ **54.** $(r \rightarrow \sim p) \wedge (q \rightarrow \sim r)$

55. $\sim[(p \vee q) \leftrightarrow (p \rightarrow \sim r)]$

56. $[(\sim r \rightarrow \sim q) \vee (p \wedge \sim r)] \rightarrow q$

In Exercises 57–64, determine the truth value for each simple statement. Then, using the truth values, determine the truth value of the compound statement.

57. If a rectangle has four sides, then a triangle has four sides.

58. If $1 + 2 = 4$, then $4 + 1 = 6$.

59. 7 and 11 are odd numbers, if and only if $7 + 11$ is an odd number.

60. If Madonna is president of the United States, then Madonna resides in the White House or Madonna does not reside in the White House.

61. If the Golden Gate Bridge is in San Francisco and Jacksonville is in Florida, then the Golden Gate Bridge is not in Florida.

62. A quart contains 32 fluid ounces, if and only if a cup contains 8 fluid ounces and four cups make a quart.

63. If this book has exactly 200 pages then this book is a mathematics book, or January 1 is New Year's Day in the United States.

64. If a dollar has the same value as 100 pennies and 100 pennies have the same value as 21 nickels, then a dollar has the same value as 21 nickels.

In Exercises 65–70, suppose that both of the following statements are true.

    p: Liz received a grade of A in mathematics.
    q: Liz made the dean's list.

Find the truth values of each compound statement.

65. If Liz received a grade of A in mathematics, then Liz made the dean's list.

66. If Liz did not receive a grade of A in mathematics, then Liz did not make the dean's list.

67. If Liz did not receive a grade of A in mathematics, then Liz made the dean's list.

68. Liz made the dean's list if and only if Liz received a grade of A in mathematics.

69. Liz did not make the dean's list if and only if Liz received a grade of A in mathematics.

70. If Liz did not make the dean's list, then Liz did not receive a grade of A in mathematics.

71. Your father makes the statement "If I get a raise in salary, then I will buy a new car." At the end of the semester you come home and there is a new car in the driveway. Can you conclude that your father got a raise? Explain.

72. Consider the statement "If your interview goes well, then you will be offered the job." If you are interviewed and then offered the job, can you conclude that your interview went well? Explain.

Problem Solving/Group Activities

In Exercises 73 and 74, construct truth tables for the symbolic statement.

73. $p \vee q \rightarrow \sim r \leftrightarrow p \wedge \sim q$

74. $[(r \rightarrow \sim q) \rightarrow \sim p] \vee (q \leftrightarrow \sim r)$

75. Is the statement "Heads I win, Tails you lose" a tautology, a self-contradiction, or neither? Explain your answer.

76. The Barr triplets have an annoying habit: Whenever a question is asked of the three of them, two tell the truth and the third lies. When I asked them which of them was born last, they replied as follows.

    Mary: Katie was born last.
    Katie: I am the youngest.
    Annie: Mary is the youngest.

Which of the Barr triplets was born last?

77. Helen Hurry shares a faculty office with three men: Sam Swift, Nick Nickelson, and Peter Perkins. One of the four teachers is a historian, another is a biologist, a third is a mathematician, and the fourth is a chemist.

I heard the following rumors when I was a student in the college. Professor Swift, who is the historian, and the biologist were fraternity brothers in college. The historian and the biologist were once legally married to each other. Professor Nickelson and the chemist are engaged to be married to each other. Helen and her boyfriend play tennis each week with the biologist and his wife. If exactly three of these four statements are true, what is each person's subject matter?

78. Construct a truth table for
    a) $(p \vee q) \rightarrow (r \wedge s)$.
    b) $(q \rightarrow \sim p) \vee (r \leftrightarrow s)$.

Research Activity

79. Select an advertisement from a newspaper or magazine that makes or implies a conditional statement. Analyze the advertisement to determine whether the consequent necessarily follows from the antecedent. Explain your answer. (See Example 3.)

3.4   EQUIVALENT STATEMENTS

Equivalent statements are an important concept in the study of logic.

> Two statements are **equivalent**, symbolized ⇔,* if both statements have exactly the same truth values in the answer columns of the truth tables.

---

*The symbol ≡ is also used to indicate equivalent statements.

Sometimes the words logically equivalent are used in place of the word equivalent.

To determine whether two statements are equivalent, construct a truth table for each statement and compare the answer columns of the truth tables. If the answer columns are identical, the statements are equivalent. If the answer columns are not identical, the statements are not equivalent.

EXAMPLE 1

Show that the following two statements are equivalent.

$$[p \vee (q \vee r)] \qquad [(p \vee q) \vee r]$$

Solution: Construct a truth table for each statement (see Table 3.21).

Table 3.21

p	*q*	*r*	**[*p**	**∨**	**(*q* ∨ *r*)]**	**[(*p* ∨ *q*)**	**∨**	**r]**
T	T	T	T	T	T	T	T	T
T	T	F	T	T	T	T	T	F
T	F	T	T	T	T	T	T	T
T	F	F	T	T	F	T	T	F
F	T	T	F	T	T	T	T	T
F	T	F	F	T	T	T	T	F
F	F	T	F	T	T	F	T	T
F	F	F	F	F	F	F	F	F
			1	3	2	1	3	2

Because the truth tables have the same answer (column 3 for both tables), the statements are equivalent. Thus we can write

$$[p \vee (q \vee r)] \Leftrightarrow [(p \vee q) \vee r].$$

▲

EXAMPLE 2

Determine whether the following statements are equivalent.

a) If you study and go to sleep by 11 P.M., then you will get an A on the test.

b) If you do not study or do not go to sleep by 11 P.M., then you will not get an A on the test.

Solution: Write each statement in symbolic form; then construct a truth table for each statement. If the answer columns of both tables are identical, the statements are equivalent. If the answer columns are not identical, the statements are not equivalent.

Let

    p: You study.
    q: You go to sleep by 11 P.M.
    r: You will get an A on the test.

In symbolic form the statements are

a) $(p \wedge q) \to r$.   **b)** $(\sim p \vee \sim q) \to \sim r$.

The truth tables for these statements are given in Tables 3.22 and 3.23, respectively. The answers in the columns labeled 5 are not identical, so the statements are not equivalent.

Table 3.22

p	q	r	$(p$	$\wedge$	$q)$	$\to$	r
T	T	T	T	T	T	T	T
T	T	F	T	T	T	F	F
T	F	T	T	F	F	T	T
T	F	F	T	F	F	T	F
F	T	T	F	F	T	T	T
F	T	F	F	F	T	T	F
F	F	T	F	F	F	T	T
F	F	F	F	F	F	T	F
			1	3	2	5	4

Table 3.23

p	q	r	$(\sim p$	$\vee$	$\sim q)$	$\to$	$\sim r$
T	T	T	F	F	F	T	F
T	T	F	F	F	F	T	T
T	F	T	F	T	T	F	F
T	F	F	F	T	T	T	T
F	T	T	T	T	F	F	F
F	T	F	T	T	F	T	T
F	F	T	T	T	T	F	F
F	F	F	T	T	T	T	T
			1	3	2	5	4

EXAMPLE 3

Select the statement that is logically equivalent to "It is not true that the plane is both overbooked and departing late."

a) If the plane is not departing late, then the plane is not overbooked.

b) The plane is not overbooked or the plane is not departing late.

c) The plane is not departing late and the plane is not overbooked.

d) If the plane is not overbooked, then the plane is not departing late.

Solution: To determine whether any of the choices are equivalent to the given statement write the given statement and the choices in symbolic form. Then construct and compare their truth tables.

Let

    p: The plane is overbooked.
    q: The plane is departing late.

The given statement may be written "It is not true that the plane is overbooked and the plane is departing late." The statement is expressed in symbolic form as $\sim(p \wedge q)$. Using p and q as indicated, choices (a)–(d) may be expressed symbolically as

a) $\sim q \to \sim p$.   **b)** $\sim p \vee \sim q$.   **c)** $\sim q \wedge \sim p$.   **d)** $\sim p \to \sim q$.

Table 3.24

p	q	$\sim$	$(p$	$\wedge$	$q)$
T	T	F	T	T	T
T	F	T	T	F	F
F	T	T	F	F	T
F	F	T	F	F	F
		4	1	3	2

Table 3.25

		(a)			(b)			(c)			(d)		
p	q	$\sim q$	$\rightarrow$	$\sim p$	$\sim p$	$\vee$	$\sim q$	$\sim q$	$\wedge$	$\sim p$	$\sim p$	$\rightarrow$	$\sim q$
T	T	F	T	F	F	F	F	F	F	F	F	T	F
T	F	T	F	F	F	T	T	T	F	F	F	T	T
F	T	F	T	T	T	T	F	F	F	T	T	F	F
F	F	T	T	T	T	T	T	T	T	T	T	T	T

Now construct a truth table for the given statement (Table 3.24) and each possible choice (Table 3.25a–d). Examining the truth tables we see that the given statement, $\sim(p \wedge q)$, is logically equivalent to choice (b), $\sim p \vee \sim q$. Therefore the correct answer is "The plane is not overbooked or the plane is not departing late." This is logically equivalent to "It is not true that the plane is both overbooked and departing late." ▲

*D*ID YOU KNOW

ADVENTURES IN LOGIC

"You should say what you mean," the March Hare went on.
"I do," Alice hastily replied; "at least—at least I mean what I say—that's the same thing, you know."
"Not the same thing a bit!" said the Hatter. "You might as well say that 'I see what I eat' is the same thing as 'I eat what I see'!"

*O*ne of the more interesting and well-known students of logic was Charles Dodgson (1832–1898), better known to us as Lewis Carroll, the author of *Alice's Adventures in Wonderland* and *Through the Looking-Glass*. Although the books have a child's point of view, many would argue that the audience best equipped to enjoy them is an adult one. Dodgson, a mathematician, logician, and photographer (among other things), uses the naïveté of a seven-year-old girl to show what can happen when the rules of logic are taken to absurd extremes. ■

De Morgan's Laws

Example 3 showed that a statement of the form $\sim(p \wedge q)$ is equivalent to a statement of the form $\sim p \vee \sim q$. Thus we may write $\sim(p \wedge q) \Leftrightarrow \sim p \vee \sim q$. This equivalent statement is one of two special laws called De Morgan's laws. The laws, named after Augustus De Morgan, an English mathematician, were first introduced in Section 2.4, where they applied to sets.

De Morgan's Laws

1. $\sim(p \wedge q) \Leftrightarrow \sim p \vee \sim q$

2. $\sim(p \vee q) \Leftrightarrow \sim p \wedge \sim q$

You can demonstrate that De Morgan's second law is true by constructing and comparing truth tables for $\sim(p \vee q)$ and $\sim p \wedge \sim q$. Do so now.

When using De Morgan's laws, if it becomes necessary to negate an already negated statement, make use of the fact that $\sim(\sim p) = p$. For example the negation of the statement ''Today is not Monday'' is ''Today is Monday.''

EXAMPLE 4

Select the statement that is logically equivalent to ''The sun is not shining but it is not raining.''

a) It is not true that the sun is shining and it is raining.

b) It is not raining or the sun is not shining.

c) The sun is shining or it is raining.

d) It is not true that the sun is shining or it is raining.

Solution: To determine which statement is equivalent, write each statement in symbolic form.

Let p: The sun is shining.
                            q: It is raining.

The statement ''The sun is not shining but it is not raining'' written symbolically is $\sim p \wedge \sim q$. Recall that the word *but* means the same as the word *and*. Write parts (a)–(d) symbolically.

a) $\sim(p \wedge q)$ **b)** $\sim q \vee \sim p$ **c)** $p \vee q$ **d)** $\sim(p \vee q)$

De Morgan's law shows that $\sim p \wedge \sim q$ is equivalent to $\sim(p \vee q)$. Therefore the answer is (d): ''It is not true that the sun is shining or it is raining.'' ▲

EXAMPLE 5

Write a statement that is logically equivalent to ''It is not true that the microwave must be set at level 6 or you will burn the food.''

Solution:
Let p: The microwave must be set at level 6.
                            q: You will burn the food.

This statement is of the form $\sim(p \vee q)$. An equivalent statement, using De Morgan's laws, is $\sim p \wedge \sim q$. Therefore an equivalent statement is "The microwave must not be set at level 6 and you will not burn the food." ▲

If we negate both sides of De Morgan's laws, we obtain the following rules.

1. $\sim(p \wedge q) \Leftrightarrow \sim p \vee \sim q$
$\sim[\sim(p \wedge q)] \Leftrightarrow \sim(\sim p \vee \sim q)$
$p \wedge q \Leftrightarrow \sim(\sim p \vee \sim q)$

2. $\sim(p \vee q) \Leftrightarrow \sim p \wedge \sim q$
$\sim[\sim(p \vee q)] \Leftrightarrow \sim(\sim p \wedge \sim q)$
$p \vee q \Leftrightarrow \sim(\sim p \wedge \sim q)$

We use rule 2 to write an equivalent statement using De Morgan's laws in Example 6.

EXAMPLE 6

Use De Morgan's laws to write an equivalent statement for "The car is too long or the garage is too short."

Solution:
Let
p: The car is too long.
q: The garage is too short.

The given statement is of the form $p \vee q$. From rule 2 we know that $p \vee q \Leftrightarrow \sim(\sim p \wedge \sim q)$. Therefore one equivalent statement is "It is false that the car is not too long and the garage is not too short." ▲

EXAMPLE 7

Use De Morgan's laws to write a statement logically equivalent to "The dog is not taken care of and the dog has fleas."

Solution:
Let
p: The dog is taken care of.
q: The dog has fleas.

The statement written symbolically is $\sim p \wedge q$. From rule 1 we see that $p \wedge q \Leftrightarrow \sim(\sim p \vee \sim q)$. If we negate p in $p \wedge q$, we obtain the statement $\sim p \wedge q$, which is what we want. If we negate p on one side of the equivalent symbol, we must also negate p on the other side of the equivalent symbol to keep the equivalence true. If we negate p on both sides of the equivalent symbol we obtain

$$p \wedge q \Leftrightarrow \sim(\sim p \vee \sim q)$$

$$\sim p \wedge q \Leftrightarrow \sim(p \vee \sim q)$$

Therefore the statement "It is false that the dog is taken care of or the dog does not have fleas" is logically equivalent to the given statement. ▲

There are strong similarities between the topics of sets and logic. We can see them by examining De Morgan's laws for sets and logic.

De Morgan's Laws: Set Theory	De Morgan's Laws: Logic
$(A \cap B)' = A' \cup B'$	$\sim(p \wedge q) \Leftrightarrow \sim p \vee \sim q$
$(A \cup B)' = A' \cap B'$	$\sim(p \vee q) \Leftrightarrow \sim p \wedge \sim q$

The complement in set theory, $'$, is similar to the negation, $\sim$, in logic. The intersection, $\cap$, is similar to the conjunction, $\wedge$; and the union, $\cup$, is similar to the disjunction, $\vee$. If we were to interchange the set symbols with the logic symbols, De Morgan's laws would remain, but in a different form.

Both $'$ and $\sim$ can be interpreted as *not*.

Both $\cap$ and $\wedge$ can be interpreted as *and*.

Both $\cup$ and $\vee$ can be interpreted as *or*.

For example, the set statement $A' \cup B$ can be written as a statement in logic as $\sim a \vee b$.

Statements containing connectives other than *and* and *or* may have equivalent statements. To illustrate this point construct truth tables for $p \rightarrow q$ and for $\sim p \vee q$. The truth tables will have the same answer columns and therefore the statements are equivalent. That is,

$$p \rightarrow q \Leftrightarrow \sim p \vee q.$$

With these equivalent statements we can write a conditional statement as a disjunction or a disjunction as a conditional statement. For example, the statement ''If the game is polo then you ride a horse'' can be equivalently stated as ''The game is not polo or you ride a horse.''

To change a conditional statement to a disjunction: Negate the antecedent, change the conditional symbol to a disjunction symbol, and keep the consequent the same. To change a disjunction statement to a conditional statement: Negate the first statement, change the disjunction symbol to a conditional symbol, and keep the second statement the same.

EXAMPLE 8

Write a conditional statement that is logically equivalent to ''The Eagles will win or the Lions will lose.'' Assume the negation of winning is losing.

Solution:

Let

p: The Eagles will win.

q: The Lions will win.

The statement may be written symbolically as $p \vee \sim q$. To write an equivalent statement negate the first statement, p, and change the disjunction symbol to a conditional symbol. Symbolically the statement is $\sim p \rightarrow \sim q$. The equivalent statement is ''If the Eagles lose then the Lions will lose.''

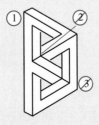
In the preceding section we showed that $p \leftrightarrow q$ has the same truth table as $(p \rightarrow q) \wedge (q \rightarrow p)$. Therefore these statements are equivalent, a useful fact for Example 9.

EXAMPLE 9

Write the following statement as an equivalent biconditional statement. "If the figure is a triangle then the figure has three angles and if the figure has three angles then the figure is a triangle."

Solution: An equivalent statement is "A figure is a triangle if and only if the figure has three angles."   ▲

Variations of the Conditional Statement

We know that $p \rightarrow q$ is equivalent to $\sim p \vee q$. Are any other statements equivalent to $p \rightarrow q$? Yes, there are many. Now let's look at the variations of the conditional statement to determine whether any are equivalent to the conditional statement. The variations of the conditional statement are made by switching and/or negating the antecedent and the consequent of a conditional statement. The variations of the conditional statement are the **converse** of the conditional, the **inverse** of the conditional, and the **contrapositive** of the conditional.

Listed here are the variations of the conditional with their symbolic form and the words we say to read each one.

Variations of the Conditional Statement

Name	Symbolic Form	Read
Conditional	$p \rightarrow q$	"If p, then q"
Converse of the conditional	$q \rightarrow p$	"If q, then p"
Inverse of the conditional	$\sim p \rightarrow \sim q$	"If not p, then not q"
Contrapositive of the conditional	$\sim q \rightarrow \sim p$	"If not q, then not p"

To write the converse of the conditional statement, switch the order of the antecedent and the consequent. To write the inverse, negate both the antecedent and the consequent. To write the contrapositive, switch the order of the antecedent and the consequent and then negate both of them.

Are any of the variations of the conditional statement equivalent? To determine the answer we can construct a truth table for each variation, as shown in Table 3.26. It reveals that the conditional statement is equivalent to the contrapositive statement and that the converse statement is equivalent to the inverse statement.

Table 3.26

p	q	Conditional $p \rightarrow q$	Contrapositive $\sim q \rightarrow \sim p$	Converse $q \rightarrow p$	Inverse $\sim p \rightarrow \sim q$
T	T	T	T	T	T
T	F	F	F	T	T
F	T	T	T	F	F
F	F	T	T	T	T

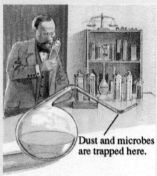

EXAMPLE 10

For the conditional statement "If there is a space station, then we will explore outer space." Write the

a) converse.   **b)** inverse.   **c)** contrapositive.

Solution:
a) Let p: There is a space station.
                 q: We will explore outer space.

The conditional statement is of the form $p \rightarrow q$, so the converse must be of the form $q \rightarrow p$. Therefore the converse is "If we explore outer space, then there is a space station."

b) The inverse is of the form $\sim p \rightarrow \sim q$. Therefore the inverse is "If there is no space station, then we will not explore outer space."

c) The contrapositive is of the form $\sim q \rightarrow \sim p$. Therefore the contrapositive is "If we do not explore outer space, then there is no space station." ▲

EXAMPLE 11

Let p: The number is divisible by 9.
           q: The number is divisible by 3.

Write the following statements and determine which are true.

a) The conditional statement, $p \rightarrow q$

b) The converse of $p \rightarrow q$

c) The inverse of $p \rightarrow q$

d) The contrapositive of $p \rightarrow q$

Solution:
a) *Conditional statement:* $(p \rightarrow q)$
If the number is divisible by 9, then the number is divisible by 3. This statement is true. A number divisible by 9 must also be divisible by 3, since 3 is a divisor of 9.

b) *Converse of the conditional:* $(q \rightarrow p)$
If the number is divisible by 3, then the number is divisible by 9. This statement is false. For instance: 6 is divisible by 3, but 6 is not divisible by 9.

c) *Inverse of the conditional:* $(\sim p \rightarrow \sim q)$
If the number is not divisible by 9, then the number is not divisible by 3. This statement is false. For instance: 6 is not divisible by 9, but 6 is divisible by 3.

d) *Contrapositive of the conditional:* $(\sim q \rightarrow \sim p)$
If the number is not divisible by 3, then the number is not divisible by 9. The statement is true, since any number that is divisible by 9 must be divisible by 3. ▲

EXAMPLE 12

Use the contrapositive to write a statement logically equivalent to "If the boat is 24 ft long, then it will not fit into the boathouse."

Solution:

Let   *p*:   The boat is 24 ft long.
                            q: The boat will fit into the boathouse.

The given statement written symbolically is

$$p \rightarrow \sim q$$

The contrapositive of the statement is

$$q \rightarrow \sim p$$

Therefore an equivalent statement is ''If the boat will fit into the boathouse, then the boat is not 24 ft long.   ▲

The contrapositive of the conditional is very important in mathematics. Consider the statement ''If a^2 is not a whole number, then *a* is not a whole number.'' Is this statement true? You may find this question difficult to answer. Writing the statement's contrapositive may enable you to answer the question. The contrapositive is ''If *a* is a whole number, then a^2 is a whole number.'' Since the contrapositive is a true statement, the original statement must also be true.

EXAMPLE 13

Determine which, if any, of the following statements are equivalent. You may use De Morgan's laws, the fact that $p \rightarrow q \Leftrightarrow \sim p \vee q$, information from the variations of the conditional, or truth tables.

a) If you leave by 9 A.M., then you will get to your destination on time.

b) You do not leave by 9 A.M. or you will get to your destination on time.

c) It is false that you get to your destination on time or you did not leave by 9 A.M.

d) If you do not get to your destination on time, then you did not leave by 9 A.M.

Solution:

Let   *p*:   You leave by 9 A.M.
                            q: You will get to your destination on time.

In symbolic form the four statements are

a) $p \rightarrow q$
b) $\sim p \vee q$
c) $\sim (q \vee \sim p)$
d) $\sim q \rightarrow \sim p$

Which of these statements are equivalent? Earlier in this section you learned that $p \rightarrow q$ is equivalent to $\sim p \vee q$. Therefore statements (a) and (b) are equivalent. Statement (d) is the contrapositive of statement (a). Therefore statement (d) is also equivalent to statement (a) and statement (b). All these statements have the same truth table (Table 3.27 on page 126).

Table 3.27

		(a)	(b)	(d)
p	q	$p \rightarrow q$	$\sim p \vee q$	$\sim q \rightarrow \sim p$
T	T	T	T	T
T	F	F	F	F
F	T	T	T	T
F	F	T	T	T

Now let's look at statement (c). If we use De Morgan's laws on statement (c), we get

$$\sim(q \vee \sim p) \Leftrightarrow \sim q \wedge p$$

If $\sim q \wedge p$ was one of the other statements, then $\sim(q \vee \sim p)$ would be equivalent to that statement. Because $\sim q \wedge p$ does not match any of the other choices, it does not necessarily mean that $\sim(q \vee \sim p)$ is not equivalent to the other statements. To determine whether $\sim(q \vee \sim p)$ is equivalent to the other statements we will construct its truth table (Table 3.28 on page 127) and compare the answer column with the answer columns in Table 3.27.

$\mathcal{D}$ID YOU KNOW

FUZZY LOGIC

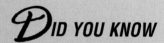

$\mathcal{M}$odern computers, like truth tables, work with only two values, 1 or 0 (equivalent to true or false in truth tables). This constraint prevents a computer from being able to reason as the human brain can and prevents the

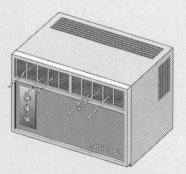

computer from being able to evaluate items involving vagueness or value judgments that so often occur in real-world situations. For example, a binary computer will have difficulty evaluating the subjective statement "the air is cool."

Fuzzy logic helps computers evaluate common-sense pictures of an uncertain world by using the key concept: Everything is a matter of degree. Fuzzy logic manipulates vague concepts such as *warm* and *fast* by assigning values between 0 and 1 to each item. For example, suppose that *warm* is assigned a value of 0.80; then *not warm* is assigned a value of $1 - 0.8 = 0.20$. As the value assigned to *warm* changes, so does the value assigned to *not warm*.

Not p is always $1 - p$, where $0 < p < 1$. Fuzzy logic is also used to manipulate values in vague sets, such as *warm* and *not warm*, to yield precise instructions that can be used to operate many devices. Fuzzy logic is used to operate air conditioners, subways, cameras, and many other devices where the change in one condition changes another condition. For example, when the air is warm the air conditioner runs faster, and when the air cools down the air conditioner runs slower. How many other devices can you name that may use fuzzy logic? See Problem-Solving Exercises 87 and 88. ∎

Table 3.28

(c)

p	q	~	(q	∨	~p)
T	T	F	T	T	F
T	F	T	F	F	F
F	T	F	T	T	T
F	F	F	F	T	T
		4	1	3	2

The answer columns of the truth tables are not the same, so $\sim(q \vee \sim p)$ is not equivalent to any of the other statements. Therefore statements (a), (b), and (d) are equivalent to each other. ▲

Section 3.4 Exercises

1. What are equivalent statements?

2. Explain how you can determine whether two statements are equivalent.

3. Suppose that two statements are connected with the biconditional and that the truth table is constructed. If the answer column of the truth table has all trues, what must be true about these two statements? Explain.

4. Write De Morgan's laws for logic.

5. For a statement of the form if p, then q, symbolically indicate the form of the
   a) converse.   **b)** inverse.   **c)** contrapositive.

6. Which of the following are equivalent statements?
   a) The converse   **b)** The contrapositive
   c) The inverse   **d)** The conditional

In Exercises 7–16, use De Morgan's laws to determine whether the two statements are equivalent.

7. $\sim p \vee \sim q, \sim(p \wedge q)$

8. $\sim(p \vee q), \sim p \wedge \sim q$

9. $\sim(p \wedge q), \sim p \wedge q$

10. $p \wedge q, \sim(\sim p \vee q)$

11. $\sim(p \wedge q), \sim p \vee \sim q$

12. $\sim(p \wedge \sim q), \sim p \vee q$

13. $\sim p \wedge q, \sim(q \vee \sim p)$

14. $\sim(p \wedge q) \rightarrow r, (\sim p \vee \sim q) \rightarrow r$

15. $q \rightarrow \sim(p \wedge \sim r), q \rightarrow \sim p \vee r$

16. $p \leftrightarrow (q \vee \sim r), p \leftrightarrow \sim(\sim q \wedge r)$

In Exercises 17–26, use a truth table to determine whether the two statements are equivalent.

17. $p \rightarrow q, \sim p \vee q$

18. $\sim p \rightarrow q, p \wedge q$

19. $p \rightarrow q, \sim q \rightarrow \sim p$

20. $(p \vee q) \vee r, p \vee (q \vee r)$

21. $(p \wedge q) \wedge r, p \wedge (q \wedge r)$

22. $\sim p \rightarrow (q \wedge r), p \vee (q \wedge r)$

23. $p \leftrightarrow (q \wedge \sim r), p \rightarrow (q \vee r)$

24. $(p \rightarrow q) \wedge (q \rightarrow r), (p \rightarrow q) \rightarrow r$

25. $\sim(q \rightarrow p) \vee r, (p \vee q) \wedge \sim r$

26. $\sim q \rightarrow (p \wedge r), \sim(p \vee r) \rightarrow q$

27. Show that $(p \rightarrow q) \wedge (q \rightarrow p) \Leftrightarrow (p \leftrightarrow q)$.

28. Determine whether $[\sim(p \rightarrow q)] \wedge [\sim(q \rightarrow p)] \Leftrightarrow \sim(p \leftrightarrow q)$.

In Exercises 29–36, use De Morgan's laws to write an equivalent statement for the sentence.

29. It is not true that the grass needs watering and the trees need fertilizer.

30. It is not true that diet food contains no salt and skim milk contains no calories.

31. Jerry subscribes to *National Geographic* or Jerry does not know geography.

32. I did not read the assignment and I did not learn the material.

33. The number is neither odd nor even.

34. If we travel to California, then we will not visit aunt Sally or we will travel to Hawaii.

35. We will sell the house, or we will not repair the roof and we will not replace the water heater.

36. It is not true that the car gets 27 miles per gallon and the car is not American made.

In Exercises 37–42, use the fact that $p \rightarrow q$ is equivalent to $\sim p \vee q$ to write an equivalent form of the statement.

37. If you do your workout, then your metabolism will increase.

38. The engine makes noise or the engine is not running.

39. Riva will watch the CNN news or the CNBC channel.

40. If the frog has legs, then it is no longer a tadpole.

41. Today is Monday or I will not go to my piano lesson.

42. I will not wear a suit today or I will not go to work.

In Exercises 43–46, use the fact that $(p \rightarrow q) \wedge (q \rightarrow p)$ is equivalent to $p \leftrightarrow q$ to write the statement in an equivalent form.

43. If a number is even then it is divisible by 2, and if a number is divisible by 2 then the number is even.

44. If you are 18 years old then you are eligible to vote, and if you are eligible to vote then you are 18 years old.

45. You need to pay taxes if and only if you receive income.

46. An animal is a mammal if and only if it is warm blooded.

In Exercises 47–56, write the converse, inverse, and contrapositive of the statement. (For Exercises 54–56, use De Morgan's laws.)

47. If the baby is a boy, then we will name him Cameron.

48. If the computer has 4 meg of memory, then we will buy the computer.

49. If the calculator is a TI 82, then the calculator can illustrate graphics.

50. If the pool water is green, then you have not added chlorine.

51. If you are hungry, then you have not eaten.

52. If the food is not healthy, then I will not go back to the restaurant.

53. If we do not stop all the crime, then our society is in trouble.

54. If Shannon is a race car driver or a bungee jumper, then Shannon has difficulty getting insurance.

55. If the figure is a parallelogram, then the opposite sides are parallel and equal.

56. If I buy a house or a condominium, then I pay real estate taxes.

In Exercises 57–64, write the contrapositive of the statement. Use the contrapositive to determine whether the conditional statement is true or false.

57. If the number is not divisible by 3, then the sum of the digits is not divisible by 3.

58. If the triangle is not isosceles, then two angles of the triangle are not equal.

59. If 2 does not divide the counting number, then 2 does not divide the units digit of the counting number.

60. If $1/n$ is not a natural number, then n is not a natural number.

61. If two lines do not intersect in at least one point, then the two lines are parallel.

62. If $\dfrac{m \cdot a}{m \cdot b} \neq \dfrac{a}{b}$, then m is not a counting number.

63. If the sum of the interior angles of a polygon do not measure 360°, then the polygon is not a quadrilateral.

64. If a and b are not both even counting numbers, then the product of a and b is not an even counting number.

In Exercises 65–80, determine which, if any, of the three statements are equivalent (see Example 13).

65. a) If you graduate from college, then you will get a college degree.
  b) If you do not get a college degree, then you did not graduate from college.
  c) You do not graduate from college or you get a college degree.

66. a) If today is Monday, then tomorrow is not Wednesday.
  b) It is false that today is Monday and tomorrow is not Wednesday.
  c) Today is not Monday or tomorrow is Wednesday.

67. a) It is false that you drive to work or you drive to school.
  b) If you do not drive to work, then you do not drive to school.
  c) You do not drive to work and you do not drive to school.

68. a) The sales tax is 7% if and only if you live in Orange County.
  b) If you live in Orange County then the sales tax is 7%, and if the sales tax is 7% then you live in Orange County.
  c) You do not live in Orange County and the sales tax is not 7%.

69. a) It is false that Garth Brooks is a country singer and Elvis is not alive.
  b) If Elvis is alive then Garth Brooks is not a country singer.
  c) Elvis is not alive or Garth Brooks is not a country singer.

70. a) It is false that if you do not leave your lights on, then your electric bill will be higher.

b) Your electric bill will be higher if and only if you leave your lights on.

c) It is false that if you leave your lights on then your electric bill will be higher.

71. a) If the car costs more than $34,000, then you need to pay a luxury tax.

b) It is false that you will need to pay a luxury tax if and only if the car costs more than $34,000.

c) You do not need to pay a luxury tax if and only if the car costs more than $34,000.

72. a) If you are fishing at 1 P.M., then you are not driving a car at 1 P.M.

b) You are not fishing at 1 P.M. or you are driving a car at 1 P.M.

c) It is false that you are fishing at 1 P.M. and you are not driving a car at 1 P.M.

73. a) Today is not Sunday or the library is open.

b) If today is Sunday, then the library is not open.

c) If the library is open, then today is not Sunday.

74. a) The grass grows and the trees are blooming.

b) If the trees are blooming then the grass does not grow.

c) The trees are not blooming or the grass does not grow.

75. a) If it is Thursday and it is 9 P.M., then we will be watching "Seinfeld."

b) If we are not watching "Seinfeld," then it is false that it is Thursday or it is 9 P.M.

c) We are watching "Seinfeld" and it is 9 P.M., or it is not Thursday.

76. a) If you are 18 years old and a citizen of the United States, then you can vote in the presidential election.

b) You can vote in the presidential election, if and only if you are a citizen of the United States and you are 18 years old.

c) You cannot vote in the presidential election, or you are 18 years old and you are not a citizen of the United States.

77. a) The package was sent by Federal Express, or the package was not sent by United Parcel Service but the package arrived on time.

b) The package arrived on time, if and only if it was sent by Federal Express or it was not sent by United Parcel Service.

c) If the package was not sent by Federal Express, then the package was not sent by United Parcel Service but the package arrived on time.

78. a) If we put the dog outside or we feed the dog, then the dog will not bark.

b) If the dog barks, then we did not put the dog outside and we did not feed the dog.

c) If the dog barks, then it is false that we put the dog outside or we feed the dog.

79. a) The car needs oil, and the car needs gas or the car is new.

b) The car needs oil, and it is false that the car does not need gas and the car is not new.

c) If the car needs oil, then the car needs gas or the car is not new.

80. a) The mortgage rate went down, if and only if Tim purchased the house and the down payment was 10%.

b) The down payment was 10%, and if Tim purchased the house then the mortgage rate went down.

c) If Tim purchased the house, then the mortgage rate went down and the down payment was not 10%.

81. Can a conditional statement and its converse both be false statements? Explain your answer with an example.

82. Can a conditional statement be true if its converse is false? Explain your answer with an example.

83. Can a conditional statement and its contrapositive both be false? Explain your answer with an example.

84. Can a conditional statement be true if its contrapositive is false? Explain.

Problem Solving/Group Activities

85. We learned that $p \rightarrow q \Leftrightarrow \sim p \vee q$. Determine a conjunctive statement that is equivalent to $p \rightarrow q$. (*Hint:* There are many answers.)

86. Determine whether $\sim[\sim(p \vee \sim q)] \Leftrightarrow p \vee \sim q$. Explain the method(s) you used to determine your answer.

87. In an appliance or device that uses fuzzy logic a change in one condition causes a change in a second condition. For example, in a camera if the brightness increases, the lens aperture automatically decreases to get the proper exposure on the film. Name at least 10 appliances or devices that make use of fuzzy logic and explain how fuzzy logic is used in each appliance or device. See the Did You Know on page 126.

88. In symbolic logic a statement is either true or false (consider true to have a value of 1 and false a value of 0). In fuzzy logic nothing is true or false, but everything is a matter of degree. For example, consider the statement "The sun is shining." In fuzzy logic this statement may have a value between 0 and 1 and may be constantly changing. For example, if the sun is partially blocked by clouds the value of this statement may be 0.25. In fuzzy logic the values of connective statements are found as follows for statements p and q.

Not p has a truth value of $1 - p$.

$p \wedge q$ has a true value equal to the lesser of p and q.

$p \vee q$ has a truth value equal to the greater of p and q.

$p \rightarrow q$ has a true value equal to the lesser of 1 and $1 - p + q$.

$p \leftrightarrow q$ has a truth value equal to $1 - |p - q|$, that is, 1 minus the absolute value of p minus q.

Suppose that the statement, p: The sun is shining, has a truth value of 0.25 and that the statement, q: Mary is getting a tan, has a truth value of 0.20. Find the truth value of

a) $\sim p$.

b) $\sim q$.

c) $p \wedge q$.

d) $p \vee q$.

e) $p \rightarrow q$.

f) $p \leftrightarrow q$.

Research Activities

89. Write a report on fuzzy logic. Many articles on the topic are available in your school library. A particularly interesting article appeared in the July 1993 issue of *Scientific American*.

90. Read one of Lewis Carroll's books and write a report on how he used logic in the book. Give at least five specific examples.

91. Do research and write a report on the life and achievements of Augustus De Morgan. Indicate in your report his contributions to sets and logic. Reference include encyclopedias and history of mathematics books.

3.5   SYMBOLIC ARGUMENTS

In the preceding sections of this chapter, we used symbolic logic to determine the truth value of a compound statement. We will now extend those basic ideas to determine whether symbolic arguments are valid or invalid.

Consider the statements:

If Jason is a singer, then he is well known.

Jason is a singer.

If you accept these two statements as true, then a conclusion that necessarily follows is that

Jason is well known.

These three statements in the following form constitute a symbolic argument.

A **symbolic argument** consists of a set of **premises** and a **conclusion**. It is called a symbolic argument because we generally write it in symbolic form to determine its validity.

Premise 1:	If Jason is a singer, then he is well known.
Premise 2:	Jason is a singer.
Conclusion:	Therefore Jason is well known.

An **argument is valid** when its conclusion necessarily follows from a given set of premises.

An **argument is invalid** or a **fallacy** when the conclusion does not necessarily follow from the given set of premises.

An argument that is not valid is invalid. The argument just presented is an example of a valid argument, as the conclusion necessarily follows from the premises. Now we will discuss a procedure to determine whether an argument is valid or invalid. We begin by writing the argument in symbolic form. To write the argument in symbolic form we let p and q be

*I*n the case of the disappearance of the racehorse Silver Blaze, Sherlock Holmes demonstrated that sometimes the absence of a clue is itself a clue. The local police inspector asked him, "Is there any point to which you would wish to draw my attention?" Holmes replied, "To the curious incident of the dog in the nighttime." The inspector, confused, asked: "The dog did nothing in the nighttime." "That was the curious incident," remarked Sherlock Holmes. From the lack of the dog's bark, Holmes concluded that the horse had been "stolen" by a stablehand. How did Holmes reach his conclusion? ■

p: Jason is a singer.

q: Jason is well known.

Symbolically, the argument is written

Premise 1: $p \rightarrow q$

Premise 2: $\underline{p}$

Conclusion: $\therefore q$ (The three-dot triangle is read "therefore.")

Write the argument in the following form.

If [*premise 1* **and** *premise 2*] **then** *conclusion*

$[(p \rightarrow q) \quad \wedge \quad p \quad] \quad \rightarrow \quad q$

Then construct a truth table for the statement $[(p \rightarrow q) \wedge p] \rightarrow q$ (Table 3.29). *If the truth table answer column is true in every case, then the statement is a tautology, and the argument is valid. If the truth table is not a tautology, then the argument is invalid.* Since the statement is a tautology the conclusion necessarily follows from the premises and the argument is valid.

Table 3.29

p	q	$[(p$	$\rightarrow$	$q)$	$\wedge$	$p]$	$\rightarrow$	q
T	T		T		T	T	T	T
T	F		F		F	T	T	F
F	T		T		F	F	T	T
F	F		T		F	F	T	F
			1		3	2	5	4

Once we have demonstrated that an argument in a particular form is valid, all arguments with exactly the same form will also be valid. In fact, many of these forms have been assigned names. The argument form just discussed,

$$p \rightarrow q$$
$$\underline{p}$$
$$\therefore q$$

is called the **law of detachment** (or *modus ponens*).

Consider the following argument.

If flowers smell sweet, then the honey is bitter.

Flowers smell sweet.

$\overline{\therefore \text{Honey is bitter.}}$

Now translate the argument into symbolic form. Let

s: Flowers smell sweet.

b: Honey is bitter.

In symbolic form the argument is

$$s \rightarrow b$$
$$\underline{s}$$
$$\therefore b$$

This argument is also the law of detachment, and therefore it is a valid argument.

Note that the argument is valid even though the conclusion, "Honey is bitter," is a false statement. It is also possible to have an invalid argument in which the conclusion is a true statement. When an argument is valid, the conclusion necessarily follows from the premises. It is not necessary for the premises or the conclusion to be true statements in an argument.

Procedure to Determine Whether an Argument is Valid

1. Write the argument in symbolic form.
2. Compare the form of the argument with forms that are known to be valid or invalid. If there are no known forms to compare it with, or you do not remember the forms, go to step 3.
3. If the argument contains two premises, write a conditional statement of the form

$$[(\text{premise 1}) \wedge (\text{premise 2})] \rightarrow \text{conclusion}$$

4. Construct a truth table for the statement in step 3.
5. If the answer column of the truth table has all trues, the statement is a tautology, and the argument is valid. If the answer column does not have all trues, the argument is invalid.

Examples 1–3 contain two premises. When an argument contains more than two premises, step 3 of the procedure will change slightly, as explained shortly.

EXAMPLE 1

Determine whether the following argument is valid or invalid.

> If Tim drives the truck, then the shipment will be delivered.
> The shipment will not be delivered.
> ∴ Tim does not drive the truck.

Solution: We first write the argument in symbolic form.

Let
p: Tim drives the truck.
q: The shipment will be delivered.

In symbolic form the argument is

$$p \rightarrow q$$
$$\sim q$$
$$\therefore \sim p$$

As we have not tested an argument in this form, we will construct a truth table to determine whether it is valid or invalid. We write the argument in the form $[(p \rightarrow q) \wedge \sim q] \rightarrow \sim p$, and construct a truth table (Table 3.30). Since the answer, column 5, has all T's, the argument is valid.

Table 3.30

p	q	$[(p \rightarrow q)$	$\wedge$	$\sim q]$	$\rightarrow$	$\sim p$
T	T	T	F	F	T	F
T	F	F	F	T	T	F
F	T	T	F	F	T	T
F	F	T	T	T	T	T
		1	3	2	5	4

The argument form in Example 1 is an example of the **law of contraposition** (or *modus tollens*).

EXAMPLE 2

Determine whether the following argument is valid or invalid.

> The car is red or the car is new.
> The car is new.
> ∴ The car is red.

Solution:

Let
p: The car is red.
q: The car is new.

In symbolic form the argument is

$$p \vee q$$
$$q$$
$$\therefore p$$

As this is not one of the forms we are familiar with, we will construct a truth table. We write the argument in the form $[(p \vee q) \wedge q] \rightarrow p$. Next, we construct a truth table, as shown in Table 3.31. The answer to the truth table, column 5, is not true in *every case*. Therefore the statement is not a tautology, and the argument is invalid, or is a fallacy.

Table 3.31

p	q	[($p \vee q$)	$\wedge$	q]	$\rightarrow$	p
T	T	T	T	T	T	T
T	F	T	F	F	T	T
F	T	T	T	T	F	F
F	F	F	F	F	T	F
		1	3	2	5	4

Standard forms of commonly used arguments are given in the following chart.

Standard Forms of Arguments

Valid Arguments	Law of Detachment	Law of Contraposition	Law of Syllogism	Disjunctive Syllogism
	$p \rightarrow q$	$p \rightarrow q$	$p \rightarrow q$	$p \vee q$
	p	$\sim q$	$q \rightarrow r$	$\sim p$
	$\therefore q$	$\therefore \sim p$	$\therefore p \rightarrow r$	$\therefore q$

Invalid Arguments	Fallacy of the Converse	Fallacy of the Inverse
	$p \rightarrow q$	$p \rightarrow q$
	q	$\sim p$
	$\therefore p$	$\therefore \sim q$

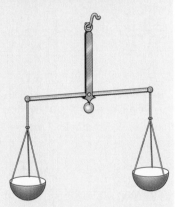

EXAMPLE 3

Determine whether the following argument is valid or invalid.

> If I drink hot milk, then I can sleep.
> If I can sleep, then I can dream.
> ∴ If I drink hot milk, then I can dream.

Solution:

Let

p:	I drink hot milk.
q:	I can sleep.
r:	I can dream.

In symbolic form the argument is

$$p \rightarrow q$$
$$q \rightarrow r$$
$$\overline{\therefore p \rightarrow r}$$

The argument is in the form of the law of syllogism. Therefore the argument is valid and there is no need to construct a truth table.   ▲

Now we consider an argument that has more than two premises. When an argument contains more than two premises, the statement we test, using a truth table, is formed by taking the conjunction of all the premises as the antecedent and the conclusion as the consequent. For example, if an argument is of the form

$$p_1$$
$$p_2$$
$$p_3$$
$$\overline{\therefore c}$$

We evaluate the truth table for $[p_1 \wedge p_2 \wedge p_3] \rightarrow c$. When we evaluate $[p_1 \wedge p_2 \wedge p_3]$, it makes no difference whether we evaluate $[(p_1 \wedge p_2) \wedge p_3]$ or $[p_1 \wedge (p_2 \wedge p_3)]$ because both give the same answer. In Example 4 we will evaluate $[p_1 \wedge p_2 \wedge p_3]$ from left to right, that is, $[(p_1 \wedge p_2) \wedge p_3]$.

EXAMPLE 4

Use a truth table to determine whether the following argument is valid or invalid.

> If the ice cream is soft, then it tastes better.
> The ice cream tastes better or it is fruit flavored.
> The ice cream is fruit flavored or the ice cream is soft.
> ∴ The ice cream is soft.

Solution: This argument contains three simple statements.

Let

p:	The ice cream is soft.
q:	The ice cream tastes better.
r:	The ice cream is fruit flavored.

In symbolic form the argument is

$$p \rightarrow q$$
$$q \vee r$$
$$r \vee p$$
$$\overline{\therefore p}$$

Write the argument in the form

$$[(p \rightarrow q) \wedge (q \vee r) \wedge (r \vee p)] \rightarrow p.$$

Now construct the truth table (Table 3.32). The answer, column 7, is not true in every case. Thus the argument is a fallacy.

Table 3.32

p	q	r	$[(p \rightarrow q)$	$\wedge$	$(q \vee r)$	$\wedge$	$(r \vee p)]$	$\rightarrow$	p
T	T	T	T	T	T	T	T	T	T
T	T	F	T	T	T	T	T	T	T
T	F	T	F	F	T	F	T	T	T
T	F	F	F	F	F	F	T	T	T
F	T	T	T	T	T	T	T	F	F
F	T	F	T	T	T	F	F	T	F
F	F	T	T	T	T	T	T	F	F
F	F	F	T	F	F	F	F	T	F
			1	3	2	5	4	7	6

Let's now investigate how we can arrive at a valid conclusion from a given set of premises.

EXAMPLE 5

Determine a logical conclusion that follows from the given statements. "If you own a house, then you will pay property tax. You own a house. Therefore, . . ."

Solution: If you recognize a specific form of an argument, you can use your knowledge of that form to draw a logical conclusion.

Let p: You own a house.
                   q: You will pay property tax.

$$\frac{\begin{array}{c} p \rightarrow q \\ p \end{array}}{\therefore ?}$$

If the question mark is replaced with a q, this argument is of the form of the law of detachment. A logical conclusion is "Therefore you will pay property tax."

Section 3.5 Exercises

1. What does it mean when an argument is valid?

2. Is it possible for an argument to be valid if its conclusion is false? Explain your answer.

3. Explain how to determine whether an argument with premises p_1 and p_2 and conclusion c is a valid or invalid argument.

4. What is another name for an invalid argument?

In Exercises 5–9, (a) indicate the form of the valid argument, and (b) write an original argument in words for each form.

5. Law of detachment

6. Law of syllogism

7. Law of contraposition

8. Disjunctive syllogism

In Exercises 9 and 10, (a) indicate the form of the fallacy, and (b) write an original argument in words for each form.

9. Fallacy of the converse

10. Fallacy of the inverse

In Exercises 11–30, determine whether the argument is valid or invalid. You may compare the argument to a standard form or use a truth table.

11. $p \rightarrow q$
$\dfrac{\sim p}{\therefore q}$

12. $p \wedge \sim q$
$\dfrac{q}{\therefore \sim p}$

13. $p \rightarrow q$
$\dfrac{p}{\therefore q}$

14. $\sim p \vee q$
$\dfrac{q}{\therefore p}$

15. $p \rightarrow q$
$\dfrac{\sim q}{\therefore \sim p}$

16. $p \vee q$
$\dfrac{\sim p}{\therefore \sim q}$

17. $p \rightarrow q$
$\dfrac{q}{\therefore p}$

18. $p \vee q$
$\dfrac{\sim q}{\therefore p}$

19. $\sim p \rightarrow q$
$\dfrac{\sim q}{\therefore \sim p}$

20. $q \wedge \sim p$
$\dfrac{\sim p}{\therefore q}$

21. $p \rightarrow q$
$\dfrac{q \rightarrow r}{\therefore p \rightarrow r}$

22. $q \wedge p$
$\dfrac{q}{\therefore \sim p}$

23. $p \leftrightarrow q$
$\dfrac{q \rightarrow r}{\therefore \sim r \rightarrow \sim p}$

24. $p \rightarrow q$
$\dfrac{q \wedge r}{\therefore p \vee r}$

25. $r \leftrightarrow p$
$\dfrac{\sim p \wedge q}{\therefore p \wedge r}$

26. $p \vee q$
$\dfrac{r \wedge p}{\therefore q}$

27. $p \rightarrow q$
$q \vee r$
$\dfrac{r \vee p}{\therefore p}$

28. $p \rightarrow q$
$r \rightarrow \sim p$
$\dfrac{p \vee r}{\therefore q \vee \sim p}$

29. $p \rightarrow q$
$q \rightarrow r$
$\dfrac{r \rightarrow p}{\therefore q \rightarrow p}$

30. $p \leftrightarrow q$
$p \vee r$
$\dfrac{q \rightarrow r}{\therefore q \vee r}$

In Exercises 31–42, translate the argument into symbolic form. Determine whether each argument is valid or invalid. You may compare the argument to a standard form or use a truth table.

31. If it is January, then we will have snow.
It is January.
∴ We will have snow.

32. I like broiled fish or I like baked chicken.
I do not like broiled fish.
∴ I like baked chicken.

33. If the speech is broadcast, then the people will listen.
The people will listen.
∴ The speech is broadcast.

34. The dog is a collie and the cat is a Siamese.
If the cat is Siamese, then the bird is a parrot.
∴ If the bird is not a parrot, then the dog is not a collie.

35. The painting is a Rembrandt or the painting is a Picasso.
If the painting is a Picasso, then it is not a Rembrandt.
∴ The painting is not a Picasso.

36. If you cook the meal, then I will vacuum the rug.
I will not vacuum the rug.
∴ You will not cook the meal.

37. The package is not more than 2 pounds.
If the package is more than 2 pounds, then we can mail the package.
∴ We can mail the package.

38. It is snowing and I am going skiing.
If I am going skiing, then I will wear a coat.
∴ If it is snowing, then I will wear a coat.

39. It is Ann's house or the house is made of brick.
If the house is not made of brick, then it is Ann's house.
∴ The house is made of brick or the house is Ann's house.

40. If the house has three bedrooms, then the Cleavers will buy the house.
If the price is not less than $90,000, then the Cleavers will not buy the house.
∴ If the house has three bedrooms, then the price is less than $90,000.

41. If the main dish is fish, then Teri will come to dinner.
If Teri comes to dinner, then we will eat at 6:00 P.M.
∴ If the main dish is fish, then we will eat at 6:00 P.M.

42. If there is an atmosphere, then there is gravity.
If an object has weight, then there is gravity.
∴ If there is an atmosphere, then an object has weight.

In Exercises 43–52, translate the argument into symbolic form. Then determine whether the argument is valid or invalid.

43. The plants are not in bloom or the plants need to be fertilized. The plants are in bloom. Therefore the plants need to be fertilized.

44. If all truck drivers are frustrated little boys, then I am a truck driver. I am not a truck driver. Therefore not all truck drivers are frustrated little boys.

45. If Bonnie passes the bar exam, then she will practice law. Bonnie will not practice law. Therefore Bonnie did not pass the bar exam.

46. The radio is on FM and tuned to station WROK. The radio is not tuned to station WROK or it is on FM. Therefore the radio is not on FM.

47. The hyenas are prowling and the zebra herd is alert. The zebra herd is not alert or the hyenas are prowling. Therefore the hyenas are not prowling.

48. If the car is new, then the car has air conditioning. The car is not new and the car has air conditioning. Therefore the car is not new.

49. If the football team wins the game, then Dave played quarterback. If Dave played quarterback, then the team is not in second place. Therefore, if the football team wins the game, then the team is in second place.

50. The engineering courses are difficult and the chemistry labs are long. If the chemistry labs are long, then the art tests are easy. Therefore the engineering courses are difficult and the art tests are not easy.

51. If the lights are on, then we can play ball. If the umpires are present, then we can play ball. Therefore if the lights are on, then the umpires are present.

52. Money is the root of all evil or the salvation of the poor. If money is the salvation of the poor, then it is not the root of all evil. Money is the root of all evil. Therefore money is not the salvation of the poor.

In Exercises 53–61, using the standard forms of arguments and other information you have learned, supply what you believe is a logical conclusion to the argument. Verify that the argument is valid for the conclusion you supplied.

53. If you drink orange juice, then you will be healthy.
You drink orange juice.
Therefore . . .

54. If you have lunch, then you must pay the bill.
You did not have lunch.
Therefore . . .

55. John reads the history assignment or he fixes the car.
John does not read the history assignment.
Therefore . . .

56. If the electric bill is too high, then I will not be able to pay the bill.
I will not be able to pay the bill.
Therefore . . .

57. If the dog is brown, then the dog is a cocker spaniel.
The dog is not a cocker spaniel.
Therefore . . .

58. If you do not read a lot, then you will not gain knowledge.
You do not read a lot.
Therefore . . .

59. If you overdraw your checking account, then the bank will send you a notice.

If the bank sends you a notice, then you will have to pay a fee.
Therefore . . .

60. If you are a waiter, then you must take orders.
If you do not get tips, then you do not take orders.
Therefore . . .

61. If the fire will burn, then the logs are dry.
If the fire will not burn, then you will have to dry the logs.
Therefore . . .

Problem Solving/Group Activities

62. Determine whether the argument is valid or invalid.

If Lynn wins the contest or strikes oil, then she will be rich.
If Lynn is rich, then she will stop working.
∴ If Lynn does not stop working, she did not win the contest.

63. It is possible for an argument to be invalid when the conclusion is a true statement. Use your own statements and the fallacy of the converse to write such an argument.

64. Is it possible for an argument to be invalid if the conjunction of the premises is false in every case of the truth table. Explain your answer.

Research Activities

65. Show how logic is used in advertising. Discuss several advertisements, and show how logic is used to persuade the reader.

66. Show how logic is used in politics. Collect articles on a particular political campaign, and show how logic is used by the candidates to gain votes.

67. Find examples of valid (or invalid) arguments in printed matter such as newspaper or magazine articles. Explain why the arguments are valid (or invalid).

3.6 EULER CIRCLES AND SYLLOGISTIC ARGUMENTS

In the preceding section we showed how to determine the validity of *symbolic arguments* using truth tables and comparing the arguments to standard forms. This section presents another form of argument called a **syllogistic argument**, better known by the shorter name **syllogism**. The validity of a syllogistic argument is determined by using Euler circles as will be explained shortly.

Syllogistic logic, a deductive process of arriving at a conclusion, was developed by Aristotle in about 350 B.C. Aristotle considered the relationships among the four types of statements that follow.

All _____ are _____.
No _____ are _____.
Some _____ are _____.
Some _____ are not _____.

Examples of these statements are: *All doctors are tall. No doctors are tall. Some doctors are tall. Some doctors are not tall.* Since Aristotle's time, other types of statements have been added to the study of syllogistic logic, two of which are

_____ is a _____.
_____ is not a _____.

Examples of these statements are: *Maria is a doctor. Maria is not a doctor.*

The difference between a symbolic argument and a syllogistic argument can be seen in the following chart. Symbolic arguments use the connectives *and, or, not, if–then,* and *if and only if.* Syllogistic arguments use the quantifiers *all, some,* and *none,* which were discussed in Section 3.1.

Symbolic Arguments vs. Syllogistic Arguments

	Words or Phrases Used	*Method of Determining Validity*
Symbolic argument	And, or, not, if–then, if and only if	Truth tables or by comparison with standard forms of arguments
Syllogistic argument	All are, some are, none are, some are not	Euler diagrams

As with symbolic logic, the premises and the conclusion together form an argument. An example of a syllogistic argument is

All German shepherds are dogs.
All dogs bark.
∴ All German shepherds bark.

This is an example of a valid argument. Recall from the previous section that an argument is *valid* when its conclusion necessarily follows from a given set of premises. Recall that an argument in which the conclusion does not necessarily follow from the given premises is said to be an *invalid argument* or a *fallacy.*

Before we give another example of a syllogism, let's review the Venn diagrams discussed in Section 2.3 in relationship with Aristotle's four statements.

All *A*s are *B*s	No *A*s are *B*s	Some *A*s are *B*s	Some *A*s are not *B*s
If an element is in set *A* then it is in set *B*.	If an element is in set *A* then it is not in set *B*.	There is at least one element that is in both set *A* and set *B*.	There is at least one element that is in set *A* that is not in set *B*.

One method used to determine whether an argument is valid or is a fallacy is by means of an **Euler diagram**, named after Leonhard Euler (1707–1783) who used circles to represent sets in syllogistic arguments. The technique of using Euler diagrams is illustrated in Example 1.

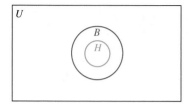

Figure 3.1

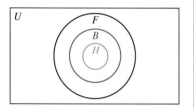

Figure 3.2

EXAMPLE 1

Determine whether the following syllogism is valid or is a fallacy.

> All horses are brown.
> All brown animals have fur.
> ∴ All horses have fur.

Solution: The statement ''All horses are brown'' may be interpreted as ''If an animal is a horse, then it is brown.'' Construct the diagram and represent the first premise ''All horses are brown,'' as shown in Fig. 3.1. The outer circle represents all brown animals and the inner circle represents all horses. Now illustrate the second premise ''All brown animals have fur,'' as shown in Fig. 3.2. The set containing all animals that have fur must contain all the brown animals, as illustrated in the diagram. Note that the premises force the set of horses to be within the set of animals that have fur. Therefore the argument is valid, since the conclusion ''All horses have fur'' necessarily follows from the set of premises. ▲

The argument in Example 1 is valid even though the conclusion ''All horses have fur'' is obviously a false statement. Similarly, an argument can be a fallacy even if the conclusion is a true statement.

When we determine the validity of an argument, we are determining whether the conclusion necessarily follows from the premises. When we say that an argument is valid, we are saying that if all the premises are true statements, then the conclusion must also be a true statement.

The form of the argument determines its validity, not the particular statements. For example, consider the syllogism

> All earth people have two heads.
> All people with two heads can fly.
> ∴ All earth people can fly.

The form of this argument is the same as that of the previous valid argument. Therefore this argument is also valid.

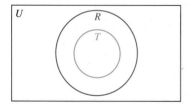

Figure 3.3

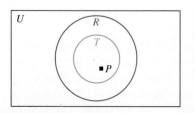

Figure 3.4

EXAMPLE 2

Determine whether the following syllogism is valid or is a fallacy.

> All teachers are rich.
> Pete is a teacher.
> ∴ Pete is rich.

Solution: The statement ''All teachers are rich'' is illustrated in Fig. 3.3. The second premise, ''Pete is a teacher,'' tells us that Pete must be placed in the inner circle (see Fig. 3.4). The Euler diagram illustrates that we must accept the conclusion ''Pete is rich'' as true (when we accept the premises as true). Therefore the argument is valid. ▲

In both Example 1 and Example 2 we had no choice as to where the second premise was to be placed in the Euler diagram. In Example 1, the set of brown animals had to be placed inside the set of animals with fur. In Example 2, Pete had to be placed inside the set of teachers. Often when determining the truth value of a syllogism a premise can be placed in more than one area in the diagram. *We always try to draw the Euler diagram so that the conclusion does not necessarily follow from the premises. If this can be done, then the conclusion does not necessarily follow from the premises and the argument is invalid.* If we cannot show that the argument is invalid, only then do we accept the argument as valid, as illustrated in Example 3.

EXAMPLE 3

Determine whether the following syllogism is valid or is a fallacy.

> All football players are strong.
> Christine is strong.
> ∴ Christine is a football player.

Solution: The statement ''All football players are strong'' is illustrated in Fig. 3.5(a). The next premise, ''Christine is strong,'' tells us that Christine must be placed in the set of strong people. Two diagrams in which both premises are satisfied are shown in Figs. 3.5(b) and 3.5(c). By examining Fig. 3.5(b),

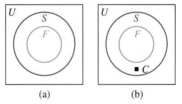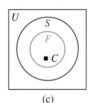

Figure 3.5   (a)   (b)   (c)

however, we see that Christine is not a football player. Therefore the conclusion ''Christine is a football player'' does not necessarily follow from the set of premises. Thus the argument is a fallacy.   ▲

EXAMPLE 4

Determine whether the following syllogism is valid or is a fallacy.

> No airplane pilots eat spinach.
> Janet does not eat spinach.
> ∴ Janet is an airplane pilot.

Solution: The diagram in Fig. 3.6 satisfies the two given premises and also shows that Janet is not an airplane pilot. Therefore the argument is invalid, or a fallacy.   ▲

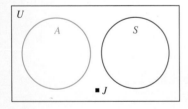

Figure 3.6

Note that in Example 4 if we placed Janet in circle *A*, the argument would appear to be valid. Remember, *whenever testing the validity of an argument, always try to show that the argument is invalid.* If there is any way of showing that the conclusion does not necessarily follow from the premises, then the argument is invalid.

EXAMPLE 5

Determine whether the following syllogism is valid or invalid.

All *A*s are *B*s.
Some *B*s are *C*s.
∴ Some *A*s are *C*s.

Solution: The statement "All *A*s are *B*s" is illustrated in Fig. 3.7. The statement "Some *B*s are *C*s" means that there is at least one *B* that is a *C*. We can illustrate this set of premises in four ways, as illustrated in Fig. 3.8.

Figure 3.7

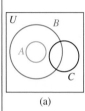 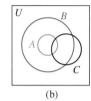 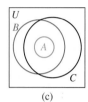

    (a)   (b)   (c)   (d)

Figure 3.8

In all four illustrations we see that (1) all *A*s are *B*s and (2) some *B*s are *C*s. The conclusion is "Some *A*s are *C*s." Since at least one of the illustrations (Fig. 3.8a) shows that the conclusion does not necessarily follow from the given premises, the argument is invalid.   ▲

EXAMPLE 6

Determine whether the following syllogism is valid or invalid.

No neurosurgeons eat licorice.
All college graduates eat licorice.
∴ No neurosurgeons are college graduates.

Solution: The first premise tells us that the neurosurgeons and the people who eat licorice are disjoint sets, as illustrated in Fig. 3.9. The second premise tells us that the set of college graduates is a subset of the people who eat licorice.

The set of college graduates and neurosurgeons cannot be made to intersect without violating a premise. Thus no neurosurgeons can be college graduates, and the syllogism is valid. Note that we did not say that this conclusion was true—only that the argument was valid.   ▲

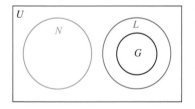

Figure 3.9

Section 3.6 Exercises

1. What does it mean when we determine that an argument is valid?

2. Explain the differences between a symbolic argument and a syllogistic argument.

3. Can an argument be valid if the conclusion is a false statement? Explain your answer.

4. Can an argument be invalid if the conclusion is a true statement? Explain.

In Exercises 5–27, use an Euler diagram to determine whether the syllogism is valid or is a fallacy.

5. All frogs are green.
   Kermit is a frog.
   ∴ Kermit is green.

6. No dogs are cats.
  All poodles are dogs.
  ∴ No poodles are cats.

7. All grass is green.
  All things that are green will grow.
  ∴ All grass will grow.

8. All *A*s are *B*s.
  All *B*s are *C*s.
  ∴ All *A*s are *C*s:

9. All trees are living things.
  Marty is a living thing.
  ∴ Marty is a tree.

10. All physicists are intelligent.
  Madame Curie was intelligent.
  ∴ Madame Curie was a physicist.

11. All football players have agents.
  Steve Young is a football player.
  ∴ Steve Young has an agent.

12. All numbers that are divisible by 2 are even.
  The number 7 is divisible by 2.
  ∴ The number 7 is even.

13. Some animals are dangerous.
  A lion is an animal.
  ∴ A lion is dangerous.

14. Some swans are not graceful swimmers.
  Candy is a swan.
  ∴ Candy is not a graceful swimmer.

15. Some medical researchers have made important discoveries.
  Some people who have made important discoveries are famous.
  ∴ Some medical researchers are famous.

16. All watermelons are edible.
  Some edible foods are fruits.
  ∴ All watermelons are fruits.

17. No plumbers are electricians.
  Lindsey is not an electrician.
  ∴ Lindsey is a plumber.

18. Some cars can float.
  All things that float are lighter than water.
  ∴ Some cars are lighter than water.

19. Some people love mathematics.
  All people who love mathematics love physics.
  ∴ Some people love physics.

20. Some desks are made of wood.
  All paper is made of wood.
  ∴ Some desks are made of paper.

21. No *x*s are *y*s.
  No *y*s are *z*s.
  ∴ No *x*s are *z*s.

22. All pilots can fly.
  All astronauts can fly.
  ∴ Some pilots are astronauts.

23. Some tall people wear glasses.
  Maria wears glasses.
  ∴ Maria is not tall.

24. All rainy days are cloudy.
  Today it is cloudy.
  ∴ Today is a rainy day.

25. All sweet things taste good.
  All things that taste good are fattening.
  All things that are fattening put on pounds.
  ∴ All sweet things put on pounds.

26. All books have red covers.
  All books that have red covers contain 200 pages.
  Some books that contain 200 pages are novels.
  ∴ All books that contain 200 pages are novels.

27. All squares are rectangles.
  All rectangles are quadrilaterals.
  All quadrilaterals are polygons.
  ∴ All squares are polygons.

Problem Solving/Group Activity

28. Statements in logic can be translated into set statements—for example, $p \wedge q$ is similar to $P \cap Q$; $p \vee q$ is similar to $P \cup Q$; $p \rightarrow q$ is equivalent to $\sim p \vee q$, which is similar to $P' \cup Q$. Euler circles can also be used to show that arguments similar to those discussed in this section are valid or invalid. Use Euler circles to show that the symbolic argument is invalid.

$$p \rightarrow q$$
$$p \vee q$$
$$\therefore \sim p$$

Research Activity

29. Leonhard Euler is considered one of the greatest mathematicians of all time. Do research and write a report on Euler's life. Include information on his contributions to sets and to logic. Also indicate other areas of mathematics in which he made important contributions. References include encyclopedias and history of mathematics books.

CHAPTER 3 SUMMARY

Key Terms

3.1

antecedent	conditional (if . . . then . . .)	consequent	inclusive *or*	simple statements
Aristotelian logic	conjunction (and)	disjunction (or)	negation (not)	statement
biconditional (if and only if)	connectives	exclusive *or*	quantifiers	symbolic logic
compound statements				

3.2

truth table

3.3

biconditional statement
conditional statement
implication
self-contradiction
tautology

3.4

contrapositive
converse
De Morgan's laws
equivalent
   statements

3.5

fallacy
invalid argument
symbolic argument
valid argument

3.6

Euler diagram
syllogism
syllogistic argument
fuzzy logic
inverse
logically equivalent

Important Facts

Quantifiers

Form of Statement	Form of Negation
All are.	Some are not.
None are.	Some are.
Some are.	None are.
Some are not.	All are.

Basic Truth Tables

Negation   Conjunction   Disjunction   Conditional   Biconditional

p	$\sim p$	p	q	$p \wedge q$	$p \vee q$	$p \to q$	$p \leftrightarrow q$
T	F	T	T	T	T	T	T
F	T	T	F	F	T	F	F
		F	T	F	T	T	F
		F	F	F	F	T	T

Summary of Connectives

Formal Name	Symbol	Read	Symbolic Form
Negation	$\sim$	not	$\sim p$
Conjunction	$\wedge$	and	$p \wedge q$
Disjunction	$\vee$	or	$p \vee q$
Conditional	$\to$	if–then	$p \to q$
Biconditional	$\leftrightarrow$	if and only if	$p \leftrightarrow q$

De Morgan's Laws

$$\sim(p \wedge q) \Leftrightarrow \sim p \vee \sim q$$
$$\sim(p \vee q) \Leftrightarrow \sim p \wedge \sim q$$

Other Equivalent Forms

$$p \to q \Leftrightarrow \sim p \vee q$$
$$p \leftrightarrow q \Leftrightarrow [(p \to q) \wedge (q \to p)]$$

Variations of the Conditional Statement

Name	Symbolic Form	Read
Conditional	$p \rightarrow q$	If p, then q.
Converse of the conditional	$q \rightarrow p$	If q, then p.
Inverse of the conditional	$\sim p \rightarrow \sim q$	If not p, then not q.
Contrapositive of the conditional	$\sim q \rightarrow \sim p$	If not q, then not p.

Symbolic Argument vs. Syllogistic Argument

	Words or Phrases Used	Method of Determining Validity
Symbolic argument	And, or, not, if–then, if and only if	Truth tables or by comparison with standard forms of arguments
Syllogistic argument	All are, some are, none are, some are not	Euler diagrams

Standard Forms of Arguments

Valid Arguments

Law of Detachment	Law of Contraposition	Law of Syllogism	Disjunctive Syllogism
$p \rightarrow q$	$p \rightarrow q$	$p \rightarrow q$	$p \vee q$
p	$\sim q$	$q \rightarrow r$	$\sim p$
$\therefore p$	$\therefore \sim p$	$\therefore p \rightarrow r$	$\therefore q$

Invalid Arguments

Fallacy of the Converse	Fallacy of the Inverse
$p \rightarrow q$	$p \rightarrow q$
q	$\sim p$
$\therefore p$	$\therefore \sim q$

Chapter 3 Review Exercises

3.1

In Exercises 1–6, write the negation of the statement.

1. All cameras use film.
2. No dogs have fleas.
3. Some apples are not red.
4. Some locks are keyless.
5. Not all people wear glasses.
6. No rabbits wear glasses.

In Exercises 7–12, write each compound statement in words.

p: I opened a can of soda.
q: I spilled some soda.
r: The soda is red.

7. $p \wedge q$

8. $\sim p \vee r$

9. $p \rightarrow (q \wedge \sim r)$

10. $p \leftrightarrow q$

11. $(p \vee \sim q) \wedge \sim r$

12. $\sim p \leftrightarrow (r \wedge \sim q)$

3.2

In Exercises 13–18, use the statements for p, q, and r as in Exercises 7–12 to write the statement in symbolic form.

13. I spilled some soda or the soda is not red.

14. I opened a can of soda and the soda is red.

15. If the soda is red, then I did not open a can of soda or I spilled some soda.

16. I spilled some soda if and only if I opened a can of soda, and the soda is not red.

17. The soda is red and I opened a can of soda, or I spilled some soda.

18. It is false that I spilled some soda or the soda is red.

In Exercises 19–26, construct a truth table for the statement.

19. $(p \vee \sim q) \wedge p$

20. $\sim p \wedge \sim q$

21. $\sim p \rightarrow (q \wedge \sim p)$

22. $\sim p \leftrightarrow q$

23. $p \wedge (\sim q \vee r)$

24. $p \rightarrow (q \wedge \sim r)$

25. $(p \vee q) \leftrightarrow (p \vee r)$

26. $(p \wedge q) \rightarrow \sim r$

3.2, 3.3

In Exercises 27–31, determine the truth value of the statement.

27. If a lion roars, then a bird has wings.

28. Super glue is a food or carrots are a food.

29. If Babe Ruth is known for his automobile designs, then Thomas Edison played professional baseball.

30. $3 + 7 = 11$ or $6 + 5 = 11$, and $7 \cdot 6 = 42$.

31. Spanish is a language, if and only if $2 + 2 = 7$ or $3 + 5 = 8$.

3.3

In Exercises 32–36, determine the truth value of the statement when p is T, q is F, and r is F.

32. $(p \vee q) \rightarrow (\sim r \wedge p)$

33. $(q \rightarrow \sim r) \vee (p \wedge q)$

34. $\sim r \leftrightarrow [(p \vee q) \leftrightarrow \sim p]$

35. $\sim[(q \wedge r) \rightarrow (\sim p \vee r)]$

36. $[\sim(q \wedge p)] \rightarrow \sim(\sim p \vee r)$

3.4

In Exercises 37–40, determine whether the pairs of statements are equivalent. You may use De Morgan's laws,

the fact that $(p \rightarrow q) \Leftrightarrow (\sim p \vee q)$, truth tables, or equivalent forms of the conditional statement.

37. $\sim p \rightarrow \sim q$ $p \vee \sim q$

38. $\sim p \vee \sim q$ $\sim p \leftrightarrow q$

39. $\sim p \vee (q \wedge r)$ $(\sim p \vee q) \wedge (\sim p \vee r)$

40. $(\sim q \rightarrow p) \wedge p$ $\sim(\sim p \leftrightarrow q) \vee p$

In Exercises 41–45, use De Morgan's laws or the fact that $(p \rightarrow q) \Leftrightarrow (\sim p \vee q)$ to write an equivalent statement for the given statement.

41. The stapler is empty or the stapler is jammed.

42. The Kings score goals and the Kings win games.

43. It is not true that New Orleans is in California or Chicago is not a city.

44. If there is not water in the vase, then the flowers will wilt.

45. I did not go to the party and I did not finish my special report.

In Exercises 46–49, write the contrapositive for the statement.

46. If the railroad crossing light is flashing red, then you must stop.

47. If John is having difficulty seeing, then John's eyes must be checked.

48. If today is not a holiday, then I will be at work.

49. If the computer has a color monitor and an extended keyboard, then it is Sam's computer.

50. Write the converse, inverse, and contrapositive of the conditional statement ''If I study, then I will get a passing grade.''

In Exercises 51–54, determine which, if any, of the three statements are equivalent.

51. a) If the temperature is over 80°, then the air conditioner will come on.
   b) The temperature is not over 80° or the air conditioner will come on.
   c) It is false that the temperature is over 80° and the air conditioner will not come on.

52. a) The screwdriver is on the workbench if and only if the screwdriver is not on the counter.
   b) If the screwdriver is not on the counter then the screwdriver is not on the workbench.
   c) It is false that the screwdriver is on the counter and the screwdriver is not on the workbench.

53. a) If $2 + 3 = 6$, then $3 + 1 = 5$.
   b) $2 + 3 = 6$ if and only if $3 + 1 \neq 5$.
   c) If $3 + 1 \neq 5$, then $2 + 3 \neq 6$.

54. a) If the painting must be done on Monday and the furniture must be delivered on Monday, then I must be home on Monday.

   b) If I must not be home on Monday, then the painting must be done on Monday and the furniture must be delivered on Monday.

   c) I must be home on Monday, or the painting must be done on Monday and the furniture must be delivered on Monday.

3.5, 3.6

In Exercises 55–60, determine whether the argument is valid.

55. $p \rightarrow q$
   $\underline{\sim p }$
   $\therefore q$

56. $p \wedge q$
   $\underline{q \rightarrow r }$
   $\therefore p \rightarrow r$

57. Nicole is in the hot tub or she is in the shower.
   $\underline{\text{Nicole is in the hot tub.} }$
   $\therefore$ Nicole is not in the shower.

58. If the car has a sound system, then Rick will buy the car. If the price is not less than $18,000, then Rick will not buy the car. Therefore if the car has a sound system, then the price is less than $18,000.

59. All grasshoppers are green.
   $\underline{\text{Some crickets are green.} }$
   $\therefore$ Some crickets are grasshoppers.

60. No mathematics books are dull.
   $\underline{\text{All textbooks are dull.} }$
   $\therefore$ Some mathematics books are textbooks.

CHAPTER 3 TEST

In Exercises 1–3, write the statement in symbolic form.

       p: Raul is a boy.
       q: Raul is 10 years old.
       r: Raul lives in New Jersey.

1. Raul lives in New Jersey or Raul is 10 years old, and Raul is a boy.

2. If Raul is a boy and Raul does not live in New Jersey, then Raul is not 10 years old.

3. It is false that Raul lives in New Jersey if and only if Raul is 10 years old.

In Exercises 4 and 5, use p, q, and r as above to write each symbolic statement in words.

4. $\sim(p \rightarrow \sim r)$ **5.** $p \leftrightarrow (q \wedge r)$

In Exercises 6 and 7, construct a truth table for the given statement.

6. $[\sim(p \rightarrow r)] \wedge q$ **7.** $(q \leftrightarrow \sim r) \vee p$

In Exercises 8 and 9, find the truth value of the statement.

8. $2 + 6 = 8$ or $7 - 12 = 5$.

9. If Germany is a country in Europe and India is a country in Asia, then Mexico is a country in Canada.

In Exercises 10 and 11, given that p is true, q is false, and r is true, determine the truth value of the statement.

10. $[\sim(r \rightarrow \sim p)] \wedge (q \rightarrow p)$

11. $(r \vee q) \leftrightarrow (p \wedge \sim q)$

12. Determine whether the pair of statements are equivalent.

       $\sim p \vee q$ $\sim(p \wedge \sim q)$

In Exercises 13 and 14, determine which, if any, of the three statements are equivalent.

13. a) If the bird is red, then it is a cardinal.
    b) The bird is not red or it is a cardinal.
    c) If the bird is not red, then it is not a cardinal.

14. a) It is not true that the test is today or the concert is tonight.
    b) The test is not today and the concert is not tonight.
    c) If the test is not today, then the concert is not tonight.

15. Translate the following argument into symbolic form. Determine whether the argument is valid or invalid by comparing the argument to a recognized form or by using a truth table.

    If the soccer team wins the game, then Sue played fullback. If Sue played fullback, then the team is in second place. Therefore, if the soccer team wins the game, then the team is in second place.

16. Use an Euler diagram to determine whether the following syllogistic argument is valid or invalid.

       All numbers divisible by 7 are odd.
       The number 28 is divisible by 7.
       ∴ The number 28 is odd.

In Exercises 17 and 18, write the negation of the statement.

17. All birds are black.

18. Some people are funny.

19. The conditional statement "If the apple is red, then it is a delicious apple" is given. Write the inverse, converse, and contrapositive of the statement.

20. Is it possible for an argument to be valid when the conclusion is a false statement? Explain your answer.

GROUP PROJECTS

Rental Agreement

1. On page 149 is a typical personal property rental agreement. It may be used for an owner renting personal property from a chain saw to a roto rooter.

   a) For each article in the agreement, list all the logic connectives that appear. Do not forget conditional statements where the word *then* is not written but is implied.

   b) For each *or* listed in part (a) indicate whether it is an *inclusive or* or an *exclusive or* (see page 89). Explain your answers.

   c) In article 5 of the agreement, represent each simple statement with a letter starting with the letter p. For example, let

   p: The owner warrants the property is free of any known faults which would affect its safe operation under normal usage.

   q: The property is in good working condition.

   Then rewrite this article, using only the letters p, q, r, and so on. Write each sentence in a vertical manner, one below the other.

   d) Fill in all the blanks in the agreement with realistic information. You may choose the item that is to be rented.

   e) If the rentee is using illegal drugs while using this property, is the rentee violating the agreement? Explain.

   f) If the rentee is using illegal drugs while using this personal property and the rentee is injured because of a proven defect in the property, do you believe that the rentee would have a case against the owner? Explain your answer.

   g) Explain in your own words the meaning of article 6 of the agreement.

   h) Suppose that the article rented is a leafblower. The owner rents the leafblower to one person for his own use. The owner finds out that before returning the blower, the renter has shared the blower with his neighbors and that every household on the street where the renter lives used it to remove leaves from their property. Has the renter violated the agreement? Explain. What, if anything, do you believe the owner should do?

Switching Circuits

2. An application of logic is *switching circuits*. There are two basic types of electric circuits: *series circuits* and *parallel circuits*. In a series circuit the current can flow in only one path; see Fig. 3.10. In a parallel circuit the current can flow in more than one path; see Fig. 3.11.

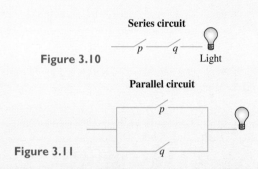

Series circuit

Figure 3.10   Light

Parallel circuit

Figure 3.11

In Figs. 3.10 and 3.11 the p and q represent switches that may be opened or closed. In the series circuit in Fig. 3.10, if both switches are closed the current will reach the bulb and the bulb will light. In the parallel circuit in Fig. 3.11 if either switch p or switch q is closed, or if both switches are closed, the current will reach the bulb and the bulb will light.

a) How many different open/closed arrangements of the two switches in Fig. 3.10 are possible? List all the possibilities.

b) Series circuits are represented using conjunctions. The circuit in Fig. 3.10 may be represented as $p \wedge q$. Construct a 4 row truth table to represent the series circuit. Construct the table with columns for p, q, and $p \wedge q$. The statement $p \wedge q$ represents the outcome of the circuit (either the bulb lighting or the bulb not lighting). Represent a closed switch with the number 1, an open switch with the number 0, the bulb lighting with the number 1, and the bulb not lighting with the number 0. For example, if both switches are closed, the bulb will light, and so we write the first row of the truth table as

p	q	$p \wedge q$
1	1	1

c) How is the truth table determined in part (b) similar to the truth table for $p \wedge q$ discussed in earlier sections of this chapter?

d) Parallel circuits are represented using disjunctions. The circuit in Fig. 3.11 may be represented as $p \vee q$. Construct a truth table to represent the parallel circuit. Construct the table with columns for p, q, and $p \vee q$. The statement $p \vee q$ represents the outcome of the circuit (either the bulb lighting or the bulb not lighting). Use 1's and 0's as indicated in part (b).

e) How is the truth table determined in part (d) similar to the truth table for $p \vee q$ discussed in earlier sections of this chapter?

Parts (f) and (g) continued on page 150.

PERSONAL PROPERTY RENTAL AGREEMENT
(Complex)

This Agreement is made on _____, 19____, between _____,
Owner, residing at _____, City of _____, State of _____, and _____.
Renter, residing at _____, City of _____, State of _____.

1. The Owner agrees to rent to the Renter and the Renter agrees to rent from the Owner the following property:

_____

2. The term of this Agreement will be from ____ o'clock ___ m. _____ 19____, until _____ o'clock ___ m., _____, 19____.

3. The rental payments will be $_____ per _____ and will be payable by the Renter to the Owner as follows: _____

4. The Renter agrees to pay a late fee of $_____ per day that the rental payment is late. If the rental payments are in default for over _____ days, the Owner may immediately demand possession of the property without advance notice to the Renter.

5. The Owner warrants that the property is free of any known faults that would affect its safe operation under normal usage and is in good working condition. The Renter states that the property has been inspected and is in good working condition. The Renter agrees to use the property in a safe manner and in normal usage and to maintain the property in good repair. The Renter further agrees not to use the property in a negligent manner or for any illegal purpose.

6. The Renter agrees to fully indemnify the Owner for any damage to or loss of the property during the term of this Agreement, unless such loss or damage is caused by a defect of the rented property. The Owner shall not be liable for any injury, loss, or damage caused by any use of the property.

7. The Renter has paid the Owner a security deposit of $_____. This security deposit will be held as security for payments of the rent and for the repair of any damages to the property by the Renter. This deposit will be returned to the Renter upon the termination of this Agreement, minus any rent still owed to the Owner and minus any amounts needed to repair the property, beyond normal wear and tear.

8. The Renter may not assign or transfer any rights under this Agreement to any other person, nor allow the property to be used by any other person, without the written consent of the Owner.

9. Renter agrees to obtain insurance coverage for the property during the term of this rental agreement in the amount of $_____. Renter agrees to provide the Owner with a copy of the insurance policy and to not cancel the policy during the term of this rental agreement.

10. This Agreement may be terminated by either party by giving 24 hours written notice to the other party. Any dispute related to this agreement will be settled by voluntary mediation. If mediation is unsuccessful, the dispute will be settled by binding arbitration using an arbitrator of the American Arbitration Association. The parties agree that this Agreement is the entire agreement between them. This Agreement binds and benefits both the Owner and Renter and any successors. Time is of the essence of this Agreement. This Agreement is governed by the laws of the State of _____.

11. The following are additional terms of this Agreement: _____

_____   _____
(Signature of owner)   (Signature of renter)

_____   _____
(Name of owner)   (Name of renter)

f) Represent the following circuit as a symbolic logic statement using parentheses. Explain how you determined your answer.

g) Draw a circuit to represent the logic statement $p \land (q \lor r)$. Explain how you determined your answer.

Computer Gates

3. Gates in computers work on the same principles as switching circuits. The three basic types of gate are the NOT gate, the AND gate, and the OR gate. Each is illustrated along with a table that indicates current flow entering and exiting the gate. If current flows into a NOT gate, then no current exists, and vice-versa. Current exits an AND gate only when both inputs have a current flow. Current exits an OR gate if current flows through either, or both, inputs. In the table a 1 represents a current flow and a 0 indicates no current flow. For example, in the AND gate, if there is a current flow in input A (I_a has a value of 1) and no current flow in input B (I_b has a value of 0), there is no current flow in the output (O has a value of 0); see row 2 of the AND Gate table.

NOT gate

Input ——▷○—— Output

AND gate　　　　　　　　　**OR gate**

Input A ——⟩　　　　　　Input A ——⟩
　　　　　　⟩—— Output　　　　　　⟩—— Output
Input B ——⟩　　　　　　Input B ——⟩

NOT Gate

I	O
1	0
0	1

AND Gate　　　　　**OR Gate**

I_a	I_b	O
1	1	1
1	0	0
0	1	0
0	0	0

I_a	I_b	O
1	1	1
1	0	1
0	1	1
0	0	0

a) If 1 is considered true and 0 is considered false, explain how these tables are similar to the *not*, *and*, and *or* truth tables.

For the inputs indicated in the following figures determine whether the output is 1 or 0.

b)

c)

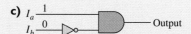

d)

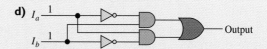

e) What values for I_a and I_b will give an output of 1 in the figure in part (d)? Explain how you determined your answer.

f) Construct a truth table using 1's and 0's for the following gate. Your truth table should have columns I_a, I_b, and O, and should indicate the four possible cases for the inputs and each corresponding output.

Logic Game

4. a) On page 151 is a photograph of a logic game at the Ontario Science Centre. There are 12 balls on top of the game board, numbered from left to right, with ball 1 on the extreme left and ball 12 on the extreme right. On the platform in front of the players are 12 buttons, one corresponding to each of the balls. When 6 buttons are pushed, the 6 respective balls are released. When 1 or 2 balls reach an *and* or *or* gate, a single ball may or may not pass through the gate. The object of the game is to select a proper combination of 6 buttons that will allow 1 ball to reach the bottom. Using your knowledge of *and* and *or*, select a combination of 6 buttons that will result in a win. (There is more than one answer.) Explain how you determined your answer.

b) Construct a game similar to this one where 15 balls are at the top and 8 must be selected to allow 1 ball to reach the bottom.

c) Indicate all solutions to the game you constructed in part (c).

Cat Puzzle

5. Solve the following puzzle. The Joneses have four cats. The parents are Tiger and Boots, and the kittens are Sam and Sue. Each cat insists on eating out of its own bowl. To complicate matters, each cat will eat only its own brand of cat food. The colors of the bowls are red, yellow, green, and blue. The different types of cat food are Puss 'n' Boots, Friskies, Nine Lives, and Meow Mix. Tiger will eat Meow Mix if and only if it is in a yellow bowl. If Boots is to eat her food, then it must be in a yellow bowl. Mrs. Jones knows that the label on the can containing Sam's food is the same color as his bowl. The name of the food that Boots will eat contains her name. Meow Mix and Nine Lives are packaged in a brown paper bag. The color of Sue's bowl is green if and only if she eats Meow Mix. The label on the Friskies can is red. Match each cat with its food and the bowl of the correct color.

Chapter 4 Systems of Numeration

*W*e are so accustomed to reading and writing that it is hard to imagine what our lives would be like without them. The number system we use—called the Hindu–Arabic system—seems to be a permanent, unchanging means of communicating quantities. However, just as languages evolve over time, so do numerical symbols that represent numbers.

Mathematics began with the practical problem of counting and record keeping. This happened so far back in human history that no one can point to a time or place and say, "It all started here." People had to count their herds, the passage of days, and objects of barter. They used physical objects—stones, shells, fingers—to represent the objects counted. For example, shepherds counted their flocks by moving one pebble from the "out" pile to the "in" pile to account for each animal.

As primitive cultures grew from villages to cities, the complexity of human activities increased. Now people needed better ways of recording and communicating. It was a

One of the earliest reasons that human beings needed numbers was for reckoning time, marking off days in the lunar month, so the seasonal changes that dictated human activity could be anticipated. The Mayans, the Egyptians, and the ancient Britons constructed monumental stone observatories that enabled them to mark the passage of the seasons, especially the summer solstice, using the alignment of the sun as a guide.

This antique print shows one person calculating with Hindu–Arabic numerals, the other with a counting board. Looking over their shoulders is the spirit Arithmeticae.

revolutionary step when people started using physical objects to represent not only specific objects like sheep and grain, but also the concept of pure quantity. The later invention of counting boards, the abacus, and wooden or ivory rods made it easier to work with large numbers.

Through the course of human history, the evolution of numeration systems has expanded our knowledge and abilities for record keeping, communication, and computation. As a society's numeration system changes, so do the capabilities of that society. Without an understanding of the binary number system, the computer as we know it today could not exist. And, without the computer, would we have had the ability to explore space or track the activities of a global marketplace? ■

Because of its flexibility, it is the Hindu–Arabic system that has become the universal language of mathematics.

Archaeologist Denise Schmandt-Besserat made a breakthrough discovery about early systems of numeration. She realized that the little clay geometric objects that had been found in many archaeological sites had actually been used by people to account for their goods. Later in history, these tokens were impressed on a clay tablet to represent quantities—the beginning of writing.

4.1 ADDITIVE, MULTIPLICATIVE, AND CIPHERED SYSTEMS OF NUMERATION

Just as the first attempts to write were made long after the development of speech, the first representation of numbers by symbols came long after people had learned to count. A tally system using physical objects, such as scratch marks in the soil or on a stone, notches on a stick, pebbles, or knots on a vine, was probably the earliest method of recording numbers.

In primitive societies such a tally system adequately served the limited need for recording livestock, agriculture, or whatever was counted. As civilization developed, however, more efficient and accurate methods of calculating and keeping records were needed. Because tally systems are impractical and inefficient, societies developed symbols to replace them. For example, the Egyptians used the symbol ∩ and the Babylonians used the symbol ◀ to represent the number we symbolize by 10.

A **number** is a quantity, and it answers the question "How many?" A **numeral** is a symbol such as ∩, ◀, and 10 used to represent the number. We think a number but write a numeral. The distinction between number and numeral will be made here only if it is helpful to the discussion.

In language, relatively few letters of the alphabet are used to construct a large number of words. Similarly, in arithmetic a small variety of numerals can be used to represent all numbers. In general, when writing a number, we use as few numerals as possible. One of the greatest accomplishments of humankind has been the development of systems of numeration, whereby all numbers are "created" from a few symbols. Without such systems, mathematics would not have developed to its present level.

> A **system of numeration** consists of a set of numerals and a scheme or rule for combining the numerals to represent numbers.

Four types of numeration systems used by different cultures are the topic of this chapter. They are additive (or repetitive), multiplicative, ciphered, and place-value systems. You do not need to memorize all the symbols, but you should understand the principles behind each system. By the end of this chapter, we hope that you better understand the system we use, the Hindu–Arabic system, and its relationship to other types of systems.

Additive Systems

An additive system is one in which the number represented by a particular set of numerals is simply the sum of the values of the numerals. The additive system of numeration is one of the oldest and most primitive types of numeration systems. One of the first additive systems, the Egyptian hieroglyphic system, dates back to about 3000 B.C. The Egyptians used symbols for the powers of 10: 10^0 or 1; 10^1 or 10; 10^2 or $10 \cdot 10$; 10^3 or $10 \cdot 10 \cdot 10$; and so on. Table 4.1 lists the Egyptian hieroglyphic numerals with the equivalent Hindu–Arabic numerals.

In order to write the number 600 in Egyptian hieroglyphics, we write the numeral for 100 six times: 999999.

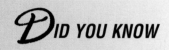
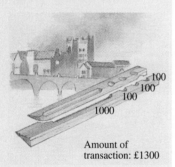

Table 4.1 Egyptian Hieroglyphics

Hindu–Arabic Numerals	Egyptian Numerals	Description
1		Staff (vertical stroke)
10	∩	Heel bone (arch)
100		Scroll (coiled rope)
1,000		Lotus flower
10,000		Pointing finger
100,000		Tadpole (or whale)
1,000,000		Astonished person

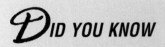

EXAMPLE 1

Write the following numeral as a Hindu–Arabic numeral.

Solution: $10{,}000 + 10{,}000 + 100 + 100 + 100 + 1 = 20{,}301$

EXAMPLE 2

Write 43,628 as an Egyptian numeral.

Solution:

$$43{,}628 = 40{,}000 + 3{,}000 + 600 + 20 + 8$$

In this system the order of the symbols is not important. For example, and both represent 100,212.

Users of additive systems easily accomplished addition and subtraction by combining or removing symbols. Multiplication and division were more difficult; they were performed by a process called *duplation and mediation* (see Section 4.5). The Egyptians had no symbol for zero, but they did have an understanding of fractions. The symbol ⌣ was used to take the reciprocal of a number; thus meant $\frac{1}{3}$ and was $\frac{1}{11}$. Writing large numbers in the Egyptian system would have taken longer than in other systems because so many symbols had to be listed. For example, 45 symbols are needed to represent the number 99,999.

The Roman numeration system, a second example of an additive system, was developed later than the Egyptian system. Roman numerals (Table 4.2) were used in most European countries until the eighteenth century. They are still commonly seen on buildings, on clocks, and in books. Roman numerals are selected letters of the Roman alphabet.

The Roman system has two advantages over the Egyptian system. The first is that it uses the subtraction principle as well as the addition principle. Starting

Table 4.2

Roman numerals	I	V	X	L	C	D	M
Hindu–Arabic numerals	1	5	10	50	100	500	1000

from the left, we add each numeral unless its value is smaller than the value of the numeral to its right. In that case, we subtract it from that numeral. Only the numbers 1, 10, 100, 1000, . . . can be subtracted, and only from the next two higher numbers. For example, C (100) can be subtracted only from D (500) or M (1000). The symbol DC represents 500 + 100, or 600, and CD represents 500 − 100, or 400. Similarly, MC represents 1000 + 100, or 1100, and CM represents 1000 − 100, or 900.

EXAMPLE 3

Write CLXII as a Hindu–Arabic numeral.

Solution: Since each numeral is larger than the one on its right, no subtraction is necessary.

$$CLXII = 100 + 50 + 10 + 1 + 1 = 162$$

▲

EXAMPLE 4

Write DCXLVI as a Hindu–Arabic numeral.

Solution: Checking from left to right, we see that X (10) has a smaller value than L (50). Therefore XL represents 50 − 10, or 40.

$$DCXLVI = 500 + 100 + (50 − 10) + 5 + 1 = 646$$

▲

EXAMPLE 5

Write 289 as a Roman numeral.

Solution:

$$289 = 200 + 80 + 9 = 100 + 100 + 50 + 10 + 10 + 10 + 9$$

(Nine is treated as 10 − 1.)

$$289 = CCLXXXIX$$

▲

In the Roman numeration system a symbol does not have to be repeated more than three consecutive times. For example, the number 646 would be written DCXLVI instead of DCXXXXVI.

The second advantage of the Roman numeration system over the Egyptian system is that it makes use of the multiplication principle for numbers over 1000. A bar above a symbol or group of symbols indicates that the symbol or symbols are to be multiplied by 1000. Thus $\overline{V} = 5 \times 1000 = 5000$, $\overline{X} = 10 \times 1000 = 10,000$, and $\overline{CD} = 400 \times 1000 = 400,000$. This greatly reduces the number of symbols needed to write large numbers. Still, it requires 19 symbols, including the bar, to write the number 33,888.

Multiplicative Systems

Multiplicative numeration systems are more similar than additive systems to our Hindu–Arabic system. The number 642 in a multiplicative system might be written (6) (100) (4) (10) (2) or

6
100
4
10
2

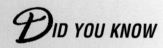

Table 4.3

Traditional Chinese numerals	一	二	三	四	五	六	七	八	九	十	百	千
Hindu–Arabic numerals	1	2	3	4	5	6	7	8	9	10	100	1000

Note that no addition signs are needed to represent the number. From this illustration, try to formulate a rule explaining how multiplicative systems work.

The principal example of a multiplicative system is the traditional Chinese system. The numerals used in this system are given in Table 4.3.

Chinese numerals are always written vertically. The number on top will be a number from 1 to 9 inclusive. This number is to be multiplied by the power of 10 below it. The number 20 is written

$$\left.\begin{matrix}二\\十\end{matrix}\right\}\ 2 \times 10 = 20$$

The number 400 is written

$$\left.\begin{matrix}四\\百\end{matrix}\right\}\ 4 \times 100 = 400$$

EXAMPLE 6

Write 428 as a Chinese numeral.

Solution:

$$428 = \begin{cases} 400 = \begin{cases} 4 & 四 \\ 100 & 百 \end{cases} \\ 20 = \begin{cases} 2 & 二 \\ 10 & 十 \end{cases} \\ 8 = \quad 8 \quad 八 \end{cases}$$

Note that in Example 6 the units digit, the 8, is not multiplied by a power of the base.

Have you noticed that there is no symbol for zero in the Chinese system? Why is a symbol for zero not needed?

The number system used today in China is different from the traditional system. The present-day system is a positional-value system rather than a multiplicative system and uses the symbol 0 for zero.

Ciphered Systems

Ciphered numeration systems require the memorization of many different symbols but have the advantage that numbers can be written in a compact form. The ciphered numeration system that we will discuss is the Ionic Greek (Table 4.4 on page 158). The Ionic Greek system was developed in about 3000 B.C. and used

Table 4.4 Ionic Greek Numerals

1	α	alpha	60	ξ	xi
2	β	beta	70	o	omicron
3	γ	gamma	80	π	pi
4	δ	delta	90	Q	koph*
5	ϵ	epsilon	100	ρ	rho
6	ζ	vau*	200	σ	sigma
7	ζ	zeta	300	τ	tau
8	η	eta	400	υ	upsilon
9	θ	theta	500	ϕ	phi
10	ι	iota	600	χ	chi
20	κ	kappa	700	ψ	psi
30	λ	lambda	800	ω	omega
40	μ	mu	900	ᴫ	sampi*
50	ν	nu			

Taken from the Phoenician alphabet.

letters of their alphabet for numerals. Other ciphered systems include the Hebrew, Coptic, Hindu, Brahmin, Syrian, Egyptian Hieratic, and early Arabic.

Since the Greek alphabet contains 24 letters but 27 symbols were needed, the Greeks borrowed the symbols ζ, Q, ᴫ from the Phoenician alphabet.

The number 24 = 20 + 4. When 24 is written as a Greek numeral, the plus sign is omitted:

$$24 = \kappa\delta.$$

The number 996 written as a Greek numeral is ᴫ Q ζ.

When a prime (′) is placed above a number, it multiplies that number by 1000. For example,

$$\beta' = 2 \times 1000 = 2000;$$

$$\sigma' = 200 \times 1000 = 200{,}000.$$

EXAMPLE 7

Write $\chi \, \nu \, \gamma$ as a Hindu–Arabic numeral.

Solution: $\chi = 600$, $\nu = 50$, and $\gamma = 3$. Adding these numbers gives 653.

▲

EXAMPLE 8

Write 8652 as an Ionic Greek numeral.

Solution:

$$
\begin{aligned}
8652 &= 8000 + 600 + 50 + 2 \\
&= (8 \times 1000) + 600 + 50 + 2 \\
&= \eta' \qquad\quad \chi \quad\;\; \nu \quad\; \beta \\
&= \eta'\chi\nu\beta
\end{aligned}
$$

▲

Section 4.1 Exercises

1. What is the difference between a number and a numeral?

2. List four numerals given in this section that may be used to represent the number ten.

3. What is a system of numeration?

4. Explain how numbers are represented in an additive numeration system.

5. Explain how numbers are represented in a multiplicative numeration system.

6. Explain how numbers are represented in a ciphered numeration system.

In Exercises 7–12, write the numeral as a Hindu–Arabic numeral.

7. 𓂥𓂥𓎆||||

8. 𓎆𓏲𓂥||

9. 𓆼𓂥𓂥𓂥𓂥𓂥𓂥𓎆𓎆𓎆||||

10. 𓍢𓍢𓍢𓍢𓆼𓂥𓂥𓎆

11. 𓂭𓂭𓍢𓍢𓍢𓆼𓆼𓂥𓂥𓎆||||

12. ✶✶✶✶𓂭𓂥𓂥𓎆𓎆𓎆𓎆|

In Exercises 13–18, write the numeral as an Egyptian numeral.

13. 321	14. 752	15. 1357
16. 1812	17. 173,845	18. 2,315,932

In Exercises 19–30, write the numeral as a Hindu–Arabic numeral.

19. XIV	20. XVI
21. CXLVII	22. CLXII
23. MCDXCII	24. MCMXVIII
25. MCMXLV	26. MMDXLVI
27. X̄MMMDCLXXV	28. L̄MCMXLIV
29. ĪX̄CDLXIV	30. V̄MCCCXXXIII

In Exercises 31–42, write the numeral as a Roman numeral.

31. 42	32. 94	33. 139
34. 269	35. 1997	36. 3564
37. 4793	38. 5842	39. 9999
40. 14,395	41. 20,644	42. 99,999

In Exercises 43–48, write the numeral as a Hindu–Arabic numeral.

43. 五十四

44. 六十二

45. 三千八十一

46. 三千二十九

47. 二千六百五十

48. 三千四百八十七

In Exercises 49–54, write the numeral as a traditional Chinese numeral.

49. 32	50. 178	51. 465
52. 1997	53. 3570	54. 6905

In Exercises 55–60, write the numeral as a Hindu–Arabic numeral.

55. $\tau\xi\delta$	56. $\chi o\eta$	57. $\mu'\beta'\phi\epsilon$
58. $\rho'\nu'\omega\iota\gamma$	59. $\theta'\zeta$	60. $\alpha'\pi Q\theta$

In Exercises 61–66, write the numeral as an Ionic Greek numeral.

61. 32	62. 178	63. 465
64. 1997	65. 35,704	66. 690,540

In Exercises 67–69, compare the advantages and disadvantages of a ciphered system of numeration with those of the named system.

67. An additive system

68. A multiplicative system

69. The Hindu–Arabic system

In Exercises 70–73, write the numeral as numerals in the indicated systems of numeration.

70. 𓆼𓎆𓎆 in Hindu–Arabic, Roman, Chinese, and Greek.

71. MCMXXXVI in Hindu–Arabic, Egyptian, Greek, and Chinese.

72. 五百二十七 in Hindu–Arabic, Egyptian, Roman, and Greek.

73. $\nu\kappa\beta$ in Hindu–Arabic, Egyptian, Roman, and Chinese.

Problem Solving/Group Activities

74. Write the Roman numeral for 999,999.

75. Write the Ionic Greek numeral for 999,999.

76. a) Create your own additive numeration system that contains at least one feature that isn't in either the Egyptian or Roman numeration system. Construct a table showing your numerals and the corresponding Hindu–Arabic numerals.
   b) Explain how to represent numbers in your system and explain the special feature not in the Egyptian or Roman numeration system.
   c) Write 458 in your system.
   d) Write 83,905 in your system.

77. a) Create your own multiplicative numeration system. Construct a table showing your numerals and the corresponding Hindu–Arabic numerals.
   b) Explain how to represent numbers in your system.
   c) Write 458 in your system.
   d) Write 83,905 in your system.

78. a) Create your own ciphered numeration system. Construct a table showing your numerals and the corresponding Hindu–Arabic numerals.
   b) Explain how to represent numbers in your system.
   c) Write 458 in your system.
   d) Write 83,905 in your system.

Research Activities

For Exercises 79–84, references include history of mathematics books and encyclopedias.

79. Write a paper on the Rhind papyrus. Indicate what information archaeologists learned from the papyrus.

80. Another example of an additive numeration system is the ancient Greek system of numeration. Write a paper describing the symbols used in the ancient Greek system of numeration and their equivalent symbols in the Hindu–Arabic numeration system. Explain how the basic operations of addition, subtraction, multiplication, and division were carried out and illustrate each with an example.

81. The Hebrew numeration system is a ciphered system similar to the Ionic Greek numeration system. Write a paper describing the symbols used in the Hebrew system of numeration and their equivalent symbols in the Hindu–Arabic numeration system. Explain how the basic operations of addition, subtraction, multiplication, and division were carried out and illustrate each with an example.

82. Do research on the Chinese numeration system used today and discuss the differences between it and the traditional Chinese numeration system.

83. Do research and write a report on how large numbers were written in the Roman numeration system. Write 2,340,000 in Roman numerals.

84. Do research and write a report on how large numbers were written in the Ionic Greek numeration system. Write 2,340,000 in Ionic Greek numerals.

4.2 PLACE-VALUE OR POSITIONAL-VALUE NUMERATION SYSTEMS

The eighteenth-century mathematician Pierre Simon, Marquis de Laplace, speaking of the positional principle, said: "The idea is so simple that this very simplicity is the reason for our not being sufficiently aware of how much attention it deserves."

Today the most common type of numeration system is the place-value system. The Hindu–Arabic numeration system, used in the United States and many other countries, is an example of a place-value system. In a place-value system the value of the symbol depends on its position in the representation of the number. For example, the 2 in 20 represents 2 tens, and the 2 in 200 represents 2 hundreds. A true positional-value system requires a base and a set of symbols, including a symbol for zero, and one for each counting number less than the base. Although any number can be written in any base, the most common positional system is the base 10 system (the decimal number system).

The Hindus in India are credited with the invention of zero and the other symbols used in our system. The Arabs, who traded regularly with the Hindus, also adopted the system—thus the name Hindu–Arabic. However, not until the middle of the fifteenth century did the Hindu–Arabic numerals take the form we know today.

The Hindu–Arabic numerals and the positional system of numeration revolutionized mathematics by making addition, subtraction, multiplication, and division much easier to learn and very practical to use. Merchants and traders no longer had to depend on the counting board or abacus. The first group of math-

Table 4.5

Babylonian numerals	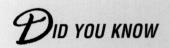	
Hindu–Arabic numerals	1	10

ematicians, who computed with the Hindu–Arabic system rather than with pebbles or beads on a wire, were known as the "algorists."

In the Hindu–Arabic system the symbols 0, 1, 2, 3, 4, 5, 6, 7, 8, and 9 are called digits. The base 10 system was developed from counting on fingers, and the word *digit* comes from the Latin word for fingers.

The positional values in the Hindu–Arabic system are

$$\ldots, (10)^5, (10)^4, (10)^3, (10)^2, 10, 1.$$

To evaluate a number in the Hindu–Arabic system, we multiply the first digit on the right by 1. We multiply the second digit from the right by the base 10. We multiply the third digit from the right by the base squared, 10^2 or 100. We multiply the fourth digit from the right by the base cubed, 10^3 or 1000, and so on. In general, we multiply the digit n places from the right by 10^{n-1}. Therefore we multiply the digit eight places from the right by 10^7. Using the place-value rule, we can write a number in expanded form. The number 1234 written in expanded form is

$$1234 = (1 \times 10^3) + (2 \times 10^2) + (3 \times 10^1) + (4 \times 1)$$

or $\qquad (1 \times 1000) + (2 \times 100) + (3 \times 10) + 4.$

The oldest known numeration system that resembled a place-value system was developed by the Babylonians in about 2500 B.C. Their system resembled a place-value system with a base of 60, a sexagesimal system. It was not a true place-value system because it lacked a symbol for zero. The lack of a symbol for zero led to a great deal of ambiguity and confusion. Table 4.5 gives the Babylonian numerals.

The positional values in the Babylonian system are

$$\ldots, (60)^3, (60)^2, 60, 1.$$

In a Babylonian numeral a gap is left between the characters to distinguish between the various place values. From right to left, the sum of the first group of numerals is multiplied by 1. The sum of the second group is multiplied by 60. The sum of the third group is multiplied by $(60)^2$, and so on.

EXAMPLE 1

Write ◄◄ ◄◄❙❙❙❙ as a Hindu–Arabic numeral.

Solution:

$$\underbrace{◄◄}_{60\text{'s}} \qquad \underbrace{◄◄❙❙❙❙}_{\text{units}}$$

$$\underbrace{10 + 10}_{60\text{'s}} \qquad \underbrace{10 + 10 + 1 + 1 + 1 + 1}_{\text{units}}$$

$$(20 \times 60) + (24 \times 1)$$
$$1200 + 24 = 1224$$

The Babylonians used the symbol ❙⃗ to indicate subtraction. The numeral ◄❙⃗❙❙ represents $10 - 2$, or 8. The numeral ◄◄◄◄❙❙❙❙❙❙❙⃗◄❙❙ represents $35 - 12$, or 23 in decimal notation.

$\mathcal{D}$ID YOU KNOW

COUNTING BOARDS

$\mathcal{O}$ne of the earliest counting devices, used in most ancient civilizations, was the counting board. On such a board each column represents a positional value. The number of times a value occurs is represented by markers (beads, stones, sticks) in the column. An empty column signifies "no value." The widespread use of counting boards meant that Europeans were already long accustomed to working with positional values when they were introduced to Hindu–Arabic numerals in the fifteenth century. People in China, Japan, the former Soviet Union, and Eastern Europe still commonly use a type of counting board known as the abacus to perform routine computations. ∎

EXAMPLE 2

Write ❶ ◀❶ ◀◀T̃❶❶ as a Hindu–Arabic numeral.

Solution: The place value of these three groups of numerals from left to right is

$$(60)^2, \quad 60, \quad 1$$

$$\text{or} \quad 3600, \quad 60, \quad 1.$$

The numeral in the group on the right has a value of $20 - 2$, or 18. The numeral in the center has a value of $10 + 1$, or 11. The numeral on the left represents 1. Multiplying each group by its positional value gives

$$(1 \times 60^2) + (11 \times 60) + (18 \times 1)$$
$$= (1 \times 3600) + (11 \times 60) + (18 \times 1)$$
$$= 3600 + 660 + 18$$
$$= 4278.$$
▲

To explain the procedure used to convert from a Hindu–Arabic numeral to a Babylonian numeral, we will consider a length of time. How can we change 9820 seconds into hours, minutes, and seconds? Since there are 3600 seconds in an hour (60 seconds to a minute and 60 minutes to an hour), we can find the number of hours by dividing 9820 by 60^2, or 3600.

$$
\begin{array}{r}
2 \leftarrow \text{Hours} \\
3600\overline{)9820} \\
7200 \\
\hline
2620 \leftarrow \text{Remaining seconds}
\end{array}
$$

Now we can determine the number of minutes by dividing the remaining seconds by 60, the number of seconds in a minute.

$$
\begin{array}{r}
43 \leftarrow \text{Minutes} \\
60\overline{)2620} \\
2400 \\
\hline
220 \\
180 \\
\hline
40 \leftarrow \text{Remaining seconds}
\end{array}
$$

Since the remaining number of seconds, 40, is less than the number of seconds in a minute, our task is complete.

$$9820 \text{ sec} = 2 \text{ hr}, 43 \text{ min}, \text{ and } 40 \text{ sec}$$

The same procedure is used to convert a decimal (base 10) number to a Babylonian number or any number in a different base.

EXAMPLE 3

Write 1602 as a Babylonian numeral.

Solution: The Babylonian numeration system has positional values of

$$\ldots, 60^3, 60^2, 60, 1,$$

which can be expressed as

$$\ldots, 216000, 3600, 60, 1.$$

The largest positional value less than or equal to 1602 is 60. To determine how many groups of 60 are in 1602, divide 1602 by 60.

$$
\begin{array}{r}
26 \\
60 \overline{)1602} \\
120 \\
\hline
402 \\
360 \\
\hline
42
\end{array}
$$

Thus $1602 \div 60 = 26$ with remainder 42. There are 26 groups of 60 and 42 units remaining. Because the remainder, 42, is less than the base, 60, no further division is necessary. The remainder represents the number of units when the number is written in expanded form. Therefore $1602 = (26 \times 60) + (42 \times 1)$. When written as a Babylonian numeral, 1602 is

$$\texttt{<<𝙸𝙸𝙸𝙸𝙸𝙸\ \ <<<<𝙸𝙸}.$$ ▲

EXAMPLE 4

Write 6270 as a Babylonian numeral.

Solution: Divide 6270 by the largest positional value less than or equal to 6270. That value is 3600.

$$6270 \div 3600 = 1 \text{ with remainder } 2670$$

There is one group of 3600 in 6270. Next divide the remainder 2670 by 60 to determine the number of groups of 60 in 2670.

$$2670 \div 60 = 44 \text{ with remainder } 30$$

There are 44 groups of 60 and 30 units remaining.

$$6270 = (1 \times 60^2) + (44 \times 60) + (30 \times 1)$$

Thus 6270 written as a Babylonian numeral is

$$\texttt{𝙸\ \ <<<<𝙸𝙸𝙸𝙸\ \ <<<}.$$ ▲

Another place-value system is the Mayan numeration system. The Mayans, who lived on the Yucatan Peninsula, developed a sophisticated numeration system based on their religious and agricultural calendar. The numbers in this system are written vertically rather than horizontally, with the units position on the bottom. In the Mayan system the number in the bottom row is to be multiplied by 1. The number in the second row from the bottom is to be multiplied by 20. The number in the third row is to be multiplied by 18×20, or 360. You probably expected the third row to be multiplied by 20^2 rather than 18×20. It is believed that the Mayans used 18×20 so that their numeration system would conform to their calendar of 360 days. The positional values above 18×20 are 18×20^2, 18×20^3, and so on.

Positional Values in the Mayan System

... $18 \times (20)^3$,	$18 \times (20)^2$,	18×20,	20,	1
or ... 144,000,	7200,	360,	20,	1

The digits 0, 1, 2, 3, ..., 19 of the Mayan system are formed by a simple grouping of dots and lines, as shown in Table 4.6 on page 164.

Table 4.6   Mayan Numerals

0	1	2	3	4	5	6	7	8	9
⊖	•	••	•••	••••	—	•—	••—	•••—	••••—

10	11	12	13	14	15	16	17	18	19
=	•=	••=	•••=	••••=	≡	•≡	••≡	•••≡	••••≡

EXAMPLE 5

Write $\begin{smallmatrix}\bullet\bullet\\\bullet\bullet\\\bullet\bullet\bullet\end{smallmatrix}$ as a Hindu–Arabic numeral.

Solution: In the Mayan numeration system the first three positional values are

$$18 \times 20$$
$$20$$
$$1$$

$$\overset{\bullet\bullet}{=} \;=\; 7 \times (18 \times 20) = 2520$$
$$\bullet\bullet \;=\; 2 \times 20 \;\;\;\;\;\;\;= \;\;\;\;40$$
$$\overset{\bullet\bullet\bullet}{=} \;=\; 13 \times 1 \;\;\;\;\;\;= \;\;\;\underline{13}$$
$$2573.$$

EXAMPLE 6

Write $\begin{smallmatrix}\bullet\bullet\bullet\\\bullet\\\bullet\bullet\bullet\bullet\end{smallmatrix}$ as a Hindu–Arabic numeral.

Solution:

$$\overset{\bullet\bullet\bullet}{=} \;=\; 8 \times (18 \times 20) = 2880$$
$$\overset{\bullet}{=} \;=\; 11 \times 20 \;\;\;\;\;\;= \;\;\;220$$
$$\bullet\bullet\bullet\bullet \;=\; 4 \times 1 \;\;\;\;\;\;\;\;= \;\;\;\;\underline{4}$$
$$3104.$$

EXAMPLE 7

Write 1023 as a Mayan numeral.

Solution: To convert from a Hindu–Arabic to a Mayan numeral, we use a procedure similar to the one used to convert to a Babylonian numeral. The Mayan positional values are . . . , 7200, 360, 20, 1. The greatest positional value less than or equal to 1023 is 360. Divide 1023 by 360.

$$1023 \div 360 = 2 \text{ with remainder } 303$$

There are two groups of 360 in 1023. Next divide the remainder, 303, by 20.

$$303 \div 20 = 15 \text{ with remainder } 3$$

There are 15 groups of 20 with three units remaining.

$$1023 = (2 \times 360) + (15 \times 20) + (3 \times 1)$$

1023 written as a Mayan numeral is

$$\left.\begin{array}{r} 2 \times 360 \\ 15 \times 20 \\ 3 \times 1 \end{array}\right\} = \begin{array}{c} \bullet\bullet \\ \equiv \\ \bullet\bullet\bullet \end{array}.$$

Section 4.2 Exercises

1. What is the most common type of numeration system used in the world today?

2. a) What is the base in the Hindu–Arabic numeration system?
  b) What are the digits in the Hindu–Arabic numeration system?

3. Explain how to write a number in expanded form in a positional-value numeration system.

4. Why was the Babylonian system not a true place-value system?

5. a) What problem did not having a symbol for zero cause in the Babylonian system?
  b) Write the numbers 133 and 72013 as Babylonian numerals.

6. a) List the first five positional values, starting with the units position, for the Mayan numeration system.
  b) Describe two ways that the Mayan place-value system differs from the Hindu–Arabic place-value system.

In Exercises 7–18, write the Hindu–Arabic numeral in expanded form.

7. 34   **8.** 75   **9.** 237

10. 896   **11.** 897   **12.** 3769

13. 4648   **14.** 23,468   **15.** 56,832

16. 125,678 **17.** 346,861 **18.** 3,765,934

In Exercises 19–24, write the numeral as a Hindu–Arabic numeral.

19. ❮❮❮❮❯❯

20. ❮❮❮❮❮❯❯❯❯

21. ❮❯❯ ❯❯❯❯

22. ❯❮ ❮❮❯❯❯

23. ❯ ❮❮❮❯❯ ❮❯❯❯

24. ❮ ❮❮❯❯❯❯ ❯❯

In Exercises 25–30, write the numeral as a Babylonian numeral.

25. 70   **26.** 95   **27.** 121

28. 512   **29.** 3878   **30.** 3030

In Exercises 31–36, write the numeral as a Hindu–Arabic numeral.

31. $\begin{array}{c}\bullet\bullet\bullet\\ \equiv\end{array}$

32. $\begin{array}{c}\equiv\\ \underline{\quad}\end{array}$

33. $\begin{array}{c}\equiv\equiv\\ \bigcirc\\ \bullet\end{array}$

34. $\begin{array}{c}\bullet\bullet\\ \bullet\bullet\\ \bullet\bullet\end{array}$

35. $\begin{array}{c}\bullet\\ \equiv\\ \bullet\bullet\\ \bigcirc\end{array}$

36. $\begin{array}{c}\bullet\bullet\bullet\bullet\\ \equiv\\ \underline{\quad}\end{array}$

In Exercises 37–42, write the numeral as a Mayan numeral.

37. 15   **38.** 227   **39.** 300

40. 406   **41.** 3060   **42.** 1978

43. Compare the advantages and disadvantages of a place-value system with those of (a) additive numeration systems, (b) multiplicative numeration systems, and (c) ciphered numeration systems.

44. Create your own place-value system. Write 1996 in your system.

In Exercises 45 and 46, write the numeral in the indicated systems of numeration.

45. ❮❮❮❯❯❯ in Hindu–Arabic and Mayan.

46. ___ in Hindu–Arabic and Babylonian.
$$\begin{array}{c}\bullet\bullet\\ \bullet\bullet\bullet\bullet\end{array}$$

Problem Solving/Group Activities

47. Is there a largest number in the Babylonian numeration system? Explain. Write the Babylonian numeral for 999,999.

48. Is there a largest number in the Mayan numeration system? Explain. Write the Mayan numeral for 999,999.

In Exercises 49–52, first convert each numeral to a Hindu–Arabic numeral and then perform the indicated operation. Finally, convert the answer back to a numeral in the original numeration system.

49. ꀪ ꀫꀫ ꀪ + ꀫꀪꀪ   **50.** ꀪ ꀫꀫꀪꀪ − ꀫꀫꀫꀪ

51.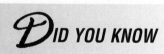

52.
$$\frac{\overset{\bullet\bullet}{\bullet}\quad\overset{\bullet}{\bullet}}{\underset{\equiv}{\bullet}}\;-\;\frac{\bullet\bullet}{\underset{\bullet\bullet\bullet}{\bullet\bullet}}$$

53. A three-digit base 10 whole number has the following properties: The hundred's digit is greater than 5; the ten's digit is an odd number; and the sum of the digits is 9. What numbers satisfy these criteria?

Research Activities

For Exercises 54–56, references include history of mathematics books and encyclopedias. In addition, for Exercises 55 and 56 references include Israh, Georges, From One to Zero: A Universal History of Numbers, Viking Press. Sizer, Walter S., "Other Numeration Systems Alive and Thriving" Mathematics Teacher (January 1990) 83:17–19.

54. Investigate and write a report on the development of the Hindu–Arabic system of numeration. Start with the earliest records of this system in India.

55. The Arabic numeration system currently in use is a base 10 positional-value system, which uses different symbols than the Hindu–Arabic numeration system. Write the symbols used in the Arabic system of numeration and their equivalent symbols in the Hindu–Arabic numeration system. Write 54, 607, and 1997 in Arabic numerals.

56. The Nepalese numeration system currently in use is a base 10 positional-value system. Write the symbols used in the Nepalese system of numeration and their equivalent symbols in the Hindu–Arabic numeration system. Write 54, 606, and 1997 in Nepalese numerals.

4.3   OTHER BASES

The positional values in the Hindu–Arabic numeration system are

$$\ldots, (10)^4, (10)^3, (10)^2, 10, 1.$$

The positional values in the Babylonian numeration system are

$$\ldots, (60)^4, (60)^3, (60)^2, 60, 1.$$

The numbers 10 and 60 are called the bases in the Hindu–Arabic and Babylonian systems, respectively.

Any counting number greater than 1 may be used as a base for a positional-value numeration system. If a positional-value system has a base b, then its positional values will be

$$\ldots, b^4, b^3, b^2, b, 1.$$

The positional values in a base 8 system are

$$\ldots, 8^4, 8^3, 8^2, 8, 1,$$

and the positional values in a base 2 system are

$$\ldots, 2^4, 2^3, 2^2, 2, 1.$$

As we indicated earlier, the Mayan numeration system is based on the number 20. However, it is not a true base 20 positional-value system. Why?

The fact that human beings have ten fingers is believed to be responsible for the almost universal acceptance of base 10 numeration systems. Even so, there are still some positional-value numeration systems that use bases other than 10. Some societies are still using a base 2 numeration system. They include some groups of people in Australia, New Guinea, Africa, and South America. Bases 3 and 4 are also used in some areas of South America. The only base 5 system in pure form at present seems to be the one used in Saraveca, a South American Arawakan language. Elsewhere, base 5 systems are combined with base 10 or base 20 systems. The pure base 6 system occurs only sparsely in Northwest Africa. Base 6 also occurs in other systems in combination with base 12, the *duodecimal system*.

We continue to see remains of other base systems in many countries. For example, there are 12 inches in a foot, 12 months in a year. Base 12 is also evident in the dozen, the 24-hour day, and the gross (12×12). English uses the word *score* to mean 20, as in ''Four score and seven years ago.'' Other traces are found in pre-English Celtic, Gaelic, Danish, and Welsh. Remains of base 60 are found in measurements of time (60 seconds to a minute, 60 minutes to an hour) and angles (60 seconds to one minute, 60 minutes to one degree).

The base 2, or **binary system**, has become very important because it is the internal language of the computer. For example, when the grocery store's cash register computer records the price of your groceries by using a scanning device, the bar codes it scans on the packages are in binary form. Computers use a two-digit ''alphabet'' that consists of the numerals 0 and 1. Every character on a standard keyboard can be represented by a combination of those two numerals. A single numeral such as 0 or 1 is called a **bit**. Other bases that computers make use of are base 8 and base 16. A group of four bits is called a **nibble**; a group of eight bits is called a **byte**. In the American Standard Code for Information Interchange (ASCII) code, used in most computers, the byte 01000001 represents the character A, 01100001 represents the character a, 00110000 represents the character 0, and 00110001 represents the character 1.

A place-value system with base b must have b distinct symbols, one for zero and one for each number less than the base. A base 6 system must have symbols for the numbers 0, 1, 2, 3, 4, and 5. All numbers in base 6 are constructed from these 6 symbols. A base 8 system must have symbols for 0, 1, 2, 3, 4, 5, 6, and 7. All numbers in base 8 are constructed from these 8 symbols, and so on.

A number in a base other than base 10 will be indicated by a subscript to the right of the number. Thus 123_5 represents a number in base 5. The number 123_6 represents a number in base 6. The value of 123_5 is not the same as the value of 123_{10} and the value of 123_6 is not the same as the value of 123_{10}. A base 10 number may be written without a subscript. For example, 123 means 123_{10} and 456 means 456_{10}. For clarity in certain problems we will use the subscript 10 to indicate a number in base 10.

Remember the symbols that represent the base itself, in any base b, are 10_b. For example, in base 5 the symbols 10_5 represent the number 5. The symbols 10_5 mean one group of 5 and no units. In base 6 the symbols 10_6 represent the number 6. The symbols 10_6 represent one group of 6 and no units, and so on.

To change a number in a base other than 10 to a base 10 number, we follow the same procedure we used in Section 4.2 to change the Babylonian and Mayan numbers to base 10 numbers. Multiply each digit in the number by its respective positional value. Then find the sum of the products.

EXAMPLE 1

Convert 234_6 to base 10.

Solution: In base 6 the positional values are . . . , 6^3, 6^2, 6, 1. In expanded form,

$$234_6 = (2 \times 6^2) + (3 \times 6) + (4 \times 1)$$
$$= (2 \times 36) + (3 \times 6) + (4 \times 1)$$
$$= \quad 72 \quad + \quad 18 \quad + \quad 4$$
$$= 94.$$

▲

EXAMPLE 2

Convert 3615_8 to base 10.

Solution:

$$3615_8 = (3 \times 8^3) \ + (6 \times 8^2) + (1 \times 8) + (5 \times 1)$$
$$= (3 \times 512) + (6 \times 64) + (1 \times 8) + (5 \times 1)$$
$$= \quad 1536 \quad + \quad 384 \quad + \quad 8 \quad + \quad 5$$
$$= 1933$$

▲

A base 12 system must have 12 distinct symbols. In this text we will use the symbols 0, 1, 2, 3, 4, 5, 6, 7, 8, 9, T, and E, where T represents ten and E represents eleven. Why will the numerals 10_{12} and 11_{12} have different meanings than 10 and 11?

EXAMPLE 3

Convert $12T6_{12}$ to base 10.

Solution:

$$12T6_{12} = (1 \times 12^3) \ + (2 \times 12^2) + (T \times 12) \ + (6 \times 1)$$
$$= (1 \times 1728) + (2 \times 144) + (10 \times 12) + (6 \times 1)$$
$$= \quad 1728 \quad + \quad 288 \quad + \quad 120 \quad + \quad 6$$
$$= 2142$$

▲

EXAMPLE 4

Convert 101101_2 to base 10.

Solution:

$$101101_2 = (1 \times 2^5) + (0 \times 2^4) + (1 \times 2^3) + (1 \times 2^2) + (0 \times 2) + (1 \times 1)$$
$$= \quad 32 \quad + \quad 0 \quad + \quad 8 \quad + \quad 4 \quad + \quad 0 \quad + \quad 1$$
$$= 45$$

▲

To change a number from a base 10 system to a different base, we will use the procedure explained in Section 4.2. Divide the number by the highest power of the base less than or equal to the given number. Record this quotient. Then divide the remainder by the next smaller power of the base and record this quotient. Repeat this procedure until the remainder is a number less than the base. The answer is the set of quotients listed from left to right, with the remainder on the far right. This procedure is illustrated in Examples 5–7.

EXAMPLE 5

Convert 406 to base 8.

Solution: The positional values in the base 8 system are . . . , 8^3, 8^2, 8, 1, or . . . , 512, 64, 8, 1. The highest power of 8 that is less than or equal to 406 is 8^2, or 64. Divide 406 by 64.

First digit in answer

$$406 \div 64 = 6 \text{ with remainder } 22$$

Therefore there are six groups of 8^2 in 406. Next divide the remainder, 22, by 8.

Second digit in answer

$$22 \div 8 = 2 \text{ with remainder } 6$$

Third digit in answer

There are two groups of 8 in 22 and 6 units remaining. Since the remainder, 6, is less than the base, 8, no further division is required.

$$406 = (6 \times 64) + (2 \times 8) + (6 \times 1)$$
$$= (6 \times 8^2) + (2 \times 8) + (6 \times 1)$$
$$= 626_8$$

EXAMPLE 6

Convert 273 to base 3.

Solution: The place values in the base 3 system are . . . , $3^6, 3^5, 3^4, 3^3, 3^2,$ 3, 1, or . . . , 729, 243, 81, 27, 9, 3, 1. The highest power of the base that is less than 273 is 3^5, or 243. Successive divisions by the powers of the base give the following result.

$$273 \div 243 = 1 \text{ with remainder of } 30$$
$$30 \div 81 = 0 \text{ with remainder } 30$$
$$30 \div 27 = 1 \text{ with remainder } 3$$
$$3 \div 9 = 0 \text{ with remainder } 3$$
$$3 \div 3 = 1 \text{ with remainder } 0$$

The remainder, 0, is less than the base, 3, so no further division is necessary. To obtain the answer, list the quotients from top to bottom followed by the remainder in the last division.

The number 273 can be represented as one group of 243, no groups of 81, one group of 27, no groups of 9, one group of 3, and no units.

$$273 = (1 \times 243) + (0 \times 81) + (1 \times 27) + (0 \times 9) + (1 \times 3) + (0 \times 1)$$
$$= (1 \times 3^5) + (0 \times 3^4) + (1 \times 3^3) + (0 \times 3^2) + (1 \times 3) + (0 \times 1)$$
$$= 101010_3$$

EXAMPLE 7

Convert 558 to base 12.

Solution: The place values in base 12 are . . . , $12^3, 12^2, 12, 1,$ or . . . , 1728, 144, 12, 1.

$$558 \div 144 = 3 \text{ with remainder } 126$$
$$126 \div 12 = T \text{ with remainder } 6$$

(Remember that T is used to represent ten in base 12.)

$$558 = (3 \times 12^2) + (T \times 12) + (6 \times 1) = 3T6_{12}$$

Section 4.3 Exercises

1. In your own words explain how to change a number in a base other than base 10 to base 10.

2. In your own words explain how to change a number in base 10 to a base other than base 10.

In Exercises 3–20, convert the numeral to a numeral in base 10.

3. 6_9

4. 50_7

5. 17_8

6. 101_2

7. 1011_2

8. 1101_2

9. 76_{12}

10. 12021_3

11. 476_9

12. 654_7

13. 20432_5

14. 101111_2

15. 3001_4

16. $123E_{12}$

17. 123_8

18. 1043_8

19. 14705_8

20. 67342_9

In Exercises 21–36, convert the base 10 numeral to a numeral in the base indicated.

21. 8 to base 2

22. 16 to base 2

23. 25 to base 2

24. 146 to base 5

25. 435 to base 8

26. 908 to base 4

27. 1695 to base 12

28. 100 to base 3

29. 230 to base 6

30. 64 to base 2

31. 2867 to base 12

32. 1234 to base 5

33. 1011 to base 2

34. 1589 to base 7

35. 2408 to base 8

36. 13469 to base 8

In Exercises 37–40, assume that a base 16 positional-value system uses the numerals 0, 1, 2, 3, 4, 5, 6, 7, 8, 9, A, B, C, D, E, and F, where A through F represent 10 through 15, respectively. Convert the numeral to a numeral in base 10.

37. 734_{16}

38. 285_{16}

39. $6D3B7_{16}$

40. $24FEA_{16}$

In Exercises 41–44, convert the numeral to a numeral in base 16.

41. 307

42. 349

43. 5478

44. 34,721

In Exercises 45–50, convert 1996 to a numeral in the base indicated.

45. 2

46. 3

47. 5

48. 7

49. 8

50. 16

In Exercises 51–54, if any numerals are written incorrectly, explain why.

51. 4063_5

52. 1203_3

53. 674_8

54. 1206_{12}

55. There is an alternative method for changing a number in base 10 to a different base. This method will be used to convert 328 to base 5. Dividing 328 by 5 gives a quotient of 65 and a remainder of 3. Write the quotient below the dividend and the remainder on the right, as shown.

$$5 \underline{|\ 328} \quad \text{remainder}$$
$$65 \qquad 3$$

Continue this process of division by 5.

$$
\begin{array}{r|l}
5 & 328 \\
5 & 65 \\
5 & 13 \\
5 & 2 \\
& 0 \\
\end{array}
\quad
\begin{array}{l}
\text{remainder} \\
3 \\
0 \\
3 \\
2 \\
\end{array}
$$

Answer

(Since the dividend, 2, is smaller than the divisor, 5, the quotient is 0 and the remainder is 2.)

Note that the division continues until the quotient is zero. The answer is read from the bottom number to the top number in the remainder column. Thus $328 = 2303_5$.

a) Explain why this procedure results in the proper answer.

b) Convert 683 to base 5 by this method.

c) Convert 763 to base 8 by this method.

Problem Solving/Group Activities

56. a) Use the numerals 0, 1, and 2 to write the first 20 numbers in the base 3 numeration system.

b) What is the next number after 222_3?

57. a) Make up your own base 20 positional-value numeration system. Indicate the 20 numerals you will use to represent the 20 numbers less than the base.

b) Write the numbers 468 and 5293 in your base 20 numeration system.

58. a) Use the numerals 0 and 1 to write the first 16 numbers in the base 2 numeration system.

b) Use the numerals 0, 1, 2, 3, 4, 5, 6, and 7 to write the first 16 numbers in the base 8 numeration system.

59. The ASCII code used by most computers utilizes the last seven positions of an eight-bit byte to represent all the characters on a standard keyboard. How many different orderings of 0s and 1s (or how many different characters) can be made by using the last seven positions of an eight-bit byte?

60. Find b if $111_b = 43$.

Research Activities

61. Investigate and write a report on how digital computers use the binary number system.

For Exercises 62 and 63, references include Flagg, Graham, Numbers Through the Ages, Macmillan, 1987; and Israh, Georges, From One to Zero: A Universal History of Numbers, Viking, 1985.

62. We mention at the beginning of this section that some societies still use a base 2 numeration system. These societies are in Australia, New Guinea, Africa, and South America. Write a report on these societies, covering the symbols they use and how they combine these symbols to represent numbers in their numeration system.

63. Societies of Tierra del Fuego and others in South America use a base 3 numeration system. Write a report on these societies, covering the symbols they use and how they combine these symbols to represent numbers in their numeration system.

4.4   COMPUTATION IN OTHER BASES

Addition

When computers perform calculations, they do so in base 2, the binary system. In this section we explain how to perform calculations in base 2 and other bases.

In a base 2 system the only digits are 0 and 1, and the place values are

$$\ldots, 2^4, 2^3, 2^2, 2, 1$$
$$\text{or} \quad \ldots, 16, 8, 4, 2, 1.$$

Suppose that we want to add $1_2 + 1_2$. The subscript 2 indicates that we are adding in base 2. Remember the answer to $1_2 + 1_2$ must be written by using only the digits 0 and 1. The sum of $1_2 + 1_2$ is 10_2 since $1_2 + 1_2$ is one group of two and no units. Recall that 10_2 means $1(2) + 0(1)$.

If we wanted to find the sum of $10_2 + 1_2$ we would add the digits in the right-hand, or units, column. Since $0_2 + 1_2 = 1_2$, the sum of $10_2 + 1_2 = 11_2$.

We are going to work additional examples and exercises in base 2, so rather than performing individual calculations in every problem, we can construct and use an addition table, Table 4.7, for base 2 (just as we used an addition table in base 10 when we first learned to add in base 10).

Table 4.7

+	0	1
0	0	1
1	1	10

EXAMPLE 1

Add 1101_2
       $\underline{111_2}$

Solution: Begin by adding the numbers in the right-hand, or units, column. From our previous discussion, and as can be seen in Table 4.7, $1_2 + 1_2 = 10_2$. Place the 0 under the units column and carry the 1 to the twos column, the second column from the right.

$$\begin{array}{c} 1\ 1\ ^{1}0\ 1 \\ \underline{1\ \ 1\ \ 1} \\ 0_2 \end{array}$$

Now add the three digits in the twos column, $1_2 + 0_2 + 1_2$. Treat this as $(1_2 + 0_2) + 1_2$. Therefore add $1_2 + 0_2$ to get 1_2, then add $1_2 + 1_2$ to get 10_2. Place the 0 under the twos column and carry the 1 to the 2^2 column (the third column from the right).

$$\begin{array}{c} 1\ ^{1}1\ ^{1}0\ 1 \\ \underline{1\ \ 1\ \ 1} \\ 0\ \ 0_2 \end{array}$$

Now add the three 1s in the 2^2 column to get $(1_2 + 1_2) + 1_2 = 10_2 + 1_2 = 11_2$. Place the 1 under the 2^2 column and carry the 1 to the 2^3 column (the fourth column from the right).

$$
\begin{array}{cccc}
{}^1 1 & {}^1 1 & {}^1 0 & 1 \\
 & 1 & 1 & 1 \\
\hline
 & 1 & 0 & 0_2 \\
\end{array}
$$

Now add the two 1s in the 2^3 column, $1_2 + 1_2 = 10_2$. Place the 10 as follows.

$$
\begin{array}{ccccc}
 & {}^1 1 & {}^1 1 & {}^1 0 & 1 \\
 & & 1 & 1 & 1 \\
\hline
1 & 0 & 1 & 0 & 0_2 \\
\end{array}
$$

Therefore the sum is 10100_2. ▲

Let's now look at addition in a base 5 system. In base 5 the only digits are 0, 1, 2, 3, and 4, and the positional values are

$$\ldots, \quad 5^4, \quad 5^3, \; 5^2, \; 5, \; 1$$

or $\quad \ldots, \; 625, \; 125, \; 25, \; 5, \; 1.$

What is the sum of $4_5 + 3_5$? We can consider this to mean $(1 + 1 + 1 + 1) + (1 + 1 + 1)$. We can regroup the seven 1s into one group of five and two units as $(1 + 1 + 1 + 1 + 1) + (1 + 1)$. Thus the sum of $4_5 + 3_5 = 12_5$ (circled in Table 4.8). Recall that 12_5 means $1(5) + 2(1)$. We can use this same procedure in obtaining the values in the base 5 addition table.

Table 4.8

+	0	1	2	3	4
0	0	1	2	3	4
1	1	2	3	4	10
2	2	3	4	10	11
3	3	4	10	11	12
4	4	10	11	⑫	13

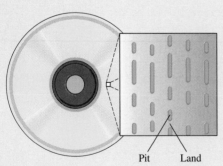

EXAMPLE 2

Add 32_5
    33_5

Solution: First determine that $2_5 + 3_5$ is 10_5 from Table 4.8. Record the 0 and carry the 1 to the fives column.

$$\begin{array}{cc} {}^1 3 & 2_5 \\ 3 & 3_5 \\ \hline & 0_5 \end{array}$$

Add the numbers in the second column, $(1_5 + 3_5) + 3_5 = 4_5 + 3_5 = 12_5$. Record the 12.

$$\begin{array}{cc} {}^1 3 & 2_5 \\ 3 & 3_5 \\ \hline 1 \quad 2 & 0_5 \end{array}$$

The sum is 120_5.   ▲

EXAMPLE 3

Add 1234_5
    2042_5

Solution:

$$\begin{array}{cccc} 1 & {}^1 2 & {}^1 3 & 4_5 \\ 2 & 0 & 4 & 2_5 \\ \hline 3 & 3 & 3 & 1_5 \end{array}$$ ▲

You can develop an addition table for any base and use it to add in that base. However, as you get more comfortable with addition in other bases, you may prefer to add numbers in other bases by using mental arithmetic. To do so, convert the sum of the numbers being added from the given base to base 10, and then convert the base 10 number back into the given base. You must clearly understand how to convert from base 10 to the given base, as discussed in Section 4.3. For example, to add $7_9 + 8_9$ add $7 + 8$ in base 10 to get 15_{10} and then mentally convert 15_{10} to 16_9 using the procedure given earlier. Remember, 16_9 when converted to base 10 becomes $1(9) + 6(1)$, or 15. Addition using this procedure is illustrated in Examples 4 and 5.

EXAMPLE 4

Add 1022_3
    2121_3

Solution: To solve this problem make the necessary conversions by using mental arithmetic. $2 + 1 = 3_{10} = 10_3$. Record the 0 and carry the 1.

$$\begin{array}{cccc} 1 & 0 & {}^1 2 & 2_3 \\ 2 & 1 & 2 & 1_3 \\ \hline & & & 0_3 \end{array}$$

$1 + 2 + 2 = 5_{10} = 12_3$. Record the 2 and carry the 1.

$$\begin{array}{cccc} 1 & {}^1 0 & {}^1 2 & 2_3 \\ 2 & 1 & 2 & 1_3 \\ \hline & & 2 & 0_3 \end{array}$$

$1 + 0 + 1 = 2_{10} = 2_3$. Record the 2.

$$
\begin{array}{cccc}
1 & {}^{1}0 & {}^{1}2 & 2_3 \\
2 & 1 & 2 & 1_3 \\
\hline
 & 2 & 2 & 0_3
\end{array}
$$

$1 + 2 = 3_{10} = 10_3$. Record the 10.

$$
\begin{array}{r}
1022_3 \\
2121_3 \\
\hline
10220_3
\end{array}
$$

▲

EXAMPLE 5

Add 444_5
$\quad\;244_5$
$\quad\;143_5$
$\quad\;214_5$

Solution: Adding the digits in the right-hand column gives $4 + 4 + 3 + 4 = 15_{10} = 30_5$. Record the 0 and carry the 3. Adding the 3 with the digits in the next column yields $3 + 4 + 4 + 4 + 1 = 16_{10} = 31_5$. Record the 1 and carry the 3. Adding the 3 with the digits in the left-hand column gives $3 + 4 + 2 + 1 + 2 = 12_{10} = 22_5$. Record both digits. The sum of these four numbers is 2210_5.

$$
\begin{array}{cccc}
{}^{3}4 & {}^{3}4 & 4_5 \\
2 & 4 & 4_5 \\
1 & 4 & 3_5 \\
2 & 1 & 4_5 \\
\hline
2 \quad 2 & 1 & 0_5
\end{array}
$$

▲

Subtraction

Subtraction can also be performed in other bases. Always remember that, when you ''borrow,'' you borrow the amount of the base. In Examples 6 and 7 we will perform the subtraction in base 10 when convenient and convert the results to the given base.

EXAMPLE 6

Subtract $\quad 3032_5$
$\qquad\; - 1004_5$

Solution: Since 4 is greater than 2, we must borrow one group of 5 from the preceding column. This action gives a sum of $5 + 2$, or 7 in base 10. Now we subtract 4 from 7; the difference is 3. We complete the problem in the usual manner. The 3 in the second column becomes a 2, $2 - 0 = 2$, $0 - 0 = 0$, and $3 - 1 = 2$.

$$
\begin{array}{r}
3032_5 \\
- 1004_5 \\
\hline
2023_5
\end{array}
$$

▲

EXAMPLE 7

Subtract 468_{12}
              -295_{12}

Solution: $8 - 5 = 3$. Next we must subtract 9 from 6. Since 9 is greater than 6, borrowing is necessary. We must borrow one group of 12 from the preceding column. We then have a sum of $12 + 6 = 18$ in base 10. Now we subtract 9 from 18, and the difference is 9. The 4 in the left column becomes 3, and $3 - 2 = 1$.

$$\begin{array}{r} 468_{12} \\ -295_{12} \\ \hline 193_{12} \end{array}$$

▲

Multiplication

Multiplication can also be performed in other bases. Doing so is helped by forming a multiplication table for the base desired. Suppose that we want to determine the product of $4_5 \times 3_5$. In base 10, 4×3 means that there are four groups of three units. Similarly in a base 5 system, $4_5 \times 3_5$ means that there are four groups of three units, or

$$(1 + 1 + 1) + (1 + 1 + 1) + (1 + 1 + 1) + (1 + 1 + 1).$$

Regrouping the 12 units into groups of five gives

$$(1 + 1 + 1 + 1 + 1) + (1 + 1 + 1 + 1 + 1) + (1 + 1),$$

or two groups of five, and two units. Thus $4_5 \times 3_5 = 22_5$.

We can construct other values in the base 5 multiplication table in the same way. However, you may find it easier to multiply the values in the base 10 system, and then change the product to base 5 by using the procedure discussed in Section 4.3. Multiplying 4×3 in base 10 gives 12, and converting 12 from base 10 to base 5 gives 22_5.

The product of $4_5 \times 3_5$ is circled in Table 4.9, the base 5 multiplication table. The other values in the table may be found by either method discussed.

Table 4.9

×	0	1	2	3	4
0	0	0	0	0	0
1	0	1	2	3	4
2	0	2	4	11	13
3	0	3	11	14	22
4	0	4	13	㉒	31

EXAMPLE 8

Multiply 13_5, using the base 5 multiplication table.
            $\times\ 3_5$

Solution: Multiply as you would in base 10, but use the base 5 multiplication table to find the products. When the product consists of two digits, record the right digit and carry the left digit. Multiplying $3_5 \times 3_5 = 14_5$. Record the 4 and carry the 1.

$$\begin{array}{r} {}^{1}13_5 \\ \times\ \ 3_5 \\ \hline 4 \end{array}$$

$(3_5 \times 1_5) + 1_5 = 4_5$. Record the 4.

$$\begin{array}{r} {}^{1}13_5 \\ \times\ \ 3_5 \\ \hline 44_5 \end{array}$$

The product is 44_5.

▲

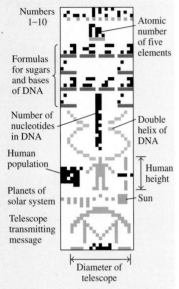

Numbers 1–10

Atomic number of five elements

Formulas for sugars and bases of DNA

Number of nucleotides in DNA

Double helix of DNA

Human population

Human height

Planets of solar system

Sun

Telescope transmitting message

Diameter of telescope

Constructing a multiplication table is often tedious, especially when the base is large. To multiply in a given base without the use of a table, multiply in base 10 and convert the products to the appropriate base number before recording them. This procedure is illustrated in Example 9.

EXAMPLE 9

Multiply $\begin{array}{r} 43_7 \\ \times\, 25_7 \end{array}$

Solution: $5 \times 3 = 15_{10} = 2(7) + 1(1) = 21_7$. Record the 1 and carry the 2.

$$\begin{array}{r} {}^{2}43_7 \\ \times\ 25_7 \\ \hline 1 \end{array}$$

$(5 \times 4) + 2 = 20 + 2 = 22_{10} = 3(7) + 1(1) = 31_7$. Record the 31.

$$\begin{array}{r} {}^{2}43_7 \\ \times\ 25_7 \\ \hline 311 \end{array}$$

$2 \times 3 = 6$. Record the 6.

$$\begin{array}{r} {}^{2}43_7 \\ \times\ 25_7 \\ \hline 311 \\ 6 \end{array}$$

$2 \times 4 = 8_{10} = 1(7) + 1(1) = 11_7$. Record the 11. Now add in base 7 to determine the answer. Remember, in base 7 there are no digits greater than 6.

$$\begin{array}{r} {}^{2}43_7 \\ \times\ 25_7 \\ \hline 311 \\ 116 \\ \hline 1501_7 \end{array}$$

Division

Division is performed in much the same manner as long division in base 10. A detailed example of a division in base 5 is illustrated in Example 10. The same procedure is used for division in any other base.

EXAMPLE 10

Divide $2_5\,\overline{\smash{\big)}\,143_5}$.

Solution: Using Table 4.9, the multiplication table for base 5, we list the multiples of the divisor, 2.

$$2_5 \times 1_5 = 2_5$$
$$2_5 \times 2_5 = 4_5$$
$$2_5 \times 3_5 = 11_5$$
$$2_5 \times 4_5 = 13_5$$

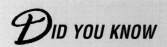
Since $2_5 \times 4_5 = 13_5$, which is less than 14_5, 2_5 goes into 14_5 four times.

$$
\begin{array}{r}
4 \\
2_5 \overline{)\ 143_5} \\
\underline{13} \\
1
\end{array}
$$

Now bring down the 3 as when dividing in base 10.

$$
\begin{array}{r}
4 \\
2_5 \overline{)\ 143_5} \\
\underline{13} \\
13
\end{array}
$$

We see that $2_5 \times 4_5 = 13_5$. Use this information to complete the problem.

$$
\begin{array}{r}
44_5 \\
2_5 \overline{)\ 143} \\
\underline{13} \\
13 \\
\underline{13} \\
0
\end{array}
$$

Therefore $143_5 \div 2_5 = 44_5$ with remainder zero. ▲

A division problem can be checked by multiplication. If the division was performed correctly, (quotient $\times$ divisor) + remainder = dividend. We can check Example 10 as follows.

$$(44_5 \times 2_5) + 0_5 = 143_5$$

$$
\begin{array}{r}
44_5 \\
\times\ 2_5 \\
\hline
143_5 \quad \text{Check}
\end{array}
$$

EXAMPLE 11

Divide $4_6 \overline{)\ 2430_6}$.

Solution: The multiples of 4 in base 6 are

$$4_6 \times 1_6 = 4_6$$
$$4_6 \times 2_6 = 12_6$$
$$4_6 \times 3_6 = 20_6$$
$$4_6 \times 4_6 = 24_6$$
$$4_6 \times 5_6 = 32_6.$$

$$
\begin{array}{r}
404_6 \\
4_6 \overline{)\ 2430_6} \\
\underline{24} \\
03 \\
\underline{00} \\
30 \\
\underline{24} \\
2
\end{array}
$$

Thus the quotient is 404_6, with remainder 2_6.

Be careful when subtracting! When you borrow, remember that you borrow 10_6, which is the same as 6 in base 10.

Check: Does $(404_6 \times 4_6) + 2_6 = 2430_6$?

$$
\begin{array}{r}
404_6 \\
\times \quad 4_6 \\
\hline
2424_6 + 2_6 = 2430_6 \quad \text{True}
\end{array}
$$

▲

Section 4.4 Exercises

1. a) What are the first five positional values, from right to left, in base b?
  b) What are the first five positional values, from right to left, in base 6?

2. In the addition

$$
\begin{array}{r}
463_7 \\
+ \ 24_7 \\
\hline
\end{array}
$$

what are the positional values of the first column on the right, the second column from the right, and the third column from the right? Explain how you determined your answer.

3. In your own words explain how to add two numbers in a given base. In your explanation answer the question, "What happens when the sum of the numbers in a column is greater than the base?"

4. In your own words explain how to subtract two numbers in a given base. Include in your explanation what you do when, in one column, you must subtract a larger number from a smaller number.

In Exercises 5–16, add in the indicated base.

5. 22_4
  103_4

6. 33_7
  65_7

7. 3412_5
  2334_5

8. 101_2
  11_2

9. 569_{12}
  238_{12}

10. 222_3
  22_3

11. 1012_3
  1011_3

12. 470_{12}
  347_{12}

13. 14631_7
  6040_7

14. 2341_5
  1341_5

15. 1011_2
  110_2

16.* $43A_{16}$
  496_{16}

In Exercises 17–28, subtract in the indicated base.

17. 312_4
  $- 103_4$

18. 536_7
  $- 124_7$

19. 2432_5
  $- 1243_5$

20. 1011_2
  $- 101_2$

21. 463_{12}
  $- 13T_{12}$

22. 1221_3
  $- 202_3$

23. 1001_2
  $- 110_2$

24. 1453_{12}
  $- 245_{12}$

25. 4223_7
  $- 304_7$

26. 4232_5
  $- 2341_5$

27. 2100_3
  $- 1012_3$

28.* $4E7_{16}$
  $- 189_{16}$

In Exercises 29–37, multiply in the indicated base.

29. 42_5
  $\times \ 2_5$

30. 123_5
  $\times \ 4_5$

31. 423_7
  $\times \ 5_7$

32. 101_2
  $\times \ 11_2$

33. 302_4
  $\times \ 23_4$

34. 124_{12}
  $\times \ 6_{12}$

35. 234_9
  $\times \ 25_9$

36. $6T3_{12}$
  $\times \ 24_{12}$

37. 111_2
  $\times 101_2$

In Exercises 38–44, divide in the indicated base.

38. $1_2 \overline{)110_2}$

39. $5_6 \overline{)221_6}$

40. $4_5 \overline{)143_5}$

41. $7_8 \overline{)466_8}$

42. $2_4 \overline{)312_4}$

43. $6_{12} \overline{)431_{12}}$

44. $3_7 \overline{)2101_7}$

Problem Solving/Group Activity

Divide in the indicated base.

45. $14_5 \overline{)242_5}$

46. $20_4 \overline{)223_4}$

47. Consider the multiplication

$$
\begin{array}{r}
462_8 \\
\times \ 35_8 \\
\hline
\end{array}
$$

  a) Multiply the numbers in base 8.
  b) Convert 462_8 and 35_8 to base 10.
  c) Multiply the base 10 numbers determined in part (b).
  d) Convert the answer obtained in base 8 in part (a) to base 10.
  e) Are the answers obtained in parts (c) and (d) the same? Why or why not?

48. If $1304_b = 204$, determine b.

For Exercises 49 and 50, study the pattern in the boxes. The number in the bottom row of each box represents the value

---

*See Exercises 37–40 in Section 4.3.

of each dot in the box directly above it. For example, the following figure represents $(3 \times 7^2) + (2 \times 7) + (4 \times 7^0)$, or the number 324_7. This number in base 10 is 165.

•••	••	••
7^2	7^1	7^0

49. Determine the base 5 number represented by the dots in the top row of the boxes. Then convert the base 5 number to a number in base 10.

••	•••		••
5^3	5^2	5^1	5^0

50. Fill in the correct amount of dots in the columns above the base values if the number represented by the dots is to equal 327 in base 10.

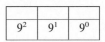

9^2	9^1	9^0

Research Activities

51. Investigate and write a report on the use of the duodecimal (base 12) system as a system of numeration. You might contact the Duodecimal Society, Nassau Community College, Garden City, NY 11530 for information.

52. One method used by computers to perform subtraction is the ''end around carry method.'' Do research in books on computers, in encyclopedias, or in other sources, and write a report explaining, with specific examples, how a computer performs subtraction by using the end around carry method.

4.5 EARLY COMPUTATIONAL METHODS

Early civilizations used various methods for multiplying and dividing. Multiplication was performed by *duplation and mediation*, by the *galley method*, and by *Napier rods*. Following is an explanation of each method.

Duplation and Mediation

EXAMPLE 1

Multiply 17×30 using duplation and mediation.

Solution: Write 17 and 30 with a dash between to separate them. Divide the number on the left, 17, in half, drop the remainder, and place the quotient, 8, under the 17. Double the number on the right, 30, obtaining 60, and place it under the 30. You will then have the following paired lines.

$$17 — 30$$
$$8 — 60$$

Continue this process, taking one-half the number in the left-hand column, disregarding the remainder, and doubling the number in the right-hand column, as shown below. When a 1 appears in the left-hand column, stop.

$$17 — 30$$
$$8 — 60$$
$$4 — 120$$
$$2 — 240$$
$$1 — 480$$

Cross out all the even numbers in the left-hand column and the corresponding numbers in the right-hand column.

$$17 — 30$$
$$\cancel{8 — 60}$$
$$\cancel{4 — 120}$$
$$\cancel{2 — 240}$$
$$1 — 480$$

Now add the remaining numbers in the right-hand column, obtaining $30 + 480 = 510$, which is the product you want. If you check, you will find that $17 \times 30 = 510$. ▲

The Galley Method

The galley method (sometimes referred to as the Gelosia method) was developed after duplation and mediation. To multiply 312×75 using the galley method, you construct a rectangle consisting of three columns (one for each digit of 312) and two rows (one for each digit of 75).

Place the digits 3, 1, 2 above the boxes and the digits 7, 5 on the right of the boxes, as shown in Fig. 4.1. Then place a diagonal in each box.

Complete each box by multiplying the number on top of the box by the number to the right of the box (Fig. 4.2). Place the units below the diagonal and the tens above.

Add the numbers along the diagonals, as shown in Fig. 4.3, starting with the bottom right diagonal. If the sum in a diagonal is 10 or greater, record the units digit below the rectangle and carry the tens digit to the next diagonal to the left.

The answer is read down the left-hand column and along the bottom, as shown by the arrow in Fig. 4.3. The answer is 23,400.

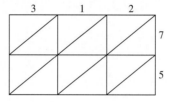

Figure 4.1

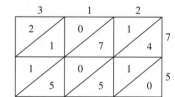

Figure 4.2

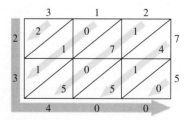

Figure 4.3

Napier Rods

The third method was developed from the galley method by John Napier in the seventeenth century. His method of multiplication, known as Napier rods, proved to be one of the forerunners of the modern-day computer. Napier developed a system of separate rods numbered from 0 through 9 and an additional strip for an index, numbered vertically 1 through 9 (Fig. 4.4). Each rod is divided into 10 blocks, and each block contains a multiple of the top number. Units are placed to the right, and tens to the left. Example 2 explains how Napier rods are used to multiply numbers.

Figure 4.4

INDEX	0	1	2	3	4	5	6	7	8	9
1	0/0	0/1	0/2	0/3	0/4	0/5	0/6	0/7	0/8	0/9
2	0/0	0/2	0/4	0/6	0/8	1/0	1/2	1/4	1/6	1/8
3	0/0	0/3	0/6	0/9	1/2	1/5	1/8	2/1	2/4	2/7
4	0/0	0/4	0/8	1/2	1/6	2/0	2/4	2/8	3/2	3/6
5	0/0	0/5	1/0	1/5	2/0	2/5	3/0	3/5	4/0	4/5
6	0/0	0/6	1/2	1/8	2/4	3/0	3/6	4/2	4/8	5/4
7	0/0	0/7	1/4	2/1	2/8	3/5	4/2	4/9	5/6	6/3
8	0/0	0/8	1/6	2/4	3/2	4/0	4/8	5/6	6/4	7/2
9	0/0	0/9	1/8	2/7	3/6	4/5	5/4	6/3	7/2	8/1

INDEX	3	6	5
1	0/3	0/6	0/5
2	0/6	1/2	1/0
3	0/9	1/8	1/5
4	1/2	2/4	2/0
5	1/5	3/0	2/5
6	1/8	3/6	3/0
7	2/1	4/2	3/5
8	2/4	4/8	4/0
9	2/7	5/4	4/5

Figure 4.5

EXAMPLE 2

Multiply 8 × 365, using Napier rods.

Solution: To multiply 8 × 365 line up the rods 3, 6, and 5 opposite the index, as shown in Fig. 4.5. Add along the diagonals as in the galley method.

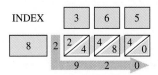

Thus 8 × 365 = 2920.

Example 3 illustrates the procedure to follow to multiply numbers containing more than one digit, using Napier rods.

EXAMPLE 3

Multiply 48 × 365, using Napier rods.

Solution: 48 × 365 = (40 + 8) × 365

Write (40 + 8) × 365 = (40 × 365) + (8 × 365). To find 40 × 365, determine 4 × 365 and multiply the product by 10. To evaluate 4 × 365, set up Napier rods for 3, 6, and 5 with index 4, and then evaluate along the diagonals, as indicated.

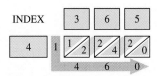

Therefore 4 × 365 = 1460. Then 40 × 365 = 10 × 1460 = 14,600.

$$48 \times 365 = (40 \times 365) + (8 \times 365)$$
$$= 14,600 + 2920$$
$$= 17,520$$

8 × 365 = 2920 from Example 2.

Section 4.5 Exercises

1. What are the three early computational methods discussed in this Section?

2. **a)** Explain in your own words how multiplication by duplation and mediation is performed.
   b) Using the procedure given in part (a), multiply 267×193.

In Exercises 3–10, multiply, using duplation and mediation.

3. 15×27 4. 35×23

5. 9×162 6. 130×87

7. 43×221 8. 96×53

9. 85×85 10. 49×124

In Exercises 11–18, multiply, using the galley method.

11. 7×365 12. 8×365

13. 3×371 14. 7×12

15. 75×12 16. 17×256

17. 244×321 18. 634×832

In Exercises 19–26, multiply, using Napier rods.

19. 3×37 20. 6×34

21. 8×75 22. 7×125

23. 5×125 24. 75×125

25. 9×6742 26. 7×3456

Problem Solving/Group Activities

In Exercises 27–29, use the method of duplation and mediation to perform the multiplication. Write the answer in the numeration system in which the exercise is given.

27. $(\cap |||) \bullet (\cap\cap ||)$

28. $(\text{XXVI}) \cdot (\text{LXVII})$

29. $(\mathbf{<} \mathbf{I I I I}) \bullet (\mathbf{<}\mathbf{<} \mathbf{I I I})$

In Exercises 30–34, use the galley method to perform the multiplication. Write the answer in the base in which the exercise is given.

30. $12_3 \times 121_3$ 31. $24_5 \times 234_5$

32. $110_2 \times 111_2$ 33. $26_7 \times 436_7$

34. Develop a set of Napier rods that can be used to multiply numbers in base 5. Illustrate how your rods can be used to multiply $3_5 \times 212_5$.

Research Activities

For Exercises 35–37, references include history of mathematics books and encyclopedias. A specific reference for Exercise 36 is: National Council of Teachers of Mathematics, Historical Topics for the Mathematics Classroom, *Thirty-first Yearbook. (Pages 93, 130 & 131).*

35. In addition to Napier rods, John Napier is credited with making other important contributions to mathematics. Write a report on John Napier and his contributions to mathematics.

36. Do research and determine which ancient societies used the duplation and mediation method.

37. Do research on the multiplication method of duplation and mediation and write a paper explaining why the technique works.

CHAPTER 4 SUMMARY

Key Terms

4.1

number
numeral
system of numeration

4.2

base
digit
expanded form
place-value system

4.3

binary system
bit
byte
nibble

Important Facts

Types of Numeration Systems

Additive (Egyptian hieroglyphics, Roman)
Multiplicative (traditional Chinese)
Ciphered (Ionic Greek)
Place-value (Babylonian, Mayan, Hindu–Arabic)

Early Computational Methods

Duplation and mediation
The galley method
Napier rods

Chapter 4 Review Exercises

4.1, 4.2

In Exercises 1–6, assume an additive numeration sytsem in which a = 1, b = 10, c = 100, and d = 1000. Find the value of the numeral.

1. dca   **2.** abcda   **3.** bccaad

4. cbdadaaa   **5.** dddccbaaaa   **6.** ccbaddac

In Exercises 7–12, assume the same additive numeration system as in Exercises 1–6. Write the numeral in terms of a, b, c, and d.

7. 43   **8.** 167   **9.** 389

10. 1996   **11.** 7542   **12.** 6314

In Exercises 13–18, assume a multiplicative numeration system in which a = 1, b = 2, c = 3, d = 4, e = 5, f = 6, g = 7, h = 8, i = 9, x = 10, y = 100, and z = 1000. Find the value of the numeral.

13. ixc   **14.** bxg   **15.** fydxh

16. dzfxh   **17.** ezfydxh   **18.** dzhyi

In Exercises 19–24, assume the same multiplicative numeration system as in Exercises 13–18. Write the Hindu–Arabic numeral in that system.

19. 57   **20.** 276   **21.** 694

22. 2020   **23.** 6004   **24.** 2001

In Exercises 25–36, use the following ciphered numeration system.

Decimal	1	2	3	4	5	6	7	8	9
Units	a	b	c	d	e	f	g	h	i
Tens	j	k	l	m	n	o	p	q	r
Hundreds	s	t	u	v	w	x	y	z	A
Thousands	B	C	D	E	F	G	H	I	J
Ten thousands	K	L	M	N	O	P	Q	R	S

Convert the numeral to a Hindu–Arabic numeral.

25. qd   **26.** wc   **27.** tmf

28. NGzqc   **29.** Qqb   **30.** Pwki

Write the numeral in the ciphered numeration system.

31. 42   **32.** 152   **33.** 493

34. 1997   **35.** 23,685   **36.** 75,496

In Exercises 37–42, convert 1462 to a numeral in the indicated numeration system.

37. Egyptian   **38.** Roman   **39.** Chinese

40. Ionic Greek   **41.** Babylonian   **42.** Mayan

In Exercises 43–48, convert the numeral to a Hindu–Arabic numeral.

43. ⌒⌾〰ᶘᶘ∩∩ⅠⅠⅠⅠⅠ

44.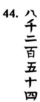

45. φ π ε

46. MCMXCI

47. ⟨⟨Ⅰ ⟨⟨⟨ⅠⅠⅠⅠ

48. ⋮
        ̄
        •
        ̄̄

4.3

In Exercises 49–54, convert the numeral to a Hindu–Arabic numeral.

49. 46_7 **50.** 101_2 **51.** 432_5

52. 2746_8 **53.** $T0E_{12}$ **54.** 20220_3

In Exercises 55–60, convert 463 to a numeral in the base indicated.

55. base 7   **56.** base 3   **57.** base 2

58. base 4   **59.** base 12   **60.** base 8

4.4

In Exercises 61–66, add in the base indicated.

61. 43_8
        $\underline{56_8}$

62. 10110_2
        $\underline{11001_2}$

63. TE_{12}
        $\underline{87_{12}}$

64. 234_7
        $\underline{456_7}$

65. 4023_5
        $\underline{4023_5}$

66. 1407_8
        $\underline{7014_8}$

In Exercises 67–72, subtract in the base indicated.

67. 4032_7
        $-\ \ 321_7$

68. 1001_2
        $-\ \ 101_2$

69. $4TE_{12}$
        $-\ \ E7_{12}$

70. 4321_5
        $-\ \ 442_5$

71. 2473_8
        -1643_8

72. 2021_3
        $-\ 212_3$

In Exercises 73–78, multiply in the base indicated.

73. 23_5
   $\times\ 3_5$

74. 23_4
   $\times 21_4$

75. 126_{12}
   $\times\ 47_{12}$

76. 202_3
   $\times\ 22_3$

77. 1011_2
   $\times\ 101_2$

78. 476_8
   $\times\ 23_8$

In Exercises 79–84, divide in the base indicated.

79. $1_2\overline{)1011_2}$

80. $2_4\overline{)330_4}$

81. $3_5\overline{)140_5}$

82. $4_6\overline{)3020_6}$

83. $4_6\overline{)2034_6}$

84. $6_8\overline{)5072_8}$

4.5

85. Multiply 142×24, using the duplation and mediation method.

86. Multiply 142×24, using the galley method.

87. Multiply 142×24, using Napier rods.

CHAPTER TEST

1. Explain the difference between a numeral and a number.

In Exercises 2–6, convert each of the numerals to a Hindu–Arabic numeral.

2. MMDCXLVII

3. ⟨⟨Ⲧ ⟨ⲦⲦⲦⲦⲦ

4. 八
   千
   九
   十

5. •••
   ⸬
   ••

6. ↄ𐡸𐡸999∩∩∩ΙΙ

7. θ′πϘθ

In Exercises 8–12, convert the number written in base 10 to a numeral in the numeration system indicated.

8. 345 to Egyptian

9. 2476 to Ionic Greek

10. 1324 to Mayan

11. 1596 to Babylonian

12. 2378 to Roman

In Exercises 13–16, describe briefly each of the systems of numeration. Explain how each type of numeration system is used to represent numbers.

13. Additive system

14. Multiplicative system

15. Ciphered system

16. Place-value system

In Exercises 17–20, convert the numeral to a numeral in base 10.

17. 24_5

18. 211_3

19. 101101_2

20. 368_9

In Exercises 21–24, convert the base 10 numeral to a numeral in the base indicated.

21. 36 to base 2

22. 79 to base 3

23. 2356 to base 12

24. 23456 to base 7

In Exercises 25–28, perform the indicated operations.

25. 124_5
    $+434_5$

26. 463_7
    -124_7

27. 45_6
    $\times 23_6$

28. $3_5 \overline{)1210_5}$

29. Multiply 14×28, using duplation and mediation.

30. Multiply 43×196, using the galley method.

GROUP PROJECTS

Adding and Subtracting with Egyptian and Roman Numerals

1. In Section 4.1 you learned that the Egyptian and the Roman numeration systems are additive systems. This information and the following references will be useful in answering parts (a) and (b). (References include history of mathematics books, encyclopedias and National Council of Teachers of Mathematics, *Historical Topics for the Mathematics Classroom*, Thirty-first Yearbook.)
   a) Develop and explain methods for adding numbers in the Egyptian and Roman numeration systems. Illustrate your methods with the following problems.
   i) 99∩∩∩∩∩∩∩ΙΙΙΙ + 9∩∩ΙΙΙΙΙΙΙ
   ii) CCCLIX + MCXLII
   b) Develop and explain methods for subtracting numbers in the Egyptian and Roman numeration systems. Illustrate your methods with the following problems.
   i) 99∩∩∩Ι − 9∩∩ΙΙΙΙΙ
   ii) CMLXXIV − DCXXI

Zip Codes

2. The U.S. Postal Service introduced a bar-coding system for zip codes in 1976. The system is known as Postnet (postal numeric encoding technique) and has been refined over the years. The zip code consists of five digits followed by a hyphen and four additional digits. This is referred to as the "zip + 4." The first digit of the zip code represents a region of the country, as shown in the accompanying map. The bar code is a series of long and short bars. The short bars represent 0 and the long bars represent 1. A bar code may have 52 or 62 bars. Each code starts and ends with a long bar that is not used in determining the zip + 4. If the code contains 52 bars, the code represents the zip + 4 and a check number. If the code contains 62 bars, it contains the zip + 4, the last two digits of the street number, and a check number. If the code contains 52 bars, the sum of the zip + 4 and the check number must equal a number that is divisible by 10. If the code contains 62 bars, the sum of the zip + 4, the last two numbers of the street address, and the check number must equal a number that is divisible by 10.

National zip-code areas

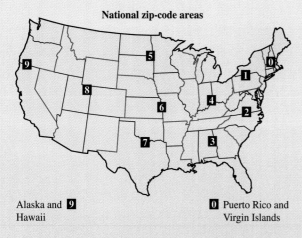

Alaska and **9**   **0** Puerto Rico and
Hawaii   Virgin Islands

Consider the following bar code with 52 bars and the zip + 4 from Pittsburgh that it corresponds to. The sum of the digits in the zip + 4 $(1 + 5 + 2 + 5 + 0 + 7 + 4 + 0 + 6)$ is 30. Because 30 is divisible by 10, the check digit must be 0. To interpret the bar code, group the bars in groups of five, remembering not to include the first and last bars. The last group of five lines on the right (which you count), must represent the digit 0.

15250-7406 (Pittsburgh, PA)

In the following box we give the patterns of zeros and ones that represent the digits 0 through 9. The number in parentheses to the right of the pattern is the number represented by the zeros and ones.

11000 (0) 00011 (1) 00101 (2) 00110 (3) 01001 (4)
01010 (5) 01100 (6) 10001 (7) 10010 (8) 10100 (9)

Use the values in the box to interpret the Postnet bar code for the Pittsburgh zip + 4. Remember that a long bar represents 1 and a short bar represents 0. Check your results against the given zip + 4.

Now let's work some problems.

a) For the bar code

determine the zip + 4 and the check number. Then check by adding the zip + 4 and the check number. Is the sum divisible by 10?

b) For each of the following Postnet codes determine the zip + 4, the last two numbers of the street address (if applicable), and the check number.

i)

ii)

c) Construct the Postnet code of long and short bars for each of the numbers. The numbers represent the zip + 4 and the last two digits of the street address.
   i) 32226-8600-34
   ii) 20794-1063-50

d) Construct the 52-bar Postnet code for your college's zip + 4. Don't forget to include the check digit.

ASCII Code and Computer

3. When we enter any letter, number, or command into a computer the computer uses a character code to interpret that instruction as a series of 0's and 1's. A character code is a standard scheme for representing characters by combinations of bits, 0's and 1's. The most widely used character code is the American Standard Code for Information Interchange, usually abbreviated to ASCII (pronounced as'key). Table 4.10 contains the ASCII Character code.

The ASCII code uses seven bits to represent each character. The seven bits can be arranged in 2^7 or 128 different ways. So ASCII can represent 128 different characters. For example the 7 bits used to represent the letter A is 1000001. Each of these numbers is a binary number (base 2). The representation of A5 in the ASCII code is 10000010110101. The first 7 digits represent the A, the second seven digits represent the 5.

a) Represent the following in ASCII code.
   i) cat
   ii) 96
   iii) Pear

The answer for part (iii) is a large number of 0's and 1's and very difficult to represent accurately. To avoid this problem of working with binary numbers,

Table 4.10   ASCII Character Code

		Leftmost Three Bits							
		000	001	010	011	100	101	110	111
	0000	NUL	DLE	Space	0	@	P	`	p
	0001	SOH	DC1	!	1	A	Q	a	q
	0010	STX	DC2	"	2	B	R	b	r
	0011	ETX	DC3	#	3	C	S	c	s
	0100	EOT	DC4	$	4	D	T	d	t
	0101	ENQ	NAK	%	5	E	U	e	u
	0110	ACK	SYN	&	6	F	V	f	v
Rightmost	0111	BEL	ETB	'	7	G	W	g	w
Four	1000	BS	CAN	(	8	H	X	h	x
Bits	1001	HT	EM	)	9	I	Y	i	y
	1010	LF	SUB	*	:	J	Z	j	z
	1011	VT	ESC	+	;	K	[	k	{
	1100	FF	FS	,	<	L	\	l	|
	1101	CR	GS	-	=	M	]	m	}
	1110	SO	RS	.	>	N	^	n	~
	1111	SI	US	/	?	O	_	o	DEL

The two- and three-letter abbreviations represent control characters. Examples are CR (carriage return), LF (line feed), and BEL (bell).

programmers sometimes use one of two alternative notations, octal (base 8) and hexadecimal (base 16).

In working with the octal notation, we divide a binary number into groups of three. For example, dividing the binary number 100111000011101001010110 into groups of three gives

100   111   000   011   101   001   010   110.

Now each of the 3-bit numbers can be represented with an octal digit 0–7. For example 100 is represented by 4 and 111 by 7. The following table shows the correspondence between 3-bit binary groups and octal digits.

3-Bit Binary Group	Octal Digit
000	0
001	1
010	2
011	3
100	4
101	5
110	6
111	7

Using the values in the table, we can convert the binary number 100111000011101001010110 to an octal number. The octal number is 47035126_8. The subscript 8 indicates that the number is in base 8.

Instead of marking off binary digits in groups of three, we can mark them off in groups of four. By doing so we are using hexadecimal notation, or base 16. The following table shows the correspondence between 4-bit binary groups and hexadecimal digits.

4-Bit Binary Group	Hexadecimal Digit
0000	0
0001	1
0010	2
0011	3
0100	4
0101	5
0110	6
0111	7
1000	8
1001	9
1010	A
1011	B
1100	C
1101	D
1110	E
1111	F

Dividing the bits in the binary number 100111000011101001010110 into groups of four, we have 1001 1100 0011 1010 0101 0110. Translating these groups into a hexadecimal number, we get $9C3A56_{16}$.

b) Determine the decimal value corresponding to the binary number.

   i) 110_2　　**ii)** 110011_2　　**iii)** 111011_2

c) Convert the octal value to binary.

   i) 105_8　　**ii)** 347_8　　**iii)** 276_8

d) Convert the hexadecimal value to binary.

   i) AD_{16}　　**ii)** $2E4_{16}$　　**iii)** $6C7F_{16}$

e) Convert 3 + 4 to the ASCII character code.

Chapter 5

Number Theory & the Real Number System

*N*umbers find their way into every aspect of life. For instance, haven't you wanted someone to tell you that "you're the only one for me"? And haven't you been in a situation where you realize that "two's company, three's a crowd"? Numbers have many uses, from describing the natural world, to communicating vast quantities of information. Most important, they can be used to model and analyze almost any phenomenon of daily life.

The ancient Greeks were among the first to think of numbers in a systematic way and

Numbers are a fact of life.

to study their properties. The Greek philosopher Pythagoras believed that the universe revealed itself in the language of numbers and looked at the shapes of nature to find the numerical relations in them. The full force of the power of numbers to describe natural phenomena would not be realized until Sir Isaac Newton (1642–1727) and Gottfried Leibniz (1646–1716) developed a branch of mathematics known to us as calculus. Calculus can describe natural phenomena, such as the motion of planets, the swing of a pendulum, and the recoil of a spring.

Interest in the mathematical description of our world continues today. Within the last 20 years, the field of chaos theory has grown in importance because it provides a model for

The artist Georges Seurat (1859–1891) believed that certain "natural proportions" exist in nature and that those proportions are the ones most likely to please the human eye. You can see this in his painting *La Parade*, which is composed of rectangles within rectangles. The sides of each rectangle are proportioned with a ratio of about 1 to 1.618.

natural phenomena, such as storms and earthquakes, which previously could not be described mathematically. Still, the words of Galileo ring true: Speaking of the universe, he said, "It is written in the language of mathematics . . . without which it is humanly impossible to understand a single word of it."

The language of mathematics now permeates almost every part of our lives. In this age of information, we are constantly being assaulted by facts, figures, and numbers big and small. In recent years, the news media have increasingly focused on a phenomenon called *innumeracy*, the inability to use and understand numerical information. Literacy of numbers, or numeracy, will be important to your career in the twenty-first century. ■

The strength of the traditional Japanese samurai sword is legendary. The master sword maker prepares the blade by heating a bar of iron until it is white hot, then folding it over and pounding it smooth. He does this 15 times. Each time the metal is folded, the layers of steel are doubled (a geometric sequence). For a sword of 15 folds, the blade contains 2^{15}, or 32,768 layers of steel!

5.1 NUMBER THEORY

This chapter introduces **number theory**, the study of numbers and their properties. The numbers we use to count are called the **counting numbers** or **natural numbers**. Since we begin counting with the number 1, the set of natural numbers begins with 1. The set of natural numbers is frequently denoted by N:

$$N = \{1, 2, 3, 4, 5, \ldots\}.$$

Any natural number can be expressed as a product of two or more natural numbers. For example $8 = 2 \times 4$, $16 = 4 \times 4$, and $19 = 1 \times 19$. The natural numbers that are multiplied together are called **factors** of the product. For example,

$$2 \times 4 = 8.$$
$$\text{Factors}$$

A natural number may have many factors. For example, what pairs of numbers have a product of 12?

$$12 \cdot 1 = 12$$
$$6 \cdot 2 = 12$$
$$4 \cdot 3 = 12$$
$$\text{Factors}$$

The numbers 1, 2, 3, 4, 6, and 12 are all factors of 12. Each of these numbers divides 12 without a remainder.

If a and b are natural numbers we say that a is a **divisor** of b or a **divides** b, symbolized $a \,|\, b$, if the quotient of b divided by a has a remainder of 0. If a divides b, then b is **divisible** by a. For example, 4 divides 12, symbolized $4 \,|\, 12$, since the quotient of 12 divided by 4 has a remainder of 0. Note that 12 is divisible by 4. The notation $7 \nmid 12$ means that 7 does not divide 12. Note that every factor of a natural number is also a divisor of the natural number. *Caution:* Do not confuse the symbols, $a \,|\, b$ and a/b; $a \,|\, b$ means "a divides b" and a/b means "a divided by b" ($a \div b$). The symbols a/b and $a \div b$ indicate that the operation of division is to be performed, and b may or may not be a divisor of a.

Prime and Composite Numbers

Every natural number greater than 1 can be classified as either a prime number or a composite number.

> A **prime number** is a natural number greater than 1 that has exactly two factors (or divisors), itself and 1.

The number 5 is a prime number because it is divisible only by the factors 1 and 5. The first eight prime numbers are 2, 3, 5, 7, 11, 13, 17, and 19. The number 2

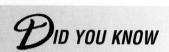
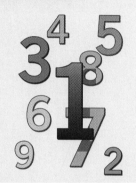

is the only even prime number. All other even numbers have at least three divisors:
1, 2, and the number itself.

> A **composite number** is a natural number that is divisible by a number
> other than itself and 1.

Any natural number greater than 1 that is not prime is composite. The first eight
composite numbers are 4, 6, 8, 9, 10, 12, 14, and 15.

The number 1 is neither prime nor composite; it is called a **unit**. The number
38 has at least three divisors, 1, 2, and 38, and, hence, is a composite number. In
contrast, the number 23 is a prime number since its only divisors are 1 and 23.

The ancient Greeks, more than 2000 years ago, developed a technique for
determining which numbers are prime numbers and which are not. This technique
is named the **sieve of Eratosthenes**, for the Greek mathematician Eratosthenes
of Cyrene who first used it.

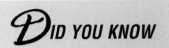

Figure 5.1

To find the prime numbers less than or equal to any natural number, say, 50,
using this method, list the first 50 counting numbers (Fig. 5.1). Cross out 1 since
it is not a prime number. Circle 2, the first prime number. Then cross out all the
multiples of 2; that is, 2, 4, 6, 8, . . . , 50. Circle the next prime number, 3. Cross
out all multiples of 3 that are not already crossed out. Continue this process until
you reach the prime number p, such that $p \cdot p$, or p^2, is greater than the last number
listed, in this case 50. Next circle 5 and cross out its multiples. Then circle 7 and
cross out its multiples. The next prime number is 11 and $11 \cdot 11$, or 121, is greater
than 50, so you are done. At this point, circle all the remaining numbers to obtain
the prime numbers less than or equal to 50. The prime numbers less than or equal
to 50 are 2, 3, 5, 7, 11, 13, 17, 19, 23, 29, 31, 37, 41, 43, and 47.

Now we turn our attention to composite numbers and their factors. The rules
of divisibility given in the chart on page 192 are helpful in finding divisors (or
factors) of composite numbers.

The test for divisibility by 6 is a particular case of the general statement that
the product of two prime divisors of a number is a divisor of the number. Thus,
for example, if both 3 and 7 divide a number, then 21 will also divide the number.

Note that the chart does not list rules of divisibility for the number 7. There
is a rule for 7, but it is difficult to remember. The easiest way to check divisibility
by 7 is just perform the division.

Rules of Divisibility

Divisible by	Test	Example
2	The number is even.	924 is divisible by 2, since 924 is even.
3	The sum of the digits of the number is divisible by 3.	924 is divisible by 3, since the sum of the digits, $9 + 2 + 4 = 15$, and 15 is divisible by 3.
4	The number formed by the last two digits of the number is divisible by 4.	924 is divisible by 4, since the number formed by the last two digits, 24, is divisible by 4.
5	The number ends in 0 or 5.	265 is divisible by 5, since the number ends in 5.
6	The number is divisible by both 2 and 3.	924 is divisible by 6, since it is divisible by both 2 and 3.
8	The number formed by the last three digits of the number is divisible by 8.	5824 is divisible by 8, since the number formed by the last three digits, 824, is divisible by 8.
9	The sum of the digits of the number is divisible by 9.	837 is divisible by 9, since the sum of the digits, 18, is divisible by 9.
10	The number ends in 0.	290 is divisible by 10, since the number ends in 0.

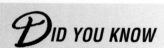

DID YOU KNOW

FRIENDLY NUMBERS

*T*he ancient Greeks often thought of numbers as having human qualities. For example the numbers 220 and 284 were considered "friendly" or "amicable" numbers because each number was the sum of the other number's proper factors. (A proper factor is any factor of a number other than the number itself.) If you sum all the proper factors of 284 (1 + 2 + 4 + 71 + 142) you get the number 220, and if you sum all the proper factors of 220 (1 + 2 + 4 + 5 + 10 + 11 + 20 + 22 + 44 + 55 + 110) you get 284. ■

EXAMPLE 1

Determine whether the number 374,832 is divisible by

a) 2   **b)** 3   **c)** 4   **d)** 5   **e)** 6   **f)** 8   **g)** 9   **h)** 10

Solution:

a) Since the number is even, it is divisible by 2.

b) Since the sum of the digits, 27, is divisible by 3, the number is divisible by 3.

c) Since the number formed by the last two digits, 32, is divisible by 4, the number is divisible by 4.

d) Since the last digit is not a 0 or a 5, the number is not divisible by 5.

e) Since the number is divisible by both 2 and 3, the number is divisible by 6.

f) Since the number formed by the last three digits, 832, is divisible by 8, the number is divisible by 8.

g) Since the sum of the digits, 27, is divisible by 9, the number is divisible by 9.

h) Since the number does not end in 0, the number is not divisible by 10.

Every composite number can be expressed as a product of prime numbers. The process of breaking a given number down into a product of prime numbers is called prime factorization. The prime factorization of 18 is $3 \times 3 \times 2$. No other natural number listed as a product of primes will have the same prime factorization as 18. The *fundamental theorem of arithmetic* states this concept formally. (A theorem is a statement or proposition that can be proven true.)

The Fundamental Theorem of Arithmetic

Every composite number can be expressed as a *unique* product of prime numbers.

In writing the prime factorization of a number, the order of the factors is immaterial. For example, we may write the prime factors of 18 as $3 \times 3 \times 2$ or $2 \times 3 \times 3$ or $3 \times 2 \times 3$.

A number of techniques can be used to find the prime factorization of a number. Two methods are illustrated.

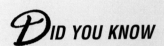

DID YOU KNOW

THE LONG LOST FACTORING MACHINE

The hunt began in 1989, when Jeffrey Shallit of the University of Waterloo in Ontario, Canada, came across an article in a 1920 French journal regarding a machine that was built in 1914 for factoring numbers (and factoring powers).

The article was written by Eugène Oliver Carissan, an officer in the French infantry and amateur mathematician who had constructed the factoring machine. The article indicated that Carissan had developed number sieve technology several years before Derrick H. Lehmer, the U.S. mathematician generally credited with being the first to design and build machines for solving number theory problems.

After considerable searching, Shallit, Hugh C. Williams of the University of Manitoba, Canada, and François Morain of École Polytechnique in Palaiseau, France, found the machine in a drawer of an astronomical observatory in Floirac, near Bordeaux. The machine, which resembles a cross between an antique music box and an old-fashioned hand-cranked phonograph, was in good condition and still worked.

Rotating the machine by a hand crank at two revolutions per minute, an operator could process 35 to 40 numbers per second. Carissan needed just 10 minutes to prove that 708,158,977 is a prime number.

The machine, which represented a significant advance at the time, is now housed at the Conservatoire National des Arts et Métiers in Paris. ▪

Method 1: Branching

To find the prime factorization of a number, select any two numbers whose product is the number to be factored. If the factors are not prime numbers, then continue factoring each composite number until all numbers are prime.

┌─ **EXAMPLE 2**

Write 300 as a product of primes.

Solution: Select any two numbers whose product is 300. Among the many choices, two possibilities are $10 \cdot 30$ and $15 \cdot 20$. Let's consider $15 \cdot 20$. Now find any two numbers whose product is 15 and any two numbers whose product is 20. Continue branching as shown in Fig. 5.2 until the numbers in the last row are all prime numbers. To determine the answer write the product of all the prime factors. The branching diagram is sometimes referred to as a *factor tree.*

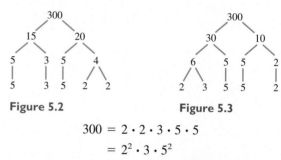

Figure 5.2   **Figure 5.3**

$$300 = 2 \cdot 2 \cdot 3 \cdot 5 \cdot 5$$
$$= 2^2 \cdot 3 \cdot 5^2$$

Thus the prime factorization of 300 is $2^2 \cdot 3 \cdot 5^2$. Note that if we had selected 30 and 10 as the initial factors, the final results would have been the same (Fig. 5.3).   ▲

Method 2: Division

To obtain the prime factorization of a number by this method, divide the given number by the smallest prime number by which it is divisible. Place the quotient under the given number. Then divide the quotient by the smallest prime number by which it is divisible, and again record the quotient. Repeat this process until the quotient is a prime number. The prime factorization is the product of all the prime divisors and the prime (or last) quotient. This procedure is illustrated in Example 3.

┌─ **EXAMPLE 3**

Write 300 as a product of prime numbers.

Solution: Because 300 is an even number, the smallest prime number that divides it is 2. Divide 300 by 2. Place the quotient, 150, below the 300. Repeat this process of dividing each quotient by the smallest prime number that divides it:

$$
\begin{array}{r|r}
2 & 300 \\ \hline
2 & 150 \\ \hline
3 & 75 \\ \hline
5 & 25 \\ \hline
 & 5
\end{array}
$$

The final quotient, 5, is a prime number, so we stop. The prime factorization is

$$300 = 2 \cdot 2 \cdot 3 \cdot 5 \cdot 5 = 2^2 \cdot 3 \cdot 5^2.$$ ▲

Note that, despite the different methods used in Examples 2 and 3, the answer is the same.

Greatest Common Divisor

The discussion, in Section 5.3, of how to reduce fractions makes use of the greatest common divisor (GCD). One technique of finding the GCD is to use prime factorization.

> The **greatest common divisor (GCD)** of a set of natural numbers is the largest natural number that divides (without remainder) every number in that set.

What is the GCD of 12 and 18? One way to determine it is to make a list of the divisors (or factors) of 12 and 18:

$$\text{Divisors of 12} \quad \{1, 2, 3, 4, 6, 12\}$$
$$\text{Divisors of 18} \quad \{1, 2, 3, 6, 9, 18\}$$

The common divisors are 1, 2, 3, and 6, and therefore the greatest common divisor is 6.

If the numbers are large, this method of finding the GCD is not practical. The GCD can be found more efficiently by using prime factorization.

> **To Find the Greatest Common Divisor of Two or More Numbers**
> 1. Determine the prime factorization of each number.
> 2. Find each prime factor with the smallest exponent that appears in each of the prime factorizations.
> 3. Determine the product of the factors found in step 2.

Example 4 illustrates this procedure.

EXAMPLE 4

Find the GCD of 24 and 60.

Solution: The branching method of finding the prime factors of 24 and 60 is illustrated in Fig. 5.4 on page 196.

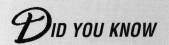

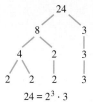

$$24 = 2^3 \cdot 3 \qquad\qquad 60 = 2^2 \cdot 3 \cdot 5$$

Figure 5.4

I. The prime factorization of 24 is $2^3 \cdot 3$ and the prime factorization of 60 is $2^2 \cdot 3 \cdot 5$.

2. The prime factors with the smallest exponents that appear in each of the factorizations of 24 and 60 are 2^2 and 3.

3. The product of the factors found in step 2 is $2^2 \cdot 3 = 4 \cdot 3 = 12$. The GCD of 24 and 60 is 12. Twelve is the largest number that divides both 24 and 60.   ▲

EXAMPLE 5

Find the GCD of 36 and 150.

Solution:

I. The prime factorization of 36 is $2^2 \cdot 3^2$ and the prime factorization of 150 is $2 \cdot 3 \cdot 5^2$.

2. The prime factors with the smallest exponents that appear in each of the factorizations of 36 and 150 are 2 and 3.

3. The product of the factors found in step 2 is $2 \cdot 3 = 6$. The GCD of 36 and 150 is 6.   ▲

Two numbers with a GCD of 1 are said to be *relatively prime*. The numbers 9 and 14 are relatively prime, since the GCD is 1.

Least Common Multiple

To perform addition and subtraction of fractions (Section 5.3) we will use the least common multiple (LCM). One technique of finding the LCM is to use prime factorization.

> The **least common multiple (LCM)** of a set of natural numbers is the smallest natural number that is divisible (without remainder) by each element of the set.

What is the least common multiple of 12 and 18? One way to determine the LCM is to make a list of the multiples of each number:

Multiples of 12 $\{12, 24, \mathbf{36}, 48, 60, \mathbf{72}, 84, 96, \mathbf{108}, 120, 132, \mathbf{144}, \ldots\}$
Multiples of 18 $\{18, \mathbf{36}, 54, \mathbf{72}, 90, \mathbf{108}, 126, \mathbf{144}, 162, \ldots\}$

Some common multiples of 12 and 18 are 36, 72, 108, and 144. The least common multiple, 36, is the smallest number that is divisible by both 12 and 18.

Usually the most efficient method of finding the LCM is to use prime factorization.

To Find the Least Common Multiple of Two or More Numbers
1. Determine the prime factorization of each number.
2. List each prime factor with the greatest exponent that appears in any of the prime factorizations.
3. Determine the product of the factors found in step 2.

Example 6 illustrates this procedure.

EXAMPLE 6

Find the LCM of 24 and 60.

Solution:
1. Find the prime factors of each number. In Example 4 we determined that

$$24 = 2^3 \cdot 3 \quad \text{and} \quad 60 = 2^2 \cdot 3 \cdot 5.$$

2. List each prime factor with the greatest exponent that appears in either of the prime factorizations: 2^3, 3, 5.
3. Determine the product of the factors found in step 2:

$$2^3 \cdot 3 \cdot 5 = 8 \cdot 3 \cdot 5 = 120.$$

Thus 120 is the LCM of 24 and 60. It is the smallest number that is divisible by both 24 and 60. ▲

EXAMPLE 7

Find the LCM of 36 and 150.

Solution:
1. Find the prime factors of each number. In Example 5 we determined that

$$36 = 2^2 \cdot 3^2 \quad \text{and} \quad 150 = 2 \cdot 3 \cdot 5^2.$$

2. List each prime factor with the greatest exponent that appears in either of the prime factorizations: 2^2, 3^2, 5^2.
3. Determine the product of the factors found in step 2:

$$2^2 \cdot 3^2 \cdot 5^2 = 4 \cdot 9 \cdot 25 = 900.$$

Thus 900 is the least common multiple of 36 and 150. It is the smallest number divisible by both 36 and 150. ▲

The Search for Larger Prime Numbers

More than 2000 years ago, the Greek mathematician Euclid proved that there is no largest prime number. However, mathematicians continue to strive to find larger and larger prime numbers.

Marin Mersenne (1588–1648), a seventeenth-century monk, found that num-

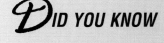

bers of the form $2^n - 1$ are often prime numbers when n is a prime number. For example,

$$2^2 - 1 = 4 - 1 = 3 \qquad 2^3 - 1 = 8 - 1 = 7$$
$$2^5 - 1 = 32 - 1 = 31 \qquad 2^7 - 1 = 128 - 1 = 127$$

Numbers of the form $2^n - 1$ that are prime are referred to as Mersenne primes. The first 10 Mersenne primes occur when $n = 2, 3, 5, 7, 13, 17, 19, 31, 61, 89$. The first time the expression $2^n - 1$ does not generate a prime number, for prime number n, is when n is 11. The number $2^{11} - 1$ is a composite number (see Exercise 94). Scientists frequently use Mersenne primes in their search for larger and larger primes.

The largest prime found to date was discovered in March 1994 by Paul Gage and David Slowinski of Cray Research, Eagan, Minnesota. The number is 2 multiplied by itself 859,433 times, minus 1, or $2^{859,433} - 1$. This new record prime, the 33rd known Mersenne prime, is 258,716 digits long. This number in print would fill eight newspaper pages. The previous record holder $2^{756,839} - 1$, which consisted of 227,832 digits, was tracked down by ABA Technology's Hartwell Laboratory in 1992. The computer used to find the new largest prime number was a Cray Y-MP M90 Series Supercomputer. The run time on the computer was about 30 minutes compared with $19\frac{1}{2}$ hours required to find the previous record holder. It is very difficult to comprehend the magnitude of this largest prime number. If you stacked $2^{859,433} - 1$ sheets of paper one on top of the other, the height of the stack would be many times the size of our universe.

More about Prime Numbers

Another mathematician who studied prime numbers was Pierre de Fermat (1601–1665). A lawyer by profession, Fermat became interested in mathematics as a hobby. He became one of the finest mathematicians of the seventeenth century. Fermat conjectured that each number of the form $2^{2^n} + 1$, now referred to as the Fermat number, was prime for each natural number n. Recall that a *conjecture* is a supposition that has not been proved nor disproved. In 1732 Leonhard Euler proved that for $n = 5$, $2^{32} + 1$ was a composite number, thus disproving Fermat's conjecture.

In the past 300 years mathematicians have only been able to evaluate the sixth, seventh, eighth, and ninth Fermat numbers to determine whether they are prime or composite. Each of these numbers has been shown to be composite. Fermat numbers are very difficult to test due to the sheer magnitude of the computations involved.

In 1772 Euler devised the formula $n^2 - n + 41$, which yields a prime number for any natural number up to 40 but fails for $n = 41$. In 1879 E. B. Escott devised the formula $n^2 - 79n + 1601$, which yields a prime number for any natural number up to 79 but fails for $n = 80$.

In 1742 Christian Goldbach conjectured in a letter to Euler that every even number greater than or equal to 6 can be represented as a sum of two prime numbers (*examples:* $6 = 3 + 3$; $8 = 3 + 5$) and that every odd number greater than or equal to 9 can be represented as the sum of three odd primes (*examples:* $9 = 3 + 3 + 3$; $11 = 3 + 3 + 5$). This conjecture is still unproven.

Another conjecture is that there are infinitely many pairs of twin primes of the form $p, p + 2$ (*examples:* 3, 5; 5, 7). What are the next two pairs of twin primes?

Section 5.1 Exercises

1. What is number theory?

2. What does "*a* and *b* are factors of *c*" mean?

3. **a)** What does "*a* divides *b*" mean?
   b) What does "*a* is divisible by *b*" mean?

4. What is a prime number?

5. What is a composite number?

6. What does the Fundamental Theorem of Arithmetic state?

7. **a)** What is the greatest common divisor of a set of natural numbers?
   b) In your own words explain how to find the GCD of a set of natural numbers by using prime factorization.
   c) Find the GCD of 16 and 40 by using the procedure given in part (b).

8. **a)** What is the least common multiple of a set of natural numbers?
   b) In your own words explain how to find the LCM of a set of natural numbers by using prime factorization.
   c) Find the LCM of 16 and 40 by using the procedure given in part (b).

9. What are Mersenne primes?

10. What are twin primes?

11. What is a conjecture?

12. Use the sieve of Eratosthenes to find the prime numbers up to and including 100.

In Exercises 13–22, determine whether the statement is true or false.

13. 8 is a factor of 56.

14. $8 \mid 56$

15. Fifty-six is a multiple of 8.

16. Eight is a divisor of 56.

17. Fifty-six is divisible by 8.

18. Eight is a multiple of 56.

19. If a number is divisible by 3, then every digit of the number is divisible by 3.

20. If every digit of a number is divisible by 3, then the number itself is divisible by 3.

21. If a number is divisible by 2 and 3, then the number is divisible by 6.

22. If a number is divisible by 3 and 4, then the number is divisible by 12.

In Exercises 23–28, determine whether the number is divisible by each of the following numbers: 2, 3, 4, 5, 6, 8, 9, and 10.

23. 48,324

24. 529,200

25. 2,763,105

26. 3,126,120

27. 1,882,320

28. 3,941,221

29. Determine a number that is divisible by 2, 3, 4, 5, and 6.

30. Determine a number that is divisible by 3, 4, 5, 9, and 10.

In Exercises 31–42, find the prime factorization of the number.

31. 52   32. 48   33. 57

34. 180   35. 312   36. 500

37. 510   38. 999   39. 1996

40. 1168   41. 2052   42. 5060

In Exercises 43–52, find the greatest common divisor of the numbers.

43. 15 and 180   44. 12 and 77

45. 42 and 56   46. 52 and 65

47. 90 and 600   48. 180 and 360

49. 96 and 212   50. 240 and 285

51. 24, 48, 128   52. 18, 78, 198

In Exercises 53–62, find the least common multiple of the numbers.

53. 15 and 180   54. 12 and 77

55. 42 and 56   56. 52 and 65

57. 90 and 600   58. 180 and 360

59. 96 and 212   60. 240 and 285

61. 24, 48, 128   62. 18, 78, 198

In Exercises 63 and 64, are the statements true or false? Explain.

63. If a number is not divisible by 5, then it is not divisible by 10.

64. If a number is not divisible by 10, then it is not divisible by 5.

65. Find the next two sets of twin primes that follow the set 5, 7.

66. The primes 2 and 3 are consecutive natural numbers. Is there another pair of consecutive natural numbers both of which are prime? Explain.

67. Show that Goldbach's conjecture is true for the even numbers 6 through 12 and the odd numbers 9 through 15.

68. Find the fifth Mersenne prime number.

69. Show that each number of the form $2^{2^n} + 1$ is prime for $n = 1, 2,$ and 3.

70. In a world championship poker match Sheila has accumulated 288 red chips and 192 blue chips. She wants to make stacks of chips on the table so that each stack contains the same number of chips. If the red and blue chips cannot be mixed and each chip is to belong to one stack, what is the largest number of chips that she can have in a stack?

71. John collects trading cards. He has 432 baseball cards and 360 football cards. He wants to make stacks of cards on a table so that each stack contains the same number of cards and each card belongs to one stack. If the baseball and football cards must not be mixed in the stacks, what is the largest number of cards that he can have in a stack?

72. Cortney has collected 168 pogs with faces on them (pogs are plastic pieces about the size and shape of a half-dollar) and 224 pogs that do not have faces on them. She wants to make stacks of pogs so that each stack contains the same number of pogs. If the two types of pogs are not to be mixed in the stacks and each pog is to belong to one stack, what is the largest number of pogs that can be placed in a stack?

73. Sears has a sale on mechanics' tool chests every 40 days and a sale on mechanics' tool sets every 60 days. If both are on sale today, how long will it be before they are on sale together again?

74. Sara and Harry both work the 3:00 P.M. to 11:00 P.M. shift. Sara has every fifth night off and Harry has every sixth night off. If they both have tonight off, how many days will pass before they have the same night off again?

75. At an all night theater one movie runs for 80 minutes and a second runs for 95 minutes. Twenty minutes is allowed for cleaning the theater after each movie. Then the movie starts again. If the movies both start at 1 P.M., when will the movies start again at the same time?

76. The United States Senate consists of 100 members. Senate committees are to be formed so that each of the committees contains the same number of senators and each senator is a member of exactly one committee. The committees are to have more than 2 members but fewer than 50 members. There are various ways that these committees can be formed.
a) What size committees are possible?
b) How many committees are there for each size?

77. Consider the first eight prime numbers greater than 3. The numbers are 5, 7, 11, 13, 17, 19, 23, and 29.
a) Determine which of these prime numbers differs by 1 from a multiple of the number 6.
b) Use inductive reasoning and the results obtained in part (a) to make a conjecture regarding prime numbers.
c) Select a few more prime numbers and determine whether your conjecture appears to be correct.

78. State a procedure that defines a divisibility test for 15.

Another method that can be used to find the greatest common divisor is known as the **Euclidean algorithm**. We will illustrate this procedure by finding the GCD of 60 and 220.

First divide 220 by 60. Disregard the quotient 3, and then divide 60 by the remainder 40. Continue this process of dividing the divisors by the remainders until you obtain a remainder of 0. The divisor in the last division, in which the remainder is 0, is the GCD.

$$
\begin{array}{ccc}
3 & 1 & 2 \\
60\overline{)220} & 40\overline{)60} & 20\overline{)40} \\
\underline{180} & \underline{40} & \underline{40} \\
40 & 20 & 0
\end{array}
$$

Since 40/20 had a remainder of 0, the GCD is 20.

In Exercises 79–84, use the Euclidean algorithm to find the GCD.

79. 35, 75

80. 20, 160

81. 18, 112

82. 96, 115

83. 150, 180

84. 210, 560

A number whose proper factors (factors other than the number itself) add up to the number is called a perfect number. For example, 6 is a perfect number because its proper factors are 1, 2, and 3, and $1 + 2 + 3 = 6$. Determine which, if any, of the following numbers are perfect.

85. 12

86. 28

87. 496

88. 48

Problem Solving/Group Activities

89. A certain procedure can be used to determine the *number of factors* (or *divisors*) of a composite number. To do so, write the number in its factored form using exponents. Add 1 to each exponent and then multiply these numbers. The product gives the number of positive divisors of the composite number.
a) Use this procedure to determine the number of divisors of 24.
b) To check your answer, list all the divisors of 24. You should obtain the same number of divisors found in part (a).

90. Recall that, if a number is divisible by both 2 and 3, then the number is divisible by 6. If a number is divisible by both 2 and 4, is the number necessarily divisible by 8? Explain your answer.

91. The product of any three consecutive natural numbers is divisible by 6. Explain why.

92. A number in which each digit except 0 appears exactly three times is divisible by 3. For example, 888,444,555 and 714,714,714 are both divisible by 3. Explain why this outcome must be true.

93. Use the fact that, if $a \mid b$ and $a \mid c$ then $a \mid (b + c)$, to determine whether 36,018 is divisible by 18. (*Hint:* Write 36,018 as 36,000 + 18.)

94. Show that $2^{11} - 1$ is a composite number.

Research Activity

95. Do research and explain what *deficient numbers* and *abundant numbers* are. Give an example of each type of number. References include history of mathematics books and encyclopedias.

5.2 THE INTEGERS

In Section 5.1 we introduced the natural or counting numbers:

$$N = \{1, 2, 3, 4, \ldots\}.$$

Another important set of numbers, the whole numbers, help to answer the question "How many?"

$$\text{Whole numbers} = \{0, 1, 2, 3, 4, \ldots\}$$

Note that the set of whole numbers contains the number 0 but that the set of counting numbers does not. If a farmer were asked how many chickens were in a coop the answer would be a whole number. If the farmer had no chickens, he or she would answer zero. Although we use the number 0 daily and take it for granted, the number 0 as we know it was not used and accepted until the sixteenth century.

If the temperature is 12°F and drops 20°, the resulting temperature is −8°F. This type of problem shows the need for negative numbers. The set of integers consists of the negative integers, 0, and the positive integers.

$$\text{Integers} = \{\ldots, -4, -3, -2, -1, 0, 1, 2, 3, \ldots\}$$
Negative integers   Positive integers

The term *positive integers* is yet another name for the natural numbers or counting numbers.

An understanding of addition, subtraction, multiplication, and division of the integers is essential in understanding algebra (Chapter 6). To aid in our explanation of addition and subtraction of integers, we will introduce the real number line (Fig. 5.5 on page 202). To construct the real number line, arbitrarily select a point for zero to serve as the starting point. Place the positive integers to the right of 0, equally spaced from one another. Place the negative integers to the left of 0, using the same spacing. The real number line contains the integers and all of the other real numbers that are not integers. Some examples of real numbers that are not integers are indicated in Fig. 5.5. We will discuss real numbers that are not integers in the next two sections.

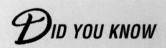

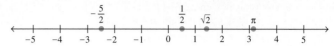

Figure 5.5

The arrows at the ends of the real number line indicate that the line continues indefinitely in both directions. Note that for any natural number, n, on the number line, the opposite of that number, $-n$, is also on the number line. This real number line was drawn horizontally, but it could just as well have been drawn vertically. In fact, in the next chapter you will see that the axes of a graph are the union of two number lines, one horizontal and the other vertical.

The number line can be used to determine the order of two integers. On the number line, the numbers increase from left to right. The number 3 is greater than 2, written $3 > 2$. Observe that 3 is to the right of 2. Similarly, we can see that $0 > -1$ by observing that 0 is to the right of -1 on the number line.

Instead of stating that 3 is greater than 2, we could state that 2 is less than 3, written $2 < 3$. Note that 2 is to the left of 3 on the number line. We can also see that $-1 < 0$ by observing that -1 is to the left of 0. The inequality symbol always points to the smaller of the two numbers when the inequality is true.

EXAMPLE 1

Insert either $>$ or $<$ in the shaded area between the paired numbers to make the statement correct.

a) -4 ■ 2 **b)** -2 ■ -4 **c)** -5 ■ -3 **d)** 0 ■ -2

Solution:

a) $-4 < 2$, since -4 is to the left of 2.

b) $-2 > -4$, since -2 is to the right of -4.

c) $-5 < -3$, since -5 is to the left of -3.

d) $0 > -2$, since 0 is to the right of -2.

Addition of Integers

Addition of integers can be represented geometrically with a number line. To do so, begin at 0 on the number line. Represent the first addend by an arrow starting at 0. Draw the arrow to the right if the addend is positive. If the addend is negative, draw the arrow to the left. From the tip of the first arrow, draw a second arrow to represent the second addend. Draw the second arrow to the right or left, as just explained. The sum of the two integers is found at the tip of the second arrow.

EXAMPLE 2

Evaluate $2 + (-4)$ using the number line.

Solution:

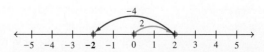

Thus $2 + (-4) = -2$.

EXAMPLE 3

Evaluate **a)** $-2 + (-5)$ and   **b)** $-5 + 2$ using the number line.

Solution:

a)

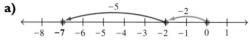

Thus $-2 + (-5) = -7$.

b)

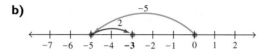

Thus $-5 + 2 = -3$.

EXAMPLE 4

Evaluate $4 + (-4)$ using the number line.

Solution:

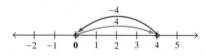

Thus $4 + (-4) = 0$.

The number -4 is said to be the additive inverse of 4, and 4 is the additive inverse of -4, because their sum is 0. In general the **additive inverse** of the number n is $-n$, since $n + (-n) = 0$. Inverses will be discussed more formally in Chapter 10.

Subtraction of Integers

Any subtraction problem can be rewritten as an addition problem. To do so, we use the following definition of subtraction.

Subtraction

$$a - b = a + (-b)$$

The rule for subtraction indicates that to subtract b from a, *add* the additive inverse of b to a. For example,

$$3 - 5 = 3 + (-5).$$

Subtraction   Addition   Additive inverse of 5

Now we can determine the value of $3 + (-5)$.

Thus $3 - 5 = 3 + (-5) = -2$.

EXAMPLE 5

Evaluate $-3 - (-2)$ using the number line.

We are subtracting -2 from -3. The additive inverse of -2 is 2; therefore we add 2 to -3.

$$-3 - (-2) = -3 + 2$$

We now add $-3 + 2$ on the number line to obtain the answer -1.

Thus $-3 - (-2) = -3 + 2 = -1$.

In Example 5 we found that $-3 - (-2) = -3 + 2$. In general, $a - (-b) = a + b$. As you get more proficient in working with integers you should be able to answer questions involving them without drawing a number line.

EXAMPLE 6

Evaluate   **a)** $-3 - 5$ **b)** $4 - (-3)$

Solution:

a) $-3 - 5 = -3 + (-5) = -8$ **b)** $4 - (-3) = 4 + 3 = 7$

EXAMPLE 7

The highest point on earth is Mount Everest, in the Himalayas, at a height of 29,028 ft above sea level. The lowest point on earth is the Mariana Trench, in the Pacific Ocean, at a depth of 36,198 ft below sea level ($-36,198$ ft). Find the vertical height difference between Mount Everest and the Mariana Trench.

Solution: We obtain the vertical difference by subtracting the lower elevation from the higher elevation.

$$29,028 - (-36,198) = 29,028 + 36,198 = 65,226$$

The vertical difference is 65,226 ft.

Multiplication of Integers

The multiplication property of zero is important in our discussion of multiplication of integers. It indicates that the product of 0 and any number is 0.

Multiplication Property of Zero

$$a \cdot 0 = 0 \cdot a = 0$$

We will develop the rules for multiplication of integers using number patterns. The four possible cases are

1. positive integer × positive integer,

2. positive integer × negative integer,

3. negative integer × positive integer, and

4. negative integer × negative integer.

Case 1: Positive integer × positive integer The product of two positive integers can be defined as repeated addition of a positive integer. Thus 3 · 2 means 2 + 2 + 2. This sum will always be positive. Thus a positive integer times a positive integer is a positive integer.

Case 2: Positive integer × negative integer Consider the following patterns.

$$4(3) = 12$$
$$4(2) = 8$$
$$4(1) = 4$$

Note that each time the second factor is reduced by 1, the product is reduced by 4. Continuing the process gives

$$4(0) = 0.$$

What comes next?

$$4(-1) = -4$$
$$4(-2) = -8$$

The pattern indicates that a positive integer times a negative integer is a negative integer.

We can confirm this result by using the number line. The expression $3(-2)$ means $(-2) + (-2) + (-2)$. Adding $(-2) + (-2) + (-2)$ on the number line, we obtain a sum of -6.

Case 3: Negative integer × positive integer A procedure similar to that used in case 2 will indicate that a negative integer times a positive integer is a negative integer.

Case 4: Negative integer × negative integer We have illustrated that a positive integer times a negative integer is a negative integer. We make use of this fact in the following pattern.

$$4(-3) = -12$$
$$3(-3) = -9$$
$$2(-3) = -6$$
$$1(-3) = -3$$

In this pattern, each time the first term is decreased by 1, the product is increased by 3. Continuing this process gives

$$0(-3) = 0;$$

$$(-1)(-3) = 3;$$

$$(-2)(-3) = 6.$$

This pattern illustrates that a negative integer times a negative integer is a positive integer.

The examples were restricted to integers. However, the rules for multiplication can be used for any numbers. We summarize them as follows.

Rules for Multiplication

1. The product of two numbers with *like signs* (positive × positive or negative × negative) is a *positive number*.

2. The product of two numbers with *unlike signs* (positive × negative or negative × positive) is a *negative number*.

┌─ **EXAMPLE 8**

Evaluate **a)** $4(-7)$ **b)** $(-3)(2)$ **c)** $(-4)(-5)$

Solution:

a) $4(-7) = -28$ **b)** $(-3)(2) = -6$ **c)** $(-4)(-5) = 20$ ▲

Division of Integers

You may already realize that a relationship exists between multiplication and division.

$$6 \div 2 = 3 \quad \text{means that} \quad 3 \cdot 2 = 6.$$

$$\frac{20}{10} = 2 \quad \text{means that} \quad 2 \cdot 10 = 20.$$

These examples demonstrate that division is the reverse process of multiplication.

Division

For any a, b, and c where $b \neq 0$, $\dfrac{a}{b} = c$ means that $c \cdot b = a$.

We will discuss the four possible cases for division, which are similar to those for multiplication.

Case 1: Positive integer ÷ positive integer A positive integer divided by a positive integer is positive.

$$\frac{6}{2} = 3, \quad \text{since} \quad 3(2) = 6$$

Case 2: Positive integer ÷ negative integer A positive integer divided by a negative integer is negative.

$$\frac{6}{-2} = -3, \quad \text{since} \quad (-3)(-2) = 6$$

Case 3: Negative integer ÷ positive integer A negative integer divided by a positive integer is negative.

$$\frac{-6}{2} = -3, \quad \text{since} \quad (-3)(2) = -6$$

Case 4: Negative integer ÷ negative integer A negative integer divided by a negative integer is positive.

$$\frac{-6}{-2} = 3, \quad \text{since} \quad 3(-2) = -6$$

The examples were restricted to integers. However, the rules for division can be used for any numbers. You should realize that division of integers does not always result in an integer. The rules for division are summarized as follows.

Rules for Division

1. The quotient of two numbers with *like signs* (positive ÷ positive or negative ÷ negative) is a *positive number*.
2. The quotient of two numbers with *unlike signs* (positive ÷ negative or negative ÷ positive) is a *negative number*.

EXAMPLE 9

Evaluate   **a)** $\dfrac{20}{-5}$ **b)** $\dfrac{-60}{4}$ **c)** $\dfrac{-25}{-5}$

Solution:

a) $\dfrac{20}{-5} = -4$ **b)** $\dfrac{-60}{4} = -15$ **c)** $\dfrac{-25}{-5} = 5$ ▲

In the definition of division we stated that the denominator could not be 0. Division by 0 is not allowed. The quotient of a number divided by zero is said to be **undefined**. For example $\frac{3}{0}$ is undefined.

Section 5.2 Exercises

1. Explain the rule for multiplication of real numbers.

2. Explain the rule for division of real numbers.

3. What is the product of 0 times any real number?

4. What do we call the quotient of 3 ÷ 0? Explain your answer.

In Exercises 5–14, evaluate the expression.

5. $-9 + 12$

6. $5 + (-12)$

7. $(-7) + 15$

8. $(-5) + (-5)$

9. $[9 + (-13)] + 0$

10. $(4 + 6) + (-3)$

11. $[(-2) + (-5)] + 8$

12. $[9 + (-4)] + (-6)$

13. $[(-24) + (-8)] + 12$

14. $(3 - 13) + (-7)$

In Exercises 15–24, evaluate the expression.

15. $6 - 9$

16. $-3 - 6$

17. $-8 - 5$

18. $6 - (-4)$

19. $-7 - (-5)$

20. $4 - (-4)$

21. $12 - 18$

22. $3 - (-8)$

23. $[5 + (-3)] - 4$

24. $6 - (6 + 8)$

End of combining

In Exercises 25–34, evaluate the expression.

25. $-4 \cdot 5$

26. $3(-6)$

27. $(-8)(-8)$

28. $(-7)(7)$

29. $[(-9)(-3)] \cdot 5$

30. $(-4)(5)(-6)$

31. $(7 \cdot 5)(-3)$

32. $(-7)(-1)(-1)$

33. $[(-3)(-6)] \cdot [(-5)(8)]$

34. $[[(-4)(8)](6)](-2)$

In Exercises 35–44, evaluate the expression.

35. $-24 \div (-6)$

36. $-24 \div 6$

37. $16 \div (-16)$

38. $\dfrac{48}{-3}$

39. $\dfrac{63}{-7}$

40. $\dfrac{-75}{25}$

41. $\dfrac{132}{-11}$

42. $\dfrac{186}{-6}$

43. $140 \div (-7)$

44. $(-600) \div (-4)$

In Exercises 45–53, determine whether the statement is true or false.

45. Every integer is a natural number.

46. Every natural number is an integer.

47. The difference of any two negative integers is a negative integer.

48. The sum of any two negative integers is a negative integer.

49. The product of any two positive integers is a positive integer.

50. The difference of a positive integer and a negative integer is always a negative integer.

51. The quotient of a negative integer and a positive integer is always a negative number.

52. The quotient of any two negative integers is a negative number.

53. The sum of a positive integer and a negative integer is always a positive integer.

54. The product of a positive integer and a negative integer is always a positive integer.

In Exercises 55–64, evaluate the expression.

55. $(7 + 5) \div 2$

56. $[5(-4)] + 3$

57. $[6(-2)] - 5$

58. $[(-5)(-6)] - 3$

59. $(4 - 8)(3)$

60. $[18 \div (-2)](-3)$

61. $[2 + (-17)] \div 3$

62. $(5 - 9) \div (-4)$

63. $[(-22)(-3)] \div (2 - 13)$

64. $[15(-4)] \div (-6)$

In Exercises 65–68, write the numbers in increasing order from left to right.

65. $-5, 4, -3, 0, 5, -1$

66. $5, -3, 2, 0, 3, -2$

67. $-5, -2, -3, -1, -4, -6$

68. $106, 33, -47, -108, 72, -76$

69. Eunice's investments gained 18%. During the same time period Betty's investments lost 12%. Find the percent difference between the performances of the two investments.

70. Mount Whitney, in the Sierra Nevada mountains of California, is the highest point in the contiguous United States. It is 14,495 ft above sea level. Death Valley, in California and Nevada, is the lowest point in the United States, 282 ft below sea level. Find the vertical height difference between Mount Whitney and Death Valley.

71. A helicopter drops a package from a height of 842 ft above sea level. The package lands in the ocean and settles at a point 927 ft below sea level. What was the vertical distance the package traveled?

72. In the first four plays of the game the Eagles gained 8 yd, lost 5 yd, gained 3 yd, and gained 4 yd. What is the total number of yards gained in the first four plays? Did the Eagles make a first down? (10 yd are needed for a first down)

73. Tito has been following the price of a cuckoo clock at a local furniture store. The changes in the price for the past 6 months were: increase of $12, decrease of $40, increase of $35, decrease of $19, decrease of $6, and increase of $37. What was the net change in the cost of the clock over the 6-month period?

74. Yvette records the closing price of her favorite stock every Friday and the change in price from the previous Friday. The changes in price for the last five Fridays were: gained 7 points, lost 2 points, lost 3 points, lost 2 points, and gained 1 point. What was the net change in the stock's value for the five Fridays?

75. Part of a World Standard Time Zones chart used by airlines and the United States Navy is shown. The scale along the bottom is just like a number line with the integers −12, −11, . . . , 11, 12 on it.
   a) Find the difference in time between Amsterdam (zone +1) and Los Angeles (zone −8).
   b) Find the difference in time between Boston (zone −5) and Puerto Vallarta (zone −7).

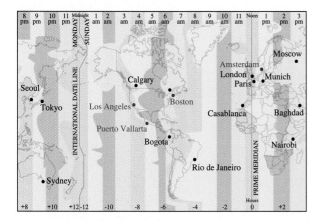

76. Explain why $\dfrac{a}{b} = \dfrac{-a}{-b}$.

77. Find the quotient:
$$\frac{-1 + 2 - 3 + 4 - 5 + \cdots - 99 + 100}{1 - 2 + 3 - 4 + 5 - \cdots + 99 - 100}$$

78. Triangular numbers and square numbers were introduced in the Section 1.1 Exercises. There are also **pentagonal numbers**, which were also studied by the Greeks. Four pentagonal numbers are 1, 5, 12, and 22.

   a) Determine the next three pentagonal numbers.
   b) Describe a procedure to determine the next five pentagonal numbers without drawing the figures.
   c) Is 72 a pentagonal number? Explain how you determined your answer.

79. Place the appropriate plus or minus signs between each digit so that the total will equal 1.

$$0 \quad 1 \quad 2 \quad 3 \quad 4 \quad 5 \quad 6 \quad 7 \quad 8 \quad 9 = 1$$

80. Do research and write a report on the history of the number 0 in the Hindu–Arabic numeration system.

5.3 THE RATIONAL NUMBERS

We introduced the number line in Section 5.1, and discussed the integers in Section 5.2. The numbers that fall between the integers on the number line are either rational or irrational numbers. In this section we will discuss the rational numbers and in Section 5.4 we will discuss the irrational numbers.

Any number that can be expressed as a quotient of two integers (denominator not 0) is a rational number.

> *"When you can measure what you are talking about and express it in numbers, you know something about it."*
> Lord Kelvin

> The set of **rational numbers**, denoted by Q, is the set of all numbers of the form p/q, where p and q are integers and $q \neq 0$.

The following numbers are examples of rational numbers:

$$\frac{3}{5}, \quad \frac{2}{7}, \quad \frac{12}{5}, \quad 2, \quad 0$$

The integers 2 and 0 are rational numbers because each can be expressed as the quotient of two integers: $2 = \frac{2}{1}$ and $0 = \frac{0}{1}$. In fact, every integer p is a rational number, since it can be written in the form of $p/1$.

Numbers such as $\frac{3}{5}$ and $-\frac{2}{7}$ are also called *fractions*. The number above the fraction line is called the *numerator*, and the number below the fraction line is called the *denominator*.

Reducing Fractions

Sometimes the numerator and denominator in a fraction have a common divisor (or common factor). For example, both the numerator and denominator of the fraction $\frac{6}{10}$ have the common divisor 2. When a numerator and denominator have a common divisor, we can *reduce the fraction to its lowest terms*.

A fraction is said to be in its lowest terms (or reduced) when the numerator and denominator are relatively prime (that is, have no common divisors other than 1). To reduce a fraction to its lowest terms, divide both the numerator and the denominator by the greatest common divisor. Recall that a procedure for finding the greatest common divisor was discussed in Section 5.1.

The fraction $\frac{6}{10}$ is reduced to its lowest terms as follows.

$$\frac{6}{10} = \frac{6 \div 2}{10 \div 2} = \frac{3}{5}$$

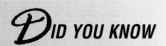
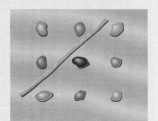
EXAMPLE I

Reduce $\frac{36}{150}$ to its lowest terms.

Solution: In Example 5 of Section 5.1 we determined that the GCD of 36 and 150 is 6. Divide the numerator and the denominator by the GCD, 6.

$$\frac{36 \div 6}{150 \div 6} = \frac{6}{25}$$

Since there are no common divisors of 6 and 25 other than 1, this fraction is in its lowest terms.   ▲

Terminating or Repeating Decimal Numbers

Note the following important property of the rational numbers.

> Every *rational number* when expressed as a decimal number will be either a terminating or a repeating decimal number.

Examples of terminating decimal numbers are 0.5, 0.75, and 4.65. Examples of repeating decimal numbers are 0.333 . . . , 0.2323 . . . , and 8.456456 . . . . One way to indicate that a number or group of numbers repeat is to place a bar above the number or group of numbers that repeat. Thus 0.333 . . . may be written $0.\overline{3}$, 0.2323 . . . may be written $0.\overline{23}$, and 8.456456 . . . may be written $8.\overline{456}$.

EXAMPLE 2

Show that the following rational numbers are terminating decimal numbers.

a) $\frac{1}{2}$ **b)** $\frac{12}{5}$ **c)** $\frac{3}{8}$

Solution: To express the rational number in decimal form, divide the numerator by the denominator. If you use a calculator, or use long division, you will obtain the following results.

a) $1 \div 2 = 0.5$ **b)** $12 \div 5 = 2.4$ **c)** $3 \div 8 = 0.375$ ▲

Note that in each part of Example 2 each quotient has a final nonzero digit. Thus each number is a terminating decimal number.

EXAMPLE 3

Show that the following rational numbers are repeating decimal numbers.

a) $\frac{1}{3}$ **b)** $\frac{23}{99}$ **c)** $\frac{2}{7}$

Solution: If you use a calculator, or use long division, you will see that each fraction results in a repeating decimal number.

a) $1 \div 3 = 0.333\ldots$ or $0.\overline{3}$

b) $23 \div 99 = 0.2323\ldots$ or $0.\overline{23}$

c) $2 \div 7 = 0.285714285714\ldots$ or $0.\overline{285714}$ ▲

Note that in each part of Example 3 the quotient has no final digit and continues indefinitely. Each number is a repeating decimal number.

When a fraction is converted to a decimal number, the maximum number of digits that can repeat is $n - 1$, where n is the denominator of the fraction. For example, when $\frac{2}{7}$ is converted to a decimal number, the maximum number of digits that can repeat is $7 - 1$, or 6.

Converting Decimal Numbers to Fractions

We can convert a terminating or repeating decimal number into a quotient of integers. The explanation of the procedure will refer to the positional values to the right of the decimal point, as illustrated here:

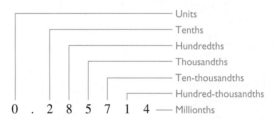

Example 4 demonstrates how to convert from a decimal number to a fraction.

EXAMPLE 4

Convert the following terminating decimal numbers to a quotient of integers.

a) 0.4   **b)** 0.62   **c)** 0.062

Solution: When converting a terminating decimal number to a quotient of integers, we observe the last digit to the right of the decimal point. The position of this digit will indicate the denominator of the quotient of integers.

a) $0.4 = \frac{4}{10}$ because the 4 is in the tenths position.

b) $0.62 = \frac{62}{100}$ because the digit on the right, 2, is in the hundredths position.

c) $0.062 = \frac{62}{1000}$ because the digit on the right, 2, is in the thousandths position.   ▲

Converting a repeating decimal number to a quotient of integers is more difficult. To do so, we must "create" another repeating decimal number with the same repeating digits so that when one repeating decimal number is subtracted

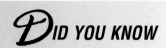

from the other repeating decimal number, the difference will be a whole number. To create a number with the same repeating digits, multiply the original repeating decimal number by 10 if one digit repeats, by 100 if two digits repeat, by 1000 if three digits repeat, and so on. Examples 5–7 demonstrate this procedure.

EXAMPLE 5

Convert $0.\overline{3}$ to a quotient of integers.

Solution: $0.\overline{3} = 0.3\overline{3} = 0.33\overline{3}$, and so on.

Let the original repeating decimal number be n; thus $n = 0.\overline{3}$. Because one digit repeats, we multiply both sides of the equation by 10, which gives $10n = 3.\overline{3}$. Then we subtract.

$$\begin{array}{r} 10n = 3.\overline{3} \\ - \quad n = 0.\overline{3} \\ \hline 9n = 3.0 \end{array}$$

Note that $10n - n = 9n$ and $3.\overline{3} - 0.\overline{3} = 3.0$.

Next, we solve for n by dividing both sides of the equation by 9.

$$\frac{9n}{9} = \frac{3.0}{9}$$

$$n = \frac{3}{9} = \frac{1}{3}$$

Therefore $0.\overline{3} = \frac{1}{3}$.

EXAMPLE 6

Convert $0.\overline{83}$ to a quotient of integers.

Solution: Let $n = 0.\overline{83}$. Since two digits repeat, multiply both sides of the equation by 100. Thus $100n = 83.\overline{83}$. Now we subtract n from $100n$.

$$\begin{array}{r} 100n = 83.\overline{83} \\ - \quad n = 0.\overline{83} \\ \hline 99n = 83 \end{array}$$

Finally, we divide both sides of the equation by 99.

$$\frac{99n}{99} = \frac{83}{99}$$

$$n = \frac{83}{99}$$

Therefore $0.\overline{83} = \frac{83}{99}$.

EXAMPLE 7

Convert $12.14\overline{2}$ to a quotient of integers.

Solution: This problem is different from the two preceding examples in that the repeating digit, 2, is not directly to the right of the decimal point. When this situation arises, move the decimal point to the right until the repeating

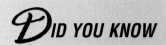

terms are directly to its right. For each place the decimal point is moved, the
number is multiplied by 10. In this example the decimal point must be moved
two places to the right. Thus the number must be multiplied by 100.

$$n = 12.14\overline{2}$$
$$100n = 100 \times 12.14\overline{2} = 1214.\overline{2}$$

Now proceed as in the previous two examples. Since one digit repeats, multiply
both sides by 10.

$$100n = 1214.\overline{2}$$
$$10 \times 100n = 10 \times 1214.\overline{2}$$
$$1000n = 12142.\overline{2}$$

Now subtract $100n$ from $1000n$ so that the repeating part will drop out.

$$\begin{array}{r} 1000n = 12142.\overline{2} \\ -\ 100n = \ \ 1214.\overline{2} \\ \hline 900n = 10928 \end{array}$$

$$n = \frac{10928}{900} = \frac{2732}{225}$$

Therefore $12.14\overline{2} = \dfrac{2732}{225}$. ▲

A set of numbers is said to be a **dense set** if between any two distinct members of the set there exists a third distinct member of the set. The set of integers
is not dense, since between any two consecutive integers, there is not another
integer. For example, between 1 and 2 there are no other integers. The set of
rational numbers is dense because between any two distinct rational numbers there
exists a third distinct rational number. Example 8 illustrates this concept.

EXAMPLE 8

Determine a rational number between the two numbers in each of the following
pairs.

a) 0.243 and 0.244 **b)** 0.0007 and 0.0008

Solution:

a) The number 0.243 can be written as 0.2430, and 0.244 can be written as
0.2440. There are many numbers between these two. Some of them are
0.2431, 0.2435, and 0.243912. How many numbers are there between these
two numbers?

b) 0.0007 can be written as 0.00070, and 0.0008 can be written as 0.00080.
One number between them is 0.00075. ▲

Consider the rational number $0.\overline{9}$. What is $0.\overline{9}$ when expressed as a quotient
of integers? We will use the procedure we used earlier in this section to answer
this question.

Let $n = 0.\overline{9}$, then

$$\begin{array}{r} 10n = 9.\overline{9} \\ -\ \ n = 0.\overline{9} \\ \hline 9n = 9 \end{array}$$

$$n = \frac{9}{9} = 1$$

We have just illustrated that $0.\overline{9} = 1$, a statement that you probably question. However, we can show that $0.\overline{9} = 1$ by using the denseness property. We stated earlier that the rational numbers are dense. Thus if $0.\overline{9}$ and 1 are two different rational numbers, we can find a rational number between the two. If we take the average of $0.\overline{9}$ and 1 by adding them and dividing the sum by 2 the result is

$$\frac{0.\overline{9} + 1}{2} = \frac{1.\overline{9}}{2} = 0.\overline{9}$$

The average of the two numbers, $0.\overline{9}$, is one of the original numbers, so the two numbers must be equal.

Example 8 illustrated a procedure used to find a rational number between any two decimal numbers. We can also find a rational number between any two fractions. One way to do so is to add the fractions and divide the sum by 2. First, we need to explain how to multiply, divide, add, and subtract fractions.

Multiplication and Division of Fractions

The product of two fractions is found by multiplying the numerators together and multiplying the denominators together.

Multiplication of Fractions

$$\frac{a}{b} \cdot \frac{c}{d} = \frac{a \cdot c}{b \cdot d} = \frac{ac}{bd}, \qquad b \neq 0, \quad d \neq 0$$

EXAMPLE 9

Evaluate the following.

a) $\dfrac{2}{3} \cdot \dfrac{5}{7}$ **b)** $\left(\dfrac{-2}{3}\right)\left(\dfrac{-4}{9}\right)$ **c)** $\left(2\dfrac{3}{4}\right)\left(1\dfrac{5}{8}\right)$

Solution:

a) $\dfrac{2}{3} \cdot \dfrac{5}{7} = \dfrac{2 \cdot 5}{3 \cdot 7} = \dfrac{10}{21}$ **b)** $\left(\dfrac{-2}{3}\right)\left(\dfrac{-4}{9}\right) = \dfrac{(-2)(-4)}{(3)(9)} = \dfrac{8}{27}$

c) First, we change each of the mixed numbers to fractions, as follows.

$$2\frac{3}{4} = \frac{(2 \cdot 4) + 3}{4} = \frac{11}{4}$$

$$1\frac{5}{8} = \frac{(1 \cdot 8) + 5}{8} = \frac{13}{8}$$

Now we multiply the fractions together.

$$\left(\frac{11}{4}\right)\left(\frac{13}{8}\right) = \frac{143}{32}$$

Note that $\frac{143}{32}$ may be written as a mixed number by dividing the numerator by the denominator.

$$
32 \overline{\smash{\big)}\, 143} \\
\underline{128} \\
15
$$

Place value: 4

$$
\frac{143}{32} = 4\frac{15}{32} \quad \begin{array}{l} \leftarrow \text{Remainder} \\ \leftarrow \text{Divisor} \end{array}
$$

$$\uparrow$$

Quotient

The **reciprocal** of any number is 1 divided by that number. The product of a number and its reciprocal must equal 1. Examples of some numbers and their reciprocals follow.

Number		Reciprocal		Product
3	$\cdot$	$\dfrac{1}{3}$	$=$	1
$\dfrac{3}{5}$	$\cdot$	$\dfrac{5}{3}$	$=$	1
-6	$\cdot$	$-\dfrac{1}{6}$	$=$	1

To find the quotient of two fractions, multiply the first fraction by the reciprocal of the second fraction.

Division of Fractions

$$
\frac{a}{b} \div \frac{c}{d} = \frac{a}{b} \cdot \frac{d}{c} = \frac{ad}{bc}, \qquad b \neq 0, \quad d \neq 0, \quad c \neq 0
$$

EXAMPLE 10

Evaluate the following.

a) $\dfrac{2}{3} \div \dfrac{5}{7}$ **b)** $\dfrac{-3}{5} \div \dfrac{5}{7}$

Solution:

a) $\dfrac{2}{3} \div \dfrac{5}{7} = \dfrac{2}{3} \cdot \dfrac{7}{5} = \dfrac{2 \cdot 7}{3 \cdot 5} = \dfrac{14}{15}$

b) $\dfrac{-3}{5} \div \dfrac{5}{7} = \dfrac{-3}{5} \cdot \dfrac{7}{5} = \dfrac{-3 \cdot 7}{5 \cdot 5} = \dfrac{-21}{25}$ or $-\dfrac{21}{25}$

Addition and Subtraction of Fractions

Before we can add or subtract fractions, the fractions must have a common denominator. A common denominator is another name for a common multiple. The least common multiple (LCM), introduced in Section 5.1, is the lowest common denominator.

To add or subtract two fractions with a common denominator, we add or subtract their numerators and retain the common denominator.

Addition and Subtraction of Fractions

$$\frac{a}{c} + \frac{b}{c} = \frac{a+b}{c}, \quad c \neq 0 \qquad \frac{a}{c} - \frac{b}{c} = \frac{a-b}{c}, \quad c \neq 0$$

EXAMPLE 11

Evaluate the following.

a) $\dfrac{3}{7} + \dfrac{2}{7}$ **b)** $\dfrac{5}{9} - \dfrac{1}{9}$

Solution:

a) $\dfrac{3}{7} + \dfrac{2}{7} = \dfrac{3+2}{7} = \dfrac{5}{7}$ **b)** $\dfrac{5}{9} - \dfrac{1}{9} = \dfrac{5-1}{9} = \dfrac{4}{9}$ ▲

Note that in Example 11 the denominators of the fractions being added or subtracted were the same; that is, they have a common denominator. *When adding or subtracting two fractions with unlike denominators, first rewrite each fraction with a common denominator. Then add or subtract the fractions.*

Writing fractions with a common denominator is accomplished with the *fundamental law of rational numbers.*

Fundamental Law of Rational Numbers

If a, b, and c are integers, with $b \neq 0$ and $c \neq 0$, then

$$\frac{a}{b} = \frac{a}{b} \cdot \frac{c}{c} = \frac{a \cdot c}{b \cdot c}$$

Examples 12 and 13 illustrate the procedure for adding and subtracting fractions with unlike denominators.

EXAMPLE 12

Evaluate $\dfrac{5}{12} - \dfrac{3}{10}$.

Solution: Using prime factorization (Section 5.1), we find that the LCM of 12 and 10 is 60. We will therefore express each fraction with a denominator of 60. Sixty divided by 12 is 5. Therefore the denominator, 12, must be multiplied by 5 to get 60. If the denominator is multiplied by 5, the numerator must also be multiplied by 5 so that the value of the fraction remains unchanged. Multiplying both numerator and denominator by 5 is the same as multiplying by 1.

We follow the same procedure for the other fraction, $\frac{3}{10}$. Sixty divided by 10 is 6. Therefore we multiply both the denominator, 10, and the numerator, 3, by 6 to obtain an equivalent fraction with a denominator of 60.

$$\frac{5}{12} - \frac{3}{10} = \left(\frac{5}{12} \cdot \frac{5}{5}\right) - \left(\frac{3}{10} \cdot \frac{6}{6}\right)$$

$$= \frac{25}{60} - \frac{18}{60}$$

$$= \frac{7}{60}$$

▲

EXAMPLE 13

Add $\dfrac{1}{36} + \dfrac{1}{150}$.

Solution: In Example 7 of Section 5.1 we determined that the LCM of 36 and 150 is 900. We rewrite both fractions using the LCM in the denominator.

$$\frac{1}{36} + \frac{1}{150} = \left(\frac{1}{36} \cdot \frac{25}{25}\right) + \left(\frac{1}{150} \cdot \frac{6}{6}\right)$$

$$= \frac{25}{900} + \frac{6}{900}$$

$$= \frac{31}{900}$$

▲

Now that you know how to add and divide fractions, you can find a fraction between any two given fractions by adding those two fractions and dividing their sum by 2.

EXAMPLE 14

Find a rational number halfway between $\frac{1}{4}$ and $\frac{1}{3}$.

Solution: First, add $\frac{1}{4}$ and $\frac{1}{3}$.

$$\frac{1}{4} + \frac{1}{3} = \frac{3}{12} + \frac{4}{12} = \frac{7}{12}$$

Next, divide the sum by 2.

$$\frac{7}{12} \div 2 = \frac{7}{12} \div \frac{2}{1} = \frac{7}{12} \cdot \frac{1}{2} = \frac{7}{24}$$

Thus $\frac{7}{24}$ is between $\frac{1}{4}$ and $\frac{1}{3}$, written

$$\frac{1}{4} < \frac{7}{24} < \frac{1}{3}.$$

▲

The procedure used in Example 14 will result in a fraction that is exactly halfway between (or the average of) the two given fractions. The result in Example 14 is more obvious if you change all the denominators to 24; that is,

$$\frac{6}{24} < \frac{7}{24} < \frac{8}{24}.$$

EXAMPLE 15

Following are the instructions given on a box of Minute Rice. Determine the amount of (a) rice and water, (b) salt, and (c) butter or margarine needed to make 3 servings of rice.

DIRECTIONS

1. Bring water, salt, and butter (or margarine) to a boil.
2. Stir in rice. Cover: remove from heat. Let stand 5 minutes. Fluff with fork.

TO MAKE	RICE & WATER (equal measures)	SALT	BUTTER OR MARGARINE (if desired)
2 servings	$\frac{2}{3}$ cup	$\frac{1}{4}$ tsp	1 tsp
4 servings	$1\frac{1}{3}$ cups	$\frac{1}{2}$ tsp	2 tsp

Solution: Since 3 is halfway between 2 and 4, we can find the amount of each ingredient by finding the average of the amount for 2 and 4 servings. To do so, we add the amounts for 2 servings and 4 servings and divide the sum by 2.

a) Rice and water: $\dfrac{\frac{2}{3} + 1\frac{1}{3}}{2} = \dfrac{\frac{2}{3} + \frac{4}{3}}{2} = \dfrac{\frac{6}{3}}{2} = \dfrac{2}{2} = 1$ cup.

b) Salt: $\dfrac{\frac{1}{4} + \frac{1}{2}}{2} = \dfrac{\frac{1}{4} + \frac{2}{4}}{2} = \dfrac{\frac{3}{4}}{2} = \dfrac{3}{4} \cdot \dfrac{1}{2} = \frac{3}{8}$ tsp.

c) Butter or margarine: $\dfrac{1 + 2}{2} = \dfrac{3}{2},$ or $1\frac{1}{2}$ tsp.

The solution to Example 15 can be found in other ways. Suggest two other procedures for solving the same problem.

Section 5.3 Exercises

1. Describe the set of rational numbers.

2. **a)** Explain how to write a terminating decimal number as a fraction.
   b) Write 0.013 as a fraction.

3. What does it mean when a set of numbers is dense?

4. **a)** Explain how to reduce a fraction to lowest terms.
   b) Reduce $\frac{12}{45}$ to lowest terms by using the procedure in part (a).

5. **a)** Explain how to multiply two fractions.
   b) Multiply $\frac{24}{15} \cdot \frac{7}{36}$ by using the procedure in part (a).

6. **a)** Explain how to determine the reciprocal of a number.
   b) Using the procedure in part (a), determine the reciprocal of -5.

7. **a)** Explain how to divide two fractions.
   b) Divide $\frac{12}{19} \div \frac{3}{8}$ by using the procedure in part (a).

8. **a)** Explain how to add or subtract two fractions having a common denominator.
   b) Add $\frac{13}{27} + \frac{8}{27}$ by using the procedure in part (a).

9. **a)** Explain how to add or subtract two fractions having unlike denominators.
   b) Using the procedure in part (a), add $\frac{5}{12} + \frac{4}{9}$.

10. In your own words state the fundamental law of rational numbers.

In Exercises 11–20, determine whether each number is a terminating or repeating decimal number.

11. 9

12. $24.\overline{8}$

13. $1.\overline{03}$

14. 0.788788

15. $0.9\overline{2}$

16. 0.7777

17. 16.52185218 . . .

18. 4.14

19. 3.132132132

20. 0.999 . . .

In Exercises 21–30, express each rational number as a terminating or repeating decimal number.

21. $\dfrac{3}{4}$ **22.** $\dfrac{2}{7}$ **23.** $\dfrac{5}{7}$

24. $\dfrac{1}{6}$ **25.** $\dfrac{5}{8}$ **26.** $\dfrac{22}{7}$

27. $\dfrac{17}{3}$ **28.** $\dfrac{70}{15}$ **29.** $\dfrac{80}{15}$

30. $\dfrac{1001}{11}$

In Exercises 31–40, express each terminating decimal number as a quotient of two integers.

31. 0.4   **32.** 0.84

33. 0.076   **34.** 0.9075

35. 6.001   **36.** 3.625

37. 1.246   **38.** 0.7913

39. 2.0678   **40.** 1.0025

In Exercises 41–50, express each repeating decimal number as a quotient of two integers.

41. $0.\overline{6}$ **42.** $0.\overline{2}$ **43.** $1.\overline{9}$

44. $0.\overline{48}$ **45.** $1.\overline{72}$ **46.** $0.\overline{357}$

47. $3.\overline{258}$ **48.** $1.6\overline{3}$ **49.** $1.2\overline{1}$

50. $4.3\overline{73}$

In Exercises 51–60, evaluate each expression.

51. $\dfrac{2}{7} \div \dfrac{5}{3}$ **52.** $\dfrac{5}{7} \cdot \dfrac{6}{11}$

53. $\left(\dfrac{-3}{8}\right)\left(\dfrac{-16}{15}\right)$ **54.** $\left(-\dfrac{3}{5}\right) \div \dfrac{10}{21}$

55. $\dfrac{5}{6} \div \dfrac{6}{5}$ **56.** $\dfrac{7}{9} \div \dfrac{7}{9}$

57. $\left(\dfrac{3}{5} \cdot \dfrac{4}{7}\right) \div \dfrac{1}{3}$ **58.** $\left(\dfrac{4}{7} \div \dfrac{4}{5}\right) \cdot \dfrac{1}{7}$

59. $\left[\left(\dfrac{-3}{4}\right)\left(\dfrac{-2}{7}\right)\right] \div \dfrac{3}{5}$ **60.** $\left(\dfrac{3}{8} \cdot \dfrac{5}{9}\right) \cdot \left(\dfrac{4}{7} \div \dfrac{5}{8}\right)$

In Exercises 61–70, reduce each fraction to lowest terms.

61. $\dfrac{24}{36}$ **62.** $\dfrac{42}{54}$ **63.** $\dfrac{48}{92}$

64. $\dfrac{160}{240}$ **65.** $\dfrac{325}{675}$ **66.** $\dfrac{11}{1001}$

67. $\dfrac{96}{212}$ **68.** $\dfrac{65}{140}$ **69.** $\dfrac{44}{121}$

70. $\dfrac{112}{196}$

In Exercises 71–80, perform the indicated operation and reduce your answer to lowest terms.

71. $\dfrac{1}{7} + \dfrac{1}{8}$ **72.** $\dfrac{1}{7} - \dfrac{4}{21}$ **73.** $\dfrac{2}{13} + \dfrac{3}{130}$

74. $\dfrac{5}{24} + \dfrac{7}{72}$ **75.** $\dfrac{5}{9} - \dfrac{7}{54}$ **76.** $\dfrac{13}{30} - \dfrac{17}{120}$

77. $\dfrac{1}{12} + \dfrac{1}{48} + \dfrac{1}{72}$ **78.** $\dfrac{3}{5} + \dfrac{7}{15} + \dfrac{9}{75}$

79. $\dfrac{1}{30} - \dfrac{3}{40} - \dfrac{7}{50}$ **80.** $\dfrac{4}{25} - \dfrac{9}{100} - \dfrac{7}{40}$

Alternative methods for adding and subtracting two fractions are shown. These methods may not result in a solution in its lowest terms.

$$\frac{a}{b} + \frac{c}{d} = \frac{ad + bc}{bd} \quad \text{and} \quad \frac{a}{b} - \frac{c}{d} = \frac{ad - bc}{bd}$$

In Exercises 81–86, use one of the two formulas to evaluate the expression.

81. $\dfrac{2}{3} + \dfrac{3}{4}$ **82.** $\dfrac{5}{7} - \dfrac{1}{12}$ **83.** $\dfrac{5}{7} + \dfrac{3}{4}$

84. $\dfrac{7}{3} - \dfrac{5}{12}$ **85.** $\dfrac{3}{8} + \dfrac{5}{12}$ **86.** $\left(\dfrac{2}{3} + \dfrac{1}{4}\right) - \dfrac{3}{5}$

In Exercises 87–92, evaluate each expression.

87. $\left(\dfrac{1}{5} \cdot \dfrac{1}{4}\right) + \dfrac{1}{3}$ **88.** $\left(\dfrac{3}{5} \div \dfrac{2}{10}\right) - \dfrac{1}{3}$

89. $\left(\dfrac{1}{2} + \dfrac{3}{10}\right) \div \left(\dfrac{1}{5} + 2\right)$ **90.** $\left(\dfrac{1}{9} \cdot \dfrac{3}{5}\right) + \left(\dfrac{2}{3} \cdot \dfrac{1}{5}\right)$

91. $\left(3 - \dfrac{4}{9}\right) \div \left(4 + \dfrac{2}{3}\right)$ **92.** $\left(\dfrac{2}{5} \div \dfrac{4}{9}\right)\left(\dfrac{3}{5} \cdot 6\right)$

In Exercises 93–100, find a rational number between the two numbers in each pair.

93. 0.25 and 0.26   **94.** 4.3 and 4.003

95. −2.176 and −2.175   **96.** 1.3457 and 1.34571

97. 3.12345 and 3.123451   **98.** 0.4105 and 0.4106

99. 4.872 and 4.873   **100.** −3.7896 and −3.7895

In Exercises 101–108, find a rational number between the two fractions in each pair.

101. $\dfrac{3}{5}$ and $\dfrac{4}{5}$ **102.** $\dfrac{1}{9}$ and $\dfrac{2}{9}$ **103.** $\dfrac{1}{20}$ and $\dfrac{1}{10}$

104. $\dfrac{7}{13}$ and $\dfrac{8}{13}$ **105.** $\dfrac{1}{4}$ and $\dfrac{1}{5}$ **106.** $\dfrac{1}{3}$ and $\dfrac{2}{3}$

107. $\dfrac{1}{10}$ and $\dfrac{1}{100}$ **108.** $\dfrac{1}{2}$ and $\dfrac{2}{3}$

109. Are the rational numbers dense? Explain.

110. Are the integers dense? Explain.

111. If a person from another planet visited you and asked what Earthlings mean by the symbol $\frac{3}{8}$, how would you explain it?

112. Mr. and Mrs. Ryckman have been increasing their children's weekly allowances each year by $\frac{1}{5}$ over the previous year's weekly allowance.
    a) If Jill's current weekly allowance is $9.00, what will it be next year?
    b) Diana's current weekly allowance is $10.00. What will her weekly allowance be in 2 years?

113. A sociology class is $\frac{2}{5}$ math majors, $\frac{1}{4}$ English majors, and $\frac{1}{10}$ chemistry majors; the remaining members of the class are art majors. What fraction of the class are art majors?

114. Wu, who lives in California, spends $\frac{11}{30}$ of her income on housing, $\frac{2}{5}$ on food and clothing, and $\frac{1}{10}$ on entertainment. If she saves the rest, what fraction of her income does she save?

115. A stairway consists of 12 stairs, each $10\frac{3}{4}$ in. high. What is the vertical height of the stairway?

116. Mary Anne gives Sam the tailor $8\frac{3}{4}$ yd of fabric. She asks him to make a skirt, a pair of slacks, and a jacket. These three items require $1\frac{7}{8}$ yd, $2\frac{5}{8}$ yd, and $2\frac{3}{8}$ yd, respectively, of fabric. How many yards of fabric are left over from the original piece?

117. If a plumber cuts three $2\frac{5}{16}$-ft lengths of pipe from an 8-ft section, how much remains?

118. In one week Disney stock rose from $53\frac{3}{8}$ to $58\frac{3}{4}$. By how many points had the stock increased?

119. A recipe for six servings of stew calls for $1\frac{1}{4}$ tsp of worcestershire sauce. How much worcestershire sauce should be used for eight servings?

120. Last year Mark's height was $46\frac{3}{4}$ in. When measured this year his height increased by $3\frac{5}{16}$ in. Find Mark's present height.

121. The instructions for assembling a computer stand include a diagram illustrating its dimensions. Find the total height of the stand.

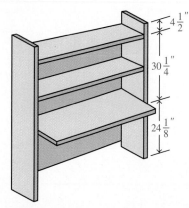

122. The width of a picture is $24\frac{7}{8}$ in., as shown in the diagram above and to the right. Find x, the distance from the edge of the frame to the center.

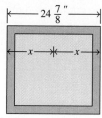

123. A piece of wood measures $15\frac{3}{8}$ in.
    a) How far from one end should you cut the wood if you want to cut the length in half?
    b) What is the length of each piece after the cut? You must allow $\frac{1}{8}$ in. for the saw cut.

124. Rafela wants to place $\frac{1}{2}$-in. molding along the floor around the perimeter of her room (excluding door openings). She finds that she needs lengths of $26\frac{1}{2}$ in., $105\frac{1}{4}$ in., $53\frac{1}{4}$ in., and $106\frac{5}{16}$ in. How much molding will she need?

Problem Solving/Group Activities

In Exercises 125–128, determine the missing number in the pair on the right. For each exercise, the second number in the pair may be found by performing one operation (either addition, subtraction, multiplication, or division) on the first number in the pair with the same rational number. For example consider the pairs.

$$\left(5, \frac{27}{5}\right), \left(\frac{3}{10}, \frac{7}{10}\right), \left(-1, -\frac{3}{5}\right), (4, \_)$$

The second number of each pair of obtained by adding $\frac{2}{5}$ to the first number:

$$5 + \frac{2}{5} = \frac{27}{5}, \quad \frac{3}{10} + \frac{2}{5} = \frac{7}{10}, \quad -1 + \frac{2}{5} = -\frac{3}{5}.$$

Therefore the blank is to be filled in with $4 + \frac{2}{5} = \frac{20}{5} + \frac{2}{5}$ or $\frac{22}{5}$.

125. $(4, 2), \left(\frac{1}{2}, -\frac{3}{2}\right), \left(\frac{25}{3}, \frac{19}{3}\right), \left(\frac{5}{9}, \_\right)$

126. $\left(\frac{1}{3}, \frac{8}{3}\right), \left(12, \frac{43}{3}\right), \left(-\frac{15}{6}, -\frac{1}{6}\right), (4, \_)$

127. $(4, 6), \left(15, \frac{45}{2}\right), \left(-\frac{16}{3}, -8\right), \left(\frac{5}{9}, \_\right)$

128. $(12, -3), \left(\frac{15}{2}, -\frac{15}{8}\right), \left(-\frac{1}{2}, \frac{1}{8}\right), (-16, \_)$

129. If a piece of wood $8\frac{3}{4}$ ft long is to be cut into four equal pieces, find the length of each piece. (Allow $\frac{1}{4}$ in. for each saw cut.)

130. The dimensions of the cover of a book have been increased from $8\frac{1}{2}$ in. by $9\frac{1}{4}$ in. to $8\frac{1}{2}$ in. by $10\frac{1}{4}$ in. By

how many square inches has the surface area increased? Use Area = length × width.

131. A rectangular room measures 8 ft 3 in. by 10 ft 8 in. by 9 ft 2 in. high.
   a) Determine the perimeter of the room in feet.
   b) Calculate the area of the floor of the room in square feet.
   c) Calculate the volume of the room in cubic feet.

132. The back of a framed picture that is to be hung is shown.

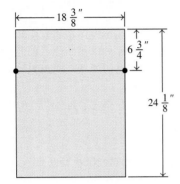

A nail is to be hammered into the wall, and the picture will be hung by the wire on the nail.
   a) If the center of the wire is to rest on the nail and a side of the picture is to be 20 in. from the window, how far from the window should the nail be placed?
   b) If the top of the frame is to be $26\frac{1}{4}$ in. from the ceiling, how far from the ceiling should the nail be placed? (Assume that the wire will not stretch.)
   c) Repeat part (b) if the wire will stretch $\frac{1}{4}$ in. when the picture is hung.

Research Activity

133. The ancient Greeks are often considered the first true mathematicians. Write a report summarizing the ancient Greeks' contributions to rational numbers. Include in your report what they learned and believed about the rational numbers. References include encyclopedias and history of mathematics books.

5.4 THE IRRATIONAL NUMBERS AND THE REAL NUMBER SYSTEM

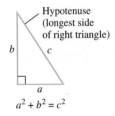

Figure 5.6

Pythagoras (ca. 585–500 B.C.), a Greek mathematician, is credited with providing a written proof that in any *right triangle* (a triangle with a 90° angle, see Fig. 5.6) the square of the length of one side (a^2) added to the square of the length of the other side (b^2) equals the square of the length of the hypotenuse (c^2). The formula $a^2 + b^2 = c^2$ is now known as the Pythagorean theorem.* Pythagoras found that the solution of the formula, where $a = 1$ and $b = 1$, is not a rational number.

$$a^2 + b^2 = c^2$$
$$1^2 + 1^2 = c^2$$
$$1 + 1 = c^2$$
$$2 = c^2$$

There is no rational number that when squared will equal 2. This prompted a need for a new set of numbers, the irrational numbers.

In Section 5.2 we introduced the real number line. The points on the real number line that are not rational numbers are referred to as irrational numbers. Recall that every rational number is either a terminating or a repeating decimal number. Therefore irrational numbers, when represented as decimal numbers, will be nonterminating, nonrepeating decimal numbers.

*The Pythagorean theorem is discussed in more detail in Section 9.3.

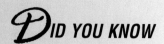

An **irrational number** is a real number whose decimal representation is a nonterminating, nonrepeating decimal number.

Nonrepeating number patterns can be used to indicate irrational numbers. For example, 6.1011011101111 . . . and 0.525225222 . . . are both irrational numbers.

The expression $\sqrt{2}$ is read "the square root of 2" or "radical 2." The symbol $\sqrt{}$ is called the radical sign and the number or expression inside the radical sign is called the radicand. In $\sqrt{2}$, 2 is the radicand.

The square roots of some numbers are rational while the square roots of other numbers are irrational. The principal (or positive) square root of a number n, written $\sqrt{n}$, is the positive number that when multiplied by itself gives n. Whenever we mention the term "square root" in this text, we mean the principal square root. For example,

$$\sqrt{25} = 5, \quad \text{since} \quad 5 \cdot 5 = 25;$$

$$\sqrt{100} = 10, \quad \text{since} \quad 10 \cdot 10 = 100.$$

Both $\sqrt{25}$ and $\sqrt{100}$ are examples of numbers that are rational numbers because their square roots, 5 and 10, are terminating decimal numbers.

Returning to the problem faced by Pythagoras, if $c^2 = 2$ then c has a value of $\sqrt{2}$. But what is $\sqrt{2}$ equal to? The $\sqrt{2}$ is an irrational number and it cannot be expressed as a terminating or repeating decimal number. It can only be approximated by a decimal number: $\sqrt{2}$ is approximately 1.4142135 (to seven decimal places).

Other irrational numbers include $\sqrt{3}$, $\sqrt{5}$, and $\sqrt{37}$. Another important irrational number used to represent the ratio of a circle's circumference to its diameter is pi, symbolized π. Pi is approximately 3.1415926.

We have discussed procedures for performing the arithmetic operations of addition, subtraction, multiplication, and division with rational numbers. We can perform the same operations with the irrational numbers. Before we can proceed, however, we must understand the numbers called perfect squares. Any number that is the square of a natural number is said to be a perfect square.

Natural numbers	1,	2,	3,	4,	5,	6, . . .
Squares of the natural numbers	1^2,	2^2,	3^3,	4^2,	5^2,	6^2, . . .
or Perfect squares	1,	4,	9,	16,	25,	36, . . .

The numbers 1, 4, 9, 16, 25, and 36 are some of the perfect square numbers. Can you determine the next two perfect square numbers? How many perfect square numbers are there? The square root of a perfect square number will be a natural number. For example, $\sqrt{1} = 1$, $\sqrt{4} = 2$, $\sqrt{9} = 3$, $\sqrt{16} = 4$, $\sqrt{25} = 5$, and so on.

The number that multiplies a radical is called the radical's coefficient. For example in $3\sqrt{5}$ the 3 is the coefficient of the radical.

Some irrational numbers can be simplified by determining whether there are any perfect square factors in the radicand. If there are, the following rule can be used to simplify the radical.

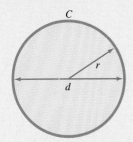
Product Rule for Radicals
$$\sqrt{a \cdot b} = \sqrt{a} \cdot \sqrt{b}, \qquad a \geq 0, \quad b \geq 0$$

To simplify a radical, write the radical as a product of two radicals. One of the radicals should contain the greatest perfect square that is a factor of the radicand in the original expression. Then simplify the radical containing the perfect square factor.

For example,
$$\sqrt{8} = \sqrt{4 \cdot 2} = \sqrt{4} \cdot \sqrt{2} = 2\sqrt{2}$$

and
$$\sqrt{50} = \sqrt{25 \cdot 2} = \sqrt{25} \cdot \sqrt{2} = 5\sqrt{2}.$$

EXAMPLE 1

Simplify   **a)** $\sqrt{45}$ **b)** $\sqrt{48}$

Solution:

a) Since 9 is a perfect square factor of 45, we write
$$\sqrt{45} = \sqrt{9 \cdot 5} = \sqrt{9} \cdot \sqrt{5} = 3\sqrt{5}.$$

Since 5 has no perfect square factors, $\sqrt{5}$ cannot be simplified.

b) Since 16 is a perfect square factor of 48 write
$$\sqrt{48} = \sqrt{16 \cdot 3} = \sqrt{16} \cdot \sqrt{3} = 4\sqrt{3}.$$

In Example 1(b) you can obtain the correct answer if you start out factoring differently:
$$\sqrt{48} = \sqrt{4 \cdot 12} = \sqrt{4} \cdot \sqrt{12} = 2\sqrt{12}.$$

Note that 12 has 4 as a perfect square factor.
$$2\sqrt{12} = 2\sqrt{4 \cdot 3} = 2\sqrt{4} \cdot \sqrt{3}$$
$$= 2 \cdot 2\sqrt{3}$$
$$= 4\sqrt{3}$$

You have to do a little more work, but you get the same answer.

Addition and Subtraction of Irrational Numbers

To add or subtract two or more square roots with the same radicand, add or subtract their coefficients. The answer is the sum or difference of the coefficients multiplied by the common radical.

EXAMPLE 2

Simplify   **a)** $2\sqrt{7} + 3\sqrt{7}$ **b)** $3\sqrt{5} - 2\sqrt{5} + \sqrt{5}$

Solution:

a) $2\sqrt{7} + 3\sqrt{7} = (2 + 3)\sqrt{7} = 5\sqrt{7}$

b) $3\sqrt{5} - 2\sqrt{5} + \sqrt{5} = (3 - 2 + 1)\sqrt{5} = 2\sqrt{5}$

(Note that $\sqrt{5} = 1\sqrt{5}$.)

EXAMPLE 3

Simplify $5\sqrt{3} - \sqrt{12}$.

Solution: These radicals cannot be added in their present form because they contain different radicands. When this occurs, determine whether one or more of the radicals can be simplified so that they have the same radicand.

$$5\sqrt{3} - \sqrt{12} = 5\sqrt{3} - \sqrt{4 \cdot 3}$$
$$= 5\sqrt{3} - \sqrt{4} \cdot \sqrt{3}$$
$$= 5\sqrt{3} - 2\sqrt{3}$$
$$= (5 - 2)\sqrt{3} = 3\sqrt{3}$$

Multiplication of Irrational Numbers

When multiplying irrational numbers, we again make use of the product rule for radicals. After the radicands are multiplied, simplify the remaining radical when possible.

EXAMPLE 4

Simplify   **a)** $\sqrt{2} \cdot \sqrt{8}$ **b)** $\sqrt{3} \cdot \sqrt{5}$ **c)** $\sqrt{8} \cdot \sqrt{3}$

Solution:

a) $\sqrt{2} \cdot \sqrt{8} = \sqrt{2 \cdot 8} = \sqrt{16} = 4$

b) $\sqrt{3} \cdot \sqrt{5} = \sqrt{3 \cdot 5} = \sqrt{15}$

c) $\sqrt{8} \cdot \sqrt{3} = \sqrt{8 \cdot 3} = \sqrt{24} = \sqrt{4} \cdot \sqrt{6} = 2\sqrt{6}$

Division of Irrational Numbers

To divide irrational numbers, use the following rule. After performing the division, simplify when possible.

Quotient Rule for Radicals

$$\frac{\sqrt{a}}{\sqrt{b}} = \sqrt{\frac{a}{b}}, \qquad a \geq 0, \quad b > 0$$

EXAMPLE 5

Divide   **a)** $\dfrac{\sqrt{8}}{\sqrt{2}}$ **b)** $\dfrac{\sqrt{96}}{\sqrt{2}}$

Solution:

a) $\dfrac{\sqrt{8}}{\sqrt{2}} = \sqrt{\dfrac{8}{2}} = \sqrt{4} = 2$

b) $\dfrac{\sqrt{96}}{\sqrt{2}} = \sqrt{\dfrac{96}{2}} = \sqrt{48} = \sqrt{16 \cdot 3} = \sqrt{16} \cdot \sqrt{3} = 4\sqrt{3}$

Rationalizing the Denominator

A denominator is *rationalized* when it contains no radical expressions. To rationalize a denominator that contains only a square root, multiply both the numerator and denominator of the fraction by a number that will result in the radicand in the denominator becoming a perfect square. (This action is the equivalent of multiplying the fraction by 1 because the value of the fraction does not change.) Then simplify the fractions when possible.

EXAMPLE 6

Rationalize the denominator of

a) $\dfrac{3}{\sqrt{2}}$ **b)** $\dfrac{5}{\sqrt{8}}$ **c)** $\dfrac{\sqrt{3}}{\sqrt{6}}$

Solution:

a) Multiply the numerator and denominator by a number that will make the radicand a perfect square.

$$\frac{3}{\sqrt{2}} = \frac{3}{\sqrt{2}} \cdot \frac{\sqrt{2}}{\sqrt{2}} = \frac{3\sqrt{2}}{\sqrt{4}} = \frac{3\sqrt{2}}{2}$$

Note that the 2s cannot be divided out because one of the 2s is a radicand and the other is not.

b) $\dfrac{5}{\sqrt{8}} = \dfrac{5}{\sqrt{8}} \cdot \dfrac{\sqrt{2}}{\sqrt{2}} = \dfrac{5\sqrt{2}}{\sqrt{16}} = \dfrac{5\sqrt{2}}{4}$

You could have obtained the same answer to this problem by multiplying both the numerator and denominator by $\sqrt{8}$, and then simplifying. Try to do so now.

c) Write $\dfrac{\sqrt{3}}{\sqrt{6}}$ as $\sqrt{\dfrac{3}{6}}$, and reduce the fraction to obtain $\sqrt{\dfrac{1}{2}}$. By the quotient rule for radicals, $\sqrt{\dfrac{1}{2}} = \dfrac{\sqrt{1}}{\sqrt{2}}$ or $\dfrac{1}{\sqrt{2}}$. Now rationalize $\dfrac{1}{\sqrt{2}}$.

$$\frac{1}{\sqrt{2}} = \frac{1}{\sqrt{2}} \cdot \frac{\sqrt{2}}{\sqrt{2}} = \frac{\sqrt{2}}{2}$$

Section 5.4 Exercises

1. Explain the difference between a rational number and an irrational number.

2. What is the principal square root of a number?

3. What is a perfect square?

4. In your own words state the Product Rule for Radicals.

5. **a)** Explain how to add square roots that have the same radicand.

   b) Using the procedure in part (a), add $7\sqrt{3} - 2\sqrt{3} + 3\sqrt{3}$.

6. In your own words state the Quotient Rule for Radicals.

7. What does it mean to rationalize a denominator?

8. a) Explain how to rationalize a denominator that contains a square root.

b) Using the procedure in part (a), rationalize $\dfrac{3}{\sqrt{5}}$.

In Exercises 9–18, determine whether the number is rational or irrational.

9. $\sqrt{25}$ **10.** $\sqrt{6}$ **11.** $\dfrac{3}{5}$

12. 0.303003000 . . .   **13.** 8.323223222 . . .

14. π **15.** $\dfrac{22}{7}$ **16.** 3.14159

17. 3.14159 . . .   **18.** $\dfrac{\sqrt{7}}{\sqrt{7}}$

In Exercises 19–28, evaluate the expression.

19. $\sqrt{81}$ **20.** $\sqrt{121}$ **21.** $\sqrt{49}$

22. $-\sqrt{144}$ **23.** $-\sqrt{169}$ **24.** $\sqrt{25}$

25. $-\sqrt{225}$ **26.** $-\sqrt{36}$ **27.** $-\sqrt{100}$

28. $\sqrt{256}$

In Exercises 29–38, classify the number as a member of one or more of the following sets: the rational numbers, the integers, the natural numbers, the irrational numbers.

29. 6   **30.** -7

31. $\sqrt{16}$ **32.** $\dfrac{3}{7}$

33. 0.131131113 . . .   **34.** 1.732

35. $-\dfrac{5}{9}$ **36.** 0.345345345 . . .

37. $0.345345\overline{345}$ **38.** 0.345334533345 . . .

In Exercises 39–48, simplify the radical.

39. $\sqrt{64}$ **40.** $\sqrt{20}$ **41.** $\sqrt{45}$

42. $\sqrt{40}$ **43.** $\sqrt{125}$ **44.** $\sqrt{28}$

45. $\sqrt{48}$ **46.** $\sqrt{50}$ **47.** $\sqrt{54}$

48. $\sqrt{98}$

In Exercises 49–58, perform the indicated operation.

49. $2\sqrt{5} + 3\sqrt{5}$ **50.** $\sqrt{7} + 5\sqrt{7}$

51. $3\sqrt{5} - 7\sqrt{5}$ **52.** $5\sqrt{3} + 4\sqrt{12}$

53. $4\sqrt{12} - 7\sqrt{27}$ **54.** $2\sqrt{7} + 5\sqrt{28}$

55. $5\sqrt{3} + 7\sqrt{12} - 3\sqrt{75}$

56. $13\sqrt{2} + 2\sqrt{18} - 5\sqrt{32}$

57. $\sqrt{8} - 3\sqrt{50} + 9\sqrt{32}$

58. $\sqrt{63} + 13\sqrt{98} - 5\sqrt{112}$

In Exercises 59–68, perform the indicated operation. Simplify the answer when possible.

59. $\sqrt{5}\sqrt{2}$ **60.** $\sqrt{8}\sqrt{6}$ **61.** $\sqrt{4}\sqrt{10}$

62. $\sqrt{5}\sqrt{10}$ **63.** $\sqrt{6}\sqrt{12}$ **64.** $\sqrt{13}\sqrt{26}$

65. $\dfrac{\sqrt{8}}{\sqrt{4}}$ **66.** $\dfrac{\sqrt{125}}{\sqrt{5}}$ **67.** $\dfrac{\sqrt{72}}{\sqrt{8}}$

68. $\dfrac{\sqrt{136}}{\sqrt{8}}$

In Exercises 69–78, rationalize the denominator.

69. $\dfrac{2}{\sqrt{3}}$ **70.** $\dfrac{6}{\sqrt{5}}$ **71.** $\dfrac{\sqrt{7}}{\sqrt{2}}$

72. $\dfrac{\sqrt{7}}{\sqrt{6}}$ **73.** $\dfrac{\sqrt{12}}{\sqrt{3}}$ **74.** $\dfrac{\sqrt{32}}{\sqrt{6}}$

75. $\dfrac{\sqrt{9}}{\sqrt{2}}$ **76.** $\dfrac{\sqrt{15}}{\sqrt{3}}$ **77.** $\dfrac{\sqrt{10}}{\sqrt{6}}$

78. $\dfrac{8}{\sqrt{8}}$

In Exercises 79–84, determine whether the statement is true or false.

79. $\sqrt{p}$ is irrational for any prime number p.

80. The sum of any two irrational numbers is a rational number.

81. The sum of any two irrational numbers is an irrational number.

82. The product of any two rational numbers is always a rational number.

83. The product of an irrational and a rational number is always an irrational number.

84. The product of any two irrational numbers is always an irrational number.

In Exercises 85–88, give an example to show that the stated case can occur.

85. The sum of two irrational numbers may be an irrational number.

86. The sum of two irrational numbers may be a rational number.

87. The product of two irrational numbers may be an irrational number.

88. The product of two irrational numbers may be a rational number.

89. Without doing any calculations, determine whether $\sqrt{3} = 1.732$. Explain your answer.

90. Without doing any calculations, determine whether $\sqrt{11} = 3.3166$. Explain.

91. Give an example to show that $\sqrt{a + b} \neq \sqrt{a} + \sqrt{b}$.

92. The number π is an irrational number. Often the values 3.14 or $\frac{22}{7}$ are used for π. Does π equal either 3.14 or $\frac{22}{7}$? Explain your answer.

93. A carpenter constructing a rectangular door 3 ft 6 in. by 6 ft 6 in. places a diagonal brace from the upper left corner to the lower right corner. What is the length of the brace? (*Hint:* Use the Pythagorean theorem.)

94. Students are decorating a rectangular dorm room that measures 12 ft by 18 ft. They want to hang a streamer diagonally across the ceiling of the room from one corner to the other. Determine the length of the streamer needed. Assume that there is no slack in the streamer. (*Hint:* Use the Pythagorean theorem.)

95. The time T required for a pendulum to swing back and forth may be found by the formula

$$T = 2\pi \sqrt{\frac{l}{g}},$$

where l is the length of the pendulum and g is the acceleration of gravity. Find the time in seconds if $l = 35$ cm and $g = 980$ cm/sec².

Problem Solving/Group Activities

96. a) If a radical expression is evaluated on a calculator, explain how you can determine whether the expression is a rational or irrational number.

b) Is $\sqrt{0.04}$ rational or irrational? Explain.

c) Is $\sqrt{0.7}$ rational or irrational? Explain.

97. One way to find a rational number between two distinct rational numbers is to add the two distinct rational numbers and divide by 2. Do you think that this method will work for finding an irrational number between two distinct irrational numbers? Explain.

98. a) Using the Pythagorean theorem, $a^2 + b^2 = c^2$, construct a right triangle and then construct a square on each side, as illustrated in Fig. 5.7. Compute the area of the square with side a, the area of the square with side b, and the area of the square with side c. What is the relationship among these areas?

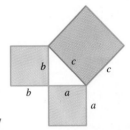

Figure 5.7

b) Construct a right triangle. Construct a semicircle on each side, with the side as the diameter of the circle. Compute the area of the three semicircles. What is the relationship between these areas?

c) Using parts (a) and (b) conjecture a rule regarding the areas of similar figures constructed on the sides of a right triangle.

Research Activities

In Exercises 99 and 100, references include history of mathematics books and encyclopedias.

99. Do the necessary research and write a two-page report on the history of the development of the irrational numbers.

100. Do the necessary research and write a two-page report on the history of pi. In your report indicate when the symbol π was first used and list the first 10 digits of π.

5.5   REAL NUMBERS AND THEIR PROPERTIES

Now that we have discussed both the rational and irrational numbers, we can discuss the real numbers and the properties of the real number system. The union of the rational numbers and the irrational numbers is the set of real numbers, symbolized by $\mathbb{R}$.

Figure 5.8 on page 228 illustrates the relationship among various sets of numbers. It shows that the natural numbers are a subset of the whole numbers, the integers, the rational numbers, and the real numbers. For example, since the number 3 is a natural or counting number, it is also a whole number, an integer, a rational number, and a real number. Since the rational number $\frac{1}{3}$ is outside the set of integers, it is not an integer, a whole number, or a natural number. The number $\frac{1}{3}$ is a real number, however, as is the irrational number $\sqrt{2}$. Note that the real numbers are the union of the rational numbers and the irrational numbers.

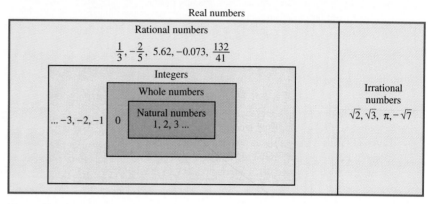

Figure 5.8

The relationship between the various sets of numbers in the real number system can also be illustrated with a tree diagram, as in Fig. 5.9.

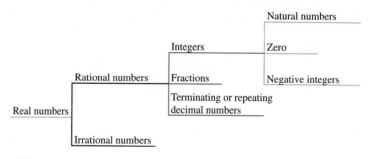

Figure 5.9

Figure 5.9 shows that, for example, the natural numbers are a subset of the integers, the rational numbers, and the real numbers. We can also see, for example, the natural numbers, zero, and the negative integers together form the integers.

Properties of the Real Number System

We are now prepared to consider the properties of the real number system. The first property that we will discuss is *closure*.

> If an operation is performed on any two elements of a set and the result is an element of the set, we say that the set is **closed** under that given operation.

Is the sum of any two natural numbers a natural number? The answer is yes. Thus we say that the natural numbers are closed under the operation of addition.

Are the natural numbers closed under the operation of subtraction? If we subtract one natural number from another natural number, must the difference

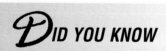

always be a natural number? The answer is no. For example, $3 - 5 = -2$, which is not a natural number. Therefore the natural numbers are not closed under the operation of subtraction.

EXAMPLE I

Determine whether the integers are closed under the operations of (a) multiplication, and (b) division.

Solution:

a) If we multiply any two integers, will the product always be an integer? The answer is yes. Thus the integers are closed under the operation of multiplication.

b) If we divide any two integers, will the quotient be an integer? The answer is no. For example, $6 \div 5 = \frac{6}{5}$, which is not an integer. Therefore the integers are not closed under the operation of division.   ▲

Next we will discuss three important properties: the commutative property, the associative property, and the distributive property. A knowledge of these properties is essential for the understanding of algebra. We begin with the commutative property.

Commutative Property

Addition	*Multiplication*
$a + b = b + a$	$a \cdot b = b \cdot a$

for any real numbers a and b.

The commutative property states that the *order* in which two numbers are added or multiplied is immaterial. For example $2 + 3 = 3 + 2 = 5$ and $5 \cdot 7 = 7 \cdot 5 = 35$. Note that the commutative property does not hold for the operations of subtraction or division. For example

$$2 - 3 \neq 3 - 2 \quad \text{and} \quad 4 \div 2 \neq 2 \div 4$$

Associative Property

Addition	*Multiplication*
$(a + b) + c = a + (b + c)$	$(a \cdot b) \cdot c = a \cdot (b \cdot c)$

for any real numbers a, b, and c.

The associative property states that when adding or multiplying three real numbers, we may place parentheses around any two adjacent numbers. For example,

$$\begin{aligned} (2 + 3) + 4 &= 2 + (3 + 4) \\ 5 + 4 &= 2 + 7 \\ 9 &= 9. \end{aligned} \qquad \begin{aligned} (2 \cdot 3) \cdot 4 &= 2 \cdot (3 \cdot 4) \\ 6 \cdot 4 &= 2 \cdot 12 \\ 24 &= 24. \end{aligned}$$

The associative property does not hold for the operations of subtraction and division. For example,

$$(4 - 2) - 3 \neq 4 - (2 - 3) \quad \text{and} \quad (8 \div 4) \div 2 \neq 8 \div (4 \div 2).$$

Note the difference between the commutative property and the associative property. The commutative property involves a change in *order*, whereas the associative property involves a change in *grouping* (or the *association* of numbers that are grouped together).

Another property of the real numbers is the distributive property of multiplication over addition.

Distributive Property of Multiplication over Addition
$$a \cdot (b + c) = a \cdot b + a \cdot c$$

for any real numbers a, b, and c.

For example, if $a = 2$, $b = 3$, and $c = 4$, then

$$2 \cdot (3 + 4) = (2 \cdot 3) + (2 \cdot 4)$$
$$2 \cdot 7 = 6 + 8$$
$$14 = 14.$$

This result indicates that, when using the distributive property, you may either add first and then multiply or multiply first and then add. Note that the distributive property involves two operations, addition and multiplication. Although positive integers were used in the example, any real numbers could have been used.

We frequently use the commutative, associative, and distributive properties without realizing that we are doing so. To add $13 + 4 + 6$, we may add the $4 + 6$ first to get 10. To this sum we then add 13 to get 23. Here we have done the equivalent of placing parentheses around the $4 + 6$. We can do so because of the associative property of addition.

To multiply 102×11 in our heads, we might multiply $100 \times 11 = 1100$ and $2 \times 11 = 22$ and add these two products to get 1122. We are permitted to do so because of the distributive property.

$$102 \times 11 = (100 + 2) \times 11 = (100 \times 11) + (2 \times 11)$$
$$= 1100 + 22 = 1122$$

EXAMPLE 2

Name the property illustrated.

Solution:

a) $4 + 11 = 11 + 4$ — Commutative property of addition

b) $(3 + 7) + 13 = 3 + (7 + 13)$ — Associative property of addition

c) $(x \cdot y) \cdot z = x \cdot (y \cdot z)$ — Associative property of multiplication

d) $13(x + z) = 13 \cdot x + 13 \cdot z$ — Distributive property of multiplication over addition

e) $3 + (x + 4) = 3 + (4 + x)$ The only change between left and right sides of the equation is the order of the x and 4. The order is changed from $x + 4$ to $4 + x$ using the commutative property of addition.

f) $5(11 \cdot 9) = (11 \cdot 9)5$ The order of 5 and $(11 \cdot 9)$ is changed by using the commutative property of multiplication. ▲

EXAMPLE 3

Use the distributive property to simplify

a) $3(5 + \sqrt{2})$ **b)** $\sqrt{3}(5 + \sqrt{2})$

Solution:

a) $3(5 + \sqrt{2}) = (3 \cdot 5) + (3\sqrt{2})$
$$= 15 + 3\sqrt{2}$$

b) $\sqrt{3}(5 + \sqrt{2}) = (\sqrt{3} \cdot 5) + (\sqrt{3} \cdot \sqrt{2})$
$$= 5\sqrt{3} + \sqrt{6}$$

Note that $\sqrt{3} \cdot 5$ is written $5\sqrt{3}$. ▲

EXAMPLE 4

Use the distributive property to multiply $3(x + 2)$. Then simplify the result.

Solution: $3(x + 2) = 3 \cdot x + 3 \cdot 2$
$$= 3x + 6$$ ▲

We summarize the properties mentioned in this section as follows, where a, b, and c are any real numbers.

Commutative property of addition	$a + b = b + a$
Commutative property of multiplication	$a \cdot b = b \cdot a$
Associative property of addition	$(a + b) + c = a + (b + c)$
Associative property of multiplication	$(a \cdot b) \cdot c = a \cdot (b \cdot c)$
Distributive property of multiplication over addition	$a \cdot (b + c) = a \cdot b + a \cdot c$

Section 5.5 Exercises

1. What are the real numbers?

2. What symbol is used to represent the set of real numbers?

3. What does it mean if a set is closed under a given operation?

4. Give the commutative property of addition, explain what it means, and give an example illustrating it.

5. Give the commutative property of multiplication, explain what it means, and give an example illustrating it.

6. Give the associative property of addition, explain what it means, and give an example illustrating it.

7. Give the associative property of multiplication, explain what it means, and give an example illustrating it.

8. Give the distributive property of multiplication over addition, explain what it means, and give an example illustrating it.

In Exercises 9–12, determine whether the natural numbers are closed under the given operation.

9. addition

10. subtraction

11. multiplication

12. division

In Exercises 13–16, determine whether the integers are closed under the given operation.

13. subtraction

14. addition

15. division

16. multiplication

In Exercises 17–20, determine whether the rational numbers are closed under the given operation.

17. addition

18. subtraction

19. multiplication

20. division

In Exercises 21–24, determine whether the irrational numbers are closed under the given operation.

21. addition

22. subtraction

23. multiplication

24. division

In Exercises 25–28, determine whether the real numbers are closed under the given operation.

25. addition

26. subtraction

27. division

28. multiplication

29. Does $3 + (4 + 5) = 3 + (5 + 4)$ illustrate the commutative property or the associative property? Explain your answer.

30. Does $(4 + 5) + 6 = 6 + (4 + 5)$ illustrate the commutative property or the associative property? Explain your answer.

31. Give an example to show that the commutative property of addition may be true for the negative integers.

32. Give an example to show that the commutative property of multiplication may be true for the negative integers.

33. Does the commutative property hold for the integers under the operation of subtraction? Give an example to support your answer.

34. Does the commutative property hold for the real numbers under the operation of subtraction? Give an example to support your answer.

35. Does the commutative property hold for the integers under the operation of division? Give an example to support your answer.

36. Does the commutative property hold for the rational numbers under the operation of division? Give an example to support your answer.

37. Give an example to show that the associative property of addition may be true for the negative integers.

38. Give an example to show that the associative property of multiplication may be true for the negative integers.

39. Does the associative property hold for the integers under the operation of subtraction? Give an example to support your answer.

40. Does the associative property hold for the integers under the operation of division? Give an example to support your answer.

41. Does the associative property hold for the real numbers under the operation of subtraction? Give an example to support your answer.

42. Does the associative property hold for the real numbers under the operation of division? Give an example to support your answer.

43. Does $a + (b \cdot c) = (a + b) \cdot (a + c)$? Give an example to support your answer.

In Exercises 44–58, state the name of the property illustrated.

44. $3 + [4 + (-2)] = (3 + 4) + (-2)$

45. $3 \cdot x = x \cdot 3$

46. $(x + 3) + 4 = x + (3 + 4)$

47. $2(3 + 6) = (2 \cdot 3) + (2 \cdot 6)$

48. $(\frac{1}{3} + \frac{1}{4}) + 2 = \frac{1}{3} + (\frac{1}{4} + 2)$

49. $x + y = y + x$

50. $x \cdot y = y \cdot x$

51. $(3 + 4) + (5 + 6) = (5 + 6) + (3 + 4)$

52. $4 + (2 + 3) = (2 + 3) + 4$

53. $2(x + y) = 2x + 2y$

54. $(2 + 3) + 4 = (3 + 2) + 4$

55. $(x + y) + z = z + (x + y)$

56. $(a + b) + c = c + (a + b)$

57. $a \cdot (x + y) = (x + y) \cdot a$

58. $(a + b) + (c + d) = (c + d) + (a + b)$

In Exercises 59–64, use the distributive property to multiply. Then simplify the resulting expression.

59. $4(x + 5)$ 60. $5(x + 4)$

61. $\sqrt{3}(2 + \sqrt{6})$ 62. $\sqrt{2}(\sqrt{6} + 3)$

63. $\sqrt{3}(x + \sqrt{3})$ 64. $x(3 + y)$

In Exercises 65 and 66, name the property used to go from step to step.

65. $3(x + 2) + 5 = (3x + 3 \cdot 2) + 5$

$\qquad\qquad\quad = (3x + 6) + 5$

66. $\qquad = 3x + (6 + 5)$

$\qquad\qquad = 3x + 11$

In Exercises 67–70, name the property used to go from step to step.

67. $5(x + 3) + 4x = (5 \cdot x + 5 \cdot 3) + 4x$

$\qquad\qquad\qquad = (5x + 15) + 4x$

68. $\qquad = 5x + (15 + 4x)$

69. $\qquad = 5x + (4x + 15)$

70. $\qquad = (5x + 4x) + 15$

$\qquad\qquad = 9x + 15$

In Exercises 71–74, name the property used to go from step to step.

71. $5 + 3(x + 2) + 4x = 5 + (3 \cdot x + 3 \cdot 2) + 4x$

$\qquad\qquad\qquad\quad = 5 + (3x + 6) + 4x$

72. $\qquad = 5 + (6 + 3x) + 4x$

73. $\qquad = (5 + 6) + 3x + 4x$

$\qquad\qquad\quad = 11 + 7x$

74. $\qquad = 7x + 11$

In Exercises 75–84, determine whether the activity can be used to illustrate the property indicated. Assume that for the property to hold the end result must be the same and when using the associative property the items within parentheses must be performed first. Explain your answer.

75. Putting on shoes and socks—commutative property.

76. Putting sugar and cream in coffee—commutative property.

77. Brushing your teeth, washing your face, and combing your hair—associative property.

78. Turning on the lamp and reading a book—commutative property.

79. Cracking an egg, pouring out the egg, and cooking the egg—associative property.

80. Removing the gas cap, putting the nozzle in the tank, and turning on the gas—associative property.

81. Writing on the blackboard and erasing the blackboard—commutative property.

82. A coffee machine dropping the cup, dispensing the coffee, and then adding the sugar—associative property.

83. In a car, putting the key in the ignition and turning the key—commutative property.

84. Starting a car, moving the stick lever to drive, and then stepping on the gas—associative property.

85. The man in the photographs is demonstrating that he can remove his sweater without removing his jacket. Can removal of the sweater and jacket be used to illustrate the commutative property? Explain your answer.

Problem Solving/Group Activities

86. a) Consider the three words *man eating tiger*. Does (man eating) tiger mean the same as man (eating tiger)?

b) Can you find three other nonassociative word triples?

87. Does $0 \div a = a \div 0$ (assume that $a \neq 0$)? Explain.

5.6 RULES OF EXPONENTS AND SCIENTIFIC NOTATION

An understanding of exponents is important in solving problems in algebra. In the expression 5^2, the 2 is referred to as the *exponent* and the 5 is referred to as the *base*. We read 5^2 as 5 to the second power, or 5 squared, which means

$$5^2 = \underbrace{5 \cdot 5}_{2 \text{ factors of } 5}.$$

The number 5 to the third power, or 5 cubed, written 5^3, means

$$5^3 = \underbrace{5 \cdot 5 \cdot 5}_{3 \text{ factors of } 5}.$$

In general, the number b to the nth power, written b^n, means

$$b^n = \underbrace{b \cdot b \cdot b \cdot \cdots \cdot b}_{n \text{ factors of } b}.$$

EXAMPLE 1

Evaluate the following.

a) 3^2 **b)** $(-3)^2$ **c)** 4^3 **d)** 1^{100} **e)** 8^1

Solution:
a) $3^2 = 3 \cdot 3 = 9$
b) $(-3)^2 = (-3)(-3) = 9$
c) $4^3 = 4 \cdot 4 \cdot 4 = 64$
d) $1^{100} = 1$. (The number 1 times itself any number of times equals 1.)
e) $8^1 = 8$. (Any number with an exponent of 1 is the number itself.)

▲

EXAMPLE 2

Evaluate the following.

a) -3^4 **b)** $(-3)^4$

Solution:
a) -3^4 means $-(3)^4$, which can also be written $-1(3)^4$.

$$-3^4 = -1(3)^4 = -1(3)(3)(3)(3) = -1(81) = -81$$

b) $(-3)^4 = (-3)(-3)(-3)(-3)$
$$= (9)(-3)(-3)$$
$$= (-27)(-3)$$
$$= 81$$

▲

From Example 2 we can see that $-x^n \neq (-x)^n$ where n is an even natural number.

EXAMPLE 3

Evaluate $-x^3$ when (a) $x = \frac{3}{2}$ and (b) $x = -\frac{3}{2}$.

Solution:

a) $-x^3$ means $-1(x)^3$. Thus

$$-x^3 = -1(x)^3 = -1\left(\frac{3}{2}\right)^3 = -1\left(\frac{27}{8}\right) = -\frac{27}{8}$$

b) $-x^3 = -1(x)^3 = -1\left(-\frac{3}{2}\right)^3 = -1\left(-\frac{27}{8}\right) = \frac{27}{8}$

Rules of Exponents

Now that we know how to evaluate powers of numbers we can discuss the rules of exponents. Consider

$$2^2 \cdot 2^3 = \underbrace{2 \cdot 2}_{2 \text{ factors}} \cdot \underbrace{2 \cdot 2 \cdot 2}_{3 \text{ factors}} = 2^5.$$

This example illustrates the product rule for exponents.

Product Rule for Exponents

$$a^m \cdot a^n = a^{m+n}$$

Therefore by using the product rule, $2^2 \cdot 2^3 = 2^{2+3} = 2^5$.

EXAMPLE 4

Use the product rule to simplify.

a) $5^2 \cdot 5^6$ **b)** $8^3 \cdot 8^5$

Solution:

a) $5^2 \cdot 5^6 = 5^{2+6} = 5^8$ **b)** $8^3 \cdot 8^5 = 8^{3+5} = 8^8$

Consider

$$\frac{2^5}{2^2} = \frac{2 \cdot 2 \cdot 2 \cdot \cancel{2} \cdot \cancel{2}}{\cancel{2} \cdot \cancel{2}} = 2 \cdot 2 \cdot 2 = 2^3.$$

This example illustrates the quotient rule for exponents.

Quotient Rule for Exponents

$$\frac{a^m}{a^n} = a^{m-n}, \qquad a \neq 0$$

Therefore $\dfrac{2^5}{2^2} = 2^{5-2} = 2^3$.

EXAMPLE 5

Use the quotient rule to simplify.

a) $\dfrac{5^8}{5^5}$ **b)** $\dfrac{8^{12}}{8^5}$

Solution:

a) $\dfrac{5^8}{5^5} = 5^{8-5} = 5^3$ **b)** $\dfrac{8^{12}}{8^5} = 8^{12-5} = 8^7$

Consider $2^3 \div 2^3$. The quotient rule gives

$$\frac{2^3}{2^3} = 2^{3-3} = 2^0.$$

But $\dfrac{2^3}{2^3} = \dfrac{8}{8} = 1$. Therefore 2^0 must equal 1. This example illustrates the zero exponent rule.

> **Zero Exponent Rule**
>
> $$a^0 = 1, \qquad a \neq 0$$

Note that 0^0 is not defined by the zero exponent rule.

EXAMPLE 6

Use the zero exponent rule to simplify.

a) 5^0 **b)** $(-3)^0$

Solution:

a) $5^0 = 1$ **b)** $(-3)^0 = 1$

Consider $2^3 \div 2^5$. The quotient rule yields

$$\frac{2^3}{2^5} = 2^{3-5} = 2^{-2}.$$

But $\dfrac{2^3}{2^5} = \dfrac{\cancel{2} \cdot \cancel{2} \cdot \cancel{2}}{\cancel{2} \cdot \cancel{2} \cdot \cancel{2} \cdot 2 \cdot 2} = \dfrac{1}{2^2}$. Since $\dfrac{2^3}{2^5}$ equals both 2^{-2} and $\dfrac{1}{2^2}$, then 2^{-2} must equal $\dfrac{1}{2^2}$. This example illustrates the negative exponent rule.

> **Negative Exponent Rule**
>
> $$a^{-m} = \frac{1}{a^m}, \qquad a \neq 0$$

EXAMPLE 7

Use the negative exponent rule to simplify.

a) 5^{-2} **b)** 8^{-1}

Solution:

a) $5^{-2} = \dfrac{1}{5^2} = \dfrac{1}{25}$ **b)** $8^{-1} = \dfrac{1}{8^1} = \dfrac{1}{8}$

Consider $(2^3)^2$:

$$(2^3)^2 = (2^3)(2^3) = 2^{3+3} = 2^6.$$

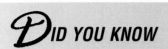
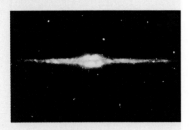

This example illustrates the power rule for exponents.

Power Rule for Exponents
$$(a^m)^n = a^{m \cdot n}$$

Thus $(2^3)^2 = 2^{3 \cdot 2} = 2^6$.

EXAMPLE 8

Use the power rule to simplify.

a) $(3^5)^4$ **b)** $(4^3)^6$

Solution:

a) $(3^5)^4 = 3^{5 \cdot 4} = 3^{20}$ **b)** $(4^3)^6 = 4^{3 \cdot 6} = 4^{18}$ ▲

Summary of the Rules of Exponents

$a^m \cdot a^n = a^{m+n}$	Product rule for exponents
$\dfrac{a^m}{a^n} = a^{m-n}, \quad a \neq 0$	Quotient rule for exponents
$a^0 = 1, \quad a \neq 0$	Zero exponent rule
$a^{-m} = \dfrac{1}{a^m}, \quad a \neq 0$	Negative exponent rule
$(a^m)^n = a^{m \cdot n}$	Power rule for exponents

Scientific Notation

Often scientific problems deal with very large and very small numbers. For example, the distance from the earth to the sun is about 93,000,000 miles. The wavelength of a yellow color of light is about 0.0000006 meter. Because working with many zeros is difficult, scientists developed a notation that expresses such numbers with exponents. For example consider the distance from the earth to the sun, 93,000,000 miles.

$$93,000,000 = 9.3 \times 10,000,000$$
$$= 9.3 \times 10^7$$

The diameter of a red blood cell may be 0.000004 inch.

$$0.000004 = 4.0 \times 0.000001$$
$$= 4.0 \times 10^{-6}$$

The numbers 9.3×10^7 and 4.0×10^{-6} are written in a form called scientific notation. Each number written in scientific notation is written as a number greater than or equal to 1 and less than 10 ($1 \leq a < 10$) multiplied by some power of 10. Some examples of numbers in scientific notation are

$$3.7 \times 10^3, \quad 2.05 \times 10^{-3}, \quad 5.6 \times 10^8, \quad \text{and} \quad 1.00 \times 10^{-5}.$$

The following is a simplified procedure for writing a number in scientific notation.

To Write a Number in Scientific Notation

1. Move the decimal point in the original number to the right or left until you obtain a number greater than or equal to 1 and less than 10.

2. Count the number of places you have moved the decimal point to obtain the number in step 1. If the decimal point was moved to the left, the count is to be considered positive. If the decimal point was moved to the right, the count is to be considered negative.

3. Multiply the number obtained in step 1 by 10 raised to the count found in step 2. (Note that the count determined in step 2 is the exponent on the base 10.)

EXAMPLE 9

Write each number in scientific notation.

a) Males have up to 5,800,000 red blood cells per cubic millimeter of blood.

b) In 1995, the world population was estimated to be approximately 5,630,000,000.

c) The probability of winning a lottery may be 0.0000018.

d) The wavelength of an X ray may be 0.000000000492 meter.

Solution:

a) $5,800,000 = 5.8 \times 10^6$

b) $5,630,000,000 = 5.63 \times 10^9$

c) $0.0000018 = 1.8 \times 10^{-6}$

d) $0.000000000492 = 4.92 \times 10^{-10}$ ▲

To convert from a number given in scientific notation to decimal notation we reverse the procedure.

To Change a Number in Scientific Notation to Decimal Notation

1. Observe the exponent on the 10.

2. a) If the exponent is positive, move the decimal point in the number to the right the same number of places as the exponent. Adding zeros to the number might be necessary.

b) If the exponent is negative, move the decimal point in the number to the left the same number of places as the exponent. Adding zeros might be necessary.

EXAMPLE 10

Write each number in decimal notation.

a) The age of the earth is estimated by some scientists to be 4.6×10^9 years.

b) The half-life of plutonium 192 is about 1×10^{15} years.

c) The average grain size in siltstone is 1.35×10^{-3} inches.

d) The quotient of 12 divided by 480,000 is 2.5×10^{-5}.

Solution:

a) $4.6 \times 10^9 = 4,600,000,000$

b) $1 \times 10^{15} = 1,000,000,000,000,000$

c) $1.35 \times 10^{-3} = 0.00135$

d) $2.5 \times 10^{-5} = 0.000025$ ▲

In scientific journals and books we occasionally see numbers like 10^{15} and 10^{-6}. We interpret these numbers as 1×10^{15} and 1×10^{-6}, respectively, when converting the numbers to decimal form.

EXAMPLE 11

Multiply $(4.3 \times 10^6)(2 \times 10^{-4})$. Write the answer in decimal notation.

Solution: $(4.3 \times 10^6)(2 \times 10^{-4}) = (4.3 \times 2)(10^6 \times 10^{-4})$
$$= 8.6 \times 10^2$$
$$= 860 \qquad ▲$$

EXAMPLE 12

Divide $\dfrac{0.0000093}{0.003}$. Write the answer in scientific notation.

Solution: First write each number in scientific notation.
$$\frac{0.0000093}{0.003} = \frac{9.3 \times 10^{-6}}{3 \times 10^{-3}} = \left(\frac{9.3}{3}\right)\left(\frac{10^{-6}}{10^{-3}}\right)$$
$$= 3.1 \times 10^{-6-(-3)}$$
$$= 3.1 \times 10^{-6+3}$$
$$= 3.1 \times 10^{-3} \qquad ▲$$

EXAMPLE 13

Multiply $(42,100,000)(0.008)$. Write the answer in (a) scientific notation and (b) decimal notation.

Solution:

a) Change each number to scientific notation.
$$(42,100,000)(0.008) = (4.21 \times 10^7)(8 \times 10^{-3})$$
$$= (4.21 \times 8)(10^7 \times 10^{-3})$$
$$= 33.68 \times 10^4$$
$$= 3.368 \times 10^5$$

b) The answer in decimal form is 336,800.   ▲

EXAMPLE 14

A personal computer that contains a pentium chip can perform 100 million operations per second. How long would such a computer take to perform 20 trillion (20,000,000,000,000) operations?

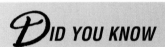

Solution: You may not be sure whether to multiply or divide to determine the solution. One approach to solving problems, as mentioned in Section 1.3, is to substitute simpler numbers in the problem. Suppose that the computer performed two operations per second and that you wanted to determine the time needed to perform 10 operations. You would use division to determine that the answer is 5 sec ($10 \div 2 = 5$ sec). Therefore you must use division to solve this problem. Divide the number of operations desired by the number of operations per second to determine the time required.

$$20 \text{ trillion} = 2.0 \times 10^{13}$$

$$100 \text{ million operations} = 1.0 \times 10^{8}$$

$$\text{Time to perform the operations} = \frac{2.0 \times 10^{13}}{1.0 \times 10^{8}}$$

$$= 2.0 \times 10^{5}, \text{ or } 200,000 \text{ sec (or } 55.\overline{5} \text{ hr)}$$

Scientific Notation on the Calculator

What will your calculator show when you multiply very large or very small numbers? The answer depends on whether your calculator has the ability to display an answer in scientific notation. On calculators without that ability you will probably get an error message because the answer will be too large or too small for the display.

For example, on a calculator without scientific notation:

$$\boxed{8000000} \; \boxed{\times} \; \boxed{600000} \; \boxed{=} \; \text{Error}$$

If your calculator has the ability to give an answer in scientific notation you will probably get

$$\boxed{8000000} \; \boxed{\times} \; \boxed{600000} \; \boxed{=} \; 4.8 \qquad 12$$

This $4.8 \qquad 12$ means 4.8×10^{12}.

On a calculator that uses scientific notation:

$$\boxed{.0000003} \; \boxed{\times} \; \boxed{.004} \; \boxed{=} \; 1.2 \qquad -9$$

This $1.2 \qquad -9$ means 1.2×10^{-9}.

Section 5.6 Exercises

1. In the expression 4^{6}, what is the name given to the 4 and what is the name given to the 6?

2. Explain the meaning of b^{n}.

3. a) Explain the Product Rule for Exponents.
   b) Use the Product Rule to simplify $3^{3} \cdot 3^{5}$.

4. a) Explain the Quotient Rule for Exponents.

b) Use the Quotient Rule to simplify $\dfrac{4^{5}}{4^{3}}$.

5. a) Explain the Negative Exponent Rule.
   b) Use the Negative Exponent Rule to simplify 3^{-5}.

6. a) Explain the Zero Exponent Rule.
   b) Use the Zero Exponent Rule to simplify 5^{0}.

7. a) Explain the Power Rule for Exponents
   b) Use the Power Rule to simplify $(4^4)^3$.

8. What is the form of a number in scientific notation?

9. a) In your own words explain how to change a number in decimal notation to scientific notation.
   b) Using the procedure in part (a) change 0.000426 to scientific notation.

10. a) In your own words explain how to change a number in scientific notation to decimal notation.
   b) Change 5.76×10^{-4} to decimal notation.

11. A number is given in scientific notation. What does it indicate about the number when the exponent on the ten is (a) positive, (b) zero, and (c) negative?

12. a) How is the number 10^5 interpreted in scientific notation?
   b) Write 10^5 as a number in decimal notation.

In Exercises 13–42, evaluate the expression.

13. 5^2 **14.** 2^3 **15.** $(-3)^2$

16. $(-2)^3$ **17.** -2^3 **18.** $(\frac{1}{2})^2$

19. $(\frac{5}{6})^2$ **20.** -3^4 **21.** $3^2 \cdot 4^3$

22. $\dfrac{6^2}{2^2}$ **23.** $\dfrac{3^5}{3^2}$ **24.** $3^2 \cdot 3^3$

25. $\dfrac{5^2}{5^5}$ **26.** $4^2 \cdot 9^0$ **27.** $(-5)^0$

28. $(-2)^4$ **29.** 2^4 **30.** $2^5 \cdot 4^0$

31. 2^{-2} **32.** 2^{-3} **33.** $(3^2)^3$

34. $(1^4)^5$ **35.** $\dfrac{5^7}{5^5}$ **36.** $7^2 \cdot 7$

37. 5^{-3} **38.** 2^{-4} **39.** $3^2 \cdot 3^2$

40. $(5^2)^3$ **41.** $\dfrac{3^5}{3^4}$ **42.** $2^{-2} \cdot 2$

43. Evaluate each expression for $x = 5$ and $x = -5$.
   a) x^2 **b)** $-x^2$ **c)** x^3 **d)** $-x^3$

44. Evaluate each expression for $x = \frac{5}{8}$ and $x = -\frac{5}{8}$.
   a) x^2 **b)** $-x^2$ **c)** x^3 **d)** $-x^3$

In Exercises 45–60, express the number in scientific notation.

45. 37,000   **46.** 5,940,000   **47.** 600

48. 0.000421   **49.** 0.053   **50.** 0.0000561

51. 19,000   **52.** 1,260,000,000   **53.** 0.00000186

54. 0.0003   **55.** 0.00000423   **56.** 54,000

57. 711   **58.** 0.02   **59.** 0.153

60. 416,000

In Exercises 61–76, express the number in decimal notation.

61. 4.2×10^3 **62.** 1.63×10^{-4} **63.** 6×10^7

64. 5.19×10^5 **65.** 2.13×10^{-5} **66.** 2.74×10^{-7}

67. 3.12×10^{-1} **68.** 4.6×10^1 **69.** 9×10^6

70. 7.3×10^4 **71.** 2.31×10^2 **72.** 1.04×10^{-2}

73. 3.5×10^4 **74.** 2.17×10^{-6} **75.** 1×10^4

76. 1×10^{-3}

In Exercises 77–86, perform the indicated operation and express each number in decimal notation.

77. $(4 \times 10^2)(3 \times 10^5)$ **78.** $(2 \times 10^{-3})(3 \times 10^2)$

79. $(5.1 \times 10^1)(3 \times 10^{-4})$

80. $(1.6 \times 10^{-2})(4 \times 10^{-3})$

81. $\dfrac{6.4 \times 10^5}{2 \times 10^3}$ **82.** $\dfrac{8 \times 10^{-3}}{2 \times 10^1}$ **83.** $\dfrac{8.4 \times 10^{-6}}{4 \times 10^{-3}}$

84. $\dfrac{25 \times 10^3}{5 \times 10^{-2}}$ **85.** $\dfrac{4 \times 10^5}{2 \times 10^4}$ **86.** $\dfrac{16 \times 10^3}{8 \times 10^{-3}}$

In Exercises 87–96, perform the indicated operation by first converting each number to scientific notation. Write the answer in scientific notation.

87. $(700,000)(6,000,000)$ **88.** $(0.0006)(5,000,000)$

89. $(0.003)(0.00015)$ **90.** $(230,000)(3000)$

91. $\dfrac{1,400,000}{700}$ **92.** $\dfrac{20,000}{0.0005}$ **93.** $\dfrac{0.00004}{200}$

94. $\dfrac{0.0012}{0.000006}$ **95.** $\dfrac{150,000}{0.0005}$ **96.** $\dfrac{24,000}{8,000,000}$

In Exercises 97–100, list the numbers from smallest to largest.

97. $5.8 \times 10^5, 3.2 \times 10^{-1}, 4.6, 8.3 \times 10^{-4}$

98. $8.5 \times 10^{-5}, 8.2 \times 10^3, 1.3 \times 10^{-1}, 6.2 \times 10^4$

99. $40,000, 4.1 \times 10^3, 0.00079, 8.3 \times 10^{-5}$

100. $267,000,000, 3.14 \times 10^7, 1,962,000, 4.79 \times 10^6$

101. The number of people in the world in 1992 was approximately 5.42×10^9, and the number of people living in China was 1.7×10^9. How many people lived outside China?

102. In the United States only 5% of the 4.0×10^9 pounds of plastic is recycled annually.
   a) How many pounds are recycled annually?
   b) How many pounds are not recycled annually?

103. The distance from the earth to the planet Jupiter is approximately 4.5×10^8 mi. If a spacecraft traveled at a speed of 25,000 mph, how many hours would the spacecraft need to travel from the earth to Jupiter? Use distance = rate × time.

104. The distance from the earth to the moon is approximately 239,000 mi. If a spacecraft travels at a speed of 20,000 mph, how many hours would the spacecraft need to travel from the earth to the moon? Use distance = rate × time.

105. If a computer can perform 4 million operations per second, how long would it take to perform 10 trillion (10,000,000,000,000) operations?

106. If a cubic millimeter of blood contains 5,800,000 red blood cells, how many red blood cells are contained in 50 cubic millimeters of blood?

107. The half-life of a radioactive isotope is the time required for half the quantity of the isotope to decompose. The half-life of uranium-238 is 4.5×10^9 years, and the half-life of uranium-234 is 2.5×10^5 years. How many times greater is the half-life of uranium-238 than uranium-234?

108. A treaty between the United States and Canada requires that during the tourist season a minimum of 100,000 cubic feet of water per second flow over Niagara Falls (another 130,000–160,000 ft³/sec are diverted for power generation). Find the minimum amount of water that will flow over the falls in a 24-hour period during the tourist season.

109. On September 30, 1992, the U.S. public debt was approximately $4,065,000,000,000 (4 trillion 65 billion dollars). The U.S. population on that date was approximately 257,000,000.
a) Find the average U.S. debt for every person in the United States (the per capita debt).
b) On September 30, 1982, the U.S. debt was approximately $1,142,000,000,000. How much larger was the debt in 1992 than in 1982?
c) How many times greater was the debt in 1992 than in 1982?

110. Laid end to end, the 18 billion disposable diapers thrown away in the United States each year would reach the moon and back 7 times.
a) Write 18 billion in scientific notation.
b) If the distance from the earth to the moon is 2.38×10^5 miles, what is the length of all these diapers placed end to end? Write the answer in decimal notation.

111. According to the U.S. Treasury, the total U.S. currency in circulation is about 3.13×10^{12} dollars. The following table and graph indicate the percent of the total circulation by currency denomination on March 31, 1993.

Denomination	Percent of Currency
1	1.7
5	1.9
10	3.8
20	22.5
50	11.9
100	57.7
*others	0.5

includes $2, $500, $1000, $5000, and $10,000 bills

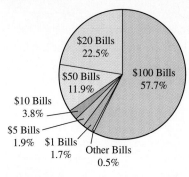

a) Estimate the amount of currency in $1 bills in circulation.
b) Estimate the amount of currency in $100 bills in circulation.
c) Estimate the number of $1 bills in circulation.
d) Estimate the number of $100 bills in circulation.
e) Are more $1 bills or $100 bills in circulation? Estimate the difference in the number of $1 bills and $100 bills in circulation.

112. The U.S. Department of Agriculture estimated the number of livestock on farms in 1890 and 1993 as follows.

	1890	1993
Total Livestock	1.676×10^8	1.8004×10^8
All Cattle	6.00×10^7	1.01×10^8
Milk Cows	1.50×10^7	9.84×10^6
Sheep	4.45×10^7	1.02×10^7
Hogs	4.81×10^7	5.90×10^7

a) In complete sentences describe the change in the number of cattle, the number of milk cows, the number of sheep, and the number of hogs from 1890 to 1993.
b) How many more cattle were raised in 1993 than in 1890?
c) How many fewer milk cows were raised in 1993 than in 1890?
d) From 1890 to 1993 did the number of cattle or the number of milk cows change more? By how much?
e) How many times greater is the total number of livestock on farms in 1993 than in 1890?

113. In the metric system 1 meter = 10^3 millimeters. How many times greater is a meter than a millimeter? Explain how you determined your answer.

114. In the metric system 1 gram = 10^3 milligrams and 1 gram = 10^{-3} kilograms. What is the relationship between milligrams and kilograms? Explain how you determined your answer.

Problem Solving/Group Activities

115. Many people have no idea of the difference in size between a million (1,000,000), a billion (1,000,000,000), and a trillion (1,000,000,000,000).

a) Write a million, a billion, and a trillion in scientific notation.

b) Determine how long it would take to spend a million dollars if you spent $1000 a day.

c) Repeat part (b) for a billion dollars.

d) Repeat part (b) for a trillion dollars.

e) How many times greater is a billion dollars than a million dollars?

116. All of human history was required for the world's population to reach 5.4×10^9 in 1992. At current rates the world population will double in about 40 years.

a) Estimate the world's population in 2032.

b) Assuming 365 days in a year, estimate the average number of additional people added to the earth's population each day between 1992 and 2032.

117. a) Light travels at a speed of 1.86×10^5 mi/sec. A *light year* is the distance that light travels in 1 year. Determine the number of miles in a light year.

b) The earth is approximately 93,000,000 mi from the sun. How long does it take light from the sun to reach the earth?

118. The exponential function, $E(t) = 2^{10} \cdot 2^t$ approximates the number of bacteria in a certain culture after t hr.

a) The initial number of bacteria is determined when $t = 0$. What is the initial number of bacteria?

b) How many bacteria are there after $\frac{1}{2}$ hr?

Research Activities

119. Dr. John Allen Paulos of Temple University wrote an entertaining book with a message about mathematical illiteracy. The book, *Innumeracy*, was published by Hill and Wang in 1988. Read *Innumeracy* and write a 500-word book report on it.

Dr. John Allen Paulos

120. Find an article in a newspaper or magazine that contains scientific notation. Write a paragraph explaining how scientific notation was used. Attach a copy of the article to your report.

5.7 ARITHMETIC AND GEOMETRIC SEQUENCES

Now that you can recognize the various sets of real numbers and know how to add, subtract, multiply, and divide real numbers, we can discuss sequences. A **sequence** is a list of numbers that are related to each other by a rule. The numbers that form the sequence are called its **terms**. If your salary increases or decreases by a fixed amount over a period of time, the listing of the amounts, over time, would form an arithmetic sequence. When interest in a savings account is compounded at regular intervals, the listing of the amounts in the account over time will be a geometric sequence.

Arithmetic Sequences

A sequence in which each term after the first term differs from the preceding term by a constant amount is called an **arithmetic sequence**. The amount by which each pair of successive terms differs is called the **common difference**, d. The common difference can be found by subtracting any term from the term that directly follows it.

Examples of Arithmetic Sequences	**Common Differences**
$1, 5, 9, 13, 17, \ldots$	$d = 5 - 1 = 4$
$-7, -5, -3, -1, 1, \ldots$	$d = -5 - (-7) = -5 + 7 = 2$
$\dfrac{5}{2}, \dfrac{3}{2}, \dfrac{1}{2}, -\dfrac{1}{2}$	$d = \dfrac{3}{2} - \dfrac{5}{2} = -\dfrac{2}{2} = -1$

EXAMPLE 1

Write the first five terms of the arithmetic sequence with first term 4 and common difference 3.

Solution: The first term is 4. The second term is $4 + 3$ or 7. The third term is $7 + 3$ or 10, and so on. Thus the five terms are 4, 7, 10, 13, and 16.   ▲

EXAMPLE 2

Write the first four terms of the arithmetic sequence with first term 3 and common difference -2.

Solution: 3, 1, -1, -3 ▲

When discussing a sequence, we often represent the first term as a_1 (read "a sub 1"), the second term as a_2, the fifteenth term as a_{15}, and so on. We use the notation a_n to represent the general or nth term of a sequence. Thus a sequence may be symbolized as

$$a_1, a_2, a_3, a_4, \ldots, a_n, \ldots$$

When we know the first term of an arithmetic sequence and the common difference, we can use the following formula to find the value of any specific term.

General or nth Term of an Arithmetic Sequence
$$a_n = a_1 + (n-1)d$$

EXAMPLE 3

Find the seventh term of the arithmetic sequence whose first term is 3 and whose common difference is -6.

Solution: To find the seventh term, or a_7, replace n in the formula with 7, a_1 with 3, and d with -6.

$$a_n = a_1 + (n - 1)d$$
$$a_7 = 3 + (7 - 1)(-6)$$
$$= 3 + (6)(-6)$$
$$= 3 - 36$$
$$= -33$$

The seventh term is -33. As a check, we have listed the first seven terms of the sequence: 3, -3, -9, -15, -21, -27, -33.   ▲

EXAMPLE 4

Write an expression for the general or nth term, a_n, for the sequence 6, 9, 12, 15, ...

Solution: In this sequence, $a_1 = 6$ and $d = 3$. We substitute these values into $a_n = a_1 + (n - 1)d$ to obtain an expression for the nth term, a_n.

$$a_n = a_1 + (n - 1)d$$
$$= 6 + (n - 1)3$$
$$= 6 + 3n - 3$$
$$= 3 + 3n$$

Note that, when $n = 1$, the first term is $3 + 3(1) = 6$. When $n = 2$, the second term is $3 + 3(2) = 9$, and so on. ▲

We can use the following formula to find the sum of the first n terms in an arithmetic sequence.

Sum of the First n Terms in an Arithmetic Sequence

$$s_n = \frac{n(a_1 + a_n)}{2}$$

In this formula, s_n represents the sum of the first n terms, a_1 is the first term, a_n is the nth term, and n is the number of terms in the sequence from a_1 to a_n.

EXAMPLE 5

Find the sum of the first 25 natural numbers.

Solution: The sequence we are discussing is

$$1, 2, 3, 4, 5, \ldots, 25.$$

In this sequence, $a_1 = 1$, $a_{25} = 25$, and $n = 25$. Thus the sum of the first 25 terms is

$$s_n = \frac{n(a_1 + a_n)}{2}$$

$$s_{25} = \frac{25(1 + 25)}{2}$$

$$= \frac{25(26)}{2} = 325.$$

Thus the sum of the terms $1 + 2 + 3 + 4 + \cdots + 25$ is 325. ▲

Geometric Sequences

The next type of sequence we will discuss is the geometric sequence. A **geometric sequence** is one in which the ratio of any term to the term that directly precedes it is a constant. This constant is called the **common ratio**. The common ratio, r, can be found by taking any term except the first and dividing that term by the preceding term.

Examples of Geometric Sequences   **Common Ratios**

$2, 4, 8, 16, 32$ $r = 4 \div 2 = 2$

$-3, 6, -12, 24, -48$ $r = 6 \div (-3) = -2$

$\dfrac{2}{3}, \dfrac{2}{9}, \dfrac{2}{27}, \dfrac{2}{81}$ $r = \dfrac{2}{9} \div \dfrac{2}{3} = \left(\dfrac{2}{9}\right)\left(\dfrac{3}{2}\right) = \dfrac{1}{3}$

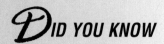

To construct a geometric sequence when the first term, a_1, and common ratio are known, multiply the first term by the common ratio to get the second term. Then multiply the second term by the common ratio to get the third term, and so on.

EXAMPLE 6

Write the first five terms of the geometric sequence with first term 5 and common ratio $\frac{1}{2}$.

Solution: The first term is 5. The second term, found by multiplying the preceding term by $\frac{1}{2}$, is $5(\frac{1}{2})$ or $\frac{5}{2}$. The third term is $(\frac{5}{2})(\frac{1}{2})$ or $\frac{5}{4}$, and so on. The sequence is

$$5, \frac{5}{2}, \frac{5}{4}, \frac{5}{8}, \frac{5}{16}, \ldots$$ ▲

When we know the first term of a geometric sequence and the common ratio we can use the following formula to find the value of the general or *n*th term, a_n.

General or *n*th Term of a Geometric Sequence

$$a_n = a_1 r^{n-1}$$

EXAMPLE 7

Find the seventh term of the geometric sequence whose first term is -3 and whose common ratio is -2.

Solution: In this sequence, $a_1 = -3$, $r = -2$, and $n = 7$. Substituting the values, we obtain

$$
\begin{aligned}
a_n &= a_1 r^{n-1} \\
a_7 &= -3(-2)^{7-1} \\
&= -3(-2)^6 \\
&= -3(64) \\
&= -192.
\end{aligned}
$$

As a check, we have listed the first seven terms of the sequence: $-3, 6, -12, 24, -48, 96, -192$.   ▲

EXAMPLE 8

Write an expression for the general or *n*th term, a_n, of the sequence 5, 15, 45, 135, ...

Solution: In this sequence, $a_1 = 5$ and $r = 3$. We substitute these values into $a_n = a_1 r^{n-1}$ to obtain an expression for the *n*th term, a_n.

$$
\begin{aligned}
a_n &= a_1 r^{n-1} \\
&= 5(3)^{n-1}
\end{aligned}
$$

Note that when $n = 1$, $a_1 = 5(3)^0 = 5(1) = 5$. When $n = 2$, $a_2 = 5(3)^1 = 15$, and so on.   ▲

EXAMPLE 9

Suppose that we form stacks of white chips such that there is one white chip in the first stack and in each successive stack we double the number of chips. Thus we have stacks of 1 chip, 2 chips, 4 chips, 8 chips, and so on. We also form stacks of red chips, starting with one red chip and then tripling the number of chips in each successive stack. Thus the stacks will contain 1 chip, 3 chips, 9 chips, 27 chips, and so on. How many more chips would the sixth stack of red chips have than the sixth stack of white chips?

Solution: The number of chips in each stack is multiplied by a constant to get the number of chips in the next stack. The number of chips in any stack may be found using the formula $a_n = a_1 r^{n-1}$.

$$a_n = a_1 r^{n-1}$$

$$\text{White} \quad a_6 = 1(2)^{6-1} = 1 \cdot 32 = 32$$

$$\text{Red} \quad a_6 = 1(3)^{6-1} = 1 \cdot 243 = 243$$

Since the sixth stack of red chips has 243 chips and the sixth stack of white chips has 32 chips, there are $243 - 32 = 211$ more chips in the sixth stack of red chips than in the sixth stack of white chips.   ▲

We can use the following formula to find the sum of the first n terms of a geometric sequence.

Sum of the First n Terms of a Geometric Sequence

$$s_n = \frac{a_1(1 - r^n)}{1 - r}, \qquad r \neq 1$$

EXAMPLE 10

Find the sum of the first five terms in the geometric sequence whose first term is 3 and whose common ratio is 4.

Solution: In this sequence, $a_1 = 3$, $r = 4$, and $n = 5$. Substituting the values, we obtain

$$s_n = \frac{a_1(1 - r^n)}{1 - r}$$

$$s_5 = \frac{3[1 - (4)^5]}{1 - 4}$$

$$= \frac{3(1 - 1024)}{-3}$$

$$= \frac{1(-1023)}{-1} = \frac{-1023}{-1} = 1023.$$

The sum of the first five terms of the sequence is 1023. The first five terms of the sequence are 3, 12, 48, 192, and 768. If you add these five numbers, you will obtain the sum 1023. ▲

EXAMPLE 11

Determine whether the following sequences are arithmetic or geometric and find the next two terms.

a) 2, 5, 8, 11, . . .   **b)** 2, 6, 18, 54, . . .
c) 2, −4, 8, −16, . . .   **d)** 7, 3, −1, −5, . . .

Solution:

a) Each term is 3 more than the preceding term. Therefore this sequence is arithmetic with $d = 3$. The next two terms are 14 and 17.

b) Each term is 3 times the preceding term. Therefore this sequence is geometric with $r = 3$. The next two terms are 162 and 486.

c) Each term is −2 times the preceding term, so the sequence is geometric with $r = -2$. The next two terms are 32 and −64.

d) Each term is 4 less than the preceding term, so the sequence is arithmetic with $d = -4$. The next two terms are −9 and −13. ▲

Section 5.7 Exercises

1. State the definition of *sequence* and give an example.

2. What are the numbers that make up a sequence called?

3. State the definition of *arithmetic sequence* and give an example.

4. Explain what the common difference in an arithmetic sequence is.

5. State the definition of *geometric sequence* and give an example.

6. Explain what the common ratio in a geometric sequence is.

In Exercises 7–16, write the first five terms of the arithmetic sequence with the first term, a_1, and common difference, d.

7. $a_1 = 4$, $d = 3$	**8.** $a_1 = 5$, $d = 2$
9. $a_1 = -4$, $d = 5$	**10.** $a_1 = -3$, $d = 1$
11. $a_1 = 6$, $d = -2$	**12.** $a_1 = 8$, $d = -5$
13. $a_1 = 24$, $d = -8$	**14.** $a_1 = \frac{1}{2}$, $d = 2$
15. $a_1 = \frac{1}{2}$, $d = \frac{1}{2}$	**16.** $a_1 = \frac{5}{2}$, $d = -\frac{3}{2}$

In Exercises 17–24, find the indicated term for the arithmetic sequence with the first term, a_1, and common difference, d.

17. Find a_5 when $a_1 = 12$, $d = 6$

18. Find a_9 when $a_1 = -8$, $d = -5$

19. Find a_8 when $a_1 = -4$, $d = 8$

20. Find a_{12} when $a_1 = 9$, $d = -7$

21. Find a_{10} when $a_1 = -30$, $d = -12$

22. Find a_{20} when $a_1 = \frac{3}{5}$, $d = -2$

23. Find a_{15} when $a_1 = -\frac{1}{2}$, $d = 5$

24. Find a_{11} when $a_1 = -20$, $d = -\frac{1}{2}$

In Exercises 25–34, write an expression for the general or nth term, a_n, for the arithmetic sequence.

25. $4, 8, 12, 16, \ldots$

26. $5, 2, -1, -4, \ldots$

27. $8, 18, 28, 38, \ldots$

28. $-3, -6, -9, -12, \ldots$

29. $-\frac{5}{2}, -\frac{3}{2}, -\frac{1}{2}, \frac{1}{2}, \ldots$

30. $-18, -9, 0, 9, \ldots$

31. $-5, -\frac{5}{2}, 0, \frac{5}{2}, \ldots$

32. $0, 4, 8, 12, \ldots$

33. $-3, -\frac{7}{2}, -4, -\frac{9}{2}, \ldots$

34. $-100, -95, -90, -85, \ldots$

In Exercises 35–44, find the sum of the terms of the arithmetic sequence.

35. $1, 2, 3, 4, \ldots, 12$

36. $5, 10, 15, \ldots, 45$

37. $20, 18, 16, 14, \ldots, 0$

38. $-1, -4, -7, -10, \ldots, -25$

39. $8, 5, 2, -1, \ldots, -16$

40. $\frac{1}{2}, \frac{5}{2}, \frac{9}{2}, \frac{13}{2}, \ldots, \frac{29}{2}$

41. $-9, -\frac{17}{2}, -8, -\frac{15}{2}, \ldots, -\frac{1}{2}$

42. $100, 110, 120, 130, \ldots, 190$

43. $10, 5, 0, -5, \ldots, -20$

44. $\frac{3}{5}, \frac{4}{5}, \frac{5}{5}, \frac{6}{5}, \ldots, 4$

In Exercises 45–54, write the first five terms of the geometric sequence with the first term, a_1, and common ratio, r.

45. $a_1 = 3$, $r = 2$

46. $a_1 = 2$, $r = 3$

47. $a_1 = 5$, $r = -2$

48. $a_1 = 3$, $r = \frac{1}{2}$

49. $a_1 = -2$, $r = -1$

50. $a_1 = 6$, $r = -3$

51. $a_1 = -2$, $r = \frac{1}{2}$

52. $a_1 = -10$, $r = -\frac{1}{2}$

53. $a_1 = 5$, $r = \frac{3}{5}$

54. $a_1 = -4$, $r = -\frac{2}{3}$

In Exercises 55–64, find the indicated term for the geometric sequence with the first term, a_1, and common ratio, r.

55. Find a_6 when $a_1 = 2$, $r = 3$

56. Find a_4 when $a_1 = 5$, $r = 2$

57. Find a_5 when $a_1 = 3$, $r = 2$

58. Find a_6 when $a_1 = -2$, $r = -2$

59. Find a_4 when $a_1 = -8$, $r = -3$

60. Find a_7 when $a_1 = 10$, $r = -3$

61. Find a_3 when $a_1 = 3$, $r = \frac{1}{2}$

62. Find a_7 when $a_1 = -3$, $r = -3$

63. Find a_5 when $a_1 = \frac{1}{2}$, $r = 2$

64. Find a_4 when $a_1 = 4$, $r = \frac{1}{3}$

In Exercises 65–73, write an expression for the general or nth term, a_n, for the geometric sequence.

65. $2, 4, 8, 16, \ldots$

66. $3, 12, 48, 192, \ldots$

67. $1, 3, 9, 27, \ldots$

68. $-6, 6, -6, 6, \ldots$

69. $-8, -4, -2, -1, \ldots$

70. $\frac{1}{3}, 1, 3, 9, \ldots$

71. $-6, 18, -54, 162, \ldots$

72. $10, \frac{10}{3}, \frac{10}{9}, \frac{10}{27}, \ldots$

73. $-4, -\frac{8}{3}, -\frac{16}{9}, -\frac{32}{27}, \ldots$

In Exercises 74–81, find the sum of the first n terms of the geometric sequence for the values of a_1 and r.

74. $n = 3$, $a_1 = 2$, $r = 5$

75. $n = 4$, $a_1 = 2$, $r = 2$

76. $n = 4$, $a_1 = 5$, $r = 3$

77. $n = 5$, $a_1 = 3$, $r = 5$

78. $n = 6$, $a_1 = 3$, $r = 4$

79. $n = 5$, $a_1 = 4$, $r = 3$

80. $n = 4$, $a_1 = -6$, $r = 2$

81. $n = 5$, $a_1 = -3$, $r = -2$

82. Find the sum of the first 50 natural numbers.

83. Find the sum of the first 50 even natural numbers.

84. Find the sum of the first 50 odd natural numbers.

85. Find the sum of the first 20 multiples of 3.

86. Donna is given a starting salary of $20,200 and promised a $1200 raise per year after each of the next eight years.

    a) Determine her salary during her eighth year of work.

    b) Determine the total salary she received over the 8 years.

87. Each swing of a pendulum (from far left to far right) is 3 in. shorter than the preceding swing. The first swing is 8 ft.
  a) Find the length of the twelfth swing.
  b) Determine the total distance traveled by the pendulum during the first 12 swings.

88. Each time a ball bounces, the height attained by the ball is 6 in. less than the previous height attained. If on the first bounce the ball reaches a height of 6 ft, find the height attained on the eleventh bounce.

89. If you are given $1 on January 1, $2 on January 2, $3 on January 3, and so on, how much money will you accumulate during the month of January (31 days)?

90. A certain substance decomposes and loses 20% of its weight each hour. If there are originally 200 g of the substance, how much remains after 6 hr?

91. If your salary were to increase at a rate of 6% per year, find your salary during your fifteenth year if your original salary is $20,000.

92. The population in the United States in 1995 was 261.5 million. If the population grows at a rate of 6% per year, find the population in 12 yr.

93. When dropped, a ball rebounds to four-fifths of its original height. How high will the ball rebound after the fourth bounce if it is dropped from a height of 30 ft?

94. A clock strikes once at 1 o'clock, twice at 2 o'clock, and so on. How many times does it strike over a 12-hr period?

Problem Solving/Group Activities

95. A geometric sequence has $a_1 = 82$ and $r = \frac{1}{2}$; find s_6.

96. The sums of the interior angles of a triangle, a quadrilateral, a pentagon, and a sextagon are 180°, 360°, 540°, and 720°, respectively. Use this pattern to find a formula for the general term, a_n, where a_n represents the sum of the interior angles of an n-sided quadrilateral.

97. Determine how many numbers between 7 and 1610 are divisible by 6.

98. Find r and a_1 for the geometric sequence with $a_2 = 24$ and $a_5 = 648$.

99. A ball is dropped from a height of 30 ft. On each bounce it attains a height four-fifths of its original height (or of the previous bounce). Find the total vertical distance traveled by the ball after it has completed its fifth bounce (therefore has hit the ground six times).

100. In calculus, an important topic is *limits*. Consider $a_n = 1/x$. Since the value of $1/x$ gets closer and closer to 0 as x gets larger and larger, we say that the limit of a_n as x approaches infinity is 0. Note that $1/x$ can never equal 0, but its value approaches 0 as x gets larger and larger. Find the limit of the following expressions as x approaches infinity.

  a) $a_n = \dfrac{1}{x - 3}$

  b) $a_n = \dfrac{x}{x + 1}$

  c) $a_n = \dfrac{1}{x^2 + 2}$

  d) $a_n = \dfrac{3x + 2}{x}$

  e) $a_n = \dfrac{6x^2 - 5}{2x - 1}$

Research Activity

101. A topic generally associated with sequences is *series*.
  a) Research *series* in several mathematics texts (intermediate algebra or college algebra texts). Explain what a series is and how it differs from a sequence. Write a formal definition. Give examples of different kinds of series.
  b) Write the arithmetic series associated with the arithmetic sequence 1, 4, 7, 10, 13, . . .
  c) Write the geometric series associated with the geometric sequence 3, 6, 12, 24, 48, . . .
  d) What is an infinite geometric series?
  e) Find the sum of the terms of the infinite geometric series $1 + \dfrac{1}{2} + \dfrac{1}{4} + \dfrac{1}{8} + \dfrac{1}{16} + \cdots$.

5.8 FIBONACCI SEQUENCE

We cannot leave the topic of sequences without discussing an interesting sequence known as the Fibonacci sequence. Leonardo of Pisa, known as Fibonacci, is often considered the most distinguished mathematician of the Middle Ages. He was born in Italy but was sent by his father to study calculations with an Arab master. His book, *Liber Abacci* (Book of the Abacus), contains interesting problems that Fibonacci liked to invent, such as: "A certain man put a pair of rabbits in a place surrounded on all sides by a wall. How many pairs of rabbits can be

produced from that pair in a year if it is assumed that every month each pair begets a new pair which from the second month becomes productive?''

The solution to this problem (Fig. 5.10) led to the development of the sequence that bears its author's name: the Fibonacci sequence. The sequence is shown in Table 5.1. The numbers in the column titled *Pairs of Adults* form the Fibonacci sequence.

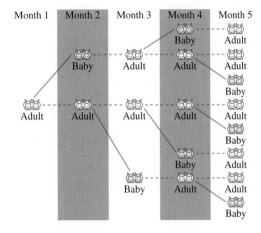

Figure 5.10

Fibonacci Sequence

1, 1, 2, 3, 5, 8, 13, 21, . . .

In the Fibonacci sequence the first and second terms are 1. The sum of these two terms is the third term. The sum of the second and third terms is the fourth term, and so on.

In the middle of the nineteenth century, mathematicians made a serious study of this sequence and found strong similarities between it and many natural phenomena. Fibonacci numbers appear in the seed arrangement of many species of plants and in the petal counts of various flowers. For example, when the flowering head of the sunflower matures to seed, the seeds' spiral arrangement becomes clearly visible. A typical count of these spirals may give 89 steeply curving to the right, 55 curving more shallowly to the left, and 34 again shallowly to the right. The largest known specimen to be examined had spiral counts of 144 right, 89 left, and 55 right. These numbers, like the other three mentioned, are consecutive terms of the Fibonacci sequence.

On the heads of many flowers, petals (or florets in composite plants) surrounding the central disk generally yield a Fibonacci number. For example, some daisies contain 21 petals, and others contain 34, 55, or 89 petals. (People who use a daisy to play the ''love me, love me not'' game will likely pluck 21, 34, 55, or 89 petals before arriving at an answer.)

Fibonacci numbers are also observed in the structure of pinecones and pineapples. The tablike or scalelike structures called bracts that make up the main body of the pinecone form a set of spirals that start from the cone's attachment to the branch. Two sets of oppositely directed spirals can be observed, one steep and the other more gradual. A count on the steep spiral will reveal a Fibonacci

Table 5.1

Month	Pairs of Adults	Pairs of Babies	Total Pairs
1	1	0	1
2	1	1	2
3	2	1	3
4	3	2	5
5	5	3	8
6	8	5	13
7	13	8	21
8	21	13	34
9	34	21	55
10	55	34	89
11	89	55	144
12	144	89	233

Many objects in nature exhibit patterns of Fibonacci numbers.

Table 5.2

Numbers	Ratio
1, 1	$\frac{1}{1} = 1$
1, 2	$\frac{2}{1} = 2$
2, 3	$\frac{3}{2} = 1.5$
3, 5	$\frac{5}{3} = 1.66\ldots$
5, 8	$\frac{8}{5} = 1.6$
8, 13	$\frac{13}{8} = 1.625$

Figure 5.11

number and a count on the gradual one will be the adjacent smaller Fibonacci number, or if not, the next smaller Fibonacci number. One investigation of 4290 pinecones from 10 species of pine trees found in California revealed that only 74 cones, or 1.7%, deviated from this Fibonacci pattern.

Like pinecone bracts, pineapple scales are patterned into spirals, and because they are roughly hexagonal in shape, three distinct sets of spirals can be counted.

Fibonacci Numbers and Divine Proportions

In 1753, while studying the Fibonacci sequence, Robert Simson, a mathematician at the University of Glasgow, noticed that when he took the ratio of any term to the term that immediately preceded it, the value he obtained remained in the vicinity of one specific number. To illustrate this, we indicate in Table 5.2 the ratio of various pairs of sequential Fibonacci numbers.

The ratio of the 50th term to the 49th term is 1.6180. Simson proved that the ratio of the $(n + 1)$ term to the nth term as n gets larger and larger is the irrational number $(\sqrt{5} + 1)/2$, which begins $1.61803\ldots$. This number was already well known to mathematicians at that time as the golden number.

Many years earlier the Bavarian astronomer and mathematician Johannes Kepler wrote that for him the golden number symbolized the Creator's intention ''to create like from like.'' The golden number $(\sqrt{5} + 1)/2$ is frequently referred to as ''phi,'' symbolized by the Greek letter Φ.

The ancient Greeks, in about the sixth century B.C., sought unifying principles of beauty and perfection, which they believed could be described by using mathematics. In their study of beauty the Greeks used the term *golden ratio*. In order to understand the golden ratio, let's consider the line segment AB in Fig. 5.11. When this line segment is divided at a point C, such that the ratio of the whole, AB, to the larger part, AC, is equal to the ratio of the larger part, AC, to the smaller part, CB, then each ratio AB/AC and AC/CB is referred to as a golden ratio. And the proportion they form, $AB/AC = AC/CB$, is called the golden proportion. Furthermore each ratio in the proportion will have a value equal to the golden number, $(\sqrt{5} + 1)/2$.

$$\frac{AB}{AC} = \frac{AC}{CB} = \frac{\sqrt{5} + 1}{2} \approx 1.618$$

The Great Pyramid of Gizeh in Egypt, built about 2600 B.C., is the earliest known example of use of the golden ratio in architecture. The ratio of any of its sides of the square base (775.75 ft) to its altitude (481.4 ft) is about 1.611. Other evidence of the use of the golden ratio appears in other Egyptian buildings and tombs.

In medieval times people referred to the golden proportion as the divine proportion, reflecting their belief in its relationship to the will of God.

The twentieth-century architect Le Corbusier developed a scale of proportions for the human body that he called the Modulor (Fig. 5.12). Note that the navel separates the entire body into golden proportions, as does the neck and knee.

From the golden proportion the golden rectangle can be formed, as shown in Fig. 5.13.

$$\frac{\text{Length}}{\text{Width}} = \frac{a + b}{a} = \frac{a}{b} = \frac{\sqrt{5} + 1}{2}$$

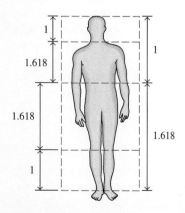

Figure 5.12

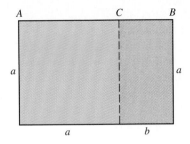

Figure 5.13

Note that when a square is cut off one end of a golden rectangle, as in Fig. 5.13, the remaining rectangle has the same properties as the original golden rectangle (creating "like from like" as Johannes Kepler had written) and is therefore itself a golden rectangle. Interestingly, the curve derived from a succession of diminishing golden rectangles, as shown in Fig. 5.14, is the same as the spiral curve of the chambered nautilus. The same curve appears on the horns of rams and some other animals. It is the same curve observed in the plant structures arranged in Fibonacci sequence mentioned earlier—in sunflowers, other flower heads, pinecones, and pineapples. The curve closely approximates what mathematicians call a *logarithmic spiral*.

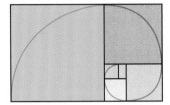 

Figure 5.14

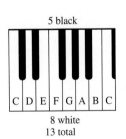

Ancient Greek civilization used the golden rectangle in art and architecture. The main measurements of many buildings of antiquity, including the Parthenon in Athens, are governed by golden ratios and rectangles. Greek statues, vases, urns, and so on also exhibit characteristics of the golden ratio. It is for Phidas, considered the greatest of Greek sculptors, that the golden ratio was named "phi." The proportions can be found abundantly in his work.

The proportions of the golden rectangle can be found in the works of many artists, from the old masters to the moderns. For example the golden rectangle can be seen in the painting *La Parade* by George Seurat, a French neoimpressionist artist, in the chapter opening display.

Fibonacci numbers are also found in another form of art, namely music. Perhaps the most obvious link between Fibonacci numbers and music can be found on the piano keyboard. An octave (Fig. 5.15) on a keyboard has 13 keys: 8 white, and 5 black (the 5 are in one group of 2 and a group of 3).

In Western music, the most complete scale, the chromatic scale, consists of 13 notes. Its predecessor, the diatonic scale, contains 8 notes (an octave). The diatonic scale was preceded by a 5-note "pentatonic scale" (*penta* is Greek for "five"). Each number is a Fibonacci number.

The visual arts deal with what is pleasing to the eye, whereas musical composition deals with what is pleasing to the ear. While art achieves some of its goals by using division of planes and area, music achieves some of its goals by a similar division of time, using notes of various duration and spacing. The musical intervals considered by many to be the most pleasing to the ear are the major sixth and minor sixth. A major sixth, for example, consists of the note C, vibrating at about 264* vibrations per second, and note A, vibrating at about 440 vibrations per second. The ratio of 440 to 264 reduces to 5 to 3, or $\frac{5}{3}$, a ratio of two consecutive Fibonacci numbers. An example of a minor sixth would be E (about 330 vibrations per second) and C (about 528 vibrations per second). The ratio 528 to

Figure 5.15

*Frequencies of notes vary in different parts of the world, and change over time.

330 reduces to 8 to 5, or $\frac{8}{5}$, the next ratio of two consecutive Fibonacci numbers. The vibrations of any sixth interval reduce to a similar ratio.

Patterns that can be expressed mathematically in terms of Fibonacci relationships have been found in Gregorian chants and works of many composers, including Bach, Beethoven, and Bartók. A number of twentieth-century works, including Ernst Krened's *Fibonacci Mobile*, have been deliberately structured by using Fibonacci proportions.

A number of studies have tried to explain why the Fibonacci series and related items are linked to so many real-life situations. It appears that the Fibonacci numbers are a part of natural harmony that is pleasing to both the eye and the ear. In the nineteenth century the German physicist and psychologist Gustav Fechner tried to determine which dimensions were most pleasing to the eye. Fechner, along with the psychologist Wilhelm Wundt, found that most people do unconsciously favor golden dimensions when purchasing greeting cards, mirrors, and other rectangular objects. This discovery has been widely used by commercial manufacturers in their packaging and labeling designs, by retailers in their store displays, and in other areas of business and advertising.

*D*ID YOU KNOW

MATHEMATICAL MUSIC

*T*he ancient Greeks are considered by many to have been the first true mathematicians. They studied mathematics not because of its applications, but because of its beauty. They believed that in nature, all harmony and everything of beauty could be explained with ratios of whole numbers (called rational numbers). This belief was reinforced by the discovery that the sound of plucked strings could be quite pleasing if the strings plucked were in the ratio of 1 to 2 (an octave), 2 to 3 (a fifth), 3 to 4 (a fourth), and so on. In other words, the secret of harmony lies in the ratio of whole numbers such as 1/2, 2/3, and 3/4.

The theory of vibrating strings has applications today that go well beyond music. How materials vibrate, and hence the stress they can absorb, is a matter important in the construction of rockets, buildings, and bridges. ■

The roadway in a suspension bridge is not unlike a string stretched between two points and so vibration must be kept to a minimum. This was the lesson learned in 1940, in the state of Washington when high winds set the Tacoma Narrows bridge into a pattern of vibration that ripped the roadway from its supports.

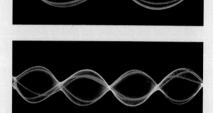

When the string of a musical instrument is plucked or bowed, it moves in a wavelike pattern like the strings shown here. The vibration this creates in the surrounding air is what your eardrums detect as sound.

Section 5.8 Exercises

1. Explain how to construct the Fibonacci sequence.

2. **a)** Find the eighth, ninth, and tenth terms of the Fibonacci sequence.
   b) Divide the ninth term by the eighth term, rounding to the nearest thousandth.
   c) Divide the tenth term by the ninth term, rounding to the nearest thousandth.
   d) Try a few more divisions and then make a conjecture about the result.

3. The ratio of the a_{n+1}/a_n terms of the Fibonacci sequence approaches 1.6180 (rounded to four decimal places) as n increases. Find the ratio of a_n/a_{n+1} rounded to four decimal places as n increases.

4. What is the golden number? Explain the relationship between the golden number and the golden proportion.

5. Select a piece of art and see if you can determine whether the artist used the golden rectangle. Write a brief description of your findings.

6. Explain what is meant by the divine proportion.

7. **a)** To what decimal value is $(\sqrt{5} + 1)/2$ approximately equal?
   b) To what decimal value is $(\sqrt{5} - 1)/2$ approximately equal?
   c) By how much do the results in part (a) and (b) differ?

8. The eleventh Fibonacci number is 89. Examine the first six terms in the decimal expression of its reciprocal, $\frac{1}{89}$. What do you find?

9. Find the ratio of the second to the first term of the Fibonacci sequence. Then find the ratio of the third to the second term of the sequence and determine whether this ratio was an increase or decrease from the first ratio. Continue this process for 10 ratios and then make a conjecture regarding the increasing or decreasing values in consecutive ratios.

10. A musical composition is described as follows. Explain why this piece is based upon the golden ratio.

Entire Composition

34 measures	55 measures	21 measures	34 measures
Theme	Fast, Loud	Slow	Repeat of theme

11. The greatest common factor of any two consecutive Fibonacci numbers is 1. Show this is true for the first 15 Fibonacci numbers.

12. The sum of any 10 consecutive Fibonacci numbers is always divisible by 11. Select any 10 consecutive Fibonacci numbers and show that for your selection this is true.

13. Twice any Fibonacci number minus the next Fibonacci number equals the second number preceding the original number. Select a number in the Fibonacci sequence and show that this pattern holds for the number selected.

14. For any four consecutive Fibonacci numbers, the difference of the squares of the middle two numbers equals the product of the smallest and largest numbers. Select four consecutive Fibonacci numbers and show that this pattern holds for the numbers you selected.

15. **a)** What are the length and width of standard size index cards?
    b) Determine the ratio of the length to the width of the index cards and compare the ratio to Φ.

16. Determine the ratio of the length to the width of this textbook and compare this ratio to Φ.

17. Determine the ratio of the length to the width of your television screen and compare this ratio to Φ.

18. Find three physical objects whose dimensions are very close to a golden rectangle.
    a) List the articles and record the dimensions.
    b) Compute the ratios of their lengths to their widths.
    c) Find the difference between the golden ratio and the ratio you obtain in part (b)—to the nearest tenth—for each object.

In Exercises 19–24, determine whether the sequence is a Fibonacci-type sequence (each term after the second term is the sum of the two preceding terms). If it is, determine the next two terms of the sequence.

19. 1, 3, 4, 7, 11, 18, . . .   **20.** 2, 3, 5, 8, 21, . . .

21. 2, 2, 2, 2, . . .   **22.** 15, 30, 45, 60, 75, . . .

23. 15, 30, 45, 75, 120, . . .   **24.** $\frac{1}{2}, \frac{1}{2}, 1, \frac{3}{2}, \frac{5}{2}, 4, \frac{13}{2}, \ldots$

25. **a)** Select any two nonzero digits and add them to obtain a third digit. Continue adding the two previous terms to get a Fibonacci-type sequence.
    b) Form ratios of successive terms to show how they will eventually approach the golden number.

26. Repeat Exercise 25 for two different nonzero numbers.

27. a) Select any three consecutive terms of a Fibonacci sequence. Subtract the product of the terms on each side of the middle term from the square of the middle term. What is the difference?
  b) Repeat part (a) with three different consecutive terms of the sequence.
  c) Make a conjecture about what will happen when you repeat this process for any three consecutive terms of a Fibonacci sequence.

28. One of the most famous number patterns involves **Pascal's triangle**. The Fibonacci sequence can be found by using Pascal's triangle. Can you explain how that can be done? A hint is shown.

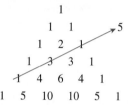

29. a) A sequence related to the Fibonacci sequence is the **Lucas sequence**. The Lucas sequence is formed in a manner similar to the Fibonacci sequence. The first two numbers of the Lucas sequence are 1 and 3. Write the first eight terms of the Lucas sequence.
  b) Complete the next two lines of the following chart.

$$1 + 2 = 3$$
$$1 + 3 = 4$$
$$2 + 5 = 7$$
$$3 + 8 = 11$$
$$5 + 13 = 18$$

  c) What do you observe about the first column in the chart in part (b)?

30. When two panes of glass are placed face to face, four interior reflective surfaces exist labeled 1, 2, 3, and 4. If light is not reflected, it has just one path through the glass. If it has one reflection, it can be reflected in two ways. If it has two reflections, it can be reflected in three ways. Use this information to answer parts (a)–(c).

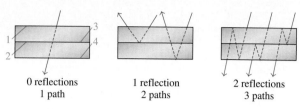

| 0 reflections | 1 reflection | 2 reflections |
| 1 path | 2 paths | 3 paths |

  a) If a ray is reflected three times, there are five paths it can follow. Show the paths.
  b) If a ray is reflected four times, there are eight paths it can follow. Show the paths.
  c) How many paths can a ray follow if it is reflected five times? Explain how you determined your answer.

Problem Solving/Group Activities

31. The divine proportion is $(a + b)/a = a/b$ (see Fig. 5.13). This can be written $1 + (b/a) = a/b$. Now let $x = a/b$, which gives $1 + (1/x) = x$. Multiply both sides of this equation by x to get a quadratic equation and then use the quadratic formula (Section 6.8) to show that one answer is $x = (1 + \sqrt{5})/2$ (the golden ratio).

32. Draw a line of length 5 in. Determine and mark the point on the line that will create the golden ratio. Explain how you determined your answer.

Research Activities

33. The digits 1 through 9 have evolved considerably since they first appeared in Fibonacci's book *Liber Abacci*. Write a report tracing the history of the evolution of the digits 1–9 since Fibonacci's time.

34. Write a report on the history and mathematical contributions of Fibonacci.

35. Write a report indicating where the golden ratio and golden rectangle have been used in art and architecture. An art teacher or a staff member of an art museum might be able to give you some information and a list of resources. You may also wish to contact the Fibonacci Association at the Department of Mathematics, Santa Clara University, Santa Clara, CA 95053.

CHAPTER 5 SUMMARY

Key Terms

5.1

composite number
counting (or natural) number
divisible
divisor
factor
Fermat number
greatest common divisor
least common multiple
Mersenne prime
number theory
prime factorization
prime number
relatively prime
theorem

5.2

additive inverse
integer
whole number

5.3

denominator
dense set
fraction
lowest terms
numerator
rational number
reciprocal

5.4

irrational number
perfect square

principal (or positive) square root
Pythagorean theorem
radical sign
radicand
rationalized denominator

5.5

associative property
closure
commutative property
distributive property
set of real numbers

5.6

base
exponent
scientific notation

5.7

arithmetic sequence
common difference
common ratio
geometric sequence
sequence
terms

5.8

divine proportion
Fibonacci sequence
golden number
golden proportion
golden ratio
golden rectangle

Important Facts

Fundamental Theorem of Arithmetic

Every composite number can be expressed as a unique product of prime numbers.

Sets of Numbers

Natural or counting numbers: $\{1, 2, 3, 4, \ldots\}$
Whole numbers: $\{0, 1, 2, 3, 4, \ldots\}$
Integers: $\{\ldots, -3, -2, -1, 0, 1, 2, 3, \ldots\}$
Rational numbers: Numbers of the form p/q where p and q are integers, $q \neq 0$. Every rational number when expressed as a decimal number will be either a terminating or repeating decimal number.
Irrational number: A real number whose representation is a nonterminating, nonrepeating decimal number (not a rational number).

Definition of Subtraction

$a - b = a + (-b)$

Fundamental Law of Rational Numbers

$$\frac{a}{b} = \frac{a}{b} \cdot \frac{c}{c} = \frac{ac}{bc}, \qquad b \neq 0, \quad c \neq 0$$

Rules of Radicals

Product rule for radicals:

$$\sqrt{a \cdot b} = \sqrt{a} \cdot \sqrt{b}, \qquad a \geq 0, \quad b \geq 0$$

Quotient rule for radicals:

$$\frac{\sqrt{a}}{\sqrt{b}} = \sqrt{\frac{a}{b}}, \qquad a \geq 0, \quad b > 0$$

Properties of Real Numbers

Commutative property of addition: $a + b = b + a$
Commutative property of multiplication: $a \cdot b = b \cdot a$
Associative property of addition:

$$(a + b) + c = a + (b + c)$$

Associative property of multiplication:

$$(a \cdot b) \cdot c = a \cdot (b \cdot c)$$

Distributive property: $a \cdot (b + c) = ab + ac$

Rules of Exponents

Product rule for exponents: $a^m \cdot a^n = a^{m+n}$

Quotient rule for exponents: $\dfrac{a^m}{a^n} = a^{m-n}, \quad a \neq 0$

Zero exponent rule: $a^0 = 1, \quad a \neq 0$

Negative exponent rule: $a^{-m} = \dfrac{1}{a^m}, \quad a \neq 0$

Power rule: $(a^m)^n = a^{m \cdot n}$

Arithmetic Sequence

$$a_n = a_1 + (n - 1)d$$

$$s_n = \frac{n(a_1 + a_n)}{2}$$

Geometric Sequence

$$a_n = a_1 r^{n-1}$$

$$s_n = \frac{a_1(1 - r^n)}{1 - r}, \quad r \neq 1$$

Fibonacci Sequence

$1, 1, 2, 3, 5, 8, 13, 21, \ldots$

Golden Number

$$\frac{\sqrt{5} + 1}{2} \approx 1.618$$

Golden Proportion

$$\frac{a + b}{a} = \frac{a}{b}$$

Chapter 5 Review Exercises

5.1

In Exercises 1 and 2, determine whether the number is divisible by each of the following numbers: 2, 3, 4, 5, 6, 8, 9, and 10.

1. 148,632   **2.** 400,644

In Exercises 3–7, find the prime factorization of the number.

3. 296   **4.** 240   **5.** 360

6. 1260   **7.** 960

In Exercises 8–13, find the GCD and the LCM of the numbers.

8. 16, 48   **9.** 75, 40   **10.** 148, 216

11. 840, 320   **12.** 60, 40, 96   **13.** 36, 108, 144

5.2

In Exercises 14–22, use the number line to evaluate the expression.

14. $-6 + 2$ **15.** $9 + (-7)$

16. $2 - 7$ **17.** $-3 + (-2)$

18. $-4 - 3$ **19.** $-4 - (-5)$

20. $(-3 + 7) - 4$ **21.** $-1 + (9 - 4)$

22. $-1 - (9 - 4)$

In Exercises 23–30, evaluate the expression.

23. $(-3)(-8)$ **24.** $-6(5)$ **25.** $6(-4)$

26. $\dfrac{-20}{-4}$ **27.** $\dfrac{8}{-2}$

28. $[8 \div (-4)](-3)$

29. $[(-4)(-3)] \div 2$

30. $[(-30) \div (10)] \div (-1)$

5.3

In Exercises 31–39, express the fraction as a terminating or repeating decimal.

31. $\frac{3}{5}$ **32.** $\frac{8}{10}$ **33.** $\frac{8}{12}$

34. $\frac{15}{4}$ **35.** $\frac{2}{7}$ **36.** $\frac{5}{12}$

37. $\frac{3}{8}$ **38.** $\frac{7}{8}$ **39.** $\frac{3}{7}$

In Exercises 40–48, express the decimal number as a quotient of two integers.

40. 0.582 **41.** $0.\overline{6}$ **42.** 2.43

43. $1.\overline{71}$ **44.** 12.083 **45.** 0.0042

46. $2.1\overline{5}$ **47.** $2.\overline{34}$ **48.** $5.0\overline{62}$

In Exercises 49–54, find a rational number between the two numbers in the pair.

49. 0.0061 and 0.0062 **50.** $\frac{3}{4}$ and $\frac{3}{5}$

51. 3.509 and 3.510 **52.** $\frac{9}{12}$ and $\frac{10}{12}$

53. $\frac{6}{12}$ and 0.51 **54.** 0.2 and $\frac{3}{9}$

In Exercises 55–63, perform the indicated operation and reduce your answer to lowest terms.

55. $\dfrac{1}{4} + \dfrac{1}{5}$ **56.** $\dfrac{3}{5} - \dfrac{2}{4}$

57. $\dfrac{5}{18} + \dfrac{3}{4}$ **58.** $\dfrac{4}{5} \cdot \dfrac{7}{9}$

59. $\dfrac{5}{9} \div \dfrac{6}{7}$ **60.** $\left(\dfrac{4}{5} + \dfrac{5}{7}\right) \div \dfrac{4}{5}$

61. $\left(\dfrac{2}{3} \cdot \dfrac{1}{7}\right) \div \dfrac{4}{7}$ **62.** $\left(\dfrac{1}{5} + \dfrac{2}{3}\right)\left(\dfrac{3}{8}\right)$

63. $\left(\dfrac{1}{5} \cdot \dfrac{2}{3}\right) + \left(\dfrac{1}{5} \div \dfrac{1}{2}\right)$

5.4

In Exercises 64–78, simplify the expression. Rationalize the denominator when necessary.

64. $\sqrt{12}$ **65.** $\sqrt{72}$ **66.** $\sqrt{2} + 3\sqrt{2}$

67. $\sqrt{3} - 4\sqrt{3}$ **68.** $\sqrt{8} + 6\sqrt{2}$ **69.** $\sqrt{3} - 7\sqrt{27}$

70. $\sqrt{75} + \sqrt{27}$ **71.** $\sqrt{3} \cdot \sqrt{6}$ **72.** $\sqrt{8} \cdot \sqrt{6}$

73. $\dfrac{\sqrt{18}}{\sqrt{2}}$ **74.** $\dfrac{\sqrt{56}}{\sqrt{2}}$ **75.** $\dfrac{3}{\sqrt{2}}$

76. $\dfrac{\sqrt{3}}{\sqrt{5}}$ **77.** $5(3 + \sqrt{5})$ **78.** $\sqrt{3}(4 + \sqrt{6})$

5.5

In Exercises 79–88, state the name of the property illustrated.

79. $x + 2 = 2 + x$

80. $3 \cdot x = x \cdot 3$

81. $(2 + 3) + 4 = 2 + (3 + 4)$

82. $5 \cdot (2 + x) = 5 \cdot 2 + 5 \cdot x$

83. $(6 + 3) + 4 = 4 + (6 + 3)$

84. $(3 + 5) + (4 + 3) = (4 + 3) + (3 + 5)$

85. $(3 \cdot a) \cdot b = 3 \cdot (a \cdot b)$

86. $a \cdot (2 + 3) = (2 + 3) \cdot a$

87. $(x + 3)2 = (x \cdot 2) + (3 \cdot 2)$

88. $x \cdot 2 + 6 = 2 \cdot x + 6$

In Exercises 89–94, determine whether the set of numbers is closed under the given operation.

89. Natural numbers, addition

90. Integers, addition

91. Integers, division

92. Real numbers, subtraction

93. Irrational numbers, multiplication

94. Rational numbers, division

5.6

In Exercises 95–102, evaluate each expression.

95. 3^3 **96.** 3^{-2} **97.** $\dfrac{5^5}{5^4}$

98. $6^2 \cdot 6$ **99.** 3^0 **100.** 5^{-3}

101. $(2^3)^2$ **102.** $(3^2)^2$

In Exercises 103–106, write each number in scientific notation.

103. 6310 **104.** 0.0000423

105. 0.00275 **106.** 4,950,000

In Exercises 107–110, express each number in decimal notation.

107. 8.7×10^2 **108.** 3.87×10^{-5}

109. 1.75×10^{-4} **110.** 1×10^5

In Exercises 111–114, perform the indicated operation and express your answer in scientific notation.

111. $(2 \times 10^6)(3.2 \times 10^{-4})$ **112.** $(3 \times 10^2)(4.6 \times 10^2)$

113. $\dfrac{8.4 \times 10^3}{4 \times 10^2}$ **114.** $\dfrac{1.5 \times 10^{-3}}{5 \times 10^{-4}}$

In Exercises 115–118, perform the indicated operation by first converting each number to scientific notation. Write your answer in decimal notation.

115. $(80{,}000)(420{,}000)$ **116.** $(75{,}000)(0.0003)$

117. $\dfrac{9{,}600{,}000}{3000}$ **118.** $\dfrac{0.000002}{0.0000004}$

119. At noon there were 12,000 bacteria in the culture. At 6:00 P.M. there were 300,000 bacteria in the culture. How many times greater is the number of bacteria at 6:00 P.M. than at noon?

5.7

In Exercises 120–125, determine whether the sequence is arithmetic or geometric. Then determine the next two terms of the sequence.

120. $3, 8, 13, 18, \ldots$ **121.** $-4, 12, -36, 108, \ldots$

122. $0, -4, -8, -12, \ldots$ **123.** $1, \frac{1}{2}, \frac{1}{4}, \frac{1}{8}, \ldots$

124. $1, 4, 7, 10, 13, \ldots$ **125.** $2, -2, 2, -2, 2, \ldots$

In Exercises 126–131, write the first five terms of the sequence with the given first term, a_1, and common difference, d, or ratio, r.

126. $a_1 = 4, d = 5$ **127.** $a_1 = -6, d = -3$

128. $a_1 = -\frac{1}{2}, d = -2$ **129.** $a_1 = 4, r = 2$

130. $a_1 = 16, r = \frac{1}{2}$ **131.** $a_1 = \frac{1}{2}, r = -\frac{1}{2}$

In Exercises 132–137, find the indicated term of the sequence with the given first term, a_1, and common difference, d, or ratio, r.

132. Find a_5, when $a_1 = 6, d = 2$

133. Find a_7, when $a_1 = -6, d = -5$

134. Find a_{10}, when $a_1 = 20, d = 8$

135. Find a_4, when $a_1 = 8, r = 3$

136. Find a_5, when $a_1 = 4, r = \frac{1}{2}$

137. Find a_4, when $a_1 = -6, r = 2$

In Exercises 138–140, find the sum of the arithmetic sequence.

138. $4, 2, 0, -2, \ldots, -18$

139. $-4, -3\frac{3}{4}, -3\frac{1}{2}, -3\frac{1}{4}, \ldots, -2\frac{1}{4}$

140. $100, 94, 88, 82, \ldots, 58$

In Exercises 141–143, find the sum of the first n terms of the geometric sequence for the values of a_1 and r.

141. $n = 3, a_1 = 4, r = 3$

142. $n = 4, a_1 = 2, r = 4$

143. $n = 5, a_1 = 3, r = -2$

In Exercises 144–149, first determine whether the sequence is arithmetic or geometric; then write an expression for the general or nth term, a_n.

144, $7, 4, 1, -2, \ldots$

145. $0, 5, 10, 15, \ldots$

146. $4, \frac{5}{2}, 1, -\frac{1}{2}, \ldots$

147. $3, 6, 12, 24, \ldots$

148. $4, -4, 4, -4, \ldots$

149. $5, \frac{5}{3}, \frac{5}{9}, \frac{5}{27}, \ldots$

5.8

In Exercises 150–153, determine whether the sequence is a Fibonacci-type sequence. If so determine its next two terms.

150. $1, 2, 3, 5, 8, 13, \ldots$

151. $2, 4, 8, 16, 32, \ldots$

152. $3, 8, 13, 18, 23, \ldots$

153. $2, 2, 4, 6, 10, 16, \ldots$

CHAPTER 5 TEST

1. Which of the numbers 2, 3, 4, 5, 6, 8, 9, and 10 divide 481,248?

2. Find the prime factorization of 280.

3. Evaluate $[(-6) + (-9)] + 8$.

4. Evaluate $-6 - 18$.

5. Evaluate $[(-70)(-5)] \div (8 - 10)$.

6. Determine a rational number between 0.873 and 0.874.

7. Determine a rational number between $\frac{1}{8}$ and $\frac{1}{9}$.

8. Write $\frac{5}{8}$ as a terminating or repeating decimal.

9. Express 6.45 as a quotient of two integers.

10. Evaluate $(\frac{5}{16} \div 3) + (\frac{4}{5} \cdot \frac{1}{2})$.

11. Perform the operation and reduce the answer to lowest terms:

$$\frac{17}{24} - \frac{3}{40}$$

12. Simplify $\sqrt{18} + \sqrt{50}$.

13. Rationalize $\dfrac{\sqrt{5}}{\sqrt{6}}$.

14. Determine whether the integers are closed under the operation of multiplication. Explain your answer.

Name the properties illustrated.

15. $(2 + x) + 3 = 2 + (x + 3)$

16. $3(x + y) = 3x + 3y$

Evaluate.

17. $\dfrac{6^7}{6^5}$ 18. $4^3 \cdot 4^2$ 19. 9^{-2}

20. Perform the operation by first converting the numerator and denominator to scientific notation. Write the answer in scientific notation.

$$\frac{64,000}{0.008}$$

21. Write an expression for the general or nth term, a_n, of the sequence $-4, -8, -12, -16, \ldots$.

22. Find the sum of the terms of the arithmetic sequence $-2, -5, -8, -11, \ldots, -32$.

23. Find a_5 when $a_1 = 3$ and $r = 3$.

24. Find the sum of the first five terms of the sequence when $a_1 = 3$ and $r = 4$.

25. Write an expression for the general or nth term, a_n, of the sequence $3, 6, 12, 24, \ldots$.

26. Write the first eight terms of the Fibonacci sequence.

GROUP PROJECTS

Making Rice

1. The amount of ingredients needed to make 3 and 5 servings of rice are:

To Make	Rice & Water	Salt	Butter
3 servings	1 cup	$\frac{3}{8}$ tsp	$1\frac{1}{2}$ tsp
5 servings	$1\frac{2}{3}$ cup	$\frac{5}{8}$ tsp	$2\frac{1}{2}$ tsp

Find the amount of each ingredient needed to make (a) 2 servings, (b) 1 serving, and (c) 29 servings. Explain how you determined your answers.

Finding Areas

2. a) Determine the area of the trapezoid shown by finding the area of the three parts indicated and finding the sum of the three areas. The necessary geometric formulas are given on pages 452–3.

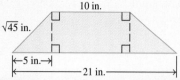

10 in.

√45 in.

←5 in.→

21 in.

b) Determine the area of the trapezoid by using the formula for the area of a trapezoid given on page 453.

c) Compare your answers from parts (a) and (b). Are they the same? If not, explain why they are different.

Medical Insurance

3. On a medical insurance policy (such as Blue Cross/ Blue Shield) the policyholder may need to make copayments for prescription drugs, office visits, and procedures until the total of all copayments reaches a specified amount. Suppose that on the Gattelaro's medical policy the copayment for prescription drugs is 50% of the cost; the copayment for office visits is $10; and the copayment for all medical tests, X-rays, and other procedures is 20% of the cost. After the family's copayment totals $500 in a calendar year, all medical and prescription bills are paid in full by the insurance company. The Gattelaros had the following medical expenses from January 1 through April 30.

Date	Reason	Cost before Copayment
January 10	Office visit	$40
	Prescription	$44
February 27	Office visit	$40
	Medical tests	$188
April 19	Office visit	$40
	X-rays	$348
	Prescription	$76

a) How much had the Gattelaros paid in copayments from January 1 through April 30?

b) How much had the medical insurance company paid?

c) What is the remaining copayment that must be paid by the Gattelaros before the $500 copayment limit is reached?

Gas and Electric Meters

4. To read an electric or gas meter follow these steps.

i) Stand directly in front of the meter.

ii) Think of the dials on the meters as clocks. It is important to notice which direction the pointers move and that each number on a dial represents 10 times as much as the number on the dial to its right.

iii) The basic rule for reading a meter is that when the pointer is between two numbers, *always read the lower number. Remember that the pointer is always moving toward the larger number,* so when the pointer is between 9 and 0, the 0 represents 10 and 9 is the lower number. When the pointer is between 0 and 1, zero is the lower number. When the pointer appears to be pointing directly to a number, read the dial to its right to determine whether the pointer is still approaching the number or has passed it. If the dial on the right has not passed 0, the dial on the left is still approaching the number. If the pointer on the right has passed 0, the dial to the left has passed the number.

iv) Read the dials from right to left, placing numbers in the same order as the dials (gas meters may have two test dials on the top or bottom, which you may ignore).

The following meter has a reading of 09846.

0   9   8   4   6

When the minute hand on a clock makes one complete revolution (60 minutes), the hour hand moves one unit (1 hour). Similarly, on the meters, when a dial makes one complete revolution (10 units), the dial to its left moves 1 unit.

a) Determine the reading of the following meter.

b) How many rotations must the dial on the far right make for the dial on the far left to move 1 unit? Explain.

c) Johanna reads her own meters and reports the readings to the Rochester Gas & Electric Company (RG&E). The electric meter on the day she is supposed to read her bill is shown. What does the meter read (in kilowatt-hours, kWh).

d) The following is the electric part of Johanna's RG&E bill. Note that the latest reading for

electricity hasn't been included on the bill and neither has any cost based on the latest reading. Use
the meter reading in part (c) and the information
provided in the bill to determine her cost of electricity for the period by filling in the shaded areas.

Your Electric Cost (SC1—Residential)

Amount of Electricity Used

Latest reading	
Previous reading	51916
Electricity used (kWh)	

Cost of Electricity Used

Basic service charge	$ 16.80
Charge for kWh used @ .099471/kWh	+
Fuel adjustment credit @ .000727/kWh	−
Gross Revenue Surcharge (on sum and difference of 3 charges directly above) @ 4.8092%	+
Electricity cost	$

e) The gas meter on the day Johanna is supposed to
read her meter is shown. What does the meter read
(in cubic feet)?

f) The following is the gas part of Johanna's RG&E
bill. Note that the latest reading for gas hasn't been
included on the bill and neither has any cost based
on the latest reading. Use the meter reading in part
(e) and the information provided in the bill to

determine her cost of gas for the period by filling in
the shaded area.

Your Gas Cost (SC1—Residential)

Amount of Gas Used

Latest reading	
Previous reading	4620
Gas used (ccf)	
Therm factor	× 1.016
Gas used (therms)	

Cost of Gas Used

Basic service charge (includes 3 therms)	$ 6.22
Next 87 therms @ .628851/therm	+
Remaining therms (rounded to the nearest therm) @ .605714/therm	+
Gas adjustment (for all therms, rounded to the nearest therm) @ .091467/therm	+
Gross Revenue Surcharge (on all above charges) @ 4.7951%	+
Gas cost	$

g) Determine Johanna's total gas and electric bill for
the period by adding the amounts obtained in parts
(d) and (f).

A Branching Plant

5. A plant grows for two months and then adds a new
branch. Each new branch grows for two months and
then adds another branch. After the second month, each
branch adds a new branch every month. Assume that
the growth begins in January.

a) How many branches will there be in February?

b) How many branches will there be in May?

c) How many branches will there be after 12 months?

d) How is this problem similar to the problem
involving rabbits that appeared in Fibonacci's book
Liber Abacci (see page 250)?

Chapter 6

Algebra, Graphs, & Functions

THE FAR SIDE

By GARY LARSON

Okay, now listen up. Nobody gets in here without answering the following question: A train leaves Philadelphia at 1:00 p.m. It's traveling at 65 miles per hour. Another train leaves Denver at 4:00... Say, you need some paper?

Math phobic's nightmare

Gary Larson, creator of *The Far Side* comic strip, touched a common fear in a cartoon titled "Math Phobic's Nightmare." In the cartoon, St. Peter guards the gates to Heaven, admitting only those who can answer an algebraic word problem. Word problems—the very mention of them is enough to frighten most people. And yet, algebra is one of the most practical tools for solving everyday problems. You probably use algebra in your daily life without realizing it.

For example, you use a coordinate system when you consult your car map to find directions to a new destination. You solve simple equations when you change a recipe to increase or decrease the number of servings. To evaluate how much interest you will earn on a savings account, or to figure out how long it will take you to travel a given distance, you use common formulas that are algebraic equations. The list of the applications of

algebra could go on and on. For this reason, the French mathematician and encyclopaedist Jean-le-Rond d'Alembert remarked, "Algebra is generous, she often gives more than is asked of her."

The symbolic language of algebra makes it an excellent tool for solving problems. Symbolism has three advantages. First, it allows us to write lengthy expressions in compact form: The eye can take in an entire statement and the mind can retain it. Thus "the product of three times a number

In the seventh century, Baghdad became the intellectual center of the world, where the mathematics and science practiced in the ancient civilizations of the East and West came together. It was here that Islamic translators brought together mathematics and the Hindu system of base 10 numeration. Not until the Crusades (1050–1250) would Europeans discover more fully the great intellectual treasures that the Islamic world possessed.

Algebraic equations can be graphed by using computer algebra systems such as Mathematica, Maple, or Derive. The results, even for relatively simple equations, are often surprisingly beautiful, and can be applied to study many fields, from physics to biology. The image here is a 3-D depiction of the equation $x^5 + y^5 = z^5$.

multiplied by a second number multiplied by itself and added to two times the product of the first, second, and third numbers" can be expressed simply as $3xy^2 + 2xyz$. Second, symbolic language is clear: Each symbol has a precise meaning. Finally, symbolism allows us to consider a large or infinite number of separate cases with a common property. For example, we can use the symbolic representation $ax + b = 0$ to represent all linear equations in one variable.

The English philosopher Alfred North Whitehead explained the power of algebra when he stated, "By relieving the brain of all unnecessary work, a good notation sets the mind free to concentrate on more advanced problems and in effect increases the mental power of the race." ■

Translating mathematical concepts into symbols allows us to discuss them and solve them more easily.

6.1 ORDER OF OPERATIONS

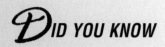

Algebra is a generalized form of arithmetic used to solve problems. It developed from the Babylonian number system as early as 1800 B.C. and spread to Greece and Europe, as well as to India, China, and Japan. The Islamic world became the center of mathematical studies between A.D. 800 and A.D. 1100. The word *algebra* is derived from the Arabic word *al-jabr* (meaning "reunion of broken parts"), which was the title of a book written by the mathematician Muhammed ibn-Musa al Khwarizmi in about A.D. 825. Algebra was reintroduced to Europe in the thirteenth and fourteenth centuries.

Why study algebra? You can solve many problems in everyday life by using arithmetic or by trial and error, but with a knowledge of algebra you can find the solutions with less effort. You can solve other problems, like some we will present in this chapter, only by using algebra.

Algebra uses letters of the alphabet called **variables** to represent numbers. Often the letters x and y are used to represent variables. However, any letter may be used as a variable. A symbol that represents a specific quantity is called a **constant**.

Multiplication of numbers and variables may be represented in several different ways in algebra. Since the "times" sign might be confused with the variable x, a dot between two numbers or variables indicates multiplication. Thus $3 \cdot 4$ means 3 times 4, and $x \cdot y$ means x times y. Placing two letters or a number and a letter next to one another, with or without parentheses, also indicates multiplication. Thus $3x$ means 3 times x, xy means x times y, and $(x)(y)$ means x times y.

An **algebraic expression** (or simply an **expression**) is a collection of variables, numbers, parentheses, and operation symbols. Some examples of algebraic expressions are

$$x, \quad x + 2, \quad 3(2x + 3), \quad \frac{3x + 1}{2x - 3}, \quad \text{and} \quad x^2 + 7x + 3.$$

Two algebraic expressions joined by an equal sign form an **equation**. Some examples of equations are

$$x + 2 = 4, \quad 3x + 4 = 1, \quad \text{and} \quad x + 3 = 2x.$$

The **solution to an equation** is the number or numbers that replace the variable to make the equation a true statement. For example, the solution to the equation $x + 3 = 4$ is $x = 1$. When we find the solution to an equation, we **solve the equation**.

We can determine if any number is a solution to an equation by **checking the solution**. To check the solution we substitute the number for the variable in the equation. If the resulting statement is a true statement, that number is a solution to the equation. If the resulting statement is a false statement, the number is not a solution to the equation. To check the number $x = 1$ in the equation $x + 3 = 4$, we do the following.

$$x + 3 = 4$$
$$1 + 3 = 4 \quad \text{Substitute 1 for } x.$$
$$4 = 4 \quad \text{True}$$

The same number is obtained on both sides of the equal sign, so the solution is correct. For the equation $x + 3 = 4$, the only solution is $x = 1$. Any other value of x would result in the check being a false statement.

To **evaluate an expression** means to find the value of the expression for a given value of the variable. To evaluate expressions and solve equations, you must have an understanding of exponents. Exponents (Section 5.6) are used to abbreviate repeated multiplication. For example, the expression 5^2 means $5 \cdot 5$. The 2 in the expression 5^2 is the **exponent**, and the 5 is the **base**. We read 5^2 as "5 to the second power" or "5 squared," and 5^2 means $5 \cdot 5$ or 25.

In general, the number b to the nth power, written b^n, means

$$\underset{\text{Base}}{\overset{\text{Exponent}}{b^n}} = \underbrace{b \cdot b \cdot b \cdot \cdots \cdot b}_{n \text{ factors of } b}.$$

An exponent refers only to its base. In the expression -5^2, the base is 5. In the expression $(-5)^2$, the base is -5.

$$-5^2 = -(5)^2 = -1(5)^2 = -1(5)(5) = -25$$

$$(-5)^2 = (-5)(-5) = 25$$

Note that $-5^2 \neq (-5)^2$ since $-25 \neq 25$.

Order of Operations

To evaluate an expression or to check the solution to an equation, we need to know the **order of operations** to follow. For example, suppose that we want to evaluate the expression $2 + 3x$ when $x = 4$. Substituting 4 for x, we obtain $2 + 3 \cdot 4$. What is the value of $2 + 3 \cdot 4$? Does it equal 20, or does it equal 14? Some standard rules called the order of operations have been developed to ensure that there is only one correct answer. In mathematics, unless parentheses indicate otherwise, always perform multiplication before addition. Thus the correct answer is 14.

$$2 + 3 \cdot 4 = 2 + (3 \cdot 4) = 2 + 12 = 14$$

The order of operations for evaluating an expression is as follows.

Order of Operations

1. First, perform all operations within parentheses or other grouping symbols (according to the following order).

2. Next, perform all exponential operations (that is, raising to powers or finding roots).

3. Next, perform all multiplications and divisions from left to right.

4. Finally, perform all additions and subtractions from left to right.

Some students use the phrase, "**P**lease **E**xcuse **M**y **D**ear **A**unt **S**ally," or the word "PEMDAS" (**P**arentheses, **E**xponents, **M**ultiplication, **D**ivision, **A**ddition, **S**ubtraction) to remind them of the order of operations. Remember: Multiplication and division are of the same order, and addition and subtraction are of the same order.

EXAMPLE 1

Evaluate the expression $-x^2 + 4x + 25$ for $x = 3$.

Solution: Substitute 3 for each x and use the order of operations to evaluate the expression.

$$-x^2 + 4x + 25$$
$$= -(3)^2 + 4(3) + 25$$
$$= -9 + 12 + 25$$
$$= 28$$

EXAMPLE 2

The temperature T, in degrees Celsius, in a sauna n minutes after the sauna is turned on can be approximated by the expression $-0.04n^2 + 1.6n + 20$ (assuming n is between 0 and 20). Find the temperature after the sauna has been on for 12 minutes.

Solution: Substitute 12 for each n.

$$-0.04n^2 + 1.6n + 20$$
$$= -0.04(12)^2 + 1.6(12) + 20$$
$$= -0.04(144) + 19.2 + 20$$
$$= -5.76 + 19.2 + 20$$
$$= 33.44$$

The temperature in the sauna is approximately 33°C (or 92°F) after 12 minutes.

EXAMPLE 3

Evaluate $-3x^2 + 4xy - y^2$ when $x = 3$ and $y = 4$.

Solution: Substitute 3 for each x and 4 for each y; then evaluate using the order of operations.

$$-3x^2 + 4xy - y^2$$
$$= -3(3)^2 + 4(3)(4) - 4^2$$
$$= -3(9) + 12(4) - 16$$
$$= -27 + 48 - 16$$
$$= 21 - 16$$
$$= 5$$

EXAMPLE 4

Determine whether 3 is a solution to the equation $2x^2 + 4x - 9 = 21$.

Solution: To determine whether 3 is a solution to the equation, substitute 3 for each x in the equation. Then evaluate the left side of the equation by using the order of operations. If this leads to a 21 on the left side of the equal sign, then both sides of the equation have the same value, and 3 is a solution.

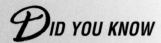

$$2x^2 + 4x - 9 = 21$$
$$2(3)^2 + 4(3) - 9 = 21$$
$$2(9) + 12 - 9 = 21$$
$$18 + 12 - 9 = 21$$
$$30 - 9 = 21$$
$$21 = 21 \quad \text{True}$$

Because 3 makes the equation a true statement, 3 is a solution to the equation.

Section 6.1 Exercises

1. What is a *variable*?

2. What is a *constant*?

3. What is an algebraic expression? Illustrate with an example.

4. What does *a number is a solution to an equation* mean?

5. **a)** For the term 4^5 identify the base and the exponent.
   b) In your own words explain how to evaluate 4^5.

6. In your own words explain the order of operations.

In Exercises 7–26, evaluate the expression for the given value(s) of the variable(s).

7. x^2, $x = 6$

8. x^2, $x = -3$

9. $-x^2$, $x = 8$

10. $-x^2$, $x = -5$

11. $-2x^3$, $x = -7$

12. $-x^3$, $x = -4$

13. $x + 9$, $x = 12$

14. $6x - 4$, $x = \frac{3}{2}$

15. $-5x + 8$, $x = -4$

16. $x^2 + 6x - 4$, $x = 5$

17. $-x^2 - 7x + 5$, $x = -2$

18. $5x^2 + 7x - 11$, $x = -1$

19. $\frac{1}{2}x^2 + 4x - 6$, $x = \frac{1}{3}$

20. $x^3 - 3x^2 + 7x + 11$, $x = 4$

21. $8x^3 - 4x^2 + 7$, $x = \frac{1}{2}$

22. $-x^2 + 4xy$, $x = 2$, $y = 3$

23. $2x^2 - 2xy + y^2$, $x = -1$, $y = -1$

24. $3x^2 + \frac{2}{5}xy - \frac{1}{2}y^2$, $x = 5$, $y = -2$

25. $4x^2 - 12xy + 9y^2$, $x = 3$, $y = 2$

26. $(x + 3y)^2$, $x = 4$, $y = -3$

In Exercises 27–36, determine whether the value(s) is (are) a solution to the equation.

27. $8x - 11 = 7$, $x = 2$

28. $3x - 6 = 4$, $x = 3$

29. $2x + y = 0$, $x = 2$, $y = -4$

30. $2x - 3y = -6$, $x = 3$, $y = 4$

31. $x^2 + 3x - 4 = 5$, $x = 2$

32. $2x^2 - x - 5 = 0$, $x = 3$

33. $2x^2 + x = 28$, $x = -4$

34. $y = x^2 + 3x - 5$, $x = 1$, $y = -1$

35. $y = -x^2 - 3x + 2$, $x = 2$, $y = -8$

36. $y = x^3 - 3x^2 + 1$, $x = 2$, $y = -3$

37. If Jon earns $5.75 per hour, the amount he earns in t hours can be found by using the expression $5.75t$. Determine Jon's pay if he works (a) 6 hr and (b) 14 hr.

38. If the sales tax on an item is 8%, the sales tax, in dollars, on an item costing d dollars can be found by using the expression $0.08d$. Determine the sales tax on a bicycle costing $875.

39. If a computer can do a calculation in 0.000002 sec, the time required to do n calculations can be found by using the expression $0.000002n$. Determine the number of seconds needed for the computer to do 8 trillion (8,000,000,000,000) calculations.

40. The total cost of an item in dollars, including a 5% sales tax, can be found by using the expression $x + 0.05x$, where x is the cost of the item before the tax. Determine the total cost of a car whose cost before the tax is $12,500.

41. The number of baskets of oranges that are produced by x trees in a small orchard can be approximated by the expression $25x - 0.2x^2$ (assuming x is no more than 100). Find the number of baskets of oranges produced by 60 trees.

42. The time, in minutes, needed for clothes hanging on a line outdoors to dry, at a specific temperature and wind

speed, depends upon the humidity, h. The time can be approximated by the expression $2h^2 + 80h + 40$, where h is the percent humidity expressed as a decimal number. Find the length of time required for clothing to dry if there is 60% humidity.

43. The rate of growth of grass in inches per week depends on a number of factors, including rainfall and temperature. For a certain area this can be approximated by the expression $0.2R^2 + 0.003RT + 0.0001T^2$, where R is the weekly rainfall, in inches, and T is the average weekly temperature, in degrees Fahrenheit. Find the amount of growth of grass for a week in which the rainfall is 2 in. and the average temperature is 70°F.

Problem Solving/Group Activities

44. Explain why $(-1)^n = 1$ for any even number n.

45. Does $(x + y)^2 = x^2 + y^2$? Complete the table and state your conclusion.

x	y	$(x + y)^2$	$x^2 + y^2$
2	3		
-2	-3		
-2	3		
2	-3		

46. Suppose that n stands for any natural number. Why does 1^n equal 1?

Research Activity

47. When were exponents first used? Write a paper explaining how exponents were first used and when mathematicians began writing them in the present form. (References include history of mathematics books and encyclopedias.)

6.2   LINEAR EQUATIONS IN ONE VARIABLE

In Section 6.1 we stated that two algebraic expressions joined by an equal sign form an equation. The solution to some equations, such as $x + 3 = 4$, can be found easily by trial and error. However, solving more complex equations, such as $2x - 3 = 4(x + 3)$, requires understanding the meaning of like terms and learning four basic properties.

The parts that are added or subtracted in an algebraic expression are called terms. The expression $4x - 3y - 5$ contains three terms, namely $4x$, $-3y$, and -5. The + and − signs that break the expression into terms are a part of the terms. When listing the terms of an expression, however, it is not necessary to include the + sign at the beginning of the term.

The numerical part of a term is called its numerical coefficient, or simply its coefficient. In the term $4x$, the 4 is the numerical coefficient. In the term $-4y$, the −4 is the numerical coefficient.

Like terms are terms that have the same variables with the same exponents on the variables. Unlike terms have different variables or different exponents on the variables.

Like Terms		Unlike Terms	
$3x$, $4x$	(Same variable, x)	$3x$, 2	(Only first term has a variable)
$4y$, $6y$	(Same variable, y)	$3x$, $4y$	(Different variables)
5, −6	(Both constants)	x, 3	(Only first term has a variable)
$-6x^2$, $7x^2$	(Same variable with same exponent)	$3x^2$, $5x^3$	(Different exponents)

To simplify an expression means to combine like terms by using the commutative, associative, and distributive properties discussed in Chapter 5. For convenience we list these properties.

> **Properties of the Real Numbers**
>
> | | |
> |---|---|
> | $a(b + c) = ab + ac$ | Distributive property |
> | $a + b = b + a$ | Commutative property of addition |
> | $ab = ba$ | Commutative property of multiplication |
> | $(a + b) + c = a + (b + c)$ | Associative property of addition |
> | $(ab)c = a(bc)$ | Associative property of multiplication |

EXAMPLE 1

Combine like terms in each expression.

a) $4x + 3x$ **b)** $5a - 7a$ **c)** $12 + x + 7$

Solution:

a) We use the distributive property (in reverse) to combine like terms.

$$4x + 3x = (4 + 3)x \quad \text{Distributive property}$$
$$= 7x$$

b) $5a - 7a = (5 - 7)a$
$$= -2a$$

c) $12 + x + 7 = x + 12 + 7$
$$= x + 19$$

We are able to rearrange the terms of an expression, as was done in Example 1(c), by the commutative and associative properties that were discussed in Section 5.5.

EXAMPLE 2

Combine like terms in each expression.

a) $3y + 4x - 3 - 2x$ **b)** $-2x + 3y - 4x + 3 - y + 5$

Solution:

a) $4x$ and $-2x$ are the only like terms. Grouping the like terms gives

$$\underbrace{4x - 2x}_{2x} + 3y - 3.$$

b) Grouping the like terms gives

$$\underbrace{-2x - 4x}_{-6x} + \underbrace{3y - y}_{+2y} + \underbrace{3 + 5}_{+8}.$$

The order of the terms in an expression is not crucial. However, when listing the terms of an expression we generally list the terms in alphabetical order with the constant, the term without a variable, last.

Solving Equations

Recall that to solve an equation means to find the value or values for the variable that make(s) the equation true. In this section we discuss solving **linear (or first degree) equations**. A linear equation in one variable is one in which the exponent on the variable is 1. The following equations are examples of linear equations: $5x - 1 = 3$ and $2x + 4 = 6x - 5$.

Equivalent equations are equations that have the same solution. The equations $2x - 5 = 1$, $2x = 6$, and $x = 3$ are all equivalent equations since they all have the same solution, 3. When we solve an equation we write the given equation as a series of simpler equivalent equations until we obtain an equation of the form $x = c$, where c is some real number.

To solve any equation, we have to **isolate the variable**. That means getting the variable by itself on one side of the equal sign. The four properties of equality that we are about to discuss will be used to isolate the variable. The first is the addition property.

Addition Property of Equality

If $a = b$, then $a + c = b + c$ for all real numbers a, b, and c.

The addition property of equality indicates that the same number can be added to both sides of an equation without changing the solution.

EXAMPLE 3

Find the solution to the equation $x - 6 = 4$.

Solution: To isolate the variable, add 6 to both sides of the equation.

$$x - 6 + 6 = 4 + 6$$
$$x + 0 = 10$$
$$x = 10$$

Check: $x - 6 = 4$
$10 - 6 = 4$ Substitute 10 for x.
$4 = 4$ True

In Example 3 we showed the step $x + 0 = 10$. Generally this step is done mentally and the step is not listed.

Subtraction Property of Equality

If $a = b$, then $a - c = b - c$ for all real numbers a, b, and c.

The subtraction property of equality indicates that the same number can be subtracted from both sides of an equation without changing the solution.

EXAMPLE 4

Find the solution to the equation $x + 5 = 7$.

Solution: To isolate the variable, subtract 5 from both sides of the equation.

$$x + 5 - 5 = 7 - 5$$
$$x = 2$$

Note that we did not subtract 7 from both sides of the equation, since this would not result in getting x on one side of the equal sign by itself. ▲

Multiplication Property of Equality

If $a = b$, then $a \cdot c = b \cdot c$ for all real numbers a, b, and c where $c \neq 0$.

The multiplication property of equality indicates that both sides of the equation can be multiplied by the same nonzero number without changing the solution.

EXAMPLE 5

Find the solution to the equation $\dfrac{x}{3} = 2$.

Solution: To solve this equation, multiply both sides of the equation by 3.

$$3\left(\frac{x}{3}\right) = 3(2)$$

$$\frac{\overset{1}{\cancel{3}}x}{\underset{1}{\cancel{3}}} = 6$$

$$1x = 6$$

$$x = 6$$

▲

In Example 5 we showed the steps $(3x)/3 = 6$ and $1x = 6$. Generally we will not illustrate these steps.

Division Property of Equality

If $a = b$, then $a/c = b/c$ for all real numbers a, b, and c, $c \neq 0$.

The division property of equality indicates that both sides of an equation can be divided by the same nonzero number without changing the solution. Note that the divisor, c, cannot be 0 because division by 0 is not permitted.

EXAMPLE 6

Find the solution to the equation $5x = 35$.

Solution: To solve this equation, divide both sides of the equation by 5.

$$\frac{5x}{5} = \frac{35}{5}$$

$$x = 7$$

▲

An **algorithm** is a general procedure for accomplishing a task. The following general procedure is an algorithm for solving linear (or first degree) equations. Sometimes the solution to an equation may be found more easily by using a variation of the general procedure. Remember that the primary objective in solving any equation is to isolate the variable.

> **A General Procedure for Solving Linear Equations**
>
> **1.** If the equation contains fractions, multiply both sides of the equation by the lowest common denominator (or least common multiple). This step will eliminate all fractions from the equation.
> **2.** Use the distributive property to remove parentheses when necessary.
> **3.** Combine like terms on the same side of the equal sign when possible.
> **4.** Use the addition or subtraction property to collect all terms with a variable on one side of the equal sign and all constants on the other side of the equal sign. It may be necessary to use the addition or subtraction property more than once. This process will eventually result in an equation of the form $ax = b$, where a and b are real numbers.
> **5.** Solve for the variable using the division or multiplication property. This will result in an answer in the form $x = c$, where c is a real number.

EXAMPLE 7

Solve the equation $3x + 4 = 13$; then check your solution.

Solution: Our goal is to isolate the variable; therefore we start by getting the term $3x$ by itself on one side of the equation.

$$3x + 4 = 13$$
$$3x + 4 - 4 = 13 - 4 \qquad \text{Subtract 4 from both sides of the equation (subtraction property) (step 4).}$$
$$3x = 9$$
$$\frac{3x}{3} = \frac{9}{3} \qquad \text{Divide both sides of the equation by 3 (division property) (step 5).}$$
$$x = 3$$

A check will show that 3 is the solution to $3x + 4 = 13$.   ▲

EXAMPLE 8

Solve the equation $2 = 3 + 5(p + 1)$ for p.

Solution: Our goal is to isolate the variable p. To do so follow the general procedure for solving equations.

$$2 = 3 + 5(p + 1)$$
$$2 = 3 + 5p + 5 \qquad \text{Distributive property (step 2)}$$
$$2 = 5p + 8 \qquad \text{Combine like terms (step 3).}$$
$$2 - 8 = 5p + 8 - 8 \qquad \text{Subtraction property (step 4)}$$
$$-6 = 5p$$
$$-\frac{6}{5} = \frac{5p}{5} \qquad \text{Division property (step 5)}$$
$$-\frac{6}{5} = p$$

▲

EXAMPLE 9

Solve the equation: $\dfrac{2x}{3} + \dfrac{1}{3} = \dfrac{3}{4}$.

Solution: When an equation contains fractions we generally begin by multiplying each term of the equation by the lowest common denominator, LCD (see Chapter 5). In this example the LCD is 12, since 12 is the smallest number that is divisible by both 3 and 4.

$$12\left(\frac{2x}{3} + \frac{1}{3}\right) = 12\left(\frac{3}{4}\right)$$ Multiply both sides of the equation by the LCD (step 1).

$$12\left(\frac{2x}{3}\right) + 12\left(\frac{1}{3}\right) = 12\left(\frac{3}{4}\right)$$ Distributive property (step 2)

$$\overset{4}{\cancel{12}}\left(\frac{2x}{\cancel{3}_1}\right) + \overset{4}{\cancel{12}}\left(\frac{1}{\cancel{3}_1}\right) = \overset{3}{\cancel{12}}\left(\frac{3}{\cancel{4}_1}\right)$$ Divide out common factors.

$$8x + 4 = 9$$

$$8x + 4 - 4 = 9 - 4$$ Subtraction property (step 4)

$$8x = 5$$

$$\frac{8x}{8} = \frac{5}{8}$$ Division property (step 5)

$$x = \frac{5}{8}$$

A check will show that $\frac{5}{8}$ is the solution to the equation. You could have worked the problem without first multiplying both sides of the equation by the LCD. Try it! ▲

EXAMPLE 10

Solve the equation: $3x + 4 = 5x + 6$.

Solution: Note that the equation has an x on both sides of the equal sign. In equations of this type you might wonder what to do first. It really does not matter as long as you don't forget the goal of isolating the variable x. Let's collect the terms containing a variable on the left-hand side of the equation.

$$3x + 4 = 5x + 6$$

$$3x + 4 - 4 = 5x + 6 - 4$$ Subtraction property (step 4)

$$3x = 5x + 2$$

$$3x - 5x = 5x - 5x + 2$$ Subtraction property (step 4)

$$-2x = 2$$

$$\frac{-2x}{-2} = \frac{2}{-2}$$ Division property (step 5)

$$x = -1$$ ▲

In the solution to Example 10 the terms containing the variable were collected on the left side of the equal sign. Now work Example 10 collecting the terms with the variable on the right-hand side of the equation. If you do so correctly you will get the same result.

EXAMPLE 11

Solve the equation and check your solution: $4x - 0.48 = 0.8x + 4$.

Solution: This problem may be solved with the decimals, or you may multiply each term by 100 and eliminate the decimals. We will solve the problem with the decimals.

$$4x - 0.48 = 0.8x + 4$$

$$4x - 0.48 + 0.48 = 0.8x + 4 + 0.48 \qquad \text{Addition property}$$

$$4x = 0.8x + 4.48$$

$$4x - 0.8x = 0.8x - 0.8x + 4.48 \qquad \text{Subtraction property}$$

$$3.2x = 4.48$$

$$\frac{3.2x}{3.2} = \frac{4.48}{3.2} \qquad \text{Division property}$$

$$x = 1.4$$

Check: $4x - 0.48 = 0.8x + 4$

$4(1.4) - 0.48 = 0.8(1.4) + 4 \qquad$ Substitute 1.4 in place of x in the equation.

$5.6 - 0.48 = 1.12 + 4$

$5.12 = 5.12 \qquad$ True   ▲

In Chapter 5 we explained that $a - b$ can be expressed as $a + (-b)$. We will use this principle in Example 12.

EXAMPLE 12

Solve $9 = -7 + 8(r - 5)$ for r.

Solution:

$$9 = -7 + 8(r - 5)$$

$$9 = -7 + 8[r + (-5)] \qquad \text{Definition of subtraction of integers}$$

$$9 = -7 + 8(r) + 8(-5) \qquad \text{Distributive property}$$

$$9 = -7 + 8r + (-40)$$

$$9 = -47 + 8r \qquad \text{Combine like terms.}$$

$$9 + 47 = -47 + 47 + 8r \qquad \text{Addition property}$$

$$56 = 8r \qquad \text{Combine like terms.}$$

$$\frac{56}{8} = \frac{8r}{8} \qquad \text{Division property}$$

$$7 = r \qquad \blacktriangle$$

So far every equation has had exactly one solution. However, some equations have no solution and others have more than one solution. Example 13 illustrates an equation that has no solution, and Example 14 illustrates an equation that has an infinite number of solutions.

EXAMPLE 13

Solve $2(x - 3) + x = 5x - 2(x + 5)$.

Solution:

$$2(x - 3) + x = 5x - 2(x + 5)$$
$$2x - 6 + x = 5x - 2x - 10 \qquad \text{Distributive property}$$
$$3x - 6 = 3x - 10 \qquad \text{Combine like terms.}$$
$$3x - 3x - 6 = 3x - 3x - 10 \qquad \text{Subtraction property}$$
$$-6 = -10 \qquad \text{False}$$

During the process of solving an equation, if you obtain a false statement like $-6 = -10$, or $4 = 0$, the equation has no solution. An equation that has no solution is called an inconsistent equation. The equation $2(x - 3) + x = 5x - 2(x + 5)$ is inconsistent and thus has no solution. ▲

EXAMPLE 14

Solve $3(x + 2) - 5(x - 3) = -2x + 21$.

Solution:

$$3(x + 2) - 5(x - 3) = -2x + 21$$
$$3x + 6 - 5x + 15 = -2x + 21 \qquad \text{Distributive property}$$
$$-2x + 21 = -2x + 21 \qquad \text{Combine like terms.}$$

Note that at this point both sides of the equation are the same. Every real number will satisfy this equation. This equation has an infinite number of solutions. An equation of this type is called an identity. When solving an equation, if you notice that the same expression appears on both sides of the equal sign, the equation is an identity. The solution to any identity is all real numbers. If you continue to solve an equation that is an identity, you will end up with $0 = 0$, as follows.

$$-2x + 21 = -2x + 21$$
$$-2x + 2x + 21 = -2x + 2x + 21 \qquad \text{Addition property}$$
$$21 = 21 \qquad \text{Combine like terms.}$$
$$21 - 21 = 21 - 21 \qquad \text{Subtraction property}$$
$$0 = 0 \qquad \text{True for any value of } x \quad ▲$$

Proportions

A ratio is a quotient of two quantities. An example is the ratio of 2 to 5, which can be written 2:5 or $\frac{2}{5}$ or 2/5.

A **proportion** is a statement of equality between two ratios.

An example of a proportion is $\dfrac{a}{b} = \dfrac{c}{d}$. Consider the proportion

$$\frac{x + 2}{5} = \frac{x + 5}{8}.$$

We can solve this proportion by first multiplying both sides of the equation by the least common denominator, 40.

$$\frac{x+2}{5} = \frac{x+5}{8}$$

$$40\left(\frac{x+2}{\overset{8}{\cancel{5}}}\right) = 40\left(\frac{x+5}{\overset{5}{\cancel{8}}}\right)$$

$$8(x+2) = 5(x+5)$$

$$8x + 16 = 5x + 25$$

$$3x + 16 = 25$$

$$3x = 9$$

$$x = 3$$

A check will show that 3 is the solution.

Proportions can often be solved more easily by using the cross product method.

Cross Product Method to Solve Proportions

$$\text{If } \frac{a}{b} = \frac{c}{d}, \text{ then } ad = bc.$$

Let's use the cross product method to solve the proportion $(x + 2)/5 = (x + 5)/8$.

$$\frac{x+2}{5} = \frac{x+5}{8}$$

$$8(x+2) = 5(x+5)$$

$$8x + 16 = 5x + 25$$

$$3x + 16 = 25$$

$$3x = 9$$

$$x = 3$$

Many practical application problems can be solved using proportions.

To Solve Application Problems Using Proportions

1. Represent the unknown quantity by a variable.

2. Set up the proportion by listing the given ratio on the left side of the equal sign, and the unknown and other given quantity on the right side of the equal sign. When setting up the right side of the proportion, the same respective quantities should occupy the same respective positions on the left and right. For example, an acceptable proportion might be

$$\frac{\text{miles}}{\text{hour}} = \frac{\text{miles}}{\text{hour}}.$$

3. Once the proportion is properly written, drop the units and use the cross product method to solve the equation.

4. Answer the question or questions asked.

─ **EXAMPLE 15**

The water rate in Monroe County is $1.93 per 1000 gallons (gal) of water used. What is the water bill if 25,000 gallons are used?

Solution: This problem may be solved by setting up a proportion. One proportion that can be used is

$$\frac{\text{cost of 1000 gal}}{1000 \text{ gal}} = \frac{\text{cost of 25,000 gal}}{25,000 \text{ gal}}.$$

We want to find the cost for 25,000 gallons of water, so we will call this quantity x. The proportion then becomes

Given ratio $\left\{\dfrac{1.93}{1000} = \dfrac{x}{25,000}\right.$.

Now we solve for x.

$$(1.93)(25,000) = 1000x$$

or $\qquad 1000x = 48,250$

$$x = \frac{48,250}{1000} = 48.25$$

The cost of 25,000 gallons of water is $48.25. ▲

─ **EXAMPLE 16**

Insulin comes in 10-cc vials labeled in the number of units of insulin per cubic centimeter (cc) of fluid. A vial of insulin marked U40 has 40 units of insulin per cubic centimeter of fluid. If a patient needs 25 units of insulin, how much fluid should be drawn into the syringe from the U40 vial?

Solution: The unknown quantity, x, is the number of cubic centimeters of fluid to be drawn into the syringe. Following is one proportion that can be used to find that quantity.

Given ratio $\left\{\dfrac{40 \text{ units}}{1 \text{ cc}} = \dfrac{25 \text{ units}}{x \text{ cc}}\right.$

$$40x = 25(1)$$

$$40x = 25$$

$$x = \frac{25}{40} = 0.625$$

The nurse or doctor putting the insulin in the syringe should draw 0.625 cc of the fluid. ▲

Section 6.2 Exercises

1. Define and give an example of *term*.

2. Define and give an example of *like terms*.

3. Define and give an example of a *numerical coefficient*.

4. Explain how to simplify an expression. Give an example.

5. Define and give an example of a *linear equation*.

6. State the addition property of equality. Give an example.

7. State the subtraction property of equality. Give an example.

8. State the multiplication property of equality. Give an example.

9. State the division property of equality. Give an example.

10. Define *algorithm*.

11. Define and given an example of a *ratio*.

12. Define and given an example of a *proportion*.

In Exercises 13–36, combine like terms.

13. $5x + 7x$

14. $-2x - 3x$

15. $2x + 3x - 7$

16. $8 - 7x + 11x$

17. $7x + 3y - 4x + 8y$

18. $x - 4x + 3$

19. $-3x + 2 - 5x$

20. $-3x + 4x - 2 + 5$

21. $2 - 3x - 2x + 1$

22. $0.4x + 1.1x - 3$

23. $7.3x + 6.4 - 4.7x$

24. $\frac{1}{2}x + \frac{3}{4}x + 5$

25. $\frac{1}{3}x - \frac{1}{4}x - 2$

26. $7x + 5y - 3 + 2x - 6y$

27. $4x - 3y - 2y + 6x + 4$ **28.** $3(p + 2) - 4(p + 3)$

29. $4(r - 5) + 6(r + 3) + 8$

30. $5(w + 5) - 2(w - 12) - 1$

31. $0.1(x + 4) - 2.3(x - 2)$

32. $\frac{1}{4}(x + 3) + \frac{1}{2}x$

33. $\frac{1}{4}x + \frac{4}{5} - \frac{3}{5}x + \frac{1}{3}$

34. $n - \frac{3}{4} + \frac{5}{9}n - \frac{1}{6}$

35. $0.5(2.6x - 4) + 2.3(1.4x - 5)$

36. $\frac{2}{3}(3x + 9) - \frac{1}{4}(2x + 5)$

In Exercises 37–62, solve the equation.

37. $y + 3 = 8$

38. $3y - 4 = 11$

39. $10 = 6x + 4$

40. $11 = 3 - 4y$

41. $\dfrac{4}{9} = \dfrac{x}{11}$

42. $\dfrac{x + 3}{3} = \dfrac{x + 7}{5}$

43. $\frac{1}{2}x + \frac{1}{3} = \frac{2}{3}$

44. $\frac{1}{2}y + \frac{1}{3} = \frac{1}{4}$

45. $0.7x - 0.3 = 1.8$

46. $5x + 0.050 = -0.732$

47. $5t + 6 = 2t + 15$

48. $\dfrac{x}{4} + 2x = -\dfrac{2}{3}$

49. $\dfrac{x - 5}{4} = \dfrac{x - 9}{3}$

50. $2r + 8 = 5 + 3r$

51. $\dfrac{x}{15} = 2 + \dfrac{x}{5}$

52. $12x - 1.2 = 3x + 1.5$

53. $2(x + 3) - 4 = 2(x - 4)$

54. $6y + 3(4 + y) = 8$

55. $6(x + 1) = 4x + 2(x + 3)$

56. $\dfrac{x}{3} + 4 = \dfrac{2x}{5} - 6$

57. $0.67x + 0.31 = 0.33x - 0.42$

58. $\frac{2}{3}(x + 5) = \frac{1}{4}(x + 2)$

59. $2x + 3 - 5x = 11 + 3x - 7x - 5$

60. $3x + 4 - 11x = 14 + 7x - 5 + 6x$

61. $2(x + 3) = 4(x - 5) + 2$

62. $5.7x - 3.1(x + 5) = 7.3$

In Exercises 63 and 64, use the Essex County water rate of $1.75 per 1000 gallons of water used.

63. What is the water bill if a resident uses 35,300 gal?

64. How many gallons of water can the customer use if the water bill is not to exceed $36.50?

65. A 30-pound bag of fertilizer will cover an area of 2500 square feet.
   a) How many pounds are needed to cover an area of 28,000 ft²?
   b) How many bags of fertilizer must be purchased to cover an area of 28,000 ft²?

66. In Northhampton County the property tax rate is $7.085 per $1000 of assessed value. If a house and lot have been assessed at $132,600, determine the amount of tax the owner will have to pay.

67. A car can travel 32 mi on 1 gal of gasoline. How far can it travel on 12 gal of gasoline?

68. A gallon of paint covers 825 ft². How much paint is needed to cover a house with a surface area of 5775 ft²?

69. A doctor asks a nurse to give a patient 250 milligrams (mg) of the drug Simethicone. The drug is available only in a solution whose concentration is 40 mg Simethicone per 0.6 millimeter (mm) of solution. How many millimeters of solution should the nurse give the patient?

70. Marisa exchanged 15 U.S. dollars for 46.75 Mexican pesos at a bank in Cancun, Mexico.
   a) What is the conversion rate per U.S. dollar (that is, what is 1 U.S. dollar worth in pesos)?
   b) At a gift shop, Marisa purchased a handmade Mayan vase for 297 pesos. What is the cost of the vase in U.S. dollars?

In Exercises 71 and 72, how much insulin (in cc) would be given for the following doses? (Refer to Example 16.)

71. 12 units of insulin from a vial marked U40

72. 35 units of insulin from a vial marked U40

73. a) In your own words summarize the procedure to use to solve an equation.
   b) Solve the equation $2(x + 3) = 4x + 3 - 5x$ with the procedure you outlined in part (a).

74. a) What is an identity?
   b) When solving an equation, how will you know if the equation is an identity?

75. a) What is an inconsistent equation?
   b) When solving an equation, how will you know if the equation is inconsistent?

Problem Solving/Group Activities

76. The pressure, P, in pounds per square inch (psi), exerted on an object x feet below the sea is given by the formula $P = 14.70 + 0.43x$. The 14.70 represents the weight in pounds of the column of air (from sea level to the top of the atmosphere) standing over a 1 in. by 1 in. square of seawater. The $0.43x$ represents the weight in pounds of a column of water 1 in. by 1 in. by x ft (see Fig. 6.1).

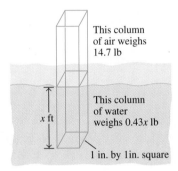

This column of air weighs 14.7 lb

This column of water weighs 0.43x lb

x ft

1 in. by 1in. square

Figure 6.1

a) A submarine is built to withstand a pressure of 148 psi. How deep can that submarine go?

b) If the pressure gauge in the submarine registers a pressure of 128.65 psi, how deep is the submarine?

77. a) If the ratio of males to females in a class is 2:3, what is the ratio of males to all the students in the class? Why?

b) If the ratio of males to females in a class is $m:n$, what is the ratio of males to all the students in the class?

Research Activities

78. Ratio and proportion are used in many different ways in everyday life. Submit two articles from newspapers or magazines in which ratio and/or proportion are used. Write a brief summary of each article explaining how ratio and/or proportion were used.

79. Write a report explaining how the Egyptians used equations. Include in your discussion the forms of the equations used. (References include history of mathematics books and encyclopedias.)

6.3 FORMULAS

A formula is an equation that typically has a real-life application. To evaluate a formula, substitute the given values for their respective variables and then evaluate using the order of operations given in the box in Section 6.1. Many of the formulas given in this section will be discussed in greater detail in other parts of the book.

EXAMPLE 1

The simple interest formula,* interest = principal × rate × time, or $i = prt$, is used to find the interest you must pay on a simple interest loan when you borrow principal, p, at simple interest rate, r, in decimal form, for time, t. Erikka borrows \$3000 at a simple interest rate of 9% for 3 years.

a) How much will Erikka pay in interest at the end of 3 years?

b) What is the total amount she will repay the bank at the end of 3 years?

Solution:

a) Substitute the values of p, r, and t into the formula; then evaluate.

$$i = prt$$
$$= 3000(0.09)(3)$$
$$= 810$$

Thus Erikka must pay \$810 interest.

b) The total she must pay at the end of 3 years is the principal, \$3000, plus the \$810 interest, for a total of \$3810.

*The simple interest formula will be discussed in Section 10.2.

EXAMPLE 2

Use the simple interest formula to find the rate on Ahmed's $6000 loan if the interest is $1080 after 2 years.

Solution: We substitute the appropriate values into the simple interest formula and solve for the desired quantity, r.

$$i = prt$$
$$1080 = (6000)r(2)$$
$$1080 = 12{,}000r$$
$$\frac{1080}{12{,}000} = \frac{12{,}000r}{12{,}000}$$
$$0.09 = r$$

Therefore the simple interest rate is 9% per year.   ▲

Many formulas contain Greek letters, such as μ (mu), σ (sigma), Σ (capital sigma), δ (delta), ϵ (epsilon), π (pi), θ (theta), and λ (lambda). Example 3 makes use of Greek letters.

EXAMPLE 3

A formula used in the study of statistics to find the standard score (or Z-score) is

$$Z = \frac{\bar{x} - \mu}{\dfrac{\sigma}{\sqrt{n}}}.$$

Find the value of Z when $\bar{x}$ (read "x bar") $= 120$, $\mu = 100$, $\sigma = 16$, and $n = 4$.

Solution:

$$Z = \frac{\bar{x} - \mu}{\dfrac{\sigma}{\sqrt{n}}} = \frac{120 - 100}{\dfrac{16}{\sqrt{4}}} = \frac{20}{\dfrac{16}{2}} = \frac{20}{8} = 2.5$$

▲

Some formulas contain subscripts. Subscripts are numbers (or letters) placed below and to the right of variables. They are used to help clarify a formula. For example, if two different amounts are used in a problem, they may be symbolized as A and A_0, or A_1 and A_2. Subscripts are read using the word "sub"; for example, A_0 is read "A sub zero" and A_1 is read "A sub one."

Exponential Equations

Many real-life problems, including population growth, growth of bacteria, decay of radioactive substances, and certain other items that increase or decrease at a very rapid rate (exponentially), can be solved by using exponential formulas. An exponential equation (or exponential formula) is of the form

$$y = a^x, \qquad a > 0, \quad a \neq 1.$$

In an exponential formula letters other than x and y may be used to represent the variables. The following equations are examples of exponential formulas: $y = 2^x$, $A = (\frac{1}{2})^x$, and $A = 2.3^t$. Note in the exponential formula the variable is the exponent of some positive constant that is not equal to 1. In many real-life applications the variable t will be used to represent time. Problems involving exponential formulas can be evaluated much more easily if you use a calculator containing a $\boxed{y^x}$ or $\boxed{x^y}$ key.

The following formula, referred to as the exponential growth or decay formula, is used to solve many real-life problems.

$$P = P_0 a^{kt}, \qquad a > 0, \quad a \neq 1$$

In the formula P_0 represents the original amount present, P represents the amount present after t years, and a and k are constants.

EXAMPLE 4

Carbon dating is used by scientists to find the age of fossils, bones, and other items. The formula used in carbon dating is

$$P = P_0 2^{-t/5600},$$

where P_0 represents the original amount of carbon 14 (C_{14}) present and P represents the amount of C_{14} present after t years. If 10 milligrams (mg) of C_{14} is present in an animal bone recently excavated, how many milligrams will be present in 5000 years?

Solution: Substituting the values in the formula gives

$$P = P_0 2^{-t/5600}$$
$$P = 10(2)^{-5000/5600}$$
$$P \approx 10(2)^{-0.89} \qquad \text{(Recall that } \approx \text{ means ``is approximately equal to'')}$$
$$P \approx 10(0.54)$$
$$P \approx 5.4 \text{ mg.}$$

Thus in 5000 years approximately 5.4 mg of the original 10 mg of C_{14} will remain.   ▲

In Example 4 we used a calculator to evaluate $(2)^{-5000/5600}$. The steps used to find this quantity on a calculator with a $\boxed{y^x}$ key are

$$2 \;\boxed{y^x}\; \boxed{(}\; 5000 \;\boxed{+/-}\; \boxed{\div}\; 5600 \;\boxed{)}\; \boxed{=}\; 0.538547.$$

To evaluate $10(2)^{-5000/5600}$ on a scientific calculator we can press the following keys.

$$10 \;\boxed{\times}\; 2 \;\boxed{y^x}\; \boxed{(}\; 5000 \;\boxed{+/-}\; \boxed{\div}\; 5600 \;\boxed{)}\; \boxed{=}\; 5.38547$$

When the a in the formula $P = P_0 a^{kt}$ is replaced with the very special letter e, we get the natural exponential formula

$$P = P_0 e^{kt}.$$

The letter e represents an irrational number whose value is approximately 2.7183. The number e plays an important role in mathematics and is used in finding the solution to many application problems.

EXAMPLE 5

Banks often credit compound interest continuously. When that is done, the principal amount in the account, P, at any time t can be calculated by the natural exponential formula $P = P_0e^{kt}$, where P_0 is the initial principal invested, k is the interest rate in decimal form, and t is the time.

Suppose $10,000 is invested in a savings account at an 8% interest rate compounded continuously. What will be the balance (or principal) in the account in 5 years?

Solution:

$$P = P_0e^{kt}$$
$$= 10,000e^{(0.08)5}$$
$$= 10,000e^{0.40}$$
$$\approx 10,000(1.4918)$$
$$\approx 14,918$$

Thus after 5 years the account's value will have grown from $10,000 to $14,918, an increase of $4918.   ▲

To evaluate $e^{(0.08)5}$ on a calculator (in Example 5) press*

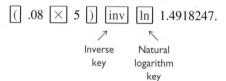

To evaluate $10,000e^{(0.08)5}$ on a calculator press

$$10,000 \quad \boxed{\times} \quad \boxed{(} \quad .08 \quad \boxed{\times} \quad 5 \quad \boxed{)} \quad \boxed{\text{inv}} \quad \boxed{\text{ln}} \quad \boxed{=} \quad 14918.247.$$

EXAMPLE 6

The world's population in 1994 was estimated to be 5.66 billion people. The world's population is continuing to grow exponentially at the rate of about 1.8% per year. The world's expected population, in billions, in t years is given by the formula $P = 5.66e^{0.018t}$. Find the expected world population in the year 2094, 100 years after 1994.

Solution:

$$P = 5.66e^{0.018t}$$
$$= 5.66e^{0.018(100)}$$
$$= 5.66e^{1.8}$$
$$\approx 5.66(6.05)$$
$$\approx 34.2$$

Thus in the year 2094 the world's population is expected to be about 34.2 billion, or about 6 times as great as it was in 1994.   ▲

*Keys to be pressed may vary on some calculators.

Solving for a Variable in a Formula or Equation

Often in mathematics and science courses, you are given a formula or an equation expressed in terms of one variable and asked to express it in terms of a different variable. For example, you may be given the formula $P = i^2 r$ and asked to solve the formula for r. To do so, treat each of the variables, except the one you are solving for, as if they were constants. Then solve for the variable desired, using the properties previously discussed. Examples 7–9 show how to do this task.

When graphing equations in Section 6.7, you will sometimes have to solve the equation for the variable y as is done in Example 7.

EXAMPLE 7

Solve the equation $2x + 3y - 6 = 0$ for y.

Solution: Isolate the term containing the y by moving the constant, -6, and the term containing the x to the right side of the equation.

$$2x + 3y - 6 = 0$$

$$2x + 3y - 6 + 6 = 0 + 6 \qquad \text{Addition property}$$

$$2x + 3y = 6$$

$$-2x + 2x + 3y = -2x + 6 \qquad \text{Subtraction property}$$

$$3y = -2x + 6$$

$$\frac{3y}{3} = \frac{-2x + 6}{3} \qquad \text{Division property}$$

$$y = \frac{-2x + 6}{3}$$

$$y = \frac{-2x}{3} + \frac{6}{3}$$

$$y = -\frac{2}{3}x + 2$$

▲

Note that, once you have found $y = (-2x + 6)/3$, you have solved the equation for y. The solution can also be expressed in the form $y = -\frac{2}{3}x + 2$. This form of the equation is convenient for graphing equations, as will be explained in Section 6.7. Example 7 can also be solved by moving the y term to the right side of the equal sign. Do so now and note that you obtain the same answer.

EXAMPLE 8

An important formula used in statistics is

$$z = \frac{x - \mu}{\sigma}.$$

Solve this formula for x.

Solution: To isolate the term x, use the general procedure for solving linear equations given in the box in Section 6.2. Treat each letter, except x, as if it were a constant.

$$z = \frac{x - \mu}{\sigma}$$

$$z \cdot \sigma = \frac{x - \mu}{\cancel{\sigma}} \cdot \cancel{\sigma} \qquad \text{Multiplication property}$$

$$z\sigma = x - \mu$$

$$z\sigma + \mu = x - \mu + \mu \qquad \text{Add } \mu \text{ to both sides of the equation.}$$

$$z\sigma + \mu = x$$

$$\text{or} \qquad x = z\sigma + \mu$$

EXAMPLE 9

A formula that may be important to you now or sometime in the future is the tax-free yield formula, $T_f = T_a(1 - F)$. This formula can be used to convert a taxable yield, T_a, into its equivalent tax-free yield, T_f, where F is the federal income tax bracket of the individual. A taxable yield is an interest rate for which income tax is paid on the interest made. A tax-free yield is an interest rate for which income tax does not have to be paid on the interest made.

a) For someone in a 28% tax bracket, find the equivalent tax-free yield of an 8% taxable investment.

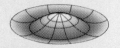 

b) Solve this formula for T_a. That is, write a formula for taxable yield in terms of tax-free yield.

Solution:

a) $T_f = T_a(1 - F)$
   $= 0.08(1 - 0.28) = 0.08(0.72) = 0.0576,$ or 5.76%

Thus a taxable investment of 8% is equivalent to a tax-free investment of 5.76% for a person in a 28% income tax bracket.

b) $T_f = T_a(1 - F)$

$$\frac{T_f}{1 - F} = \frac{T_a(1 - F)}{1 - F}$$ Divide both sides of the equation by $1 - F$.

$$\frac{T_f}{1 - F} = T_a,   \text{or}   T_a = \frac{T_f}{1 - F}$$

Section 6.3 Exercises

1. What is a formula?

2. Explain how to evaluate a formula.

3. What are subscripts?

4. What is an exponential equation?

In Exercises 5–36, use the formula to find the value of the indicated variable for the values given. Use a calculator when one is needed.

5. $d = rt$;   find d when $r = 55$ and $t = 6$ (physics).

6. $P = a + b + c$;   find P when $a = 32$, $b = 42$,   and $c = 52$ (geometry).

7. $P = 2l + 2w$;   find P when $l = 6$ and $w = 10$ (geometry).

8. $E = IR$;   find I when $E = 120$ and $R = 720$ (electronics).

9. $C = 2\pi r$;   find r when $C = 25.12$ and $\pi = 3.14$ (geometry).

10. $p = i^2 r$;   find r when $p = 62{,}500$ and $i = 5$ (electronics).

11. $m = \dfrac{a + b}{2}$;   find b when $m = 55$ and $a = 27$ (statistics).

12. $z = \dfrac{x - \mu}{\sigma}$;   find z when $x = 100$, $\mu = 110$,   and $\sigma = 5$ (statistics).

13. $z = \dfrac{x - \mu}{\sigma}$;   find μ when $z = 2.5$, $x = 42.1$,   and $\sigma = 2$ (statistics).

14. $A = \frac{1}{2}h(b_1 + b_2)$;   find h when $A = 60$, $b_1 = 10$,   and $b_2 = 30$ (geometry).

15. $T = \dfrac{PV}{k}$;   find P when $T = 80$, $V = 20$,   and $k = 0.5$ (physics).

16. $m = \dfrac{a + b + c}{3}$;   find a when $m = 70$, $b = 60$,   and $c = 90$ (statistics).

17. $A = P(1 + rt)$;   find P when $A = 2360$, $r = 0.06$,   and $t = 3$ (economics).

18. $K = \frac{1}{2}mv^2$;   find m when $K = 6000$ and $v = 20$ (physics).

19. $V = \pi r^2 h$;   find h when $V = 437.25$, $\pi = 3.14$,   and $r = 5$ (geometry).

20. $F = \frac{9}{5}C + 32$;   find F when $C = 7$ (temperature conversion).

21. $C = \frac{5}{9}(F - 32)$;   find C when $F = 50$ (temperature conversion).

22. $K = \dfrac{F - 32}{1.8} + 273.1$;   find K when $F = 100$ (chemistry).

23. $z = \dfrac{\bar{x} - \mu}{\dfrac{\sigma}{\sqrt{n}}}$;   find z when $\bar{x} = 66$, $\mu = 60$, $\sigma = 15$,   and $n = 25$ (statistics).

24. $m = \dfrac{y_2 - y_1}{x_2 - x_1}$;   find m when $y_2 = 8$, $y_1 = -4$, $x_2 = -3$,   and $x_1 = -5$ (mathematics).

25. $C = a + by_D$;   find a when $C = 1100$, $b = 10$,   and $y_D = 60$ (economics).

26. $R_T = \dfrac{R_1 R_2}{R_1 + R_2}$;   find R_T when $R_1 = 100$ and
$R_2 = 200$ (electronics).

27. $E = a_1 p_1 + a_2 p_2 + a_3 p_3$;   find E when $a_1 = 10$,
$p_1 = 0.3$, $a_2 = 100$, $p_2 = 0.3$, $a_3 = 500$,
$p_3 = 0.4$ (probability).

28. $x = \dfrac{-b + \sqrt{b^2 - 4ac}}{2a}$;   find x when $a = 2$,
$b = -5$,   and $c = -12$ (mathematics).

29. $x = \dfrac{-b - \sqrt{b^2 - 4ac}}{2a}$;   find x when $a = 2$,
$b = -5$,   and $c = -12$ (mathematics).

30. $R = O + (V - D)r$;   find O when $R = 670$,
$V = 100$, $D = 10$,   and $r = 4$ (economics).

31. $S = \dfrac{n}{2}(f + l)$;   find n when $S = 52$, $f = 6$,   and
$l = 20$ (mathematics).

32. $V = \sqrt{V_x^2 + V_y^2}$;   find V when $V_x = 3$ and
$V_y = 4$ (physics).

33. $F = \dfrac{G m_1 m_2}{r^2}$;   find G when $F = 625$, $m_1 = 100$,
$m_2 = 200$,   and $r = 4$ (physics).

34. $P = \dfrac{nRT}{V}$;   find V if $P = 12$, $n = 10$,
$R = 60$,   and $T = 8$ (chemistry).

35. $S_n = \dfrac{a_1(1 - r^n)}{1 - r}$;   find S_n when $a_1 = 8$,
$r = \frac{1}{2}$,   and $n = 3$ (mathematics).

36. $A = P\left(1 + \dfrac{r}{n}\right)^{nt}$;   find A when $P = 100$, $r = 6\%$,
$n = 1$,   and $t = 3$ (banking).

In Exercises 37–46, solve for y in the equation.

37. $5x - 3y = 7$

38. $2x - 3y = 6$

39. $4x + 7y = 14$

40. $-2x + 4y = 9$

41. $2x - 3y + 6 = 0$

42. $3x + 4y = 0$

43. $-5x - 7y = 14$

44. $4x + 3y - 2z = 4$

45. $9x + 4z = 7 + 8y$

46. $2x - 3y + 5z = 0$

In Exercises 47–66, solve for the variable indicated.

47. $d = rt$ for t

48. $E = IR$ for I

49. $p = a + b + c$ for c

50. $p = a + b + s_1 + s_2$ for s_1

51. $V = lwh$ for h

52. $p = irt$ for t

53. $C = \pi d$ for d

54. $PV = KT$ for T

55. $y = mx + b$ for b

56. $y = mx + b$ for x

57. $P = 2l + 2w$ for w

58. $A = \dfrac{m + d}{2}$ for d

59. $A = \dfrac{a + b + c}{3}$ for c

60. $A = \frac{1}{2}bh$ for b

61. $V = \frac{1}{3}Bh$ for h

62. $y = \dfrac{kX}{Z}$ for Z

63. $F = \frac{9}{5}C + 32$ for C

64. $C = \frac{5}{9}(F - 32)$ for F

65. $y_2 - y_1 = m(x_2 - x_1)$ for x_1

66. $a_n = a_1 + (n - 1)d$ for n

67. Carmen deposited \$2500 in a savings account that paid 4% simple interest per year. If the interest is added to the account at the end of each year, how much is in her account at the end of 1 year?

68. Victor lent his brother \$500 for 3 years. At the end of 3 years, his brother repaid the \$500 plus \$105 interest. What simple interest rate did his brother pay?

69. Determine the volume of an ice cream cone (cone only) if its diameter is 2.5 in. and its height is 4 in. (The formula for the volume of a cone is $V = \frac{1}{3}\pi r^2 h$.)

70. The formula $A = P(1 + r)^n$ is the **compound interest formula**. When interest is compounded periodically (yearly, monthly, daily), the formula can be used to find the amount, A, in the account after n periods. In the formula, P is the principal, r is the interest rate per compounding period, and n is the number of compounding periods. Suppose that Natasha invests \$10,000 at 8% annual interest compounded quarterly for 6 years. Then $P = 10{,}000$, $r = \dfrac{8\%}{4} = 2\%$, and $n = 6 \cdot 4 = 24$. Find the amount in the account after 6 years.

71. If 20 mg of carbon 14 are originally present in an animal bone, how much will remain at the end of 500 years? (See Example 4.)

72. If P is the price of an item today, the price of the same item n years from today, P_n, is $P_n = P(1 + r)^n$, where r is the constant rate of inflation. Determine the price of a pizza 10 years from today if the price today is \$8.50 and the annual rate of inflation is constant at 3%.

73. Assume that the value of the island of Manhattan has grown at an exponential rate of 8% per year since 1626 when Peter Minuit of the Dutch West India Company purchased the island for \$24. The value of the island, V, at any time, t, in years, can be found by the formula $V = 24e^{0.08t}$. What is the value of the island in 1996, 370 years after Minuit purchased it?

74. Plutonium, a radioactive material used in most nuclear reactors, decays exponentially at a rate of 0.003% per year. The amount of plutonium, P, left after t years can be found by the formula $P = P_0 e^{-0.00003t}$, where P_0 is the original amount of plutonium present. If there are originally 2000 grams of plutonium, find the amount of plutonium left after 50 years.

Problem Solving/Group Activities

75. Determine the volume of the block shown in Fig. 6.2, excluding the hole.

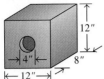

Figure 6.2

76. On Sunday, December 8, 1991, A. J. Kitt of Rochester, New York, was the first American to win the World Cup men's ski title since Bill Johnson won the title in 1984. Kitt skied the 2.1-mi course in 1 min 55.69 sec. Use the formula, distance = rate · time, to determine Kitt's average speed (or rate), in miles per hour, for the course.

Research Activities

77. The formula $R = LF\left(\dfrac{1 - C}{2}\right) + M$ may be used to describe an innocent person's reaction to a lie detector test. Write a report describing the variables in this formula. See Cadidy, Margie, "Formula for Fibbing," *Psychology Today*, March 1975.

78. Research the topic of loans and interest and write a paper that answers the following questions. When did people start paying interest? What rates were they charged? Did the borrower pay simple or compound interest? (References include history of mathematics books and encyclopedias.)

6.4   APPLICATIONS OF LINEAR EQUATIONS IN ONE VARIABLE

One of the main reasons for studying algebra is that it can be used to solve everyday problems. In this section we will do two things: (1) show how to translate a written problem into a mathematical equation and (2) show how linear equations can be used in solving everyday problems. We begin by illustrating how English phrases can be written as mathematical expressions. When writing a mathematical expression we may use any letter to represent the variable. In the following illustrations we use the letter x.

Phrase	Mathematical Expression
Six more than a number	$x + 6$
A number increased by 3	$x + 3$
Four less than a number	$x - 4$
A number decreased by 9	$x - 9$
Twice a number	$2x$
Four times a number	$4x$

Sometimes the phrase that must be converted to a mathematical expression involves more than one operation.

Phrase	Mathematical Expression
Four less than 3 times a number	$3x - 4$
Ten more than twice a number	$2x + 10$
The sum of 5 times a number and 3	$5x + 3$
Eight times a number decreased by 7	$8x - 7$

The word *is* often represents the equal sign.

Phrase	Mathematical Equation
Six more than a number is 10.	$x + 6 = 10$
Five less than a number is 20.	$x - 5 = 20$
Twice a number decreased by 6 is 12.	$2x - 6 = 12$
A number decreased by 13 is 6 times the number.	$x - 13 = 6x$

The following is a general procedure for solving word problems.

To Solve a Word Problem

1. Read the problem carefully at least twice to be sure that you understand it.

2. If possible, draw a sketch to help visualize the problem.

3. Determine which quantity you are being asked to find. Choose a letter to represent this unknown quantity. Write down exactly what this letter represents.

4. Write the word problem as an equation.

5. Solve the equation for the unknown quantity.

6. Answer the question or questions asked.

7. Check the solution.

This general procedure for solving word problems is illustrated in Examples 1–4.

EXAMPLE I

Smiley Bob's Truck Rental Agency charges $210 per week plus $0.18 per mile for a small truck. How far can you travel, to the nearest mile, on a maximum budget of $400?

Solution: In this problem the unknown quantity is the number of miles you can travel. Let's select m to represent the number of miles you can travel. Then we construct an equation using the given information that will allow us to solve for m.

Let m = number of miles you can travel;

then $0.18m$ = amount spent for m miles traveled at 18¢ per mile.

$$\text{Rental fee} + \text{mileage charge} = \text{total amount spent}$$
$$\$210 \quad + \quad \$0.18m \quad = \quad \$400$$

Now solve the equation.

$$210 + 0.18m = 400$$
$$210 - 210 + 0.18m = 400 - 210$$
$$0.18m = 190$$
$$\frac{0.18m}{0.18} = \frac{190}{0.18}$$
$$m = 1055.\overline{5}$$

Therefore you can travel 1055 mi. (If you go 1056 mi, you will have to spend more than $400.)

Check: The check is made with the information given in the original problem.

$\mathcal{D}$ID YOU KNOW

AHA!

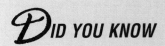

Aha:

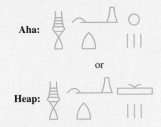

or

Heap:

$\mathcal{T}$he Egyptians as far back as 1650 B.C. had a knowledge of linear equations. They used the words "aha" or "heap" in place of the variable. Problems involving linear equations can be found in the Rhind Papyrus (Chapter 4). ∎

$$\text{Total amount spent} = \text{rental fee} + \text{mileage charge}$$
$$= 210 + 0.18(1055)$$
$$= 210 + 189.90$$
$$= 399.90$$

The check shows that the solution is correct.   ▲

EXAMPLE 2

Forty hours of overtime must be split among three workers. One worker will be assigned twice the number of hours as each of the other two. How many hours of overtime will be assigned to each worker?

Solution: Two workers receive the same amount of overtime, and the third worker receives twice that amount.

Let $x =$ number of hours of overtime for the first worker;

$x =$ number of hours of overtime for the second worker;

$2x =$ number of hours of overtime for the third worker.

Then $x + x + 2x =$ total amount of overtime

$$x + x + 2x = 40$$
$$4x = 40$$
$$x = 10.$$

Thus two workers are assigned 10 hours, and the third worker is assigned $2(10)$, or 20, hours of overtime. A check in the original problem will verify that this answer is correct.   ▲

EXAMPLE 3

Mrs. Chang wants to build a sandbox for her daughter. She has only 26 feet of wood to build the perimeter of the sandbox. What should be the dimensions of the sandbox if she wants the length to be 3 feet greater than the width?

Solution: The formula for finding the perimeter of a rectangle is $P = 2l + 2w$, where P is the perimeter, l is the length, and w is the width. A diagram such as Fig. 6.3 is often helpful in solving problems of this type.

Let w equal the width of the sandbox. The length is 3 feet more than the width, so $l = w + 3$. The total distance around the sandbox, P, is 26 feet.

Substitute the known quantities in the formula.

$$P = 2w + 2l$$
$$26 = 2w + 2(w + 3)$$
$$26 = 2w + 2w + 6$$
$$26 = 4w + 6$$
$$20 = 4w$$
$$5 = w$$

The width of the sandbox is 5 ft, and the length of the sandbox is $5 + 3$ or 8 ft.   ▲

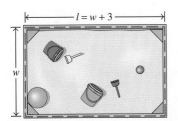

Figure 6.3

In shopping and other daily activities we are occasionally asked to solve problems using percents. The word *percent* means "per hundred." Thus, for example, 7% means 7 per hundred, or $\frac{7}{100}$. When $\frac{7}{100}$ is converted to a decimal number, we obtain 0.07. Thus 7% = 0.07.

Let's look at one example involving percent. (See Section 11.1 for a more detailed discussion of percent.)

EXAMPLE 4

Mr. Ramesh is planning to open a hot dog stand in New York City. What should be the price of the hot dog before tax if the total cost of the hot dog, including a 7% sales tax, is to be $2.50?

Solution: We are asked to find the cost of the hot dog before sales tax.

Let x = the cost of the hot dog before sales tax.

Then $0.07x$ = 7% of the cost of the hot dog (the sales tax).

Cost of hot dog before tax + tax on hot dog = 2.50

$$x + 0.07x = 2.50$$
$$1.07x = 2.50$$
$$\frac{1.07x}{1.07} = \frac{2.50}{1.07}$$
$$x = \frac{2.50}{1.07}$$
$$= 2.336, \quad \text{or} \quad 2.34$$

Thus the cost of the hot dog before tax is $2.34 to the nearest cent. ▲

Section 6.4 Exercises

1. What is the difference between a mathematical expression and an equation?

2. Give an example of a mathematical expression and an example of a mathematical equation.

In Exercises 3–24, write the phrase as a mathematical expression.

3. 4 times x increased by 9

4. 8 decreased by 5 times r

5. $5r$ decreased by 13

6. 7 more than 5 times p

7. 11 decreased by twice p

8. 4 times n decreased by 15

9. 7 more than x

10. 4 decreased by 3 times y

11. The sum of 7 and r, divided by 9

12. 11 decreased by x, divided by 7

13. 6 less than the product of 5 times y, increased by 3

14. The quotient of 8 and y, decreased by 3 times x

In Exercises 15–26, write an equation and solve.

15. A number decreased by 5 is 22.

16. The sum of a number and 12 is 37.

17. A number multiplied by 5 is 95.

18. A number decreased by 9 is 21.

19. Thirteen decreased by 3 times a number is 40.

20. Five more than 7 times a number is 82.

21. The product of a number and 3 is the number decreased by 12.

22. Six more than five times a number is 7 times the number decreased by 18.

23. A number increased by 11 is 1 more than 3 times the number.

24. A number divided by 3 is 4 less than the number.

25. Three more than a number is 5 times the sum of the number and 7.

26. The product of 3 and a number, decreased by 4, is 4 times the number.

In Exercises 27–46, set up an equation that can be used to solve the problem. Solve the equation and find the desired value(s).

27. Two families, the Eagles and Stones, pick peaches. How can they divide the 240 peaches picked so that one family has four times as many as the other family?

28. Scratch Miback purchases two shirts for $87.50. If one of the shirts sells for $14.00 more than the other, how much is the more expensive shirt?

29. The village of Montgomery, which has a population of 6200, is growing by 500 each year. In how many years will the population reach 12,200?

30. Budget Warehouse has a plan whereby for a yearly fee of $80 you save 8% of the price of all items purchased in the store. What is the total Vito will need to spend during the year for his savings to equal the yearly fee?

31. A store owner sells a toy for $3.00, including a 7% sales tax. Determine the cost of the toy before the tax.

32. Paula pays 8 cents to make a copy of a page at a copy shop. She is considering purchasing a photocopy machine that is on sale for $250, including tax. How many copies would Paula have to make in the copy shop for her cost to equal the purchase price of the machine she is considering buying?

33. A town recycles 28 tons of newspaper and cardboard each week. The amount of newspaper is 3 times the amount of cardboard. Determine the number of tons of newspaper and the number of tons of cardboard recycled each week.

34. The U.S. Bureau of Labor Statistics reported a 65% increase in the number of employed mothers from January 1970 to January 1990. If the number of employed mothers reported in 1990 was 16.8 million, how many mothers were employed in 1970?

35. Yekcim Esuom budgeted $100 for renting a truck. How far can he travel in one day if the charges are $60 a day plus 20¢ per mile?

36. The Connells would like to build a rectangular sandbox for their children. They have a total of 20 ft of wood for the perimeter of the sandbox. What will be the dimensions of the box if the width of the box is to be two-thirds its length?

37. The total floor space in three barns is 45,000 ft². The two smaller ones have the same area, and the largest one is 3 times the area of the smaller ones.

a) Determine the floor space for each barn.
b) Can merchandise that takes up 8500 ft² of floor space fit into either of the smaller barns?

38. Mr. Appelton worked a 55-hr week last week. He is paid $1\frac{1}{2}$ times his regular hourly rate for all hours over a 40-hr week. His pay last week was $500. What is his hourly rate?

39. A person's weight on the moon is $\frac{1}{6}$ of that on earth. Assume that the sum of a person's weight on earth and the moon is 203 lb. What does the person weigh on earth?

40. The Peretskys purchased a car. If the total cost, including a 5% sales tax, was $12,836.25, find the cost of the car before tax.

41. The Goullets will need 186 ft of fence to enclose their rectangular yard. If the length of the yard is 9 ft more than the width what are the dimensions of the yard?

42. A bookcase with 3 shelves is to be built by a woodworking student. If the height of the bookcase is to be 2 ft longer than the length of a shelf, and the total amount of wood to be used is 32 ft, find the dimensions of the bookcase.

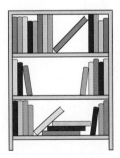

43. The cost of doing the family laundry for a month at a local laundromat is $70. A new washer and dryer cost a total of $760. How many months would it take for the cost of doing the laundry at the laundromat to equal the cost of a new washer and dryer?

44. Ms. Saplin is being hired as a salesperson. She is given a choice of two salary plans. Plan *A* is $100 a week plus 5% commission on sales. Plan *B* is a straight 15% commission on sales. How many dollars in sales are necessary per week for Plan *B* to result in the same salary as Plan *A*?

45. Florence has been told that with her half-off airfare coupon her airfare from New York to San Diego will be $227.00. The $227.00 includes a 7% tax *on the regular fare*. On the way to the airport, Florence realizes that she has lost her coupon. What will her regular fare be, including tax?

46. The cost of renting a small truck at the U-Haul rental agency is $35 per day plus 20 cents a mile. The cost of renting the same truck at the Ryder rental agency is $25 per day plus 32 cents a mile. How far would you have to drive in one day for the cost of renting from U-Haul to equal the cost of renting from Ryder?

Problem Solving/Group Activities

47. Some states allow a husband and wife to file individual tax returns (on a single form) even though they have filed a joint federal tax return. It is usually to the taxpayers' advantage to do so when both husband and wife work. The smallest amount of tax owed (or the largest refund) will occur when the husband's and wife's taxable incomes are the same.

Mr. Neuland's 1996 taxable income was $24,200, and Mrs. Neuland's taxable income for that year was $26,400. The Neulands' total tax deduction for the year was $3640. This deduction can be divided between Mr.

and Mrs. Neuland any way they wish. How should the $3640 be divided between them to result in each individual's having the same taxable income and therefore the greatest tax refund?

48. Write each equation as a sentence. There are many correct answers.
   a) $x + 3 = 13$
   b) $3x + 5 = 8$
   c) $3x - 8 = 7$

49. Show that the sum of any three consecutive integers is 3 less than 3 times the largest.

50. A driver education course at the East Lake School of Driving costs $45 but saves those under 25 years of age 10% of their annual insurance premiums until they are 25. Dan has just turned 18, and his insurance costs $600.00 per year.
   a) When will the amount saved from insurance equal the price of the course?
   b) When Dan turns 25, how much will he have saved?

6.5 VARIATION

In Sections 6.3 and 6.4 we presented many applications of algebra. In this section we introduce variation, which is an important tool in solving applied problems.

Direct Variation

Many scientific formulas are expressed as variations. A variation is an equation that relates one variable to one or more other variables through the operations of multiplication or division (or both operations). Essentially there are three types of variation problems: direct, inverse, and joint variation.

In direct variation the two related variables increase together or decrease together; that is, as one increases so does the other, and as one decreases so does the other.

Consider a car traveling at 40 miles an hour. The car travels 40 miles in 1 hour, 80 miles in 2 hours, and 120 miles in 3 hours. Note that, as the time increases, the distance traveled increases, and, as the time decreases, the distance traveled decreases.

The formula used to calculate distance traveled is

$$\text{Distance} = \text{rate} \cdot \text{time}.$$

Since the rate is a constant, 40 miles per hour, the formula can be written

$$d = 40t.$$

We say that distance varies directly as time or that distance is directly proportional to time.

The preceding equation is an example of direct variation.

Direct Variation

If a variable y varies directly with a variable x, then

$$y = kx,$$

where k is the **constant of proportionality** (or the variation constant).

Examples 1–4 illustrate direct variation.

EXAMPLE 1

The circumference of a circle, C, is directly proportional to (or varies directly as) its radius, r. Write the equation for the circumference of a circle if the constant of proportionality, k, is 2π.

Solution:

$$C = kr \qquad \text{C varies directly as } r.$$
$$= 2\pi r \qquad \text{Constant of proportionality is } 2\pi.$$

▲

EXAMPLE 2

The resistance, R, of a wire varies directly as its length, L.

a) Write this variation as an equation.

b) Find the resistance (measured in ohms) of a 30-foot length of wire, assuming that the constant of proportionality for the wire is 0.008.

Solution:

a) $R = kL$

b) $R = 0.008(30) = 0.24$

The resistance of the wire is 0.24 ohm. ▲

In certain variation problems the constant of proportionality, k, may not be known. In such cases we can often find it by substituting the given values in the variation formula and solving for k.

EXAMPLE 3

The gravitational force of attraction, F, between an object and the earth is directly proportional to the mass, m, of the object. If the force of attraction is 832 when the object's mass is 26, find the constant of proportionality.

Solution:

$$F = km$$
$$832 = k26$$
$$\frac{832}{26} = \frac{26k}{26}$$
$$32 = k$$

Thus the constant of proportionality is 32. ▲

EXAMPLE 4

w varies directly as the square of *y*. If *w* is 60 when *y* is 20, find *w* when *y* is 80.

Solution: Since *w* varies directly as the *square of y*, we begin with the formula $w = ky^2$. Since the constant of proportionality is not given, we must first find *k* using the given information.

$$w = ky^2$$
$$60 = k(20)^2$$
$$60 = 400k$$
$$\frac{60}{400} = \frac{400k}{400}$$
$$0.15 = k$$

We now use $k = 0.15$ to find *w* when *y* is 80.

$$w = ky^2$$
$$= 0.15(80)^2$$
$$= 960$$

Thus, when *y* equals 80, *w* equals 960.   ▲

Inverse Variation

A second type of variation is **inverse variation**. When two quantities vary inversely, as one quantity increases, the other quantity decreases, and vice versa.

To explain inverse variation, we use the formula, distance = rate · time. If we solve for time, we get time = distance/rate. Assume that the distance is fixed at 100 miles; then

$$\text{Time} = \frac{100}{\text{rate}}.$$

At 100 miles per hour it would take 1 hour to cover this distance. At 50 miles an hour, it would take 2 hours. At 25 miles an hour, it would take 4 hours. Note that, as the rate (or speed) decreases, the time increases and vice versa.

The preceding equation can be written

$$t = \frac{100}{r}.$$

This equation is an example of an inverse variation. The time and rate are inversely proportional. The constant of proportionality is 100.

Inverse Variation

If a variable *y* varies inversely with a variable *x*, then

$$y = \frac{k}{x}, \quad \text{or} \quad xy = k,$$

where *k* is the constant of proportionality.

Two quantities vary inversely, or are inversely proportional, when as one quantity increases the other quantity decreases and vice versa. Examples 5 and 6 illustrate inverse variation.

EXAMPLE 5

The illuminance, I, of a light source varies inversely as the square of the distance, d, from the source. Assuming that the illuminance is 75 units at a distance of 6 meters, find the equation that expresses the relationship between the illuminance and the distance.

Solution: Since the illuminance varies inversely as the *square* of the distance the general form of the equation is

$$I = \frac{k}{d^2}, \quad \text{or} \quad Id^2 = k.$$

To find k, we substitute the given values for I and d.

$$75 = \frac{k}{6^2}$$

$$75 = \frac{k}{36}$$

$$(75)(36) = k$$

$$2700 = k$$

Thus the formula is $I = \dfrac{2700}{d^2}$.

▲

EXAMPLE 6

y varies inversely as x. If $y = 8$ when $x = 15$, find y when $x = 18$.

Solution: First write the inverse variation, then solve for k.

$$y = \frac{k}{x}$$

$$8 = \frac{k}{15}$$

$$120 = k$$

Now substitute 120 for k in $y = \dfrac{k}{x}$ and find y when $x = 18$.

$$y = \frac{120}{x} = \frac{120}{18} = 6.7 \quad \text{(to the nearest tenth)}$$

▲

Joint Variation

One quantity may vary directly as a product of two or more other quantities. This type of variation is called **joint variation**.

> ### Joint Variation
> The general form of a joint variation, where y varies directly as x and z, is
>
> $$y = kxz,$$
>
> where k is the constant of proportionality.

EXAMPLE 7

The area, A, of a triangle varies jointly as its base, b, and height, h. If the area of a triangle is 48 square inches when its base is 12 inches and its height is 8 inches, find the area of a triangle whose base is 15 inches and whose height is 20 inches.

Solution: First write the joint variation; then substitute the known values and solve for k.

$$A = kbh$$
$$48 = k(12)(8)$$
$$48 = k(96)$$
$$\frac{48}{96} = k$$
$$k = \frac{1}{2}$$

Now solve for the area of the given triangle.

$$A = kbh$$
$$= \frac{1}{2}(15)(20)$$
$$= 150 \text{ square inches}$$

Summary of Variations

Direct	Inverse	Joint
$y = kx$	$y = \dfrac{k}{x}$	$y = kxz$

Combined Variation

Often in real-life situations one variable varies as a combination of variables. The following examples illustrate the use of combined variations.

EXAMPLE 8

The owners of the Colonel Mustard Pretzel Shop find that their weekly sales of pretzels, S, varies directly with their advertising budget, A, and inversely with their pretzel price, P. When their advertising budget is $500 and the price is $1, they sell 6500 pretzels.

a) Write an equation of variation expressing S in terms of A and P. Include the value of the proportionality constant.

b) Find the expected sales if the advertising budget is $720 and the price is $1.20.

Solution:

a) We begin with the equation

$$S = \frac{kA}{P}.$$

We now find k using the known values.

$$6500 = \frac{k(500)}{1}$$

$$6500 = 500k$$

$$13 = k$$

Therefore the equation for the sales of pretzels is $S = \frac{13A}{P}.$

b) $S = \frac{13A}{P}$

$$= \frac{13(720)}{1.20} = 7800$$

They can expect to sell 7800 pretzels.

EXAMPLE 9

The electrostatic force, F, of repulsion between two positive electrical charges is jointly proportional to the two charges, q_1 and q_2, and inversely proportional to the square of the distance, d, between the two charges. Express F in terms of q_1, q_2, and d.

Solution: $F = \frac{kq_1q_2}{d^2}$

EXAMPLE 10

A varies jointly as B and C and inversely as the square of D. If $A = 1$ when $B = 9$, $C = 4$, and $D = 6$, find A when $B = 8$, $C = 12$, and $D = 5$.

Solution: We begin with the equation

$$A = \frac{kBC}{D^2}.$$

We must first find the constant of proportionality, k, by substituting the known values for A, B, C, and D and solving for k.

$$1 = \frac{k(9)(4)}{6^2}$$

$$1 = \frac{36k}{36}$$

$$1 = k$$

Thus the constant of proportionality equals 1. Now we find A for the corresponding values of B, C, and D.

$$A = \frac{kBC}{D^2}$$

$$A = \frac{(1)(8)(12)}{5^2} = \frac{96}{25} = 3.84$$

▲

Section 6.5 Exercises

In Exercises 1–4, use complete sentences to answer the question.

1. Describe direct variation.

2. Describe inverse variation.

3. Describe joint variation.

4. Describe combined variation.

In Exercises 5–20, use your intuition to determine whether the variation between the indicated quantities is direct or inverse.

5. The diameter of a hose and the volume of water coming from the hose.

6. The speed and distance traveled by a car in a specified time period.

7. The distance between two cities on a map and the actual distance between the two cities.

8. A weight and the force needed to lift that weight.

9. The light illuminating an object and the distance the light is from the object.

10. The volume of a balloon and its radius.

11. The cubic-inch displacement in liters and the horsepower of the engine.

12. The length of a board and the force needed to break the board at the center.

13. The shutter opening of a camera and the amount of sunlight that reaches the film.

14. The number of pages a person can read in a fixed period of time and his reading speed.

15. The time required for an ice cube to melt in water and the temperature of the water.

16. A person's weight (due to the earth's gravity) and her distance from the earth.

17. The time needed to properly expose a film and the size of the opening of the camera lens.

18. The time to reach a certain point for a plane flying with the wind and the speed of the wind.

19. The time needed to heat a casserole in a microwave oven and the power level selected (1 is lowest and 10 is highest).

20. The number of calories eaten and the amount of exercise required to burn off those calories.

In Exercises 21 and 22, use Exercises 5–20 as a guide.

21. Name two items that have not been mentioned in this section that have a direct variation.

22. Name two items that have not been mentioned in this section that have an inverse variation.

In Exercises 23–40, (a) write the variation and (b) find the quantity indicated.

23. x varies directly as y. Find x when $y = 12$ and $k = 6$.

24. C varies directly as the square of Z. Find C when $Z = 9$ and $k = \frac{3}{4}$.

25. y varies directly as R. Find y when $R = 180$ and $k = 1.7$.

26. x varies inversely as y. Find x when $y = 25$ and $k = 5$.

27. R varies inversely as W. Find R when $W = 160$ and $k = 8$.

28. L varies inversely as the square of P. Find L when $P = 4$ and $k = 100$.

29. A varies directly as B and inversely as C. Find A when $B = 12$, $C = 4$, and $k = 3$.

30. A varies jointly as R_1 and R_2 and inversely as the square of L. Find A when $R_1 = 120$, $R_2 = 8$, $L = 5$, and $k = \frac{3}{2}$.

31. T varies directly as the square of D and inversely as F. Find T when $D = 8$, $F = 15$, and $k = 12$.

32. x varies jointly as y and z. If x is 72 when y is 18 and $z = 2$, find x when y is 36 and $z = 1$.

33. Z varies jointly as W and Y. If Z is 28 when W is 28 and $Y = 2$, find Z when W is 140 and $Y = 6$.

34. y varies directly as the square of R. If y is 5 when $R = 5$, find y when R is 10.

35. *S* varies inversely as *G*. If *S* is 12 when *G* is 0.4, find *S* when *G* is 5.

36. *C* varies inversely as *J*. If *C* is 7 when *J* is 0.7, find *C* when *J* is 12.

37. *x* varies inversely as the square of *P*. If $x = 10$ when *P* is 6, find *x* when $P = 20$.

38. *F* varies jointly as M_1 and M_2 and inversely as *d*. If *F* is 20 when $M_1 = 5$, $M_2 = 10$, and $d = 0.2$, find *F* when $M_1 = 10$, $M_2 = 20$, and $d = 0.4$.

39. *F* varies jointly as q_1 and q_2 and inversely as the square of *d*. If *F* is 8 when $q_1 = 2$, $q_2 = 8$, and $d = 4$, find *F* when $q_1 = 28$, $q_2 = 12$, and $d = 2$.

40. *S* varies jointly as *I* and the square of *T*. If *S* is 8 when $I = 20$ and $T = 4$, find *S* when $I = 2$ and $T = 2$.

41. The volume of a gas, *V*, varies inversely as its pressure, *P*. If the volume, *V*, is 800 cc when the pressure is 200 millimeters (mm) of mercury, find the volume when the pressure is 25 mm of mercury.

42. The length that a spring will stretch, *S*, varies directly with the force (or weight), *F*, attached to the spring. If a spring stretches 1.4 in. when 20 lb is attached, how far will it stretch when 10 lb is attached?

43. The intensity, *I*, of light received at a source varies inversely as the square of the distance, *d*, from the source. If the light intensity is 20 footcandles at 15 ft, find the light intensity at 12 ft.

44. On earth the weight of an object varies directly with its mass. If an object with a weight of 256 lb has a mass of 8 slugs, find the mass of an object weighing 120 lb.

45. The weekly videotape rentals, *R*, at Busterblock Video vary directly with their advertising budget, *A*, and inversely with the daily rental price, *P*. When their advertising budget is $400 and the rental price is $2 per day, they rent 4600 tapes per week. How many tapes would they rent per week if they increased their advertising budget to $500 and raised their rental price to $2.50?

46. The weight, *W*, of an object in the earth's atmosphere varies inversely with the square of the distance, *d*, between the object and the center of the earth. A 140-lb person standing on earth is approximately 4000 mi from the earth's center. Find the weight (or gravitational force of attraction) of this person at a distance 100 mi from the earth's surface.

47. The wattage rating of an appliance, *W*, varies jointly as the square of the current, *I*, and the resistance, *R*. If the wattage is 1 watt when the current is 0.1 ampere and the resistance is 100 ohms, find the wattage when the current is 0.4 ampere and the resistance is 250 ohms.

48. The electrical resistance of a wire, *R*, varies directly as its length, *L*, and inversely as its cross-sectional area, *A*. If the resistance of a wire is 0.2 ohm when the length is 200 ft and its cross-sectional area is 0.05 in^2, find the resistance of a wire whose length is 5000 ft with a cross-sectional area of 0.01 in^2.

49. The number of phone calls between two cities during a given time period, *N*, varies directly as the populations p_1 and p_2 of the two cities and inversely to the distance, *d*, between them. If 100,000 calls are made between two cities 300 miles apart and the populations of the cities are 60,000 and 200,000, how many calls are made between two cities with populations of 125,000 and 175,000 that are 450 miles apart?

50. a) If *y* varies directly as *x* and the constant of proportionality is 2, does *x* vary directly or inversely as *y*? Explain.
 b) Give the new constant of proportionality for *x* as a variation of *y*.

51. a) If *y* varies inversely as *x* and the constant of proportionality is 0.3, does *x* vary directly or inversely as *y*? Explain.
 b) Give the new constant of proportionality for *x* as a variation of *y*.

Problem Solving/Group Activities

52. An article in the magazine *Outdoor and Travel Photography* states, ''If a surface is illuminated by a point-source of light, the intensity of illumination produced is inversely proportional to the square of the distance separating them. In practical terms, this means that foreground objects will be grossly overexposed if your background subject is properly exposed with a flash. Thus direct flash will not offer pleasing results if there are any intervening objects between the foreground and the subject.''

 If the subject you are photographing is 4 ft from the flash, and the illumination on this subject is 1/16 of the light of the flash, what is the intensity of illumination on an intervening object that is 3 ft from the flash?

53. In a specific region of the country, the amount of a customer's water bill, *W*, is directly proportional to the average daily temperature for the month, *T*, the lawn area, *A*, and the square root of *F*, where *F* is the family size, and inversely proportional to the number of inches of rain, *R*.

 In one month, the average daily temperature is 78° and the number of inches of rain is 5.6. If the average family of four who has a thousand square feet of lawn pays $68.00 for water for that month, estimate the water bill in the same month for the average family of six who has fifteen-hundred square feet of lawn.

6.6   LINEAR INEQUALITIES

The first four sections of this chapter have dealt with equations. However, we often encounter statements of inequality. The symbols of inequality are as follows.

Symbols of Inequality

$a < b$ means that a is less than b.

$a \le b$ means that a is less than or equal to b.

$a > b$ means that a is greater than b.

$a \ge b$ means that a is greater than or equal to b.

An *inequality* consists of two (or more) expressions joined by an inequality sign.

Examples of Inequalities

$$3 < 5, \qquad x < 2, \qquad 3x - 2 \ge 5$$

A statement of inequality can be used to indicate a set of real numbers. For example, $x < 2$ represents the set of all real numbers less than 2. Listing all these numbers is impossible, but some are $-\frac{1}{2}, -1, 0, -2, \frac{97}{163}, -1.234$.

A method of picturing all real numbers less than 2 is to graph the solution on the number line. The number line was discussed in Chapter 5.

To indicate the solution set of $x < 2$ on the number line, we draw an open circle at 2 and a line to the left of 2 with an arrow at its end. This technique indicates that all points to the left of 2 are part of the solution set. The open circle indicates that the solution set does not include the number 2.

Graph the solution set of $x \ge -1$, where x is a real number, on the number line.

Solution: The numbers greater than or equal to -1 are all the points on the number line to the right of -1 and -1 itself. The solid dot shows that -1 is included in the solution.

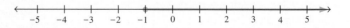

The inequality statements $x < 2$ and $2 > x$ have the same meaning. Note that the inequality symbol points to the x in both cases. Thus one may be written in place of the other. Likewise, $x > 2$ and $2 < x$ have the same meaning. Note that the inequality symbol points to the 2 in both cases. We make use of this fact in Example 2.

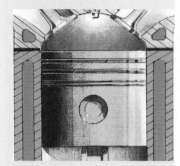
EXAMPLE 2

Graph $-4 \leq x$, where x is a real number, on the number line.

Solution: We can restate $-4 \leq x$ as $x \geq -4$. Both statements have identical solutions. Any number that is greater than or equal to -4 satisfies the inequality $x \geq -4$. Thus the graph includes -4 and all the points to the right of -4 on the number line.

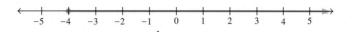

We can find the solution to an inequality by adding, subtracting, multiplying, or dividing both sides of the inequality by the same number or expression. We use the procedure discussed in Section 6.2 to isolate the variable, with one important exception: *When both sides of an inequality are multiplied or divided by a negative number, the direction of the inequality symbol is reversed.* When we change the direction of the inequality symbol we say we change the **order** (or sense) of the inequality.

EXAMPLE 3

Solve the inequality $-x > 5$ and graph the solution set on the number line.

Solution: To solve this inequality, we must eliminate the negative sign in front of the x. To do so, we multiply both sides of the inequality by -1 and change the order of the inequality.

$$-x > 5$$
$$-1(-x) < -1(5) \quad \text{Multiply both sides of the inequality by } -1 \text{ and}$$
$$x < -5 \qquad\qquad \text{change the order of the inequality.}$$

The solution set is graphed on the number line as follows.

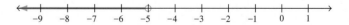

EXAMPLE 4

Solve the inequality $-2x \geq 6$ and graph the solution set on the number line.

Solution: Solving this inequality requires making the coefficient of the x term 1. To do so, divide both sides of the inequality by -2 and change the order of the inequality.

$$-2x \geq 6$$
$$\frac{-2x}{-2} \leq \frac{6}{-2} \quad \text{Divide both sides of the inequality by } -2 \text{ and change}$$
$$x \leq -3 \qquad\qquad \text{the order of the inequality.}$$

The solution set is graphed on the number line as follows.

EXAMPLE 5

Solve the inequality $2x - 4 < 6$ and graph the solution set on the number line.

Solution: To find the solution set, isolate x on one side of the inequality symbol.

$$2x - 4 < 6$$
$$2x - 4 + 4 < 6 + 4 \qquad \text{Add 4 to both sides of the inequality.}$$
$$2x < 10$$
$$\frac{2x}{2} < \frac{10}{2} \qquad \text{Divide both sides of the inequality by 2.}$$
$$x < 5$$

Thus the solution set to $2x - 4 < 6$ is all real numbers less than 5.

Note that in Example 5 the *order of the inequality* did not change when both sides of the inequality were divided by the positive number 2.

EXAMPLE 6

Solve the inequality $x + 4 < 7$, where x is an integer, and graph the solution set on the number line.

Solution:

$$x + 4 < 7$$
$$x + 4 - 4 < 7 - 4$$
$$x < 3$$

Since x is an integer and is less than 3, the solution set is the set of integers less than 3, or $\{\ldots -3, -2, -1, 0, 1, 2\}$. To graph the solution, we make solid dots at the corresponding points on the number line. The three smaller dots to the left of -3 indicate that all the integers to the left of -3 are included.

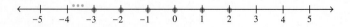

An inequality of the form $a < x < b$ is called a **compound inequality**. Consider the compound inequality $-3 < x \leq 2$, which means that $-3 < x$ *and* $x \leq 2$.

EXAMPLE 7

Graph the solution set of the inequality $-3 < x \le 2$,

a) where x is an integer.

b) where x is a real number.

Solution:

a) The solution set is all the integers between -3 and 2, including the 2 but not including the -3, or $\{-2, -1, 0, 1, 2\}$.

b) The solution set consists of all the real numbers between -3 and 2, including the 2 but not including the -3.

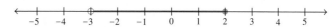

EXAMPLE 8

Solve the compound inequality for x and graph the solution set.

$$-4 < \frac{x+3}{2} \le 5$$

Solution: To solve a compound inequality, we must isolate the x as the middle term. To do so, we use the same principles used to solve inequalities.

$$-4 < \frac{x+3}{2} \le 5$$

$$2(-4) < 2\left(\frac{x+3}{2}\right) \le 2(5) \qquad \text{Multiply the three terms by 2.}$$

$$-8 < x + 3 \le 10$$

$$-8 - 3 < x + 3 - 3 \le 10 - 3 \qquad \text{Subtract 3 from all three terms.}$$

$$-11 < x \le 7$$

The solution set is graphed on the number line as follows.

EXAMPLE 9

A student must have an average (the mean) on five tests that is greater than or equal to 80% but less than 90% to receive a final grade of B. Darrell's grades on the first four tests were 98%, 76%, 86%, and 92%. What range of grades on the fifth test would give him a B in the course?

Solution: The unknown quantity is the range of grades on the fifth test. First construct an inequality that can be used to find the range of grades on the fifth exam. The average (mean) is found by adding the grades and dividing the sum by the number of exams.

Let x = the fifth grade. Then

$$\text{Average} = \frac{98 + 76 + 86 + 92 + x}{5}.$$

For Darrell to obtain a B, his average must be greater than or equal to 80 but less than 90.

$$80 \le \frac{98 + 76 + 86 + 92 + x}{5} < 90$$

$$80 \le \frac{352 + x}{5} < 90$$

$$5(80) \le 5\left(\frac{352 + x}{5}\right) < 5(90) \qquad \text{Multiply the three terms of the inequality by 5.}$$

$$400 \le 352 + x < 450$$

$$400 - 352 \le 352 - 352 + x < 450 - 352 \qquad \text{Subtract 352 from all three terms.}$$

$$48 \le x < 98$$

"In mathematics the art of posing problems is easier than that of solving them."

George Cantor

Thus a grade of 48% up to but not including a grade of 98% on the fifth test will result in a grade of B.   ▲

Section 6.6 Exercises

1. When solving an inequality, under what conditions do you need to change the order of the inequality symbol?

2. Does $x > -3$ have the same meaning as $-3 < x$? Explain.

3. Does $x < 2$ have the same meaning as $2 > x$? Explain.

4. When graphing the solution set to an inequality on the number line, when should you use an open circle and when should you use a closed circle?

In Exercises 5–22, graph the solution set of the inequality, where x is a real number, on the real number line.

5. $x > 3$

6. $x \le 4$

7. $x + 3 < 7$

8. $4x \ge 2$

9. $-3x \le 18$

10. $-4x < 12$

11. $\dfrac{x}{6} < -2$

12. $\dfrac{x}{2} > 4$

13. $\dfrac{x}{3} \ge -2$

14. $\dfrac{-x}{5} \ge 2$

15. $2x - 5 \ge 17$

16. $2x - 7 < 4x + 9$

17. $2(x + 6) \le 15$

18. $-4(x + 2) - 2x > -6x + 2$

19. $3(x + 4) - 2 < 3x + 10$

20. $-2 \le x \le 1$

21. $3 < x - 7 \le 6$

22. $\dfrac{1}{2} < \dfrac{x + 4}{2} \le 4$

In Exercises 23–42, graph the solution set of the inequality, where x is an integer, on the number line.

23. $x \ge 1$

24. $-5 < x$

25. $4x \ge 16$

26. $-4x \le 16$

27. $-5x < 30$

28. $2 + x > 7$

29. $\dfrac{x}{6} < -2$

30. $\dfrac{x}{3} > -3$

31. $-\dfrac{x}{6} \ge 3$

32. $\dfrac{2x}{3} \le 4$

33. $-11 < -5x + 4$

34. $2x + 5 < -3 + 6x$

35. $2(x + 8) \le 3x + 17$

36. $-2(x - 1) < 3(x - 4) + 5$

37. $5(x + 4) - 6 \le 2x + 8$

38. $-2 \le x \le 10$

39. $1 > -x > -5$

40. $-2 < 2x + 3 < 6$

41. $0.2 < \dfrac{x - 3}{10} \le 0.4$

42. $-\dfrac{1}{3} \le \dfrac{x - 3}{6} < \dfrac{1}{2}$

In Exercises 43–51, write a statement of inequality and solve the inequality.

43. You can rent a compact car from Hungry Howie's Auto Rental for $190 per week with no charge for mileage or from Sleepy Sam's Auto Rental for $110 per week plus 20 cents for each mile driven. At what mileage does renting from Hungry Howie's Auto Rental cost less?

44. For a small plane to take off safely, it must have a maximum load of no more than 1500 lb. The gasoline weighs 400 lb and the passengers weigh a total of 670 lb.
   a) Write an inequality that can be used to determine the maximum weight of the luggage that the plane can safely carry.
   b) Determine the maximum weight of the luggage.

45. The janitor must move a large shipment of books from the first floor to the fifth floor. Each box of books weighs 60 lb, and the janitor weighs 180 lb. The sign on the elevator reads, ''Maximum weight 1200 lb.''
   a) Write a statement of inequality to determine the maximum number of boxes of books the janitor can place on the elevator at one time. (The janitor must ride in the elevator with the books.)
   b) Determine the maximum number of boxes that can be moved in one trip.

46. After Mrs. Franklin is seated in a restaurant, she realizes that she has only $19.00. If she must pay 7% tax and wants to leave a 15% tip, what is the price range of meals that she can order?

47. For a business to realize a profit, its revenue, R, must be greater than its costs, C; that is, a profit will result only if $R > C$ (the company breaks even when $R = C$). A record manufacturer has a weekly cost equation $C = 300 + 1.5x$ and a weekly revenue equation $R = 2x$, where x is the number of records produced and sold in a week. How many records must be sold weekly for the company to make a profit?

48. The velocity v, in ft/sec, t sec after an object is projected directly upward is given by the formula $v = 92 - 32t$. How many seconds after being projected upward will the velocity be between 19 ft/sec and 32 ft/sec?

49. In Example 9, what range of grades on the fifth test would result in Darrell receiving a grade of B if his grades on the first four tests were 78%, 64%, 88%, and 76%?

50. The minimum speed for vehicles on a highway is 40 mph, and the maximum speed is 55 mph. If Philip has been driving nonstop along the highway for 4 hr, what range in miles could he have legally traveled?

51. The cost of running a day camp for one week is $8000, plus $175 per person enrolled. If $18,000 is the minimum amount and $24,000 is the maximum amount that will be spent running the camp for 1 week, what is the minimum number and the maximum number of people that can be enrolled in the day camp?

Problem Solving/Group Activities

52. J. B. Davis is painting the exterior of his house. The instructions on the paint can indicate that 1 gal covers from 250 to 400 ft². The total surface of the house to be painted is 2750 ft². Determine the number of gallons of paint he could use, and express the answer as an inequality.

53. Teresa's five test grades for the semester are 86%, 74%, 68%, 96%, and 72%. Her final exam counts one-third of her final grade. What range of grades on her final exam would result in Teresa receiving a final grade of B in the course? (See Example 9.)

54. A student multiplied both sides of the inequality $-\frac{1}{3}x \le 4$ by -3 and forgot to reverse the inequality symbol. What is the relation between the student's incorrect solution set and the correct solution set? Is any number in both the correct solution set and the student's incorrect solution set? If so, what is it?

Research Activity

55. Find a newspaper or a magazine article that contains the mathematical concept of inequality.
   a) From the information in the article write a statement of inequality.
   b) Summarize the article and explain how you arrived at the inequality statement in part (a).

6.7 GRAPHING LINEAR EQUATIONS

In Section 6.2 we solved equations with a single variable. However, real-world problems often involve two or more unknowns. For example the profit, p, of a company may depend on the sales, s; or the cost, c, of mailing a package may depend on the weight, w, of the package. Thus it is helpful to be able to work with equations with two variables (for example, $x + 2y = 6$). Doing so requires understanding the Cartesian (or rectangular) coordinate system, named after the French mathematician René Descartes (1596–1650).

The rectangular coordinate system consists of two perpendicular number lines (Fig. 6.4). The horizontal line is the *x*-axis and the vertical line is the *y*-axis. The point of intersection of the *x*-axis and *y*-axis is called the origin. The numbers on the axes to the right and above the origin are positive. The numbers on the axes on the left and below the origin are negative. The axes divide the plane into four parts: the first, second, third, and fourth quadrants.

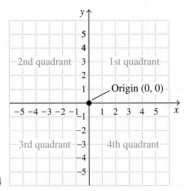

Figure 6.4

We indicate the location of a point in the rectangular coordinate system by means of an ordered pair of the form (*x*, *y*). The *x*-coordinate is always placed first and the *y*-coordinate is always placed second in the ordered pair. Consider the point illustrated in Fig. 6.5. Since the *x*-coordinate of the point is 5 and the *y*-coordinate is 3, the ordered pair that represents this point is (5, 3).

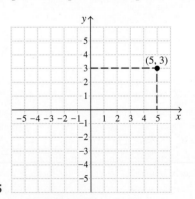

Figure 6.5

The origin is represented by the ordered pair (0, 0). Every point on the plane can be represented by one and only one ordered pair (*x*, *y*), and every ordered pair (*x*, *y*) represents one and only one point on the plane.

EXAMPLE I

Plot the points $A(-2, 4)$, $B(3, -4)$, $C(6, 0)$, $D(4, 1)$,   and $E(0, 3)$.

Solution: Point *A* has an *x*-coordinate of -2 and a *y*-coordinate of 4. Project a vertical line up from -2 on the *x*-axis and a horizontal line to the left from 4 on the *y*-axis. The two lines intersect at the point denoted *A* (Fig. 6.6). The other points are plotted in a similar manner.

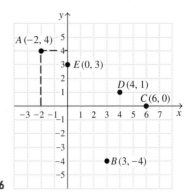

Figure 6.6

EXAMPLE 2

The points A, B, and C are three vertices of a parallelogram with a side parallel to the x-axis. Plot the three points and find the coordinates of the fourth vertex, D.

$$A(1, 2) \qquad B(2, 4) \qquad C(7, 4)$$

Solution: A parallelogram is a figure that has opposite sides that are of equal length and are parallel. (Parallel lines are two lines in the same plane that do not intersect.) The horizontal distance between points B and C is 5 units (see Fig. 6.7). Therefore the horizontal distance between points A and D must also be 5 units. This problem has two possible solutions, as illustrated in Fig. 6.7(a) and 6.7(b).

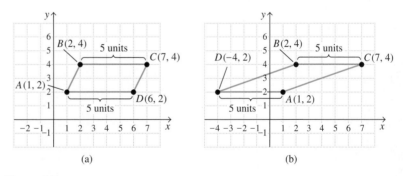

(a)   (b)

Figure 6.7

The solutions are the points $(6, 2)$ and $(-4, 2)$. ▲

Graphing Linear Equations by Plotting Points

Consider the following equation in two variables: $y = x + 1$. Every ordered pair that makes the equation a true statement is a solution to, or satisfies, the equation. We can mentally find some ordered pairs that satisfy the equation $y = x + 1$ by picking some values of x and solving the equation for y. For example, suppose that we let $x = 1$; then $y = 1 + 1 = 2$. The ordered pair $(1, 2)$ is a solution to the equation $y = x + 1$. We can make a chart of other ordered pairs that are solutions to the equation.

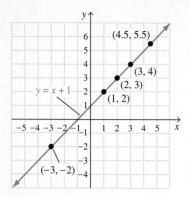

Figure 6.8

x	y	Ordered Pair
1	2	(1, 2)
2	3	(2, 3)
3	4	(3, 4)
4.5	5.5	(4.5, 5.5)
−3	−2	(−3, −2)

How many other ordered pairs satisfy the equation? Infinitely many ordered pairs satisfy the equation. Since we cannot list all of the solutions, we show them by means of a graph. A **graph** is an illustration of all of the points whose coordinates satisfy an equation.

The points (1, 2), (2, 3), (3, 4), (4.5, 5.5), and (−3, −2) are plotted in Fig. 6.8. With a straightedge we can draw one line that contains all these points. This line, when extended indefinitely in either direction, passes through all the points in the plane that satisfy the equation $y = x + 1$. The arrows on the ends of the line indicate that the line extends indefinitely.

All equations of the form $ax + by = c$, $a \neq 0$, $b \neq 0$, will be straight lines when graphed. Thus such equations are called **linear equations in two variables**. The exponents on the variables x and y must be 1 for the equation to be linear. Since only two points are needed to draw a line, only two points are needed to graph a linear equation. It is always a good idea to graph a third point as a checkpoint. If no error has been made, all three points will be in a line, or **collinear**. One method that can be used to obtain points is to solve the equation for y, substitute values for x, and find the corresponding values of y.

EXAMPLE 3

Graph $y = 3x + 6$.

Solution: Since the equation is already solved for y, select values for x and find the corresponding values for y. The table indicates values arbitrarily selected for x and the corresponding values for y. The ordered pairs are (0, 6), (1, 9), and (−2, 0). The graph is shown in Fig. 6.9.

x	y
0	6
1	9
−2	0

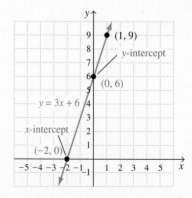

Figure 6.9

To Graph Equations by Plotting Points

1. Solve the equation for y.
2. Select at least three values for x and find their corresponding values of y.
3. Plot the points.
4. If the points are in a straight line, draw a line through the set of points and place arrow tips at both ends of the line.

In step 4 of the procedure, if the points are not in a straight line, recheck your calculations and find your error.

Graphing by Using Intercepts

Example 3 contained two special points on the graph, $(-2, 0)$ and $(0, 6)$. At these points the line crosses the x-axis and the y-axis, respectively. The ordered pairs $(-2, 0)$ and $(0, 6)$ represent the *x-intercept* and the *y-intercept*, respectively. Another method that can be used to graph linear equations is to find the x- and y-intercepts of the graph.

Finding the x- and y-Intercepts

To find the y-intercept, set $x = 0$ and solve the equation for y.
To find the x-intercept, set $y = 0$ and solve the equation for x.

An equation may be graphed by finding the x- and y-intercepts, plotting the intercepts, and drawing a straight line through the intercepts. When graphing by this method, you should always plot a checkpoint before drawing your graph. To obtain a checkpoint, select a nonzero value for x and find the corresponding value of y. The checkpoint should be collinear with the x- and y-intercepts.

EXAMPLE 4

Graph $2x + 3y = 6$ by using the x- and y-intercepts.

Solution: To find the x-intercept, set $y = 0$ and solve for x.

$$2x + 3(0) = 6$$
$$2x = 6$$
$$x = 3$$

The x-intercept is $(3, 0)$. To find the y-intercept, set $x = 0$ and solve for y.

$$2(0) + 3y = 6$$
$$3y = 6$$
$$y = 2$$

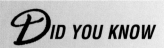

The y-intercept is $(0, 2)$. As a checkpoint try $x = 1$ and find the corresponding value for y.

$$2x + 3y = 6$$
$$2(1) + 3y = 6$$
$$2 + 3y = 6$$
$$3y = 4$$
$$y = \tfrac{4}{3}$$

The checkpoint is the ordered pair $(1, \tfrac{4}{3})$ or $(1, 1\tfrac{1}{3})$.

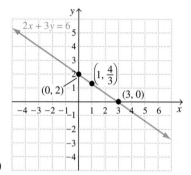

Figure 6.10

Since all three points are collinear in Fig. 6.10, draw a line through the three points to obtain the graph. ▲

Slope

Another useful concept when you are working with straight lines is slope, which is a measure of the "steepness" of a line. The slope of a line is a ratio of the vertical change to the horizontal change for any two points on the line. Consider Fig. 6.11. Point A has coordinates (x_1, y_1), and point B has coordinates (x_2, y_2). The vertical change between points A and B is $y_2 - y_1$, and the horizontal change between points A and B is $x_2 - x_1$. Thus the slope, which is often symbolized with the letter m, can be found as follows.

$$\textbf{Slope} = \frac{\text{vertical change}}{\text{horizontal change}}$$

$$m = \frac{y_2 - y_1}{x_2 - x_1}$$

The Greek capital letter delta, Δ, is used to represent the words "the change in." Therefore slope may be defined as

$$m = \frac{\Delta y}{\Delta x}.$$

A line may have a positive slope, or a negative slope, or zero slope, as indicated in Fig. 6.12. A line with a positive slope rises from left to right (Fig. 6.12a). A line with a negative slope falls from left to right (Fig. 6.12b). A horizontal line, which neither rises nor falls, has a slope of zero (Fig. 6.12c). Since a vertical line does not have any horizontal change (the x value remains constant)

Figure 6.11

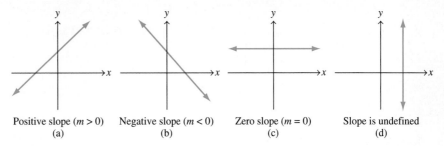

Positive slope ($m > 0$) Negative slope ($m < 0$) Zero slope ($m = 0$) Slope is undefined
(a) (b) (c) (d)

Figure 6.12

and since we cannot divide by 0, the slope of a vertical line is undefined (Fig. 6.12d).

EXAMPLE 5

Determine the slope of the line that passes through the points $(-1, -3)$ and $(1, 5)$.

Solution: Let's begin by drawing a sketch, illustrating the points and the line (Fig. 6.13a).

We will let (x_1, y_1) be $(-1, -3)$ and (x_2, y_2) be $(1, 5)$. Then

$$\text{Slope} = \frac{y_2 - y_1}{x_2 - x_1} = \frac{5 - (-3)}{1 - (-1)} = \frac{5 + 3}{1 + 1} = \frac{8}{2} = \frac{4}{1} \quad \text{or} \quad 4.$$

The slope of 4 means that there is a vertical change of 4 units for each horizontal change of 1 unit (see Fig. 6.13b). The slope is positive, and the line rises from left to right. Note that we would have obtained the same results if we let (x_1, y_1) be $(1, 5)$ and (x_2, y_2) be $(-1, -3)$. Try this now and see.

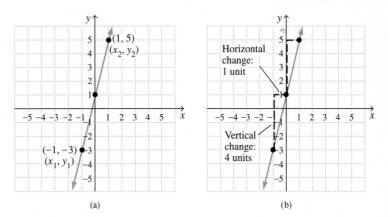

(a) (b)

Figure 6.13

Graphing Equations by Using the Slope and y-Intercept

A linear equation given in the form $y = mx + b$ is said to be in slope–intercept form.

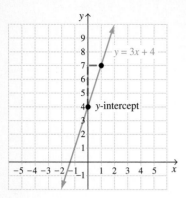

Figure 6.14

Slope–Intercept Form of a Line

$$y = mx + b$$

where m is the slope of the line and b is the y-intercept of the line.

Consider the graph of the equation $y = 3x + 4$, which appears in Fig. 6.14. By examining the graph we can see that the y-intercept is 4. We can also see that the graph has a positive slope, since it rises from left to right. Since the vertical change is 3 units for every 1 unit of horizontal change, the slope must be $\frac{3}{1}$ or 3.

We could graph this equation by marking the y-intercept at 4 and then moving *up* 3 units and to the *right* 1 unit to get another point. If the slope were -3, which means $\frac{-3}{1}$, we could start at the y-intercept and move *down* 3 units and to the *right* 1 unit. Thus, if we know the slope and y-intercept of a line, we can graph the line.

To Graph Equations by Using the Slope and y-Intercept

1. Solve the equation for y to place the equation in slope–intercept form.
2. Determine the slope and y-intercept from the equation.
3. Plot the y-intercept.
4. Obtain a second point using the slope.
5. Draw a straight line through the points.

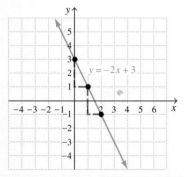

Figure 6.15

EXAMPLE 6

Graph $y = -2x + 3$ by using the slope and y-intercept.

Solution: The slope is -2 and the y-intercept is 3. Plot $(0, 3)$ on the y-axis. Then plot the next point by moving *down* two units and to the *right* 1 unit (see Fig. 6.15). A third point has been plotted in the same way. The graph of $y = -2x + 3$ is the line drawn through these three points.   ▲

EXAMPLE 7

Graph $y = \dfrac{5}{3}x - 2$ using the slope and y-intercept.

Solution: Plot $(0, -2)$ on the y-axis. Then plot the next point by moving *up* 5 units and to the *right* 3 units (see Fig. 6.16). The graph of $y = \frac{5}{3}x - 2$ is the line drawn through these two points.

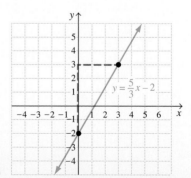

Figure 6.16

EXAMPLE 8

a) Write $3x + 4y = 4$ in slope–intercept form.

b) Graph the equation.

Solution:

a) Solve the given equation for y.

$$3x + 4y = 4$$
$$3x - 3x + 4y = -3x + 4$$
$$4y = -3x + 4$$
$$\frac{4y}{4} = \frac{-3x + 4}{4}$$
$$y = -\frac{3x}{4} + \frac{4}{4}$$
$$y = -\frac{3}{4}x + 1$$

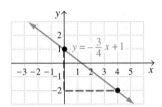

Figure 6.17

Thus in slope–intercept form the equation is $y = -\frac{3}{4}x + 1$.

b) The y-intercept is 1, and the slope is $-\frac{3}{4}$. Place a dot at $(0, 1)$ on the y-axis, then move *down* 3 units and to the *right* 4 units to obtain the second point (see Fig. 6.17). Draw a line through the two points.   ▲

EXAMPLE 9

Determine the equation of the line in Fig. 6.18.

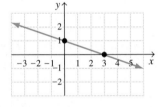

Figure 6.18

Solution: If we determine the slope and the y-intercept of the line, then we can write the equation using slope–intercept form, $y = mx + b$. We see from the graph that the y-intercept is 1; thus $b = 1$. The slope of the line is negative because the graph falls from left to right. The change in y is one unit for each three-unit change in x. Thus m, the slope of the line, is $-\frac{1}{3}$.

$$y = mx + b$$
$$y = -\frac{1}{3}x + 1$$

The equation of the line is $y = -\frac{1}{3}x + 1$.   ▲

EXAMPLE 10

In the Cartesian coordinate system, graph (a) $y = 2$ and (b) $x = -3$.

Solution:

a) For any value of x, the value of y is 2. Therefore the graph will be a horizontal line through $y = 2$ (Fig. 6.19).

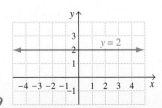

Figure 6.19

b) For any value of y, the value of x is -3. Therefore the graph will be a vertical line through $x = -3$ (Fig. 6.20).

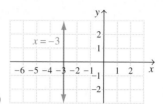

Figure 6.20

Note that the graph of $y = 2$ has a slope of 0. The slope of $x = -3$ is undefined.

▲

We will discuss labeling of the axes of a graph before looking at a couple of applications. In graphing the equations in this section, we labeled the horizontal axis the x-axis and the vertical axis the y-axis. For each equation we can determine values for y by substituting values for x. Since the value of y depends on the value of x, we refer to y as the **dependent variable** and x as the **independent variable**. We label the *vertical axis* with the *dependent variable*, and the *horizontal axis* with the *independent variable*. For the equation $C = 3n + 5$, the C is the dependent variable and n is the independent variable. Thus to graph this equation we label the vertical axis C and the horizontal axis n.

In many graphs the values to be plotted on one axis are much greater than the values to be plotted on the other axis. When that occurs we can use different scales on the horizontal and the vertical axes, as illustrated in Examples 11 and 12.

EXAMPLE 11

The Paul Presley Peterson Paving Company has a contract from the state to pave 20 miles of highway. The distance, d, in miles they can pave in t hours can be approximated by the formula $d = 0.2t$.

a) Graph $d = 0.2t$, for $t \le 300$ hours.

b) Estimate the distance paved in 150 hours.

Solution:

a) Since $d = 0.2t$ is a linear equation, its graph will be a straight line. Select three values for t, find the corresponding values for d, and then draw the graph (Fig. 6.21).

$$d = 0.2t$$

t	d
0	0
100	20
300	60

Let $t = 0$, $d = 0.2(0) = 0$

Let $t = 100$, $d = 0.2(100) = 20$

Let $t = 300$, $d = 0.2(300) = 60$

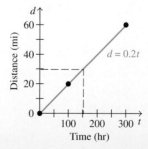

Figure 6.21

b) By drawing a vertical line up to the graph at $t = 150$ and a horizontal line across to the distance axis, we can determine that the distance covered is about 30 miles. ▲

EXAMPLE 12

Tishia Jones owns a small business that manufactures compact discs. She believes that the profit (or loss) from each CD produced can be estimated by the formula $P = 3.5S - 200,000$, where S is the number of copies of the CD sold.

a) Graph $P = 3.5S - 200,000$, for $S \leq 500,000$ copies.

b) From the graph estimate the number of copies that must be sold for the company to break even on a CD.

c) If the profit from a CD is $1 million, estimate the number of copies sold.

Solution:

a) Select values for S and find the corresponding values of P.

S	P
0	−200,000
100,000	150,000
500,000	1,550,000

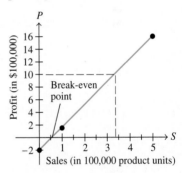

Figure 6.22

b) On the graph (Fig. 6.22) note that the break-even point is about 0.6, or 60,000 copies.

c) We can obtain the answer by drawing a horizontal line from 10 on the profit axis. Since the horizontal line cuts the graph at about 3.4 on the S axis, approximately 340,000 copies were sold. ▲

Section 6.7 Exercises

1. What is a graph?

2. Explain how to find the x-intercept of a linear equation.

3. Explain how to find the y-intercept of a linear equation.

4. What is the slope of a line?

5. Describe the three methods used to graph a linear equation in this section.

6. a) Explain in your own words how to find the slope of a line between two points.

    b) Based on your explanation in part (a), find the slope of the line through the points (6, 2) and (−3, 5).

In Exercises 7–14, plot all the points on the same axes.

7. (2, 3)

8. (−4, 1)

9. (2, −3)

10. (0, 0)

11. (−4, 0)

12. (0, 4)

13. (0, −7)

14. $(3\frac{1}{2}, 4\frac{1}{2})$

In Exercises 15–22, plot all the points on the same axes.

15. (1, 3)

16. (0, 3)

17. (−2, −3)

18. (0, −4)

19. (−3, 1)

20. (−3, 0)

21. (4, −1)

22. (4.5, 3.5)

In Exercises 23–32 (indicated on Fig. 6.23), write the coordinates of the corresponding point.

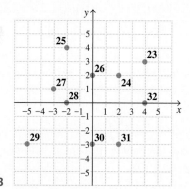

Figure 6.23

In Exercises 33 and 34, the points A, B, and C are three vertices (the points where two lines meet) of a rectangle. Plot the three points. (a) Find the coordinates of the fourth point, D, to complete the rectangle. (b) Find the area of the rectangle; use $A = lw$.

33. $A(-3, 4)$, $B(4, 4)$, $C(4, -2)$
34. $A(-4, 2)$, $B(7, 2)$, $C(7, 8)$

In Exercises 35 and 36, the points A, B, and C are three vertices of a parallelogram with sides parallel to the x-axis. Plot the three points. Find the coordinates of the fourth point, D, to complete the parallelogram. Note: There are two possible answers for point D.

35. $A(3, 2)$, $B(5, 5)$, $C(9, 5)$
36. $A(-2, 2)$, $B(3, 2)$, $C(6, -1)$

In Exercises 37–40, for what value of b will the line joining the points P and Q be parallel to the indicated axis?

37. $P(-4, 3)$, $Q(b, 1)$; y-axis
38. $P(-5, 2)$, $Q(7, b)$; x-axis
39. $P(3b - 1, 5)$, $Q(8, 4)$; y-axis
40. $P(-6, 2b + 1)$, $Q(2, 7)$; x-axis

In Exercises 41–48, determine which ordered pairs satisfy the given equation.

41. $2x - y = 0$ $(2, 4)$, $(2, 3)$, $(-2, 4)$
42. $x - 4y = 6$ $(2, -3)$, $(0, -\frac{3}{2})$, $(6, 0)$
43. $4x - 2y = -10$ $(0, 5)$, $(-\frac{5}{2}, 0)$, $(2, -3)$
44. $3x = 5y + 2$ $(0, -\frac{2}{5})$, $(\frac{1}{3}, -\frac{1}{5})$, $(-2, -1)$
45. $\frac{x}{2} + 3y = 4$ $(0, \frac{4}{3})$, $(8, 0)$, $(10, -2)$
46. $7y = 3x - 5$ $(1, -1)$, $(-3, -2)$, $(2, 5)$
47. $2x - 5y = -7$ $(2, 1)$, $(-1, 1)$, $(4, 3)$
48. $\frac{x}{2} + \frac{3y}{4} = 2$ $(0, \frac{8}{3})$, $(1, \frac{11}{4})$, $(4, 0)$

In Exercises 49–52, graph the equation and state the slope if it exists (see Example 10).

49. $x = 4$ **50.** $x = -2$
51. $y = 3$ **52.** $y = -5$

In Exercises 53–62, graph the equation by plotting points as in Example 3.

53. $y = x + 3$ **54.** $y = x - 5$
55. $y = 4x - 3$ **56.** $y = -x + 5$
57. $y - 4x = 8$ **58.** $y + 3x = 6$
59. $y = \frac{1}{2}x + 4$ **60.** $3y = 2x - 3$
61. $2y = -x + 6$ **62.** $y = -\frac{3}{4}x$

In Exercises 63–72, graph the equation, using the x- and y-intercepts as in Example 4.

63. $x + y = 5$ **64.** $x - y = 2$
65. $2x + y = 8$ **66.** $3x - 2y = 6$
67. $3x = 5y - 15$ **68.** $y = 4x + 4$
69. $y = -2x - 5$ **70.** $2x + 4y = 6$
71. $-3x + y = 4$ **72.** $5y = 3x + 10$

In Exercises 73–82, find the slope of the line through the given points. If the slope is undefined, so state.

73. $(4, 5)$ and $(2, 7)$ **74.** $(5, 8)$ and $(7, 12)$
75. $(-2, -4)$ and $(3, 2)$ **76.** $(2, -2)$ and $(-3, 5)$
77. $(5, 2)$ and $(-3, 2)$
78. $(-3, -5)$ and $(-1, -2)$
79. $(8, -3)$ and $(8, 3)$ **80.** $(2, 6)$ and $(2, -3)$
81. $(-2, 3)$ and $(1, -1)$ **82.** $(-7, -5)$ and $(5, -6)$

In Exercises 83–92, graph the equation, using the slope and y-intercept as in Examples 6–8.

83. $y = x + 4$ **84.** $y = 2x + 1$
85. $y = -x - 2$ **86.** $y = -3x + 1$
87. $y = -\frac{3}{5}x + 3$ **88.** $y = -x - 2$
89. $7y = 4x - 7$ **90.** $3x + 2y = 6$
91. $3x - 2y + 6 = 0$ **92.** $3x + 4y - 8 = 0$

In Exercises 93–96, determine the equation of the graph.

93.

94.

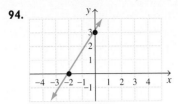

95.

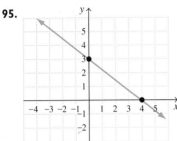

96.

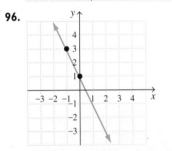

97. The charge, C, for processing a roll of 35-millimeter (mm) film at the George Eastman Drug Store is \$2.50, plus \$0.25 per picture, or $C = 2.50 + 0.25n$, where n is the number of pictures printed.

   a) Draw a graph of the cost of processing film for up to and including 36 pictures.

   b) From the graph estimate the cost of processing a role of 35-mm film containing 20 pictures.

   c) If the total cost of processing a roll of 35-mm film is \$11.50, estimate the number of pictures.

98. The monthly profit, p in dollars, at a computer repair shop can be estimated by the equation $p = 40n - 500$, where n is the number of computers repaired in a week.

   a) Graph $p = 40n - 500$, for $n \leq 100$.

   b) Estimate the profit for a week if 50 computers are repaired.

   c) How many computers would need to be repaired in a week for the shop to break even?

99. The weekly cost of operating a taxi, C in dollars, can be approximated by the equation $C = 80 + 0.25n$, where n is the number of miles driven.

   a) Graph $C = 80 + 0.25n$, for $n \leq 1000$.

   b) Estimate the cost if 600 mi are driven in a week.

   c) If the weekly cost is \$180, how many miles were driven in a week?

100. When \$1000 is invested in a savings account paying simple interest for a year, the interest, i in dollars, earned can be found by the formula $i = 1000r$, where r is the rate in decimal form.

   a) Graph $i = 1000r$, for r up to and including a rate of 15%.

   b) If the rate is 6%, what is the simple interest?

   c) If the rate is 12%, what is the simple interest?

Problem Solving/Group Activities

101. a) Two lines are parallel when they do not intersect no matter how far they are extended. Explain how you can determine, without graphing the equations, whether two equations will be parallel lines when graphed.

   b) Determine whether the graphs of the equations $2x - 3y = 6$ and $4x = 6y + 6$ are parallel lines.

102. In which quadrants will the set of points that satisfy the equation $x + y = 1$ lie? Explain.

Research Activity

103. René Descartes is known for his contributions to algebra. Write a paper on his life and his contributions to algebra.

6.8   LINEAR INEQUALITIES IN TWO VARIABLES

In Section 6.6 we introduced linear inequalities in one variable. Now we will introduce linear inequalities in two variables. Some examples of linear inequalities in two variables are $2x + 3y \leq 7$, $x + 7y \geq 5$, and $x - 3y < 6$.

The solution set of a linear inequality in one variable may be indicated on a number line. The solution set of a linear inequality in two variables is indicated on a coordinate plane.

An inequality that is strictly less than ($<$) or greater than ($>$) will have as its solution set a **half-plane**. A half-plane is the set of all the points on one side of a line. An inequality that is less than or equal to ($\leq$) or greater than or equal to ($\geq$) will have as its solution set the set of points that consists of a half-plane and a line. To indicate that the line is part of the solution set, we draw a solid line. To indicate that the line is not part of the solution set, we draw a dashed line.

To Graph Inequalities in Two Variables

1. Mentally substitute the equal sign for the inequality sign and plot points as if you were graphing the equation.

2. If the inequality is $<$ or $>$ draw a dashed line through the points. If the inequality is $\leq$ or $\geq$ draw a solid line through the points.

3. Select a test point not on the line and substitute the x- and y-coordinates into the inequality. If the substitution results in a true statement, shade in the area on the same side of the line as the test point. If the test point results in a false statement, shade in the area on the opposite side of the line as the test point.

EXAMPLE 1

Draw the graph of $x + 3y < 9$.

Solution: To obtain the solution set, start by graphing $x + 3y = 9$. Since the original inequality is strictly "less than," draw a dashed line (Fig. 6.24). The dashed line indicates that the points on the line are not part of the solution set.

The line $x + 3y = 9$ divides the plane into three parts—the line itself and two half-planes. The line is the boundary between the two half-planes. The points in one half-plane will satisfy the inequality $x + 3y < 9$. The points in the other half-plane will satisfy the inequality $x + 3y > 9$.

To determine the solution set of the inequality $x + 3y < 9$, pick any point on the plane that is not on the line. The simplest point to work with is the origin, $(0, 0)$. Substitute $x = 0$ and $y = 0$ in $x + 3y < 9$.

$$x + 3y < 9$$
$$\text{Is } 0 + 3(0) < 9?$$
$$0 + 0 < 9$$
$$0 < 9 \quad \text{True}$$

Since 0 is less than 9, the point $(0, 0)$ is part of the solution set. All the points on the same side of the line $x + 3y = 9$ as the point $(0, 0)$ are members of the solution set. Indicate this condition by shading the half-plane that contains $(0, 0)$. The graph is shown in Fig. 6.25.   ▲

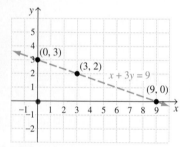

Figure 6.24

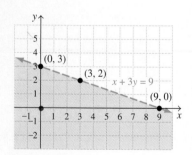

Figure 6.25

EXAMPLE 2

Draw the graph of $4x - 2y \geq 12$.

Solution: First draw the graph of the equation $4x - 2y = 12$. Use a solid line because the points on the boundary are included in the solution set. Now pick a point that is not on the line. Take $(0, 0)$ as the test point.

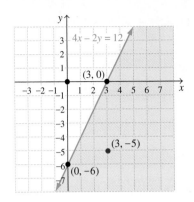

Figure 6.26

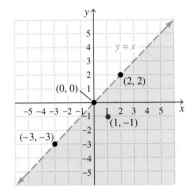

Figure 6.27

$$4x - 2y \geq 12$$
$$\text{Is } 4(0) - 2(0) \geq 12?$$
$$0 \geq 12 \quad \text{False}$$

Since 0 is not greater than or equal to 12 ($0 \not\geq 12$), the solution set is the line and the half-plane that does not contain the point (0, 0). The graph is shown in Fig. 6.26.

If you had arbitrarily selected the test point $(3, -5)$, you would have found that the inequality would be true: $4(3) - 2(-5) \geq 12$, or $22 \geq 12$. Thus the point $(3, -5)$ would be in the half-plane containing the solution set. ▲

EXAMPLE 3

Draw the graph of $y < x$.

Solution: The inequality is strictly "less than," so the boundary is not part of the solution set. In graphing the equation $y = x$, draw a dashed line (Fig. 6.27). Since (0, 0) is *on* the line, it cannot serve as a test point. Let's pick the point $(1, -1)$.

$$y < x$$
$$-1 < 1 \quad \text{True}$$

Since $-1 < 1$ is true, the solution set is the half-plane containing the point $(1, -1)$. ▲

Section 6.8 Exercises

In Exercises 1–20, draw the graph of the inequality.

1. $x > 3$

2. $y \leq 2$

3. $y > x + 2$

4. $y \leq x - 4$

5. $y \geq 2x - 6$

6. $y < -2x + 2$

7. $3x - 4y > 12$

8. $x + 2y > 4$

9. $5x - 2y \leq 10$

10. $2y - 5x \geq 10$

11. $3x + 2y < 6$

12. $-x + 2y < 2$

13. $x + y > 0$

14. $x + 2y \leq 0$

15. $3y - 2x \leq 0$

16. $y \geq 2x + 7$

17. $3x + 2y > 12$

18. $y \leq 3x - 4$

19. $\frac{2}{5}x - \frac{1}{2}y \leq 1$

20. $0.1x + 0.3y \leq 0.4$

21. $0.2x + 0.5y \leq 0.3$

22. Ann is allowed to talk a maximum of 15 min on the telephone. She has two friends (x and y) with whom she wants to share the time.

a) State this problem as an inequality in two variables.
b) Graph the inequality.

23. Jose has 30 ft of fencing to place around a sandbox he plans to build for his children.
a) Write an inequality illustrating all possible dimensions of the rectangular sandbox. $P = 2l + 2w$ is the formula for the perimeter of a rectangle.
b) Graph the inequality.

Problem Solving/Group Activity

24. Yolanda has \$150,000 to spend on purchasing land and building a new house in the country. She wants at least 1 acre of land but less than 10 acres. If land costs \$1500 per acre and building costs are \$75 per square foot, the inequality $1500x + 75y \leq 150,000$, where $1 \leq x < 10$, describes the restriction on her purchase.
a) What quantities do x and y represent in the inequality?
b) Graph the inequality.

c) If Yolanda decides that her house must be at least 1950 ft^2 in size, how many acres of land can she buy?

d) If Yolanda decides that she wants to own at least 5 acres of land, what size house can she afford?

25. The Toys 'b' We Company must ship x toy cars to one outlet and y toy cars to a second outlet. The maximum number of toy cars that the manufacturer can produce and ship is 200. We can represent this situation with the inequality $x + y \leq 200$.

a) Can x or y be negative? Explain.

b) Graph the inequality.

c) Write one or two paragraphs interpreting the information that the graph provides.

6.9   SOLVING QUADRATIC EQUATIONS BY USING FACTORING AND BY USING THE QUADRATIC FORMULA

In Section 6.2 we solved linear, or first degree, equations. In those equations the exponent on all variables was 1. Now we deal with the quadratic equation. The standard form of a quadratic equation in one variable is shown in the box.

Standard Form of a Quadratic Equation

$$ax^2 + bx + c = 0, \qquad a \neq 0$$

Note that in the standard form of a quadratic equation, the greatest exponent on x is 2 and the right side of the equation is equal to zero. *To solve a quadratic equation* means to find the value or values that make the equation true. In this section we will solve quadratic equations by factoring and by the quadratic formula.

Before we examine factoring, we will look at the FOIL method of multiplying two binomials. A binomial is an expression that contains two terms, in which each exponent that appears on a variable is a whole number.

Examples of Binomials

$$x + 3 \qquad x - 5$$
$$3x + 5 \qquad 4x - 2$$

To multiply two binomials, we can use the FOIL method. The name of the method, FOIL, is an acronym to help its users remember it as a method that obtains the products of the First, Outer, Inner, and Last terms of the binomials.

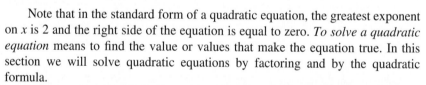

$$(a + b)(c + d) = a \cdot c + a \cdot d + b \cdot c + b \cdot d$$

First   Outer   Inner   Last

After multiplying the first, outer, inner, and last terms, combine all like terms.

EXAMPLE 1

Multiply $(x + 2)(x + 4)$.

Solution: The FOIL method of multiplication yields

$$\begin{array}{cccc} \text{F} & \text{O} & \text{I} & \text{L} \end{array}$$
$$(x + 2)(x + 4) = x \cdot x + (x)(4) + 2 \cdot x + 2 \cdot 4$$
$$= x^2 + 4x + 2x + 8$$
$$= x^2 + 6x + 8.$$

Note that $4x$ and $2x$ were combined to get $6x$. ▲

EXAMPLE 2

Multiply $(3x + 2)(x - 5)$.

Solution:

$$(3x + 2)(x - 5) = 3x \cdot x + 3x(-5) + 2 \cdot x + 2(-5)$$
$$= 3x^2 - 15x + 2x - 10$$
$$= 3x^2 - 13x - 10$$ ▲

Factoring Trinomials of the Form $x^2 + bx + c$

The expression $x^2 + 6x + 8$ is an example of a trinomial. A **trinomial** is an expression containing three terms in which each exponent that appears on a variable is a whole number.

Example 1 showed that

$$(x + 2)(x + 4) = x^2 + 6x + 8.$$

Since the product of $x + 2$ and $x + 4$ is $x^2 + 6x + 8$, we say that $x + 2$ and $x + 4$ are **factors** of $x^2 + 6x + 8$. **To factor an expression** means to write the expression as a product of its factors. For example, to factor $x^2 + 6x + 8$, we would write

$$x^2 + 6x + 8 = (x + 2)(x + 4).$$

Let's look at the factors more closely.

$$2 + 4 = 6$$
$$2 \cdot 4 = 8$$
$$x^2 + 6x + 8 = (x + 2)(x + 4)$$

Note that the sum of the two numbers in the factors is $2 + 4$ or 6. The 6 is the coefficient of the x-term. Also note that the product of the numbers in the two factors is $2 \cdot 4$, or 8. The 8 is the constant. In general, when factoring an expression of the form $x^2 + bx + c$, we need to find two numbers whose product is c and whose sum is b. When we determine the two numbers, the factors will be of the form

$$(x + \boxed{}) \; (x + \boxed{}).$$
$$\uparrow \qquad\qquad \uparrow$$
$$\text{One} \qquad \text{Other}$$
$$\text{number} \qquad \text{number}$$

EXAMPLE 3

Factor $x^2 + 5x + 6$.

Solution: We need to find two numbers whose product is 6 and whose sum is 5. Since the product is $+6$, the two numbers must both be positive or both be negative. Because the coefficient of the x-term is positive, only the positive factors of 6 need to be considered. Can you explain why? We begin by listing the positive numbers whose product is 6.

Factors of 6	Sum of Factors
1(6)	$1 + 6 = 7$
2(3)	$2 + 3 = 5$

Since $2 \cdot 3 = 6$ and $2 + 3 = 5$, 2 and 3 are the numbers we are seeking. Thus we write

$$x^2 + 5x + 6 = (x + 2)(x + 3).$$

Note $(x + 3)(x + 2)$ is also an acceptable answer.   ▲

To Factor Trinomial Expressions of the Form $x^2 + bx + c$

1. Find two numbers whose product is c and whose sum is b.

2. Write factors in the form

$$(x + \boxed{})(x + \boxed{}).$$

$\uparrow$ $\uparrow$
One number   Other number
from step 1   from step 1

If, for example, the numbers found in step 1 were 6 and -4, the factors would be written $(x + 6)(x - 4)$.

EXAMPLE 4

Factor $x^2 - x - 12$.

Solution: We must find two numbers whose product is -12 and whose sum is -1. (Remember: $-x$ means $-1x$.) Begin by listing the factors of -12.

Factors of -12	Sum of Factors
$-12(1)$	$-12 + 1 = -11$
$-6(2)$	$-6 + 2 = -4$
$-4(3)$	$-4 + 3 = -1$
$-3(4)$	$-3 + 4 = 1$
$-2(6)$	$-2 + 6 = 4$
$-1(12)$	$-1 + 12 = 11$

The table lists all the factors of -12. The only numbers with a product of -12 and a sum of -1 are -4 and 3. We listed all factors in this example so that you would see that, for example, $-4(3)$ is a different set of factors than $-3(4)$.

Once you find the factors you are seeking, there is no reason to go any further. The trinomial can be written in factored form as

$$x^2 - x - 12 = (x - 4)(x + 3).$$ ▲

Factoring Trinomials of the Form $ax^2 + bx + c$, $a \neq 1$

Now we discuss how to factor an expression of the form $ax^2 + bx + c$, where a, the coefficient of the squared term, is not equal to 1.

Consider the multiplication problem $(2x + 1)(x + 3)$.

$$\begin{aligned} (2x + 1)(x + 3) &= 2x \cdot x + 2x \cdot 3 + 1 \cdot x + 1 \cdot 3 \\ &= 2x^2 + 6x + x + 3 \\ &= 2x^2 + 7x + 3 \end{aligned}$$

Since $(2x + 1)(x + 3) = 2x^2 + 7x + 3$, the factors of $2x^2 + 7x + 3$ are $2x + 1$ and $x + 3$.

Let's study these factors more closely.

$$\mathbf{F} = 2 \cdot 1 = 2 \qquad \mathbf{O + I} = (2 \cdot 3) + (1 \cdot 1) = 7 \qquad \mathbf{L} = 1 \cdot 3 = 3$$

Note that the product of the coefficient of the first terms in the multiplication of the binomials equals 2, the coefficient of the squared term. The sum of the products of the coefficients of the outer and inner terms equals 7, the coefficient of the x-term. The product of the last terms equals 3, the constant.

A procedure to factor expressions of the form $ax^2 + bx + c, a \neq 1$, follows.

To Factor Trinomial Expressions of the Form $ax^2 + bx + c$, $a \neq 1$

1. Write all pairs of factors of the coefficient of the squared term, a.
2. Write all pairs of factors of the constant, c.
3. Try various combinations of these factors until the sum of the products of the outer and inner terms is bx.

EXAMPLE 5

Factor $3x^2 + 17x + 10$.

Solution: The only factors of 3 are 1 and 3. Therefore we write

$$3x^2 + 17x + 10 = (3x \qquad)(x \qquad).$$

The number 10 has both positive and negative factors. However, since both the constant, 10, and the sum of the products of the outer and inner terms, 17, are positive, the two factors must be positive. Why? The positive factors of 10 are 1(10) and 2(5). The following is a list of the possible factors.

Possible Factors	Sum of Products of Outer and Inner Terms
$(3x + 1)(x + 10)$	$31x$
$(3x + 10)(x + 1)$	$13x$
$(3x + 2)(x + 5)$	$17x$ ← Correct middle term
$(3x + 5)(x + 2)$	$11x$

Thus $3x^2 + 17x + 10 = (3x + 2)(x + 5)$. ▲

Note that factoring problems of this type may be checked by using the FOIL method of multiplication. We will check the results to Example 5:

$$(3x + 2)(x + 5) = 3x \cdot x + 3x \cdot 5 + 2 \cdot x + 2 \cdot 5$$
$$= 3x^2 + 15x + 2x + 10$$
$$= 3x^2 + 17x + 10.$$

Since we obtained the expression we started with, our factoring is correct.

EXAMPLE 6

Factor $6x^2 - 11x - 10$.

Solution: The factors of 6 will be either $6 \cdot 1$ or $2 \cdot 3$. Therefore the factors may be of the form $(6x \quad)(x \quad)$ or $(2x \quad)(3x \quad)$. When there is more than one set of factors for the first term, we generally try the medium-sized factors first. If this does not work, we try the other factors. Thus we write

$$6x^2 - 11x - 10 = (2x \quad)(3x \quad).$$

The factors of -10 are $(-1)(10)$, $(1)(-10)$, $(-2)(5)$, and $(2)(-5)$. There will be eight different pairs of possible factors of the trinomial $6x^2 - 11x - 10$. Can you list them?

The correct factoring is $6x^2 - 11x - 10 = (2x - 5)(3x + 2)$. ▲

Note that in Example 6 we first tried factors of the form $(2x \quad)(3x \quad)$. If we had not found the correct factors using them, we would have tried $(6x \quad)(x \quad)$.

Solving Quadratic Equations by Factoring

To solve a quadratic equation by factoring set one side of the equation equal to 0 and then use the *zero-factor* property.

Zero-Factor Property
If $a \cdot b = 0$, then $a = 0$ or $b = 0$.

The zero-factor property indicates that, if the product of two factors is 0, then one (or both) of the factors must have a value of 0.

EXAMPLE 7

Solve the equation $(x - 2)(x + 4) = 0$.

Solution: When we use the zero-factor property, either $x - 2$ or $x + 4$ must equal 0 for the product to equal 0. Thus we set each individual factor equal to 0 and solve each resulting equation for x.

$$(x - 2)(x + 4) = 0$$

$$x - 2 = 0 \quad \text{or} \quad x + 4 = 0$$

$$x = 2 \qquad\qquad x = -4$$

Thus the solutions are 2 and -4.

Check:

$x = 2$	$x = -4$
$(x - 2)(x + 4) = 0$	$(x - 2)(x + 4) = 0$
$(2 - 2)(2 + 4) = 0$	$(-4 - 2)(-4 + 4) = 0$
$0(6) = 0$	$(-6)(0) = 0$
$0 = 0$ True	$0 = 0$ True

To Solve a Quadratic Equation by Factoring

1. Use the addition or subtraction property to make one side of the equation equal to 0.

2. Factor the side of the equation not equal to 0.

3. Use the zero-factor property to solve the equation.

Examples 8 and 9 illustrate this procedure.

EXAMPLE 8

Solve the equation $x^2 - 8x = -15$.

Solution: First add 15 to both sides of the equation to make the right side of the equation equal to 0.

$$x^2 - 8x = -15$$

$$x^2 - 8x + 15 = -15 + 15$$

$$x^2 - 8x + 15 = 0$$

Factor the left side of the equation. The object is to find two numbers whose product is 15 and whose sum is -8. Since the product of the numbers is positive and the sum of the numbers is negative, the two numbers must both be negative. The numbers are -3 and -5. Note that $(-3)(-5) = 15$ and $-3 + (-5) = -8$.

$$x^2 - 8x + 15 = 0$$

$$(x - 3)(x - 5) = 0$$

Now use the zero-factor property to find the solution.

$$x - 3 = 0 \quad \text{or} \quad x - 5 = 0$$
$$x = 3 \qquad\qquad x = 5$$

The solutions are 3 and 5. ▲

EXAMPLE 9

Solve the equation $2x^2 - 11x + 12 = 0$.

Solution: $2x^2 - 11x + 12$ factors into $(2x - 3)(x - 4)$. Thus we write

$$2x^2 - 11x + 12 = 0$$
$$(2x - 3)(x - 4) = 0$$
$$2x - 3 = 0 \quad \text{or} \quad x - 4 = 0$$
$$2x = 3 \qquad\qquad x = 4$$
$$x = \frac{3}{2}.$$

The solutions are $\frac{3}{2}$ and 4. ▲

Solving Quadratic Equations by Using the Quadratic Formula

Not all quadratic equations can be solved by factoring. When a quadratic equation cannot be easily solved by factoring, we can solve the equation with the quadratic formula. The quadratic formula can be used to solve any quadratic equation.

> **Quadratic Formula**
> For a quadratic equation in standard form, $ax^2 + bx + c = 0$, $a \neq 0$, the quadratic formula is
> $$x = \frac{-b \pm \sqrt{b^2 - 4ac}}{2a}.$$

To use the quadratic formula, first write the quadratic equation in standard form. Then determine the values for a (the coefficient of the squared term), b (the coefficient of the x term), and c (the constant). Finally substitute the values of a, b, and c into the quadratic formula and evaluate the expression.

EXAMPLE 10

Solve the equation $x^2 + 6x - 16 = 0$ using the quadratic formula.

Solution: In this equation, $a = 1$, $b = 6$, and $c = -16$. Substituting these values into the quadratic formula gives

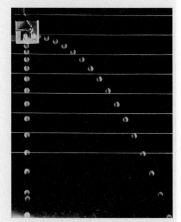
$$x = \frac{-b \pm \sqrt{b^2 - 4ac}}{2a} = \frac{-6 \pm \sqrt{6^2 - 4(1)(-16)}}{2(1)}$$

$$= \frac{-6 \pm \sqrt{36 + 64}}{2}$$

$$= \frac{-6 \pm \sqrt{100}}{2}$$

$$= \frac{-6 \pm 10}{2}$$

$$\frac{-6 + 10}{2} = \frac{4}{2} = 2 \qquad\qquad \frac{-6 - 10}{2} = \frac{-16}{2} = -8.$$

The solutions are 2 and −8.   ▲

Note that Example 10 can also be solved by factoring. We suggest that you do so now.

EXAMPLE 11

Solve $3x^2 - 6x = 5$, using the quadratic formula.

Solution: Begin by writing the equation in standard form by subtracting 5 from both sides of the equation.

$$3x^2 - 6x - 5 = 0$$

$$a = 3, \qquad b = -6, \qquad c = -5$$

$$x = \frac{-b \pm \sqrt{b^2 - 4ac}}{2a} = \frac{-(-6) \pm \sqrt{(-6)^2 - 4(3)(-5)}}{2(3)}$$

$$= \frac{6 \pm \sqrt{36 + 60}}{6}$$

$$= \frac{6 \pm \sqrt{96}}{6}$$

Since $\sqrt{96} = \sqrt{16}\sqrt{6} = 4\sqrt{6}$ (see Section 5.4), we write

$$\frac{6 \pm \sqrt{96}}{6} = \frac{6 \pm 4\sqrt{6}}{6} = \frac{\overset{1}{\cancel{2}}(3 \pm 2\sqrt{6})}{\underset{3}{\cancel{6}}} = \frac{3 \pm 2\sqrt{6}}{3}.$$

The solutions are $\dfrac{3 + 2\sqrt{6}}{3}$ and $\dfrac{3 - 2\sqrt{6}}{3}$.   ▲

Note that the solutions to Example 11 are irrational numbers. It is also possible for a quadratic equation to have no real solution. In solving an equation, if the radicand (the expression inside the square root) is a negative number, then the quadratic equation has **no real solution**.

Section 6.9 Exercises

1. Define *trinomial*. Give three examples of trinomials.

2. In your own words explain the *FOIL* method used to multiply two binomials.

3. In your own words state the zero-factor property.

4. Have you memorized the quadratic formula? If not, you need to do so. Without looking at the book write the quadratic formula.

In Exercises 5–20, factor the trinomial. If the trinomial cannot be factored, so state.

5. $x^2 + 8x + 12$

6. $x^2 + 5x + 6$

7. $x^2 + 6x - 7$

8. $x^2 - 6x - 7$

9. $x^2 + 2x - 24$

10. $x^2 - 6x + 8$

11. $x^2 - 2x - 3$

12. $x^2 - 5x - 6$

13. $x^2 - 10x + 21$

14. $x^2 - 25$

15. $x^2 - 49$

16. $x^2 - x - 30$

17. $x^2 + 3x - 28$

18. $x^2 + 4x - 32$

19. $x^2 + 2x - 63$

20. $x^2 - 2x - 48$

In Exercises 21–32, factor the trinomial. If the trinomial cannot be factored, so state.

21. $2x^2 + 3x - 2$

22. $2x^2 - 7x - 15$

23. $3x^2 - 13x - 10$

24. $2x^2 + 7x - 15$

25. $5x^2 + 12x + 4$

26. $2x^2 - 9x + 10$

27. $5x^2 - 7x - 6$

28. $4x^2 + 16x + 15$

29. $5x^2 - 13x + 6$

30. $6x^2 - 11x + 4$

31. $3x^2 - 14x - 24$

32. $6x^2 + 5x + 1$

In Exercises 33–36, solve each equation, using the zero-factor property.

33. $(x - 6)(x + 3) = 0$

34. $(2x - 3)(3x + 1) = 0$

35. $(3x + 5)(4x - 3) = 0$

36. $(x - 6)(5x - 4) = 0$

In Exercises 37–56, solve each equation by factoring.

37. $x^2 + 6x + 8 = 0$

38. $x^2 + 2x - 3 = 0$

39. $x^2 - 8x + 15 = 0$

40. $x^2 + 2x - 8 = 0$

41. $x^2 - 15 = 2x$

42. $x^2 - 7x = -6$

43. $x^2 = 4x - 3$

44. $x^2 - 13x + 40 = 0$

45. $x^2 - 81 = 0$

46. $x^2 - 64 = 0$

47. $x^2 + 5x - 36 = 0$

48. $x^2 + 12x + 20 = 0$

49. $3x^2 + x = 2$

50. $4x^2 + 11x = 3$

51. $5x^2 + 11x = -2$

52. $2x^2 = -5x + 3$

53. $3x^2 - 4x = -1$

54. $5x^2 + 16x + 12 = 0$

55. $4x^2 - 9x + 2 = 0$

56. $6x^2 + x - 2 = 0$

In Exercises 57–76, solve the equation, using the quadratic formula. If the equation has no real solution, so state.

57. $x^2 - 9x + 20 = 0$

58. $x^2 - 4x - 21 = 0$

59. $x^2 - 3x - 10 = 0$

60. $x^2 - 5x - 14 = 0$

61. $x^2 - 8x = 9$

62. $x^2 = -8x + 15$

63. $x^2 - 2x + 3 = 0$

64. $2x^2 - x - 3 = 0$

65. $x^2 - 5x = 4$

66. $2x^2 - 7x + 2 = 0$

67. $3x^2 - 13x = 12$

68. $3x^2 - 13x + 15 = 0$

69. $4x^2 - x - 1 = 0$

70. $4x^2 - 5x - 3 = 0$

71. $2x^2 + 7x + 5 = 0$

72. $3x^2 = 9x - 5$

73. $3x^2 - 10x + 7 = 0$

74. $4x^2 + 7x - 1 = 0$

75. $4x^2 - 11x + 13 = 0$

76. $5x^2 + 9x - 2 = 0$

Problem Solving/Group Activities

77. The weekly profit p (in thousands of dollars) of The Red Dog Refrigerator Company is given by $p = x^2 + 16x - 24$, where x is the number of refrigerators produced and sold. How many refrigerators must be produced and sold to have a weekly profit of \$40,000?

78. **a)** Explain why solving $(x - 4)(x - 7) = 6$ by setting each factor equal to 6 is not correct.
    b) Determine the correct solution to $(x - 4)(x - 7) = 6$.

79. The radicand in the quadratic formula, $b^2 - 4ac$, is called the **discriminant**. How many solutions will the quadratic equation have if the discriminant is (a) greater than 0, (b) equal to 0, or (c) less than zero? Explain your answer.

Research Activity

80. The Italian mathematician Girolamo Cardano (1501–1576) is recognized for his skill in solving equations. Write a paper about his life and his contributions to mathematics, in particular his contribution to solving equations.

6.10   FUNCTIONS AND THEIR GRAPHS

The concepts of relations and functions are extremely important in mathematics. A **relation** is any set of ordered pairs. Therefore every graph will be a relation. A function is a special type of relation. Suppose that you are purchasing oranges at a supermarket where each orange costs $0.20. Then one orange would cost $0.20, two oranges $2 \times \$0.20 = \$0.40$, three oranges $0.60, and so on. We can indicate this relation in a table of values.

Number of Oranges	Cost
0	0.00
1	0.20
2	0.40
3	0.60
⋮	⋮
10	2.00
⋮	⋮

In general the cost for purchasing n oranges will be 20 cents times the number of oranges, or $0.20n$. We can represent the cost, c, of n items by the equation $c = 0.20n$. Since the value of c depends on the value of n we refer to n as the *independent variable* and c as the *dependent variable. Note for each value of the independent variable, n, there is one and only one value of the dependent variable, c.* Such an equation is called a **function**. In the equation $c = 0.20n$, the value of c depends on the value of n, so we say that "c is a function of n."

> A **function** is a special type of relation where each value of the independent variable corresponds to a unique value of the dependent variable.

The set of values that can be used for the independent variable is called the **domain** of the function and the resulting set of values obtained for the dependent variable is called the **range**. The domain and range for the function $c = 0.20n$ are illustrated in Fig. 6.28.

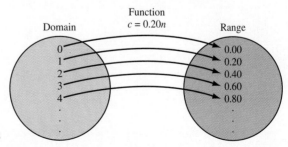

Figure 6.28

When we graphed equations of the form $ax + by = c$ in Section 6.7 we found that they were straight lines. For example, the graph of $y = 2x - 1$ is illustrated in Fig. 6.29 on page 332.

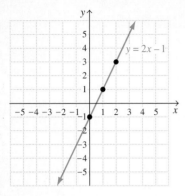

Figure 6.29

Is the equation $y = 2x - 1$ a function? To answer this question we must ask, "Does each value of x correspond to a unique value of y?" The answer is "yes"; therefore this equation is a function.

For the equation $y = 2x - 1$, we say that "y is a function of x" and write $y = f(x)$. The notation $f(x)$ is read "f of x." When we are given an equation that is a function, we may replace the y in the equation with $f(x)$, since $f(x)$ represents y. Thus $y = 2x - 1$ may be written $f(x) = 2x - 1$.

To evaluate a function for a specific value of x, replace each x in the function with the given value, then evaluate. For example, to evaluate $f(x) = 2x - 1$ when $x = 8$, we do the following.

$$f(x) = 2x - 1$$
$$f(8) = 2(8) - 1 = 16 - 1 = 15$$

Thus $f(8) = 15$. Since $f(x) = y$, when $x = 8$, $y = 15$. What is the domain and range of $f(x) = 2x - 1$? Because x can be any real number, the domain is the set of real numbers, symbolized $\mathbb{R}$. The range is also $\mathbb{R}$.

We can determine whether a graph is a function by using the **vertical line test**: If a vertical line can be drawn so that it intersects the graph at more than one point, then each x does not have a unique y and the graph is not a function. If a vertical line cannot be made to intersect the graph in at least two different places, then the graph is a function.

EXAMPLE I

Use the vertical line test to determine which of the following graphs are functions.

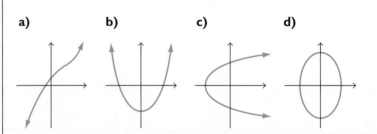

Solution: (a) and (b) are functions but (c) and (d) are not.

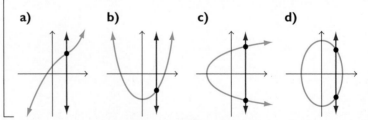

There are many real-life applications of functions—in fact, all the applications illustrated in Sections 6.2–6.4 were functions. Let's consider two more applications.

EXAMPLE 2

The average stopping distance, d, in meters, for a car traveling v kilometers per hour is given by the function $d = f(v) = 0.16v + 0.01v^2$. Find the stopping distance for a car traveling at 60 km/hr.

Solution: Substitute 60 for v in the function.

$$f(v) = 0.16v + 0.01v^2$$
$$f(60) = 0.16(60) + 0.01(60)^2$$
$$= 9.6 + 36 = 45.6$$

Thus the average stopping distance of a car traveling at 60 km/hr is 45.6 meters.

▲

EXAMPLE 3

On July 20, 1969, Neil Armstrong became the first person to walk on the moon. The velocity, v, of his spacecraft, the *Eagle*, in meters per second, was a function of time before touchdown, t, given by

$$v = f(t) = 3.2t + 0.45.$$

The height of the spacecraft, h, above the moon's surface, in meters, was also a function of time before touchdown, given by

$$h = g(t) = 1.6t^2 + 0.45t.$$

What was the velocity of the spacecraft and its distance from the surface of the moon

a) at 3 seconds before touchdown.   **b)** at touchdown (0 sec).

Solution:

a) $v = f(t) = 3.2t + 0.45,$ $h = g(t) = 1.6t^2 + 0.45t$

$\qquad f(3) = 3.2(3) + 0.45 \qquad\qquad g(3) = 1.6(3)^2 + 0.45(3)$

$\qquad\qquad = 9.6 + 0.45 \qquad\qquad\qquad = 1.6(9) + 1.35$

$\qquad\qquad = 10.05 \text{ m/sec} \qquad\qquad\quad = 14.4 + 1.35$

$\qquad\qquad\qquad\qquad\qquad\qquad\qquad\qquad = 15.75 \text{ m}$

The velocity 3 seconds before touchdown was 10.05 meters per second and the height 3 seconds before touchdown was 15.75 meters.

b) $v = f(t) = 3.2t + 0.45,$ $h = g(t) = 1.6t^2 + 0.45t$

$\qquad f(0) = 3.2(0) + 0.45 \qquad\qquad g(0) = 1.6(0)^2 + 0.45(0)$

$\qquad\qquad = 0 + 0.45 \qquad\qquad\qquad\quad = 0 + 0$

$\qquad\qquad = 0.45 \text{ m/sec} \qquad\qquad\qquad = 0 \text{ m}$

The touchdown velocity was 0.45 meters per second. At touchdown the *Eagle* is on the moon, and therefore the distance from the moon is 0 meters.

▲

In Section 6.3 we discussed exponential equations. All exponential equations are also functions.

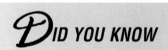

SEA OF TRANQUILLITY

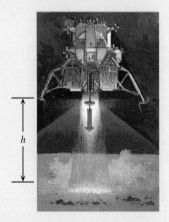

*A*pollo II touched down at Mare Tranqillitatis, the Sea of Tranquillity. The rock samples taken there placed the age of the rock at 3.5 billion years old—as old as the oldest known earth rocks. ■

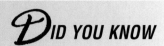

EXAMPLE 4

The power supply of a satellite is a radioisotope. The power output, p, in watts (w), remaining in the power supply is a function of the time the satellite is in space. If there are originally 100 grams of the isotope, the power remaining after t days is $p = 100e^{-0.001t}$. What will be the remaining power after 1 year in space?

Solution: Substitute 365 days for t in the function, then evaluate using a calculator as described in Section 6.3.

$$p = 100e^{-0.001t}$$
$$= 100e^{-0.001(365)}$$
$$= 100e^{-0.365}$$
$$\approx 100(0.694)$$
$$\approx 69.4 \text{ watts}$$

Graphs of Quadratic Functions

Equations of the form $y = ax + b$ are **linear functions**. The graphs of linear functions are straight lines that will pass the vertical line test.

Equations of the form $y = ax^2 + bx + c$, $a \neq 0$, are **quadratic functions**. Examples of quadratic functions are $y = 2x^2 + 6x - 3$ and $y = -\frac{1}{2}x^2 + 6$.

The graph of every quadratic function is a **parabola**. Two parabolas are illustrated in Fig. 6.30. A parabola opens upward when the coefficient of the squared term is greater than 0, Fig. 6.30(a). A parabola opens downward when the coefficient of the squared term is less than 0, Fig. 6.30(b).

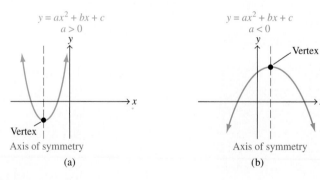

Figure 6.30

The *vertex* of a parabola is the lowest point on a parabola that opens upward and the highest point on a parabola that opens downward. Every parabola is *symmetric* with respect to a vertical line through its vertex. This line is called the *axis of symmetry* of the parabola. The x-coordinate of the vertex and the equation of the axis of symmetry can be found by using the following equation.

Axis of Symmetry of a Parabola

$$x = \frac{-b}{2a}$$

Once the x-coordinate of the vertex has been determined, the y-coordinate can be found by substituting the value found for the x-coordinate into the quadratic equation and evaluating the equation. This procedure is illustrated in Example 5.

EXAMPLE 5

Consider the equation $y = 2x^2 - 4x - 6$.

a) Determine whether the graph will be a parabola that opens upward or downward.

b) Find the equation of the axis of symmetry of the parabola.

c) Find the vertex of the parabola.

Solution:

a) Since $a = 2$, which is greater than 0, the parabola opens upward.

b) To find the axis of symmetry, we use the equation $x = -b/2a$. In the equation $y = 2x^2 - 4x - 6$, $a = 2$, $b = -4$, and $c = -6$, so

$$x = \frac{-b}{2a} = \frac{-(-4)}{2(2)} = \frac{4}{4} = 1.$$

The equation of the axis of symmetry is $x = 1$.

c) The x-coordinate of the vertex is 1 (from part b). To find the y-coordinate, we substitute 1 for x in the equation and then evaluate.

$$\begin{aligned}
y &= 2x^2 - 4x - 6 \\
&= 2(1)^2 - 4(1) - 6 \\
&= 2(1) - 4 - 6 \\
&= 2 - 4 - 6 \\
&= -8
\end{aligned}$$

Therefore the vertex of the parabola is located at the point $(1, -8)$ on the graph.   ▲

$\mathcal{D}$ID YOU KNOW

GRAVITY AND THE PARABOLA

$\mathcal{A}$ny football fan knows that a football will arch in the same path going down as it did going up. Early gunners knew it too. To hit a distant target, the cannon barrel was pointed skyward, not directly at the target. A cannon fired at a 45° angle will travel the greatest horizontal distance. What the football player and gunner alike were allowing for is the effect of gravity on a projectile. Projectile motion follows a parabolic path. Galileo was neither a gunner nor a football player, but he gave us the formula that effectively describes that motion, and the distance traveled by an object if it is projected at a specific angle with a specific initial velocity. ■

General Procedure to Sketch the Graph of a Quadratic Equation

1. Determine whether the parabola opens upward or downward.
2. Determine the equation of the axis of symmetry.
3. Determine the vertex of the parabola.
4. Determine the y-intercept by substituting $x = 0$ into the equation.
5. Determine the x-intercepts (if they exist) by substituting $y = 0$ into the equation and solving for x.
6. Draw the graph, making use of the information gained in steps 1–5. Remember the parabola will be symmetric with respect to the axis of symmetry.

In step 5, to determine the x-intercepts, you may use either factoring or the quadratic formula.

EXAMPLE 6

Sketch the graph of the equation $y = x^2 - 6x + 8$.

Solution: We will follow the steps outlined in the general procedure.

1. Because $a = 1$, which is greater than 0, the parabola opens upward.

2. $x = \dfrac{-b}{2a} = \dfrac{-(-6)}{2(1)} = \dfrac{6}{2} = 3$

Thus the axis of symmetry is $x = 3$.

3. $y = x^2 - 6x + 8$
$y = (3)^2 - 6(3) + 8 = 9 - 18 + 8 = -1$
Thus the vertex is at $(3, -1)$.

4. $y = x^2 - 6x + 8$
$y = 0^2 - 6(0) + 8 = 8$
Thus the y-intercept is at $(0, 8)$.

5. $0 = x^2 - 6x + 8$, or $x^2 - 6x + 8 = 0$
We can solve this equation by factoring.

$$(x - 4)(x - 2) = 0$$
$$x - 4 = 0 \quad \text{or} \quad x - 2 = 0$$
$$x = 4 \qquad\qquad x = 2$$

Thus the x-intercepts are at $(4, 0)$ and $(2, 0)$.

6. The graph is shown in Fig. 6.31.

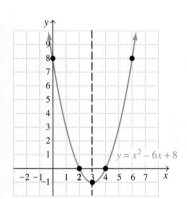

Figure 6.31

Note that the domain of the graph in Example 6, the possible x-values, is the set of all real numbers, $\mathbb{R}$. The range, the possible y-values, is the set of all real numbers greater than or equal to -1. When graphing parabolas, if you feel that you need additional points to graph, you can always substitute values for x and find the corresponding values of y and plot those points. For example, if you substituted 1 for x, the corresponding value of y is 3. Thus you could plot the point $(1, 3)$.

EXAMPLE 7

a) Sketch the graph of the equation $y = -2x^2 + 3x + 4$.

b) Determine the domain and range of the function.

Solution:

a) Follow the steps outlined in the general procedure.
   1. Since $a = -2$, which is less than 0, the parabola opens downward.

   2. Axis of symmetry: $x = \dfrac{-b}{2a} = \dfrac{-(3)}{2(-2)} = \dfrac{-3}{-4} = \dfrac{3}{4}$.

3. y-coordinate of vertex: $y = -2x^2 + 3x + 4$

$$= -2\left(\frac{3}{4}\right)^2 + 3\left(\frac{3}{4}\right) + 4$$

$$= -2\left(\frac{9}{16}\right) + \frac{9}{4} + 4$$

$$= -\frac{9}{8} + \frac{9}{4} + 4$$

$$= -\frac{9}{8} + \frac{18}{8} + \frac{32}{8} = \frac{41}{8} \quad \text{or} \quad 5\frac{1}{8}.$$

Thus the vertex is at $\left(\frac{3}{4}, 5\frac{1}{8}\right)$.

4. y-intercept: $y = -2x^2 + 3x + 4$

$$= -2(0)^2 + 3(0) + 4 = 4.$$

The y-intercept is $(0, 4)$.

5. x-intercepts: $y = -2x^2 + 3x + 4$

$$0 = -2x^2 + 3x + 4 \quad \text{or} \quad -2x^2 + 3x + 4 = 0.$$

This equation cannot be factored, so we will use the quadratic formula to solve it.

$$a = -2, \quad b = 3, \quad c = 4$$

$$x = \frac{-b \pm \sqrt{b^2 - 4ac}}{2a}$$

$$= \frac{-3 \pm \sqrt{3^2 - 4(-2)(4)}}{2(-2)}$$

$$= \frac{-3 \pm \sqrt{9 + 32}}{-4}$$

$$= \frac{-3 \pm \sqrt{41}}{-4}$$

Since $\sqrt{41} \approx 6.4$,

$$x \approx \frac{-3 + 6.4}{-4} \approx \frac{3.4}{-4} \approx -0.85 \quad \text{or} \quad x \approx \frac{-3 - 6.4}{-4} \approx \frac{-9.4}{-4} \approx 2.35.$$

6. Plot the vertex $(\frac{3}{4}, 5\frac{1}{8})$, the y-intercept $(0, 4)$, and the x-intercepts $(-0.85, 0)$ and $(2.35, 0)$. Then sketch the graph (Fig. 6.32).

b) The domain, the values that can be used for x, is the set of all real numbers, $\mathbb{R}$. The range, the values of y, is $y \leq 5\frac{1}{8}$.   ▲

When we use the quadratic formula to find the x-intercepts of a graph, if the radicand, $b^2 - 4ac$, is a negative number, the graph has no x-intercepts. The graph will lie totally above or below the x-axis.

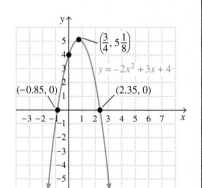

Figure 6.32

Exponential Functions

What does the graph of an **exponential function** of the form $y = a^x$, $a > 0$, $a \neq 1$, look like? Examples 8 and 9 illustrate graphs of exponential functions.

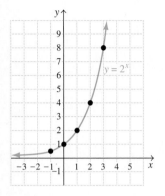

Figure 6.33

EXAMPLE 8

a) Graph $y = 2^x$.

b) Determine the domain and range of the function.

Solution:

a) Substitute values for x and find the corresponding values of y. The graph is shown in Fig. 6.33.

$$y = 2^x$$

$x = -3, \quad y = 2^{-3} = \dfrac{1}{2^3} = \dfrac{1}{8}$

$x = -2, \quad y = 2^{-2} = \dfrac{1}{2^2} = \dfrac{1}{4}$

$x = -1, \quad y = 2^{-1} = \dfrac{1}{2^1} = \dfrac{1}{2}$

$x = 0, \quad y = 2^0 = 1$

$x = 1, \quad y = 2^1 = 2$

$x = 2, \quad y = 2^2 = 4$

$x = 3, \quad y = 2^3 = 8$

x	y
-3	$\frac{1}{8}$
-2	$\frac{1}{4}$
-1	$\frac{1}{2}$
0	1
1	2
2	4
3	8

b) The domain is all real numbers, $\mathbb{R}$. The range is $y > 0$. Note that y can never have a value of 0. ▲

All exponential functions of the form $y = a^x$, $a > 1$, will have the general shape of the graph illustrated in Example 8 (Fig. 6.33).

EXAMPLE 9

a) Graph $y = (\frac{1}{2})^x$.

b) Determine the domain and range of the function.

Solution:

a) We begin by substituting values for x and calculating values for y. We then plot the ordered pairs and use these points to sketch the graph. To evaluate a fraction with a negative exponent, we use the fact that

$$\left(\frac{a}{b}\right)^{-x} = \left(\frac{b}{a}\right)^x$$

For example,

$$\left(\frac{1}{2}\right)^{-3} = \left(\frac{2}{1}\right)^{3} = 8.$$

Then

$$y = \left(\frac{1}{2}\right)^{x}.$$

$$x = -3, \quad y = \left(\frac{1}{2}\right)^{-3} = 2^3 = 8$$

$$x = -2, \quad y = \left(\frac{1}{2}\right)^{-2} = 2^2 = 4$$

$$x = -1, \quad y = \left(\frac{1}{2}\right)^{-1} = 2^1 = 2$$

$$x = 0, \quad y = \left(\frac{1}{2}\right)^{0} = 1$$

$$x = 1, \quad y = \left(\frac{1}{2}\right)^{1} = \frac{1}{2}$$

$$x = 2, \quad y = \left(\frac{1}{2}\right)^{2} = \frac{1}{4}$$

$$x = 3, \quad y = \left(\frac{1}{2}\right)^{3} = \frac{1}{8}$$

x	y
−3	8
−2	4
−1	2
0	1
1	$\frac{1}{2}$
2	$\frac{1}{4}$
3	$\frac{1}{8}$

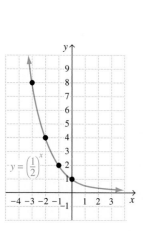

Figure 6.34

The graph is illustrated in Fig. 6.34.

b) The domain is the set of all real numbers, $\mathbb{R}$. The range is $y > 0$.

All exponential functions of the form $y = a^x$, $0 < a < 1$, will have the general shape of the graph illustrated in Example 9 (Fig. 6.34). Can you now predict the shape of the graph of $y = e^x$? Remember: e has a value of about 2.7183.

Section 6.10 Exercises

1. What is a relation?

2. What is a function?

3. What is the domain of a function?

4. What is the range of a function?

5. Explain how and why the vertical line test can be used to determine whether a graph is a function.

6. Give three examples of one quantity being a function of another quantity.

In Exercises 7–22, determine whether the graph is a function. For each function give the domain and range.

7.

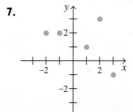

8.

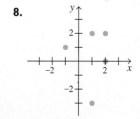

9.
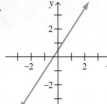

10.
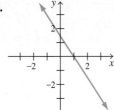

21.
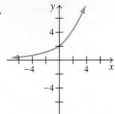

22.
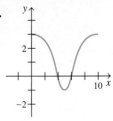

11.
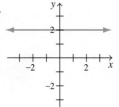

12.
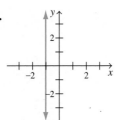

In Exercises 23–28, determine whether the set of ordered pairs is a function.

23. $\{(1, 2), (2, 3), (3, 4)\}$

24. $\{(3, 3), (4, 4), (5, 5)\}$

25. $\{(3, 3), (3, 4), (2, 1)\}$

26. $\{(1, 4), (2, 1), (1, 1)\}$

27. $\{(7, 1), (6, 1), (5, 1)\}$

28. $\{(1, 7), (1, 6), (1, 5)\}$

13.
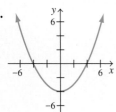

14.
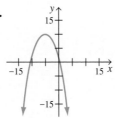

In Exercises 29–42, evaluate the function for the given value of x.

29. $f(x) = x - 5, \quad x = 8$

30. $f(x) = 2x + 7, \quad x = 3$

31. $f(x) = -5x + 8, \quad x = -2$

32. $f(x) = 4x - 6, \quad x = 0$

33. $f(x) = -4x + 7, \quad x = -2$

34. $f(x) = 9x - 8, \quad x = 4$

15.
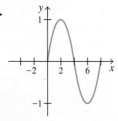

16.
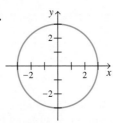

35. $f(x) = x^2 + 2x + 4, \quad x = 6$

36. $f(x) = x^2 - 12, \quad x = 6$

37. $f(x) = 2x^2 - 2x - 8, \quad x = -2$

38. $f(x) = -x^2 + 3x + 7, \quad x = 2$

39. $f(x) = -3x^2 + 5x + 4, \quad x = -3$

40. $f(x) = 5x^2 + 2x + 5, \quad x = 4$

41. $f(x) = -6x^2 - 6x - 12, \quad x = -3$

42. $f(x) = -3x^2 + 5x - 9, \quad x = -2$

17.
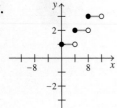

18.
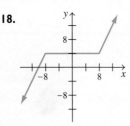

In Exercises 43–58,

   a) *determine whether the parabola will open upward or downward.*

   b) *find the equation of the axis of symmetry.*

   c) *find the vertex.*

   d) *find the y-intercept.*

   e) *find the x-intercepts if they exist.*

   f) *sketch the graph.*

   g) *find the domain and range of the function.*

19.
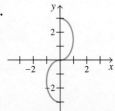

20.
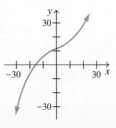

43. $y = x^2 - 9$

44. $y = x^2 - 16$

45. $y = -x^2 + 9$

46. $y = -x^2 + 25$

47. $y = -x^2 - 4$

48. $y = -2x^2 - 8$

49. $y = 2x^2 - 3$

50. $y = -3x^2 - 6$

51. $y = x^2 + 2x - 15$

52. $y = x^2 + 4x - 5$

53. $y = x^2 + 5x + 6$

54. $y = x^2 - 7x - 8$

55. $y = -x^2 + 4x - 6$

56. $y = -x^2 + 8x - 8$

57. $y = -3x^2 + 14x - 8$

58. $y = 2x^2 - x - 6$

In Exercises 59–70, draw the graph of the function and state the domain and range.

59. $y = 3^x$

60. $y = 4^x$

61. $y = (\frac{1}{3})^x$

62. $y = (\frac{1}{4})^x$

63. $y = 2^x + 1$

64. $y = 3^x - 1$

65. $y = 4^x + 1$

66. $y = 2^x - 1$

67. $y = 3^{x-1}$

68. $y = 3^{x+1}$

69. $y = 4^{x+1}$

70. $y = 4^{x-1}$

71. In biology there is a function called the *bioclimatic rule*. The rule states that in spring and early summer annual events—such as blossoming for a given species of plants, the appearance of certain insects, and ripening of fruit—usually occur about 4 days later for each 500 feet of altitude above a specified altitude. This rule may be represented by the function $d = 4(a/500)$, where d is the delay in days, and a is the change in altitude. Determine the delay in days for the following altitudes.
a) 800 ft
b) 1200 ft
c) 1500 ft

72. The annual growth, g, in inches, of a certain type of maple tree is a function of its age. The younger a tree is, the greater its growth. The annual growth may be approximated by the function $g = 5.2 + 0.01t - 0.008t^2$, where t is years and $1 \le t \le 20$. Estimate the annual growth of the maple tree when the tree is (a) 4 years old and (b) 20 years old.

73. The stopping distance, d, in meters, for a car traveling v kilometers per hour is given by the function $d = f(v) = 0.18v + 0.01v^2$. Determine the stopping distance for the following speeds.
a) 88 km/hr
b) 72 km/hr

74. A ball is thrown downward from the top of a 90-story building. The distance, d, in feet, the ball is above the ground after t seconds is given by the function $d = f(t) = -16t^2 - 80t + 800$. Determine $f(2)$ and explain what it means.

75. A checkerboard contains 64 squares. If two pennies are placed on the first square, 4 pennies on the second square, 8 pennies on the third square, 16 pennies on the fourth square, and so on, the number of pennies on the nth square can be found by the function $A = 2^n$.
a) Find the number of pennies placed on the 30th square.
b) What is this amount in dollars?

76. The spacing of the frets on the neck of a classical guitar is determined from the equation $d = (21.9)(2)^{(20-x)/12}$, where $x = $ the fret number and $d = $ the distance in centimeters of the xth fret from the bridge.
a) Determine how far the 19th fret should be from the bridge (rounded to one decimal place).
b) Determine how far the 4th fret should be from the bridge (rounded to one decimal place).
c) The distance of the nut from the bridge can be found by letting $x = 0$ in the given exponential equation. Find the distance (rounded to one decimal place).

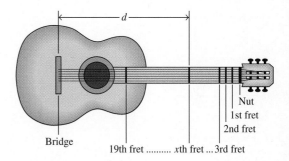

Problem Solving/Group Activities

77. A house initially cost $85,000. The value, V, of the house after n years if it appreciates at a constant rate of 4% per year can be determined by the function $V = f(n) = \$85,000(1.04)^n$.
a) Determine $f(8)$ and explain its meaning.
b) After how many years is the value of the house greater than $153,000? (Find by trial and error.)

78. While exercising, a person's recommended target heart rate is a function of age. The recommended number of beats per minute, y, is given by the function $y = f(x) = -0.85x + 187$, where x represents a person's age in years. Determine the number of recommended heart beats per minute for the following ages and explain the results.
a) 20
b) 30
c) 50
d) 60
e) How long would a person have to live to have a recommended target rate of 85?

79. Light travels at about 186,000 miles per second through space. The distance, d, in miles that light travels in t seconds can be determined by the function $d = 186,000t$.

a) Light reaches the moon from the earth in about 1.3 sec. Determine the approximate distance from the earth to the moon.

b) Express the distance in miles, d, traveled by light in t *minutes* as a function of time, t.

c) Light travels from the sun to the earth in about 8.3 min. Determine the approximate distance from the sun to the earth.

Research Activity

80. The idea of using variables in algebraic equations was introduced by the French mathematician François Viète (1540–1603). Write a paper about his life and his contributions to mathematics. In particular discuss his work with algebraic equations. (References include history of mathematics books and encyclopedias.)

CHAPTER 6 SUMMARY

Key Terms

6.1

algebra
algebraic expression
base
check the solution
constant
equation
evaluate an expression
exponent
order of operations
solution to an equation
solve the equation
variable

6.2

algorithm
identity
inconsistent equation
like terms
linear (or first degree) equation
numerical coefficient
proportion
ratio
simplify an expression
term
unlike terms

6.3

evaluate a formula
exponential equation (formula)
formula
subscript

6.5

combined variation
constant of proportionality
direct variation
inverse variation
joint variation

6.6

compound inequality
inequality
order (or sense) of inequality

6.7

Cartesian (or rectangular) coordinate system
collinear
dependent variable
graph
independent variable
linear equations in two variables
ordered pair
origin
quadrants
slope of a line
x- and y-axes
x- and y-intercepts

6.8

half-plane

6.9

binomial
factor
FOIL method
quadratic equation
quadratic formula
trinomial

6.10

domain
exponential function
function
linear function
parabola
quadratic function
range
relation
vertical line test

Important Facts

Properties used to solve equations

If $a = b$, then $a + c = b + c$
                      Addition property of equality

If $a = b$, then $a - c = b - c$
                    Subtraction property of equality

If $a = b$, then $ac = bc$
                    Multiplication property of equality

If $a = b$, then $a/c = b/c$, $c \neq 0$
                    Division property of equality

Variation

Direct: $y = kx$

Inverse: $y = \dfrac{k}{x}$, or $xy = k$

Joint: $y = kxz$

Inequality symbols

$a < b$ means that a is less than b.
$a \leq b$ means that a is less than or equal to b.
$a > b$ means that a is greater than b.
$a \geq b$ means that a is greater than or equal to b.

Intercepts

To find the x-intercept, set $y = 0$ and solve the resulting equation for x.
To find the y-intercept, set $x = 0$ and solve the resulting equation for y.

Slope

Slope (m): $m = \dfrac{y_2 - y_1}{x_2 - x_1}$

Equations and Formulas

Linear equation in two variables:
$$ax + by = c, \quad a \neq 0, \quad b \neq 0$$
Quadratic equation in one variable:
$$ax^2 + bx + c = 0, \quad a \neq 0$$
Quadratic equation (or function) in two variables:
$$y = ax^2 + bx + c, \quad a \neq 0$$
Exponential equation (or function):
$$y = a^x, \quad a \neq 1, \quad a > 0$$
Exponential growth or decay formula:
$$P = P_0 a^{kt}, \quad a \neq 1, \quad a > 0$$
Quadratic formula:
$$x = \frac{-b \pm \sqrt{b^2 - 4ac}}{2a}$$
Slope–intercept form of a line:
$$y = mx + b$$
Axis of symmetry of a parabola:
$$x = \frac{-b}{2a}$$

Zero-factor property

If $a \cdot b = 0$, then $a = 0$ or $b = 0$.

Chapter 6 Review Exercises

6.1

In Exercises 1–6, evaluate the expression for the given value(s) of the variable.

1. $x^2 + 11$, $x = 3$

2. $-x^2 + 8$, $x = -3$

3. $5x^2 - 3x + 1$, $x = 4$

4. $-x^2 + 7x - 3$, $x = \frac{1}{2}$

5. $4x^3 - 7x^2 + 3x + 1$, $x = -2$

6. $4x^2 - 2xy + 3y^2$, $x = 2$, $y = -1$

6.2

In Exercises 7–9, combine like terms.

7. $3x + 4 - 6 - x$

8. $2x + 5(x - 3) + 4x$

9. $2(x - 4) + \frac{1}{2}(2x + 3)$

In Exercises 10–14, solve the equation for the given variable.

10. $3s + 7 = 22$

11. $3t + 8 = 6t - 13$

12. $\dfrac{x + 5}{6} = \dfrac{x - 3}{3}$

13. $4(x - 2) = 3 + 5(x + 4)$

14. $\dfrac{x}{3} + \dfrac{2}{5} = 4$

15. A recipe for Hot Oats Cereal calls for 2 cups of water and for $\frac{1}{3}$ cup of dry oats. How many cups of dry oats would be used with 3 cups of water?

16. A mason lays 120 blocks in 1 hr 40 min. How long will it take her to lay 450 blocks?

6.3

In Exercises 17–20, use the formula to find the value of the indicated variable for the values given.

17. $A = \frac{1}{2}bh$

Find A when $b = 6$ and $h = 12$ (geometry).

18. $V = 2\pi R^2 r^2$

Find V when $R = 3$, $r = 1\frac{3}{4}$, and $\pi = 3.14$ (geometry).

19. $Z = \dfrac{\bar{x} - \mu}{\dfrac{\sigma}{\sqrt{n}}}$

Find $\bar{x}$ when $Z = 2$, $\mu = 100$, $\sigma = 3$, and $n = 16$ (statistics).

20. $V = \frac{1}{3}\pi r^2 h$

Find h when $r = 3$, $\pi = 3.14$, and $V = 56.52$ (geometry).

In Exercises 21–24, solve for y.

21. $3x - 2y = 6$ **22.** $3x + 4y = 8$

23. $2x - 3y + 52 = 30$ **24.** $-3x - 4y + 5z = 4$

In Exercises 25–28, solve for the variable indicated.

25. $A = lw$,   for l

26. $P = 2l + 2w$,   for w

27. $L = 2(wh + lh)$,   for l

28. $a_n = a_1 + (n-1)d$,   for d

6.4

In Exercises 29–32, write the phrase in mathematical terms.

29. 8 decreased by 7 times x

30. 5 times x decreased by 3

31. 7 more than 6 times r

32. 11 less than 8 divided by q

In Exercises 33–36, write an equation that can be used to solve the problem. Solve the equation and find the desired value(s).

33. Twelve decreased by 3 times a number is 21.

34. The product of 3 and a number, increased by 8, is 6 less than the number.

35. Five times the difference of a number and 4 is 45.

36. Fourteen more than 10 times a number is 8 times the sum of the number and 12.

In Exercises 37–40, write the equation and then find the solution.

37. On a joint income tax return Wesley's income was $\frac{1}{3}$ of Marie's income. Their total income was $48,000. Determine Wesley's income.

38. The Merry Mailman Mailbox Company has fixed costs of $10,000 per month and variable costs of $8.50 per mailbox manufactured. The company has $85,000 available to meet its total monthly expenditures. What is the maximum number of mailboxes the company can manufacture in a month? (Fixed costs, like rent and insurance, are those that occur regardless of the level of production. Variable costs depend on the level of production.)

39. The systolic blood pressure of a particular patient is 40 units higher than his diastolic blood pressure. The sum of his diastolic and systolic blood pressures is 200. Determine both blood pressures.

40. A tiller can be rented for $30 an hour and purchased for $480. How many hours would the tiller have to be rented for the rental cost to equal the cost of purchasing a tiller?

6.5

In Exercises 41–44, find the quantity indicated.

41. A is directly proportional to the square of C. If A is 5 when C is 5, find A when $C = 10$.

42. x is inversely proportional to y. If x is 20 when $y = 5$, find x when $y = 100$.

43. W is directly proportional to L and inversely proportional to A. If $W = 80$ when $L = 100$ and $A = 20$, find W when $L = 50$ and $A = 40$.

44. z is jointly proportional to x and y and inversely proportional to the square of r. If z is 12 when x is 20, $y = 8$, and $r = 8$, find z when $x = 10$, $y = 80$, and $r = 3$.

45. The scale of a map is 1 in. to 60 mi. What distance on the map represents 300 mi?

46. An electric company charges $0.162 per kWh. What is the electric bill if 740 kWh are used in a month?

47. The distance, d, an object drops in free fall is directly proportional to the square of the time, t. If an object falls 16 ft in 1 sec, how far will an object fall in 5 sec?

48. The area, A, of a circle varies directly with the square of its radius, r. If the area is 78.5 when the radius is 5, find the area when the radius is 8.

6.6

In Exercises 49–52, graph the solution set for the set of real numbers.

49. $6 + 7x \geq -3x - 4$ **50.** $3x + 7 \geq 5x + 9$

51. $3(x + 9) \leq 4x + 11$ **52.** $-3 \leq x + 1 < 7$

In Exercises 53–56, graph the solution set for the set of integers.

53. $2 + 7x > -12$ **54.** $5x + 13 \geq -22$

55. $-1 < x \leq 7$ **56.** $-8 \leq x + 2 \leq 7$

6.7

In Exercises 57–60, graph the ordered pair in the Cartesian coordinate system.

57. $(3, 5)$ **58.** $(-3, 2)$

59. $(-3, -4)$ **60.** $(6, -7)$

In Exercises 61 and 62, points A, B, and C are vertices of a rectangle. Plot the points. Find the coordinates of the fourth point, D, to complete the rectangle. Find the area of the rectangle.

61. $A(-3, 3)$, $B(2, 3)$, $C(2, -1)$

62. $A(-3, 1)$, $B(-3, -2)$, $C(4, -2)$

In Exercises 63–66, graph the equation by plotting points.

63. $x - y = 4$ **64.** $2x + 3y = 12$

65. $x = y$ **66.** $x = 3$

In Exercises 67–70, graph the equation, using the x- and y-intercepts.

67. $x + 2y = 4$ **68.** $3x - 2y = 6$

69. $3x - 5y = 15$ **70.** $2x + 3y = 9$

In Exercises 71–74, find the slope of the line through the given points.

71. $(2, 4)$, $(6, 3)$ **72.** $(4, 2)$, $(6, -3)$

73. $(-1, -4)$, $(5, 3)$ **74.** $(6, 2)$, $(6, -2)$

In Exercises 75–78, graph the equation by plotting the y-intercept and then plotting a second point by making use of the slope.

75. $y = 3x - 2$ **76.** $2y - 4 = 3x$

77. $2y + x = 8$ **78.** $y = -x - 1$

In Exercises 79 and 80, determine the equation of the graph.

79. 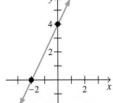   **80.**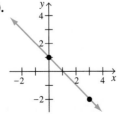

81. The monthly disability income, I, that Nadja receives is $I = 460 - 0.5e$, where e is her monthly earnings for her part-time job for the previous month.

a) Draw a graph of disability income versus earnings for earnings up to and including \$920.

b) If Nadja earns \$600 in January, how much disability income will she receive in February?

c) If she received \$380 disability income in November, how much did she earn in October?

82. The monthly rental cost, C, in dollars, for space in the Galleria Mall can be approximated by the equation $C = 1.70A + 3000$, where A is the area, in square feet, of space rented.

a) Draw a graph of monthly rental cost versus square feet for up to and including 12,000 ft².

b) Determine the monthly rental cost if 2000 ft² are rented.

c) If the rental cost is \$10,000 per month, how many square feet are rented?

6.8

In Exercises 83–86, graph the inequality.

83. $6x + 9y \leq 54$ **84.** $3x + 2y \geq 12$

85. $2x - 3y > 12$ **86.** $-7x - 2y < 14$

6.9

In Exercises 87–92, factor the trinomial. If the trinomial cannot be factored, so state.

87. $x^2 + 10x + 21$ **88.** $x^2 + x - 6$

89. $x^2 - 10x + 24$ **90.** $x^2 - 9x + 20$

91. $2x^2 + x - 21$ **92.** $3x^2 + 5x - 2$

In Exercises 93–96, solve the equation by factoring.

93. $x^2 + 3x + 2 = 0$ **94.** $x^2 - 6x = -5$

95. $3x^2 - 17x + 10 = 0$ **96.** $3x^2 = -7x - 2$

In Exercises 97–100, solve the equation, using the quadratic formula. If the equation has no real solution, so state.

97. $x^2 - 5x + 1 = 0$ **98.** $x^2 - 3x + 2 = 0$

99. $2x^2 - 3x + 4 = 0$ **100.** $2x^2 - x - 3 = 0$

6.10

In Exercises 101–104, determine whether the graph is a function. If it is a function give the domain and range.

101. 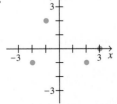   **102.**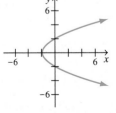

103. 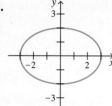   **104.**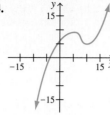

In Exercises 105–108, evaluate f(x) for the given value of x.

105. $f(x) = 5x + 13, \quad x = -2$

106. $f(x) = -4x + 5, \quad x = -3$

107. $f(x) = 2x^2 - 3x + 4, \quad x = 5$

108. $f(x) = -4x^2 + 7x + 9, \quad x = 4$

In Exercises 109 and 110, for each function

    a) *determine whether the parabola will open upward or downward.*

    b) *find the equation of the axis of symmetry.*

    c) *find the vertex.*

    d) *find the y-intercept.*

    e) *find the x-intercepts if they exist.*

    f) *sketch the graph.*

    g) *determine the domain and range.*

109. $y = -x^2 - 4x + 21$ **110.** $y = 3x^2 - 24x - 30$

In Exercises 111 and 112, draw the graph of the function and state the domain and range.

111. $y = 2^{2x}$ **112.** $y = (\frac{1}{2})^x$

6.2, 6.3, 6.10

113. The gas mileage, m, of a specific car can be estimated by the equation (or function)

$$m = 30 - 0.002n^2, \quad 20 \le n \le 80,$$

where n is the speed of the car in miles per hour. Estimate the gas mileage when the car travels at 60 mph.

114. The approximate number of accidents in one month, n, involving drivers between 16 and 30 years of age inclusive can be approximated by the equation (or function)

$$n = 2a^2 - 80a + 5000, \quad 16 \le a \le 30,$$

where a is the age of the driver. Approximate the number of accidents in one month that involved

    a) 18-year-olds.   **b)** 25-year-olds.

115. The percent of light filtering through Swan Lake, p, can be approximated by the equation (or function) $P = 100(0.92)^x$, where x is the depth in feet. Find the percent of light filtering through at a depth of 4.5 ft.

CHAPTER 6 TEST

1. Evaluate $-4x^2 + 7x + 11$, when $x = 2$.

In Exercises 2 and 3, solve the equation.

2. $3x + 5 = 2(4x - 7)$

3. $-2(x - 3) + 6x = 2x + 3(x - 4)$

In Exercises 4 and 5, write an equation to represent the problem. Then solve the equation.

4. The product of a number and 5, decreased by 4, is 26.

5. Ross can rent a stall in a marketplace for one day for $60. He will sell crafts that cost him $4.35 apiece for $7.75 each. How many items must he sell in 1 day to break even?

6. Evaluate $L = ah + bh + ch$, when $a = 3$, $b = 4$, $c = 5$, and $h = 7$.

7. Solve $5x - 8y = 17$, for y.

8. W varies jointly as P and Q and inversely as the square of T. If $W = 6$ when $P = 20$, $Q = 8$, and $T = 4$, find W if $P = 30$, $Q = 4$, and $T = 8$.

9. The time, t, for an ice cube to melt is inversely proportional to the temperature of the water it is in. If an ice cube takes 1.7 min to melt in 70°F water temperature, how long will an ice cube of the same size take to melt in 50°F water?

10. Graph the solution set of $-3x + 11 \leq 5x + 35$ on the real number line.

11. Determine the slope of the line through the points $(-4, 8)$ and $(11, -14)$.

In Exercises 12 and 13, graph the equation.

12. $y = 3x - 4$

13. $2x - 3y = 15$

14. Graph the inequality $3y \geq 5x - 12$.

15. Solve the equation $x^2 + 2x = 35$ by factoring.

16. Solve the equation $3x^2 + 2x = 8$ by using the quadratic formula.

17. Determine whether the following graph is a function. Explain your answer.

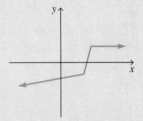

18. Evaluate $f(x) = -4x^2 - 11x + 5$, when $x = -2$.

19. For the equation $y = x^2 - 2x + 4$,
a) determine whether the parabola will open upward or downward.
b) find the equation of the axis of symmetry.
c) find the vertex.
d) find the y-intercept.
e) find the x-intercepts if they exist.
f) sketch the graph.
g) determine the domain and range.

GROUP PROJECTS

Archeology

1. Archeologists have developed formulas to predict the height and, in some cases, the age at death of the deceased by knowing the lengths of certain bones in the body. The long bones of the body grow at approximately the same rate. Thus a linear relationship exists between the length of the bones and the person's height. If the length of one of these major bones—the femur (F), the tibia (T), the humerus (H), and the radius (R)—is known, the height, h, of a person can be calculated with one of the following formulas. The relationship between bone length and height is different for males and females.

Male	Female
$h = 2.24F + 69.09$	$h = 2.23F + 61.41$
$h = 2.39T + 81.68$	$h = 2.53T + 72.57$
$h = 2.97H + 73.57$	$h = 3.14H + 64.98$
$h = 3.65R + 80.41$	$h = 3.88R + 73.51$

All measurements are in centimeters.

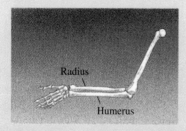

a) Measure your humerus and use the appropriate formula to predict your height in centimeters. How close is this predicted height to your actual height? (The result is an approximation because measuring a bone covered with flesh and muscle is difficult.)

b) Determine and describe where the femur and tibia bones are located.

c) Dr. Juarez, an archeologist, had one female humerus that was 29.42 cm in length. He concluded that the height of the entire skeleton would have been 157.36 cm. Was his conclusion correct?

d) If a 21-year-old female is 167.64 cm tall, about how long should her tibia be?

e) Sometimes the age of a person may be determined by using the fact that the height of a person, and the length of his or her long bones, decreases at the rate of 0.06 cm per year after the age of 30.

  i) At age 30 Jolene is 168 cm tall. Estimate the length of her humerus.

  ii) Estimate the length of Jolene's humerus when she is 60 years old.

f) Select six people of the same sex and measure their height and one of the bones for which an equation is given (the same bone on each person). Each measurement should be made to the nearest 0.5 cm. For each person you will have two measurements, which can be considered an ordered pair (bone length, height). Plot the ordered pairs on a piece of graph paper, with the bone length on the horizontal axis and the height on the vertical axis. Start the scale on both axes at zero. Draw a straight line that you feel is the best approximation, or best fit, through these points. Determine where the line crosses the y-axis and the slope of the line. Your y-intercept and slope should be close to the values in the given equation for that bone. (Reference: Trotter, M., and G. C. Gleser. "Estimation of Stature from Long Bones of American Whites and Negroes." *American Journal of Physical Anthropology*, 1952, 10:463–514.)

Graphing Calculator

2. The functions that we graphed in this chapter can be easily graphed with a graphing calculator (or grapher). If you do not have a graphing calculator, borrow one from your instructor or a friend.

a) Explain how you would set the domain and range. The calculator key to set the domain and range may be labeled *range* or *window*. Set the grapher with the following range or window settings: Xmin = −12, Xmax = 12, Xscl = 1, Ymin = −13, Ymax = 6, and Yscl = 1.

b) Explain how to enter a function in the graphing calculator. Enter the function $y = 3x^2 - 7x - 8$ in the calculator.

c) Graph the function you entered in part (b).

d) Learn how to use the trace feature. Then use it to estimate the x-intercepts. Record the estimated values for the x-intercepts.

e) Learn how to use the zoom feature to obtain a better approximation of the x-intercepts. Use the zoom feature twice and record the x-intercepts each time.

Writing Equations with Sets of Points

3. Suppose that we have a set of ordered pairs that we know are in a straight line. Can we determine the equation of that line? The answer is *yes*. We do so by using the point–slope form of a linear equation:

$$y - y_1 = m(x - x_1).$$

In the equation m is the slope of the line and (x_1, y_1) is a point on the line. Consider a line that contains the points (1, 6), (3, 10), and (8, 20). We can determine the slope by using any two points. Regardless which two points we select, the slope, m, is 2. If we select (1, 6) to represent (x_1, y_1), the equation of the line is

$$y - 6 = 2(x - 1)$$
$$y - 6 = 2x - 2$$
$$y = 2x + 4.$$

If we select either of the other two points for (x_1, y_1), we would obtain the same equation for the line. Therefore the equation of the line that contains the three points is $y = 2x + 4$.

Ms. James, a salesperson, earns a monthly salary plus a commission on sales. In January, on sales of $1000 she earned a total of $1600. In February, on sales of $3000 she earned a total of $2000. In March, on sales of $6000 she earned a total of $2600.

a) Which item, sales or total salary, is the independent variable and which is the dependent variable? (*Hint:* Does total salary depend on sales, or does sales depend on total salary?) Explain your answer.

b) Write the information for sales and total salary as a set of three ordered pairs.

c) Determine the slope of the line that passes through the three points.

d) Determine the equation of the line that passes through these three points by using the point–slope form of a linear equation.

e) Use the equation in part (d) to determine the monthly salary and the commission rate.

f) Determine Ms. James's total salary in a month in which she sells $8000 worth of merchandise.

g) Graph the equation in part (d) for sales up to and including $10,000.

h) Use the graph in part (g) to determine whether your answer to part (f) appears to be correct. Explain why.

World Population

4. The following graph shows the world's population estimated to the year 2016. The first coordinate in each ordered pair represents the year and the second coordinate represents the world population, in billions. For example, the ordered pair (1650, 0.5) indicates that in 1650 the world's population was 0.5 billion, or 500,000,000.

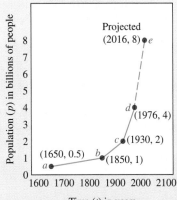

World population growth

a) Would you say that this graph represents a linear equation? Explain.

b) Would you say that this graph represents a function? Explain.

c) In a paragraph or two explain what this graph shows.

d) Estimate the amount of time required for the world population to double, called the *doubling time*, starting in 1650.

e) Estimate the world population doubling time starting in 1976.

f) Find the slope of the line segment between each pair of points, that is, *ab*, *bc*, and so on.

g) On the graph draw a red line from point *a* (1650, 0.5) to point *e* (2016, 8) and calculate the slope of this line.

h) Do you believe that the slope of the line from *a* to *e* is the same as the average of the slopes for *ab*, *bc*, *cd*, and *de*? Explain.

i) Find the average of the slopes for *ab*, *bc*, *cd*, and *de* by adding the four slopes and dividing the sum by 4.

j) Compare your answers to parts (g) and (i) to determine whether you answered part (h) correctly.

Chapter 7

Systems of Linear Equations & Inequalities

*I*magine that you and a friend have hit upon a great idea for a new T-shirt. First you make a few to give away to friends, using your own money. Then other students see the shirt, and soon everyone on campus wants one. Suddenly, you have entered the T-shirt business. To make a profit in your business, you need to keep track of the cost of your materials, the quantity of shirts sold, and the price at which you sell them—a relatively straightforward calculation.

But now suppose that you come up with three other designs, and you want to put them on sweatshirts as well as T-shirts, and you want to offer a variety of colors: black, white, blue, and maroon. The equation for finding the profitability of your venture becomes more complicated because there are more variables. To track your profits, you may need to develop and solve systems of equations.

In fact, student entrepreneurs often face these problems. Consider the case of David Mays and Jon Shecter, who met during their first week at Harvard in 1986. They started writing a free newsletter, *The Source*, to promote their rap music radio program. By their senior year, it had turned into a magazine with a circulation of 10,000.

Following their graduation in 1990, they

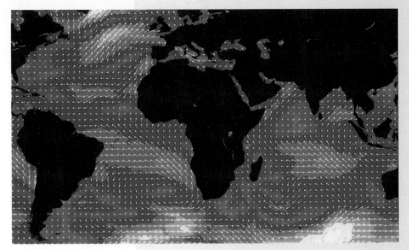

Mathematics can provide the means to handle and systematize a large body of numerical data. Here a matrix is used to display "vectors" that show wind speed and direction for the world's oceans. NASA used radar to collect 350,000 wind measurements over a three-day period.

moved to New York, borrowed money, and increased subscription rates and advertising income. The growth of their business required juggling many financial details, sophisticated accounting, and projecting profitability. Mays and Shecter hope that it will be the *Rolling Stone* of the next generation.

For most businesses, numerous factors must be considered to determine not only whether the business is profitable, but also how much they should charge their customers, which production method is most efficient, what return they can expect by placing advertisements, and so on. Many small-business owners routinely make these calculations based on their own experience, mathematics, and sometimes computer programs. Larger companies often employ

inventory analysts, quality control engineers, and efficiency experts to aid them, along with computers, to keep track of vast quantities of data. ◼

Linear programming provides businesses and governments with a mathematical form of decision making that makes the most efficient use of time and resources. Telecommunications companies use it to route calls through regional switching centers so few of their customers will reach a "no circuits available" message.

7.1 SYSTEMS OF LINEAR EQUATIONS

In Chapter 6 we discussed linear equations in two variables. In algebra it is often necessary to find the common solution to two or more such equations. We refer to the equations in this type of problem as a *system of linear equations* or as *simultaneous linear equations*. A *solution to a system of equations* is the ordered pair or pairs that satisfy all equations in the system. A system of linear equations may have exactly one solution, no solution, or infinitely many solutions.

The solution to a system of linear equations may be found by a number of different techniques. In this section we illustrate how a system of equations may be solved by graphing. In Section 7.2 we illustrate two algebraic methods, substitution and the addition method, for solving a system of linear equations.

EXAMPLE 1

Determine which of the ordered pairs is a solution to the following system of equations.

$$2x - 5y = 10$$
$$3x + y = 15$$

a) $(0, -2)$ **b)** $(5, 0)$ **c)** $(4, 3)$

Solution: For the ordered pair to be a solution to the system it must satisfy each equation in the system.

a) $2x - 5y = 10$ $3x + y = 15$

$2(0) - 5(-2) = 10$ $3(0) + (-2) = 15$

$10 = 10$ True $-2 = 15$ False

Since $(0, -2)$ does not satisfy both equations, it is not a solution to the system.

b) $2x - 5y = 10$ $3x + y = 15$

$2(5) - 5(0) = 10$ $3(5) + 0 = 15$

$10 = 10$ True $15 = 15$ True

Since $(5, 0)$ satisfies both equations, it is a solution to the system.

c) $2x - 5y = 10$ $3x + y = 15$

$2(4) - 5(3) = 10$ $3(4) + 3 = 15$

$-7 = 10$ False $15 = 15$ True

Since $(4, 3)$ does not satisfy both equations in the system, it is not a solution to the system.   ▲

To find the solution to a system of linear equations graphically, we graph both of the equations on the same axes. The coordinates of the point or points of intersection of the graphs are the solution or solutions to the system of equations. When two linear equations are graphed, three situations are possible. The two lines may intersect at one point, as in Example 2; or the two lines may be parallel and not intersect, as in Example 3; or the two equations may represent the same line, as in Example 4.

> **Procedure for Solving a System of Equations by Graphing**
> **1.** Determine three ordered pairs that satisfy each equation.
> **2.** Plot the ordered pairs and sketch the graphs of both equations on the same axes.
> **3.** The coordinates of the point or points of intersection of the graphs are the solution or solutions to the system of equations.

Since the solution to a system of equations may not be integer values, you may not be able to obtain the exact solution by graphing.

EXAMPLE 2

Find the solution to the following system of linear equations graphically.

$$x + 2y = 4$$

$$2x - 3y = 1$$

Solution: To find the solution, graph both $x + 2y = 4$ and $2x - 3y = 1$ on the same axes (Fig. 7.1). Three points that satisfy each equation are shown in the table.

$x + 2y = 4$

x	y
0	2
2	1
4	0

$2x - 3y = 1$

x	y
0	$-\frac{1}{3}$
$\frac{1}{2}$	0
2	1

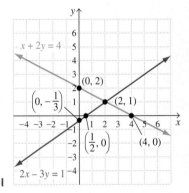

Figure 7.1

The graphs intersect at $(2, 1)$, which is the solution. This is the only point that satisfies *both* equations.

Check:

$x + 2y = 4$	$2x - 3y = 1$
$2 + 2(1) = 4$	$2(2) - 3(1) = 1$
$2 + 2 = 4$	$4 - 3 = 1$
$4 = 4$ True	$1 = 1$ True

The system of equations in Example 2 is an example of a **consistent system of equations**. A consistent system of equations is one that has a solution.

EXAMPLE 3

Find the solution to the following system of equations graphically.

$$2x + y = 3$$

$$2x + y + 5 = 0$$

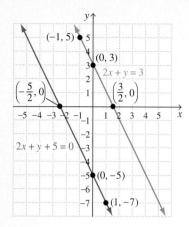

Figure 7.2

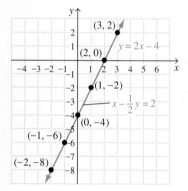

Figure 7.3

Solution: Three ordered pairs that satisfy the equation $2x + y = 3$ are $(0, 3)$, $(\frac{3}{2}, 0)$, and $(-1, 5)$. Three ordered pairs that satisfy the equation $2x + y + 5 = 0$ are $(0, -5)$, $(-\frac{5}{2}, 0)$, and $(1, -7)$. The graphs of both equations are given in Fig. 7.2. Since the two lines are parallel, they do not intersect; therefore the system has *no solution*.   ▲

The system of equations in Example 3 has no solution. A system of equations that has no solution is called an **inconsistent system**.

EXAMPLE 4

Find the solution to the following system of equations graphically.

$$x - \frac{1}{2} y = 2$$

$$y = 2x - 4$$

Solution: Three ordered pairs that satisfy the equation $x - \frac{1}{2}y = 2$ are $(1, -2)$, $(2, 0)$, and $(-1, -6)$. Three ordered pairs that satisfy the equation $y = 2x - 4$ are $(0, -4)$, $(-2, -8)$, and $(3, 2)$. Graph the equations on the same axes (Fig. 7.3). Because all six points are on the same line, the two equations represent the same line. Therefore every ordered pair that is a solution to one equation is also a solution to the other equation. Every point on the line satisfies both equations; thus this system has an *infinite number of solutions*. Solving the first equation for y reveals that the equations are equivalent.   ▲

When a system of equations has an infinite number of solutions, as in Example 4, it is called a **dependent system**. Note that a dependent system is also a consistent system, since it has a solution.

Figure 7.4 summarizes the three possibilities for a system of linear equations.

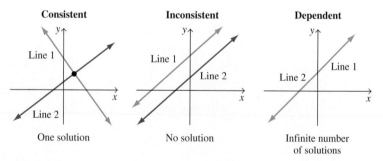

Figure 7.4

EXAMPLE 5

Ulernalot College is considering purchasing one of two types of computer systems. System 1 is a minicomputer that costs $12,000 with terminals that cost $800 each. System 2 is a networking system in which the networking device costs $2200 and the terminals cost $1500 each.

a) Write a system of equations to represent the cost of the two types of computer systems each with n terminals.

b) Graph both equations on the same axes and determine the number of terminals needed for both systems to have the same cost.

c) If 20 terminals are to be used, which system is the least expensive?

Solution: Let n = the number of terminals used. The total cost of each computer system is the initial cost plus the cost of the terminals.

a) Minicomputer system: $C = 12,000 + 800n$.
     Network system: $C = 2200 + 1500n$.

b) We graphed the cost, C, versus the number of terminals, n, for 0–26 terminals (Fig. 7.5). On the graph the lines intersect at the point (14, 23,200). Thus for 14 terminals, both systems would have the same cost, $23,200.

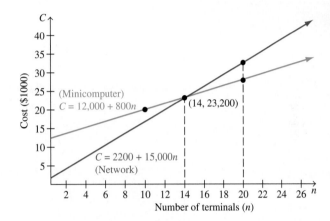

Figure 7.5

c) The graph shows that for more than 14 terminals the minicomputer is the least expensive system. Thus for 20 terminals the minicomputer is the least expensive. ▲

Manufacturers use a technique called **break-even analysis** to determine how many units of an item must be sold for the business to "break even"—that is, for its total revenue to equal its total cost. Suppose that we let the horizontal axis represent the number of units manufactured and sold and the vertical axis represent dollars. Then linear equations for cost, C, and revenue, R, can both be sketched on the same axes (Fig. 7.6). Both are expressed in dollars and both are a function of the number of units.

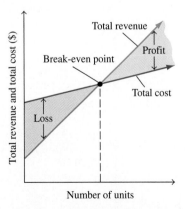

Figure 7.6

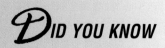

Initially the cost curve is higher than the revenue curve because of fixed (overhead) costs. During low levels of production the manufacturer suffers a loss (the cost curve is greater). During higher levels of production the manufacturer realizes a profit (the profit curve is greater). The point at which the two curves intersect is called the **break-even point**. At that number of units sold revenue equals cost, and the manufacturer breaks even.

EXAMPLE 6

Charlie's Little Box company can sell a particular box for $20 per unit. The costs for making the box are a fixed cost of $100.00 and a production cost of $10 per unit.

a) How many boxes must Charlie sell to break even?

b) Determine whether Charlie makes a profit or loss if he sells 12 boxes. What is the profit or loss?

c) How many units must Charlie sell to make a profit of $450?

Solution: Let x denote the number of boxes made and sold. The revenue is given by the equation

$$R = 20x, \text{($20 times the number of units)}$$

and the cost is given by the equation

$$C = 100 + 10x. \text{($100 plus $10 times the number of units)}$$

a) The break-even point is the point at which the revenue and cost graphs intersect. In Fig. 7.7 the graphs intersect at the point (10, 200), which is the break-even point. Thus for Charlie to break even he must manufacture 10 boxes. When 10 boxes are made and sold the cost and revenue are both $200.

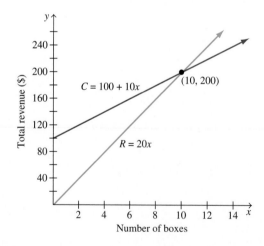

Figure 7.7

b) Examining the graph we can see that if Charlie sells 12 boxes he will have a profit, P, which is the revenue minus the cost. The profit formula is

$$P = R - C$$
$$= 20x - (100 + 10x)$$
$$= 20x - 100 - 10x$$
$$= 10x - 100.$$

Thus for 12 boxes

$$P = 10x - 100$$
$$= 10(12) - 100 = 20.$$

Charlie has a profit of $20 if he sells 12 boxes.

c) We can determine the number of boxes that Charlie must sell to have a profit of $450 by using the profit formula. Substituting 450 for P we have

$$P = 10x - 100$$
$$450 = 10x - 100$$
$$550 = 10x$$
$$55 = x.$$

Thus Charlie must sell 55 boxes to make a profit of $450.   ▲

Section 7.1 Exercises

1. What is a system of linear equations?

2. What is the solution to a system of linear equations?

3. Define a *consistent system of equations.*

4. Define an *inconsistent system of equations.*

5. Define a *dependent system of equations.*

6. Define *break-even point.*

In Exercises 7–10, solve the system of equations graphically.

7. $x = 3$
$y = 5$

8. $x = -4$
$y = 3$

9. $x = 4$
$y = -3$

10. $x = -5$
$y = -3$

In Exercises 11–26, solve the system of equations graphically. If the system does not have a single ordered pair as a solution, state whether the system is inconsistent or dependent.

11. $y = x - 3$
$y = -x - 5$

12. $x - y = 2$
$x + y = 0$

13. $2x - y = 6$
$x + 2y = -2$

14. $3x - y = 3$
$3y - 4x = 6$

15. $y + x = 2$
$x + y = -3$

16. $x - y = 1$
$x + y = 5$

17. $y = 2x - 4$
$2x + y = 0$

18. $2x + y = 3$
$2y = 6 - 4x$

19. $2x - y = -3$
$2x + y = -9$

20. $y = x + 3$
$y = -1$

21. $x = 1$
$x + y + 3 = 0$

22. $3x + 2y = 6$
$6x + 4y = 12$

23. $3x - 2y = 6$
$2y - 3x = 3$

24. $y = \frac{1}{2}x - 3$
$2y - x = 3$

25. $y = \frac{1}{3}x - 2$
$x - 3y = 6$

26. $2(x - 1) + 2y = 0$
$3x + 2(y + 2) = 0$

27. **a)** If the two lines in a system of equations have different slopes, how many solutions will the system have? Explain your answer.

b) If the two lines in a system of equations have the same slope but different *y*-intercepts, how many solutions will the system have? Explain.

c) If the two lines in a system of equations have the same slope and the same *y*-intercept, how many solutions will the system have? Explain.

In Exercises 28–39, determine without graphing whether the system of equations has exactly one solution, no solution, or an infinite number of solutions. (Consider your answers to Exercise 27.)

28. $2x + y = 5$
$y = -2x + 5$

29. $3x + 2y = 4$
$4y = -6x + 6$

30. $x - y - 7 = 0$
$4x + 4y = 5$

31. $5x - y = 4$
$5y - x = 4$

32. $3x + y = 7$
$y = -3x + 9$

33. $2x - 3y = 6$
$x - \frac{3}{2}y = 3$

34. $x + 4y = 12$
$x = 4y + 3$

35. $3x = 6y + 5$
$y = \frac{1}{2}x - 3$

36. $3y = 6x + 4$
$-2x + y = \frac{4}{3}$

37. $x - 2y = 6$
$x + 2y = 4$

38. $12x - 5y = 4$
$3x + 4y = 6$

39. $4x + 7y = 2$
$4x = 6 + 7y$

Two lines are perpendicular when they meet at a right angle (90° angle). Two lines are perpendicular to each other when their slopes are negative reciprocals of each other. The negative reciprocal of 2 is $-\frac{1}{2}$, the negative reciprocal of $\frac{3}{5}$ is $-1/(\frac{3}{5})$ or $-\frac{5}{3}$, and so on. If a represents *any real number, except 0, its negative reciprocal is $-1/a$. Note that the product of a number and its negative reciprocal is -1. In Exercises 40–43, determine, by finding the slope of each line, whether the lines will be perpendicular to each other when graphed.*

40. $3y + 2x = 21$
$2y - 3x = 15$

41. $2x + 5y = 8$
$-5x + 2y = 6$

42. $4y - x = 6$
$y = x + 8$

43. $6x + 5y = 3$
$-10x = 2 + 12y$

44. In Example 5, if the minicomputer costs $10,000 with terminals that cost $3000 and the networking device costs $5000 with terminals that cost $4000,
   a) write the system of equations to represent the cost of the two types of computer systems.
   b) graph both equations (for 0–8 terminals) on the same axes.
   c) determine the number of terminals that must be used for both systems to have the same cost.

45. The total cost of printing a book consists of a setup charge and an additional fee for material for each book printed. The Sivle Printing Company charges a $1600 setup fee plus $6 per book it prints. The Yelserp Printing Company has a setup fee of $1200, plus $8 per book it prints.
   a) Write a system of equations to represent the cost of printing the books with each company.
   b) Graph both equations (for up to and including 300 books) on the same axes.
   c) Determine the number of books that need to be printed for both companies to have the same cost.
   d) If 100 books are to be printed, which is the less expensive printer?

46. An Ly, a salesperson, has the option of selecting the method by which her weekly salary will be determined. Option 1 is a straight 8% of her total dollar sales, and option 2 is a salary of $200 plus 4% of her total dollar sales.
   a) Write a system of equations to represent her salary by each option.
   b) Graph both equations (for $0–$10,000 in sales) on the same axes.
   c) Find the total dollar sales needed for her salary from option 1 to equal her salary from option 2.
   d) If she expects to average $8000 per week in sales, which option should she choose?

47. Talking Ed's Little Horse Harness company can sell a simple halter for $20 per unit. The costs for making the halter are a fixed cost of $300 and a production cost of $10 per unit (see Example 6).
   a) Write the cost and revenue equations.
   b) Graph both equations, for 0–50 units, on the same axes.

   c) How many halters must Talking Ed's company sell to break even?
   d) Determine whether the company makes a profit or loss if it sells 10 halters. What is the profit or loss?
   e) How many halters must Ed's company sell to realize a profit of $1000?

48. A manufacturer can sell a certain speaker for $165 per unit. Manufacturing costs consist of a fixed cost of $8400 and a production cost of $95 per unit.
   a) Write the cost and revenue equations.
   b) Graph both equations, for 0–150 units, on the same axes.
   c) How many units must the manufacturer sell to break even?
   d) What is the manufacturer's profit or loss if 100 units are sold?
   e) How many units must the manufacturer sell to make a profit of $1250?

49. Explain how you can determine whether a system of two equations will be consistent, dependent, or inconsistent without graphing the equations.

Problem Solving/Group Activities

50. Hubert had two job offers for sales positions. One pays a salary of $300 per week plus a 15% commission on his dollar sales volume. The second position pays a salary of $450 per week with no commission.
   a) For each offer, write an equation that expresses the weekly pay.
   b) Graph the system of equations and determine the solution.
   c) For what dollar sales volume will the two offers result in the same pay?

51. a) The Hokuspokus Telephone Company charges 50 cents for the first minute and 30 cents for each additional minute or part thereof for a long-distance call from Happytown to Pleasantville. For the same call the Pacific Edison Telephone Company charges 60 cents for the first minute and 25 cents for each additional minute or part thereof. Write equations to determine the cost of a long-distance call for the Hokuspokus Telephone Company and for the Pacific Edison Telephone Company. Let x represent the number of additional minutes after the first minute and y the cost of the call in cents.
   b) Graph the system of equations and determine the solution.
   c) After how many minutes will the cost from the two telephone companies be the same?

52. a) If two lines have different slopes, what is the maximum possible number of points of intersection?

b) If three lines all have different slopes, what is the maximum possible number of points of intersection?

c) If four lines all have different slopes, what is the maximum possible number of points of intersection?

d) If five lines all have different slopes, what is the maximum possible number of points of intersection?

e) Is there a pattern in the number of points of intersection? If so, use it to determine the maximum possible number of points of intersection for six lines.

53. The Rhind Papyrus indicates that the early Egyptians used linear equations. Do research and write a paper on the use of the symbols used in linear equations and the use of the linear equations by the early Egyptians. (References include history of mathematics books and encyclopedias.)

7.2 SOLVING SYSTEMS OF EQUATIONS BY THE SUBSTITUTION AND ADDITION METHODS

Having solved systems of equations by graphing in Section 7.1, we are ready for two other methods used to solve systems of linear equations: the substitution method and the addition method.

Substitution Method

> **Procedure for Solving a System of Equations Using the Substitution Method**
>
> **1.** Solve one of the equations for one of the variables. If possible solve for a variable with a numerical coefficient of 1. By doing so you may avoid working with fractions.
> **2.** Substitute the expression found in step 1 into the other equation. This step yields an equation in terms of a single variable.
> **3.** Solve the equation found in step 2 for the variable.
> **4.** Substitute the value found in step 3 into the equation you rewrote in step 1, and solve for the remaining variable.

Examples 1, 2, and 3 illustrate the *substitution method*. These systems of equations are the same as in Examples 2, 3, and 4 in Section 7.1.

EXAMPLE 1

Solve the following system of equations by substitution.

$$x + 2y = 4$$

$$2x - 3y = 1$$

Solution: The numerical coefficient of the x term in the equation $x + 2y = 4$ is 1. Therefore solve this equation for x.

Step 1.
$$x + 2y = 4$$
$$x + 2y - 2y = 4 - 2y$$
$$x = 4 - 2y$$

Step 2. Substitute $4 - 2y$ for x in the other equation.

$$2x - 3y = 1$$
$$2(4 - 2y) - 3y = 1$$

Step 3. Now solve the equation for y.

$$8 - 4y - 3y = 1$$
$$8 - 7y = 1$$
$$8 - 8 - 7y = 1 - 8$$
$$-7y = -7$$
$$\frac{-7y}{-7} = \frac{-7}{-7}$$
$$y = 1$$

Step 4. Substitute $y = 1$ in the equation solved for x and determine the value of x:

$$x = 4 - 2y$$
$$x = 4 - 2(1)$$
$$x = 2.$$

Thus the solution is the ordered pair (2, 1). This answer checks with the solution obtained graphically in Section 7.1, Example 2.　▲

EXAMPLE 2

Solve the following system of equations by substitution.

$$2x + y = 3$$
$$2x + y + 5 = 0$$

Solution:　Solve for y in the first equation.

$$2x + y = 3$$
$$2x - 2x + y = 3 - 2x$$
$$y = 3 - 2x$$

Now substitute $3 - 2x$ in place of y in the second equation.

$$2x + y + 5 = 0$$
$$2x + (3 - 2x) + 5 = 0$$
$$2x + 3 - 2x + 5 = 0$$
$$8 = 0 \quad \text{False}$$

Since 8 cannot be equal to 0, there is no solution to the system of equations. Thus the system of equations is inconsistent. This answer checks with the solution obtained graphically in Section 7.1, Example 3.　▲

When solving Example 2 we obtained $8 = 0$ and indicated that the system was inconsistent and that there was no solution. When solving a system of equations, if you obtain a false statement, like $4 = 0$ or $-2 = 0$, the system is *inconsistent* and has *no solution*.

EXAMPLE 3

Solve the following system of equations by substitution.

$$x - \frac{1}{2}y = 2$$

$$y = 2x - 4$$

Solution: The second equation is already solved for y, so we will substitute $2x - 4$ for y in the first equation.

$$x - \frac{1}{2}y = 2$$

$$x - \frac{1}{2}(2x - 4) = 2$$

$$x - x + 2 = 2$$

$$2 = 2 \quad \text{True}$$

Since 2 equals 2, the system has an infinite number of solutions. Thus the system of equations is dependent. This answer checks with the solution obtained in Section 7.1, Example 4. ▲

When solving Example 3 we obtained $2 = 2$ and indicated that the system was dependent and had an infinite number of solutions. When solving a system of equations, if you obtain a true statement, such as $0 = 0$ or $2 = 2$, the system is *dependent* and has an *infinite number of solutions*.

An application in economics involving systems of equations is the law of supply and demand. The market price p of a commodity is a determining factor in the number of units of the commodity that manufacturers are willing to supply and consumers are willing to buy. Generally, as the market price increases, manufacturers' supply increases and consumers' demand decreases. An example of a pair of linear supply and demand curves is sketched in Fig. 7.8. The vertical axis represents price, p. The horizontal axis represents quantity, q. Quantity depends on price, making price the independent variable. However, for this type of graph economists display the dependent variable on the horizontal axis. This method is the opposite of what we have learned, but it best serves their purpose.

The point of intersection of the supply and demand curve is called the market equilibrium. The price coordinate of this point, p_e, called the equilibrium price, is the market price at which supply equals demand. When supply equals demand, the market price is such that there will not be a surplus or a shortage of the commodity.

Although we will not graph supply and demand curves, we can still find the market equilibrium price. To do so we use the substitution method to solve a system of equations, as is illustrated in Example 4.

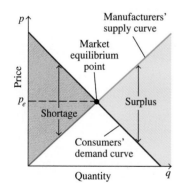

Figure 7.8

EXAMPLE 4

The supply, S, and demand, D, equations for a certain commodity are given by $S = 2p - 5$ and $D = 20 - 3p$, respectively, where p represents the price of the commodity in dollars. Find the value of p when the supply equals the demand (the equilibrium price).

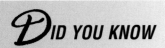

Solution: When supply equals demand, $S = D$, we can substitute $2p - 5$ for D in the equation $D = 20 - 3p$ to obtain

$$2p - 5 = 20 - 3p$$
$$2p + 3p - 5 = 20 - 3p + 3p$$
$$5p - 5 = 20$$
$$5p - 5 + 5 = 20 + 5$$
$$5p = 25$$
$$p = 5.$$

Thus supply equals demand when the price of the commodity is $5.   ▲

If neither of the equations in a system of linear equations has a variable with a coefficient of 1, it is generally easier to solve the system by using the **addition** (or **elimination**) **method**.

To solve a system of linear equations by the addition method it is necessary to obtain two equations whose sum will be a single equation containing only one variable. To achieve this goal, we rewrite the system of equations as two equations where the coefficients of one of the variables are opposites (or additive inverses) of each other. For example, if one equation has a term of $2x$, we might rewrite the other equation so that its x term will be $-2x$. To obtain the desired equations, it might be necessary to multiply one or both equations in the original system by a number. When an equation is to be multiplied by a number, we will place brackets around the equation and place the number that is to multiply the equation before the brackets. For example, $4[2x + 3y = 6]$ means that each term in the equation $2x + 3y = 6$ is to be multiplied by 4:

$$4[2x + 3y = 6] \quad \text{gives} \quad 8x + 12y = 24.$$

This notation will make our explanations much more efficient and easier for you to follow.

**Procedure for Solving a System of Equations
by the Addition Method**

1. If necessary, rewrite the equations so that the variables appear on one side of the equal sign and the constants on the other side of the equal sign.
2. If necessary, multiply one or both equations by a constant(s), so that when you add them, the result will be an equation containing only one variable.
3. Add the equations to obtain a single equation in one variable.
4. Solve the equation in step 3 for the variable.
5. Substitute the value found in step 4 into either of the original equations and solve for the other variable.

EXAMPLE 5

Solve the following system of equations by the addition method.

$$x + 2y = 4$$
$$-x + y = 2$$

Solution: Since the coefficients of the x terms, 1 and -1, are additive inverses, the sum of the x terms will be zero. Thus the sum of the two equations will contain only one variable, y. Add the two equations to obtain one equation in one variable. Then solve for the remaining variable.

$$\begin{array}{r} x + 2y = 4 \\ -x + y = 2 \\ \hline 3y = 6 \end{array}$$

$$\frac{3y}{3} = \frac{6}{3}$$

$$y = 2$$

Now substitute 2 for y in either of the original equations to find the value of x.

$$x + 2y = 4$$
$$x + 2(2) = 4$$
$$x + 4 = 4$$
$$x = 0$$

The solution to the system is $(0, 2)$.   ▲

EXAMPLE 6

Solve the following system of equations by the addition method.

$$2x + y = 8$$
$$3x + y = 5$$

Solution: We want the sum of the two equations to have only one variable. We can eliminate the variable y by multiplying either equation by -1 and then adding. We will multiply the first equation by -1.

$$-1[2x + y = 8] \quad \text{gives} \quad -2x - y = -8$$
$$3x + y = 5 \qquad\qquad\qquad 3x + y = 5$$

$$\begin{array}{r} -2x - y = -8 \\ 3x + y = 5 \\ \hline x = -3 \end{array}$$

Now we solve for y by substituting -3 for x in either of the original equations.

$$2x + y = 8$$
$$2(-3) + y = 8$$
$$-6 + y = 8$$
$$y = 14$$

The solution is $(-3, 14)$.   ▲

EXAMPLE 7

Solve the following system of equations by the addition method.

$$2x + y = 6$$
$$3x + 3y = 9$$

Solution: We can multiply the top equation by -3 and then add to eliminate the variable y.

$$-3[2x + y = 6] \quad \text{gives} \quad -6x - 3y = -18$$
$$3x + 3y = 9 \qquad\qquad 3x + 3y = 9$$

$$\begin{array}{r} -6x - 3y = -18 \\ 3x + 3y = 9 \\ \hline -3x = -9 \end{array}$$

$$x = 3$$

Now we find y.

$$2x + y = 6$$
$$2(3) + y = 6$$
$$6 + y = 6$$
$$y = 0$$

The solution is (3, 0). ▲

Note that in Example 7 we could have eliminated the variable x by multiplying the top equation by 3 and the bottom equation by -2, then adding. Try this method now.

EXAMPLE 8

Solve the following system of equations by the addition method.

$$3x - 4y = 8$$
$$2x + 3y = 9$$

Solution: In this system we cannot eliminate a variable by multiplying only one equation by an integer value and then adding. To eliminate a variable, we can multiply each equation by a different number. To eliminate the variable x, we can multiply the top equation by 2 and the bottom by -3 (or the top by -2 and the bottom by 3) and then add the two equations. If we want, we can instead eliminate the variable y by multiplying the top equation by 3 and the bottom by 4 and then adding the two equations. Let's eliminate the variable x.

$$2[3x - 4y = 8] \quad \text{gives} \quad 6x - 8y = 16$$
$$-3[2x + 3y = 9] \quad \text{gives} \quad -6x - 9y = -27$$

$$\begin{array}{r} 6x - 8y = 16 \\ -6x - 9y = -27 \\ \hline -17y = -11 \end{array}$$

$$y = \frac{11}{17}$$

We could now find x by substituting $\frac{11}{17}$ for y in either of the original equations. Although it can be done, it gets messy. Instead, let's solve for x by eliminating

the variable y from the two original equations. To do so we multiply the first equation by 3 and the second equation by 4.

$$3[3x - 4y = 8] \quad \text{gives} \quad 9x - 12y = 24$$

$$4[2x + 3y = 9] \quad \text{gives} \quad 8x + 12y = 36$$

$$\begin{array}{r} 9x - 12y = 24 \\ 8x + 12y = 36 \\ \hline 17x \qquad\;\; = 60 \end{array}$$

$$x = \frac{60}{17}$$

The solution to the system is $(\frac{60}{17}, \frac{11}{17})$.   ▲

When solving a system of linear equations by the addition method, if you obtain the equation $0 = 0$, it indicates that the system is *dependent* (both equations represent the same line; see Fig. 7.4), and there are an infinite number of solutions. When solving, if you obtain an equation such as $0 = 6$, or any other equation that is false, it means that the system is *inconsistent* (the two equations represent parallel lines; see Fig. 7.4 on page 354), and there is no solution.

EXAMPLE 9

U Hall rental agency charges $26 a day plus 20 cents a mile for a specific truck. For the same truck, Carol's charges $18 a day plus 26 cents a mile. How far would you have to drive in one day for the total cost of a U Hall truck to equal the total cost of a Carol's truck?

Solution: We are asked to find the number of miles that must be driven for each company to have the same total cost, C. First, write a system of equations to represent the total cost for each of the companies in terms of the daily fee and the mileage charge.

Let x = number of miles driven.

$$\text{Total cost} = \text{daily fee} + \text{mileage charge}$$

$$\text{U Hall: } C = 26 + 0.20x$$

$$\text{Carol's: } C = 18 + 0.25x$$

We want to determine when the cost will be the same, so we set the two costs equal to each other (substitution method) and solve the resulting equation:

$$26 + 0.20x = 18 + 0.25x$$

$$26 - 18 + 0.20x = 18 - 18 + 0.25x$$

$$8 + 0.20x = 0.25x$$

$$8 + 0.20x - 0.20x = 0.25x - 0.20x$$

$$8 = 0.05x$$

$$\frac{8}{0.05} = \frac{0.05x}{0.05}$$

$$160 = x.$$

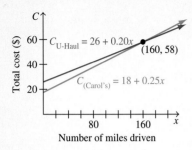

Figure 7.9

Thus if you travel 160 mi, the two companies' costs will be the same. If we construct a graph (Fig. 7.9) of the two cost equations, the point of intersection is (160, 58). This means that if you were to drive 160 mi in either truck, the cost would be $58. ▲

EXAMPLE 10

Florence Nightcap, a druggist, needs 100 milliliters (ml) of a 10% phenobarbital solution. She has only a 5% solution and a 25% solution available. How many milliliters of each solution should she mix to obtain the desired solution?

Solution: First we set up a system of equations. The unknown quantities are the amount of the 5% solution and the amount of the 25% solution that must be used. Let

$$x = \text{number of ml of 5\% solution};$$

$$y = \text{number of ml of 25\% solution}.$$

We know that 100 ml of solution are needed. Thus

$$x + y = 100.$$

The total amount of pure phenobarbital in a solution is determined by multiplying the percent of phenobarbital by the number of milliliters of solution. The second equation comes from the fact that

$$\begin{pmatrix} \text{total amount of} \\ \text{phenobarbital in} \\ \text{5\% solution} \end{pmatrix} + \begin{pmatrix} \text{total amount of} \\ \text{phenobarbital in} \\ \text{25\% solution} \end{pmatrix} = \begin{pmatrix} \text{total amount of} \\ \text{phenobarbital} \\ \text{in 10\% mixture} \end{pmatrix}$$

$$0.05x \quad + \quad 0.25y \quad = \quad 0.10(100)$$

or $\qquad 0.05x + 0.25y = 10.$

The system of equations is

$$x + y = 100$$

$$0.05x + 0.25y = 10.$$

Let's solve this system of equations by using the addition method.

$$-5[x + y = 100] \quad \text{gives} \quad -5x - 5y = -500$$

$$100[0.05x + 0.25y = 10] \quad \text{gives} \quad 5x + 25y = 1000$$

$$\begin{array}{r} -5x - 5y = -500 \\ 5x + 25y = 1000 \\ \hline 20y = 500 \end{array}$$

$$\frac{20y}{20} = \frac{500}{20}$$

$$y = 25$$

Now we find x.

$$x + y = 100$$

$$x + 25 = 100$$

$$x = 75$$

Therefore 75 ml of a 5% phenobarbital solution must be mixed with 25 ml of a 25% phenobarbital solution to obtain 100 ml of a 10% phenobarbital solution.   ▲

Example 10 can also be solved by using substitution. Try to do so now.

Section 7.2 Exercises

1. In your own words explain how to solve a system of equations by using the substitution method.

2. In your own words explain how to solve a system of equations by using the addition method.

3. How will you know, when solving a system of equations by either the substitution or the addition method, whether the system is dependent?

4. How will you know, when solving a system of equations by either the substitution or the addition method, whether the system is inconsistent?

In Exercises 5–22, solve the system of equations by the substitution method. If the system does not have a single ordered pair as a solution, state whether the system is inconsistent or dependent.

5. $y = x + 3$
$y = -x - 5$

6. $y = 4x - 6$
$y = x - 3$

7. $x - 2y = 6$
$2x + y = -3$

8. $3x + y = 3$
$4y + 3x = 6$

9. $y - x = 4$
$x - y = 3$

10. $x + y = 3$
$y + x = 5$

11. $x = 5y - 12$
$x - y = 0$

12. $3y + 2x = 4$
$3y = 6 - x$

13. $y - 2x = 3$
$2y = 4x + 6$

14. $x = y + 3$
$x = -3$

15. $y = 2$
$y + x + 3 = 0$

16. $x + 2y = 6$
$y = 2x + 3$

17. $y + 3x - 4 = 0$
$2x - y = 7$

18. $x + 4y = 7$
$2x + 3y = 5$

19. $x = 3y + 1$
$y = 2x - 1$

20. $x + 3y = 8$
$2x - y - 4 = 0$

21. $6x - y = 5$
$y = 6x - 3$

22. $x + 3y = 6$
$y = -\frac{1}{3}x + 2$

In Exercises 23–36, solve the system of equations by the addition method. If the system does not have a single ordered pair as a solution, state whether the system is inconsistent or dependent.

23. $5x + y = 4$
$3x - y = 4$

24. $5x - y = 6$
$x + y = 6$

25. $3x + 2y = 6$
$-3x + y = 0$

26. $x - 3y = 4$
$-x + 2y = -5$

27. $2x - y = -4$
$-3x - y = 6$

28. $x + y = 6$
$-2x + y = -3$

29. $2x + y = 6$
$3x + y = 5$

30. $2x + y = 11$
$x + 3y = 18$

31. $2x - 3y = 4$
$2x + y + 4 = 0$

32. $5x - 2y = 11$
$-3x + 2y = 1$

33. $3x - 4y = 11$
$3x + 5y = -7$

34. $4x - 2y = 6$
$4y = 8x - 12$

35. $4x + y = 6$
$-8x - 2y = 13$

36. $2x + 3y = 6$
$5x - 4y = -8$

37. $7x + 8y = 11$
$5x + 6y = 7$

38. $8x + 3y = 7$
$3x + 2y = 9$

In Exercises 39–51, write a system of equations that can be used to solve the problem. Then solve the system and determine the answer.

39. A plane can travel 280 mph with the wind and 240 mph against the wind. Find the speed of the plane in still air and the speed of the wind.

40. Jill can make a weekly salary of $200 plus 5% commission on sales or a weekly salary consisting of a straight 15% commission on sales. Determine the amount of sales necessary for the 15% straight commission salary to equal the $200 plus 5% commission salary.

41. The Syracuse women's basketball team made 45 field goals in a recent game; some were 2-pointers and some were 3-pointers. How many 2-point baskets were made and how many 3-point baskets were made if Syracuse scored 101 points?

42. Hockey teams receive 2 points when they win a game and 1 point when they tie a game. The Flyers won a division title with 58 points. They won 23 more games than they tied. How many wins and how many ties did the Flyers have?

43. The speed limit is 55 miles per hour in Oxon Hill–Grassmanner, Maryland. The fine for exceeding the speed limit carries a fixed charge plus a charge for each mile per hour over the speed limit. The fine for driving 70 mph is $175, and the fine for driving 75 mph is $200. Determine the fixed charge and the charge per mile.

44. Soybean meal is 16% protein and corn meal is 7% protein. How many pounds of each should be used to get a 300-lb mixture that is 10% protein?

45. Ramon wishes to mix 30 lb of coffee to sell for a total cost of $200. To obtain the mixture, he will mix coffee that sells for $6 per pound with coffee that sells for $8 per pound. How many pounds of each type of coffee should he use?

46. The Guidas own a dairy. They have milk that is 5% butterfat and skim milk without butterfat. How much of the 5% milk and how much of the skim milk should they mix to make 100 gal of milk that is 3.5% butterfat?

47. In chemistry class, Mark has an 80% acid solution and a 50% acid solution. How much of each solution should he mix to get 100 liters of a 75% acid solution?

48. An airplane has a total of 159 seats. The number of coach-class seats is 7 more than 18 times the number of first-class seats. How many of each type of seat are there on the plane?

49. Animals in an experiment are to be kept on a strict diet. Each animal is to receive, among other things, 20 g of protein and 6 g of carbohydrates. The scientist has only two food mixes of the following compositions available.

	Protein (%)	Carbohydrates (%)
Mix A	10	6
Mix B	20	2

How many grams of each mix should she use to obtain the right diet for a single animal?

50. Membership in Chippers Country Club costs $3000 per year and entitles a member to play a round of golf for a greens fee of $18.00. At Birdies Country Club, membership costs $2500 per year and the greens fee is $20.00.
 a) How many rounds must a golfer play in a year in order for the costs in the two clubs to be the same?

 b) If Yvonne Passaro planned to play 30 rounds of golf in a year, which club would be the least expensive?

51. The charge for maintaining a checking account at Union Bank is $6 per month plus 10 cents for each check that is written. The charge at Citrus Bank is $2 per month and 20 cents per check.
 a) How many checks would a customer have to write in a month for the total charges to be the same at both banks?
 b) If Eb Scrooge planned to write 14 checks per month, which bank would be the least expensive?

In Exercises 52 and 53, supply and demand equations are given. Find the equilibrium price. See Example 4.

52. $S = 1000 + 5x$
$D = 200 + 20x$

53. $S = 1200 - 2.8x$
$D = 800 + 3.2x$

Problem Solving/Group Activities

54. Solve the following system of equations for u and v by first substituting x for $1/u$ and y for $1/v$.

$$\frac{1}{u} + \frac{2}{v} = 8$$

$$\frac{3}{u} - \frac{1}{v} = 3$$

55. Develop a system of equations that has $(6, 5)$ as its solution. Explain how you developed your system of equations.

56. The substitution or addition methods can also be used to solve a system of three equations in three variables. Consider the following system.

$$x + y + z = 7$$
$$x - y + 2z = 9$$
$$-x + 2y + z = 4$$

The ordered triple (x, y, z) is the solution to the system if it satisfies all three equations.
 a) Show that the ordered triple $(2, 1, 4)$ is a solution to the system.
 b) Use the substitution or addition method to determine the solution to the system. (*Hint:* Eliminate one variable by using two equations. Then eliminate the same variable by using two different equations.)

7.3 MATRICES

We have discussed solving systems of equations by graphing, using substitution, and using the addition method. In Section 7.4 we will discuss solving systems of linear equations by using matrices. So that you will become familiar with matrices, in this section we explain how to add, subtract, and multiply matrices. We

also explain how to multiply a matrix by a real number. Matrix techniques are easily adapted to computers.

A **matrix** is a rectangular array of elements. An array is a systematic arrangement of numbers or symbols in rows and columns. In this text we use brackets to indicate a matrix. Matrices (the plural of matrix) may be used to display information and to solve systems of linear equations.

The following matrix displays information about a survey of 500 registered voters.

	Democrats	Republicans	Conservatives	Liberals	Others
Men	100	93	20	35	4
Women	80	92	21	47	8

The numbers inside the brackets are called **elements** of the matrix. The preceding matrix contains 10 elements. Because it has two rows and five columns, it is referred to as a 2 by 5, written 2×5, matrix. The **dimensions** of a matrix may be indicated with the symbol $r \times s$, where r is the number of rows and s is the number of columns. A matrix that contains the same number of rows and columns is called a **square matrix**. Following are examples of 2×2 and 3×3 square matrices.

$$\begin{bmatrix} 2 & 3 \\ 5 & 2 \end{bmatrix} \qquad \begin{bmatrix} 4 & 6 & -1 \\ 2 & 3 & 0 \\ 5 & 2 & 1 \end{bmatrix}$$

Two matrices are equal if and only if they have the same elements in the same relative positions.

EXAMPLE I

Given $A = B$, find x and y.

$$A = \begin{bmatrix} 1 & 4 \\ 6 & 3 \end{bmatrix}, \qquad B = \begin{bmatrix} x & 4 \\ 6 & y \end{bmatrix}$$

Solution: The corresponding elements must be the same, so $x = 1$ and $y = 3$. ▲

Addition of Matrices

Two matrices can be added only if they have the same dimensions (same number of rows and same number of columns). To obtain the sum of two matrices with the same dimensions, add the corresponding elements of the two matrices.

EXAMPLE 2

$$A = \begin{bmatrix} 1 & 4 \\ -2 & 6 \end{bmatrix}, \qquad B = \begin{bmatrix} 3 & 8 \\ 6 & 0 \end{bmatrix}. \quad \text{Find } A + B.$$

Solution: $A + B = \begin{bmatrix} 1 & 4 \\ -2 & 6 \end{bmatrix} + \begin{bmatrix} 3 & 8 \\ 6 & 0 \end{bmatrix}$

$$= \begin{bmatrix} 1 + 3 & 4 + 8 \\ -2 + 6 & 6 + 0 \end{bmatrix} = \begin{bmatrix} 4 & 12 \\ 4 & 6 \end{bmatrix}$$ ▲

The following example illustrates an application of addition of matrices.

EXAMPLE 3

The Ski Swap Corporation owns and operates two stores, one in Idaho and the other in Wyoming. The number of pairs of downhill skis and cross-country skis sold in each store for the months of January through June and July through December are indicated in the matrices that follow.

$$\begin{array}{cc} & \text{Idaho} \\ & \begin{array}{cc} \text{CC} & \text{DH} \end{array} \end{array} \qquad \begin{array}{cc} & \text{Wyoming} \\ & \begin{array}{cc} \text{CC} & \text{DH} \end{array} \end{array}$$

$$\begin{array}{c} \text{Jan.–June} \\ \text{July–Dec.} \end{array} \begin{bmatrix} 200 & 230 \\ 452 & 500 \end{bmatrix} = A \qquad \begin{bmatrix} 230 & 190 \\ 377 & 502 \end{bmatrix} = B$$

Find the total number of each type of ski sold by the corporation during each time period.

Solution: To solve the problem, we add matrices A and B.

$$\begin{array}{c} \\ \text{Jan.–June} \\ \text{July–Dec.} \end{array} \begin{array}{cc} \text{CC} \qquad\qquad \text{DH} \\ \begin{bmatrix} 200 + 230 & 230 + 190 \\ 452 + 377 & 500 + 502 \end{bmatrix} \end{array} = \begin{array}{cc} \text{CC} \quad \text{DH} \\ \begin{bmatrix} 430 & 420 \\ 829 & 1002 \end{bmatrix} \end{array}$$

We can see from the sum matrix that during the period from January through June, 430 pairs of cross-country skis and 420 pairs of downhill skis were sold. During the period from July through December, a total of 829 pairs of cross-country skis and 1002 pairs of downhill skis were sold. ▲

The matrix

$$I = \begin{bmatrix} 0 & 0 \\ 0 & 0 \end{bmatrix}$$

is the **additive identity matrix** for 2×2 matrices. We denote this matrix with the letter I. Note that for any 2×2 matrix A, $A + I = I + A = A$.

Subtraction of Matrices

Only matrices with the same dimensions may be subtracted. To do so, we subtract each entry in one matrix from the corresponding entry in the other matrix.

EXAMPLE 4

Find $A - B$ if

$$A = \begin{bmatrix} 4 & 6 \\ 5 & -1 \end{bmatrix} \quad \text{and} \quad B = \begin{bmatrix} 3 & -4 \\ 7 & -3 \end{bmatrix}.$$

Solution:

$$A - B = \begin{bmatrix} 4 & 6 \\ 5 & -1 \end{bmatrix} - \begin{bmatrix} 3 & -4 \\ 7 & -3 \end{bmatrix}$$

$$= \begin{bmatrix} 4 - 3 & 6 - (-4) \\ 5 - 7 & -1 - (-3) \end{bmatrix} = \begin{bmatrix} 1 & 10 \\ -2 & 2 \end{bmatrix}$$

▲

Multiplying a Matrix by a Real Number

A matrix may be multiplied by a real number by multiplying each entry in the matrix by the real number. Sometimes when we multiply a matrix by a real number, we call that real number a *scalar*.

EXAMPLE 5

For matrices A and B, find (a) $3A$ and (b) $3A - 2B$.

$$A = \begin{bmatrix} 4 & 6 \\ -3 & 5 \end{bmatrix}, \quad B = \begin{bmatrix} -1 & 5 \\ 2 & 6 \end{bmatrix}$$

Solution:

a) $3A = 3 \begin{bmatrix} 4 & 6 \\ -3 & 5 \end{bmatrix} = \begin{bmatrix} 3(4) & 3(6) \\ 3(-3) & 3(5) \end{bmatrix} = \begin{bmatrix} 12 & 18 \\ -9 & 15 \end{bmatrix}$

b) We found $3A$ in part (a). Now we find $2B$.

$$2B = 2 \begin{bmatrix} -1 & 5 \\ 2 & 6 \end{bmatrix} = \begin{bmatrix} 2(-1) & 2(5) \\ 2(2) & 2(6) \end{bmatrix} = \begin{bmatrix} -2 & 10 \\ 4 & 12 \end{bmatrix}$$

$$3A - 2B = \begin{bmatrix} 12 & 18 \\ -9 & 15 \end{bmatrix} - \begin{bmatrix} -2 & 10 \\ 4 & 12 \end{bmatrix}$$

$$= \begin{bmatrix} 12 -(-2) & 18 - 10 \\ -9 - 4 & 15 - 12 \end{bmatrix} = \begin{bmatrix} 14 & 8 \\ -13 & 3 \end{bmatrix} \quad \blacktriangle$$

Multiplication of Matrices

Multiplication of matrices is slightly more difficult than addition of matrices. Multiplication of matrices is possible only when the number of *columns* of the first matrix, A, is the same as the number of *rows* of the second matrix, B. We use the notation

$$A$$
$$3 \times 4$$

to indicate that matrix A has three rows and four columns. Suppose that matrix A is a 3×4 matrix and matrix B is a 4×5 matrix. Then

$$\begin{array}{cc} A & B \\ 3 \times 4 & 4 \times 5. \end{array}$$

Same

Product matrix 3×5

This indicates that matrix A has four columns and matrix B has four rows. Therefore we can multiply these two matrices. The product matrix will have the same number of rows as matrix A and the same number of columns as matrix B. Thus the dimensions of the product matrix are 3×5.

EXAMPLE 6

Determine which of the following pairs of matrices can be multiplied.

a) $A = \begin{bmatrix} 3 & 2 \\ 5 & 7 \end{bmatrix}$, $\qquad B = \begin{bmatrix} 0 & 6 \\ 4 & 1 \end{bmatrix}$

b) $A = \begin{bmatrix} 2 & 3 \\ 5 & 6 \end{bmatrix}$, $\qquad B = \begin{bmatrix} 2 & 4 & -1 \\ 6 & 8 & 0 \end{bmatrix}$

c) $A = \begin{bmatrix} 2 & 1 & 4 \\ 3 & 2 & 8 \end{bmatrix}$, $\qquad B = \begin{bmatrix} 2 & 1 & 3 \\ 1 & 0 & -2 \end{bmatrix}$

Solution:

a)

$$\begin{array}{cc} A & B \\ 2 \times 2 & 2 \times 2 \end{array}$$

$$\underbrace{\qquad\qquad}_{\text{Same}}$$

Because matrix A has two columns and matrix B has two rows, the two matrices can be multiplied. The product is a 2×2 matrix.

b)

$$\begin{array}{cc} A & B \\ 2 \times 2 & 2 \times 3 \end{array}$$

$$\underbrace{\qquad\qquad}_{\text{Same}}$$

Because matrix A has two columns and matrix B has two rows, the two matrices can be multiplied. The product is a 2×3 matrix.

c)

$$\begin{array}{cc} A & B \\ 2 \times 3 & 2 \times 3 \end{array}$$

$$\underbrace{\qquad\qquad}_{\text{Not same}}$$

Because matrix A has three columns and matrix B has two rows, the two matrices cannot be multiplied. ▲

To explain matrix multiplication let's use matrices A and B that follow.

$$A = \begin{bmatrix} 3 & 2 \\ 5 & 7 \end{bmatrix} \quad \text{and} \quad B = \begin{bmatrix} 0 & 6 \\ 4 & 1 \end{bmatrix}$$

Since A contains two rows and B contains two columns, the product matrix will contain two rows and two columns. To multiply two matrices, we use a row–column scheme of multiplying. The numbers in the first row of matrix A are multiplied by the numbers in the first column of matrix B.

$$A \times B = \begin{bmatrix} 3 & 2 \\ 5 & 7 \end{bmatrix} \begin{bmatrix} 0 & 6 \\ 4 & 1 \end{bmatrix}$$

$$\text{First row} \qquad \text{First column}$$

$$\begin{bmatrix} 3 & 2 \\ 5 & 7 \end{bmatrix} \qquad \begin{bmatrix} 0 & 6 \\ 4 & 1 \end{bmatrix}$$

$$(3 \times 0) + (2 \times 4) = 0 + 8$$
$$= 8$$

The 8 is placed in the first-row, first-column position of the product matrix. The other numbers in the product matrix are obtained similarly, as illustrated in the matrix that follows.

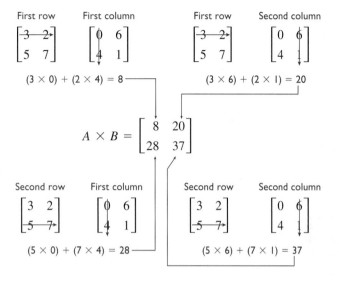

We can shorten the procedure as follows.

$$A \times B = \begin{bmatrix} 3 & 2 \\ 5 & 7 \end{bmatrix} \begin{bmatrix} 0 & 6 \\ 4 & 1 \end{bmatrix}$$

$$= \begin{bmatrix} 3(0) + 2(4) & 3(6) + 2(1) \\ 5(0) + 7(4) & 5(6) + 7(1) \end{bmatrix}$$

$$= \begin{bmatrix} 8 & 20 \\ 28 & 37 \end{bmatrix}$$

In general, if

$$A = \begin{bmatrix} a & b \\ c & d \end{bmatrix} \quad \text{and} \quad B = \begin{bmatrix} e & f \\ g & h \end{bmatrix},$$

then

$$A \times B = \begin{bmatrix} a & b \\ c & d \end{bmatrix} \begin{bmatrix} e & f \\ g & h \end{bmatrix} = \begin{bmatrix} ae + bg & af + bh \\ ce + dg & cf + dh \end{bmatrix}.$$

Let's do one more multiplication.

EXAMPLE 7

Find $A \times B$, given

$$A = \begin{bmatrix} 2 & 3 \\ 5 & 6 \end{bmatrix} \quad \text{and} \quad B = \begin{bmatrix} 2 & 4 & -1 \\ 6 & 8 & 0 \end{bmatrix}.$$

Solution: Matrix A contains two columns, and matrix B contains two rows. Thus the matrices can be multiplied. Since A contains two rows and B con-

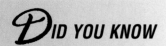

tains three columns, the product matrix will contain two rows and three columns.

$$A \times B = \begin{bmatrix} 2 & 3 \\ 5 & 6 \end{bmatrix}\begin{bmatrix} 2 & 4 & -1 \\ 6 & 8 & 0 \end{bmatrix}$$

$$= \begin{bmatrix} 2(2) + 3(6) & 2(4) + 3(8) & 2(-1) + 3(0) \\ 5(2) + 6(6) & 5(4) + 6(8) & 5(-1) + 6(0) \end{bmatrix}$$

$$= \begin{bmatrix} 22 & 32 & -2 \\ 46 & 68 & -5 \end{bmatrix}$$

It should be noted that multiplication of matrices *is not* commutative; that is, $A \times B \neq B \times A$, except in special instances.

We previously discussed the additive identity matrix. *Square matrices* also have a **multiplicative identity matrix**. The multiplicative identity matrices for a 2×2 and a 3×3 matrix, denoted I, follow. Note that in any multiplicative identity matrix 1s go diagonally from top left to bottom right, and all other elements in the matrix are 0s.

$$I = \begin{bmatrix} 1 & 0 \\ 0 & 1 \end{bmatrix} \qquad I = \begin{bmatrix} 1 & 0 & 0 \\ 0 & 1 & 0 \\ 0 & 0 & 1 \end{bmatrix}$$

For any square matrix, A, $A \times I = I \times A = A$.

EXAMPLE 8

Use the multiplicative identity matrix for a 2×2 matrix and matrix A to show that $A \times I = A$.

$$A = \begin{bmatrix} 4 & 3 \\ 2 & 1 \end{bmatrix}$$

Solution: The identity matrix is $I = \begin{bmatrix} 1 & 0 \\ 0 & 1 \end{bmatrix}$.

$$A \times I = \begin{bmatrix} 4 & 3 \\ 2 & 1 \end{bmatrix}\begin{bmatrix} 1 & 0 \\ 0 & 1 \end{bmatrix}$$

$$= \begin{bmatrix} 4(1) + 3(0) & 4(0) + 3(1) \\ 2(1) + 1(0) & 2(0) + 1(1) \end{bmatrix}$$

$$= \begin{bmatrix} 4 & 3 \\ 2 & 1 \end{bmatrix} = A$$

Example 9 illustrates an application of multiplication of matrices.

EXAMPLE 9

The Retrop Bathing Suit Company manufactures three types of bathing suits: a one-piece suit, a two-piece suit, and a three-piece suit (a two-piece suit with a matching hat). On a particular day the firm produces 20 one-piece suits, 30 two-piece suits, and 50 three-piece suits. Each one-piece suit requires 4 units of material and 1 hour of work to produce; each two-piece suit requires 3 units of material and 2 hours of work to produce; each three-piece suit requires 5 units of material and 3 hours to produce. Use matrix multiplication to determine the total number of units of material and the total number of hours needed for that day's production.

Solution: Let matrix *A* represent the number of each type of bathing suit produced.

$$
\begin{array}{c}
\begin{array}{cccc}
 & \text{One} & \text{Two} & \text{Three} \\
\text{Type} & \text{piece} & \text{piece} & \text{piece}
\end{array} \\
A = [\,20 \quad 30 \quad 50\,]
\end{array}
$$

The units of material and time requirements for each type are indicated in matrix *B*.

$$
B = \begin{array}{c}
\begin{array}{cc}
\text{Material} & \text{Hours}
\end{array} \\
\begin{bmatrix} 4 & 1 \\ 3 & 2 \\ 5 & 3 \end{bmatrix}
\end{array}
\begin{array}{l}
\text{One piece} \\ \text{Two piece} \\ \text{Three piece}
\end{array}
$$

The product of *A* and *B*, or $A \times B$, will give the total number of units of material and the total number of hours of work needed for that day's production.

$$
A \times B = [20 \quad 30 \quad 50] \begin{bmatrix} 4 & 1 \\ 3 & 2 \\ 5 & 3 \end{bmatrix}
$$

$$
= [20(4) + 30(3) + 50(5) \qquad 20(1) + 30(2) + 50(3)]
$$

$$
= [420 \quad 230]
$$

Thus a total of 420 units of material and a total of 230 hours of work were required that day. ▲

Section 7.3 Exercises

1. What is a matrix?

2. Explain how to determine the dimensions of a matrix.

3. What is a square matrix?

4. **a)** In your own words, explain the procedure used to add matrices.
   b) Use the procedure given in part (a) to add

   $$\begin{bmatrix} 5 & 3 & -1 \\ 0 & 2 & 4 \end{bmatrix} \text{ and } \begin{bmatrix} 4 & 5 & 6 \\ -1 & 3 & 2 \end{bmatrix}.$$

5. **a)** In your own words, explain the procedure used to subtract matrices.
   b) Use the procedure given in part (a) to subtract

   $$\begin{bmatrix} 5 & 3 & -1 \\ 0 & 2 & 4 \end{bmatrix} \text{ from } \begin{bmatrix} 4 & 5 & 6 \\ -1 & 3 & 2 \end{bmatrix}.$$

6. In order to multiply two matrices, what must be true about the dimensions of those matrices?

7. **a)** In your own words, explain the procedure used to multiply matrices.
   b) Use the procedure given in part (a) to multiply

   $$\begin{bmatrix} 6 & -1 \\ 5 & 0 \end{bmatrix} \text{ by } \begin{bmatrix} 2 & -3 \\ 1 & -4 \end{bmatrix}.$$

8. Records are kept each day for a month at a local cinema that houses three movie theaters *A*, *B*, and *C*. The daily average Monday through Sunday receipts for the three theaters are as follows.

   A: $654, $785, $458, $345, $1478, $2109, $543
   B: $764, $778, $568, $451, $1024, $1689, $853
   C: $567, $764, $873, $407, $2034, $2432, $567

   Express this information in the form of a 3 × 7 matrix.

In Exercises 9–12, find A + B.

9. $A = \begin{bmatrix} 8 & 5 \\ 9 & 1 \end{bmatrix}$, $B = \begin{bmatrix} -3 & -2 \\ 0 & 3 \end{bmatrix}$

10. $A = \begin{bmatrix} 1 & 0 & 2 \\ 0 & 2 & 2 \end{bmatrix}$, $B = \begin{bmatrix} 7 & 4 & -1 \\ 5 & -3 & 6 \end{bmatrix}$

11. $A = \begin{bmatrix} 4 & 1 \\ -6 & 4 \\ 2 & 7 \end{bmatrix}$, $B = \begin{bmatrix} -2 & 3 \\ 0 & 5 \\ 2 & 1 \end{bmatrix}$

12. $A = \begin{bmatrix} 4 & 0 & 2 \\ 5 & 1 & -3 \\ -1 & 4 & -6 \end{bmatrix}$, $B = \begin{bmatrix} -4 & 2 & -1 \\ 5 & -3 & 6 \\ -2 & 0 & 6 \end{bmatrix}$

In Exercises 13–16, find A − B.

13. $A = \begin{bmatrix} 8 & 6 \\ 4 & -2 \end{bmatrix}$, $B = \begin{bmatrix} 4 & -2 \\ -3 & 5 \end{bmatrix}$

14. $A = \begin{bmatrix} 5 & -3 & 6 \\ 2 & 4 & 2 \end{bmatrix}$, $B = \begin{bmatrix} 1 & 3 & 8 \\ 0 & -4 & -5 \end{bmatrix}$

15. $A = \begin{bmatrix} 5 & 3 & -1 \\ 7 & 4 & 2 \\ 6 & -1 & -5 \end{bmatrix}$, $B = \begin{bmatrix} 4 & 3 & 6 \\ -2 & -4 & 9 \\ 0 & -2 & 4 \end{bmatrix}$

16. $A = \begin{bmatrix} -4 & 3 \\ 6 & 2 \\ 1 & -5 \end{bmatrix}$, $B = \begin{bmatrix} -6 & -8 \\ -10 & -11 \\ 3 & -7 \end{bmatrix}$

In Exercises 17–22, let

$A = \begin{bmatrix} 1 & 2 \\ 0 & 5 \end{bmatrix}$, $B = \begin{bmatrix} 3 & 2 \\ 5 & 0 \end{bmatrix}$, and $C = \begin{bmatrix} -2 & 3 \\ 4 & 0 \end{bmatrix}$.

Find

17. $2B$.

18. $-3B$.

19. $2B + 3C$.

20. $2B + 3A$.

21. $3B - 2C$.

22. $4C - 2A$.

In Exercises 23–28, find A × B.

23. $A = \begin{bmatrix} 3 & 5 \\ 2 & 6 \end{bmatrix}$, $B = \begin{bmatrix} 4 & 2 \\ 0 & 3 \end{bmatrix}$

24. $A = \begin{bmatrix} -1 & 1 \\ 0 & 3 \end{bmatrix}$, $B = \begin{bmatrix} 1 & 4 \\ -1 & 0 \end{bmatrix}$

25. $A = \begin{bmatrix} 2 & 3 & -1 \\ 0 & 4 & 6 \end{bmatrix}$, $B = \begin{bmatrix} 2 \\ 4 \\ 1 \end{bmatrix}$

26. $A = \begin{bmatrix} 1 & 1 \\ 1 & 1 \end{bmatrix}$, $B = \begin{bmatrix} 1 & -1 \\ -1 & 2 \end{bmatrix}$

27. $A = \begin{bmatrix} 2 & 5 \\ 1 & 3 \end{bmatrix}$, $B = \begin{bmatrix} 5 & 1 \\ -3 & -2 \end{bmatrix}$

28. $A = \begin{bmatrix} 2 & 3 & 1 \\ -2 & -1 & 0 \\ 4 & 5 & 6 \end{bmatrix}$, $B = \begin{bmatrix} 1 & 0 & 0 \\ 0 & 1 & 0 \\ 0 & 0 & 1 \end{bmatrix}$

In Exercises 29–34, if possible, find A + B and A × B. If an operation cannot be performed, explain why.

29. $A = \begin{bmatrix} 0 & 3 & 1 \\ 2 & 4 & 5 \end{bmatrix}$, $B = \begin{bmatrix} 1 & 2 & 3 \\ 2 & -2 & 3 \end{bmatrix}$

30. $A = \begin{bmatrix} 2 & 3 & 4 \\ 4 & 6 & -2 \end{bmatrix}$, $B = \begin{bmatrix} 2 & -3 \\ 0 & 4 \end{bmatrix}$

31. $A = \begin{bmatrix} 4 & 5 & 3 \\ 6 & 2 & 1 \end{bmatrix}$, $B = \begin{bmatrix} 3 & 2 \\ 4 & 6 \\ -2 & 0 \end{bmatrix}$

32. $A = \begin{bmatrix} 1 & 2 \\ 3 & 4 \\ 5 & 6 \end{bmatrix}$, $B = \begin{bmatrix} 1 & 2 \\ 3 & 4 \\ 5 & 6 \end{bmatrix}$

33. $A = \begin{bmatrix} 1 & 2 \\ 3 & 4 \end{bmatrix}$, $B = \begin{bmatrix} -3 \\ 2 \end{bmatrix}$

34. $A = \begin{bmatrix} 5 & -1 \\ 6 & -2 \end{bmatrix}$, $B = \begin{bmatrix} 1 & 2 \\ 3 & 4 \end{bmatrix}$

In Exercises 35–37, show that the commutative property of addition, A + B = B + A, holds for matrices A and B.

35. $A = \begin{bmatrix} 4 & 2 \\ -3 & 1 \end{bmatrix}$, $B = \begin{bmatrix} 8 & 5 \\ 6 & 1 \end{bmatrix}$

36. $A = \begin{bmatrix} -4 & 3 \\ 5 & 7 \end{bmatrix}$, $B = \begin{bmatrix} -5 & -8 \\ 0 & -7 \end{bmatrix}$

37. $A = \begin{bmatrix} 0 & -1 \\ 3 & -4 \end{bmatrix}$, $B = \begin{bmatrix} 8 & 1 \\ 3 & -4 \end{bmatrix}$

38. Make up two matrices with the same dimensions, A and B, and show that $A + B = B + A$.

In Exercises 39–41, show that the associative property of addition, (A + B) + C = A + (B + C), holds for the matrices given.

39. $A = \begin{bmatrix} 4 & -3 \\ -1 & 5 \end{bmatrix}$, $B = \begin{bmatrix} 2 & 3 \\ 7 & 1 \end{bmatrix}$,
$C = \begin{bmatrix} 0 & -4 \\ 5 & 1 \end{bmatrix}$

40. $A = \begin{bmatrix} -9 & -8 \\ -7 & -6 \end{bmatrix}$, $B = \begin{bmatrix} -5 & -4 \\ -3 & -2 \end{bmatrix}$,
$C = \begin{bmatrix} -1 & 0 \\ 1 & 2 \end{bmatrix}$

41. $A = \begin{bmatrix} 7 & 4 \\ 9 & -36 \end{bmatrix}$, $B = \begin{bmatrix} 5 & 6 \\ -1 & -4 \end{bmatrix}$,
$C = \begin{bmatrix} -7 & -5 \\ -1 & 3 \end{bmatrix}$

42. Make up three matrices with the same dimensions, A, B, and C, and show that $(A + B) + C = A + (B + C)$.

In Exercises 43–47, determine whether the commutative property of multiplication, $A \times B = B \times A$, holds for the matrices given.

43. $A = \begin{bmatrix} 1 & 2 \\ 3 & -1 \end{bmatrix}$, $B = \begin{bmatrix} 1 & 3 \\ 2 & -1 \end{bmatrix}$

44. $A = \begin{bmatrix} 1 & -1 \\ 3 & 5 \end{bmatrix}$, $B = \begin{bmatrix} 1 & -1 \\ 3 & 5 \end{bmatrix}$

45. $A = \begin{bmatrix} 4 & 2 \\ 1 & -3 \end{bmatrix}$, $B = \begin{bmatrix} 2 & 4 \\ -3 & 1 \end{bmatrix}$

46. $A = \begin{bmatrix} -3 & 2 \\ 6 & -5 \end{bmatrix}$, $B = \begin{bmatrix} -\frac{5}{3} & -\frac{2}{3} \\ -2 & -1 \end{bmatrix}$

47. $A = \begin{bmatrix} 3 & 2 & 1 \\ 4 & 2 & 0 \\ 0 & -2 & 5 \end{bmatrix}$, $B = \begin{bmatrix} 1 & 0 & 0 \\ 0 & 1 & 0 \\ 0 & 0 & 1 \end{bmatrix}$

48. Make up two matrices A and B with the same dimensions, and determine whether $A \times B = B \times A$.

In Exercises 49–53, show that the associative property of multiplication, $(A \times B) \times C = A \times (B \times C)$, holds for the matrices given.

49. $A = \begin{bmatrix} 1 & 2 \\ 4 & 0 \end{bmatrix}$, $B = \begin{bmatrix} 2 & 1 \\ 3 & 0 \end{bmatrix}$, $C = \begin{bmatrix} 4 & 2 \\ 3 & 1 \end{bmatrix}$

50. $A = \begin{bmatrix} -2 & 3 \\ 0 & 4 \end{bmatrix}$, $B = \begin{bmatrix} 4 & 0 \\ 3 & 5 \end{bmatrix}$, $C = \begin{bmatrix} 3 & 4 \\ -2 & 5 \end{bmatrix}$

51. $A = \begin{bmatrix} 4 & 3 \\ -6 & 2 \end{bmatrix}$, $B = \begin{bmatrix} 1 & 2 \\ 0 & 1 \end{bmatrix}$, $C = \begin{bmatrix} 4 & 3 \\ 0 & -2 \end{bmatrix}$

52. $A = \begin{bmatrix} -1 & -2 \\ -3 & -4 \end{bmatrix}$, $B = \begin{bmatrix} 1 & 0 \\ 0 & 1 \end{bmatrix}$, $C = \begin{bmatrix} 0 & 0 \\ 0 & 0 \end{bmatrix}$

53. $A = \begin{bmatrix} 3 & 4 \\ -1 & -2 \end{bmatrix}$, $B = \begin{bmatrix} 0 & 1 \\ 1 & 0 \end{bmatrix}$, $C = \begin{bmatrix} 2 & 0 \\ 3 & 0 \end{bmatrix}$

54. Make up three matrices with the same dimensions, A, B, and C, and show that $(A \times B) \times C = A \times (B \times C)$.

55. The Original Cookie Factory bakes and sells four types of cookies: chocolate chip, sugar, molasses, and peanut butter. Matrix A shows the number of units of various ingredients used in baking a dozen of each type of cookie.

	Sugar	Flour	Milk	Eggs	
	2	2	$\frac{1}{2}$	1	Chocolate chip
$A =$	3	2	1	2	Sugar
	0	1	0	3	Molasses
	$\frac{1}{2}$	1	0	0	Peanut butter

The cost, in cents per cup or per egg, for each ingredient when purchased in small quantities and in large quantities is given in matrix B.

	Small quantities	Large quantities	
	10	12	Sugar
$B =$	5	8	Flour
	8	8	Milk
	4	6	Eggs

Use matrix multiplication to find a matrix representing the comparative cost per item for small and large quantities purchased.

In Exercises 56 and 57, use the information given in Exercise 55. Suppose that a typical day's order consists of 40 dozen chocolate chip cookies, 30 dozen sugar cookies, 12 dozen molasses cookies, and 20 dozen peanut butter cookies.

56. a) Express these orders as a 1×4 matrix.
   b) Use matrix multiplication to determine the amount of each ingredient needed to fill the day's order.

57. Use matrix multiplication to determine the cost under the two purchase options (small and large quantities) to fill the day's order.

58. Food orders from the Geology Club and the Rugby Club are summarized in matrix A.

		Burger	Fries	Cola
$A =$	Geology Club	28	32	25
	Rugby Club	33	26	31

The prices (in dollars) of a burger, fries, and a cola at three fast-food restaurants are summarized in matrix B.

		McDougal's	Burger Prince	Mendy's
	Burger	2.45	2.95	3.15
$B =$	Fries	1.35	0.99	1.00
	Cola	1.40	0.92	1.20

a) Multiply the two matrices to form a 2×3 matrix that shows the amount each club would be charged by the fast-food restaurant.
b) Determine which fast-food restaurant offers each club the best total price.

*Two matrices whose sum is the additive identity matrix are said to be **additive inverses**. That is, if $A + B = B + A = I$, where I is the additive identity matrix, then A and B are additive inverses. In Exercises 59 and 60, determine whether A and B are additive inverses.*

59. $A = \begin{bmatrix} 6 & 3 \\ 4 & -2 \end{bmatrix}$, $B = \begin{bmatrix} -6 & -3 \\ -2 & 4 \end{bmatrix}$

60. $A = \begin{bmatrix} 4 & 6 & 3 \\ 2 & 3 & -1 \\ -1 & 0 & 6 \end{bmatrix}$, $B = \begin{bmatrix} -4 & -6 & -3 \\ -2 & -3 & 1 \\ 1 & 0 & -6 \end{bmatrix}$

Two matrices whose product is the multiplicative identity matrix are said to be *multiplicative inverses. That is, if $A \times B = B \times A = I$, where I is the multiplicative identity matrix, then A and B are multiplicative inverses. In Exercises 61 and 62, determine whether A and B are multiplicative inverses.*

61. $A = \begin{bmatrix} 5 & -2 \\ -2 & 1 \end{bmatrix}$, $B = \begin{bmatrix} 1 & 2 \\ 2 & 5 \end{bmatrix}$

62. $A = \begin{bmatrix} 7 & 3 \\ 2 & 1 \end{bmatrix}$, $B = \begin{bmatrix} 1 & -3 \\ -2 & 7 \end{bmatrix}$

Problem Solving/Group Activities

In Exercises 63 and 64, determine whether the statement is true or false. Give an example to support your answer.

63. $A - B = B - A$, where A and B are any matrices.

64. For scalar a and matrices B and C, $a(B + C) = aB + aC$.

65. The number of hours of labor required to manufacture one boat of various sizes is summarized in matrix L.

$$
\begin{array}{c}
\text{Department} \\
\overbrace{\text{Cutting \quad Assembly \quad Packing}}
\end{array}
$$

$$
L = \begin{bmatrix} 1.4 \text{ hr} & 0.7 \text{ hr} & 0.3 \text{ hr} \\ 1.8 \text{ hr} & 1.4 \text{ hr} & 0.3 \text{ hr} \\ 2.7 \text{ hr} & 2.8 \text{ hr} & 0.5 \text{ hr} \end{bmatrix} \begin{array}{l} \text{Small} \\ \text{Medium} \\ \text{Large} \end{array} \Big\} \begin{array}{l} \text{Boat} \\ \text{size} \end{array}
$$

The hourly labor rates for cutting, assembly, and packing at the Ames City Plant and at the Bay City Plant are given in matrix C.

$$
\begin{array}{c}
\text{Plant} \\
\overbrace{\begin{array}{cc} \text{Ames} & \text{Bay} \\ \text{City} & \text{City} \end{array}}
\end{array}
$$

$$
C = \begin{bmatrix} \$14 & \$12 \\ \$10 & \$9 \\ \$7 & \$5 \end{bmatrix} \begin{array}{l} \text{Cutting} \\ \text{Assembly} \\ \text{Packaging} \end{array} \Big\} \text{Department}
$$

a) What is the total labor cost for manufacturing a small-sized boat at the Ames City plant?

b) What is the total cost for manufacturing a large-sized boat at the Bay City plant?

c) Determine the product $L \times C$ and explain the meaning of the results.

Research Activities

66. Find an article that shows information illustrated in matrix form. Write a short paper explaining how to interpret the information provided by the matrix. Include the article with your report.

67. Do research and write a paper on the development of matrices. In your paper cover the contributions of James Joseph Sylvester, Arthur Cayley, and William Rowan Hamilton. (References include history of mathematics books and encyclopedias.)

7.4 SOLVING SYSTEMS OF EQUATIONS BY USING MATRICES

In Section 7.3 we introduced matrices. Now we will discuss the procedure to solve a system of linear equations using matrices. We will illustrate how to solve a system of two equations and two unknowns. Systems of equations containing three equations and three unknowns (called third-order systems) and higher-order systems can also be solved by using matrices.

The first step in solving a system of equations using matrices is to write an *augmented matrix*. An augmented matrix consists of two smaller matrices—one for the coefficients of the variables in the equations and one for the constants in the equations. To determine the augmented matrix, first write each equation in standard form, $ax + by = c$. For the system of equations on the left, its augmented matrix is shown to its right.

System of Equations

$$a_1 x + b_1 y = c_1$$
$$a_2 x + b_2 y = c_2$$

Augmented Matrix

$$\left[\begin{array}{cc|c} a_1 & b_1 & c_1 \\ a_2 & b_2 & c_2 \end{array}\right]$$

Following is another example.

System of Equations

$$x + 2y = 8$$
$$3x - y = 7$$

Augmented Matrix

$$\begin{bmatrix} 1 & 2 & | & 8 \\ 3 & -1 & | & 7 \end{bmatrix}$$

Note that the bar in the augmented matrix separates the numerical coefficients from the constants. The matrix is just a shortened way of writing the system of equations. Thus we can solve a system of equations by using matrices in a manner very similar to solving a system of equations with the addition method.

To solve a system of equations by using matrices, we use *row transformations* to obtain new matrices that have the same solution as the original system. We will discuss three row transformation procedures.

Procedures for Row Transformations

1. Any two rows of a matrix may be interchanged (this is the same as interchanging any two equations in the system of equations).

2. All the numbers in any row may be multiplied by any nonzero real number. (This is the same as multiplying both sides of an equation by any nonzero real number.)

3. All the numbers in any row may be multiplied by any nonzero real number, and these products may be added to the corresponding numbers in any other row of numbers.

We use row transformations to obtain an augmented matrix whose numbers to the left of the vertical bar are the same as in the *multiplicative identity matrix*. From this type of augmented matrix we can determine the solution to the system of equations. For example, if we get

$$\begin{bmatrix} 1 & 0 & | & 3 \\ 0 & 1 & | & -2 \end{bmatrix},$$

it tells us that $1x = 3$ or $x = 3$, and $1y = -2$ or $y = -2$. Thus the solution to the system of equations that yielded this augmented matrix is $(3, -2)$. Now let's work an example.

EXAMPLE 1

Solve the following system of equations by using matrices.

$$x + 2y = 5$$
$$3x - y = 8$$

Solution: First, we write the augmented matrix.

$$\begin{bmatrix} 1 & 2 & | & 5 \\ 3 & -1 & | & 8 \end{bmatrix}$$

Our goal is to obtain a matrix of the form

$$\begin{bmatrix} 1 & 0 & | & c_1 \\ 0 & 1 & | & c_2 \end{bmatrix},$$

where c_1 and c_2 may represent any real numbers. It is generally easier to work by columns. Therefore we will try to get the first column of the augmented matrix to be $\begin{smallmatrix}1\\0\end{smallmatrix}$ and the second column to be $\begin{smallmatrix}0\\1\end{smallmatrix}$. Since the element in the top left position is already a 1, we must work to change the 3 into a 0. We use row transformation procedure 3 to change the 3 into a 0. If we multiply the top row of numbers by -3 and add these products to the second row of numbers, the element in the first column, second row will become a 0:

$$\begin{bmatrix} 1 & 2 & | & 5 \\ 3 & -1 & | & 8 \end{bmatrix}.$$

The top row of numbers multiplied by -3 gives

$$1(-3), \qquad 2(-3), \quad \text{and} \quad 5(-3).$$

Now add these products to their respective numbers in row 2.

$$\begin{bmatrix} 1 & 2 & | & 5 \\ 3 + 1(-3) & -1 + 2(-3) & | & 8 + 5(-3) \end{bmatrix} = \begin{bmatrix} 1 & 2 & | & 5 \\ 0 & -7 & | & -7 \end{bmatrix}$$

The next step is to obtain a 1 in the second column, second row. At present, -7 is in this position. To change the -7 to a 1, we use row transformation procedure 2. If we multiply -7 by $-\frac{1}{7}$, the product will be 1. Therefore we multiply all the numbers in the second row by $-\frac{1}{7}$ to get

$$\begin{bmatrix} 1 & 2 & | & 5 \\ 0(-\frac{1}{7}) & -7(-\frac{1}{7}) & | & -7(-\frac{1}{7}) \end{bmatrix} = \begin{bmatrix} 1 & 2 & | & 5 \\ 0 & 1 & | & 1 \end{bmatrix}.$$

The next step is to obtain a 0 in the second column, first row. At present a 2 is in this position. Multiplying the numbers in the second row by -2 and adding the products to the corresponding numbers in the first row gives a 0 in the desired position.

$$\begin{bmatrix} 1 + 0(-2) & 2 + 1(-2) & | & 5 + 1(-2) \\ 0 & 1 & | & 1 \end{bmatrix} = \begin{bmatrix} 1 & 0 & | & 3 \\ 0 & 1 & | & 1 \end{bmatrix}$$

We now have the desired augmented matrix:

$$\begin{bmatrix} 1 & 0 & | & 3 \\ 0 & 1 & | & 1 \end{bmatrix}.$$

With this matrix we see that $1x = 3$, or $x = 3$, and $1y = 1$, or $y = 1$. The solution to the system is (3, 1).

Check: $x + 2y = 5$ $3x - y = 8$

$3 + 2(1) = 5$ $3(3) - 1 = 8$

$5 = 5$ True $8 = 8$ True

To Change an Augmented Matrix to the Form $\begin{bmatrix} 1 & 0 & | & c_1 \\ 0 & 1 & | & c_2 \end{bmatrix}$

1. Change the element in the first column, first row, to a 1.
2. Change the element in the first column, second row, to a 0.
3. Change the element in the second column, second row, to a 1.
4. Change the element in the second column, first row, to a 0.

Generally, when changing an element in the augmented matrix to a 1, we use step 2 in the row transformation box. When changing an element to a 0, we use step 3 in the row transformation box.

> **EXAMPLE 2**
> Solve the following system of equations using matrices.
>
> $$2x - 3y = 10$$
> $$2x + 2y = 5$$
>
> *Solution:*   First, write the augmented matrix.
>
> $$\begin{bmatrix} 2 & -3 & | & 10 \\ 2 & 2 & | & 5 \end{bmatrix}$$
>
> To obtain a 1 in the first column, first row, multiply the numbers in the first row by $\frac{1}{2}$.
>
> $$\begin{bmatrix} 1 & -\frac{3}{2} & | & 5 \\ 2 & 2 & | & 5 \end{bmatrix}$$
>
> To get a 0 in the first column, second row, multiply the numbers in the first row by -2, and add the products to the corresponding numbers in the second row.
>
> $$\begin{bmatrix} 1 & -\frac{3}{2} & | & 5 \\ 2 + 1(-2) & 2 + (-\frac{3}{2})(-2) & | & 5 + 5(-2) \end{bmatrix} = \begin{bmatrix} 1 & -\frac{3}{2} & | & 5 \\ 0 & 5 & | & -5 \end{bmatrix}$$
>
> To obtain a 1 in the second column, second row, multiply the numbers in the second row by $\frac{1}{5}$.
>
> $$\begin{bmatrix} 1 & -\frac{3}{2} & | & 5 \\ 0(\frac{1}{5}) & 5(\frac{1}{5}) & | & -5(\frac{1}{5}) \end{bmatrix} = \begin{bmatrix} 1 & -\frac{3}{2} & | & 5 \\ 0 & 1 & | & -1 \end{bmatrix}$$
>
> To obtain a 0 in the second column, first row, multiply the numbers in the second row by $\frac{3}{2}$, and add the products to the corresponding numbers in the first row.
>
> $$\begin{bmatrix} 1 + \frac{3}{2}(0) & -\frac{3}{2} + \frac{3}{2}(1) & | & 5 + (-1)(\frac{3}{2}) \\ 0 & 1 & | & -1 \end{bmatrix} = \begin{bmatrix} 1 & 0 & | & \frac{7}{2} \\ 0 & 1 & | & -1 \end{bmatrix}$$
>
> The solution to the system of equations is $(\frac{7}{2}, -1)$.   ▲

Inconsistent and Dependent Systems

When you solve a system of two equations, obtaining an augmented matrix in which one row of numbers on one side of the vertical line are all zeros but a zero does not appear in the same row on the other side of the vertical line means that the system is inconsistent and has no solution. For example, a system of equations that yields the following augmented matrix is an inconsistent system.

$$\begin{bmatrix} 1 & 2 & | & 5 \\ 0 & 0 & | & 4 \end{bmatrix} \quad \text{Inconsistent system}$$

The second row of the matrix represents the equation

$$0x + 0y = 4,$$

which is never true.

If you obtain a matrix in which a 0 appears across an entire row, the system of equations is dependent. For example, a system of equations that yields the following matrix is a dependent system.

$$\begin{bmatrix} 4 & 5 & | & -6 \\ 0 & 0 & | & 0 \end{bmatrix} \quad \text{Dependent system}$$

The second row of the matrix represents the equation

$$0x + 0y = 0,$$

which is always true.

Triangularization Method

Another procedure to solve a system of two equations is to use row transformation procedures to obtain an augmented matrix of the form

$$\begin{bmatrix} 1 & a & | & b \\ 0 & 1 & | & c \end{bmatrix}$$

where a, b, and c represent real numbers. This procedure is called the **Triangularization Method**, for the ones and zeros form a triangle.

When the matrix is in this form we can write the following system of equations.

$$\begin{array}{ccc} 1x + ay = b & & x + ay = b \\ & \text{or} & \\ 0x + 1y = c & & y = c \end{array}$$

Using substitution we can easily solve the system.

For example, in Example 2 we obtained the augmented matrix

$$\begin{bmatrix} 1 & -\frac{3}{2} & | & 5 \\ 0 & 1 & | & -1 \end{bmatrix}$$

This matrix represents the following system of equations.

$$x - \frac{3}{2}y = 5$$

$$y = -1$$

To solve for x, we substitute -1 for y in the equation

$$x - \frac{3}{2}y = 5.$$

$$x - \frac{3}{2}(-1) = 5$$

$$x = \frac{7}{2}$$

Thus the solution to the system is $(\frac{7}{2}, -1)$, as was obtained in Example 2. You may use either method when solving a system of equations with matrices unless your instructor specifies otherwise.

Section 7.4 Exercises

1. What is an augmented matrix?

2. In your own words, write the three row transformation procedures.

In Exercises 3–16, use matrices to solve the system of equations.

3. $x + y = 1$
$2x - y = 5$

4. $x - y = 1$
$2x - y = 3$

5. $x + 2y = -4$
$2x - y = -3$

6. $x - y = 3$
$2x + 5y = -1$

7. $2x - 5y = -6$
$-4x + 10y = 12$

8. $x + y = 5$
$3x - y = 3$

9. $x + 3y = 1$
$-2x + y = 5$

10. $2x + 4y = 6$
$4x - 2y = -8$

11. $2x - 4y = 0$
$x - 3y = -1$

12. $4x + 2y = 6$
$5x + 4y = 9$

13. $-3x + 6y = 5$
$2x - 4y = 8$

14. $2x - 5y = 10$
$3x + y = 15$

15. $2x + y = 11$
$x + 3y = 18$

16. $4x - 3y = 7$
$-2x + 5y = 14$

In Exercises 17–20, use matrices to solve the problem.

17. If Max buys 2 lb of chocolate-covered cherries and 3 lb of chocolate-covered mints, his total cost is $23. If he buys 1 lb of chocolate-covered cherries and 2 lb of chocolate-covered mints, his total cost is $14. Find the cost of 1 lb of chocolate-covered cherries and 1 lb of chocolate-covered mints.

18. The length of a rectangle is 3 ft greater than its width. If its perimeter is 14 ft, find the length and width.

19. The Chemistry Club and the German Club had their meetings at the local pizza parlor. The Chemistry Club had three large pizzas and four pitchers of beer for a cost of $61.00. The German Club had two large pizzas and three pitchers of beer for a cost of $42.00. Find the cost of a pizza and the cost of a pitcher of beer.

20. Peoplepower, Inc., a daily employment agency, charges $10 per hour for a truck driver and $8.00 per hour for a laborer. On a certain job the laborer worked two more hours than the truck driver, and together they cost $144. How many hours did each work?

Problem Solving/Group Activity

21. Pencil World sells two types of mechanical pencils to stationery stores. The nonrefillable pencil that contains 6 pieces of lead sells for $1.50 each, and the refillable pencil sells for $2.00 each. Pencil World received an order for 200 pencils and a check for $337.50 for the pencils. When placing the order, the stationery store clerk failed to specify the number of each type of pencil being ordered. Can Pencil World fill the order with the information given? If so, determine the number of nonrefillable and the number of refillable pencils the clerk ordered.

7.5 SYSTEMS OF LINEAR INEQUALITIES

In earlier sections we showed how to find the solution to a system of linear equations in two variables. Now we are going to explore the techniques of finding the solution set to a system of linear inequalities in two variables.

The solution set of a system of linear inequalities is the set of points that satisfy all inequalities in the system. The solution set of a system of linear inequalities may consist of infinitely many ordered pairs. To determine the solution set to a system of linear inequalities, graph each inequality on the same axes. The ordered pairs common to all of the inequalities are the solution set to the system.

Procedure for Solving a System of Linear Inequalities

1. Select one of the inequalities. Replace the inequality symbol with an equality symbol and draw the graph of the equation. Draw the graph with a dashed line if the inequality is $<$ or $>$ and with a solid line if the inequality is $\leq$ or $\geq$.

2. Select a test point on one side of the line and determine whether the point is a solution to the inequality. If so, shade the area on the side of the line containing the point. If the point is not a solution shade the area on the other side of the line.

3. Repeat steps 1 and 2 for the other inequality.

4. The intersection of the two shaded areas and any solid line common to both inequalities form the solution set to the system of inequalities.

EXAMPLE I

Graph the following system of inequalities and indicate the solution set.

$$x + y < 1$$

$$x - y < 5$$

Solution: Graph both inequalities on the same axes. First, draw the graph of $x + y < 1$. When drawing the graph, remember to use a dashed line, since the inequality is "less than" (see Fig. 7.10a). If you have forgotten how to graph inequalities, review Section 6.8.

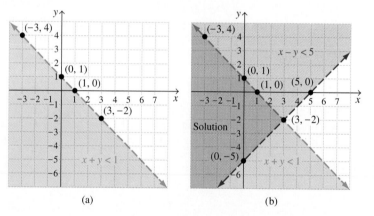

(a)   (b)

Figure 7.10

Now, on the same axes, shade the half-plane determined by the inequality $x - y < 5$ (see Fig. 7.10b). The solution set consists of all the points common to the two shaded half-planes. These are the points in the region on the graph containing both color shadings. Figure 7.10(b) shows that the two lines intersect at $(3, -2)$. This ordered pair can also be found by any of the algebraic methods discussed in Sections 7.2 and 7.3.   ▲

EXAMPLE 2

Graph the following system of inequalities and indicate the solution set.

$$2x + 3y \geq 4$$

$$2x - y > -6$$

Solution: Graph the inequality $2x + 3y \geq 4$. Remember to use a solid line because the inequality is "greater than or equal to" (see Fig. 7.11a).

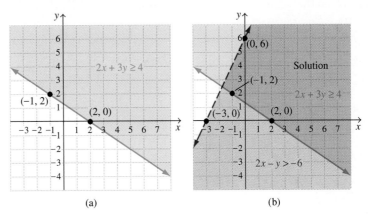

(a)   (b)

Figure 7.11

On the same set of axes, draw the graph of $2x - y > -6$. Use a dashed line, since the inequality is "greater than" (see Fig. 7.11b). The solution is the region of the graph that contains both color shadings and the part of the solid line that satisfies the inequality $2x - y > -6$. Note that the point of intersection of the two lines is not a part of the solution set.   ▲

EXAMPLE 3

Graph the following system of inequalities and indicate the solution set.

$$x \leq 5$$

$$y > -1$$

Solution: Graph the inequality $x \leq 5$ (Fig. 7.12a). On the same axes, graph $y > -1$ (Fig. 7.12b). The solution set is that region of the graph that is shaded

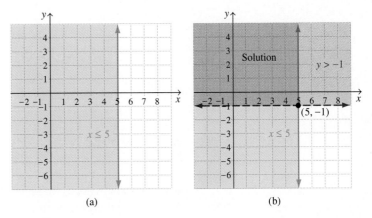

(a)   (b)

Figure 7.12

in both colors and the part of the solid line that satisfies the inequality $y > -1$. The point of intersection of the two lines, $(5, -1)$, is not part of the solution because it does not satisfy the inequality $y > -1$.   ▲

Section 7.5 Exercises

1. What is the solution set of a system of linear inequalities?

2. In your own words give the four-step procedure for solving a system of linear inequalities.

In Exercises 3–18, graph the system of linear inequalities and indicate the solution set.

3. $y - 2x > 0$
   $y + x > 3$

4. $y \leq 3x - 1$
   $y > -2x + 1$

5. $x + y < 3$
   $x - y < 7$

6. $3x + 2y > 6$
   $x + 2y < 4$

7. $x - y < 4$
   $x + y < 5$

8. $3x - y \leq 6$
   $x - y > 4$

9. $x - 3y \leq 3$
   $x + 2y \geq 4$

10. $x + 2y \geq 4$
    $3x - y \geq -6$

11. $y \leq 3x$
    $x \geq 3y$

12. $y \leq 4$
    $x - y < 1$

13. $x \geq 1$
    $y \leq 1$

14. $x \leq 0$
    $y \leq 0$

15. $3x \leq 5y + 15$
    $x \geq y - 5$

16. $3x - 4y < 8$
    $x > 3y - 1$

17. $3x \geq 2y + 6$
    $x \leq y + 7$

18. $2x - 5y < 7$
    $x > 2y + 1$

Problem Solving/Group Activities

19. Reuben Ricardo's on a diet. He must consume less than 500 calories at a meal that consists of a serving of chicken and a serving of rice. The meal must contain at least 150 calories from each source.
    a) Translate the problem into a system of linear inequalities.
    b) Solve the system graphically. Graph calories from chicken on the horizontal axis and calories from rice on the vertical axis.
    c) There are about 200 calories in 8 oz of rice and about 180 calories in 3 oz of chicken. Select a point in the solution set in part (b). This point represents the number of ounces of chicken and rice to be served at a meal. For the point selected, determine the number of ounces of chicken and rice to be served.

20. Write a system of linear inequalities whose solution is the second quadrant, including the axes.

21. a) Do all systems of linear inequalities have solutions? Explain.
    b) Write a system of inequalities that has no solution.

22. Can a system of linear inequalities have a solution set consisting of a single point? Explain.

23. Can a system of linear inequalities have as its solution set all the points on the coordinate plane? Explain your answer, giving an example to support it.

7.6 LINEAR PROGRAMMING

Government, business, and industry often require decision makers to find cost-effective solutions to a variety of problems. Linear programming often serves as a method of expressing the relationships in many of these problems and utilizes a system of linear inequalities.

The typical linear programming problem has many variables and is generally so lengthy that it is solved on a computer by a technique called the simplex algorithm. The simplex method was developed in the 1940s by George B. Dantzig. Linear programming is used to solve problems in the social sciences, health care, land development, nutrition, and many other fields.

We will not discuss the simplex method in this textbook. We will merely give a brief introduction to how linear programming works. You can find a detailed explanation in books on finite mathematics.

In a linear programming problem, there are restrictions called constraints. Each constraint is represented as a linear inequality. The list of constraints forms a system of linear inequalities. When the system of inequalities is graphed, we often obtain a region bounded on all sides by line segments (Fig. 7.13). Since the boundaries of the region form a polygon, the region is called a polygonal region.

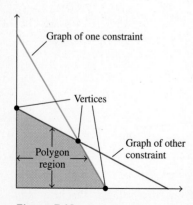

Graph of one constraint

Vertices

Graph of other constraint

Polygon region

Figure 7.13

The points where two or more boundaries intersect are called the **vertices** of the polygon. The points on the boundary of the region and the points inside the polygonal region are the solution set for the system of inequalities.

For each linear programming problem we will obtain a formula of the form $K = Ax + By$. In this formula, K (or some other variable) will be the quantity we want to maximize or minimize. The values we substitute for x and y determine the value of K. From the information given in the problem we determine the real number constants A and B. A typical equation that might be used to find the maximum profit, P, in a particular linear programming problem is $P = 3x + 7y$. We would find the maximum profit by substituting the ordered pairs (x, y) of the vertices of the polygonal region into the formula to see which ordered pair yields the greatest value of P and therefore the maximum profit. The ordered pair that yields the smallest value of P determines the minimum profit.

Linear programming is used to determine which ordered pair will yield the maximum (or minimum) value of the variable that is being maximized (or minimized). The fundamental principle of linear programming provides a rule for finding those maximum and minimum values.

Fundamental Principle of Linear Programming

If a linear equation of the form $K = Ax + By$ is evaluated at each point in a closed polygonal region, the maximum and minimum values of the equation occur at vertices of the region.

$\mathcal{D}$ID YOU KNOW

THE LOGISTICS OF D-DAY

$\mathcal{L}$inear programming was first used to deal with the age-old military problem of logistics: obtaining, maintaining, and transporting military equipment and personnel. George Dantzig developed the simplex method for the Allies of World War II to do just that. Consider the logistics of the Allied invasion of Normandy. Meteorologic experts had settled on three possible dates in June 1944. It had to be a

Winston Churchill called the invasion of Normandy "the most difficult and complicated operation that has ever taken place."

day when low tide and first light would coincide, the winds should not exceed 8–13 mph, and visibility had to be not less than 3 miles. A force of 170,000 assault troops were to be assembled and moved to 22 airfields in England where 1200 air transports and 700 gliders would then take them to the coast of France to converge with 5000 ships of the D-Day armada. The code name for the invasion was Operation Overlord, but it is known to most as D-Day. ■

Example 1 illustrates how the fundamental principle is used to solve a linear programming problem.

EXAMPLE I

The Ric Shaw Chair company makes two types of rocking chairs: a plain chair and a fancy chair. Each rocking chair must be assembled and then finished. The plain chair takes 4 hours to assemble and 4 hours to finish. The fancy chair takes 8 hours to assemble and 12 hours to finish. The company can provide at most 160 worker-hours of assembling and 180 worker-hours of finishing a day. If the profit on the plain chair is $10.00 and the profit on the fancy chair is $18.00, how many rocking chairs of each type should the company make per day to maximize profits? What is the maximum profit?

Solution: From the information given, we know the following facts.

	Assembly Time (hr)	Finishing Time (hr)	Profit ($)
Plain chair	4	4	10.00
Fancy chair	8	12	18.00

Let

$$x = \text{the number of plain chairs.}$$

$$y = \text{the number of fancy chairs.}$$

$$10x = \text{profit on the plain chairs.}$$

$$18y = \text{profit on the fancy chairs.}$$

$$P = \text{the total profit.}$$

The total profit is the sum of the profit on the plain chairs and the profit on the fancy chairs. Since $10x$ is the profit on the plain chairs and $18y$ is the profit on the fancy chairs, the profit formula is $P = 10x + 18y$.

The maximum profit, P, is dependent on several conditions, called *constraints*. The number of chairs manufactured each day cannot be a negative amount. This condition gives us the constraints $x \geq 0$ and $y \geq 0$. Another constraint is determined by the total number of hours allocated for assembling. Four hours are needed to assemble the plain chair, so the total number of hours per day to assemble x plain chairs is $4x$. Eight hours are required to assemble a fancy chair, so the total number of hours needed to assemble y fancy chairs is $8y$. The maximum number of hours allocated for assembling is 160. Thus the third constraint is $4x + 8y \leq 160$. The final constraint is determined by the number of hours allotted for finishing. Finishing a plain chair takes 4 hours, or $4x$ hours to finish x plain chairs. Finishing a fancy chair takes 12 hours, or $12y$ hours to finish y fancy chairs. The total number of hours allotted for finishing is 180. Therefore the fourth constraint is $4x + 12y \leq 180$. Thus the four constraints are

$$x \geq 0$$

$$y \geq 0$$

$$4x + 8y \leq 160$$

$$4x + 12y \leq 180.$$

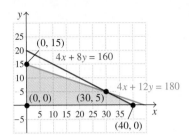

Figure 7.14

The list of constraints is a system of linear inequalities in two variables. The solution to the system of inequalities is the set of ordered pairs that satisfies all the constraints. These points are plotted in Fig. 7.14. Note that the solution to the system consists of the colored region and the solid boundaries. The points (0, 0), (0, 15), (30, 5), and (40, 0) are the points at which the boundaries intersect. These points can also be found by the substitution method described in Section 7.2.

The object in this example is to maximize the profit. The total profit is given by the formula $P = 10x + 18y$. According to the fundamental principle, the maximum profit will be found at one of the vertices of the polygonal region.

Calculate P for each one of the vertices.

$$P = 10x + 18y$$

At (0, 0), $P = 10(0) + 18(0) = 0$

At (40, 0), $P = 10(40) + 18(0) = 400$

At (30, 5), $P = 10(30) + 18(5) = 390$

At (0, 15), $P = 10(0) + 18(15) = 270$

The maximum profit is at (40, 0), which means that the company should manufacture 40 plain rocking chairs and no fancy rocking chairs. The maximum profit would be $400. The minimum profit would be at (0, 0), when no rocking chairs of either style were manufactured. ▲

A variation of the original problem could be that the company knows that it cannot sell more than 15 plain rocking chairs per day. With this additional constraint, we now have the following set of constraints.

$$x \geq 0$$

$$x \leq 15$$

$$y \geq 0$$

$$4x + 8y \leq 160$$

$$4x + 12y \leq 180$$

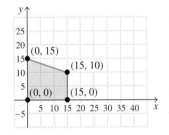

Figure 7.15

The graph of these constraints is shown in Fig. 7.15.

The vertices of the polygonal region are (0, 0), (0, 15), (15, 10), and (15, 0). To determine the maximum profit, we calculate P for each of these vertices:

$$P = 10x + 18y$$

At (0, 0), $P = 10(0) + 18(0) = 0$

At (0, 15), $P = 10(0) + 18(15) = 270$

At (15, 10), $P = 10(15) + 18(10) = 330$

At (15, 0), $P = 10(15) + 18(0) = 150.$

This set of constraints gives the maximum profit of $330.00 when the company manufactures 15 plain rocking chairs and 10 fancy rocking chairs.

Section 7.6 Exercises

1. What are constraints in a linear programming problem? How are they represented?

2. In a linear programming problem how is a polygonal region formed?

3. What are the points of intersection of the boundaries of the polygonal region called?

4. In your own words, state the fundamental principle of linear programming.

In Exercises 5–10, a set of constraints and a profit formula are given.

    a) *Draw the graph of the constraints, and find the points of intersection of the boundaries.*

    b) *Use these points of intersection to determine the maximum and minimum profit.*

5. $x + y \leq 6$
$x + 2y \leq 10$
$x \geq 0$
$y \geq 0$
$P = 3x + 4y$

6. $x + y \leq 4$
$x + 2y \geq 2$
$x \geq 0$
$y \geq 0$
$P = 2x + 3y$

7. $x + y \leq 4$
$x + 3y \leq 6$
$x \geq 0$
$y \geq 0$
$P = 6x + 7y$

8. $x + y \leq 50$
$x + 3y \leq 90$
$x \geq 0$
$y \geq 0$
$P = 20x + 40y$

9. $4x + 3y \geq 12$
$3x + 4y \leq 36$
$x \geq 2$
$y \leq 5$
$y \geq 1$
$P = 2.20x + 1.65y$

10. $x + 2y \leq 14$
$7x + 4y \geq 28$
$x \geq 2$
$x \leq 10$
$y \geq 1$
$P = 15.13x + 9.35y$

11. The P³ Company manufactures two types of power lawn mowers. Type *A* is self-propelled, and type *B* must be pushed by the operator. The company can produce a maximum of 18 mowers per week. It can make a profit of $20 on mower *A* and a profit of $25 on mower *B*. The company's planners want to make at least 2 mowers of type *A* but not more than 5. To keep certain customers happy, they must make at least 2 mowers of type *B*.
    a) List the constraints.
    b) List the profit formula.
    c) Graph the set of constraints.
    d) Find the vertices of the polygonal region.
    e) How many of each type should be made to maximize the profit?
    f) Find the maximum profit.

12. A craftsman makes wooden toy cars and trucks. He wants to make at least 2 cars and 2 trucks but no more than 5 trucks. He can make a total of 15 toys in his spare time in a month. At craft fairs he makes a profit of $8.00 per car and $10.00 per truck sold.
    a) List the constraints.
    b) List the profit formula.
    c) Graph the set of constraints.
    d) Find the vertices of the polygonal region.
    e) How many of each type should he make to maximize his profit?
    f) Find the maximum profit.

13. A liquid portion of a diet is to provide at least 300 calories, 80 units of vitamin A, and 90 units of vitamin C daily. A cup (8 oz) of dietary drink Trimfit provides 60 calories, 8 units of vitamin A, and 6 units of vitamin C. A cup (8 oz) of dietary drink Usave provides 50 calories, 20 units of vitamin A, and 30 units of vitamin C. Assume that Trimfit dietary drink costs $0.25 per cup and that Usave dietary drink costs $0.32 per cup.
    a) List the constraints.
    b) List the cost formula.
    c) Graph the set of constraints. Place cups of Trimfit on the horizontal axis and cups of Usave on the vertical axis.
    d) Find the vertices of the polygonal region.
    e) How many cups of each drink should a person consume each day to minimize the cost and still meet the stated daily requirements?
    f) Find the minimum cost.

Problem Solving/Group Activities

14. To make one package of all-beef hot dogs, a manufacturer uses 1 lb of beef; to make one package of regular hot dogs, the manufacturer uses $\frac{1}{2}$ lb each of beef and pork. The profit on regular hot dogs is 30 cents per pack and for the all-beef hot dogs is 40 cents per pack. If there are 200 lb of beef and 150 lb of pork available, how many packs of all-beef and regular hot dogs should the manufacturer make to maximize the profit? What is the profit?

15. Ross's engine reconditioning company works on 4- and 6-cylinder automobile engines. Each 4-cylinder engine requires 1 hr for cleaning, 5 hr for overhauling, and 3 hr for testing. Each 6-cylinder engine requires 1 hr for cleaning, 10 hr for overhauling, and 2 hr for testing. The cleaning station is available for at most 9 hr per week, the overhauling equipment is available for at most 80 hr per week, and the testing equipment is available for at most 24 hr per week. For each reconditioned 4-cylinder engine, the company makes a profit of $150, and for each 6-cylinder engine it makes a profit of $250. The company can sell all the

reconditioned engines it produces. How many of each type should be reconditioned to maximize profit? What is the maximum profit?

Research Activity

16. *Operations research* draws on several disciplines, including mathematics, probability theory, statistics, and economics. George Dantzig was one of the key people in developing operations research. Write a paper on Dantzig and his contributions to operations research and linear programming. (Reference: *College Mathematics Journal*, September 1986. Your instructor or a statistics or economics instructor may be able to provide additional sources.)

CHAPTER 7 SUMMARY

Key Terms

7.1

break-even analysis
break-even point
consistent system of equations
dependent system of equations
inconsistent system of equations
simultaneous linear equations
system of linear equations

7.2

addition (or elimination) method
equilibrium price
law of supply and demand
market equilibrium
substitution method

7.3

additive identity matrix
matrix
multiplicative identity matrix
scalar
square matrix

7.4

augmented matrix

7.6

constraints
polygonal region
vertices

Important Facts

Systems of Equations

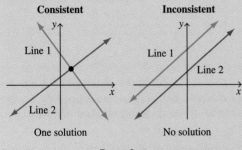

Consistent

Line 1
Line 2

One solution

Inconsistent

Line 1
Line 2

No solution

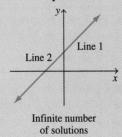

Dependent

Line 1
Line 2

Infinite number
of solutions

Methods of Solving Systems of Equations

1. Graphing
2. Substitution
3. Addition (or elimination) method

Additive Identity Matrix

$$\begin{bmatrix} 0 & 0 \\ 0 & 0 \end{bmatrix}$$

Multiplicative Identity Matrix

$$\begin{bmatrix} 1 & 0 \\ 0 & 1 \end{bmatrix}$$

Fundamental Principle of Linear Programming

If a linear equation of the form $K = Ax + By$ is evaluated at each point in a closed polygonal region, the maximum and minimum values of the equation occur at vertices of the region.

Chapter 7 Review Exercises

7.1

In Exercises 1–4, solve the system of equations graphically. If the system does not have a single ordered pair as a solution, state whether the system is inconsistent or dependent.

1. $x = 3$
$y = 7$

2. $2x + y = 1$
$-2x + y = 1$

3. $x = 3$
$x + y = 5$

4. $x + 2y = 5$
$2x + 4y = 4$

In Exercises 5–8, determine without graphing whether the system of equations has exactly one solution, no solution, or an infinite number of solutions.

5. $y = \frac{2}{3}x + 5$
$3y - 2x = 15$

6. $3y - 2x = 15$
$3y = 2x - 18$

7. $4y - x = 12$
$3y + x = 15$

8. $2x - 4y = 8$
$-2x + y = 6$

7.2

In Exercises 9–12, solve the system of equations by the substitution method. If the system does not have a single ordered pair as a solution, state whether the system is inconsistent or dependent.

9. $x - 3y = 2$
$2x - y = 9$

10. $x + 2y = 4$
$y = 3x + 2$

11. $2x - y = 4$
$3x - y = 2$

12. $3x + y = 1$
$3y = -9x - 4$

In Exercises 13–18, solve the system of equations by the addition method. If the system does not have a single ordered pair as a solution, state whether the system is inconsistent or dependent.

13. $-x + y = 12$
$x + 2y = -3$

14. $2x + y = 2$
$-3x - y = 5$

15. $2x - 3y = 1$
$x + 2y = -3$

16. $3x + 4y = 6$
$2x - 3y = 4$

17. $3x + 5y = 15$
$2x + 4y = 0$

18. $3x + y = 6$
$-6x - 2y = -12$

7.3

$Given\ A = \begin{bmatrix} -2 & 3 \\ 4 & -5 \end{bmatrix}\ and\ B = \begin{bmatrix} 1 & -6 \\ -7 & 8 \end{bmatrix},\ find$

19. $A + B$.

20. $A - B$.

21. $3A$.

22. $2A - 3B$.

23. $A \times B$.

24. $B \times A$.

7.4

In Exercises 25–32, use matrices to solve the system of equations.

25. $x + 2y = 4$
$x + y = 2$

26. $-x + y = 4$
$x + 2y = 2$

27. $2x + y = 4$
$4x - y = 2$

28. $2x + 3y = 2$
$4x - 9y = 4$

29. $x + 3y = 3$
$3x - 2y = 2$

30. $3x - 6y = 3$
$4x + 5y = 17$

7.1–7.4

31. The MacPhersons plan to install a security system in their home. They have narrowed down their choices to two security dealers: Moneywell and Doile security systems. Moneywell's system costs $3660 to install, and its monitoring fee is $17 per month. Doile's equivalent system costs only $2560 to install, but its monitoring fee is $28 per month.

   a) Assuming that the monthly monitoring fees do not change, in how many months would the total cost of Moneywell's and Doile's system be the same?

   b) If both dealers guarantee not to raise monthly fees for 10 years and the MacPhersons plan to use the system for 10 years, which system would be the least expensive?

32. More than half of all the water used in the United States is used in livestock production. The number of gallons of water used to produce one pound of meat is 1250 gallons, or more than 50 times the number of gallons of water needed to produce one pound of wheat. To produce both one pound of meat and one pound of wheat requires 2524 gallons of water. Determine the number of gallons of water needed to produce one pound of meat and the number of gallons of water needed to produce one pound of wheat.

33. Steve Trinter, an electronics salesman, earns a weekly salary plus a commission on sales. In one week his salary on sales of $4000 was $660. The next week his salary on sales of $6000 was $740. Find his weekly salary and his commission rate.

34. Cheryl has a total of 60 coins in her piggy bank. If the coins consist of only nickels and dimes and the total value of the coins is $5.75, what is the number of nickels and the number of dimes?

35. Emily needs to purchase a new air conditioner for the office. Model 1600A costs $950 to purchase and $32 per month to operate. Model 6070B, a more efficient unit, costs $1275 to purchase and $22 per month to operate.

a) After how many months will the total cost of both units be equal?

b) Which model will be the more cost effective if the life of both units is guaranteed for 10 years?

7.5

In Exercises 37–40, graph the system of linear inequalities and indicate the solution set.

36. $2x + y < 8$
$y \geq 2x - 1$

37. $y < -2x + 3$
$y \geq 4x - 2$

38. $x + 3y \leq 6$
$2x - 7y \geq 14$

39. $x - y > 5$
$6x + 5y \leq 30$

7.6

40. For the following set of constraints and profit formula, graph the constraints and find the points of intersection of the boundaries. Use these points of intersection to determine the maximum profit.

$$x + y \leq 10$$
$$2x + 1.8y \leq 18$$
$$x \geq 0$$
$$y \geq 0$$
$$P = 6x + 5y$$

CHAPTER 7 TEST

1. From a graph explain how you would identify a consistent system of equations, an inconsistent system of equations, and a dependent system of equations.

2. Solve the system of equations graphically.
$$y = 2x + 5$$
$$-2x + 3y = 3$$

3. Determine without graphing whether the system of equations has exactly one solution, no solution, or an infinite number of solutions.
$$4x + 5y = 6$$
$$-3x + 5y = 13$$

Solve the system of equations by the method indicated.

4. $3x + 5y = 3$
   $x - y = -7$
   (substitution)

5. $y = 4x + 6$
   $y = 2x + 18$
   (substitution)

6. $x - y = 4$
   $2x + y = 5$
   (addition)

7. $4x + 3y = 5$
   $2x + 4y = 10$
   (addition)

8. $2x + 3y = 4$
   $6x + 4y = 7$
   (addition)

9. $x + 3y = 4$
   $5x + 7y = 4$
   (matrices)

For $A = \begin{bmatrix} 6 & -2 \\ 3 & -5 \end{bmatrix}$ and $B = \begin{bmatrix} 1 & -5 \\ 4 & 2 \end{bmatrix}$, *find*

10. $A + B$.

11. $3A - B$.

12. $A \times B$.

13. Graph the system of linear inequalities and indicate the solution set.
$$y \le -2x + 4$$
$$2x - y > 8$$

Solve Exercises 14 and 15 by using a system of equations.

14. Ms. Chardoney, a grocer, plans to mix candy that sells for $5.40 per pound with candy that sells for $8.40 per pound to get 10 lb of a mixture to sell for $6.00 per pound. How many pounds of each type of candy should she mix to obtain the desired mixture?

15. The Crossroads apartment building has 20 rental units with one- and two-bedroom units. The rental price for a one-bedroom unit is $350 per month and for a two-bedroom unit is $425 per month. If the total rental income from all units is $7600 per month, determine the number of each type of rental unit in the building.

16. The set of constraints and profit formula for a linear programing problem are:
$$x + 3y \le 6$$
$$4x + 3y \le 15$$
$$x \ge 0$$
$$y \ge 0$$
$$p = 2x + 3y$$

   a) Draw the graph of the constraints and determine the vertices of the polygonal region.
   b) Use the vertices to determine the maximum and minimum profit.

GROUP PROJECTS

Supply and Demand Curves

1. Supply and Demand curves are discussed in Section 7.2. Recall that economists plot price on the vertical axis and supply and demand on the horizontal axis when graphing supply and demand curves.

   The distributor Fast Julie supplied the following information on price, supply, and demand for a new computer game. The supply represents the number of games the seller is willing and able to sell at a particular time and price. The demand refers to the number of games the buyers are willing and able to buy at a particular time and price. The graph relating price and supply is called the *supply curve*. The graph relating price and demand is called the *demand curve*.

Price of Game (dollars)	Supply of Game (units)	Demand of Game (units)
$30	150	500
40	250	400
60	450	200

   a) Use the data provided in the table to construct supply and demand curves on the same axes. Remember to place price on the vertical axis and quantity (supply and demand) on the horizontal axis.

   b) From the graph, estimate the price and the quantity at which the two curves intersect.

   Using the data in the table and the point–slope form of a linear equation discussed in Group Project Exercise 3 at the end of Chapter 6, we can determine the following supply and demand equations.

   $$S = 10p - 150$$
   $$D = -10p + 800$$

   c) Algebraically determine the value of the price, p, when supply equals demand.

   d) Algebraically determine the quantity when supply equals demand.

   e) Based on your results in parts (c) and (d), does your answer in part (b) appear to be correct?

Linear Programming

2. The Bookholder Company manufactures two types of bookcases out of oak and walnut. Model 01 requires 5 board feet of oak and 2 board feet of walnut. Model 02 requires 4 board feet of oak and 3 board feet of walnut. A profit of $75 is made on each Model 01 bookcase, and a profit of $125 is made on each Model 02 bookcase. The company has a supply of 1000 board feet of oak and 600 board feet of walnut. The company has orders for 40 Model 01 bookcases and 50 Model 02 bookcases. These orders indicate the minimum number the company must manufacture of each model.

   a) Write the set of constraints.

   b) Write the profit formula.

   c) Graph the set of constraints.

   d) Determine the number of bookcases of each type the company should manufacture in order to maximize profits.

   e) Determine the maximum profit.

Create Your Own Word Problem

3. a) Write a word problem that can be solved by using a system of two equations with two unknowns.

   b) For the problem in part (a), write the system of equations and find the answer.

   c) Explain how you developed the problem in part (a).

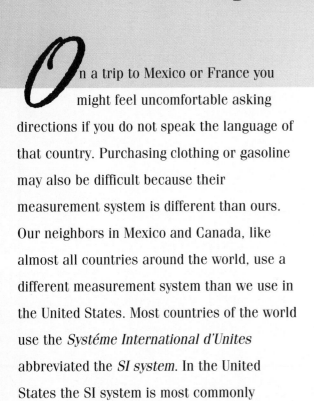

Chapter **8** **The Metric System**

Evidence of metric measurements are all around us. The numbers illustrated on the tires shown represent metric measurements. And the bottle of soda is a liter bottle, which is slightly larger than a quart. If you are working in any industry that exports products you may need to learn to use the metric system.

On a trip to Mexico or France you might feel uncomfortable asking directions if you do not speak the language of that country. Purchasing clothing or gasoline may also be difficult because their measurement system is different than ours. Our neighbors in Mexico and Canada, like almost all countries around the world, use a different measurement system than we use in the United States. Most countries of the world use the *Systéme International d'Unites* abbreviated the *SI system*. In the United States the SI system is most commonly referred to as the metric system.

The metric system originated in France during the French Revolution because of dissatisfaction with the lack of standards of measurement in Europe. Most countries have successfully made the change to the metric system, but the United States continues to have a difficult time doing so, even though it has tried on numerous occasions. In 1866, Congress passed the following law: "It shall be lawful throughout the United States of America to employ the weights and measures of the metric system; and no contract or dealing, or pleading in any court, shall be deemed invalid or liable to objection because the weights or measures expressed or referred to therein are weights or measures of the metric system."

The system of measurement most commonly used in the United States is called the U.S. customary system. However, metric units are used in many ways in the United States. You can purchase metric tools at your local hardware store. Some clothing is measured in metric units. Are your car's tires

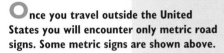

Once you travel outside the United States you will encounter only metric road signs. Some metric signs are shown above.

Years ago in the United States, automobile engines were advertised in horsepower. Now, if you look at the sales stickers of new automobiles you will find that they are measured in liters.

The 1996 Ford Mustang has a standard engine of 3.6 liters, although you can select the optional 4.6 liter engine. The 1996 Chevy Corvette has a standard 5.7 liter engine. The 1996 Dodge Caravan has a standard 2.4 liter engine but you can purchase an optional 3.0, 3.3, or 3.8 liter engine.

size 227-75? These numbers represent metric measurements. Even some road signs now give metric measurements.

Why has the United States had such a difficult time converting to the international standard of measurement? Is it because of inertia or the fear of trying something new or the cost? Some English-speaking countries adopted the SI system in the 1960s and 1970s. The United States prepared to do so in 1975 with the Metric Conversion Act. Citizens objected to the act, and Congress made the conversion voluntary, rendering it ineffective.

As you study this chapter you will see the many advantages of the metric system. You eventually may even support the movement to change from the U.S. customary system to the metric system. ■

As soon as you cross either of our borders to the north or to the south, you will find that the metric system is used. Gas is measured in liters, letters are weighed in grams, and dresses and pants are measured in centimeters. Learning the metric system will be helpful if you plan to visit our neighbors to the north or south, or across the oceans.

8.1   BASIC TERMS AND CONVERSIONS WITHIN THE METRIC SYSTEM

Most countries of the world use the Systéme International d'Unites or SI system. The SI system is generally referred to as the metric system in the United States. The metric system was named for the Greek word *metron*, meaning ''measure.'' The standard units in the metric system have gone through many changes since the system was first developed in France during the French Revolution. For example, one unit of measure, the meter, was first defined as one ten-millionth of the distance between the North Pole and the Equator. Later the meter was defined as 1,650,763.73 wavelengths of the orange–red line of krypton-86. Since 1893, the meter has been defined as the distance traveled by light in a vacuum in $\frac{1}{299,792,458}$ of a second. In this chapter we will discuss metric measurements of length, area, volume, mass, and temperature.

Two systems of weights and measures exist side by side in the United States today, the U.S. customary system and the SI system. The SI system is used predominantly in the automotive, construction, farm equipment, computer, bottling industries and in health-related professions. Many other industries are now in the process of converting to the SI system.

Using the metric system has many advantages. Some of them are summarized here.

1. The metric system is the worldwide accepted standard measurement system. All industrial nations that trade internationally, except the United States, use the metric system as the official system of measurement.

2. There is only one basic unit of measurement for each physical quantity. In the customary system many units are often used to represent the same physical quantity. For example, when discussing length, we use inches, feet, yards, miles, and so on. Converting from one of these units to the other is often a tedious task (consider changing 12 miles to inches). In the metric system we can make many conversions by simply moving the decimal point.

3. The SI system is based on the number 10, and there is little need for fractions, because most quantities can be expressed as decimals.

Basic Terms

Because the official definitions of many metric terms are quite technical, we will present them informally.

The meter (m) is the base unit of *length* in the metric system. The meter is a little more than a yard. A door is about 2 meters high.

The kilogram (kg) is the base unit of *mass*. (The difference between mass and weight will be discussed in Section 8.3.) The kilogram is a little more than 2 pounds. A newborn baby may have a mass of about 3 kilograms. The gram (g), a unit of mass derived from the kilogram, is used to measure small amounts. A nickel has a mass of about 5 grams.

The liter (ℓ) is commonly used to measure *volume*. A liter is a little more than a quart. The gas tank of a compact car may hold 50 liters of gasoline.

Thus

$$1 \text{ m} \approx 1 \text{ yd}$$

$$1 \text{ kg} \approx 2.2 \text{ lb}$$

$$1 \ \ell \approx 1 \text{ qt.}$$

The term degree Celsius (°C) is used to measure temperature. The freezing point of water is 0°C, and the boiling point of water is 100°C. The temperature on a warm day may be 30°C.

$$0°C = 32°F \qquad \text{Water freezes}$$

$$37°C = 98.6°F \qquad \text{Body temperature}$$

$$100°C = 212°F \qquad \text{Water boils}$$

Prefixes

The metric system is based on the number 10 and therefore is a decimal system. Prefixes are used to denote a multiple or part of a base unit. For example, a *deka*meter represents 10 meters, and a *centi*meter represents $\frac{1}{100}$ of a meter. In the metric system, as used outside the United States, groups of three digits in large numbers are separated by a space, not a comma. For example, the number for thirty thousand is 30 000, and the number for nine million is 9 000 000. Groups of three digits to the right of the decimal point are also separated by spaces. Commas are not used in the SI system because many countries use the comma as we use the decimal point. For example, 16 millionths is written 0,000 016 in many countries of the world. We will use the decimal point in this book and write 0.000 016. In this section we will separate groups of three digits using spaces as done outside the United States. Note, however, that the space between groups of three digits is usually omitted if there are only four digits to the left or right of the decimal point. Thus we will write three thousand as 3000 and five ten-thousandths as 0.0005.

Table 8.1 summarizes the more commonly used prefixes and their meanings. These prefixes pertain to all the units of measurement in the metric system. Thus 1 kiloliter = 1000 liters, 1 kilogram = 1000 grams, 1 centimeter = $\frac{1}{100}$ meter, and 1 milliliter = $\frac{1}{1000}$ liter.

The abbreviations or symbols for units of measure are never pluralized, but full names are. For example, 5 milliliters is symbolized as 5 mℓ not 5 mℓs.

Table 8.1

Prefix	Symbol	Meaning
kilo	k	1000 × base unit
hecto	h	100 × base unit
deka	da	10 × base unit
		base unit
deci	d	$\frac{1}{10}$ of base unit
centi	c	$\frac{1}{100}$ of base unit
milli	m	$\frac{1}{1000}$ of base unit

For scientific work, which involves very large and very small quantities, the following prefixes are also used: *mega* (M) is one million times the base unit, *giga* (G) is one billion times the base unit, *tera* (T) is one trillion times the base unit, *micro* (m) is one millionth of the base unit, *nano* (n) is one billionth of the base unit, and *pico* (p) is one trillionth of the base unit.

Conversions within the Metric System

We will use Table 8.2 to help demonstrate how to change from one metric unit to another metric unit (meters to kilometers and so on).

The meters in Table 8.2 can be replaced by grams, liters, or any other base unit of the metric system. Regardless of which unit we choose, the procedure is the same. For purposes of explanation we have used the meter.

Table 8.2

kilometer	hectometer	dekameter	meter	decimeter	centimeter	millimeter
km	hm	dam	m	dm	cm	mm
1000 m	100 m	10 m	1 m	0.1 m	0.01 m	0.001 m

The table shows that 1 hectometer equals 100 meters and 1 millimeter is 0.001 (or $\frac{1}{1000}$) meter. The millimeter is the smallest unit in the table. A centimeter is 10 times as large as a millimeter, a decimeter is 10 times as large as a centimeter, a meter is 10 times as large as a decimeter, and so on. Because each unit is 10 times as large as the unit on its right, converting from one unit to another is simply a matter of multiplying or dividing by powers of 10.

Changing Units within the Metric System

1. To change from a smaller unit to a larger unit (for example from meters to kilometers), move the decimal point in the original quantity one place to the left for each larger unit of measurement until you obtain the desired unit of measurement.

2. To change from a larger unit to a smaller unit (for example from kilometers to meters), move the decimal point in the original quantity one place to the right for each smaller unit of measurement until you obtain the desired unit of measurement.

┌─ **EXAMPLE 1**

a) Convert 403.2 meters to kilometers.

b) Convert 14 grams to centigrams.

c) Convert 0.18 liter to milliliters.

d) Convert 240 dekaliters to kiloliters.

Solution:

a) Table 8.2 shows that dekameters, hectometers, and kilometers are all larger units of measurements than meters. Kilometers are three places to the left of meters in the table. Therefore, to change a measure from meters to kilometers, we must move the decimal point in the number three places to the left, or

$$403.2\text{m} = 0.403\,2 \text{ km}.$$

Note that, since we are changing from a smaller unit of measurement (meter) to a larger unit of measurement (kilometer), the answer will be a smaller number of units.

b) Grams are a larger unit of measurement than centigrams. To convert grams to centigrams, we move the decimal point two places to the right, or

$$14 \text{ g} = 1400 \text{ cg}.$$

Note that, since we are changing from a larger unit of measurement (gram) to a smaller unit of measurement (centigram), the answer will be a larger number of units.

c) $0.18 \ \ell = 180 \ m\ell$

d) $240 \ \text{da}\ell = 2.40 \ \text{k}\ell$ ▲

EXAMPLE 2

a) Convert 305 mm to hm.

b) Convert 6.34 dam to dm.

Solution:

a) Table 8.2 shows that hm is five places to the left of mm. Therefore, to make the conversion, we must move the decimal point in the number five places to the left, or

$$305 \text{ mm} = 0.003\,05 \text{ hm}.$$

b) Table 8.2 shows that dam is two places to the left of dm. Therefore, to make the conversion, we must move the decimal point in the number two places to the right, or

$$6.34 \text{ dam} = 634 \text{ dm}.$$ ▲

EXAMPLE 3

Arrange in order from the smallest to largest length: 3.4 m, 3421 mm, and 104 cm.

Solution: To be compared, these lengths should all be in the same units of measure. Let's convert all the measures to millimeters, the smallest units of the lengths being compared.

$$3.4 \text{ m} = 3400 \text{ mm} \qquad 3421 \text{ mm} \qquad 104 \text{ cm} = 1040 \text{ mm}$$

The lengths arranged in order from smallest to largest are 104 cm, 3.4 m, and 3421 mm. ▲

Section 8.1 Exercises

1. What is the name commonly used for the Systéme International d'Unites in the United States?

2. What is the name of the system of measurement primarily used in the United States today?

3. List three advantages of the metric system.

4. What is the base metric unit commonly used to measure
   a) length?
   b) mass?
   c) volume?
   d) temperature?

5. a) In your own words explain how to convert from one metric unit of length to a different metric unit of length. Then use this procedure to convert
   b) 209 cm to km.
   c) 92.4 hm to dm.

6. What is the name of the prefix that is
   a) a million times the base unit?
   b) one millionth of the base unit?

7. Without referring to any table, name as many of the metric system prefixes as you can and give their meanings. If you don't already know all the prefixes in Table 8.1, memorize them now.

8. a) How many times greater is 1 hectometer than 1 centimeter?
   b) Write 1 hm as centimeters.
   c) Write 1 cm as hectometers.

In Exercises 9–12, fill in the blank.

9. A meter is a little longer than a _____ .

10. A kilogram is a little more than _____ pounds.

11. A nickel has a mass of about _____ grams.

12. The temperature on a warm day may be _____ °C.

In Exercises 13–18, match the prefix with the one letter, a–f, that gives the meaning of the prefix.

13. deci   a) 1000 times base unit

14. milli   b) $\dfrac{1}{100}$ of base unit

15. hecto   c) $\dfrac{1}{1000}$ of base unit

16. kilo   d) 100 times base unit

17. centi   e) $\dfrac{1}{10}$ of base unit

18. deka   f) 10 times base unit

In Exercises 19–28, unscramble the word to make a metric unit of measurement.

19. rteli   20. migradec   21. terem

22. leritililm   23. magr   24. timenceret

25. reketolim   26. raktileed   27. greeed sulesic

28. togmeharc

29. Complete the following.
   a) 1 hectogram = _____ grams
   b) 1 milligram = _____ gram
   c) 1 kilogram = _____ grams
   d) 1 centigram = _____ gram
   e) 1 dekagram = _____ grams
   f) 1 decigram = _____ gram

30. Complete the following.
   a) 1 dekaliter = _____ liters
   b) 1 centiliter = _____ liter
   c) 1 milliliter = _____ liter
   d) 1 deciliter = _____ liter
   e) 1 kiloliter = _____ liters
   f) 1 hectoliter = _____ liters

In Exercises 31–37, without referring to any of the tables or your notes, give the symbol and the equivalent in grams for the unit.

31. milligram   32. centigram   33. decigram

34. gram   35. dekagram   36. hectogram

37. kilogram

In Exercises 38–48, fill in the missing values.

38. 2 m = _____ cm

39. 3 dam = _____ m

40. 15.7 hg = _____ g

41. 0.024 hℓ = _____ ℓ

42. 242.6 cm = _____ hm

43. 1.34 mℓ = _____ ℓ

44. 417 g = _____ kg

45. 14.27 kℓ = _____ ℓ

46. 1.34 hm = _____ cm

47. 0.076 mm = _____ m

48. 0.000 52 kg = _____ cg

In Exercises 49–56, convert the given unit to the unit indicated.

49. 130 cm to hm   50. 8.3 m to cm

51. 1 049 mm to m   52. 94.5 kg to g

53. 3 472 mℓ to dℓ   54. 895 ℓ to mℓ

55. 14 000 mℓ to ℓ   56. 24 dm to km

In Exercises 57–62, arrange the quantities in order from smallest to largest.

57. 64 dm, 67 cm, 680 mm

58. 5.6 dam, 0.47 km, 620 cm

59. 4.3 ℓ, 420 cℓ, 0.045 kℓ

60. 2.2 kg, 2 400 g, 24 300 dg

61. 0.045 kℓ, 460 dℓ, 48 000 cℓ

62. 193 000 mm, 1.6 km, 52.6 hm

63. Cindy ran 100 m, and Chuck ran 100 yd in the same length of time. Who ran faster? Explain.

64. If 5 kg are placed on one side of a balance and a 10-lb weight is placed on the other side, which way would the balance tip? Explain.

65. Would you be walking faster if you walked 1 dam in 10 min or 1 hm in 10 min? Explain.

66. One pump removes 1 daℓ of water in 1 min, and another pump removes 1 dℓ of water in 1 min. Which pump removes water faster? Explain.

67. Charlie cut a piece of lumber 7.8 m in length into six equal pieces. What was the length of each section
   a) in meters?
   b) in centimeters?

68. A baseball diamond is 27 m along each side.
   a) How many meters does a batter run if he hits a home run?
   b) How many kilometers?
   c) How many millimeters?

69. How many centimeters of picture molding should be purchased to frame two pictures? One picture is 33 cm by 440 mm, and the other is 3.4 dm by 44 cm. Molding is sold in centimeter lengths.

70. At 70 cents a meter, what will it cost Karen to sew a lace border around a bedspread that measures 190 cm by 114 cm?

71. The filter pump on an aquarium circulates 360 mℓ of water every minute. If the aquarium holds 30 ℓ of water, how long will it take to circulate all the water?

72. Dale drove 1200 km and used 187 ℓ of gasoline. What was her average rate of gas use for the trip
   a) in kilometers/liter?
   b) in meters/liter?

73. The high school track is 400 m long. If Patty runs around the track 8 times, how many kilometers has she traveled?

74. If a 2 ℓ bottle of soda is divided equally among 6 people, how many milliliters will each person receive?

75. A bottle of soda contains 360 mℓ.
   a) How many milliliters are contained in a six-bottle carton?

b) How many liters does the amount in part (a) equal?
c) At $2.45 for the carton of soda, what is its cost per liter?

76. Mr. Smith ordered a beef roast large enough to serve 12 people. The roast weighs 4.8 kg and costs $6.30/kg.
   a) What is the cost of the roast?
   b) How many grams of meat are available for each person?

Problem Solving/Group Activities

In Exercises 77–79, fill in the blank to make a true statement.

77. 1 gigameter = _____ megameters

78. 1 nanogram = _____ micrograms

79. 1 teraliter = _____ picoliters

The recommended daily amount of calcium for an American adult is 0.8 g. In Exercises 80–84, how much of the food indicated must an adult eat in order to satisfy the entire daily allowance using only that food?

80. Milk—1 cup contains 288 mg calcium.

81. Eggs—1 egg contains 27 mg calcium.

82. Raisin bran—49 g contains 1.6 mg calcium.

83. Cottage cheese—1 cup contains 207 mg calcium.

84. Broccoli—1 cup (cooked) contains 195 mg calcium.

One advantage of the metric system is that by using the proper prefix you can write large and small numbers without large groups of zeros. For example, you can write 3000 m without zeros as 3 km and 0.0003 hm as 3 cm.

   In Exercises 85–90, write an equivalent statement that does not contain any zeros.

85. 5000 mm

86. 7000 cm

87. 0.0004 km

88. 3000 dm

89. 500 m

90. 0.001 hm

Research Activities

91. Write a report on the development of the metric system in Europe. Indicate who had the most influence in its development.

92. Write a report on why you believe many people oppose switching to the metric system. Give your opinion about whether the United States will eventually switch to the metric system, and if so when it might do so. Explain your answer.

8.2 LENGTH, AREA, AND VOLUME

This section and the next section are designed to help you *think metric*, that is, to become acquainted with day-to-day usage of metric units. In this section we consider length, area, and volume.

Length

The basic unit of length is the meter. In all English-speaking countries except the United States, *meter* is spelled "metre." Until 1960 the meter was officially defined by the length of a platinum bar kept in a vault in France. The modern definition of the meter is based on the speed of light, a constant that has been defined with great precision. Other commonly used units of length are the kilometer, centimeter, and millimeter. The meter, which is a little longer than 1 yard, is used to measure things that we normally measure in yards and feet. A man whose height is about 2 meters is a tall man. A tractor trailer unit (an 18-wheeler) is about 18 meters long.

The kilometer is used to measure what we normally measure in miles. For example, the distance from New York to Seattle is about 5120 km. One kilometer is about 0.6 mile, and one mile is about 1.6 kilometers.

Centimeters and millimeters are used to measure what we normally measure in inches. The centimeter is a little less than $\frac{1}{2}$ inch (see Fig. 8.1), and the millimeter is a little less than $\frac{1}{20}$ inch. A millimeter is about the thickness of a dime. A book may measure 20 cm by 25 cm with a thickness of about 3 cm. Millimeters are often used in scientific work and other areas in which small quantities must be measured. The length of a small insect may be measured in millimeters.

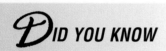
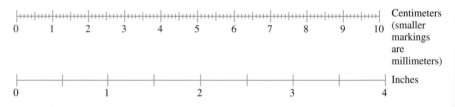

Figure 8.1

EXAMPLE 1

Determine which unit of length you would use to express

a) the length of a football field.

b) the height of a door.

c) the diameter of a round kitchen table.

d) the length of a butterfly.

e) the diameter of a door knob.

f) the distance between your house and your school.

g) the diameter of the lead in a pencil.

h) the diameter of a quarter.

i) the thickness of a quarter.

j) your height.

Solution:

a) Meters **b)** Meters or centimeters

c) Meters or centimeters **d)** Centimeters or millimeters

e) Centimeters **f)** Kilometers

g) Millimeters **h)** Centimeters or millimeters

i) Millimeters **j)** Meters or centimeters

In some parts of this solution more than one possible answer is listed. Measurements can often be made by using more than one unit. For example, if someone asks your height, you might answer $5\frac{1}{2}$ feet or 66 inches. Both answers are correct. ▲

Area

The area enclosed in a square with 1-centimeter sides (Fig. 8.2) is 1 cm × 1 cm = 1 cm². A square whose sides are 2 cm (Fig. 8.3) has an area of 2 cm × 2 cm = 2² cm² = 4 cm².

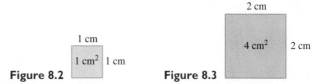

Figure 8.2 **Figure 8.3**

Areas are always expressed in square units, such as square centimeters, square kilometers, or square meters. When finding areas, be careful that all the numbers being multiplied are expressed in the same units.

In the metric system the square centimeter replaces the square inch. The square meter replaces the square foot and square yard. In the future you might purchase carpet or other floor covering by the square meter instead of by the square yard.

For measuring large land areas the metric system uses a square unit 100 meters on each side (a square hectometer). This unit is called a **hectare** (pronounced "hectair" and symbolized ha). A hectare is about 2.5 acres. One square mile of land contains about 260 hectares. Very large units of area are measured in square kilometers. One square kilometer is about $\frac{4}{10}$ square mile.

EXAMPLE 2

Determine which unit of area you would use to measure

a) the top of a kitchen table.

b) the floor of the classroom.

c) a football field.

d) a person's property with an average-sized lot.

e) the face of a quarter.

f) a national forest.

g) the space covered by a blanket on the beach.

h) a 15-cm ruler.

i) a page of this book.

j) a dollar bill.

Solution:

a) Square meters   **b)** Square meters

c) Hectares or square meters   **d)** Square meters or hectares

e) Square centimeters   **f)** Square kilometers

g) Square meters   **h)** Square centimeters

i) Square centimeters   **j)** Square centimeters ▲

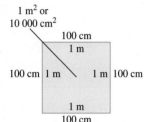

1 m² or
10 000 cm²

Figure 8.4

EXAMPLE 3

A square meter is how many times as large as a square centimeter?

Solution: Since 1 m equals 100 cm, we can replace 1 m with 100 cm (see Fig. 8.4). The area of 1 m² = 1 m × 1 m = 100 cm × 100 cm = 10 000 cm². Thus the area of one square meter is 10 000 times the area of one square centimeter. This technique can be used to convert from any square unit to a different square unit. ▲

Volume

When a figure has only two dimensions—length and width—we can find its area. When a figure has three dimensions—length, width, and height—we can find its volume. The volume of an item can be considered the space occupied by the item.

In the metric system, volume may be expressed in terms of liters or cubic meters, depending on what is being measured. In all English-speaking countries except the United States *liter* is spelled "litre."

The volume of liquids is expressed in liters. A liter is a little larger than a quart. Liters are used in place of pints, quarts, and gallons. A liter can be divided into 1000 equal parts, each of which is called a milliliter. Figure 8.5 illustrates a type of liter container (a 1000 mℓ graduated cylinder) that is often used in chemistry. Milliliters are used to express the volume of very small amounts of liquid. Drug dosages are often expressed in milliliters. An 8-oz cup will hold about 240 mℓ of liquid.

The kiloliter, 1000 liters, is used to represent the volume of large amounts of liquid. Tank trucks carrying gasoline to service stations hold about 10.5 kℓ of gasoline.

Cubic meters are used to express the volume of large amounts of solid material. The volume of a dump truck's load of topsoil is measured in cubic meters. The volume of natural gas used to heat a house may soon be measured in cubic meters instead of cubic feet.

The liquid in a liter container will fit exactly in a cubic decimeter (Fig. 8.6). Note that 1 ℓ = 1000 mℓ and that 1 dm³ = 1000 cm³. Because 1 ℓ = 1 dm³,

Figure 8.5

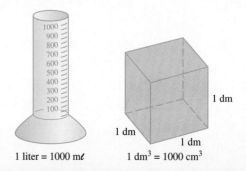

1 liter = 1000 mℓ   1 dm³ = 1000 cm³

Figure 8.6

1 mℓ must equal 1 cm³. Other useful facts are illustrated in Table 8.3. Thus, within the metric system, conversions are much simpler than in the U.S. customary system. For example, how would you change cubic feet of water into gallons of water?

Table 8.3

Volume in Cubic Units		Volume in Liters
1 cm³	=	1 mℓ
1 dm³	=	1 ℓ
1 m³	=	1 kℓ

EXAMPLE 4

Determine which unit of volume you would use to measure

a) a quart of milk.

b) a truckload of concrete.

c) a drug dosage.

d) the space available in your refrigerator.

e) the space in your classroom.

f) water in a drinking glass.

g) air in a large helium balloon.

h) water in a water bed.

i) water in an Olympic-sized swimming pool.

j) dirt removed to lay the foundation for a basement.

Solution:

a) Liters   **b)** Cubic meters

c) Milliliters   **d)** Cubic meters

e) Cubic meters   **f)** Milliliters

g) Cubic meters   **h)** Kiloliters or liters

i) Kiloliters   **j)** Cubic meters

EXAMPLE 5

A swimming pool is 18 m long and 9 m wide, and it has a uniform depth of 3 m (Fig. 8.7). Find (a) the volume of the pool in cubic meters and (b) the volume of water in the pool in kiloliters.

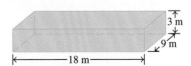

Figure 8.7

Solution:

a) To find the volume in cubic meters we use the formula

$$V = l \times w \times h.$$

Substituting, we have

$$V = 18 \text{ m} \times 9 \text{ m} \times 3 \text{ m}$$
$$= 486 \text{ m}^3.$$

b) Since $1 \text{ m}^3 = 1 \text{ k}\ell$, the pool will hold 486 kiloliters of water.   ▲

EXAMPLE 6

Select the most appropriate answer. The volume of this textbook is approximately

a) 1500 mm^3.   b) $1500 \ \ell$.   c) 1500 cm^3.

Solution: This textbook is not a liquid, so its volume is not expressed in liters. Thus (b) is not the answer. The cube in Fig. 8.8 is approximately 1500 mm^3, so (a) is not an appropriate answer. This textbook measures about 26 cm $\times$ 21 cm $\times$ 3 cm, or 1638 cm^3. Therefore 1500 cm^3 or (c) is the most appropriate answer.

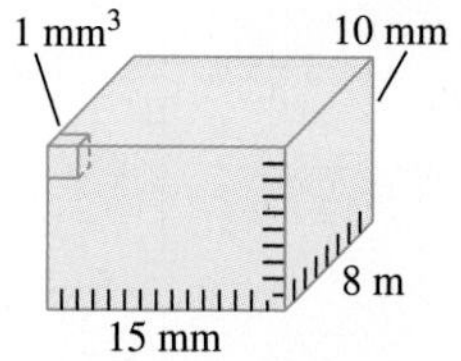

Figure 8.8

When the volume of a liquid is measured, the abbreviation cc is often used instead of cm^3 to represent cubic centimeters. For example, a nurse may give a patient an injection of 3 cc or 3 $\text{m}\ell$ of the drug ampicillin.

EXAMPLE 7

A nurse must give a patient 3 cc of the drug Gentamycin mixed in 100 cc of a normal saline solution.

a) How many $\text{m}\ell$ of the drug will the nurse administer?

b) What is the total volume of the drug and water in milliliters?

Solution:

a) Because 1 cc is equal in volume to 1 $\text{m}\ell$, the nurse will administer 3 $\text{m}\ell$ of the drug.

b) The total volume is $3 + 100$, or 103, $\text{m}\ell$.   ▲

EXAMPLE 8

A barrel in the shape of a right circular cylinder has a radius of 30 cm and a height of 80 cm. If the barrel is filled with water, how many liters of water does it contain?

Solution: The barrel is illustrated in Fig. 8.9. The formula for the volume of a right circular cylinder is $V = \pi r^2 h$, where π is approximately 3.14. If we express all the measurements in meters, the volume will be given in cubic meters. Thus 30 cm = 0.3 m, and 80 cm = 0.8 m.

$$V = \pi r^2 h$$
$$= 3.14(0.3)^2(0.8)$$
$$= 3.14(0.09)(0.8) = 0.226 \text{ m}^3$$

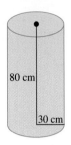

Figure 8.9

We want the volume in liters, so we must change the answer from cubic meters to liters.

$$1 \text{ m}^3 = 1000 \ \ell$$

so, $0.226 \text{ m}^3 = 0.226 \times 1000 = 226 \ \ell$ ▲

EXAMPLE 9

a) How many times larger is a cubic meter than a cubic centimeter?

b) How many times larger is a cubic dekameter than a cubic meter?

Solution:

a) The procedure used to determine the answer is similar to that used in Example 3 in this section. First, we draw a cubic meter, which is a cube 1 m long by 1 m wide by 1 m high. In Fig. 8.10 we represent each meter as 100 centimeters. The volume of the cube is its length times its width times its height, or

$$V = l \times w \times h$$
$$= 100 \text{ cm} \times 100 \text{ cm} \times 100 \text{ cm} = 1\,000\,000 \text{ cm}^3.$$

Since $1 \text{ m}^3 = 1\,000\,000 \text{ cm}^3$, a cubic meter is one million times larger than a cubic centimeter.

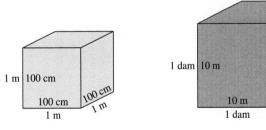

Figure 8.10   **Figure 8.11**

b) Work part (b) in a similar manner (see Fig. 8.11),

$$V = l \times w \times h$$
$$= 10 \text{ m} \times 10 \text{ m} \times 10 \text{ m} = 1000 \text{ m}^3.$$

Since $1 \text{ dam}^3 = 1000 \text{ m}^3$, a cubic dekameter is one thousand times larger than a cubic meter.   ▲

Section 8.2 Exercises

In Exercises 1–14, indicate the metric unit of measurement that you would use to express

1. the height of a window.
2. the length of a new pencil.
3. the distance between cities.
4. the length of a paper clip.
5. the width of a football field.
6. the diameter of a pencil.
7. the waist size of a person.
8. the diameter of a nickel.
9. the width of a tablecloth.
10. the length of a car.
11. the length of a fly.
12. the width of a piano key.
13. the distance to the moon.
14. the length of a newborn infant.
15. Starting with a straight piece of wood of sufficient size, construct a meter stick. Indicate decimeters, centimeters, and millimeters on the meter stick. Use the centimeter measure in Fig. 8.1 as a guide.
16. Construct a metric tape measure from a piece of tape or rope and then determine your waist measurement.

In Exercises 17–24, choose the best answer.

17. A U.S. 32-cent postage stamp is about how wide and how long?
    a) 2 cm × 3 cm
    b) 2 mm × 3 mm
    c) 2 hm × 3 hm
18. A football field is about how long?
    a) 90 km
    b) 90 m
    c) 90 cm
19. A grown woman is about how tall?
    a) 160 cm
    b) 160 mm
    c) 160 dm
20. The distance between freeway exits could be how long?
    a) 5 mm   b) 5 m   c) 5 km
21. Your thumbnail is about how wide?
    a) 1.5 cm   b) 1.5 mm   c) 1.5 dm
22. A kitchen stove is about how wide?
    a) 75 mm   b) 75 cm   c) 75 dm
23. The Empire State Building is about how tall?
    a) 375 m   b) 375 km   c) 375 cm

24. A full-length gown may be how long?
    a) 150 m   b) 150 mm   c) 150 cm

In Exercises 25–34, (a) estimate the item in metric units and (b) measure it with a metric ruler. Record your result.

25. The width of a door
26. The length of your shoe
27. The thickness of 10 sheets of paper placed one on top of the other
28. The width of a card from a deck of cards
29. The length of a car
30. The diameter of a can of soda
31. The length of a needle
32. The length of your classroom
33. The height of a milk carton
34. The thickness of a quarter

In Exercises 35–40, replace the customary measure (shown in parentheses) with the appropriate metric measure.

35. There was a crooked man and he walked a crooked _____ (mile).
36. Give him a _____ (inch), and he will take a _____ (mile).
37. One hundred _____ (yard) dash.
38. I wouldn't touch a skunk with a 10- _____ (foot) pole.
39. This is a _____ (mile)stone in my life.
40. I found _____ (an inch) worm.

In Exercises 41–50, indicate the metric unit of measurement you would use to express the area of

41. the top of a desk.
42. Sequoia National Park.
43. the face of a dime.
44. a baseball field.
45. the living room floor.
46. a building lot for a house.
47. a postage stamp.
48. Washington, D.C.
49. a ceiling tile.
50. a professional basketball court.

In Exercises 51–58, choose the best answer.

51. A 32-cent U.S. postage stamp has an area of about
    a) 5 cm²   b) 5 mm²   c) 5 dm²
52. The area of the face of a dime is about
    a) 2.5 cm²   b) 2.5 m²   c) 2.5 mm²

53. The area of a living room floor is about
   a) 20 cm^2 **b)** 20 m^2 **c)** 20 km^2

54. The area of a city lot is about
   a) $\frac{1}{8}$ m^2 **b)** $\frac{1}{8}$ ha   **c)** $\frac{1}{8}$ km^2

55. The area of a city lot is about
   a) 800 m^2 **b)** 800 hm^2 **c)** 800 cm^2

56. The area of a ceiling tile is about
   a) 360 m^2 **b)** 360 km^2 **c)** 360 cm^2

57. The area of Dinosaur National Park is about
   a) 15 km^2 **b)** 15 ha   **c)** 15 m^2

58. The area of a table model TV screen is about
   a) 2080 dm^2 **b)** 2080 mm^2 **c)** 2080 cm^2

In Exercises 59–64, replace the question mark with the appropriate value.

59. 1 m^2 = ? mm^2

60. 1 hm^2 = ? cm^2

61. 1 km^2 = ? hm^2

62. 1 cm^2 = ? mm^2

63. 1 cm^2 = ? m^2

64. 1 mm^2 = ? hm^2

In Exercises 65–70, (a) estimate the area of the item in metric units and (b) measure it in metric units and compute its area.

65. A teacher's desk top

66. A bedroom floor

67. A five-dollar bill

68. The lot your house is built on

69. The end of a 12-ounce soda can (Use $A = \pi r^2$.)

70. The face of a penny

71. Use a metric ruler to find the radius of the circle. Then compute the circumference and area of the circle. Give your answers in metric units.

72. Use a metric ruler to measure the length of the sides and the height of the parallelogram. Then compute its perimeter and area. Give your answers in metric units.

73. A rectangular building 40 m by 64 m is surrounded by a walk 1.5 m wide.
   a) Find the area of the region covered by the building and the walk.
   b) Find the area of the walk.

74. A framed picture is shown. Find the matted area.

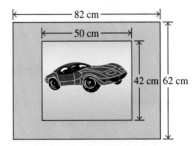

75. Herman's garden is 13.4 m by 16.5 m. How large is his garden in hectares?

76. Mr. Jones has purchased a farm that is in the shape of a rectangle. The dimensions of the piece of land are 1.4 km by 3.75 km. How many hectares of land did he purchase?

In Exercises 77–86, determine the metric unit that would best be used to measure the volume of

77. oil in a barrel.

78. water flowing over Niagara Falls per minute.

79. liquid in an eye dropper.

80. space in a garbage can.

81. water in a hot-water heater.

82. a truckload of ready-mix concrete.

83. oil needed to change oil in your car.

84. asphalt needed to pave a driveway.

85. medication in a syringe.

86. alcohol in a jigger.

In Exercises 87–94, choose the best answer to indicate the volume of

87. a quarter.
   a) 1 cm^3 **b)** 1 mm^3 **c)** 1 dm^3

88. water in an Olympic-sized swimming pool.
   a) 455 ℓ **b)** 455 mℓ **c)** 455 kℓ

89. blood in an adult.
   a) 6.4 ℓ **b)** 6.4 kℓ **c)** 6.4 mℓ

90. juice that can be squeezed out of an orange.
   a) 120 kℓ **b)** 120 mℓ **c)** 120 ℓ

91. space occupied by a textbook.
   a) 1370 mm^3 **b)** 1370 dm^3 **c)** 1370 cm^3

92. air in a basketball.
   a) 14 000 m^3 **b)** 14 000 cm^3
   c) 14 000 mm^3

93. air in a balloon with a diameter of 4 meters.
a) 30 m³   **b)** 30 cm³   **c)** 30 km³

94. a can of vegetables.
a) 550 cm³   **b)** 550 mm³   **c)** 550 dm³

In Exercises 95–100, (a) estimate the volume in metric units and (b) compute the actual volume of the item.

95. Water in a water bed that is 2 m long, 1.5 m wide, and 25 cm deep (Use $V = lwh$.)

96. Oil in a barrel that has a height of 1 m and a diameter of 0.5 m (Use $V = \pi r^2 h$.)

97. Water in a cylindrical tank that is 40 cm in diameter and 2 m high

98. The room you are now in

99. A new piece of chalk

100. An ice cream cone with a height of 10 cm and a radius of 3 cm (Use $V = \frac{1}{3}\pi r^2 h$.)

In Exercises 101–104, fill in the blank.

101. 620 cm³ = _____ mℓ

102. 620 cm³ = _____ ℓ

103. 76 kℓ = _____ m³

104. 4.2 ℓ _____ cm³

105. How many liters of water will a rectangular fish tank hold if the tank is 60 cm long, 45 cm wide, and 20 cm high?

106. a) What is the volume of water in a swimming pool that is 18 m long and 10 m wide and has an average depth of 2.5 m? Give your answer in cubic meters.
b) How many kiloliters of water will the pool hold?

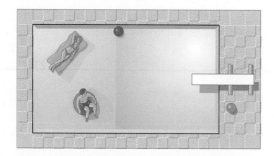

107. The first coat of paint for the outside of a building requires 1 ℓ of paint for each 10 square meters. The second coat requires 1 ℓ for every 15 square meters. If the paint costs $4.75 per liter, what will be the cost of two coats of paint for the four outside walls of a building 20 m long, 12 m wide, and 6 m high?

108. A milk container has a square base with 9.5 cm sides. Its height is 19 cm. Determine the volume of milk that the container holds in liters.

109. A circular cylinder containing glass cleaner has a base with a radius of 6 cm. Its height is 24 cm. Determine the number of liters of glass cleaner in the container.

110. How many times larger is a square kilometer than a square dekameter?

111. How many times larger is a square dekameter than a square meter?

112. How many times larger is a cubic meter than a cubic decimeter?

113. How many times larger is a cubic centimeter than a cubic millimeter?

Problem Solving/Group Activities

In Exercises 114 and 115, fill in the blank to make a true statement.

114. 6.7 kℓ = _____ dm³

115. 1.4 ha = _____ cm²

116. In Example 3 we illustrated how to change an area in a metric unit to an area in a different metric unit
a) Using Example 3 as a guide, change 1 square mile to square inches.
b) Is converting from one unit of area to a different unit of area generally easier in the metric system or the customary system? Explain.

117. In Example 9 we illustrated how to change a volume in one metric unit to a volume in a different metric unit.
a) Using Example 9 as a guide, change 6 cubic yards (a volume 1 yard by 2 yards by 3 yards) into cubic inches.
b) Is converting from one unit of volume to a different unit of volume generally easier in the metric system or the customary system? Explain.

Research Activities

118. The definition of the meter has changed several times throughout history. Use an encyclopedia or other reference book and write a one- to two-page report on the history of the meter, from when it was first named to the present.

119. Over time the metric system replaced the measurement systems of nearly all countries. The history of the development of measurements before the metric and English systems were developed is quite interesting. Read the section on Weights and Measures in the *Encyclopedia Brittanica*, or use another reference book, and answer the following questions, using complete sentences.
a) What did the *kite* measure in the ancient Egyptian system?
b) What were the *mina*, the *kus*, and the *ka* in the Babylonian system?

c) Indicate the relationship that existed between the *sacred mina* and the *shekel* in the ancient Hebrew system.

d) In the ancient Greek system, what was the basic unit of length?

e) Explain how the ancient Romans subdivided a foot, and what the basic unit of volume was in the Roman system.

f) In the ancient Chinese system, what unit was the basic unit of land measure?

8.3   MASS AND TEMPERATURE

In this section we discuss the metric measurements of mass and temperature. As with Section 8.2 the focus of this section is on thinking metric.

Mass

Weight and mass are not the same. **Mass** is a measure of the amount of matter in an object. It is determined by the molecular structure of the object, and it will not change from place to place. Weight is a measure of the gravitational pull on an object. For example, the gravitational pull of the earth is about six times as great as the gravitational pull of the moon. Thus a person on the moon weighs about $\frac{1}{6}$ as much as on earth, even though the person's mass remains the same. In space, where there is no gravity, a person has no weight.

Even on the earth the gravitational pull varies from point to point. The closer you are to the center of the earth, the greater the gravitational pull. Thus a person weighs very slightly less on a mountain than in a nearby valley. Because the mass of an object does not vary with location, scientists use mass rather than weight.

Although weight and mass are not the same, on the earth they are proportional to each other (the greater the weight, the greater the mass). Therefore for our purposes we can treat weight and mass as the same.

Figure 8.12

Weight is measured on a scale (Fig. 8.12). Mass is measured on a balance, and the object under consideration is compared with a known mass (Fig. 8.13).

The kilogram is the basic unit of mass in the metric system. It is a little more than 2 pounds. Recall that the official kilogram is a cylinder of platinum–iridium alloy kept by the International Bureau of Weights and Measures, located in Sèvres, near Paris.

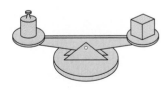

Figure 8.13

Items that we normally measure in pounds are usually measured in kilograms in other parts of the world. For example, an average-sized man has a mass of about 75 kilograms.

The gram (a unit that is 0.001 kilogram) is relatively small and is used in place of the ounce. A nickel has a mass of 5 g, a cube of sugar has a mass of about 2 g, and a large paper clip has a mass of about 1 g.

The milligram is used extensively in the medical and scientific fields, as well as in the pharmaceutical industry. Nearly all bottles of tablets are now labeled in either milligrams or grams.

The metric tonne (t) is used to express the mass of heavy items. A metric tonne equals 1000 kg. It is a little larger than our customary ton of 2000 pounds. The mass of a large truck may be expressed in metric tonnes.

EXAMPLE 1

Determine the appropriate unit(s) used to express the mass of

a) a teaspoon of sugar.   **b)** an adult.

c) a newborn child.   **d)** a car.

e) a blue whale.   **f)** a box of cereal.

g) a quarter.   **h)** a fly.

i) a steak.   **j)** a refrigerator.

Solution:

a) Grams   **b)** Kilograms

c) Kilograms   **d)** Kilograms or metric tonnes

e) Metric tonnes   **f)** Kilograms or grams

g) Grams   **h)** Milligrams

i) Kilograms or grams   **j)** Kilograms   ▲

A kilogram of water has a volume of exactly 1 liter. In fact, a liter is defined to be the volume of 1 kilogram of water at a specified temperature and pressure. Thus mass and volume are easily interchangeable in the metric system. Converting from weight to volume is not nearly as convenient in our customary system. For example, how would you change pounds of water to cubic feet or gallons of water in our customary system?

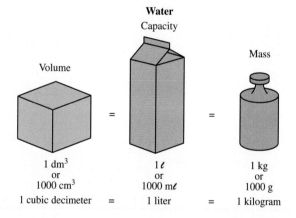

Water
Capacity

Mass

Volume

$1\ \mathrm{dm}^3$
or
$1000\ \mathrm{cm}^3$
1 cubic decimeter

$1\ \ell$
or
$1000\ \mathrm{m}\ell$
1 liter

$1\ \mathrm{kg}$
or
$1000\ \mathrm{g}$
1 kilogram

= =

= =

Figure 8.14

Figure 8.14 illustrates the relationship between volume of water in cubic decimeters, capacity in liters, and mass in kilograms. Table 8.4 expands on this relationship between volume and mass of water.

Table 8.4

Volume in Cubic Units		Volume in Liters		Mass of Water
$1\ \mathrm{cm}^3$	=	$1\ \mathrm{m}\ell$	=	1 g
$1\ \mathrm{dm}^3$	=	$1\ \ell$	=	1 kg
$1\ \mathrm{m}^3$	=	$1\ \mathrm{k}\ell$	=	1 t (1000 kg)

EXAMPLE 2

A fish tank is 1 m long, 50 cm high, and 250 mm wide (Fig. 8.15).

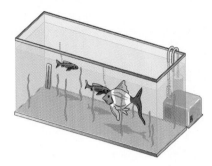

Figure 8.15

a) Determine the number of liters of water the tank holds.

b) What is the mass of the water in kilograms?

Solution:

a) We must convert all the measurements to the same units. Let's convert them all to meters; 50 cm is 0.5 m, and 250 mm is 0.25 m.

$$V = l \times w \times h$$
$$= 1 \times 0.25 \times 0.5$$
$$= 0.125 \text{ m}^3$$

Since 1 m³ of water = 1 kℓ of water,

$$0.125 \text{ m}^3 = 0.125 \text{ k}\ell, \text{ or } 125 \ \ell \text{ of water.}$$

b) Since 1 ℓ = 1 kg, 125 ℓ = 125 kg of water. ▲

To convince yourself of the advantages of the metric system, do a similar problem involving the customary system of measurement, such as Problem Solving/Group Activity Exercise 54 at the end of this section.

Temperature

The Celsius scale is used to measure temperatures in the metric system. Figure 8.16 on page 416 shows a thermometer with the Fahrenheit scale on the left and the Celsius scale on the right.

The Celsius scale was named for the Swedish astronomer Anders Celsius (1701–1744), who first devised it in 1742. On the Celsius scale, water freezes at 0°C and boils at 100°C. The Celsius thermometer in the past was called a "centigrade thermometer." Recall that *centi* means $\frac{1}{100}$, and there are 100 degrees between the freezing point of water and the boiling point of water. Thus 1°C is $\frac{1}{100}$ of this interval.

Table 8.5 on page 416 gives some common temperatures in both Celsius (°C) and Fahrenheit (°F).

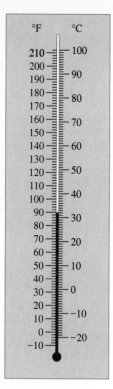

°F   °C

210 — 100
200 —
190 — 90
180 — 80
170 —
160 — 70
150 —
140 — 60
130 —
120 — 50
110 —
100 — 40
90 —
80 — 30
70 — 20
60 —
50 — 10
40 —
30 — 0
20 —
10 — −10
0 —
−10 — −20

Figure 8.16

Table 8.5

Celsius Temperature		Farenheit Temperatures
−18°C	A very cold day	0°F
0°C	Freezing point of water	32°F
10°C	A warm winter day	50°F
20°C	A mild spring day	68°F
30°C	A warm summer day	86°F
37°C	Body temperature	98.6°F
100°C	Boiling point of water	212°F
177°C	Oven temperature for baking	350°F

EXAMPLE 3

Choose the best answer. (Refer to the dual-scale thermometer in Fig. 8.16.)

a) Toronto, Canada, on New Year's Day might have a temperature of
i) −15°C.   **ii)** 15°C.   **iii)** 40°C.

b) Washington, D.C., on July 4, might have a temperature of
i) 20°C.   **ii)** 30°C.   **iii)** 40°C.

c) The oven temperature for baking a cake might be
i) 60°C.   **ii)** 100°C.   **iii)** 175°C.

Solution:

a) A temperature of 15°C is possible if it is a very mild winter, but 40°C is much too hot. The best answer for a normal winter is −15°C.

b) The best estimate is 30°C. A temperature of 20°C is too chilly, and 40°C is too hot for July 4.

c) A cake bakes at temperatures well above boiling, so the only reasonable answer is 175°C.

▲

Comparing the temperature scales in Fig. 8.16, we see that the Celsius scale has 100° from boiling to freezing, and the Fahrenheit scale has 180°. Therefore one Celsius degree represents a greater change in temperature than one Fahrenheit degree does. In fact, one Celsius degree is the same as $\frac{180}{100}$, or $\frac{9}{5}$ Fahrenheit degrees. When converting from one system to the other system, use the following formulas.

From Celsius to Fahrenheit	From Fahrenheit to Celsius
$$F = \frac{9}{5} C + 32$$	$$C = \frac{5}{9} (F - 32)$$

EXAMPLE 4

A typical setting for home thermostats is 68°F. What is the equivalent temperature on the Celsius thermometer?

Solution: We use the formula $C = \frac{5}{9}(F - 32)$ to convert from °F to °C. Substituting $F = 68$ gives

$$C = \frac{5}{9}(68 - 32)$$

$$= \frac{5}{9}(36)$$

$$= 20°.$$

EXAMPLE 5

If the temperature outdoors is 22°C, will you need to wear a sweater if going outdoors?

Solution: We use the formula $F = \frac{9}{5}C + 32$ to convert from °C to °F. Substituting $C = 22$ yields

$$F = \frac{9}{5}(22) + 32$$

$$= 39.6 + 32$$

$$= 71.6°.$$

Since the temperature is about 72°F, you will not need a sweater.

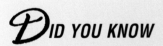

*D*ID YOU KNOW

IT'S A METRIC WORLD

*T*he United States is the only westernized country not presently using the metric system as its primary system of measurement. The only countries in the world besides the United States not presently using or committed to using the metric system are Yemen, Brunei, and a few small islands, see Fig. 8.17.

Will the United States ever convert to the metric system? If so, when? The European Community (EC) adopted a directive that requires all exporters to EC nations to indicate the dimensions of their products in metric units. Currently, U.S. manufacturers who export goods are doing so. Little by little the United States is becoming more metric. For example, soft drinks now come in liter bottles and prescription drug dosages are given in metric units. Maybe in the not too distant future gasoline will be measured in liters, not gallons, as it is in Canada and Mexico. ∎

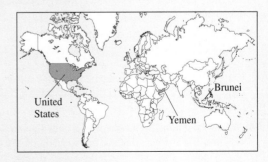

Figure 8.17

Section 8.3 Exercises

In Exercises 1–10, what metric unit of measurement would you use to express the mass of

1. a refrigerator.

2. a man.

3. a spoon.

4. a box of cereal.

5. a new pencil.

6. a compact car.

7. an oil tanker.

8. a mosquito.

9. a quarter.

10. a calculator.

In Exercises 11–16, select the best answer.

11. The mass of a woman is about how much?
a) 50 kg   **b)** 50 g   **c)** 50 dag

12. The mass of a 5-lb bag of flour is about how much?
a) 2.26 g   **b)** 2.26 kg   **c)** 2.26 dag

13. The mass of a box of bran flakes is about how much?
a) 0.45 t   **b)** 0.45 g   **c)** 0.45 kg

14. The mass of a can of tomato sauce is about how much?
a) 425 g   **b)** 425 kg   **c)** 425 mg

15. The mass of a steer is about how much?
a) 530 g   **b)** 530 kg   **c)** 530 dag

16. The mass of a full-size car is about how much?
a) 1 962 000 hg   **b)** 380 kg   **c)** 1.6 t

In Exercises 17–22, estimate the mass of the item. If a scale with metric measure is available, find the mass.

17. A quarter

18. Your body

19. A textbook

20. A pencil

21. A gallon of water

22. Your shoe

23. A mixture of 15 g of salt and 16 g of baking soda is poured into 250 mℓ of water. What is the total mass of the mixture in grams?

24. A jet can travel about 1 km on 15 kg of fuel. How many metric tonnes of fuel will the jet use flying nonstop between New York and Denver, a distance of 2600 km?

25. The dimensions of a storage tank are length 16 m, width 12 m, and height 12 m. If the tank is filled with water, determine
a) the volume of water in the tank in cubic meters.
b) the number of kiloliters of water the tank will hold.
c) the mass of the water in metric tonnes.

26. A hot-water heater in the shape of a right circular cylinder has a radius of 30 cm and a height of 150 cm. If the tank is filled with water, determine
a) the volume of water in the tank in cubic meters.
b) the number of liters of water the tank will hold.
c) the mass of the water in kilograms.

In Exercises 27–30, convert as indicated

27. 1.2 kg = _____ t

28. 3.62 t = _____ kg

29. 42.6 t = _____ g

30. 1 460 000 mg = _____ t

In Exercises 31–38, choose the best answer. Use Table 8.5 and Fig. 8.16 to help select your answers.

31. A cup of coffee might have a temperature of
a) 15°C   **b)** 50°C   **c)** 98°C

32. The thermostat for an air conditioner was set for 80°F. This setting is closest to
a) 2°C   **b)** 27°C   **c)** 57°C

33. The temperature of the water in a certain lake is 5°C. You could
a) ice fish
b) dress warmly and walk along the lake
c) swim in the lake

34. Freezing rain is most likely to occur at a temperature of
a) −25°C   **b)** 32°C   **c)** 0°C

35. The weather forecast calls for a high of 15°C. You should plan to wear
a) a down-lined jacket
b) a sweater
c) a bathing suit

36. What might be the temperature of milk from a refrigerator?
a) 30°C   **b)** 5°C   **c)** 0°C

37. What might be the temperature of an apple pie baking in the oven?
a) 90°C   **b)** 100°C   **c)** 177°C

38. Upon reentry into the earth's atmosphere, the space shuttle's tiles protect the shuttle from temperatures as high as 2300°F, or about
a) 195°C   **b)** 620°C   **c)** 1260°C

In Exercises 39–52, convert each temperature as indicated. Give your answer to the nearest tenth of a degree.

39. 30°C = _____ °F   **40.** −15°C = _____ °F

41. 72°F = _____ °C   **42.** −45°F = _____ °C

43. 350°F = _____ °C   **44.** 98°F = _____ °C

45. 37°C = _____ °F   **46.** −4°C = _____ °F

47. 13°F = _____ °C   **48.** 65°F = _____ °C

49. 20°C = _____ °F   **50.** 10°C = _____ °F

51. 113°F = _____ °C   **52.** 425°F = _____ °C

53. Maria's body temperature is 38.2°C. Should she take an aspirin or put on a sweater? Explain.

Problem Solving/Group Activities

In Example 2 we showed how to find the volume and mass of water in a fish tank. Exercise 54 demonstrates how much more complicated solving a similar problem is in our customary system.

54. A fish tank is 1 yd long by 1.5 ft high by 15 in. wide.
   a) Determine the volume of water in the fish tank in cubic feet.
   b) Determine the weight of the water in pounds. One cubic foot of water weighs about 62.5 lb.
   c) If one gallon of water weighs about 8.3 lb, how many gallons will the tank hold?

55. Let x represent a real number. Find the value of x when $x°F = x°C$.

Research Activities

56. The SI system is built on a foundation of seven basic units.
   a) Do research and list the seven basic units.
   b) Which, if any, of these basic units is defined by an actual physical object?
   c) For each base unit not defined by a physical object, give its official definition.

57. Do industries in your area export goods? If so, are they training employees to use and understand the metric system? Contact local industries that export goods and write a report on your findings.

8.4 DIMENSIONAL ANALYSIS AND CONVERSIONS TO AND FROM THE METRIC SYSTEM

You may sometimes need to change units of measurement in the metric system to equivalent units in the customary system. To do so, use **dimensional analysis** which is a procedure used to convert from one unit of measurement to a different unit of measurement. To perform dimensional analysis you must first understand what is meant by a unit fraction. A **unit fraction** is any fraction in which the numerator and denominator contain different units, and the value of the fraction is 1. From Table 8.6 we can obtain many unit fractions involving customary units.

Table 8.6   Customary Units

1 foot = 12 inches
1 yard = 3 feet
1 mile = 5280 feet
1 pound = 16 ounces
1 ton = 2000 pounds
1 cup (liquid) = 8 fluid ounces
1 pint = 2 cups
1 quart = 2 pints
1 gallon = 4 quarts
1 minute = 60 seconds
1 hour = 60 minutes
1 day = 24 hours
1 year = 365 days

Examples of Unit Fractions

$$\frac{12 \text{ in.}}{1 \text{ ft}} \qquad \frac{1 \text{ ft}}{12 \text{ in.}} \qquad \frac{16 \text{ oz}}{1 \text{ lb}} \qquad \frac{1 \text{ lb}}{16 \text{ oz}} \qquad \frac{60 \text{ min}}{1 \text{ hr}} \qquad \frac{1 \text{ hr}}{60 \text{ min}}$$

In each of these examples the numerator equals the denominator, so the value of the fraction is 1.

To convert an expression from one unit of measurement to a different unit, multiply the given expression by the unit fraction (or fractions) that will result in the answer having the units you are seeking. When two fractions are being multiplied, and the same unit appears in the numerator of one fraction and the denominator of the other fraction, then that common unit may be divided out. For example suppose that we want to convert 30 inches to feet. We consider the following:

$$30 \text{ in.} = ? \text{ ft}$$

Since inches are given we will need to eliminate them. Thus inches will need to appear in the denominator of the unit fraction. We need to convert to feet, so feet will need to appear in the numerator of the unit fraction. If we multiply a quantity in inches by a unit fraction containing feet/inches, the units will divide out as follows.

$$(\cancel{\text{in.}})\left(\frac{\text{ft}}{\cancel{\text{in.}}}\right) = \text{ft}$$

Thus the solution to the problem is

$$30 \text{ in.} = (30 \, \cancel{\text{in.}})\left(\frac{1 \text{ ft}}{12 \, \cancel{\text{in.}}}\right) = \frac{30}{12} \text{ ft} = 2.5 \text{ ft.}$$

In Examples 1–3, we will give examples that do not involve the metric system. After that we will use dimensional analysis to make conversion to and from the metric system.

EXAMPLE 1

Convert 26 ounces to pounds.

Solution: One pound is 16 ounces. Therefore we write

$$26 \text{ oz} = (26 \, \cancel{\text{oz}})\left(\frac{1 \text{ lb}}{16 \, \cancel{\text{oz}}}\right) = \frac{26}{16} \text{ lb} = 1.625 \text{ lb.}$$

Thus 26 ounces equals 1.625 pounds.   ▲

EXAMPLE 2

If $1 = 7.2 pesos, what is the amount in dollars of 1000 pesos?

Solution:

$$1000 \text{ pesos} = (1000 \, \cancel{\text{pesos}})\left(\frac{\$1}{7.2 \, \cancel{\text{pesos}}}\right) = \frac{\$1000}{7.2}$$

$$= \$138.89$$

Thus 1000 pesos have a value of about $139.00.   ▲

If more than one unit needs to be changed, more than one multiplication may be needed as illustrated in Example 3.

EXAMPLE 3

Convert 80 miles per hour to feet per second.

Solution: Let's consider the units given and where we want to end up. We are given $\frac{\text{mi}}{\text{hr}}$ and wish to end with $\frac{\text{ft}}{\text{sec}}$. Thus we need to change miles into feet and hours into seconds. Because two units need to be changed, we will need to multiply the given quantity by two unit fractions, one for each conversion. First, we show how to convert the units of measurement from miles per hour to feet per second:

$$\left(\frac{\cancel{mi}}{\cancel{hr}}\right)\left(\frac{ft}{\cancel{mi}}\right)\left(\frac{\cancel{hr}}{sec}\right) \quad \text{gives an answer in} \quad \frac{ft}{sec}.$$

Now we multiply the given quantity by the appropriate unit fractions to obtain the answer:

$$80\,\frac{mi}{hr} = \left(80\,\frac{\cancel{mi}}{\cancel{hr}}\right)\left(\frac{5280\ ft}{1\ \cancel{mi}}\right)\left(\frac{1\ \cancel{hr}}{3600\ sec}\right) = \frac{(80)(5280)}{(1)(3600)}\,\frac{ft}{sec}$$

$$\approx 117.3\,\frac{ft}{sec}.$$

Note that $\left(\dfrac{80\ mi}{hr}\right)\left(\dfrac{1\ hr}{3600\ sec}\right)\left(\dfrac{5280\ ft}{1\ mi}\right)$ will give the same answer. ▲

Conversions to and from the Metric System

Now we will apply dimensional analysis to the metric system.

Table 8.7 is used in making conversions to and from the metric system. A more exact table of conversion factors may be found in many science books at

Table 8.7   Conversions Table

Length

1 inch (in.)	= 2.54	centimeters (cm)
1 foot (ft)	= 30	centimeters (cm)
1 yard (yd)	= 0.9	meter (m)
1 mile (mi)	= 1.6	kilometers (km)

Area

1 square inch (in^2)	= 6.5	square centimeters (cm^2)
1 square foot (ft^2)	= 0.09	square meter (m^2)
1 square yard (yd^2)	= 0.8	square meter (m^2)
1 square mile (mi^2)	= 2.6	square kilometers (km^2)
1 acre	= 0.4	hectare (ha)

Volume

1 teaspoon (tsp)	= 5	milliliters (mℓ)
1 tablespoon (tbsp)	= 15	milliliters (mℓ)
1 fluid ounce (fl oz)	= 30	milliliters (mℓ)
1 cup (c)	= 0.24	liter (ℓ)
1 pint (pt)	= 0.47	liter (ℓ)
1 quart (qt)	= 0.95	liter (ℓ)
1 gallon (gal)	= 3.8	liters (ℓ)
1 cubic foot (ft^3)	= 0.03	cubic meter (m^3)
1 cubic yard (yd^3)	= 0.76	cubic meter (m^3)

Weight (Mass)

1 ounce (oz)	= 28	grams (g)
1 pound (lb)	= 0.45	kilogram (kg)
1 ton (T)	= 0.9	tonne (t)

your college's library. However, we can use this table to obtain many unit fractions.

Table 8.7 shows that 1 in. = 2.54 cm. From this equality we can write the two unit fractions

$$\frac{1 \text{ in.}}{2.54 \text{ cm}} \quad \text{or} \quad \frac{2.54 \text{ cm}}{1 \text{ in.}}.$$

Examples of other unit fractions from Table 8.7 are

$$\frac{1 \text{ yd}}{0.9 \text{ m}}, \quad \frac{0.9 \text{ m}}{1 \text{ yd}}, \quad \frac{1 \text{ gal}}{3.8 \ \ell}, \quad \frac{3.8 \ \ell}{1 \text{ gal}}, \quad \frac{1 \text{ lb}}{0.45 \text{ kg}}, \quad \text{and} \quad \frac{0.45 \text{ kg}}{1 \text{ lb}}.$$

To change from a metric unit to a customary unit or vice versa, multiply the given quantity by the unit fraction whose product will result in the units you are seeking. For example, to convert 5 inches to centimeters, multiply 5 in. by a unit fraction with centimeters in the numerator and inches in the denominator.

$$5 \text{ in.} = (5 \text{ in.})\left(\frac{2.54 \text{ cm}}{1 \text{ in.}}\right)$$
$$= 5(2.54) \text{ cm}$$
$$= 12.7 \text{ cm}$$

EXAMPLE 4

a) A recipe for chicken soup requires $4\frac{1}{2}$ cups of water. How many liters does this amount equal?

b) On a trip from Des Moines, Iowa, to Tulsa, Oklahoma, Daryl used 72.9 liters of gasoline. How many gallons of gasoline does this amount equal?

Solution:

a) In Table 8.7, under the heading of volume, we see that 1 cup = 0.24 liter. Thus the unit fractions involving cups and liters are

$$\frac{1 \text{ cup}}{0.24 \ \ell} \quad \text{or} \quad \frac{0.24 \ \ell}{1 \text{ cup}}.$$

We need to convert from cups to liters. Since $4\frac{1}{2}$ is 4.5 in decimal form, we write

$$4.5 \text{ cups} = 4.5 \text{ cups}\left(\frac{0.24 \ \ell}{1 \text{ cup}}\right) = 1.08 \ \ell.$$

b) Table 8.7 indicates that 1 gallon = 3.8 liters. Thus the unit fractions are

$$\frac{1 \text{ gal}}{3.8 \ \ell} \quad \text{or} \quad \frac{3.8 \ \ell}{1 \text{ gal}}.$$

We need to convert from liters to gallons, so we write

$$72.9 \ \ell = 72.9 \ \ell\left(\frac{1 \text{ gal}}{3.8 \ \ell}\right) = \frac{72.9}{3.8} \text{ gal} \approx 19.18 \text{ gal.}$$

EXAMPLE 5

A farm in Ontario, Canada, has an area of 37.4 hectares. Find the area in acres.

Solution: From Table 8.7 we determine that 0.4 hectare = 1 acre. Thus

$$37.4 \text{ ha} = (37.4 \text{ ha})\left(\frac{1 \text{ acre}}{0.4 \text{ ha}}\right) = \frac{37.4}{0.4} \text{ acres} = 93.5 \text{ acres.}$$ ▲

EXAMPLE 6

The distance from Boston, Massachusetts, to New Orleans, Louisiana, is about 1359 miles. Express the distance (a) in kilometers and (b) in meters.

Solution: From Table 8.7 we can determine the unit fractions needed to work this problem.

a) $1359 \text{ mi} = (1359 \text{ mi})\left(\frac{1.6 \text{ km}}{1 \text{ mi}}\right) = (1359)(1.6) \text{ km} = 2174.4 \text{ km}$

b) Since 1 km = 1000 m, 2174.4 km = 2 174 400 m. ▲

EXAMPLE 7

Brian, a nurse, must administer 4 cc of codeine elixir to a patient.

a) How many milliliters of the drug will he administer?

b) How many ounces is this dosage equivalent to?

Solution:

a) Since 1 cc = 1 mℓ, he will administer 4 mℓ of the drug.

b) Since 1 fl oz = 30 mℓ,

$$4 \text{ mℓ} = (4 \text{ mℓ})\left(\frac{1 \text{ fl oz}}{30 \text{ mℓ}}\right) = \frac{4}{30} \text{ fl oz} \approx 0.13 \text{ fl oz.}$$ ▲

Suppose that we want to convert 150 millimeters to inches. Table 8.7 does not have a conversion factor from millimeters to inches, but it does have one for inches to centimeters. Because 1 inch = 2.54 centimeters and 1 centimeter = 10 millimeters, we can reason that 1 inch = 25.4 millimeters.

$$1 \text{ cm} = 10 \text{ mm}$$
$$\text{therefore, } 2.54 \text{ cm} = 10(2.54) = 25.4 \text{ mm}$$

We can solve the problem as follows.

$$150 \text{ mm} = (150 \text{ mm})\left(\frac{1 \text{ in.}}{25.4 \text{ mm}}\right) = \frac{150}{25.4} \text{ in.}$$
$$\approx 5.91 \text{ in.}$$

If we wish we can use dimensional analysis using two unit fractions to make the conversion. The procedure follows:

$$150 \text{ mm} = (150 \text{ mm})\left(\frac{1 \text{ cm}}{10 \text{ mm}}\right)\left(\frac{1 \text{ in.}}{2.54 \text{ cm}}\right) = \frac{150}{(10)(2.54)} \text{ in.}$$
$$\approx 5.91 \text{ in.}$$

EXAMPLE 8

The manager of a perfume shop wants to order the equivalent of 60 fluid ounces of Lancone perfume from the manufacturer in Paris. How many liters will the shop order?

Solution: Table 8.7 shows no direct conversion factors for fluid ounces to liters. However, it does show a conversion from fluid ounces to milliliters. We can perform the conversion by first changing 60 fluid ounces to milliliters and then changing the answers to liters.

$$60 \text{ fl oz} = (60 \text{ fl oz})\left(\frac{30 \text{ m}\ell}{1 \text{ fl oz}}\right) = 1800 \text{ m}\ell$$

Since $1000 \text{ m}\ell = 1 \text{ }\ell$, $1800 \text{ m}\ell = 1.8 \text{ }\ell$. Thus the answer is 1.8 liters.

We could have also worked the problem by multiplying 60 fl oz by two unit fractions as follows.

$$60 \text{ fl oz} = (60 \text{ fl oz})\left(\frac{30 \text{ m}\ell}{1 \text{ fl oz}}\right)\left(\frac{1 \text{ }\ell}{1000 \text{ m}\ell}\right)$$

$$= \frac{(60)(30)}{1000} \text{ }\ell = 1.8 \text{ }\ell$$

▲

EXAMPLE 9

The label on a bottle of Vicks 44D Cough Syrup indicates that the active ingredient is dextromethorphan hydrobromide, and that 5 mℓ (1 teaspoon) contains 10 mg of this ingredient. If the recommended dosage for adults is 3 teaspoons determine the following.

a) How many milliliters of cough medicine will be taken?

b) How many milligrams of the active ingredient will be taken?

c) If the bottle contains 8 fluid ounces of medicine, how many milligrams of the active ingredient are in the bottle?

Solution:

a) Since each teaspoon contains 5 mℓ and 3 teaspoons will be taken, 15 mℓ of the cough medicine will be taken.

$$3 \text{ tsp} = (3 \text{ tsp})\left(\frac{5 \text{ m}\ell}{1 \text{ tsp}}\right) = 15 \text{ m}\ell$$

b) Since each teaspoon contains 10 mg of the active ingredient, 30 mg of the active ingredient will be taken.

$$3 \text{ tsp} = (3 \text{ tsp})\left(\frac{10 \text{ mg}}{1 \text{ tsp}}\right) = 30 \text{ mg}$$

c) Table 8.7 shows that each fluid ounce contains 30 mℓ. Since each 5 mℓ contains 10 mg of the active ingredient, we can work the problem as follows.

$$8 \text{ fl oz} = (8 \text{ fl oz})\left(\frac{30 \text{ m}\ell}{1 \text{ fl oz}}\right)\left(\frac{10 \text{ mg}}{5 \text{ m}\ell}\right) = \frac{8(30)(10)}{5} \text{ mg} = 480 \text{ mg}$$

Therefore there are 480 mg (or 0.48 g) of the active ingredient in the bottle of cough syrup.

▲

EXAMPLE 10

Drug dosage is often administered according to a patient's weight. For example, 30 mg of the drug Vancomicin is to be given for each kilogram of a person's weight. If Martha, who weighs 136 pounds, is to be given the drug, what dosage should she be given?

Solution: First, we need to convert Martha's weight into kilograms. From Table 8.7, we see that 1 lb = 0.45 kg. We need to determine the number of milligrams of the drug for Martha's weight in kilograms. The answer may be found as follows.

$$136 \text{ lb} = (136 \cancel{\text{ lb}})\left(\frac{0.45 \cancel{\text{ kg}}}{1 \cancel{\text{ lb}}}\right)\left(\frac{30 \text{ mg}}{1 \cancel{\text{ kg}}}\right) = (136)(0.45)(30) \text{ mg} = 1836 \text{ mg}$$

Thus 1836 mg, or 1.836 g, of the drug should be given.   ▲

Section 8.4 Exercises

In Exercises 1–16, convert the quantity to the indicated units.

1. 80 lb to kg   **2.** 43 km to mi   **3.** 4.2 ft to m

4. 11 in. to cm   **5.** 15 yd^2 to m^2 **6.** 28 mi to km

7. 150 kg to lb   **8.** 765 mm to in.

9. 675 ha to acres   **10.** 346 g to oz

11. 14.5 gal to ℓ   **12.** 27.4 m^3 to ft^3

13. 45.6 mℓ to fl oz   **14.** 1.6 km^2 to mi^2

15. 120 lb to kg   **16.** 2.8 lb to kg

17. Tom's new car traveled 125 mi on 5 gal of gasoline. How many kilometers can Tom travel with the same amount of gasoline?

18. The Smiths are buying a new carpet for their living room, which is 12 ft by 18 ft. The carpeting is sold in square meters. How many square meters of carpeting will they need?

19. The distance from Los Angeles to the nearest border of Maine is about 3300 mi. What is the distance in kilometers?

20. The speed limit on a certain road is 55 mph. What is the speed limit in kilometers per hour?

21. A road in England has a speed limit sign of 60 kmh. What is the speed limit in miles per hour?

22. A car's gas tank holds 23 gal. How many liters will it hold?

23. The distance between New York City and Washington, D.C., is 640 km. What is the distance in miles between the two cities?

24. Louise's property has an area of 1.4 acres. How many hectares does she own?

25. A basement is to be 50 ft long, 30 ft wide, and 8 ft high. How much dirt will have to be removed when this basement is built? Answer in cubic meters.

26. Yosemite National Park has an area of 1189 mi^2. What is its area in square kilometers?

27. A German-made car has a weight of 1.3 t.
   a) How many tons does this weight equal?
   b) How many pounds?

28. If 1.3 kg of steak costs $10.27, how much does it cost per pound?

29. A tank truck holds 34.5 kℓ of gasoline. How many gallons does it hold?

30. A 0.25-oz bottle of Chanel perfume costs $80. What is the cost per gram?

In Exercises 31–38, replace the metric measure(s) in the phrase with the appropriate customary measure(s).

31. 28 g of prevention are worth 0.45 kg of cure.

32. Give him 2.54 cm and he'll take 1600 m.

33. More bounce to the 28 g.

34. He demanded his 0.45 kg of flesh.

35. 157 cm and eyes of blue.

36. A miss is as good as 1.6 km.

37. First down and 9.1 m to go.

38. The longest 0.9 m.

39. One gram is the same as five carats. Joe's new ring contains a precious stone that is $\frac{1}{8}$ carat. Find the weight of the stone in grams.

40. One meter is about 3.3 ft. Use this information to determine
   a) the equivalent of one square meter in square feet.
   b) the equivalent of one cubic meter in cubic feet.

41. One foot is about 30 cm. Use this information to determine
   a) the equivalent of one square foot in square centimeters.
   b) the equivalent of one cubic foot in cubic centimeters.

42. The area of a football field including end zones is about 5100 square meters. Find the area in (a) square yards, (b) square feet, and (c) hectares.

43. The recommended dosage of the drug codeine for pediatric patients is 1 mg per kilogram of a child's weight. What dosage of codeine should be given to April, who weighs 56 lb?

44. For each kilogram of a person's weight, 1.5 mg of the antibiotic drug gentamycin is to be administered. If Ron Gigliotti weighs 170 lb, how much of the drug should he receive?

45. The recommended dosage of the drug ampicillin for pediatric patients is 200 mg per kilogram of a patient's weight. If Janine weighs 76 lb, how much ampicillin should she receive?

46. The recommended dosage of the drug theophylline is 5 mg per kilogram of weight. If Alex weighs 178 lb, how much of the drug should he receive?

47. The recommended dosage of the generic antibiotic vancomycin is 40 mg per kilogram of a person's weight. If Maria weighs 125 lb, how much of the drug should she receive?

48. For each kilogram of weight of a dog, 5 mg of the drug bretylium is to be given. If Blaster, an Irish setter, weighs 82 lb, how much of the drug should be given?

49. A taxi costs $1.75 for the first half mile and $1.25 for each additional half mile. Fractional parts of a half mile will be rounded up to a half mile. What is the cost of a 10-km taxi trip?

50. The postal service charges 32¢ for first-class mail weighing up to 1 oz and 23¢ for each additional ounce or fraction thereof. How much postage will Jeanne need to use on a letter that weighs 65 g?

51. The label on the bottle of Triaminic Expectorant indicates that each teaspoon (5 mℓ) contains 12.5 mg of the active ingredient phenylpropanolamine hydrochloride.
   a) Determine the amount of the active ingredient in the recommended adult dosage of 2 teaspoons.
   b) Determine the quantity of the active ingredient in a 12-oz bottle.

52. The label on the bottle of Maximum Strength Pepto-Bismol indicates that each tablespoon contains 236 mg of the active ingredient bismuth subsalicyate.
   a) Determine the amount of the active ingredient in the recommended dosage of 2 tablespoons.
   b) If the bottle contains 8 fl oz, determine the quantity of the active ingredient in the bottle.

53. The label on a bottle of Robitussin Cough Syrup indicates that each teaspoon (5 mℓ) contains 100 mg of guaifenesin in a pleasant-tasting syrup with 3.5% alcohol. If the recommended dosage is 2 teaspoons determine
   a) the amount of the active ingredient guaifenesin in the recommended dosage.
   b) the amount of alcohol, in milliliters, in the recommended dosage.
   c) the amount of guaifenesin in the 8-fl-oz bottle.

54. On January 1, 1996, the price of gasoline in El Paso, Texas, was $1.19 per gallon. Across the border in Juarez, Mexico, the cost of the same octane gasoline was 3.64 pesos per liter.
   a) How much would it cost (in the country's own currency) to fill a 20-gal tank on that day in El Paso and in Juarez?
   b) On the same date $1 of American currency could be exchanged in a bank in Juarez for 7.5750 pesos. How much, in terms of U.S. dollars, would it cost to fill the tank in Juarez?

55. Change all the measurements in the cookie recipe to metric units. Do not forget pan size, temperature, and size of cookies.

   Magic Cookie Bar
   $\frac{1}{2}$ c graham cracker crumbs
   12 oz nuts
   8 oz chocolate pieces
   $1\frac{1}{3}$ c flaked coconut
   $1\frac{1}{3}$ c condensed milk

   Coat the bottom of a 9 in. × 13 in. pan with melted margarine. Add rest of ingredients one by one—crumbs, nuts, chocolate, and coconut. Pour condensed milk over all. Bake at 350°F for 25 minutes. Allow to cool 15 minutes before cutting. Makes about two dozen $1\frac{1}{2}$ in. by 3 in. bars.

56. Write each of the metric units, labeled (*a*) through (*n*), in customary units.
   The first human flight, December 17, 1903, was (*a*) *37 meters.* Just 66 years later Neil Armstrong stepped on the moon after journeying (*b*) *370 140 km.*

On April 12, 1981, a new era in space flight began when the space shuttle embarked on its maiden voyage.

Here are some characteristics of and facts about the space shuttle. The two solid rocket boosters are jettisoned at (*c*) *44 km*. During reentry, portions of Orbiter's exterior reach temperatures up to (*d*) *1260°C*. The orbiter lands at a speed of (*e*) *335 km/hr*. It can deliver to orbit up to (*f*) *29 484 kg* of payload in its huge (*g*) *4.5 × 18-m* cargo bay. Propellants can be supplied to the engines at a rate of about (*h*) *171 396 ℓ/min* of hydrogen and (*i*) *63 588 ℓ/min* of oxygen. The external tank is (*j*) *46.89 m* long and (*k*) *8.4 m* in diameter. When fully loaded the tank contains (*l*) *632 772 kg* of liquid oxygen and (*m*) *106 142 kg* of cold liquid hydrogen at about (*n*) *−251°C*.

Problem Solving/Group Activities

57. a) The measuring cup in the figure is supposed to be calibrated in cups, ounces, and milliliters. However, the milliliter marks were left off. At appropriate locations, add horizontal marks to the blue vertical line for 100, 200, 300, 400, and 500 milliliters.

b) Estimate, in ounces, the amount of liquid in the cup.

c) Estimate, in milliliters, the amount of liquid in the cup.

58. a) The graduated cylinder shows milliliter markings from 10 to 100 mℓ. Indicate alongside the cylinder where markings for 0.5, 1.0, 1.5, 2.0, 2.5, 3.0, and 3.5 ounces would be placed if they were to be added to the cylinder.

b) Estimate the amount of liquid in the cylinder in milliliters.

c) Estimate the amount of liquid in the cylinder in ounces.

59. The following question was selected from a nursing exam. Can you answer it?

In caring for a patient after delivery, you are to give 0.2 mg Ergotrate Maleate. The ampule is labeled $\frac{1}{300}$ grain/mℓ. How much would you draw and give? (60 mg = 1 grain)

a) 15 cc   **b)** 1.0 cc   **c)** 0.5 cc   **d)** 0.01 cc

60. Phil is planning a picnic and plans on purchasing 0.18 kg of ground beef for each 100 lb of weight of guests who will be in attendance. If he expects 15 people whose average weight is 130 lb, how many pounds of beef should he purchase?

61. The displacement of automobile engines is measured in liters. A certain model of 1996 automobile has a 2.2-ℓ engine.

a) Determine the displacement of the engine in cubic centimeters.

b) Determine the displacement of the engine in cubic inches.

Research Activity

62. Write the following phrase in metric units of measurement. (Conversion factors are not given in Table 8.7.)

I love you a bushel and a peck.

CHAPTER 8 SUMMARY

Key Terms

8.1

Systéme International d'Unites (SI system)
degree Celsius
kilogram
liter
meter
metric system
U.S. customary system

8.2

hectare

8.3

mass

8.4

dimensional analysis
unit fraction

Important Facts

Metric Units

Prefix	Symbol	Meaning
kilo	k	$1000 \times$ base unit
hecto	h	$100 \times$ base unit
deka	da	$10 \times$ base unit
		base unit
deci	d	$\frac{1}{10}$ of base unit
centi	c	$\frac{1}{100}$ of base unit
milli	m	$\frac{1}{1000}$ of base unit

Water

Volume in cubic units		Volume in liters		Mass of water
1 cm^3	=	$1 \text{ m}\ell$	=	1g
1 dm^3	=	1ℓ	=	1 kg
1 m^3	=	$1 \text{ k}\ell$	=	$1 \text{ t } (1000 \text{ kg})$

Temperature

$$°C = \frac{5}{9} (°F - 32)$$

$$°F = \frac{9}{5} °C + 32$$

Chapter 8 Review Exercises

8.1

In Exercises 1–6, indicate the meaning of the prefix.

1. centi **2.** kilo **3.** milli

4. hecto **5.** deka **6.** deci

In Exercises 7–12, change the given quantity to that indicated.

7. 72 mg to g **8.** 24.2 ℓ to cℓ

9. 0.997 cm to mm **10.** 1 000 000 mg to kg

11. 4.62 kℓ to ℓ **12.** 192.6 dag to dg

In Exercises 13 and 14, arrange the quantities from smallest to largest.

13. 2.67 kℓ, 3000 mℓ, 14 630 cℓ

14. 0.047 km, 4700 m, 47 000 cm

8.2, 8.3

In Exercises 15–24, indicate the metric unit of measurement that would best express

15. the length of a telephone.

16. the mass of a telephone.

17. the temperature of the water in the ocean.

18. the diameter of a quarter.

19. the area of a room of a house.

20. the volume of a glass of milk.

21. the length of an ant.

22. the mass of a car.

23. the distance from Denver, Colorado, to San Diego, California.

24. the volume of water in a small fish tank.

In Exercises 25 and 26, (a) first estimate the following in metric units and then (b) measure with a metric ruler. Record your results.

25. Your height

26. The length of a new pencil

In Exercises 27–32, select the best answer.

27. The length of the distance between Los Angeles and San Francisco is
a) 8000 m   **b)** 1000 km   **c)** 650 km

28. The mass of a full-grown raccoon is about
a) 600 g   **b)** 20 kg   **c)** 100 kg

29. The volume of a quart of grape juice is about
a) 0.1 kℓ   **b)** 0.5 ℓ   **c)** 1 ℓ

30. The area of a vegetable garden in a person's yard may be
a) 200 m²   **b)** 0.5 ha   **c)** 0.02 km²

31. The temperature on a hot summer day in Florida may be about
a) 34°C   **b)** 45°C   **c)** 25°C

32. The height of a giant sequoia tree is about
a) 300 m   **b)** 3000 cm   **c)** 0.3 m

33. Convert 1640 kg to tonnes.

34. Convert 6.3 t to grams.

35. If the temperature outside is 28°C, what is the Fahrenheit temperature?

36. If the room temperature is 68°F, what is the Celsius temperature?

37. If your outdoor thermometer shows a temperature of −6°F, what is the Celsius temperature?

38. If Ron's body temperature is 39°C, what is his Fahrenheit temperature?

In Exercises 39 and 40, (a) measure, in centimeters, each of the line segments and then (b) compute the area of the figure.

39.   **40.**

41. a) What is the volume of water in a full swimming pool that is 8 m long and 4 m wide and has an average depth of 2 m? Answer in m³.
b) What is the mass of the water in kilograms?

42. A rectangular yard measures 22 m by 30 m. Find
a) the area in square meters.
b) the area in square kilometers.

43. A small fish tank measures 80 cm long, 40 cm wide, and 30 cm high.
a) What is its volume in cubic centimeters?
b) What is its volume in cubic meters?
c) How many mℓ of water will the tank hold?
d) How many kℓ of water will the tank hold?

44. A square kilometer is a square with length and width both 1 km. How many times larger is a square kilometer than a square dekameter?

8.4

In Exercises 45–58, change the given quantity to the indicated quantity.

45. 15 in. = _____ cm

46. 105 kg = _____ lb

47. 220 yd = _____ m

48. 100 m = _____ yd

49. 55 m/h = _____ km/h

50. 180 lb = _____ kg

51. 15 gal = _____ ℓ

52. 40 m³ = _____ yd³

53. 5 lb = _____ kg

54. 4 qt = _____ ℓ

55. 15 yd³ = _____ m³

56. 150 mi = _____ km

57. 27 cm = _____ ft

58. $3\frac{1}{4}$ in. = _____ mm

59. John bought 700 bricks to build a chimney. Each brick has a mass of 1.5 kg.
a) What is the total mass of the bricks in kilograms?
b) What is the total weight of the bricks in pounds?

60. The distance between Rochester, New York, and Reading, Pennsylvania, is 320 mi. What is this distance in kilometers?

61. The Holcombs are buying new carpet for their family room. The room is 15 ft wide and 24 ft long. The carpeting is sold only in square meters. How many square meters of carpeting will they need? Round your answer up to the nearest square meter.

62. A plane flies from Chicago to New York in 2.5 hours at an average speed of 461.6 kilometers per hour. What is the distance from Chicago to New York in (a) kilometers, and (b) miles? (Use distance = rate · time)

63. The speed limit on a certain road is 35 mph. What is the speed limit in (a) kilometers per hour? (b) meters per hour?

64. A rectangular tank used to test leaks in tires is 90 cm by 70 cm by 40 cm deep.
a) Determine the number of liters of water the tank holds.
b) What is the mass of the water in kilograms?

CHAPTER 8 TEST

1. Change 406 mm to hm.

2. Change 14.8 kg to mg.

3. How many times greater is a kilometer than a dekameter?

4. If John's high school track is 300 m long, and John jogs around the track six times, how many kilometers has he gone?

In Exercises 5–9, choose the best answer.

5. The length of this page is about
   a) 10 cm
   b) 25 cm
   c) 60 cm

6. The surface area of the top of a kitchen table is about
   a) $2 m^2$
   b) $200 cm^2$
   c) $2000 cm^2$

7. The volume of a car's gasoline tank is about
   a) 200ℓ
   b) 20ℓ
   c) 75ℓ

8. The mass of a refrigerator is about
   a) 1400 g
   b) 0.6 t
   c) 80 kg

9. The outside temperature on a snowy day is about
   a) 18°C
   b) −2°C
   c) −40°C

10. How many times greater is a square meter than a square centimeter?

11. How many times greater is a cubic meter than a cubic millimeter?

12. Convert 862 cm to inches.

13. Change 65 mph to kilometers per hour.

14. Change 62°F to °C.

15. Change 30°C to °F.

16. A giraffe may be 12 ft tall. How many centimeters is this?

17. A fish tank at an aquarium is 20 m long by 20 m wide by 8 m deep.
    a) Determine the volume of the tank in cubic meters.
    b) Determine the number of liters of water the tank holds.
    c) Determine the weight of the water in kilograms.

18. The first coat of paint for the outside walls of a building requires 1ℓ of paint for each $10 m^2$ of wall surface. The second coat requires 1ℓ for every $15 m^2$. If the paint costs $3.50 per liter, what will be the cost of two coats of paint for the four outside walls of a building 20 m long, 15 m wide, and 6 m high?

19. a) What is a unit fraction?
    b) Explain how to use unit fractions to make conversions.
    c) If 1 gelot = 9.47 keruts, change 32.5 gelots to keruts. Use a unit fraction and show your work.
    d) Change 4.6 keruts to gelots. Use a unit fraction and show your work.

GROUP PROJECTS

Health and Medicine

Throughout this chapter we have shown the importance of the metric system in the medical professions. The following 3 questions involve applications of metrics to medicine.

1. a) Twenty milligrams of the drug Lincomycin is to be given for each kilogram of a person's weight. The drug is to be mixed with 250 cc of a normal saline solution and the mixture is to be administered intravenously (I.V.) over a 1-hr period. Clyde Dexter, who weighs 196 lb, is to be given the drug. Determine the dosage of the drug he will be given.

b) At what rate per minute should the 250-cc solution be administered?

2. a) At a pharmacy a parent asks a pharmacist why her child needs such a small dosage of a certain medicine. The pharmacist explains that a general formula may be used to estimate a child's dosage of certain medicines. The formula is

$$\text{Child's dose} = \frac{\substack{\text{child's weight} \\ \text{in kilograms}}}{67.5 \text{ kg}} \times \text{adult dose}.$$

What is the amount you would give a 60-lb child if the adult dosage of the medicine is 70 mg?

b) At what weight, in pounds, would the child receive an adult dose?

3. a) At blood banks, routine blood-sample analysis is done to compute the mean corpuscular volume (MCV). The formula used is

$$MCV = \frac{H}{RBC} \times 10^7,$$

where H = the hematocrit value (the percentage of red blood cells in the blood) and RBC = the red blood count per cubic centimeter. If H = 38% and RBC = 4.8×10^3 cells per cc, find the MVC.

b) There are about 2.7×10^3 hemoglobin molecules in one red blood cell. There are about 5.0×10^6 red blood cells in one cc of blood. How many hemoglobin molecules are there in one cc of blood and in one cubic millimeter of blood? (Write your answers in scientific notation; see Section 5.6.)

Traveling to Other Countries

4. Dale Pollinger is a buyer at Xerox Corporation and travels frequently on business to foreign countries. He always plans ahead and does his holiday shopping overseas where he can purchase items not easily found in the United States.

a) On a trip to Tokyo he decides to buy a kimono for his sister. To determine the length of a kimono, one measures, in centimeters, the distance from the bottom of a person's neck to 5 cm above the floor. If the distance from the bottom of Dale's sister's neck to the floor is 5 ft 2 in., calculate the length of the kimono that Dale should purchase.

b) If the conversion rate at the time is 1 U.S. dollar = 102.30 yen, and the kimono cost 8695.5 yen, determine the cost of the kimono in U.S. dollars.

c) On a trip to Mexico City, Mexico, Dale finds a small replica of a Mayan castle that he wants to purchase for his wife. He is going directly from Mexico to Rome, so he wants to mail the castle back to the United States. The mailing rate from Mexico to the United States is 6 pesos per hundred grams. Determine the mailing cost, in U.S. dollars, if the castle weighs 6 lb and the exchange rate is 1 peso = 0.1320 U.S. dollars.

d) Dale takes a personal trip from Rochester, New York, to the Canadian side of Niagara Falls. He notices that in Niagara Falls, New York, the cost of regular gasoline is $1.39 per gallon in U.S. dollars and that in Niagara Falls, Canada, regular gas costs $0.57 per liter in Canadian dollars. His gas tank has a capacity of 18 gallons. If the exchange rate is 1 Canadian dollar = 0.7339 U.S. dollar, is it more expensive to fill the car's empty gas tank in Niagara Falls, Canada, or Niagara Falls, New York, and by how much?

Chapter 9 Geometry

*A*s a child, you probably learned to draw a house as a rectangle with a triangle on top of it, and to draw the sun as a circle with rays radiating from it. The simple geometry of this child's-eye view powerfully illustrates the fact that human beings universally recognize that the world is composed of planes, angles, straight lines, and smooth curves. People perceive most objects to be three-dimensional in nature.

The geometry of fractals now provides mathematicians with a means of describing objects in nature in which a pattern endlessly repeats itself in smaller and smaller versions. This mountain scene is a computer-generated fractal.

Indeed, most everyday objects can be described in terms of Euclidian geometry: the frisbee you toss to your dog or the soccer ball you kick across a rectangular field. Geometric patterns have been used since the beginning of human history to decorate everyday objects such as quilts, floor-tiling, and decorative containers. Patterns have endured because they can be varied nearly endlessly to provide a rhythm that pleases the eye and comforts the mind.

In the twentieth century, those who study mathematics, physics, and other disciplines have discovered that other geometries shape the universe. Einstein's theory of relativity is based on a non-Euclidean geometry, and adds a fourth dimension—time—to the three dimensions of space. Benoit Mandelbrot's work opened new fields of mathematics with

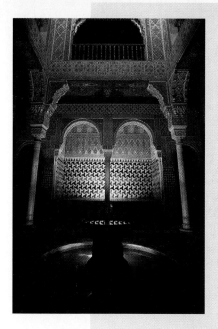

Islamic artists perfected the art of ornamentation and mosaic by using geometric forms because Islamic law strictly prohibited the use of the human form in art. Mosaic patterns at the Alhambra, a palace and fortress built in Grenada, Spain, illustrate such mathematical concepts as symmetries, tessellations, reflections, and rotations.

Nature displays geometric forms at even the most microscopic level. A recently discovered molecule called the buckminsterfullerene, is formed from 12 pentagons and 20 hexagons. Because of the molecule's resiliency, scientists anticipate that it will have applications in conductors and insulators. The molecule was named after Buckminster Fuller, the famous inventor of geodesic domes such as Disney World's Epcot Center.

the discovery of fractal geometry, which has been used to describe the shape of items such as snowflakes, coastlines, and tree branches.

Wherever human beings go, geometry is sure to follow. The patterns and variations of geometry act almost as a rhythm to human life, comparable to music or poetry. Sonya Kovalevskaya, a mathematician whose many contributions included studies of geometric sequences, remarked on this connection between mathematics and poetry. She explained, ". . . it is impossible to be a mathematician without being a poet in soul. . . . It seems to me that the poet has only to perceive that which others do not perceive, to look deeper than others look. And the mathematician must do the same thing." ■

Architects make use of a variety of polygons in designing both the basic structure and the ornamentation of a building. Frank Lloyd Wright, the American architect known for his command of space, pioneered twentieth-century residential design by creating homes where rooms flowed together in uninterrupted space that was, in effect, continued outside.

9.1   POINTS, LINES, PLANES, AND ANGLES

"It is the glory of geometry that from so few principles, fetched from without, ... it is able to accomplish so much."
I. Newton

Human beings recognized shapes, sizes, and physical forms long before geometry was developed. Geometry as a science is said to have begun in the Nile Valley of ancient Egypt. The Egyptians used geometry to measure land and to build pyramids and other structures.

The word *geometry* is derived from two Greek words, *ge*, meaning earth, and *metron*, meaning measure. Thus geometry means ''earth measure'' or ''measurement of the earth.''

Unlike the Egyptians, the Greeks were interested in more than just the applied aspects of geometry. The Greeks attempted to apply their knowledge of logic to geometry. Thales of Miletus in about 600 B.C. was the first to be credited with using deductive methods to develop geometric concepts. Another outstanding Greek geometer, Pythagoras, continued the systematic development of geometry that Thales had begun.

In about 300 B.C., Euclid collected and summarized much of the Greek mathematics of his time. In a set of 13 books called *Elements*, Euclid laid the foundation for plane geometry, which is also called Euclidean geometry.

Euclid is credited with being the first mathematician to use the axiomatic method in developing a branch of mathematics. First, Euclid introduced undefined terms, such as point, line, plane, and angle. He related these to physical space by such statements as ''A line is length without breadth,'' so that we may intuitively understand them. Because such statements play no further role in his system, they constitute primitive or undefined terms.

Second, Euclid introduced certain definitions. The definitions are introduced when needed and are often based on the undefined terms. Some terms that Euclid introduced and defined include triangle, right angle, and hypotenuse.

Third, Euclid asserted certain primitive propositions called postulates (now called axioms*) about the undefined terms and definitions. The reader is asked to accept these statements as true on the basis of their ''obviousness'' and their relationship with the physical world. For example, the Greeks accepted all right angles as being equal, which is Euclid's fourth postulate.

Fourth, Euclid proved, using deductive reasoning (Section 1.1), other propositions called theorems. One theorem that Euclid proved is known as the Pythagorean theorem: ''The sum of the areas of the squares constructed on the arms of a right triangle is equal to the area of the square constructed on the hypotenuse.'' He also proved that the sum of the angles of a triangle is 180°.

Using only 10 axioms, Euclid deduced 465 propositions (or theorems) in plane and solid geometry, number theory, and Greek geometrical algebra.

Point and Line

Three basic terms in geometry are point, line, and plane. These three terms are not given a formal definition, but we recognize points, lines, and planes when we see them.

Let's consider some properties of a line. Assume that a line means a straight line unless otherwise stated.

---

*The concept of the axiom has changed significantly since Euclid's time. Now any statement may be designated as an axiom, whether it is self-evident or not. All axioms are *accepted* as true. A set of axioms forms the foundation for a mathematical system.

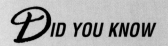

1. A line is a set of points. Each point is on the line and the line passes through each point.

2. Any two distinct points determine a unique line. Figure 9.1(a) illustrates a line containing three points. The arrows at both ends of the line indicate that the line continues in each direction. The line in Fig. 9.1(a) may be symbolized with any two points on the line—for example, $\overleftrightarrow{AB}$, $\overleftrightarrow{BA}$, $\overleftrightarrow{AC}$, or $\overleftrightarrow{CA}$.

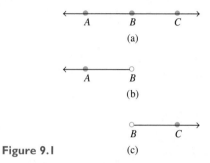

Figure 9.1

3. Any point on a line separates the line into three parts: the point itself and two *half lines* (neither of which includes the point). For example, in Fig. 9.1(a) point B separates the line into the point B and two half lines. Half line BA, symbolized $\overset{\circ}{\longleftarrow} BA$, is illustrated in Fig. 9.1(b). The open circle above the B indicates that point B is not included in the half line. Figure 9.1(c) illustrates half line BC, symbolized $\overset{\circ}{\longrightarrow} BC$.

Look at the half line $\overset{\circ}{\longrightarrow} AB$ in Fig. 9.2(b). If the *end point*, A, is included with the set of points on the half line, the result is called a *ray*. Ray AB, symbolized $\overrightarrow{AB}$, is illustrated in Fig. 9.2(c). Ray BA is illustrated in Fig. 9.2(d).

A *line segment* is that part of a line between two points, including the end points. Line segment AB, symbolized $\overline{AB}$, is illustrated in Fig. 9.2(e).

An open line segment is the set of points on a line between two points, excluding the end points. Open line segment AB, symbolized $\overset{\circ\;\circ}{AB}$, is illustrated in Fig. 9.2(f).

Description	Diagram	Symbol
(a) Line AB		$\overleftrightarrow{AB}$
(b) Half line AB		$\overset{\circ}{\longrightarrow}{AB}$
(c) Ray AB		$\overrightarrow{AB}$
(d) Ray BA		$\overleftarrow{BA}$
(e) Line segment AB		$\overline{AB}$
(f) Open line segment AB		$\overset{\circ\;\circ}{AB}$
(g) Half open line segements AB		$\overline{AB}$ $\overset{\circ}{\underline{AB}}$

Figure 9.2

Intersection

Solution: $\overrightarrow{AB}$

(a)

Union

Solution: $\overleftrightarrow{AB}$

(b)

Figure 9.3

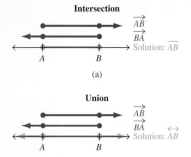
Figure 9.2(g) illustrates two half open line segments, symbolized $\overrightarrow{AB}$ and $\overset{\circ}{A}\overline{B}$.

In Chapter 2 we discussed intersection of sets. Recall that the intersection (∩) of two sets is the set of elements (points in this case) common to both sets. Consider the lines in Fig. 9.3(a). The intersection of $\overrightarrow{AB}$ and $\overrightarrow{BA}$ is line segment $\overline{AB}$. Thus $\overrightarrow{AB} \cap \overrightarrow{BA} = \overline{AB}$.

We also discussed the union of two sets in Chapter 2. The union (∪) of two sets is the set of elements (points in this case) that belong to either of the sets or both sets. The union of $\overrightarrow{AB}$ and $\overrightarrow{BA}$ is $\overleftrightarrow{AB}$ (Fig. 9.3b). Thus $\overrightarrow{AB} \cup \overrightarrow{BA} = \overleftrightarrow{AB}$.

EXAMPLE 1

Use a line similar to line AD to determine the solution to each part.

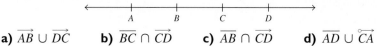

a) $\overrightarrow{AB} \cup \overrightarrow{DC}$ **b)** $\overline{BC} \cap \overrightarrow{CD}$ **c)** $\overline{AB} \cap \overrightarrow{CD}$ **d)** $\overline{AD} \cup \overset{\circ}{C}\overrightarrow{A}$

Solution:

a) $\overrightarrow{AB} \cup \overrightarrow{DC}$

Ray AB and ray DC are shown here. The union of these two rays is the entire line AD.

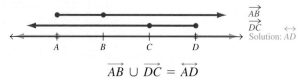

Solution: $\overleftrightarrow{AD}$

$$\overrightarrow{AB} \cup \overrightarrow{DC} = \overleftrightarrow{AD}$$

b) $\overline{BC} \cap \overrightarrow{CD}$

The intersection of line segment BC and ray CD is point C.

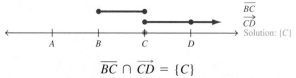

Solution: $\{C\}$

$$\overline{BC} \cap \overrightarrow{CD} = \{C\}$$

c) $\overline{AB} \cap \overrightarrow{CD}$

Line segment AB and ray CD have no points in common, so their intersection is empty.

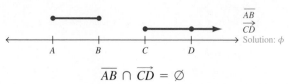

Solution: ϕ

$$\overline{AB} \cap \overrightarrow{CD} = \varnothing$$

d) $\overline{AD} \cup \overset{\circ}{C}\overrightarrow{A}$

The union of line segment AD and half line CA is ray DA (or $\overrightarrow{DB}$ or $\overrightarrow{DC}$).

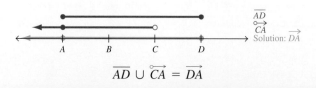

Solution: $\overrightarrow{DA}$

$$\overline{AD} \cup \overset{\circ}{C}\overrightarrow{A} = \overrightarrow{DA}$$

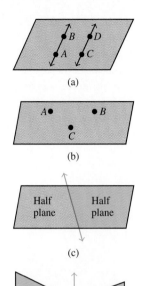

Figure 9.4

Plane

Even though the term *plane* is not formally defined, we can think of it as a two-dimensional surface that extends infinitely in both directions, like an infinitely large blackboard. Euclidean geometry is called plane geometry because it is the study of two-dimensional figures in a plane.

Two lines in the same plane that do not intersect are called parallel lines. Figure 9.4(a) illustrates two parallel lines in a plane ($\overleftrightarrow{AB}$ is parallel to $\overleftrightarrow{CD}$).

Properties of planes include the following.

1. Any three points that are not on the same line (noncollinear points) determine a unique plane (Fig. 9.4b).

2. A line in a plane divides the plane into three parts—the line and two half planes (Fig. 9.4c).

3. Any line and a point not on the line determine a unique plane.

4. The intersection of two planes is a line (Fig. 9.4d).

Two planes that do not intersect are said to be parallel planes. For example, in Fig. 9.5, plane ABE is parallel to plane GHF.

Two lines that do not lie in the same plane and do not intersect are called skewed lines. Figure 9.5 illustrates skewed lines ($\overleftrightarrow{AB}$ and $\overleftrightarrow{CD}$).

Angles

An angle, denoted $\angle$, is the union of two rays with a common end point (Fig. 9.6),

$$\overrightarrow{BA} \cup \overrightarrow{BC} = \angle ABC \text{ (or } \angle CBA).$$

An angle can be formed by the rotation of a ray about a point. An angle has an initial side and a terminal side: The initial side indicates the position of the ray prior to rotation; the terminal side indicates the position of the ray after rotation. The point common to both rays is called the vertex of the angle. The letter designating the vertex is always the middle one of the three letters designating an angle. The rays that make up the angle are called its sides.

There are several ways to name an angle. The angle in Fig. 9.6 may be denoted

$$\angle ABC, \qquad \angle CBA, \quad \text{or} \quad \angle B.$$

An angle divides a plane into three distinct parts: the angle itself, its interior, and its exterior. In Fig. 9.6 the angle is represented by the blue lines, the interior of the angle is shaded red, and the exterior is shaded green.

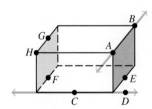

Figure 9.5

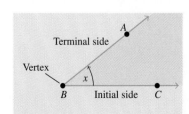

Figure 9.6

┌─ **EXAMPLE 2**

Refer to Fig. 9.7. Find the following.

a) $\overrightarrow{BF} \cup \overrightarrow{BD}$ **b)** $\angle ABF \cap \angle FBE$ **c)** $\overleftrightarrow{AC} \cap \overleftrightarrow{DE}$ **d)** $\overrightarrow{BA} \cup \overrightarrow{BC}$

Solution:
a) $\overrightarrow{BF} \cup \overrightarrow{BD} = \angle FBD$
b) $\angle ABF \cap \angle FBE = \overrightarrow{BF}$
c) $\overleftrightarrow{AC} \cap \overleftrightarrow{DE} = \{B\}$
d) $\overrightarrow{BA} \cup \overrightarrow{BC} = \overleftrightarrow{AC}$

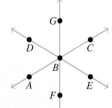

Figure 9.7

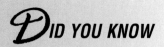
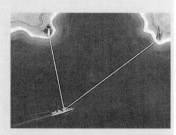

The measure of an angle, symbolized *m*, is the amount of rotation from its initial side to its terminal side. In Fig. 9.6, the letter *x* represents the measure of $\angle ABC$; therefore we may write $m\angle ABC = x$.

Angles can be measured in degrees, radians, or gradients. In this text we will discuss only the degree unit of measurement. The symbol for degrees is the same as the symbol for temperature degrees. An angle of 45 degrees is written 45°. A *protractor* is used to measure angles. The angle shown being measured by the protractor in Fig. 9.8 is 50°.

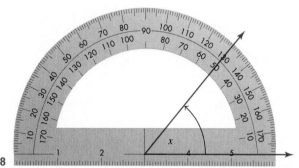

Figure 9.8

Consider a circle whose circumference is divided into 360 equal parts. If we draw a line from each mark on the circumference to the center of the circle, we get 360 wedge-shaped pieces. The measure of an angle formed by the straight sides of each wedge-shaped piece is defined to be 1 degree.

Angles are classified by their degree measurement, as shown in the following summary. A right angle is 90°; an acute angle is less than 90°; an obtuse angle is greater than 90° but less than 180°; a straight angle is 180°.

Right Angle	**Acute Angle**	**Obtuse Angle**	**Straight Angle**
$x = 90°$. The symbol ∟ is used to indicate right angles.	$0° < x < 90°$	$90° < x < 180°$	$x = 180°$

Two angles in the same plane are adjacent angles when they have a common vertex and a common side but no common interior points. $\angle DBC$ and $\angle CBA$ in Fig. 9.9 are adjacent angles. However, $\angle DBA$ and $\angle CBA$ are not adjacent angles.

Two angles the sum of whose measures is 90° are called complementary angles. Each angle is called a complement of the other.

EXAMPLE 3

If $m\angle ABC = 25°$, and $\angle ABC$ and $\angle CBD$ are complementary angles, determine $m\angle CBD$ (Fig. 9.10).

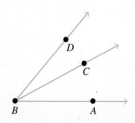

Figure 9.9

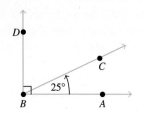

Figure 9.10

Solution: The sum of the two angles must be 90°, so

$$m\angle CBD = 90° - m\angle ABC$$
$$= 90° - 25°$$
$$= 65°$$

EXAMPLE 4

If $\angle ABC$ and $\angle DEF$ are complementary angles and $\angle DEF$ is twice as large as $\angle ABC$, determine the measure of each angle (Fig. 9.11).

Solution: If $m\angle ABC = x$, then $m\angle DEF = 2 \cdot x$ or $2x$, since it is twice as large. The sum of the measures of these angles must be 90°, so

$$m\angle ABC + m\angle DEF = 90°$$
$$x + 2x = 90°$$
$$3x = 90°$$
$$x = 30°$$

Therefore $m\angle ABC = 30°$ and $m\angle DEF = 2 \cdot 30°$, or 60°.

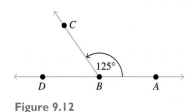

Figure 9.11

Two angles the sum of whose measures is 180° are said to be **supplementary angles**. Each angle is called a supplement of the other.

EXAMPLE 5

If $\angle ABC$ and $\angle CBD$ are supplementary angles and $m\angle ABC = 125°$, find $m\angle CBD$ (Fig. 9.12).

Solution: The sum of the two angles must be 180°, so

$$m\angle CBD = 180° - m\angle ABC$$
$$= 180° - 125°$$
$$= 55°$$

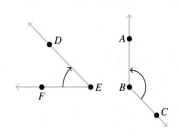

Figure 9.12

EXAMPLE 6

If $\angle ABC$ and $\angle DEF$ are supplementary angles and $\angle ABC$ is three times as large as $\angle DEF$, determine $m\angle ABC$ and $m\angle DEF$ (Fig. 9.13).

Solution: If $m\angle DEF = x$, then $m\angle ABC = 3x$.

$$m\angle ABC + m\angle DEF = 180°$$
$$3x + x = 180°$$
$$4x = 180°$$
$$x = 45°$$

Thus $m\angle DEF = 45°$, and $m\angle ABC = (3)(45°) = 135°$.

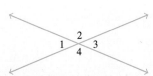

Figure 9.13

When two straight lines intersect, the nonadjacent angles formed are called **vertical angles**. In Fig. 9.14, $\angle 1$ and $\angle 3$ are vertical angles, and $\angle 2$ and $\angle 4$ are vertical angles. We can show that vertical angles have the same measure; that is, they are equal. For example, Fig. 9.14 shows that

$$m\angle 1 + m\angle 2 = 180°. \quad \text{Why?}$$
$$m\angle 2 + m\angle 3 = 180°. \quad \text{Why?}$$

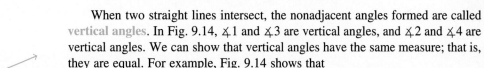

Figure 9.14

Since $\angle 2$ has the same measure in both cases, $m\angle 1$ must equal $m\angle 3$.

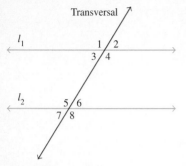

Transversal

Figure 9.15

A line that intersects two different lines, l_1 and l_2, at two different points is called a **transversal**. Figure 9.15 illustrates that when two parallel lines are cut by a transversal, eight angles are formed. Angles 3, 4, 5, and 6 are called **interior angles**, and angles 1, 2, 7, and 8 are called **exterior angles**. Eight pairs of supplementary angles are formed. Can you list them?

Special names are given to the angles formed by a transversal crossing two parallel lines.

Name	Description	Illustration	Pairs of Angles Meeting Criteria
Alternate interior angles	Interior angles on opposite sides of the transversal		∡3 and ∡6 ∡4 and ∡5
Alternate exterior angles	Exterior angles on opposite sides of the transversal		∡1 and ∡8 ∡2 and ∡7
Corresponding angles	One interior and one exterior angle on the same side of the transversal		∡1 and ∡5 ∡2 and ∡6 ∡3 and ∡7 ∡4 and ∡8

When two parallel lines are cut by a transversal, alternate interior angles have the same measure, alternate exterior angles have the same measure, and corresponding angles have the same measure.

EXAMPLE 7

Figure 9.16 shows two parallel lines cut by a transversal. Determine the measure of ∡1 through ∡7.

Solution:

$m∡6 = 53°$ ∡8 and ∡6 are vertical angles.

$m∡5 = 127°$ ∡8 and ∡5 are supplementary angles.

$m∡7 = 127°$ ∡5 and ∡7 are vertical angles.

$m∡1 = 127°$ ∡1 and ∡7 are alternate exterior angles.

$m∡4 = 53°$ ∡4 and ∡6 are alternate interior angles.

$m∡2 = 53°$ ∡6 and ∡2 are corresponding angles.

$m∡3 = 127°$ ∡3 and ∡1 are vertical angles.

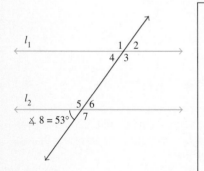

Figure 9.16

Section 9.1 Exercises

1. What is the difference between an axiom and a theorem?

2. a) What are the four key parts in the axiomatic method used by Euclid?
   b) Discuss each of the four parts.

3. What is the difference between a half line and a ray?

4. What are parallel lines?

5. What are skewed lines?

6. What are adjacent angles?

7. What are complementary angles?

8. What are supplementary angles?

9. What is a right angle?

10. What is an acute angle?

11. What is an obtuse angle?

12. What is a straight angle?

In Exercises 13–18, identify the figure as a line, half line, ray, line segment, open line segment, or half open line segment. Denote it by its appropriate symbol.

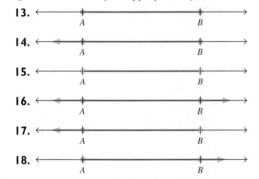

13.

14.

15.

16.

17.

18.

In Exercises 19–30, use the figure to find the following.

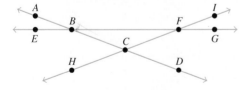

19. $\overrightarrow{FB} \cup \overrightarrow{FC}$

20. $\overset{\circ\circ}{FE} \cup \overrightarrow{FG}$

21. $\overrightarrow{BC} \cup \overset{\circ\circ}{CD}$

22. $\overleftrightarrow{AD} \cup \overline{BC}$

23. $\overleftrightarrow{AB} \cap \overline{HC}$

24. $\overrightarrow{BA} \cap \overrightarrow{BE}$

25. $\overleftrightarrow{AD} \cap \overline{BC}$

26. $\angle HCD \cap \angle ACF$

27. $\overleftrightarrow{BD} \cup \overset{\circ\circ}{CB}$

28. $\overleftrightarrow{BD} \cap \overleftrightarrow{CB}$

29. $\{C\} \cap \overset{\circ\circ}{CH}$

30. $\overrightarrow{BC} \cup \overline{CF} \cup \overline{FB}$

In Exercises 31–42, use the figure to find each of the following.

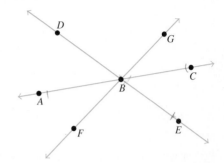

31. $\overset{\circ\circ}{BF} \cap \overrightarrow{BE}$

32. $\overrightarrow{BD} \cup \overset{\circ\circ}{BE}$

33. $\angle GBC \cap \angle CBE$

34. $\overleftrightarrow{DE} \cup \overrightarrow{BE}$

35. $\angle ABE \cup \overrightarrow{AB}$

36. $\overrightarrow{BF} \cup \overset{\circ\circ}{BE}$

37. $\angle CBE \cap \angle EBC$

38. $\{B\} \cap \overrightarrow{BA}$

39. $\overset{\circ\circ}{AC} \cap \overline{AC}$

40. $\overleftrightarrow{AC} \cap \overset{\circ\circ}{BE}$

41. $\overrightarrow{EB} \cap \overset{\circ\circ}{BE}$

42. $\overleftrightarrow{GF} \cap \overline{AB}$

In Exercises 43–46, classify the angle as acute, right, straight, or obtuse.

43.

44.

45.

46.

In Exercises 47–52, find the complementary angle of the given angle.

47. $43°$

48. $55°$

49. $67\frac{1}{2}°$

50. $36.8°$

51. $1°$

52. $2\frac{1}{3}°$

In Exercises 53–58, find the supplementary angle of the given angle.

53. $94°$

54. $14°$

55. $115°$

56. $13\frac{3}{4}°$

57. $89\frac{1}{4}°$

58. $114.7°$

In Exercises 59–64, match the name of the angles with angles 1 and 2 shown in (a)–(f).

a)

b)

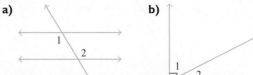

c)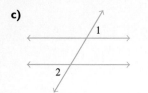

d)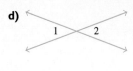

e)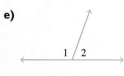

f)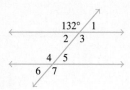

59. Vertical angles

60. Supplementary angles

61. Complementary angles

62. Alternate interior angles

63. Alternate exterior angles

64. Corresponding angles

65. The difference between the measures of two complementary angles is 24°. Determine the measures of the two angles.

66. If $\angle 1$ and $\angle 2$ are complementary angles and $\angle 1$ is seven times as large as $\angle 2$, find the measures of $\angle 1$ and $\angle 2$.

67. If $\angle 1$ and $\angle 2$ are supplementary angles and $\angle 1$ is eight times as large as $\angle 2$, find the measures of $\angle 1$ and $\angle 2$.

68. The difference between the measures of two supplementary angles is 74°. Determine the measures of the two angles.

69. From the parallel lines cut by a transversal shown here, determine the measures of $\angle 1$ through $\angle 7$.

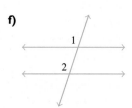

In Exercises 70–73, the angles are complementary angles. Find the measures of $\angle 1$ and $\angle 2$.

70.

71.

72.

73.

In Exercises 74–77, the angles are supplementary angles. Find the measures of $\angle 1$ and $\angle 2$.

74.

75.

76.

77.

78. a) How many lines can be drawn through a given point?
b) How many planes can be drawn through a given point?

79. What is the intersection of two distinct nonparallel planes?

80. How many planes can be drawn through a given line?

81. What is the intersection of a plane and a line (on a nonparallel plane)?

82. a) Will three noncollinear points A, B, and C always determine a plane? Explain.
b) Is it possible to determine more than one plane with three noncollinear points? Explain.
c) How many planes can be constructed through three collinear points?

83. Mark three noncollinear points A, B, and C. Draw the ray $\overrightarrow{AB}$ and the ray $\overrightarrow{AC}$. What geometric figure is produced?

The figure suggests a number of lines and planes. The lines may be described by naming two points, and the planes may be described by naming three points. In Exercises 84–93, use the figure to name the following.

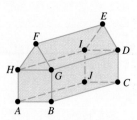

84. A pair of planes whose intersection is a line

85. Three planes whose intersection is a single point

86. Three planes whose intersection is a line

87. A line and a plane whose intersection is a point

88. A line and a plane whose intersection is a line

89. A pair of skewed lines

90. Three lines that intersect at a single point

91. Four planes whose intersection is a single point

92. What geometric advantage is there in building chairs with three legs rather than four legs?

93. Why is Euclid's *Elements* important?

In Exercises 94–99, determine whether the statement is always true, sometimes true, or never true. Explain your answer.

94. Two lines that are both parallel to a third line must be parallel to each other.

95. A triangle contains two acute angles.

96. Vertical angles are complementary angles.

97. Alternate exterior angles are supplementary angles.

98. Alternate interior angles are complementary angles.

99. A triangle contains two obtuse angles.

Problem Solving/Group Activity

100. Suppose that you have three distinct lines, all lying in the same plane. Find all the possible ways in which the three lines can be related. Sketch each case (four cases).

101. Suppose that you have three distinct planes in space. Find all the possible ways these planes can be related. Sketch each case (five cases).

102. Suppose that you have three distinct lines in space (not necessarily in the same plane). Find all the possible ways in which these three lines can be related. Sketch each case (nine cases).

Research Activity

103. Write a paper on Euclid's contributions to geometry.

9.2   POLYGONS

A polygon is a closed figure in a plane determined by three or more straight line segments. Examples of polygons are given in Fig. 9.17.

The straight line segments that form the polygon are called its sides and a point where two sides meet is called a vertex (plural vertices). The union of the sides of a polygon and its interior is called a polygonal region. A regular polygon is one whose sides are all the same length and whose interior angles all have the same measure. Figures 9.17(b) and (d) are regular polygons.

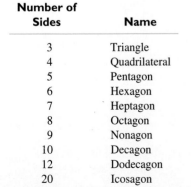

(a)   (b)   (c)   (d)

Figure 9.17

Polygons are named according to their number of sides. The names of some polygons are given in Table 9.1.

One of the most important polygons is the triangle. The sum of the measures of the interior angles of a triangle is 180°. To illustrate, consider triangle *ABC* given in Fig. 9.18. The triangle is formed by drawing two transversals through two parallel lines l_1 and l_2 with the two transversals intersecting at a point on l_1.

Suppose that the transversals make angles of 130° and 110° with l_2, as illustrated in Fig. 9.18.

Table 9.1

Number of Sides	Name
3	Triangle
4	Quadrilateral
5	Pentagon
6	Hexagon
7	Heptagon
8	Octagon
9	Nonagon
10	Decagon
12	Dodecagon
20	Icosagon

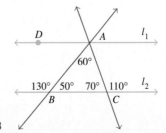

Figure 9.18

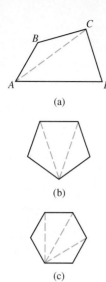

(a)

(b)

(c)

Figure 9.19

Table 9.2

Sides	Triangles	Sum of the Measures of the Interior Angles
3	1	$1(180°) = 180°$
4	2	$2(180°) = 360°$
5	3	$3(180°) = 540°$
6	4	$4(180°) = 720°$

Then $\angle ABC$ must measure 50° and $\angle ACB$ must measure 70°. By alternate interior angles, $\angle DAC$ must measure 110° and $\angle DAB$ must measure 50°. If we subtract 50° from 110°, we get 60° for the measure of $\angle BAC$. Now by adding the measures of $\angle ABC$, $\angle BCA$, and $\angle CAB$, we see that the sum of the measures of the interior angles of triangle ABC is 180°.

Consider the quadrilateral $ABCD$ (Fig. 9.19a). Drawing a straight line segment between any two vertices forms two triangles. Since the sum of the measures of the angles of a triangle is 180°, the sum of the measures of the interior angles of a quadrilateral is $2 \cdot 180°$, or 360°.

Now let's examine a pentagon (Fig. 9.19b). We can draw two straight line segments to form three triangles. Thus the sum of the measures of the interior angles of a five-sided figure is $3 \cdot 180°$, or 540°. Figure 9.19(c) shows that four triangles can be drawn in a six-sided figure. Table 9.2 summarizes this information.

If we continue this procedure, we can see that for an n-sided polygon the sum of the measures of the interior angles is $(n - 2)180°$.

> The *sum* of the measures of the interior angles of an n-sided polygon is $(n - 2)180°$.

EXAMPLE I

A gazebo is in the shape of a regular octagon (Fig. 9.20). Determine (a) the measure of an interior angle and (b) the measure of exterior angle 1.

Solution:

a) Using the formula $(n - 2)180°$, we can determine the sum of the measures of the interior angles of an octagon.

The sum of the measures of the interior angles is found as follows:

$$\text{Sum} = (8 - 2)180°$$
$$= 6(180°)$$
$$= 1080°$$

The measure of an interior angle of a regular polygon can be determined by dividing the sum of the interior angles by the number of angles.

The measure of an interior angle of a regular octagon is found as follows:

$$\text{Measure} = \frac{1080°}{8} = 135°$$

b) Since $\angle 1$ is the supplement of an interior angle,

$$m\angle 1 = 180° - 135° = 45°.$$ ▲

Figure 9.20

In order to discuss area in the next section we must be able to identify various types of triangles and quadrilaterals. The following is a summary of certain types of triangles and their characteristics.

Acute Triangle	**Obtuse Triangle**	**Right Triangle**
All angles are acute angles.	One angle is obtuse.	One angle is a right angle. hypotenuse
Isosceles Triangle	**Equilateral Triangle**	**Scalene Triangle**
Two equal sides Two equal angles	Three equal sides Three equal angles (60° each)	No two sides are equal in length.

Similar Figures

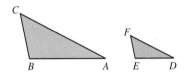

Figure 9.21

In everyday living we often have to deal with geometric figures that have the "same shape" but are of different sizes. For example, an architect will make a small-scale drawing of a floor plan, or a photographer will make an enlargement of a photograph. Figures that have the same shape but may be of different sizes are called similar figures. Two similar figures are illustrated in Fig. 9.21.

Similar figures have *corresponding angles* and *corresponding sides*. In Fig. 9.21 triangle *ABC* has angles *A*, *B*, and *C*. Their respective corresponding angles in triangle *DEF* are angles *D*, *E*, and *F*. Sides *AB*, *BC*, and *AC* in triangle *ABC* have corresponding sides *DE*, *EF*, and *DF*, respectively, in triangle *DEF*.

> Two polygons are **similar** if their corresponding angles have the same measure and their corresponding sides are in proportion.

In Fig. 9.21, $\angle A$ and $\angle D$ have the same measure, as do $\angle B$ and $\angle E$ and $\angle C$ and $\angle F$. Also, the corresponding sides of similar triangles are in proportion:

$$\frac{\overline{AB}}{\overline{DE}} = \frac{\overline{CB}}{\overline{FE}} = \frac{\overline{AC}}{\overline{DF}}.$$

EXAMPLE 2

Consider the similar figures in Fig. 9.22.

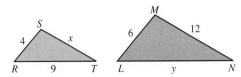

Figure 9.22

Determine the following:

a) The length of side $\overline{ST}$. **b)** The length of side $\overline{LN}$.

Solution:

a) We represent side $\overline{ST}$ with the letter x. Because the corresponding sides must be in proportion, we can write a proportion (as explained in Section 6.2) to find side $\overline{ST}$. Corresponding sides $\overline{RS}$ and $\overline{LM}$ are known, so we use them as one ratio in the proportion.

$$\frac{\overline{RS}}{\overline{LM}} = \frac{\overline{ST}}{\overline{MN}}$$

$$\frac{4}{6} = \frac{x}{12}$$

Now we solve for x.

$$4 \cdot 12 = 6 \cdot x$$
$$48 = 6x$$
$$8 = x$$

Thus the length of $\overline{ST} = 8$.

b) We represent $\overline{LN}$ with the letter y.

$$\frac{\overline{RS}}{\overline{LM}} = \frac{\overline{RT}}{\overline{LN}}$$

$$\frac{4}{6} = \frac{9}{y}$$

$$4 \cdot y = 6 \cdot 9$$
$$4y = 54$$
$$y = \frac{54}{4} = \frac{27}{2}, \text{ or } 13.5$$

Thus the length of $\overline{LN} = 13.5$.

▲

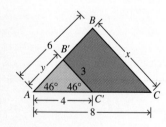

Figure 9.23

EXAMPLE 3

Consider the similar triangles ABC and $AB'C'$ (Fig. 9.23). Determine

a) the length of side $\overline{BC}$.   **b)** the length of side $\overline{AB'}$.

c) $m\angle AB'C'$.   **d)** $m\angle ACB$.

e) $m\angle ABC$.

Solution:

a) We set up a proportion to find the length of side $\overline{BC}$. Since corresponding sides $\overline{AC}$ and $\overline{AC'}$ are known, we use them as one ratio in the proportion.

$$\frac{\overline{AC}}{\overline{AC'}} = \frac{\overline{BC}}{\overline{B'C'}}$$

$$\frac{8}{4} = \frac{x}{3}$$

$$24 = 4x$$
$$6 = x$$

Therefore the length of $\overline{BC} = 6$.

b)
$$\frac{\overline{AC}}{\overline{AC'}} = \frac{\overline{AB}}{\overline{AB'}}$$

$$\frac{8}{4} = \frac{6}{y}$$

$$8y = 24$$

$$y = 3$$

Therefore the length of $\overline{AB'} = 3$.

c) The sum of the angles in a triangle must measure 180°, so $m\sphericalangle AB'C'$ must be 180° − (46° + 46°), or 88°.

d) Triangles ABC and $AB'C'$ are similar, so their corresponding angles must have the same measure. Therefore $\sphericalangle ACB$ must have the same measure as its corresponding angle $\sphericalangle AC'B'$. Thus $m\sphericalangle ACB = 46°$.

e) Because $\sphericalangle ABC$ and $\sphericalangle AB'C'$ are corresponding angles of similar triangles, $m\sphericalangle ABC = m\sphericalangle AB'C'$. Using the results from part (c), we determine that $m\sphericalangle ABC = 88°$.   ▲

Congruent Figures

If the corresponding sides of two similar triangles are the same length, the figures are called congruent figures. Corresponding angles of congruent figures have the same measure, and the corresponding sides are equal in length. Two congruent figures coincide when placed one upon the other.

Figure 9.24

EXAMPLE 4

Triangles ABC and DEF in Fig. 9.24 are congruent. Find

a) the length of side $\overline{DF}$.   **b)** the length of side $\overline{AB}$.

c) $m\sphericalangle FDE$.   **d)** $m\sphericalangle ACB$.

e) $m\sphericalangle ABC$.

Solution: Because $\triangle ABC$ is congruent to $\triangle DEF$, we know that the corresponding sides and angles are equal.

a) $\overline{DF} = \overline{AC} = 12$

b) $\overline{AB} = \overline{DE} = 7$

c) $m\sphericalangle FDE = m\sphericalangle CAB = 65°$

d) $m\sphericalangle ACB = m\sphericalangle DFE = 34°$

e) The sum of the angles of a triangle is 180°. Since $m\sphericalangle BAC = 65°$ and $m\sphericalangle ACB = 34°$, $m\sphericalangle ABC = 180° - 65° - 34° = 81°$.   ▲

Quadrilaterals

Quadrilaterals are four-sided polygons. Quadrilaterals may be classified according to their characteristics, as illustrated in the summary box. The sum of the angles in any quadrilateral measures 360°.

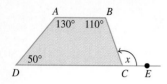

Figure 9.25

EXAMPLE 5

Find the measure of the exterior angle, x, of the trapezoid in Fig. 9.25.

Solution: Since sides AB and CD are parallel, BC may be considered a transversal. Then by alternate interior angles, $m\angle x = 110°$. ▲

A second method to solve Example 5 is to find the measure of the unknown interior angle of the trapezoid by subtracting the sum of the measures of the three given angles from 360°: $m\angle BCD = 360° - (50° + 130° + 110°) = 360° - 290° = 70°$. Since $\angle DCE$ is a straight angle, $m\angle x$ must be $180° - 70° = 110°$.

Trapezoid	**Parallelogram**	**Rhombus**
Two sides are parallel.	Both pairs of opposite sides are parallel.	Both pairs of opposite sides are parallel. The four sides are equal in length.

Rectangle	**Square**
Both pairs of opposite sides are parallel. The angles are right angles.	Both pairs of opposite sides are parallel. The four sides are equal in length. The angles are right angles.

Section 9.2 Exercises

1. What are similar figures?

2. What are congruent figures?

In Exercises 3–7, name the polygon.

3.

4.

5.

6.

7.

In Exercises 8–12, identify the triangle as scalene, isosceles, or equilateral.

8.

9.

10.

11.

12.

In Exercises 13–17, identify the triangle as acute, obtuse, or right.

13.   **14.**   **15.**

16.   **17.**

In Exercises 18–22, identify the quadrilateral.

18.    **19.**

20.    **21.**   **22.**

In Exercises 23–25, find the measure of ∡x.

23.

24.

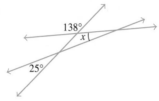

25.

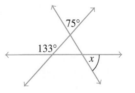

26. Lines l_1 and l_2 are parallel. Determine the measures of ∡1 through ∡12.

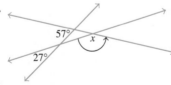

27. Sketch an octagon.
   a) Show that the figure can be divided into six triangles.
   b) Use the figure to determine the sum of the interior angles.
   c) Use the formula, sum = $(n - 2)\,180°$, to verify the results obtained in part (b).

28. Sketch a decagon. Answer similar questions for this figure as asked in Exercise 27.

In Exercises 29–32, find the sum of the interior angles of the polygon.

29. Nonagon   **30.** Hexagon

31. Pentagon   **32.** Icosagon

In Exercises 33–38, find the measure of an interior angle of the regular polygon. If a side of the polygon is extended, find the supplement of the interior angle. See Example 1.

33. Triangle   **34.** Quadrilateral   **35.** Heptagon

36. Decagon   **37.** Pentagon   **38.** Octagon

In Exercises 39–41, determine the measure of the angle. In the figure, ∡ABC makes an angle of 125° with the floor and l_1 and l_2 are parallel.

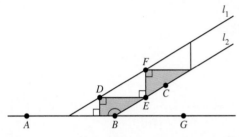

39. ∡GBC   **40.** ∡EDF   **41.** ∡DFE

42. The legs of a picnic table form an isosceles triangle as indicated in the figure. If ∡ABC = 80°, find ∡x and ∡y so that the top of the table will be parallel to the ground.

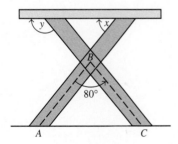

For the shopping cart illustrated, $\overline{wz}$ is parallel to $\overline{xy}$. Find

43. m∡w.   **44.** m∡x.

45. m∡y.   **46.** m∡z.

In Exercises 47–49, the figures are similar. Find the length of all sides colored red.

47.

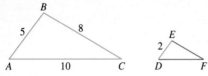

48.

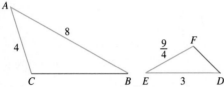

49.

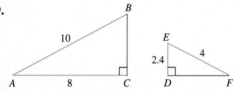

In Exercises 50–52, triangles ABC and A'B'C' are similar figures. Find the length of

50. side $\overline{A'C}$.　　**51.** side $\overline{BC}$.　　**52.** side $\overline{BB'}$.

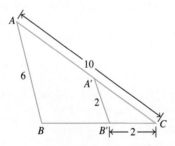

In Exercises 53–55, quadrilaterals ABCD and A'B'C'D' are similar figures. Find the length of

53. side $\overline{AB}$.　　**54.** side $\overline{BC}$.　　**55.** side $\overline{D'C'}$.

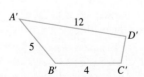

In Exercises 56–60, find the measures of the angles for similar hexagons ABCDEF and A'B'C'D'E'F'.

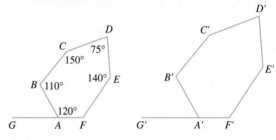

56. ∡EFA　　　　　　　　**57.** ∡F'A'B'

58. ∡B'C'D'　　　　　　**59.** ∡GAB

60. ∡F'E'D'

In Exercises 61–66, find the length of the sides and the measures of the angles for the congruent triangles ABC and A'B'C'.

61. The length of side $\overline{A'B'}$

62. The length of side $\overline{A'C'}$

63. The length of side $\overline{BC}$

64. ∡B'A'C'

65. ∡ACB

66. ∡ABC

In Exercises 67–72, find the length of the sides and the measures of the angles for the congruent quadrilaterals ABCD and A'B'C'D'.

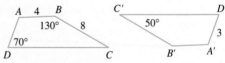

67. The length of side $\overline{A'B'}$

68. The length of side $\overline{AD}$

69. The length of side $\overline{B'C'}$

70. ∡BCD

71. ∡A'D'C'

72. ∡DAB

Problem Solving/Group Activity

73. As shown, Emily holds a stick 1 ft long at a distance of 2.0 ft from her eye. Her eye is 192 ft from a line that passes through the center of the tree. She holds the stick so that it is parallel to the tree as she sights along the top of the stick to a point *A* (the top of the tree) and along the bottom of the stick to point *B* (the base of the tree).

a) Explain why triangles *CDE* and *CAB* are similar figures.

b) Determine the height of the tree.

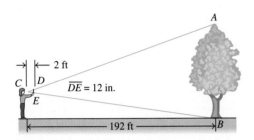

74. a) In the figure $m \angle CED = m \angle ABC$. Explain why triangles *ABC* and *DEC* must be similar.

b) Determine the distance across the lake, $\overline{DE}$.

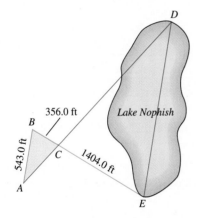

75. You are asked to measure the height of an inside wall of a warehouse. No ladder tall enough to measure the height is available. You borrow a mirror from a salesclerk and place it on the floor. You then move away from the mirror until you can see the reflection of the top of the wall in it, as shown in the figure.

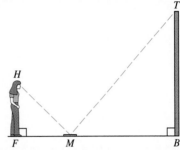

a) Explain why triangle *HFM* is similar to triangle *TBM*. (*Hint:* In the reflection of light the angle of incidence equals the angle of reflection. Thus $\angle HMF = \angle TMB$.)

b) If your eyes are $5\frac{1}{2}$ ft above the floor, you are $2\frac{1}{2}$ ft from the mirror, and the mirror is 20 ft from the wall, how high is the wall?

76. In this book the discussion of congruence is limited to polygons. Can three-dimensional objects be congruent? List the criteria that you believe would be necessary for two three-dimensional figures to be congruent. Give several examples of three-dimensional figures that you believe are congruent.

Research Activities

77. Write a paper on transformation geometry. Include a discussion and examples of rotation, reflection, and translation.

78. Write a paper on the history and use of a theodolite, a surveying instrument.

Figure 9.26

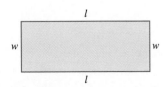

Figure 9.27

9.3 PERIMETER AND AREA

Perimeter and Area

Geometric shapes abound in the natural world and the world made by human beings. For example, a basketball court is a rectangle, a basketball is a sphere, a can of food is a cylinder, and a ream of paper is a rectangular solid.

The *perimeter*, *P*, of a two-dimensional figure, is the sum of the lengths of the sides of the figure. In Figs. 9.26 and 9.27 the sums of the red lines are the perimeters. Perimeters are measured in the same units as the sides. For example, if the sides of a figure are measured in feet, the perimeter will be measured in feet.

The *area*, *A*, is the region within the boundaries of the figure. The blue color in Figs. 9.26 and 9.27 indicate the areas of the figures. Area is measured in square units. For example, if the sides of a figure are measured in inches, the area of the figure will be measured in square inches (in^2). (See Table 8.7 on page 421 for common units of area in the U.S. customary and metric systems.)

Consider the rectangle in Fig. 9.27. Two sides of the rectangle have length l, and two sides of the rectangle have width w. Thus, if we add the lengths of the four sides to get the perimeter, we find $P = l + w + l + w = 2l + 2w$.

Perimeter of a Rectangle

$$P = 2l + 2w$$

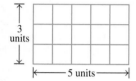

Consider a rectangle of length 5 units and width 3 units (Fig. 9.28). Counting the number of 1-unit by 1-unit squares within the figure we obtain the area of the rectangle, 15 square units. The area can also be obtained by multiplying the number of units of length by the number of units of width, or 5 units $\times$ 3 units = 15 square units. We can find the area by the formula Area = length $\times$ width.

Figure 9.28

Area of a Rectangle

$$A = l \times w$$

Using the formula for the area of a rectangle, we can determine the formulas for the areas of other figures.

A square (Fig. 9.29) is a rectangle that contains four equal sides. Therefore the length equals the width. If we call both the length and the width of the square s,

Figure 9.29

$$A = l \times w, \quad \text{so} \quad A = s \times s = s^2.$$

Area of a Square

$$A = s^2$$

A parallelogram with height h and base b is shown in Fig. 9.30(a).

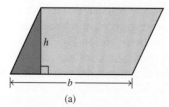
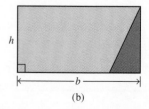

(a)   (b)

Figure 9.30

If we were to cut off the red portion of the parallelogram on the left, Fig. 9.30(a), and attach it to the right side of the figure, the resulting figure would be a rectangle, Fig. 9.30(b). Since the area of the rectangle is $b \times h$, the area of the parallelogram is also $b \times h$.

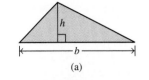

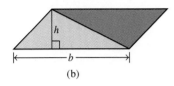

Figure 9.31

Area of a Parallelogram

$$A = b \times h$$

Consider the triangle with height, h, and base, b, shown in Fig. 9.31(a). Using this triangle and a second identical triangle we can construct a parallelogram, Fig. 9.31(b). The area of the parallelogram is bh. The area of the triangle is one-half that of the parallelogram. Therefore the area of the triangle is $\frac{1}{2}$(base)(height).

Area of a Triangle

$$A = \tfrac{1}{2}bh$$

Now consider the trapezoid shown in Fig. 9.32(a). We can partition the trapezoid into two triangles by drawing diagonal $\overline{DB}$, as in Fig. 9.32(b). One triangle has base $\overline{AB}$ (called b_2) with height $\overline{DE}$, and the other triangle has base $\overline{DC}$ (called b_1) with height $\overline{FB}$. Note the line used to measure the height of the triangle need not be inside the triangle. Because heights $\overline{DE}$ and $\overline{FB}$ are equal, both triangles have the same height, h. The area of triangles DCB and ADB are $\frac{1}{2}b_1h$ and $\frac{1}{2}b_2h$, respectively. The area of the trapezoid is the sum of the areas of the triangles, $\frac{1}{2}b_1h + \frac{1}{2}b_2h$, which can be written $\frac{1}{2}h(b_1 + b_2)$.

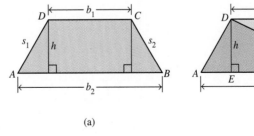

(a)   (b)

Figure 9.32

Area of a Trapezoid

$$A = \tfrac{1}{2}h(b_1 + b_2)$$

Following is a summary of the areas and perimeters of selected figures.

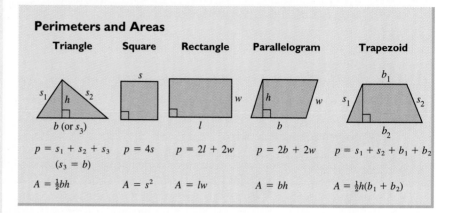

Perimeters and Areas

Triangle	Square	Rectangle	Parallelogram	Trapezoid
$p = s_1 + s_2 + s_3$ ($s_3 = b$)	$p = 4s$	$p = 2l + 2w$	$p = 2b + 2w$	$p = s_1 + s_2 + b_1 + b_2$
$A = \frac{1}{2}bh$	$A = s^2$	$A = lw$	$A = bh$	$A = \frac{1}{2}h(b_1 + b_2)$

EXAMPLE 1

You are coating your driveway with blacktop sealer. One can of sealer costs \$9.49 and covers 300 ft². If your driveway is 50 ft long and 10 ft wide, **(a)** find the area of your driveway, **(b)** determine how many cans of blacktop sealer you need, and **(c)** determine the cost of sealing the driveway.

Solution:

a) The area of your driveway in square feet is

$$A = l \cdot w = 50 \cdot 10 = 500 \text{ ft}^2.$$

The area of the driveway is in square feet because both the length and width are measured in feet.

b) To determine the number of cans of sealer you need, divide the area of the driveway by the area covered by one can of sealer.

$$\frac{500}{300} = 1\frac{2}{3}$$

You cannot ordinarily purchase two-thirds of a can of sealer, so you need two cans of sealer.

c) The cost of the sealer is $2 \times \$9.49$, or \$18.98.   ▲

Pythagorean Theorem

We introduced the Pythagorean theorem in Chapter 5. This theorem is an important tool for finding the perimeter and area of triangles. For this reason we restate it here.

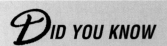
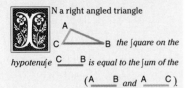
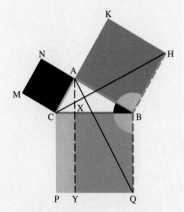
Pythagorean Theorem

The sum of the squares of the lengths of the legs of a right triangle equals the square of the length of the hypotenuse.

Symbolically, if a and b represent the lengths of the legs and c represents the length of the hypotenuse (the side opposite the right angle), then $a^2 + b^2 = c^2$.

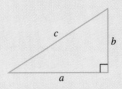

EXAMPLE 2

The bases on a baseball field are 90 ft apart (Fig. 9.33). Find the distance from home plate to second base.

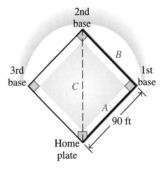

Figure 9.33

Solution: The line from home plate to second base is the hypotenuse of a right triangle with sides of 90 ft. By the Pythagorean theorem,

$$c^2 = a^2 + b^2$$
$$c^2 = (90)^2 + (90)^2$$
$$c^2 = 8100 + 8100 = 16{,}200$$
$$c = \sqrt{16{,}200} \approx 127.3 \text{ ft.}$$

Therefore the distance from home plate to second base is about 127 ft. ▲

Circles

A commonly used plane figure that is not a polygon is a *circle*. A circle is a set of points equidistant from a fixed point called the center. A radius, r, of a circle is a line segment from the center of the circle to any point on the circle (Fig. 9.34 on page 456). A diameter, d, of a circle is a line segment through the center of a circle with both end points on the circle. Note that the diameter of the circle is

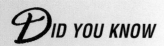

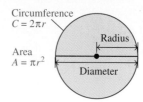

Figure 9.34

twice its radius. The circumference is the length of the simple closed curve that forms the circle. The formulas for the area and circumference of a circle are given in Fig. 9.34. We introduced the symbol pi, π, in Chapter 5. Recall that π is approximately 3.14.

EXAMPLE 3

At the local pizza shop the price for a medium plain pizza is $7.50, and the price for a large plain pizza is $14.00. The diameters of the large and medium pizzas are 18 inches and 12 inches, respectively, and both pizzas are the same thickness. To get the most for your money should you buy two medium pizzas or one large pizza?

Solution: We begin by finding the surface area of the two pizzas. The radius of the larger pizza is 9 in. and the radius of the smaller pizza is 6 in.

Area of a Large Pizza	Area of a Medium Pizza
$A = \pi r^2$	$A = \pi r^2$
$= 3.14(9)^2$	$= 3.14(6)^2$
$= 3.14(81) = 254.34 \text{ in}^2$	$= 3.14(36) = 113.04 \text{ in}^2$

The two medium pizzas have an area of $2(113.04) = 226.08 \text{ in}^2$.

	One Large Pizza	Two Medium Pizzas
Area	254.34 in^2	226.08 in^2
Cost	$14	$15

One large pizza has a greater area and a lower cost. Therefore one large pizza should be purchased.   ▲

EXAMPLE 4

You plan to fertilize your lawn. The shapes and dimensions of your lot, house, pool, and shed are shown in Fig. 9.35. One bag of fertilizer costs $25.95 and

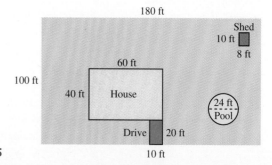

Figure 9.35

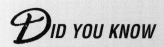

DID YOU KNOW

CITY PLANNING

$\mathcal{T}$he Roman poet Virgil tells the story of Queen Dido, who fled to Africa after her brother murdered her husband. There, she begged some land from King Iarbus, telling him she only needed as much land as the hide of an ox would enclose. Being very clever, she decided that the greatest area would be enclosed if she tore the hide into thin strips and formed them into a circle. On this land she founded the city of Byrsa (the Greek word for "hide"), later known as Carthage. ∎

covers 5000 ft^2. Determine how many bags of fertilizer you need and the total cost of the fertilizer.

Solution: The total area of the lot is $100 \cdot 180$, or 18,000 ft^2. From this total subtract the area of the house, driveway, pool, and shed.

$$\text{Area of house} = 60 \cdot 40 = 2400 \text{ ft}^2$$

$$\text{Area of driveway} = 20 \cdot 10 = 200 \text{ ft}^2$$

$$\text{Area of shed} = 8 \cdot 10 = 80 \text{ ft}^2$$

The diameter of the pool is 24 ft, so its radius is 12 ft.

$$\text{Area of pool} = \pi r^2 = 3.14(12)^2 = 3.14(144) = 452.16 \text{ ft}^2$$

The total area of the house, driveway, shed, and pool is approximately $2400 + 200 + 80 + 452$, or 3132 ft^2. The area to be fertilized is $18,000 - 3132$, or 14,868 ft^2. The number of bags of fertilizer is found by dividing the total area to be fertilized by the number of square feet covered per bag.

The number of bags of fertilizer is 14,868/5000, or 2.97 bags. Therefore you need three bags. At $25.95 per bag, the total cost is $77.85. ▲

Sometimes converting an area from one square unit of measure to a different square unit of measure may be necessary. For example, you may have to change square feet of carpet to square yards of carpet before placing an order with a dealer. Examples 5 and 6 show how to do conversions.

EXAMPLE 5

a) Convert one square foot to square inches.

b) Convert 7.5 square feet to square inches.

Solution:

a) 1 ft = 12 in.
  Therefore 1 ft^2 = 12 in. × 12 in. = 144 in^2.

b) Since 1 ft^2 = 144 in^2, 2 ft^2 = 2(144 in^2) = 288 in^2, and 7.5 ft^2 is

$$7.5 \text{ ft}^2 = 7.5 \times 144 = 1080 \text{ in}^2. \qquad ▲$$

EXAMPLE 6

The area of a four-wall racquetball court is 115,200 in^2. How large is this area in square feet?

Solution: Since 1 ft^2 = 144 in^2, divide 115,200 in^2 by 144 in^2 to obtain the number of square feet.

$$115,200 \text{ in}^2 = \frac{115,200}{144} = 800 \text{ ft}^2 \qquad ▲$$

When multiplying units of length, be sure that the units are the same. You can multiply feet by feet to get square feet or yards by yards to get square yards. However, you cannot get a valid answer if you multiply numbers expressed in feet by numbers expressed in yards.

Section 9.3 Exercises

In Exercises 1–3, find the area of the triangle.

1.

6 in.
9 in.

2.

3 cm
5 cm

3.

5 yd
3 ft

In Exercises 4–10, find the area and perimeter of the quadrilateral.

4.
5 ft
12 ft

5.

4 in.
5 in.
8 in.

6.

14 in.
3 ft   2 ft   4 ft
30 in.

7.
6 ft
6 ft

8.
2 ft
2 yd

9.
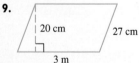
20 cm   27 cm
3 m

10.
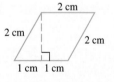
2 cm
2 cm
2 cm
1 cm   1 cm

In Exercises 11–14, find the area and circumference of the circle.

11.

4 in.

12.
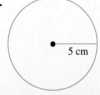
5 cm

13.
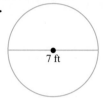
7 ft

14.
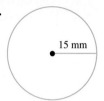
15 mm

In Exercises 15–18, use the Pythagorean theorem to determine the third side of the triangle.

15.
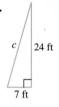
c 24 ft
7 ft

16.

15 in.   13 in.
a

17.
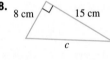
b 18 m
15 m

18.
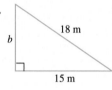
8 cm   15 cm
c

In Exercises 19–26, find the shaded area.

19.
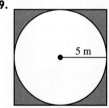
5 m

20.
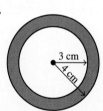
3 cm
4 cm

21.
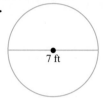
4 in.
4 in.

22.
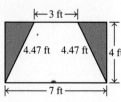
3 ft
4.47 ft   4.47 ft   4 ft
7 ft

23.
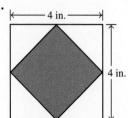
8 m   6 m
10 m

24.

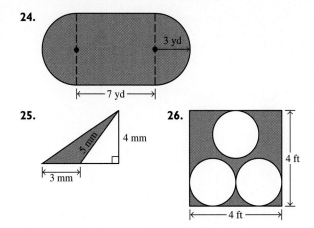

25.   **26.**

27. 11.3 ft² to square yards.   **28.** 302 ft² to square yards.
29. 18.3 yd² to square feet.   **30.** 14.7 yd² to square feet.

One square yard equals 9 sq ft. Use this information to convert

One square meter equals 10,000 square centimeters. Use this information to convert

31. 15.6 m² to cm²   **32.** 19.2 m² to cm²
33. 608 cm² to m²   **34.** 1075 cm² to m²

In Exercises 35–37, use the measurements given on the floor plans to obtain an answer.

35. One method of estimating the cost of building a new house is to use square footage of living space (using exterior dimensions). If the current rate is $60 per square foot, find the cost of building the house.

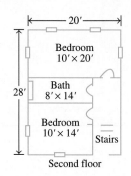

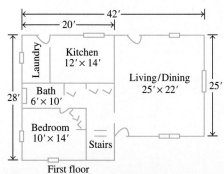

36. Determine the cost of putting a hardwood floor in the living/dining area of the house. The cost of the hardwood flooring selected is $4.50 per square foot. Add 5% to the cost for waste.

37. Determine the cost of covering the three bedroom floors in the house with carpet. The price of the carpet selected is $14.95 per square yard. (Add 5% for waste.)

38. A building 40 m by 64 m is surrounded by a walk 1.5 m wide. Find the dimensions of the region covered by the building and the walk. Find the area of the walk.

39. A gallon of wood sealer covers 300 ft². Golda Zenna plans to seal the floor of a gymnasium that is 100 ft by 160 ft. How many gallons of sealer are needed?

40. Thong Seng has selected carpet that costs $19.99 per square yard to use in the living room and family room. The living room measures 15 ft by 21 ft, and the family room measures 18 ft by 16 ft. How many square yards of carpeting will she need? What will it cost?

41. The Pellas' lot is illustrated here.

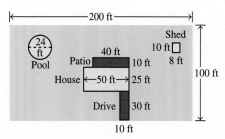

The Pellas' son Tony is taking a college mathematics course and so Carmine Pella asks him the following questions. Can you answer them?
a) How many square feet of lawn do they have?
b) If a bag of fertilizer covers 5000 ft², how many bags of fertilizer will they need to cover the lawn?
c) If a bag of fertilizer costs $29.95, what will be the total cost of the fertilizer?

In Exercises 42–49, explain your answer.

42. What is the effect on the area of a square if a side is doubled?

43. What is the effect on the area of a square if a side is tripled?

44. What is the effect on the area of a square if a side is cut in half?

45. What is the effect on the area of a rectangle if its length is doubled and its width is left unchanged?

46. What is the effect on the area of a rectangle if its length and width are both doubled?

47. What is the effect on the area of a rectangle if its length is doubled and its width is cut in half?

48. What is the effect on the area of a triangle if its base is doubled and its altitude is left unchanged?

49. What is the effect on the area of a triangle if its base is doubled and its altitude is cut in half?

50. An isosceles right triangle has an area of 72 cm². What are the lengths of the sides?

51. Herman's garden is 11.5 m by 15.4 m.
   a) How large is his garden in square meters?
   b) If 1 ha equals 10 000 m², how large is his garden in hectares?

52. The shape and measurements of the base of a free-standing fireplace are shown. The base will be covered with a special stone. Determine the number of square yards of stone the mason must order.

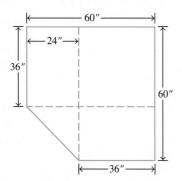

53. Which hamburger has the larger surface area: a square hamburger 3 in. on a side from Wendy's or a $3\frac{1}{2}$-in.-diameter round hamburger from Burger King? Explain your answer and give the difference in their surface areas.

54. Can two figures have the same perimeters and areas and not be congruent figures? Explain, or show how you determined your answer.

Problem Solving/Group Activity

55. The Kents are painting their house this summer. The dimensions of the house are shown.

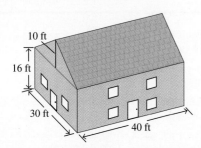

   a) Find the total number of square feet to be painted. (Assume that there are 12 windows, each 2 ft 3 in. × 3 ft 4 in., and two doors, each 80 in. × 36 in. The roof will not be painted.)
   b) Determine the number of gallons needed if two coats of paint are required. For each coat assume that 1 gal of paint covers 500 ft².

   c) Determine the cost of the paint if paint sells for $24.95/gal.

56. Find the surface area of the right circular cylinder shown. Use $s = 2\pi rh + 2\pi r^2$.

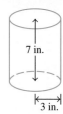

57. Find the surface area of the rectangular solid shown.

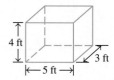

In Exercises 58–60, write an expression that represents the shaded area.

58.

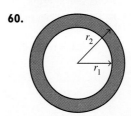

59.

60.

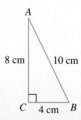

61. A second formula for determining the area of a triangle (called Heron's formula) is

$$A = \sqrt{s(s - a)(s - b)(s - c)},$$

where $s = \frac{1}{2}(a + b + c)$ and a, b, and c are the lengths of the sides of the triangle. Use Heron's formula to determine the area of right triangle ABC and check your answer using the formula $A = \frac{1}{2}ab$.

62. A developer is offered two triangular pieces of land. The dimensions of the first piece are 15 ft × 25 ft × 35 ft. The dimensions of the second piece are 60 ft × 100 ft × 140 ft. Since the dimensions of the second piece are four times the dimensions of the first, the two triangles are similar. The price of the larger piece of land is eight times the price of the smaller piece of land. Determine which piece of land is the least expensive per square foot.

Research Activities

For Exercises 63 and 64, references include history of mathematics books and encyclopedias.

63. The early Babylonians and Egyptians did not know about π and had to devise techniques to approximate the area of a circle. Do research and write a paper on the techniques these societies used to determine the area of a circle.

64. Write a paper on the contributions of Heron of Alexandria to geometry.

9.4 VOLUME

When discussing a one-dimensional figure, such as a line, we can find its length. When discussing a two-dimensional figure, such as a rectangle, we can find its area. When discussing a three-dimensional figure, such as a sphere, we can find its volume. **Volume** is a measure of the capacity of a figure. The measure of volume may be confusing because we use different units to measure different types of volumes. For example water and other liquids may be measured in ounces, quarts, or gallons. A volume of topsoil may be measured in cubic yards. In the metric system, liquid may be measured in liters or milliliters, and topsoil may be measured in cubic meters.

Solid geometry is the study of three-dimensional solid figures (also called space figures). Volumes of three-dimensional figures are measured in cubic units, such as cubic feet or cubic meters.

We will begin our discussion with the *rectangular solid*. If the length of the solid is 5 units, the width is 2 units, and the height is 3 units, the total number of cubes is 30 (Fig. 9.36). Thus the volume is 30 cubic units. The volume of a rectangular solid can also be found by multiplying its length times width times height; in this case, 5 units × 2 units × 3 units = 30 cubic units. In general, the volume of any rectangular solid, as shown in part (a) of the summary box on page 462, is $V = l \times w \times h$.

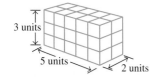

Figure 9.36

> ### Volume of a Rectangular Solid
> $$V = l \times w \times h$$

A *cube* is a rectangular solid having the same length, width, and height (part b of the box on page 462). If we call the side of a cube *s* and use the formula for a rectangular solid, substituting *s* for *l*, *w*, and *h*, we obtain $V = s \cdot s \cdot s = s^3$.

> ### Volume of a Cube
> $$V = s^3$$

Now consider the right circular cylinder (part c of the box on page 462). The base is a circle with area πr^2. When we add height, *h*, the figure becomes a cylinder. For the same circular base, the greater the height, the greater is the volume. The volume of the right circular cylinder is found by multiplying the area of the base, πr^2, by the height h.

Volume of a Cylinder

$$V = \pi r^2 h$$

A cone is illustrated in part (d) of the box. Imagine a cone inside a cylinder, sharing the same circle as base. The volume of the cone is less than the volume of the cylinder that has the same base and height (Fig. 9.37). In fact, the volume of the cone is one-third the volume of the cylinder.

Volume of a Cone

$$V = \tfrac{1}{3}\pi r^2 h$$

The next shape we will discuss in this section is the sphere (part e of the box). Basketballs, golf balls, and so on have the shape of a sphere. The formula for the volume of a sphere is as follows.

Volume of a Sphere

$$V = \tfrac{4}{3}\pi r^3$$

The following is a summary of the volumes of selected three-dimensional figures.

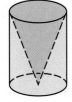

Figure 9.37

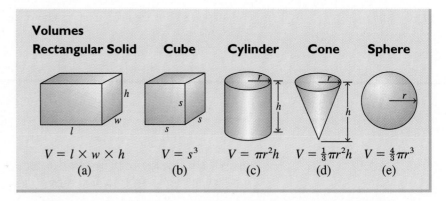

Volumes

Rectangular Solid	Cube	Cylinder	Cone	Sphere
$V = l \times w \times h$	$V = s^3$	$V = \pi r^2 h$	$V = \tfrac{1}{3}\pi r^2 h$	$V = \tfrac{4}{3}\pi r^3$
(a)	(b)	(c)	(d)	(e)

EXAMPLE I

The base of a patio is to be a slab of concrete 12 ft long by 9 ft wide by 6 in. thick (Fig. 9.38).

a) How many cubic yards of concrete are needed?

b) If one cubic yard of concrete costs \$54.50, how much will the concrete for the patio cost?

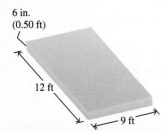

6 in.
(0.50 ft)

12 ft

9 ft

Figure 9.38

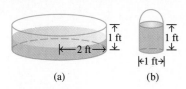

(a)   (b)

Figure 9.39

Solution:

a) When the calculations are performed, all the measurements must be in the same units. We will therefore convert all units to yards. There are 36 in. in a yard. Therefore 6 in. is $\frac{6}{36}$ or $\frac{1}{6}$ yd. The length of 12 ft equals 4 yd, and the width of 9 ft equals 3 yd.

$$V = l \cdot w \cdot h$$
$$= 4 \cdot 3 \cdot \tfrac{1}{6}$$
$$= 2 \text{ yd}^3$$

Note that since all three measurements are in terms of yards, the answer is in cubic yards.

b) One cubic yard of concrete costs $54.50, so two cubic yards will cost 2 × $54.50, or $109.00.   ▲

EXAMPLE 2

Myra and Belinda and their friends are at a picnic at the town park. They have brought a children's wading pool in the shape of a right circular cylinder with a radius of 2 ft and a height of 1 ft into which they will put cold water to keep the soda cold (Fig. 9.39a). They carry the water from the faucet to the pool in a bucket that is also in the shape of a right circular cylinder, with a diameter of 1 ft and a height of 1 ft (Fig. 9.39b).

a) How many buckets of water are needed to fill the pool to a height of $\frac{1}{2}$ ft?

b) If one cubic foot of water weighs 62.5 lb, what is the weight of the water in the pool?

c) If there are 7.5 gal of water per cubic foot, how many gallons of water are in the pool?

Solution:

a) Volume of Water in Pool   Volume of Water in Bucket

$V = \pi r^2 h$ $V = \pi r^2 h$

$\quad = 3.14(2)^2(0.5)$ $\quad = 3.14(0.5)^2(1)$

$\quad = 3.14(4)(0.5) = 6.28 \text{ ft}^3$ $\quad = 3.14(0.25)(1) = 0.785 \text{ ft}^3$

To find the number of buckets needed, we divide

$$\frac{6.28}{0.785} = 8.0.$$

Thus eight bucketfuls are needed.

b) Since 1 ft^3 of water weighs 62.5 lb, 6.28 ft^3 of water weighs 6.28 × 62.5, or 392.5 lb.

c) Since 1 ft^3 of water contains 7.5 gal, 6.28 ft^3 of water contains 6.28 × 7.5, or 47.1 gal of water. Thus there are 47.1 gal of water in the pool.   ▲

A **polyhedron** is a closed surface formed by the union of polygonal regions. Figure 9.40 illustrates some polyhedrons.

Each polygonal region is called a *face* of the polyhedron. The line segment formed by the intersection of two faces is called an *edge*. The point at which two

Polyhedrons

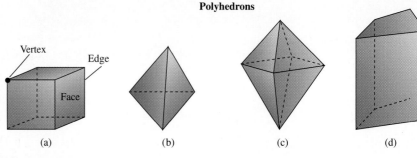

(a)   (b)   (c)   (d)

Figure 9.40

or more edges intersect is called a *vertex*. In Fig. 9.40(a) there are 6 faces, 12 edges, and 8 vertices. Note that

$$\text{Number of vertices} - \text{number of edges} + \text{number of faces} = 2$$
$$8 \quad - \quad 12 \quad + \quad 6 \quad = 2.$$

This formula, credited to Leonhard Euler, is true for any polyhedron.

Euler's Polyhedron Formula

Number of vertices − number of edges + number of faces = 2

We suggest that you verify that this formula holds for Fig. 9.40(b), (c), and (d).

A **regular polyhedron** is one whose faces are all regular polygons of the same size and shape. Figure 9.40(a) and (b) are regular polyhedrons.

A **prism** is a special type of polyhedron whose bases are congruent polygons and whose sides are parallelograms. These parallelogram regions are called the *lateral faces* of the prism. If all the lateral faces are rectangles, the prism is said to be a **right prism**. Some right prisms are illustrated in Fig. 9.41.

Prisms

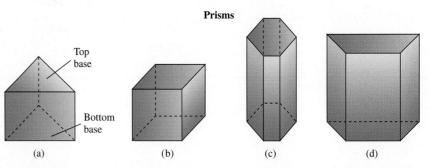

Top base

Bottom base

(a)   (b)   (c)   (d)

Figure 9.41

The volume of any prism can be found by multiplying the area of the base B by the height h of the prism.

Volume of a Prism

$$V = Bh$$

where B is the area of a base and h is the height.

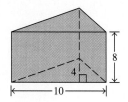

Figure 9.42

EXAMPLE 3

Find the volume of the triangular prism shown in Fig. 9.42.

Solution: First find the area of the triangular base.

$$\text{Area of base} = \tfrac{1}{2}bh$$
$$= \tfrac{1}{2}(10)(4) = 20 \text{ square units}$$

Now find the volume by multiplying the area of the triangular base by the height of the prism.

$$V = B \cdot h$$
$$= 20 \cdot 8 = 160 \text{ cubic units} \quad \blacktriangle$$

EXAMPLE 4

Find the volume of the remaining solid after the cylinder, triangular prism, and square prism have been cut from the solid (Fig. 9.43).

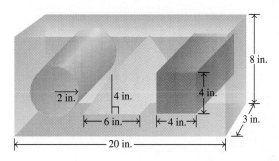

Figure 9.43

Solution: To find the volume of the remaining solid, find the volume of the rectangular solid. Then subtract the volume of the two prisms and the cylinder that were cut out.

$$\text{Volume of rectangular solid} = l \cdot w \cdot h$$
$$= 20 \cdot 3 \cdot 8 = 480 \text{ in}^3$$
$$\text{Volume of circular cylinder} = \pi r^2 h$$
$$= (3.14)(2^2)(3)$$
$$= (3.14)(4)(3) = 37.68 \text{ in}^3$$
$$\text{Volume of triangular prism} = \text{area of the base} \cdot \text{height}$$
$$= \tfrac{1}{2}(6)(4)(3) = 36 \text{ in}^3$$
$$\text{Volume of square prism} = s^2 \cdot h$$
$$= 4^2 \cdot 3 = 48 \text{ in}^3$$
$$\text{Volume of solid} = 480 - 37.68 - 36 - 48$$
$$= 358.32 \text{ in}^3 \quad \blacktriangle$$

Another special category of polyhedrons is the **pyramid**. Unlike prisms, pyramids have only one base. Some pyramids are illustrated in Fig. 9.44. Note that all but one face of a pyramid intersect at a common vertex.

Pyramids

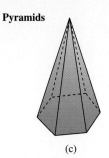

(a)   (b)   (c)   (d)

Figure 9.44

Figure 9.45

If a pyramid is drawn inside a prism, as shown in Fig. 9.45, the volume of the pyramid is less than that of the prism. In fact, the volume of the pyramid is one-third the volume of the prism.

> **Volume of a Pyramid**
> $$V = \tfrac{1}{3}Bh$$
> where B is the area of the base, and h is the height.

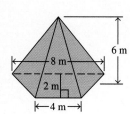

Figure 9.46

EXAMPLE 5

Find the volume of the pyramid shown in Fig. 9.46.

Solution: First find the area of the base of the pyramid. Since the base of the pyramid is a trapezoid,

$$\text{Area of base} = \tfrac{1}{2}h(b_1 + b_2)$$
$$= \tfrac{1}{2}(2)(4 + 8)$$
$$= 12 \text{ m}^2.$$

Now find the volume:

$$V = \tfrac{1}{3}B \cdot h$$
$$= \tfrac{1}{3}(12)(6)$$
$$= 24 \text{ m}^3.$$

In certain situations converting volume from one cubic unit to a different cubic unit might be necessary. For example, when purchasing topsoil you might have to change the amount of topsoil from cubic feet to cubic yards prior to placing your order.

EXAMPLE 6

a) Convert 1 yd^3 to cubic feet.

b) Convert 8.3 yd^3 to cubic feet.

Solution:

a) 1 yd = 3 ft
   1 yd^3 = 3 ft · 3 ft · 3 ft = 27 ft^3

b) 1 yd^3 = 27 ft^3
   Thus 8.3 yd^3 = 8.3 × 27 = 224.1 ft^3.

EXAMPLE 7

Corinne built a flower box with the dimensions shown in Fig. 9.47. How many cubic yards of topsoil does she need to fill the box?

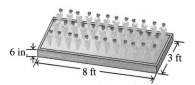

Figure 9.47

Solution: To find the volume, we first have to convert all the measurements to the same units. Since 6 in. = 0.5 ft,

$$V = lwh$$
$$= 8(3)(0.5) = 12 \text{ ft}^3.$$

Now we must convert 12 ft^3 to cubic yards. How shall we do this? We know from Example 6 that 1 yd^3 = 27 ft^3. Thus 12 ft^3 must be less than 1 yd^3. If we divide the total number of cubic feet, 12, by the number of cubic feet in one cubic yard, 27, we will get a number less than 1.

$$\frac{12}{27} \approx 0.44$$

Thus Corinne needs approximately 0.44 yd^3 of topsoil.

Section 9.4 Exercises

1. In your own words, define *volume*.

2. What is solid geometry?

3. In your own words, explain the difference between a prism and a pyramid.

4. In your own words, state Euler's polyhedron formula.

In Exercises 5–10, find the volume of the solid.

5.

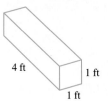

4 ft   1 ft   1 ft

6.

3 ft   3 ft   3 ft

7.

6 in.   2 in.

8.

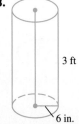

3 ft   6 in.

9.

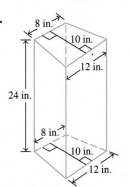

8 in.   10 in.   12 in.   24 in.   8 in.   10 in.   12 in.

10.

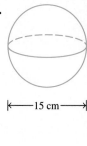

15 cm

In Exercises 11–14, find the volume of the pyramid.

11.

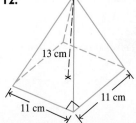

13 ft   15 ft   9 ft

12.

13 cm   11 cm   11 cm

13. **14.**

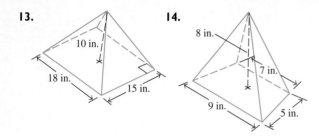

In Exercises 15–20, find the volume of the shaded area.

15. **16.**

17. **18.**

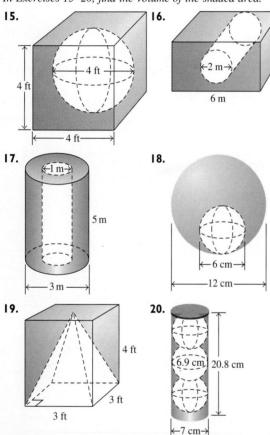

19. **20.**

In Exercises 21–24, use the fact that one cubic yard (yd³) equals 27 cubic feet (ft³) to make the conversion.

21. 4 yd³ to ft³   **22.** 3.2 yd³ to ft³

23. 212 ft³ to yd³   **24.** 84 ft³ to yd³

In Exercises 25–28, use the fact that one cubic meter (m³) equals 1 000 000 cubic centimeters (cm³) to make the conversion.

25. 1.5 m³ to cm³   **26.** 6.4 m³ to cm³

27. 4 000 000 cm³ to m³   **28.** 6 800 000 cm³ to m³

29. The dimensions of the interior of an upright freezer are height 46 in., width 25 in., and depth 25 in. Find its volume
   a) in cubic inches.   **b)** in cubic feet.

30. The dimensions of a square Wendy's hamburger are length and width 4 in. and thickness $\frac{3}{16}$ in. The

dimensions of a Ground Round circular hamburger are diameter $4\frac{1}{2}$ in. and thickness $\frac{1}{4}$ in. Which hamburger has the greater volume? What is the difference in their volumes?

31. A bread pan is 12 in. × 4 in. × 3 in. How many quarts does it hold, if 1 in³ ≈ 0.01736 qt?

32. Rob has two cylindrical containers for storing gasoline. One has a diameter of 10 in. and a height of 12 in. The other has a diameter of 12 in. and a height of 10 in.
   a) Which container holds the greater amount of gasoline, the taller one or the one with the greater diameter?
   b) What is the difference in volume?

33. **a)** How many cubic centimeters of water will a rectangular fish tank hold if the tank is 80 cm long, 50 cm wide, and 30 cm high?
   b) If one cubic centimeter holds one milliliter of liquid, how many milliliters will the tank hold?
   c) If $1\ell = 1000$ mℓ, how many liters will the tank hold?

34. **a)** What is the volume of water in a swimming pool that is 18 m long and 10 m wide and has an average depth of 3 m? Give your answer in cubic meters.
   b) If 1 m³ = 1 kℓ, how many kiloliters of water will the pool hold?

35. **a)** How many cubic feet of dirt are needed to fill a ditch 5 ft wide by 7 ft long by 30 in. deep?
   b) How many cubic yards of dirt are needed?

36. George Lockhart's driveway is 80 ft long by 12.5 ft wide. He plans to lay 4 in. of blacktop on the driveway.
   a) How many cubic feet of blacktop does he need?
   b) How many cubic yards?

37. The Pyramid of Cheops in Egypt has a square base measuring 720 ft on a side. Its height is 480 ft. What is its volume?

38. A gallon is a measure of liquid volume. It is equivalent to 231 in³. Measure the base and height *of the liquid* in a half-gallon paper carton of milk and compute its volume in cubic inches. Do your results check with the value given?

39. The ball at the end of a ballpoint pen has a diameter of 1.2 mm. Find the volume of the ball.

40. What is the total piston displacement of a two-cylinder engine if each piston has a diameter of $2\frac{1}{4}$ in. and a $3\frac{1}{2}$-in. stroke?

41. In baking a cake you have to choose between a round pan with a 9-in. diameter and a 7-in. × 9-in. rectangular pan.
a) Determine the area of the base of each pan.
b) If both pans are 2 in. deep, determine the volume of each pan.
c) Which pan has the larger volume?

42. A flower box is 4 ft long, and its ends are in the shape of a trapezoid. The upper and lower bases of the trapezoid measure 12 in. and 8 in., respectively, and the height is 9 in. Find the volume of the flower box
a) in cubic inches.   **b)** in cubic feet.

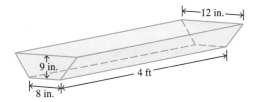

43. What is the smallest number of cuts that would divide a cube of wood whose edge measures 3 in. into cubes, each of whose edges measures 1 in.?

44. A soda straw is 30 cm long and 4 mm in diameter. How much soda, in milliliters, can the straw hold? Note that 1 cc = 1 mℓ.

In Exercises 45–50, find the missing value indicated by the question mark. Use the following formula.

$$\left(\begin{array}{c}\text{Number of}\\\text{vertices}\end{array}\right) - \left(\begin{array}{c}\text{number of}\\\text{edges}\end{array}\right) + \left(\begin{array}{c}\text{number}\\\text{of faces}\end{array}\right) = 2$$

	Number of Vertices	Number of Edges	Number of Faces
45.	12	14	?
46.	8	?	2
47.	?	8	4
48.	7	12	?
49.	11	?	5
50.	?	10	4

51. What effect does doubling the length of each edge of a cube have on its volume? Explain.

52. What effect does doubling the radius of the base of a cylinder have on the volume? Explain.

53. What effect does doubling the radius of a sphere have on its volume? Explain.

54. A half-inch cube is a cube measuring $\frac{1}{2}$ in. on a side. A half cubic inch is half the volume of a cube measuring 1 in. on a side. Is the volume of a half-inch cube the same as the volume of a half cubic inch? Explain.

55. A king-sized waterbed mattress is 7 ft long, 7 ft wide, and 9 in. high.
a) How many cubic feet of water are needed to fill the waterbed mattress?
b) If 1 ft^3 of water weighs 62.5 lb, what is the weight of the water in the mattress?
c) If 1 gal of water weighs 8.3 lb, how many gallons of water does the mattress hold?

56. A hot-water heater is in the shape of a circular cylinder. It has a height of 5 ft 6 in. and a diameter of 18 in.
a) How many cubic feet of water will the tank hold?
b) How many gallons of water does the tank hold? (1 ft^3 ≈ 7.5 gal.)

Problem Solving/Group Activity

57. The general procedure for determining the number of rolls of wallpaper needed to wallpaper a given rectangular room is as follows.
a) Determine the room's overall perimeter, in feet, by measuring the distance around the room where the wall meets the floor. Ignore windows and doors at this time.
b) Multiply the perimeter by the height in feet of the room. This result gives the number of square feet of wall space.
c) Divide this number by 30, the approximate number of usable square feet in a roll of wallpaper.
d) From this number, deduct one roll for every two averaged-sized openings (windows, doors, fireplace, and so on).
Helen Latvis plans to wallpaper her kitchen. It is 12 ft long, 10 ft wide, and 8 ft high, and it contains two doorways and two windows. How many rolls of wallpaper will she need for her kitchen?

58. A theater decides to change the shape of its popcorn container from a regular cone to a right circular cylinder and charge twice as much. If the containers have the same radius and height, is this new circular cylinder container a better buy for the customer? Explain.

59. Which is a better buy per cubic centimeter, a grapefruit 6 cm in radius that costs $0.33 or a grapefruit 7 cm in radius that costs $0.49? Explain.

60. A box is packed with six cans of orange juice. The cans are touching each other and the sides of the box, as shown. What percent of the volume of the interior of the box is not occupied by the cans?

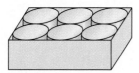

61. A faucet is dripping at the rate of 12 drops per minute. There are 20 drops per milliliter. How many liters of water are wasted in a 30-day period? (1000 mℓ = 1 ℓ.)

62. a) Explain how to demonstrate, using the cube shown, that

$$(a + b)^3 = a^3 + 3a^2b + 3ab^2 + b^3.$$

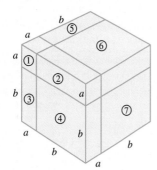

b) What is the volume in terms of a and b of each numbered piece in the figure?

c) An eighth piece is not illustrated. What is its volume?

Research Activities

63. Calculate the volume of the room in which you sleep or study. Go to a store that sells room air conditioners and find out how many cubic feet can be cooled by the different models available. Describe the model that would be the proper size for your room. What is the initial cost? How much does that model cost to operate? If you moved to a room that had twice the amount of floor space and the same height, would the air conditioner you selected still be adequate? Explain.

64. Pappus of Alexandria (ca. A.D. 350) was the last of the well-known ancient Greek mathematicians. Write a paper on his life and his contributions to mathematics.

65. Read the book *SPHERELAND: A Fantasy About Curved Spaces and an Expanding Universe* by Dionys Burger, translated by Cornelie J. Rheinboldt. Present a review of the book to the class.

9.5   GRAPH THEORY

In the eighteenth-century Prussian town of Königsberg (now the city of Kaliningrad, Russia near the Baltic Sea) seven bridges crossed the Prigel River (Fig. 9.48a). Individuals in the area tried to determine whether it was possible to walk a path that would cross each of the seven bridges exactly once. They found that they ended up either not crossing one of the bridges, or crossing one of the bridges more than once. The problem was brought to the attention of the Swiss mathe-

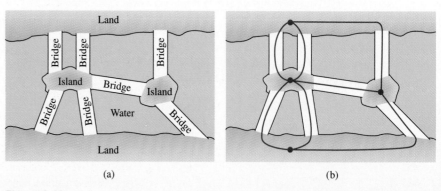

(a)   (b)

Figure 9.48

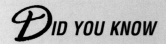
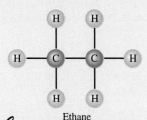
matician Leonhard Euler (pronounced "öiler," 1707–1783). His study of this problem, now known as the Königsberg bridge problem, laid the groundwork for a modern branch of mathematics called *graph theory*, a topic in a more general area called topology. To solve the problem, Euler drew figures called graphs (or *networks*) like that shown in red in Fig. 9.48(b). Each dot represented a plot of land and each line a bridge or path. Can you find a path that crosses each bridge exactly once?

Before we determine whether there is a path that will cross each bridge exactly once, let's consider Example 1.

EXAMPLE I

In Fig. 9.49 start at any point and try to trace each figure without retracing a line and without removing your pencil from the paper. If you succeed, indicate your starting point and ending point.

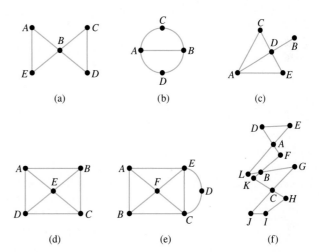

Figure 9.49

Solution:

a) The figure can be traced if you start at any point. You will end at the point at which you started.

b) The figure can be traced, but only if you start at point *A* or point *B*. If you start at point *A*, you will end at point *B*, and vice versa.

c) The figure can be traced, but only if you start at point *A* or point *B*. If you start at point *A*, you will end at point *B*, and vice versa.

d) The figure cannot be traced without retracing a line.

e) The figure can be traced, but only if you start at point *A* or point *B*. If you start at point *A*, you will end at point *B*, and vice versa.

f) The figure can be traced if you start at any point. You will end at the point at which you started. ▲

In order for you to be able to answer the Königsberg bridge problem and understand why we were able to trace all but one of the figures in Example 1 without retracing a line, we must introduce some new terms and concepts.

A vertex is any designated point. An arc (or an edge) is any line, either straight or curved, that begins and ends at a vertex. Figure 9.50(a) has three

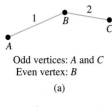

Odd vertices: *A* and *C*
Even vertex: *B*

(a)

Odd vertices: *A* and *D*
Even vertices: *B*, *C*, and *E*

(b)

Figure 9.50

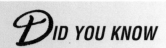

NETWORKS

*T*oday, graph (network) theory is a flourishing field, useful for dealing with a variety of every-day problems. It is used by the airline industry to arrange the complex network of flights connecting cities and by sports officials planning tournaments and playoff schedules. City planners use it to work out new roads and traffic patterns. Tests are underway in traffic-clogged cities such as Los Angeles to link cars, via an on-board navigational computer, to the city's central traffic-control computer. In this way motorists can be rerouted to less congested roadways. ∎

designated vertices, *A*, *B*, and *C*, and two arcs, 1 and 2. Figure 9.50(b) has five designated vertices, *A*, *B*, *C*, *D*, and *E*, and six arcs.

A vertex with an odd number of attached arcs is called an odd vertex. A vertex with an even number of attached arcs is called an even vertex. Figure 9.50(a) has two odd vertices and one even vertex. Figure 9.50(b) has two odd vertices and three even vertices.

A network is any continuous (not broken) system of arcs and vertices.

> A network is said to be **traversable** if it can be traced without removing the pencil from the paper and without tracing an arc more than once.

After completing Example 1, do you have an intuitive feeling as to when a figure is traversable? Think about odd and even vertices. Leonhard Euler discovered an important scientific principle concealed in the Königsberg bridge problem. He presented his simple and ingenious solution of that problem to the Russian Academy at St. Petersburg in 1735. Euler developed the following rules of traversability in solving the problem.

> **Rules of Traversability**
> 1. A network with no odd (all even) vertices is traversable; you may start from any vertex, and you will end where you began.
> 2. A network with exactly two odd vertices is traversable; you must start at either of the odd vertices and finish at the other.
> 3. A network with more than two odd vertices is not traversable.

The network in the Königsberg bridge problem (Fig. 9.48b) has four odd vertices, so it cannot be traversed. Therefore crossing each bridge only once is impossible. Note that it is impossible for a network to contain an odd number of odd vertices. (If you don't believe this statement, try to construct such a network.)

Now go back to Example 1 and determine which figures are traversable, using these rules. Note that Figs. 9.49(a) and 9.49(f) are traversable from any point because they contain only even vertices. Figures 9.49(b), (c), and (e) have exactly two odd vertices and can be traversed but only by starting at one of the odd vertices, either point *A* or point *B*. Figure 9.49(d) contains more than two odd vertices and therefore cannot be traversed.

EXAMPLE 2

The floor plan of a six-gallery art museum is shown in Fig. 9.51(a). The openings represent doors, and the letters represent galleries.

a) Determine the galleries that contain an odd number of doors; an even number of doors.

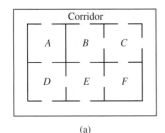

(a)

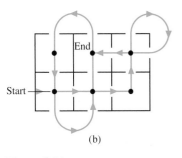

(b)

Figure 9.51

b) Each gallery can be represented as an odd or even vertex. Use this information to determine whether it is possible to walk through each gallery by using each door only once.

c) Determine a path to walk through each gallery by using each door only once.

Solution:

a) Galleries B and D contain three doors each. Galleries A and F contain two doors each. Galleries C and E have four doors.

b) There are only two odd vertices, B and D, so the figure is traversable, and you can walk through the museum by using each door only once.

c) You must start in either Gallery B or D (see Fig. 9.51b). If you start in B, you will end in D, and vice versa. When you leave Gallery E, you can leave by either door. ▲

The floor plan in Example 2 can be reduced to a map, where the rooms are the vertices and the doors are the paths. Construct a map of that floor plan now.

Section 9.5 Exercises

1. What is a vertex?

2. What is an arc (or an edge)?

3. Explain how to determine whether a vertex is odd or even.

4. In your own words, explain the rules for determining whether a graph is traversable.

In Exercises 5–8, determine the number of vertices and the number of arcs.

5.

6.

7.

8.

9. Explain why these two figures represent the same graph.

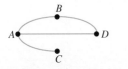

10. For the graph, list the vertices that are odd and the vertices that are even.

In Exercises 11–18, determine whether the network is traversable. If it is, state the points from which you may start and end.

11.

12.

13.

14.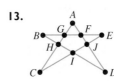

15.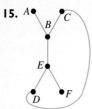

16.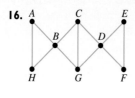

17.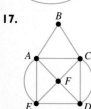

18.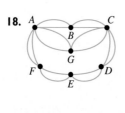

In Exercises 19–26, the floor plan of a building is shown.

a) *Determine the rooms that contain an odd number of doors; an even number of doors.*

b) *Use the rules of traversability to determine whether it is possible to walk through the building using each door only once.*

c) *If the answer to part (b) is yes, determine a path through the building, using each door exactly once.*

19.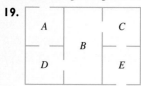

20.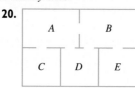

21.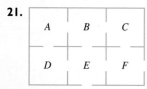

22.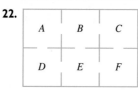

23.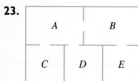

24.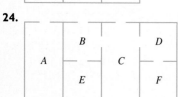

25.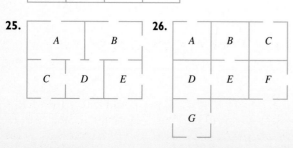

26.

27. The floor plan for a suite of rooms is shown. Add an exit door in one of the rooms so that the security guard can enter through the door marked enter, pass through each door only once locking it behind him, and then exit by the door you added. Explain why there is only one possible room in which the door may be placed.

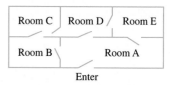

In Exercises 28 and 29, is it possible to cross each bridge exactly once? Explain your answer, and show the path if it is possible.

28.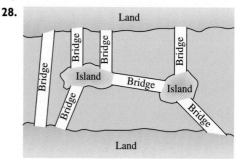

29.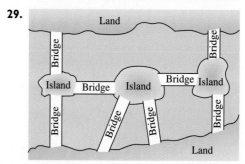

30. Draw a graph that contains four vertices and is traversable from exactly two points.

31. Can you draw a graph that contains an odd number of odd vertices? If you answer yes draw such a graph.

32. Draw a graph that contains five vertices that is traversable from exactly two points.

33. In the figure, lines connecting two states indicate that the two states share a common border. Which of these states share a common border with (a) Tennessee? (b) Missouri?

34. In the figure, a line connecting two countries indicates that they share a common border. Which of these countries share a common border with (a) Brazil? (b) Bolivia?

35. The remaining games in a soccer league are illustrated in the graph.

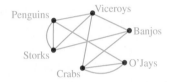

a) How many games do the Viceroys have remaining?
b) How many games do the Penguins have remaining?
c) How many games still need to be played?

36. Dawn, Jessica, Pam, Bill, Ed, and Scott go to a dance. Dawn dances with Bill and Scott. Jessica dances with all the boys and Pam dances only with Scott. Draw a graph that displays this information.

37. France has common borders with Belgium, Germany, Switzerland, Italy, and Spain. Belgium has common borders with the Netherlands, France, and Germany. Germany has common borders with Belgium, Poland, the Czech Republic, Austria, Switzerland, and France. Switzerland has common borders with France, Italy, Austria, and Germany. Draw a graph that displays this information.

38. Gretchen's Delivery service, located in town C, makes deliveries every day to each town shown on the map. The map also shows the highways connecting the towns and the distances between towns. The delivery service's expenses are lowest when the driver's route is the shortest.

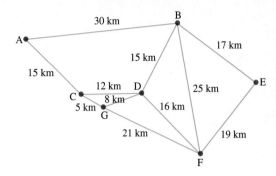

a) Is the map traversable? Explain.
b) Plan the delivery route for which the company will have the lowest expense. The driver must start at C and end at C. However the driver need not travel on every road.
c) What is the length of the shortest route?

39. Networks (a) and (b) illustrate relationships of arcs, regions, and vertices. Network (a) has three vertices (*A*, *B*, and *C*), four arcs (1, 2, 3, and 4), and three regions (*x*, *y*, and *z*). Network (b) has five vertices (*A*, *B*, *C*, *D*, and *E*), seven arcs (1, 2, 3, 4, 5, 6, 7), and four regions (*w*, *x*, *y*, and *z*). Using these networks and others that you can make up yourself, develop a formula expressing the number of arcs in terms of the number of vertices and regions. This formula is known as Euler's formula for networks.

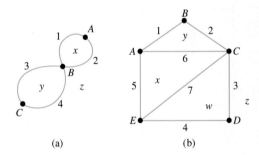

(a)   (b)

40. In 1856 William Rowan Hamilton (1805–1865) introduced a problem similar to the Königsberg bridge problem. Hamilton turned his problem into a game that was marketed in 1856. Write a short paper explaining the game. (References include encyclopedias and history of mathematics books.)

9.6 THE MÖBIUS STRIP, KLEIN BOTTLE, AND MAPS

One of the first pioneers of topology was the German astronomer and mathematician August Ferdinand Möbius (1790–1866). A student of Gauss, Möbius was the director of the University of Leipzig's observatory. He spent a great deal

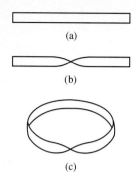

Figure 9.52

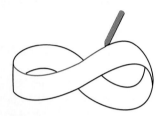

Figure 9.53

Figure 9.54

Figure 9.55

Figure 9.56

of time studying geometry and played an essential part in the systematic development of projective geometry. He is best known for his studies of the properties of one-sided surfaces, including the one called the Möbius strip.

Möbius Strip

If you place a pencil on one surface of a sheet of paper and do not remove it from the sheet, you must cross the edge to get to the other surface. Thus a sheet of paper has one edge and two surfaces. The sheet retains these properties even when crumpled into a ball. The Möbius strip is a one-sided, one-edged surface. You can construct one, as shown in Fig. 9.52, by (a) taking a strip of paper, (b) giving one end a half twist, and (c) taping the ends together.

The Möbius strip has some very interesting properties. In order to better understand these properties, perform the following experiments.

Experiment 1 Make a Möbius strip using a strip of paper and tape as illustrated in Fig. 9.52. Place the point of a felt tip pen on the edge of the strip (Fig. 9.53). Pull the strip slowly so that the pen marks the edge—do not remove the pen from the edge. Continue pulling the strip and observe what happens.

Experiment 2 Make a Möbius strip. Place the tip of a felt tip pen on the surface of the strip (Fig. 9.54). Pull the strip slowly so that the pen marks the surface. Continue and observe what happens.

Experiment 3 Make a Möbius strip. With a scissors make a small slit in the middle of the strip. Starting at the slit, cut along the strip, keeping the scissors in the middle of the strip (Fig. 9.55). Continue cutting and observe what happens.

Experiment 4 Make a Möbius strip. Make a small slit at a point about one-third of the width of the strip. Cut along the strip, keeping the scissors the same distance from the edge (Fig. 9.56). Continue cutting and observe what happens.

Klein Bottle

Another topological object is the punctured Klein bottle. This object, named after Felix Klein (1849–1925), resembles a bottle but only has one side.

A punctured Klein bottle can be made by stretching a hollow piece of glass tubing. The neck is then passed through a hole and joined to the base (Fig. 9.57).

Figure 9.57

Look closely at the model of the Klein bottle shown in Fig. 9.57. Note that the bottle could not hold any water. The punctured Klein bottle has only one edge (the hole) and no outside or inside because it has just one side.

Maps

Maps have fascinated topologists for years because of the many challenging problems they present. Mapmakers have known for a long time that, regardless of the complexity of the map and whether drawn on a flat surface or a sphere, only four colors are needed to differentiate each country (or state) from its immediate neighbors. Thus every map can be drawn by using only four colors, and no two countries with a common border will have the same color. Regions that meet at only one point (such as the states of Arizona, Colorado, Utah, and New Mexico) are not considered to have a common border. In Fig. 9.58(a) no two states with a common border are marked with the same color.

In 1976 Kenneth Appel and Wolfgang Haken of the University of Illinois—using their ingenuity, logic, and 1200 hours of computer time—succeeded in proving that only four colors are needed to draw a map. They solved the four-color map problem by reducing any map to a graph as Euler did with the Königsberg bridge problem. They replaced each country with a point. They connected two countries having a common border with a straight line (Fig. 9.58b). They then showed that the points of any graph in the plane could be colored by using only four colors in such a way that no two points connected by the same line were the same color.

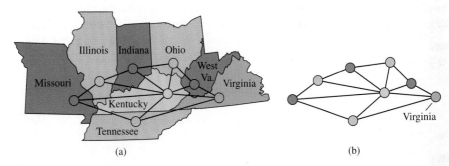

(a)   (b)

Figure 9.58

Mathematicians have shown that, in certain cases, more than four colors may be needed to draw a map. For example, a map drawn on a Möbius strip requires a maximum of six colors and a map drawn on a torus (the shape of a doughnut) requires a maximum of seven colors.

Jordan Curves

A Jordan curve is a topological object that can be thought of as a circle twisted out of shape (Fig. 9.59a–d). Like a circle, it has an inside and an outside. To get from one side to the other at least one line must be crossed. Consider the Jordan curve in Fig. 9.59(d). Are points A and B inside or outside the curve?

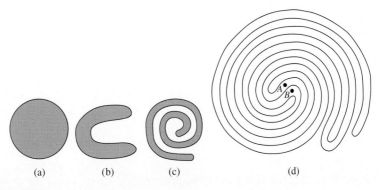

(a)   (b)   (c)   (d)

Figure 9.59

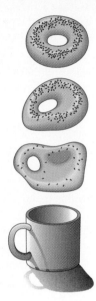

Figure 9.60

A quick way to tell whether the two dots are inside or outside the curve is to draw a straight line from each dot to a point that is clearly outside the curve. If the straight line crosses the curve an even number of times, the dot is outside. If the straight line crosses the curve an odd number of times, the dot is inside the curve. Can you explain why this procedure works? Determine whether point A and point B are inside or outside the curve (see Exercises 20 and 21 at the end of this section).

Topological Equivalence

Someone once said that a topologist is a person who does not know the difference between a doughnut and a coffee cup. Two geometric figures are said to be **topologically equivalent** if one figure can be elastically twisted, stretched, bent, or shrunk into the other figure without puncturing or ripping the original figure. If a doughnut is made of elastic material, it can be stretched, twisted, bent, shrunk, and distorted until it resembles a coffee cup with a handle, as shown in Fig. 9.60. Thus the doughnut and coffee cup are topologically equivalent.

In topology, figures are classified according to their *genus*. The **genus** of an object is determined by the number of holes in the object. A cup and a doughnut each have one hole and are of genus 1. A kettle and a double box end wrench each have two holes and are of genus 2. Figure 9.61 illustrates this type of classification.

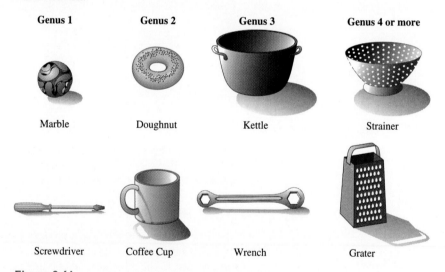

Genus 1	Genus 2	Genus 3	Genus 4 or more
Marble	Doughnut	Kettle	Strainer
Screwdriver	Coffee Cup	Wrench	Grater

Figure 9.61

Section 9.6 Exercises

1. What is a Möbius strip?

2. Explain how to make a Möbius strip.

3. What is a Klein bottle?

4. What is a Jordan curve?

5. When are two figures topologically equivalent?

6. How is the genus of a figure determined?

7. Use the result of Experiment 1 to find the number of edges on a Möbius strip.

8. Use the result of Experiment 2 to find the number of surfaces on a Möbius strip.

9. How many separate strips are obtained in Experiment 3?

10. How many separate strips are obtained in Experiment 4?

11. **a)** Take a strip of paper, give it one full twist, and connect the ends. Is this a Möbius strip with only one side? Explain.

b) Determine the number of edges, as in Experiment 1.
c) Determine the number of surfaces, as in Experiment 2.
d) Cut the strip down the middle. What is the result?

12. Make a Möbius strip. Cut it one-third of the way from the edge, as in Experiment 4. You should get two loops, one going through the other. Determine whether either (or both) of these loops is itself a Möbius strip.

13. Take a strip of paper, make one whole twist and another half twist, and then tape the ends together. Test by a method of your choice to determine whether this has the same properties as a Möbius strip.

14. Can you see any advantage in a Möbius conveyor belt? Explain.

In Exercises 15–19,

a) *color the map by using a maximum of four colors so that no two regions with a common border have the same color.*
b) *show how each map can be reduced to a graph.*

15.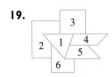

16.

17.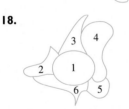

18.

19.

20. Determine whether point *A* in Fig. 9.59(d) is inside or outside the Jordan curve.

21. Determine whether point *B* in Fig. 9.59(d) is inside or outside the Jordan curve.

22. When testing to determine whether a point is inside or outside a Jordan curve, explain why if you count an odd number of lines the point is inside the curve, and if you count an even number of lines the point is outside the curve.

In Exercises 23–30, give the genus of the object.

23.

24.

25.

26.

27.

28.

29.

30.
> Euler's Konigsberg bridge problem laid the groundwork for a branch of modern mathematics called topology, the study of the features of shapes and objects.

31. Name at least three objects not mentioned in this section that have
a) genus 0.
b) genus 1.
c) genus 2.
d) genus 3 or more.

Problem Solving/Group Activity

32. Using playdough (or glazing compound) make a doughnut. Without puncturing or tearing the doughnut, reshape it into a topologically equivalent figure, a cup with a handle.

33. Using at most four colors, try to color the map of the Middle East. Do not use the same color for any two countries that share a common border.

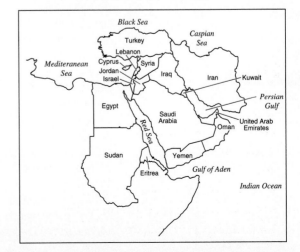

34. Write a paper on the history of the four-color problem. Include in the paper the names of the individuals who

solved the problem and the technique they used to do so.

35. Write a paper on the major contributors to the development of topology, starting with Euler.

9.7   NON-EUCLIDEAN GEOMETRY AND FRACTAL GEOMETRY

Non-Euclidean Geometry

In Section 9.1 we stated that postulates or axioms are statements that are to be accepted as true. In his book *Elements*, Euclid's fifth postulate was "If a straight line falling on two straight lines makes the interior angles on the same side less than two right angles, the two straight lines, if produced indefinitely, meet on that side on which the angles are less than the two right angles."

Euclid's fifth axiom may be better understood by observing Fig. 9.62. The sum of angles *A* and *B* is less than the sum of two right angles (180°). Therefore the two lines will meet if extended.

John Playfair (1748–1819), a Scottish physicist and mathematician, wrote a geometry book that was published in 1795. In his book, Playfair gave a logically equivalent interpretation of Euclid's fifth postulate. This version is often referred to as Playfair's postulate or the Euclidean parallel postulate.

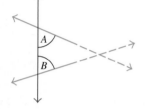

Figure 9.62

> **The Euclidean Parallel Postulate**
>
> Given a line and a point not on the line, one and only one line can be drawn through the given point parallel to the given line (Fig. 9.63).

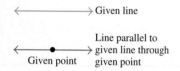

Figure 9.63

The Euclidean parallel postulate may be better understood by looking at Fig. 9.63.

Many mathematicians after Euclid believed that this postulate was not as self-evident as the other nine. Others believed that this postulate could be proved from the other nine postulates and therefore was not needed at all. Of the many attempts to prove that the fifth postulate was not needed, the most noteworthy one was presented by Girolamo Saccheri (1667–1733), a Jesuit in Italy. In the course of his elaborate chain of deductions, Saccheri proved many of the theorems of what is now called hyperbolic geometry. However, Saccheri did not realize what he had done. He believed that Euclid's geometry was the only "true" geometry and concluded that his own work was in error. Thus Saccheri narrowly missed receiving credit for a great achievement—the founding of non-Euclidean geometry.

In 1763 the German mathematician Georg Simon Klügel listed nearly 30 attempts to prove axiom 5 from the remaining axioms, and he rightly concluded that all the alleged proofs were unsound. Fifty years later a new generation of geometers were becoming more and more frustrated at their inability to prove Euclid's fifth postulate. One of them, a Hungarian named Farkos Bolyai, wrote a letter to his son, Janos Bolyai. "I entreat you leave the science of parallels alone. . . . I have traveled past all reefs of this infernal dead sea and have always come back with a broken mast and torn sail.'' The son, refusing to heed his father's advice, continued to think about parallels until, in 1823, he saw the whole truth and enthusiastically declared, ''I have created a new universe from nothing.'' He recognized that geometry branches in two directions, depending on whether Euclid's fifth postulate is applied. He recognized two different geometries and published his discovery as a 24-page appendix to a textbook written by his father. The famous mathematician George Bruce Halsted called it ''the most extraordinary two dozen pages in the whole history of thought.'' Farkos Bolyai proudly presented a copy of his son's work to his friend Carl Friedrich Gauss, then Germany's greatest mathematician, whose reply to the father had a devastating effect on the son. Gauss wrote, ''I am unable to praise this work. . . . To praise it would be to praise myself. Indeed, the whole content of the work, the path taken by your son, the results to which he is led, coincides almost entirely with my meditations which occupied my mind partly for the last thirty or thirty-five years.'' We now know from his earlier correspondence that Gauss had indeed been familiar with **hyperbolic geometry** even before Janos was born. In his letter, Gauss also indicated that it was his intention not to let his theory be published during his lifetime, but to record it so that the theory would not perish with him. It is believed that the reason Gauss did not publish his work was that he feared being ridiculed by other prominent mathematicians of his time.

Ivanovich Lobachevsky

At about the same time as Bolyai's publication, the Russian Nikolay Ivanovich Lobachevsky published a paper that was remarkably like Bolyai's, although it was quite independent of it. Lobachevsky made a deeper investigation and wrote several books. In marked contrast to Bolyai, who received no recognition during his lifetime, Lobachevsky received great praise and became a professor at the University of Kazan.

After the initial discovery, little attention was paid to the subject until 1854, when G. F. Bernhard Riemann (1826–1866), a student of Gauss, suggested a second type of non-Euclidean geometry, which is now called **spherical**, **elliptical**, or **Riemannian geometry**. The hyperbolic geometry of his predecessors was synthetic; that is, it was not based on or related to any concrete model when it was developed. Riemann's geometry was closely related to the theory of surfaces. A **model** may be considered a physical interpretation of the undefined terms that satisfies the axioms. A model may be a picture or an actual physical object.

G. F. Bernhard Riemann

The two types of non-Euclidean geometries that we have mentioned are elliptical geometry and hyperbolic geometry. The major difference among the three geometries lies in the fifth axiom. The fifth axiom of the three geometries is summarized here.

The Fifth Axiom of Geometry

Euclidean	**Elliptical**	**Hyperbolic**
Given a line and a point not on the line, one and only one line can be drawn parallel to the given line through the given point.	Given a line and a point not on the line, no line can be drawn through the given point parallel to the given line.	Given a line and a point not on the line, two or more lines can be drawn through the given point parallel to the given line.

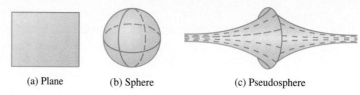

(a) Plane   (b) Sphere   (c) Pseudosphere

Figure 9.64

To understand the fifth axiom of the two non-Euclidean geometries, remember that the term *line* is undefined. Thus a line can be interpreted differently in different geometries. A model for Euclidean geometry is a plane, such as a blackboard (Fig. 9.64a). A model for elliptical geometry is a sphere (Fig. 9.64b). A model for hyperbolic geometry is a pseudosphere (Fig. 9.64c). A pseudosphere is similar to two trumpets placed bell to bell. Obviously, a line on a plane cannot be the same as a line on either of the other two figures. The curved red lines in Fig. 9.65(a) and both colored lines in Fig. 9.66 are examples of lines in elliptical and hyperbolic geometry, respectively.

Elliptical Geometry

A circle on the surface of a sphere is called a great circle if it divides the sphere into two equal parts. If we were to cut through a sphere along a great circle, we would have two identical pieces. If we interpret a line to be a great circle, we can see that the fifth axiom of elliptical geometry is true. Two great circles on a sphere must intersect, hence there can be no parallel lines (Fig. 9.65a).

If we were to construct a triangle on a sphere, the sum of its angles would be greater than 180° (Fig. 9.65b). The theorem, "The sum of the measures of the angles of a triangle is greater than 180°," has been proved by means of the axioms of elliptical geometry. The sum of the measures of the angles varies with the area of the triangle and gets closer to 180° as the area decreases.

Hyperbolic Geometry

The lines in hyperbolic geometry are represented by geodesics on the surface of the pseudosphere. A *geodesic* is the shortest and least-curved arc between two points on a surface. Figure 9.66 illustrates two lines on the surface of a pseudosphere.

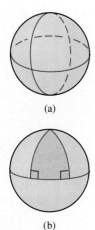

(a)

(b)

Figure 9.65

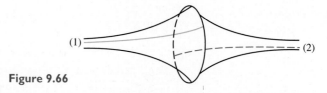

(1)   (2)

Figure 9.66

Figure 9.67(a) illustrates the fifth axiom of hyperbolic geometry.* Note that, through the given point, two lines are drawn parallel to the given line. If we were to construct a triangle on a pseudosphere, the sum of the measures of the angles would be less than 180° (Fig. 9.67b). The theorem, "The sum of the measures

---

*A formal discussion of hyperbolic geometry is beyond the scope of this text.

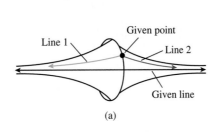

(b)

Figure 9.67

of the angles of a triangle is less than 180°,'' has been proved by means of the axioms of hyperbolic geometry.

We have stated that the sum of the measures of the angles of a triangle is 180°, is greater than 180°, and is less than 180°. Which statement is correct? Each statement is correct *in its own system*. Many theorems hold true for all three

$\mathcal{D}$ID YOU KNOW

IT'S ALL RELATIVE

$\mathcal{E}$instein's general theory of relativity, published in 1916, approached space and time differently from our everyday understanding of them. Einstein's theory unites the three dimensions of space with one of time in a four-dimensional space–time continuum. His theory dealt with the path that light and objects take while moving through space under the force of gravity. Einstein conjectured that mass (such as stars and planets) caused space to be curved. The greater the mass, the greater the curvature. Also, in the region nearer to the mass, the curvature of the space is greater.

Einstein's theory was confirmed by the solar eclipses of 1919 and 1922.

To prove his conjecture, Einstein exposed himself to Riemann's non-Euclidean geometry. Einstein felt that the trajectory of a particle in space represents not a straight line but the straightest curve possible, a geodesic.

Space–time is now thought to be a combination of three different types of curvature: spherical (described by Riemannian geometry), flat (described by Euclidean geometry), and saddle-shaped (described by hyperbolic geometry). ■

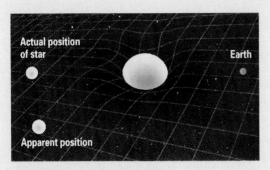

You can visualize Einstein's theory by thinking of space as a rubber sheet pulled taut on which a mass is placed, causing the rubber sheet to bend.

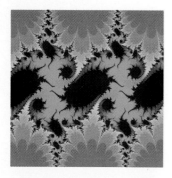

Fractal Images

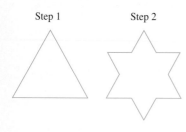

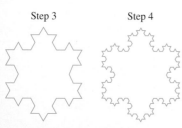

Figure 9.68

geometries—vertical angles still have the same measure, we can uniquely bisect a line segment with a straightedge and compass alone, and so on.

The many theorems based on the fifth postulate may differ in each geometry. It is important for you to realize that each theorem proved is true *in its own system* because each is logically deduced from the given set of axioms of the system. No one system is the "best" system. Euclidean geometry may appear to be the one to use in the classroom, where the blackboard is flat. However, in discussions involving the earth as a whole, elliptical geometry may be the most useful, since the earth is a sphere. If the object under consideration has the shape of a saddle or pseudosphere, hyperbolic geometry may be the most useful.

Fractal Geometry

We are familiar with one-, two-, and three-dimensional figures. However, there are many objects that are difficult to categorize as one-, two-, or three-dimensional. For example, how would you classify the irregular shapes we see in nature such as a coast line, or the bark on a tree, or a mountain, or a path followed by lightning? For a long time mathematicians assumed that making realistic geometric models of natural shapes and figures was almost impossible. However, the development of **fractal geometry** now makes it possible. Both color photos on this page were made by using fractal geometry. The discovery and study of fractal geometry has been one of the hottest mathematical topics of the 1980s and 1990s.

The word *fractal* (from the Latin word *fractus*, "broken up, fragmented") was first used in the mid-1970s by the mathematician Benoit Mandelbrot to describe shapes that had several common characteristics, including some form of "self-similarity," as will be seen shortly in the Koch snowflake.

Typical fractals are extremely irregular curves or surfaces that wiggle enough so that they are not considered one-dimensional. Fractals do not have integer dimensions; their dimensions are between 1 and 2. For example, a fractal may have a dimension of 1.26. Fractals are developed by applying the same rule over and over again, with the end point of each simple step becoming the starting point for the next step, a process called **recursion**.

Using the recursive process, we will develop a famous fractal called the **Koch snowflake** after Helga von Koch, the Swedish mathematician who first discovered its remarkable characteristics. The Koch snowflake illustrates a property of all fractals called *self-similarity*—that is, each smaller piece of the curve resembles the whole curve.

To develop the Koch snowflake:

1. start with an equilateral triangle (Step 1, Fig. 9.68), and

2. whenever you see an edge —— replace it with ⌁ (Steps 2–4).

What is the perimeter of the snowflake in Fig. 9.68, and what is its area? A portion of the boundary of the Koch snowflake known as the Koch curve or the snowflake curve is represented in Fig. 9.69.

Because the Koch curve consists of infinitely many pieces of the form ⌁, the perimeter is also infinite. The area of the Koch snowflake is 1.6 times the area of the starting equilateral triangle. Thus the area of the snowflake is finite. The Koch snowflake has a finite area enclosed by an infinite boundary! This fact may seem difficult to accept, but it is true. However, the Koch snowflake, like other fractals, is not an everyday run-of-the-mill geometric shape.

Figure 9.69

Fractals provide a way to study natural forms such as coastlines, trees, mountains, galaxies, polymers, rivers, weather patterns, brains, lungs, and blood supplies. Fractals also help explain that which appears chaotic. The blood supply in the body is one example. The branching of arteries and veins appear chaotic, but closer inspection reveals that the same type of branching occurs for smaller and smaller blood vessels, down to the capillaries. Thus fractal geometry provides a geometric structure for chaotic processes in nature. The study of chaotic processes is called chaos theory.

Section 9.7 Exercises

In Exercises 1–5, list the accomplishments of the mathematician.

1. Janos Bolyai

2. Carl Friedrich Gauss

3. Nikolay Ivanovich Lobachevsky

4. Girolamo Saccheri

5. G. F. Bernhard Riemann

6. State the fifth axiom of
   a) Euclidean geometry. b) hyperbolic geometry.
   c) elliptical geometry.

7. State the theorem concerning the sum of the measures of the angles of a triangle in
   a) Euclidean geometry. b) hyperbolic geometry.
   c) elliptical geometry.

8. What model is often used in describing and explaining Euclidean geometry?

9. What model is often used in describing and explaining elliptical geometry?

10. What model is often used in describing and explaining hyperbolic geometry?

11. What do we mean when we say that no one axiomatic system of geometry is "best"?

12. What did Einstein conjecture about the relationship between mass and space?

13. List the three types of curvature of space and the types of geometry that correspond to them.

14. In our discussion of the Koch snowflake we indicated that it had a finite area enclosed in an infinite perimeter. Explain how this can be true.

15. a) Develop a fractal by beginning with a square and replacing each side ⎯ with a ⌐⌐ . Repeat this process twice.
    b) If you continue this process will the fractal's perimeter be finite or infinite? Explain.
    c) Will the fractal's area be finite or infinite? Explain.

Problem Solving/Group Activity

16. In forming the Koch snowflake in Figure 9.68 the perimeter becomes greater at each step in the process. If each side of the original triangle is 1 unit, a general formula for the perimeter, L, of the snowflake at any step, n, may be found by the formula

$$L = 3\left(\frac{4}{3}\right)^{n-1}$$

For example, at the first step when $n = 1$, the perimeter is 3 units, which can be verified by the formula as follows:

$$L = 3\left(\frac{4}{3}\right)^{1-1} = 3\left(\frac{4}{3}\right)^{0} = 3 \cdot 1 = 3.$$

At the second step, when $n = 2$, we find the perimeter as follows:

$$L = 3\left(\frac{4}{3}\right)^{2-1} = 3\left(\frac{4}{3}\right) = 4$$

Thus at the second step the perimeter of the snowflake is 4 units.

a) Use the formula to complete the following table.

Step	Perimeter
1	
2	
3	
4	
5	
6	

b) Use the results of your calculations to explain why the perimeter of the Koch snowflake is infinite.

17. We can create another model for hyperbolic geometry with the following construction.

i) Draw a large circle, as shown in Fig. 9.70; call it the great circle.

ii) Draw a smaller circle intersecting the great circle at right angles, as shown in Fig. 9.70. Two curves that intersect at right angles are said to be *orthogonal*.

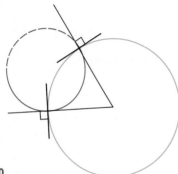

Figure 9.70

iii) Draw additional orthogonal circles to the great circle. All the lines and points inside the great circle are lines and points in hyperbolic geometry. Two lines that share a common point on the

circumference of the great circle are parallel, as shown in Fig. 9.71. For example, l_1 is parallel to l_2.

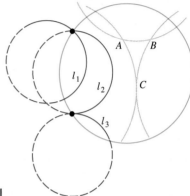

Figure 9.71

a) Name another pair of parallel lines.

b) In our model, ABC is a triangle. Measure its interior angles and determine whether the sum of the measures of the angles is less than, equal to, or greater than 180°.

Research Activity

18. The Renaissance painters, for the first time perhaps, worked at giving visual depth to their paintings. To accomplish visual depth the painter had to understand *perspective*. That is, in a painting they had to determine how large a tree or building should be to give the impression that the object was in the background. The branch of geometry that provides this information is called *projective geometry*. Write a paper on the use of projective geometry in art, including specific examples of applications. (References include books on art, encyclopedias, and history of mathematics books.)

CHAPTER 9 SUMMARY

Key Terms

9.1

acute angle
adjacent angles
alternate exterior angles
alternate interior angles
angle
axiomatic method
axioms
complementary angles
corresponding angles
definitions

end point
Euclidean (or plane) geometry
exterior angles
half lines
interior angles
line
line segment
measure of an angle
obtuse angle
postulates
parallel lines
parallel planes

plane
point
ray
right angle
skewed lines
straight angle
supplementary angles
theorems
transversal
undefined terms
vertex
vertical angles

9.2

congruent figures
polygon
quadrilaterals
regular polygon
similar figures

9.3

area
circumference
diameter
perimeter
radius

9.4

polyhedron
prism

pyramid
solid geometry
volume

9.5

arc (or edge)
even vertex
graphs
network
odd vertex
topology
traversable network
vertex

9.6

genus
Jordan curve

Klein bottle
Möbius strip

9.7

chaos theory
fractal geometry
hyperbolic geometry
Koch snowflake
model
non-Euclidean geometry
parallel postulate
recursion
Riemannian (or spherical or
    elliptical) geometry

Important Facts

The sum of the measures of the angles of a triangle is
180°.
The sum of the measures of the angles of a quadrilateral is
360°.
The sum of the measures of the interior angles of an
n-sided polygon is $(n - 2)180°$.

Triangle

$A = \frac{1}{2}bh$

$p = s_1 + s_2 + s_3$

Rectangle

$A = lw$

$p = 2l + 2w$

Trapezoid

$A = \frac{1}{2}h(b_1 + b_2)$

$p = s_1 + s_2 + b_1 + b_2$

Pythagorean Theorem

$a^2 + b^2 = c^2$

Square

$A = s^2$

$p = 4s$

Parallelogram

$A = bh$

$p = 2b + 2w$

Circle

$A = \pi r^2$; $C = 2\pi r$ or $C = \pi d$

Cube

$V = s^3$

Rectangular Solid

$V = lwh$

Cylinder

$V = \pi r^2 h$

Cone

$V = \frac{1}{3}\pi r^2 h$

Sphere

$v = \frac{4}{3}\pi r^3$

Prism

$V = Bh$, where B is the area of the base

Pyramid

$V = \frac{1}{3}Bh$, where B is the area of the base

Rules of Traversability

1. A network with no odd (all even) vertices is traversable; you may start from any vertex, and you will end where you began.
2. A network with exactly two odd vertices is traversable; you must start at either of the odd vertices and finish at the other.
3. A network with more than two odd vertices is not traversable.

Fifth Postulate in Euclidean Geometry

Given a line and a point not on the line, only one line can be drawn through the given point parallel to the given line.

Fifth Postulate in Elliptical Geometry

Given a line and a point not on the line, no line can be drawn through the given point parallel to the given line.

Fifth Postulate in Hyperbolic Geometry

Given a line and a point not on the line, two or more lines can be drawn through the given point parallel to the given line.

Chapter 9 Review Exercises

9.1

In Exercises 1–6, use the figure shown to find the following.

1. $\angle ABH \cap \angle FBC$
2. $\overleftrightarrow{AB} \cap \overleftrightarrow{BC}$
3. $\overline{BF} \cup \overline{FC} \cup \overline{BC}$
4. $\overleftrightarrow{HI} \cap \overleftrightarrow{EG}$
5. $\overrightarrow{BH} \cup \overrightarrow{HB}$
6. $\overrightarrow{CF} \cap \overrightarrow{CG}$

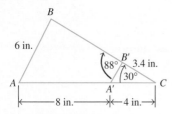

9.2

In Exercises 7–10, use the similar triangles ABC and A′B′C shown to find the following.

7. The length of $\overline{A'B'}$
8. The length of $\overline{BC}$
9. The measure of $\angle ABC$
10. The measure of $\angle BAC$

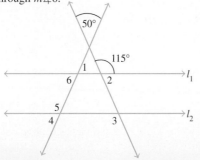

11. In the following figure, l_1 and l_2 are parallel lines. Find $m\angle 1$ through $m\angle 6$.

9.3

In Exercises 12–16, find the area of the figure.

12. 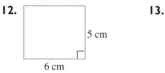 13.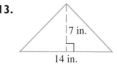

14.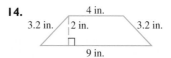

15.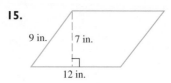

16.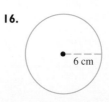

9.4

In Exercises 17 and 18, find the volume of the figure.

17.

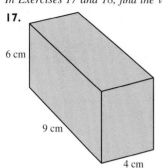

6 cm

9 cm

4 cm

18.

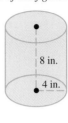

8 in.

4 in.

19. Find the area of the isosceles triangle.

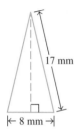

17 mm

8 mm

In Exercises 20–23, find the volume of the solid.

20.

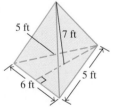

5 ft 7 ft

5 ft

6 ft

21.

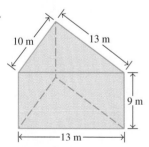

10 m 13 m

9 m

13 m

22.

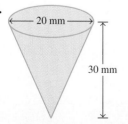

20 mm

30 mm

23.

10 ft

10 ft

9 ft

8 ft

24. Determine the total cost of covering a 14-ft × 16-ft kitchen floor with tile. The cost of the tile selected is $18.50 per square yard. (Add 5% of cost for waste.)

25. Farmer Jones has a water trough whose ends are trapezoids and whose sides are rectangles, as illustrated. He is afraid that the base it is sitting on will not support the weight of the trough when it is filled with water. He knows that the base will support 4800 lb.

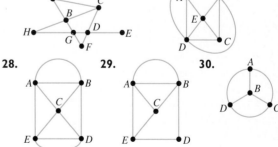

3 ft   4 ft   8 ft

2 ft   8 ft

3 ft

a) Find the number of cubic feet of water contained in the trough.
b) Find the total weight, assuming that the trough weighs 375 lb and the water weighs 62.5 lb per cubic foot. Is the base strong enough to support the trough filled with water?
c) If 1 gal of water weighs 8.3 lb, how many gallons of water will the trough hold?

9.5

Determine whether the networks in Exercises 26–30 are traversable. If they are, indicate from which points you may start.

26.

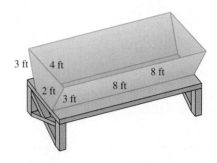

A
C
B
H
D
G F
E

27.

A B
E
D C

28.

A B
C
E D

29.

A B
C
E D

30.

A
B
D C

9.7

31. Summarize the history of non-Euclidean geometry.

32. State the fifth axiom of Euclidean, elliptical, and hyperbolic geometry.

CHAPTER 9 TEST

In Exercises 1–4, use the figure to describe the following sets of points:

1. $\overset{\circ}{AF} \cap \overset{\circ}{EF}$

2. $\overline{BC} \cup \overline{CD} \cup \overline{BD}$

3. $\measuredangle EDF \cap \measuredangle BDC$

4. $\overrightarrow{AC} \cup \overrightarrow{BA}$

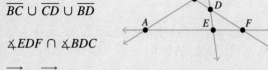

5. $m\measuredangle A = 53.2°$. Find the measure of the complement of $\measuredangle A$.

6. $m\measuredangle B = 79.8°$. Find the measure of the supplement of $\measuredangle B$.

7. In the figure, find the measure of $\measuredangle x$.

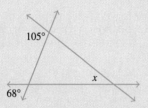

8. Find the sum of the interior angles of a hexagon.

9. Triangles ABC and $A'B'C'$ are similar figures. Find the length of side $\overline{B'C'}$.

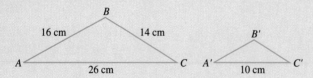

10. Right triangle ABC has one leg of length 15 in. and a hypotenuse of length 25 in.
    a) Find the length of the other leg.
    b) Find the perimeter of the triangle.
    c) Find the area of the triangle.

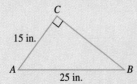

11. Find the volume of a sphere of diameter 15 cm.

12. The sketch shows the dimensions of the base of a pier for a bridge with semicircular ends. How many cubic yards of concrete are needed to build a pier 6 ft high?

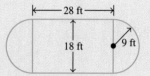

13. Find the volume of the pyramid.

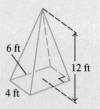

In Exercises 14 and 15, determine whether the network is traversable. If it is traversable, state at which points you may start.

14.
15.

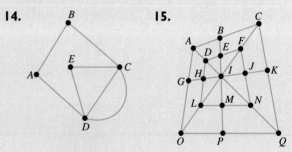

16. Explain how Euclidean geometry, hyperbolic geometry, and elliptical geometry differ.

17. What is a Möbius strip?

18. a) Sketch an object that is of genus 1.
    b) Sketch an object that is of genus 2.

GROUP PROJECTS

Supporting a Jacuzzi

1. You are thinking of buying a circular Jacuzzi 12 ft in diameter, 4 ft deep, and weighing 475 lb. You want to place the Jacuzzi on a deck built to support 30,000 lb.

    a) Determine the volume of the water in the Jacuzzi in cubic feet.

    b) Determine the number of gallons of water the Jacuzzi will hold. Note 1 ft$^3 \approx$ 7.5 gal.

    c) Determine the weight of the water in the Jacuzzi. (*Hint:* fresh water weighs about 52.4 lb/ft^3.)

    d) Will the deck support the weight of the Jacuzzi and water?

    e) Will the deck support the weight of the Jacuzzi, water, and four people, whose average weight is 115 lb?

Designing a Ramp

2. The Troutmans are planning to build a ramp so that their front entrance is wheel-chair accessible. The ramp will be 36 in. wide. It will rise two inches for each foot of length of horizontal distance. Where the ramp meets the porch, the ramp must be 2 ft high. To provide stability for the ramp, the Troutmans will install a slab of concrete 4 in. thick and 6 in. longer and wider than the ramp (see accompanying figure). The top of the slab will be level with the ground. The ramp may be constructed of concrete or pressure-treated lumber. You are to estimate the cost of materials for constructing the slab, the ramp of concrete, and the ramp of pressure-treated lumber.

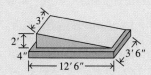

Slab

    a) Determine the length of the base of the ramp.

    b) Determine the dimensions of the concrete slab on which the ramp will set.

    c) Determine the volume of the concrete in cubic yards needed to construct the slab.

    d) If ready-mix concrete costs $45 per cubic yard, determine the cost of the concrete needed to construct the slab.

Concrete Ramp

    e) To build the ramp of concrete a form in the shape of the ramp must be framed. The two sides of the form are triangular and the shape of the end, which is against the porch, is rectangular. The form will be framed from $\frac{3}{4}$-in. plywood, which comes in 4-ft × 8-ft sheets. Determine the number of sheets of plywood needed. Assume that the entire sheet(s) will be used to make the sides and the end of the form and that there is no waste.

    f) If the plywood costs $18.95 for a 4-ft × 8-ft sheet, determine the cost of the plywood.

    g) To brace the form the Troutmans will need two boards 2 in. × 4 in. × 8 ft (referred to as 8 ft 2 × 4's) and six pieces of 2 in. × 4 in. × 3 ft. These six pieces of lumber will be cut from an 8 ft 2 × 4. Determine the number of 8 ft 2 × 4's needed.

    h) Determine the cost of the 8 ft 2 × 4's needed in part (g) if one board costs $2.14.

    i) Determine the volume, in cubic yards, of concrete needed to fill the form.

    j) Determine the cost of the concrete needed to fill the form.

    k) Determine the total cost of materials for building the ramp of concrete by adding the results in parts (d), (f), (h), and (j).

Wooden Ramp

    l) Determine the length of the top of the ramp.

    m) The top of the ramp will be constructed of $\frac{5}{4}$-in. × 6-in. × 10-ft pressure-treated lumber. The boards will be butted end to end to make the necessary length and will be supported from underneath by a wooden frame. Determine the number of boards needed to cover the top of the ramp. The boards are laid lengthwise on the ramp.

    n) Determine the cost of the boards to cover the top of the ramp if the price of a 10-ft length is $6.47.

    o) To support the top of the ramp the Troutmans will need 10 pieces of 8 ft 2 × 4's. The price of a pressure-treated 8 ft 2 × 4 is $2.44. Determine the cost of the supports.

    p) Determine the cost of materials for building a wooden ramp by adding the amounts from parts (d), (n), and (p).

    q) Are the materials for constructing a concrete ramp or a wooden ramp less expensive?

Chapter 10 Mathematical Systems

*W*e have, throughout human history, devised different ways of making sense out of our world. For primitive societies, explanations of why things happened the way they did took the form of mythology. With the philosophers of ancient Greece we see a more reasoned, scientific approach used to discover the essence of things. Today, education is highly specialized and fragmented into different disciplines. Yet the driving force behind all these disciplines is still the same: to understand the most basic laws and patterns of the subject.

In the last 200 years, much of the focus of scientific study of the fundamental laws of nature has shifted from what things are to how they change in space and time. The mathematics used for this purpose belongs to a branch of mathematics known as group theory. A group is a collection of

fundamentally basic elements: It could be numbers, it could be the elementary particles of physics, or it could be a pattern of repeating geometric designs. The way in which the elements change or remain the same when acted upon by some operation or transformation defines membership in the group. A certain underlying pattern and symmetry exists in all this.

To appreciate why group theory is so useful to scientists, consider the physicists who are trying to piece together the history of the universe. They have a clear sense of what the universe is like today, but what about 20 billion years ago when scientists believe the

Physicists use group theory to categorize the different elementary particles that spin off when subatomic particles such as protons and neutrons are smashed together. This photo shows a computer simulation of the tracks made by such a collision.

The concepts of group theory can be found in music. Consider the musical form known as the fugue. A single theme is repeated over time, in several voices, but changed in key and pitch to create a densely layered piece of music that eventually returns to the original theme.

universe exploded into existence in what is called the Big Bang? Group theory may enable them to derive what the initial conditions of the universe may have been from knowledge of what matter is like today. Group theory can also be used to describe number systems as well as the basic building blocks of crystalline solids. It can even be used to define the symmetries that appear in the artifacts of a given culture. No wonder the physicist Sir Arthur Stanley Eddington called group theory the "super-mathematics." ■

Despite the seemingly endless variety of patterns the human imagination can devise, group theory can be used to catalog and define patterns by the way in which the design elements are transformed and positioned.

10.1 GROUPS

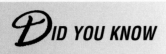

Many different types of mathematical systems have been developed. Some systems are used in solving everyday problems, such as planning work schedules. Others are more abstract and are used primarily in research, in chemistry, physical structure, matter, the nature of genes, and other scientific fields.

When you learned how to add integers, you were introduced to a mathematical system. When you learned how to multiply integers, you became familiar with a second mathematical system. The set of integers with the operation of subtraction and the set of integers with the operation of division are two other examples of mathematical systems. Addition, subtraction, multiplication, and division are called binary operations. A binary operation is an operation, or rule, that can be performed on two and only two elements of a set. The result is a single element. When we add *two* integers, the sum is *one* integer. When we multiply *two* integers, the product is *one* integer. Is finding the reciprocal of a number a binary operation? No, it is an operation on a single element of a set.

A **mathematical system** consists of a set of elements and at least one binary operation.

Commutative and Associative Properties

Once a mathematical system is defined, its structure may display certain properties. Consider the set of integers:

$$I = \{\ldots, -3, -2, -1, 0, 1, 2, 3, \ldots\}.$$

Recall that the ellipsis, the three dots, at each end of the set indicates that the set continues in the same manner.

The set of integers can be studied with the operations of addition, subtraction, multiplication, or division as separate mathematical systems. For example, when we study the set of integers under the operations of addition or multiplication, we see that the commutative and associative properties hold. The general forms of the properties are shown here.

For Any Elements *a*, *b*, and *c*		
	Addition	**Multiplication**
Commutative property	$a + b = b + a$	$a \cdot b = b \cdot a$
Associative property	$(a + b) + c = a + (b + c)$	$(a \cdot b) \cdot c = a \cdot (b \cdot c)$

The integers *are commutative* under the operations of *addition and multiplication*. For example,

$$2 + 4 = 4 + 2 \quad \text{and} \quad 2 \cdot 4 = 4 \cdot 2$$
$$6 = 6 \qquad\qquad\qquad 8 = 8$$

However, the integers *are not commutative* under the operations of *subtraction and division*. For example,

$$4 - 2 \neq 2 - 4 \quad \text{and} \quad 4 \div 2 \neq 2 \div 4$$
$$2 \neq -2 \qquad\qquad 2 \neq \tfrac{1}{2}$$

The integers *are associative* under the operations of *addition and multiplication*. For example,

$$(1 + 2) + 3 = 1 + (2 + 3) \quad \text{and} \quad (1 \cdot 2) \cdot 3 = 1 \cdot (2 \cdot 3)$$
$$3 + 3 = 1 + 5 \qquad\qquad 2 \cdot 3 = 1 \cdot 6$$
$$6 = 6 \qquad\qquad 6 = 6$$

However, the integers *are not associative* under the operations of *subtraction and division*. See Exercises 14 and 15 at the end of this section.

To say that a set of elements is commutative under a given operation means that the commutative property holds for *any* elements a and b in the set. Similarly, to say that a set of elements is associative under a given operation means that the associative property holds for *any* elements a, b, and c in the set.

Consider the mathematical system consisting of the set of integers under the operation of addition. Because the set of integers is infinite, this mathematical system is an example of an *infinite mathematical system*. We will study certain properties of this mathematical system. The first property that we will examine is closure.

Closure

The sum of any two integers is an integer. Therefore the set of integers is said to be *closed*, or to satisfy the *closure property*, under the operation of addition.

> If a binary operation is performed on any two elements of a set and the result is an element of the set, then that set is **closed** (or has **closure**) under the given binary operation.

Is the set of integers closed under the operation of multiplication? The answer is yes. When any two integers are multiplied, the product will be an integer.

Is the set of integers closed under the operation of subtraction? Again the answer is yes. The difference of any two integers is an integer.

Is the set of integers closed under the operation of division? The answer is no because two integers may have a quotient that is not an integer. For example, if we select the integers 2 and 3, the quotient of 2 divided by 3 is $\tfrac{2}{3}$, which is not an integer. Thus the integers are not closed under the operation of division.

We showed that the set of integers was not closed under the operation of division by finding two integers whose quotient was not an integer. A specific example illustrating that a specific property is not true is called a *counter-example*. Mathematicians and scientists often try to find a counterexample to confirm that a specific property is not always true.

Identity Element

Now we will discuss the identity element for the set of integers under the operation of addition. Is there an element in the set that, when added to any given integer, results in a sum that is the given integer? The answer is yes. The sum of 0 and any integer is the given integer. For example, $1 + 0 = 0 + 1 = 1$, $2 + 0 = 0 + 2 = 2$, and so on. For this reason we call 0 the additive identity element for the set of integers. Note that for any integer a, $a + 0 = 0 + a = a$.

> An **identity element** is an element in a set such that when a binary operation is performed on it and any given element in the set, the result is the given element.

Is there an identity element for the set of integers under the operation of multiplication? The answer is yes; it is the number 1. Note that $2 \cdot 1 = 1 \cdot 2 = 2$, $3 \cdot 1 = 1 \cdot 3 = 3$, and so on. For any integer a, $a \cdot 1 = 1 \cdot a = a$. For this reason, 1 is called the multiplicative identity element for the set of integers.

Inverses

What integer when added to 4 gives a sum of 0; that is, $4 + \blacksquare = 0$? The shaded area is to be filled in with the integer -4: $4 + (-4) = 0$. We say that -4 is the additive inverse of 4, and 4 is the additive inverse of -4. Note that the sum of the element and its additive inverse gives the additive identity element 0. What is the additive inverse of 12? Since $12 + (-12) = 0$, -12 is the additive inverse of 12.

Other examples of integers and their additive inverses are

Element + Additive Inverse = Identity Element

0	+	0	=	0
2	+	(−2)	=	0
−5	+	5	=	0

Note that for the operation of addition, every integer a has a unique inverse, $-a$, such that $a + (-a) = -a + a = 0$.

> When a binary operation is performed on two elements in a set and the result is the identity element for the binary operation, then each element is said to be the **inverse** of the other.

Does every integer have an inverse under the operation of multiplication? For multiplication the product of an integer and its inverse must yield the multiplicative identity element, 1. What is the multiplicative inverse of 2? That is, 2 times what number gives 1?

$$2 \cdot ? = 1 \qquad 2 \cdot \tfrac{1}{2} = 1$$

However, since $\tfrac{1}{2}$ is not an integer, 2 does not have a multiplicative inverse in the set of integers.

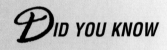

*P*ROFILE IN MATHEMATICS

NIELS ABEL (1802–1829)
ÉVARISTE GALOIS
(1811–1832)

s Abel   Évariste Galois

Untimely ends: Important contributions to the development of
group theory were made by two
young men who would not live
to see their work gain acceptance: *Niels Abel*, a Norwegian,
and *Évariste Galois*, a Frenchman.
While their work showed bril-
liance, neither was noticed in his
lifetime by the mathematics
community. Abel lived in poverty and died of malnutrition
and tuberculosis at the age of
26, just two days before a letter
arrived with the offer of a
teaching position. Galois died of
complications from a gunshot
wound he received in a duel
over a love affair. The 20-year-
old man had spent the night before summarizing his theories in
the hope that eventually someone would find it "profitable to
decipher this mess." After their
deaths, both men had important
elements of group theory
named after them. ■

Group

Let's review what we have learned about the mathematical system consisting of the set of integers under the operation of addition.

1. The set of integers is *closed* under the operation of addition.
2. The set of integers has an *identity element* under the operation of addition.
3. Each element in the set of integers has an *inverse* under the operation of addition.
4. The *associative property* holds for the set of integers under the operation of addition.

The set of integers under the operation of addition is an example of a group. The properties of a group can be summarized as follows.

> **Properties of a Group**
> Any mathematical system that meets the following four requirements is called a group.
>
> **1.** The set of elements is *closed* under the given operation.
> **2.** An *identity element* exists for the set.
> **3.** Every element in the set has an *inverse*.
> **4.** The set of elements is *associative* under the given operation.

It is often very time consuming to show that the associative property holds for all cases. In many of the examples that follow we will state that the associative property holds for the given set of elements under the given operation.

Commutative Group

The commutative property does not need to hold for a mathematical system to be a group. However, if a mathematical system meets the four requirements of a group and is also commutative under the given operation, the mathematical system is a *commutative (or abelian) group*. The abelian group is named after Niels Abel (see Profile in Mathematics).

> A group that satisfies the commutative property is called a **commutative group** (or **abelian group**).

Because the commutative property holds for the set of integers under the operation of addition, the set of integers under the operation of addition is not only a group, but it is a commutative group.

> **Properties of a Commutative Group**
> A mathematical system is a commutative group if all five conditions hold.
>
> **1.** The set of elements is *closed* under the given operation.
> **2.** An *identity element* exists for the set.

Continued on page 498.

3. Every element in the set has an *inverse*.
4. The set of elements is *associative* under the given operation.
5. The set of elements is *commutative* under the given operation.

To determine whether a mathematical system is a group under a given operation, check, in the following order, to determine whether (a) the system is closed under the given operation, (b) there is an identity element in the set for the given operation, (c) every element in the set has an inverse under the given operation, and (d) the associative property holds under the given operation. If any of these four requirements is *not* met, stop and state that the mathematical system is not a group. If asked to determine whether the mathematical system is a commutative group, then check to determine whether the commutative property holds for the given operation.

--- **EXAMPLE 1**

Determine whether the set of rational numbers under the operation of multiplication forms a group.

Solution: Recall from Chapter 5 that the rational numbers are the set of numbers of the form p/q where p and q are integers, $q \neq 0$. All fractions and integers are rational numbers.

1. *Closure:* The product of two rational numbers is a rational number. Therefore the rational numbers are closed under the operation of multiplication.

2. *Identity Element:* The multiplicative identity element for the set of rational numbers is 1. Note, for example, $3 \cdot 1 = 1 \cdot 3 = 3$, and $\frac{3}{8} \cdot 1 = 1 \cdot \frac{3}{8} = \frac{3}{8}$. For any rational number a, $a \cdot 1 = 1 \cdot a = a$.

3. *Inverse Elements:* For the mathematical system to be a group under the operation of multiplication, *each and every* rational number must have a multiplicative inverse in the set of rational numbers. Remember that for the operation of multiplication, the product of a number and its inverse must give the multiplicative identity element, 1. Let's check a few rational numbers:

$$\text{Rational number} \cdot \text{inverse} = \text{identity element}$$

$$3 \quad \cdot \quad \frac{1}{3} \quad = \quad 1$$

$$\frac{2}{3} \quad \cdot \quad \frac{3}{2} \quad = \quad 1$$

$$-\frac{1}{5} \quad \cdot \quad -5 \quad = \quad 1$$

Looking at these examples you might deduce that each rational number does have an inverse. However, one rational number, 0, does not have an inverse.

$$0 \cdot ? = 1$$

Because there is no rational number that, when multiplied by 0, gives 1, 0 does not have a multiplicative inverse. Since every rational number does not have an inverse, this mathematical system is not a group.

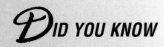

There is no need at this point to check the associative property because we have already shown that the mathematical system of rational numbers under the operation of multiplication is not a group.   ▲

Section 10.1 Exercises

1. What is a binary operation?

2. What are the parts of a mathematical system?

3. Explain why each of the following is a binary operation. Give an example to illustrate each binary operation.
   a) Addition
   b) Subtraction
   c) Multiplication
   d) Division

4. Explain the closure property. Give an example of the property.

5. What is an identity element? Explain. Give the additive and multiplicative identity elements for the set of integers.

6. What is an inverse element? Explain. Give the additive and multiplicative inverse of the number 2 for the set of rational numbers.

7. Give the associative property of addition and illustrate the property with an example.

8. Give the associative property of multiplication and illustrate the property with an example.

9. Give the commutative property of addition and illustrate the property with an example.

10. Give the commutative property of multiplication and illustrate the property with an example.

11. What is a counterexample?

12. Give an example to show that the commutative property does not hold for the set of integers under the operation of subtraction.

13. Give an example to show that the commutative property does not hold for the set of integers under the operation of division.

14. Give an example to show that the associative property does not hold for the set of integers under the operation of subtraction.

15. Give an example to show that the associative property does not hold for the set of integers under the operation of division.

16. What properties are required for a mathematical system to be a group?

17. What properties are required for a mathematical system to be a commutative group?

18. What is another name for a commutative group?

In Exercises 19–23, explain your answer.

19. Is the set of positive integers a commutative group under the operation of addition?

20. Is the set of integers a group under the operation of multiplication?

21. Is the set of rational numbers a commutative group under the operation of addition?

22. Is the set of positive integers a group under the operation of subtraction?

23. Is the set of integers a group under the operation of subtraction?

24. For the set of integers list two operations that are not binary. Explain your answer.

Problem Solving/Group Activity

25. Is the set of irrational numbers a group under the operation of addition?

26. Is the set of irrational numbers a group under the operation of multiplication?

27. Is the set of real numbers a group under the operation of addition?

28. Is the set of real numbers a group under the operation of multiplication?

29. Create a mathematical system with two binary operations. Select a set of elements and the binary operations so that one binary operation with the set of elements will meet the requirements for a group and the other binary operation will not. Explain why the one binary operation with the set of elements is a group. For the other binary operation and the set of elements find counterexamples to show that it is not a group.

Research Activity

30. There are other classifications of mathematical systems besides groups. For example, there are *rings* and *fields*. Do research to determine the requirements that must be met for a mathematical system to be (a) a ring and (b) a field. (c) Is the set of real numbers, under the operations of addition and multiplication, a field? (References include history of mathematics books.)

10.2 FINITE MATHEMATICAL SYSTEMS

In the preceding section we presented infinite mathematical systems. In this section we present some finite mathematical systems. A **finite mathematical system** is one whose set contains a finite number of elements.

Clock Arithmetic

Figure 10.1

Let's develop a finite mathematical system called **clock arithmetic**. The set of elements in this system will be the hours on a clock: {1, 2, 3, 4, 5, 6, 7, 8, 9, 10, 11, 12}. The binary operation that we will use is addition, which we define as movement of the hour hand in a clockwise direction. Assume that it is 4 o'clock: What time will it be in nine hours? (See Fig 10.1.) If we add 9 hours to 4 o'clock, the clock will read 1 o'clock. Thus $4 + 9 = 1$ in clock arithmetic. Would $9 + 4$ be the same as $4 + 9$? Yes, $4 + 9 = 9 + 4 = 1$.

Table 10.1 is the addition table for clock arithmetic. Its elements are based on the definition of addition as previously illustrated. For example, the sum of 4 and 9 is 1, so we put a 1 in the table where the row to the right of the 4 intersects the column below the 9. Likewise, the sum of 11 and 10 is 9, so we put a 9 in the table where the row to the right of the 11 intersects the column below the 10.

The binary operation of this system is defined by the table. It is denoted by the symbol $+$. To determine the value of $a + b$, where a and b are any two numbers in the set, find a in the left-hand column and find b along the top row. Assume that there is a horizontal line through a and a vertical line through b; the point of intersection of these two lines is where you find the value of $a + b$. For example, $10 + 4 = 2$ has been circled in Table 10.1. Note that $4 + 10$ also equals 2, but this result will not necessarily hold for all examples in this chapter.

Table 10.1

+	1	2	3	4	5	6	7	8	9	10	11	12
1	2	3	4	5	6	7	8	9	10	11	12	1
2	3	4	5	6	7	8	9	10	11	12	1	2
3	4	5	6	7	8	9	10	11	12	1	2	3
4	5	6	7	8	9	10	11	12	1	2	3	4
5	6	7	8	9	10	11	12	1	2	3	4	5
6	7	8	9	10	11	12	1	2	3	4	5	6
7	8	9	10	11	12	1	2	3	4	5	6	7
8	9	10	11	12	1	2	3	4	5	6	7	8
9	10	11	12	1	2	3	4	5	6	7	8	9
10	11	12	1	②	3	4	5	6	7	8	9	10
11	12	1	2	3	4	5	6	7	8	9	10	11
12	1	2	3	4	5	6	7	8	9	10	11	12

EXAMPLE 1

Determine whether the clock arithmetic system under the operation of addition is a commutative group.

Solution: Check the five requirements that must be satisfied for a commutative group.

1. *Closure:* Is the set of elements in clock arithmetic closed under the operation of addition? Yes, since Table 10.1 contains only the elements in the set {1, 2, 3, 4, 5, 6, 7, 8, 9, 10, 11, 12}. If Table 10.1 had contained an element other than the numbers 1 through 12, the set would not have been closed under addition.

2. *Identity element:* Is there an identity element for clock arithmetic? If it is currently 4 o'clock, how many hours have to pass before it is 4 o'clock again? Twelve hours: $4 + 12 = 12 + 4 = 4$. In fact, given any hour, in 12 hours the clock will return to the starting point. Therefore 12 is the additive identity element in clock arithmetic.

In examining Table 10.1 we see that the row of numbers next to the 12 in the left-hand column is identical to the row of numbers along the top. We also see that the column of numbers under the 12 in the top row is identical to the column of numbers on the left. The search for such a column and row is one technique for determining whether an identity element exists for a system defined by a table.

3. *Inverse elements:* Is there an inverse for the number 4 in clock arithmetic for the operation of addition? Recall that the identity element in clock arithmetic is 12. What number when added to 4 gives 12; that is, $4 + $ 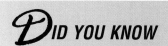 $ = 12$? Table 10.1 shows that $4 + 8 = 12$ and also that $8 + 4 = 12$. Thus 8 is the additive inverse of 4, and 4 is the additive inverse of 8.

To find the additive inverse of 7, find 7 in the left-hand column. Look to the right of the 7 until you come to the identity element 12. Determine the number at the top of this column. The number is 5. Since $7 + 5 = 5 + 7 = 12$, 5 is the inverse of 7, and 7 is the inverse of 5. The other inverses can be found in the same way, as shown in Table 10.2. Note that each element in the set has an *inverse*.

Table 10.2

Element	+	Inverse	=	Identity Element
1	+	11	=	12
2	+	10	=	12
3	+	9	=	12
4	+	8	=	12
5	+	7	=	12
6	+	6	=	12
7	+	5	=	12
8	+	4	=	12
9	+	3	=	12
10	+	2	=	12
11	+	1	=	12
12	+	12	=	12

4. *Associative property:* Now consider the associative property. Does $(a + b) + c = a + (b + c)$ for all values a, b, and c of the set? Remember to always evaluate the values within the parentheses first. Let's select some values for a, b, and c. Let $a = 2$, $b = 6$, and $c = 8$. Then

$$(2 + 6) + 8 = 2 + (6 + 8)$$
$$8 + 8 = 2 + 2$$
$$4 = 4 \quad \text{True}$$

Let $a = 5$, $b = 12$, and $c = 9$. Then

$$(5 + 12) + 9 = 5 + (12 + 9)$$
$$5 + 9 = 5 + 9$$
$$2 = 2 \quad \text{True}$$

Randomly selecting *any* elements a, b, and c of the set reveals that $(a + b) + c = a + (b + c)$. Thus the system of clock arithmetic is associative under the operation of addition. Note that if there is just one set of values a, b, and c such that $(a + b) + c \neq a + (b + c)$, the system is not associative. Normally you will not be asked to check every case to determine whether the associative property holds. However, if every element in the set does not appear in every row and column of the table, you need to check the associative property carefully.

5. *Commutative property:* Does the commutative property hold under the given operation? Does $a + b = b + a$ for all elements a and b of the set? Let's randomly select some values for a and b to determine whether the commutative property appears to hold. Let $a = 5$ and $b = 8$; then Table 10.1 shows that

$$5 + 8 = 8 + 5$$
$$1 = 1 \quad \text{True}$$

Let $a = 9$ and $b = 6$; then

$$9 + 6 = 6 + 9$$
$$3 = 3 \quad \text{True}$$

The commutative property holds for these two specific cases. In fact, if we were to randomly select *any* values for a and b, we would find that $a + b = b + a$. Thus the commutative property of addition is true in clock arithmetic. Note that, if there is just one set of values a and b such that $a + b \neq b + a$, the system is not commutative.

This system satisfies the five properties required for a mathematical system to be a commutative group. Thus clock arithmetic under the operation of addition is a commutative group. ▲

One method that can be used to determine whether a system defined by a table is commutative under the given operation is to determine whether the elements in the table are symmetric about the main diagonal. The main diagonal is the diagonal from the upper left-hand corner to the lower right-hand corner of the table. In Table 10.3 the main diagonal is shaded in color.

If the elements are symmetric about the main diagonal, then the system is commutative. If the elements are not symmetric about the main diagonal, then the system is not commutative. If you examine the system in Table 10.3 you will see that its elements are symmetric about the main diagonal. Therefore this mathematical system is commutative.

It is possible to have groups that are not commutative. Such groups are called noncommutative or nonabelian groups. However, a *noncommutative group defined by a table must be at least a six-element by six-element table.* Nonabelian groups are illustrated in Exercises 53 and 56 at the end of this section. We suggest that you do at least Exercise 53.

Table 10.3

+	0	1	2	3	4
0	0	1	2	3	4
1	1	2	3	4	0
2	2	3	4	0	1
3	3	4	0	1	2
4	4	0	1	2	3

Now we will look at another finite mathematical system.

EXAMPLE 2

Table 10.4

$\odot$	1	3	5	7
1	5	7	1	3
3	7	1	3	5
5	1	3	5	7
7	3	5	7	1

Consider the mathematical system defined by Table 10.4. Assume that the associative property holds for the given operation.

a) List the elements in the set of this mathematical system.

b) Identify the binary operation.

c) Determine whether this mathematical system is a commutative group.

Solution:

a) The set of elements for this mathematical system consists of the elements found on the top (or left-hand side) of the table: $\{1, 3, 5, 7\}$.

b) The binary operation is $\odot$.

c) We must determine whether the five requirements for a commutative group are satisfied.

1. *Closure:* All the elements in the table are in the original set of elements, $\{1, 3, 5, 7\}$, so the system is closed.

2. *Identity element:* The identity element is 5. Note that the column of elements under the 5 is identical to the left-hand column *and* the row of elements to the right of the 5 is identical to the top row.

3. *Inverse elements:* When an element operates on its inverse, the result is the identity element. For this example the identity element is 5. To determine the inverse of 1, find the element to replace the question mark:

$$1 \odot ? = 5$$

Since $1 \odot 1 = 5$, 1 is the inverse of 1. Thus 1 is its own inverse.

To find the inverse of 3, find the element to replace the question mark:

$$3 \odot ? = 5$$

Table 10.5

Element	$\odot$	Inverse	=	Identity Element
1	$\odot$	1	=	5
3	$\odot$	7	=	5
5	$\odot$	5	=	5
7	$\odot$	3	=	5

Since $3 \odot 7 = 7 \odot 3 = 5$, 7 is the inverse of 3 (and 3 is the inverse of 7). The elements and their inverses are shown in Table 10.5. Every element has a unique inverse.

4. *Associative property:* It is given that the associative property holds for this operation. One example of the associative property is

$$(7 \odot 3) \odot 1 = 7 \odot (3 \odot 1)$$
$$5 \odot 1 = 7 \odot 7$$
$$1 = 1 \quad \text{True}$$

5. *Commutative property:* The elements in the table are symmetric about the main diagonal, so the commutative property holds for the operation of $\odot$. One example of the commutative property is

$$3 \odot 5 = 5 \odot 3$$
$$3 = 3 \quad \text{True}$$

The five necessary properties hold. Thus the mathematical system is a commutative group.

Mathematical Systems Without Numbers

Thus far all the systems we have discussed have been based on sets of numbers. Example 3 illustrates a mathematical system of symbols rather than numbers.

--- EXAMPLE 3

Use the mathematical system defined by Table 10.6 and determine

Table 10.6

$\otimes$	#	$\triangle$	L
#	$\triangle$	L	#
$\triangle$	L	#	$\triangle$
L	#	$\triangle$	L

a) the set of elements.

b) the binary operation.

c) closure or nonclosure of the system.

d) the identity element.

e) the inverse of #.

f) # $\otimes$ $\triangle$ and L $\otimes$ $\triangle$.

g) (# $\otimes$ $\triangle$) $\otimes$ $\triangle$ and # $\otimes$ ($\triangle$ $\otimes$ $\triangle$).

Solution:

a) The set of elements of this mathematical system is {#, $\triangle$, L}.

b) The binary operation is $\otimes$.

c) Because the table does not contain any symbols other than #, $\triangle$, and L, the system is closed under $\otimes$.

d) The identity element is L, since the row next to L in the left-hand column is the same as the top row and the column under L is identical to the left-hand column. Note that

$$\# \otimes L = L \otimes \# = \#$$
$$\triangle \otimes L = L \otimes \triangle = \triangle$$
$$L \otimes L = L$$

e) Element $\otimes$ inverse element = identity element, or

$$\# \otimes ? = L$$

To find the inverse of #, we must determine the element to replace the question mark. Since # $\otimes$ $\triangle$ = L and $\triangle$ $\otimes$ # = L, $\triangle$ is the inverse of #.

f) # $\otimes$ $\triangle$ = L   and   L $\otimes$ $\triangle$ = $\triangle$

g) (# $\otimes$ $\triangle$) $\otimes$ $\triangle$ = L $\otimes$ $\triangle$ and   # $\otimes$ ($\triangle$ $\otimes$ $\triangle$) = # $\otimes$ #

   = $\triangle$ = $\triangle$

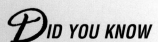

DID YOU KNOW

GROUP THEORY CAN BE FUN

*G*roup theory can be fun as well as educational, at least so thought Erno Rubik, a Hungarian teacher of architecture and design. In 1976 he presented the world with a puzzle in the form of a cube with six faces, each of a different color. Each face is divided into nine squares; each row and column of each face can rotate so that in no time at all six faces are transformed into a mix of colors. Your job, should you choose to accept it, is to get the cube back to its original condition. ∎

Table 10.7

*	A	B	C	D
A	D	A	B	C
B	A	B	C	D
C	B	C	D	A
D	C	D	A	B

Table 10.8

Element	*	Inverse	=	Identity Element
A	*	C	=	B
B	*	B	=	B
C	*	A	=	B
D	*	D	=	B

Table 10.9

⊖	x	y	z
x	x	z	y
y	z	y	x
z	y	x	z

EXAMPLE 4

Determine whether the mathematical system in Table 10.7 in which $*$ is an associative operation is a commutative group.

Solution:

1. *Closure:* The system is closed.

2. *Identity element:* The identity element is B.

3. *Inverse elements:* Each element has an inverse as illustrated in Table 10.8.

4. *Associative property:* It is given that the associative property holds. An example illustrating the associative property is

$$(D * A) * C = D * (A * C)$$
$$C * C = D * B$$
$$D = D \quad \text{True}$$

5. *Commutative property:* By examining the table we can see that it is symmetric about the main diagonal. Thus the system is commutative under the given operation. One example of the commutative property is

$$D * C = C * D$$
$$A = A \quad \text{True}$$

All five properties are satisfied. Thus the system is a commutative group.

▲

EXAMPLE 5

Determine whether the mathematical system in Table 10.9 is a commutative group under the operation of $\ominus$.

Solution:

1. The system is closed.

2. No row is identical to the top row, so there is no identity element. Therefore this mathematical system is *not a group*. There is no need to go any further, but for practice, let's look at a few more items.

3. Since there is no identity element, there can be no inverses.

4. The associative property does not hold. The following counterexample illustrates that the associative property does not hold for every case.

$$(x \ominus y) \ominus z \neq x \ominus (y \ominus z)$$
$$z \ominus z \neq x \ominus x$$
$$z \neq x$$

5. The table is symmetric about the main diagonal. Therefore the commutative property does hold for the operation of $\sim$.

Note that the associative property does not hold even though the commutative property does hold. This outcome can occur when there is no identity element and every element does not have an inverse, as in this example.

▲

Table 10.10

*	a	b	c
a	a	b	c
b	b	b	a
c	c	a	c

EXAMPLE 6

Determine whether the mathematical system in Table 10.10 is a commutative group under the operation of *.

Solution:

1. The system is closed.

2. There is an identity element, *a*.

3. Each element has an inverse; *a* is the inverse of *a*, *b* is the inverse of *c*, and *c* is the inverse of *b*.

4. Every element in the set does not appear in every row and every column of the table, so we need to check the associative property carefully. There are many specific cases where the associative property does hold. However, the following counterexample illustrates that the associative property does not hold for every case.

$$(b * b) * c \neq b * (b * c)$$
$$b * c \neq b * a$$
$$a \neq b$$

𝒟ID YOU KNOW

CREATING PATTERNS BY DESIGN

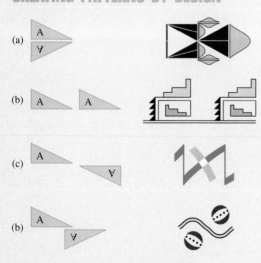

(a)

(b)

(c)

(b)

Patterns from *Symmetries of Culture* by Dorothy K. Washburn and Donald W. Crowe (University of Washington Press, 1988)

𝒲hat makes group theory such a powerful tool is that it can be used to reveal the underlying structure of just about any physical phenomenon that involves symmetry and patterning, such as wallpaper or quilt patterns. Interest in the formal study of symmetry in design came out of the Industrial Revolution in the late nineteenth century. The new machines of the Industrial Revolution could vary any given pattern almost indefinitely. Designers needed a way to describe and manipulate patterns systematically. At the same time, explorers were discovering artifacts of other cultures, which stimulated interest in categorizing patterns. Shown here are the four geometric motions that generate all two-dimensional patterns: (a) reflection, (b) translation, (c) rotation, and (d) glide reflection. How these motions are applied, or not applied, is the basis of pattern analysis. ∎

5. The commutative property holds because there is symmetry about the main diagonal.

Since we have shown that the associative property does not hold under the operation of ∗, this system is not a group. Therefore it cannot be a commutative group. ▲

Section 10.2 Exercises

1. Explain how the clock 12 addition table is formed.

2. Explain one method of determining whether a system defined by a table is commutative under the given operation.

In Exercises 3–14, use Table 10.1 to determine the sum in clock arithmetic.

3. 7 + 5

4. 9 + 5

5. 11 + 10

6. 8 + 4

7. 12 + 11

8. 12 + 12

9. 3 + (7 + 9)

10. (11 + 7) + 6

11. (4 + 4) + 8

12. (11 + 1) + 12

13. (5 + 7) + (9 + 6)

14. (7 + 11) + (9 + 5)

In Exercises 15–20, determine the difference in clock arithmetic by starting at the first number and counting counterclockwise on the clock the number of units given by the second number.

15. 11 − 7

16. 10 − 5

17. 5 − 8

18. 3 − 9

19. 5 − 11

20. 3 − 10

21. Can you find another way to obtain the answers for Exercises 15–20? Explain.

22. Develop an addition table for the seven-hour clock in Fig. 10.2.

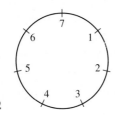

Figure 10.2

In Exercises 23–34, determine the sum or difference in seven-hour clock arithmetic.

23. 3 + 5

24. 4 + 6

25. 5 + 2

26. 4 + 4

27. 7 + 4

28. 3 + 7

29. 6 − 2

30. 2 − 5

31. 3 − 6

32. 1 − 7

33. (3 − 5) − 6

34. 4 + (2 − 6)

35. Determine whether seven-hour clock arithmetic under the operation of addition is a commutative group. Explain.

36. A mathematical system is defined by a 3-element by 3-element table where every element in the set appears in each row and each column. Must the mathematical system be a commutative group? Explain.

37. Consider the mathematical system indicated by the following table in which the operation ∗ is associative.

∗	0	1	2	3
0	0	1	2	3
1	1	2	3	0
2	2	3	0	1
3	3	0	1	2

a) What are the elements of the set in this mathematical system?

b) What is the binary operation?

c) Is the system closed? Explain.

d) Is there an identity element for the system under the given operation? If so, what is it?

e) Does every element in the system have an inverse? If so, give each element and its corresponding inverse.

f) Give an example to illustrate the associative property.

g) Is the system commutative? Give an example to verify your answer.

h) Is the mathematical system a commutative group? Explain.

In Exercises 38–40, repeat parts (a)–(h) of Exercise 37 for the mathematical system in the table. Assume that the operations are associative.

38.

§	16	19	23
16	19	23	16
19	23	16	19
23	16	19	23

39.

∞	2	3	7	11
2	7	11	2	3
3	11	2	3	7
7	2	3	7	11
11	3	7	11	2

40.

~	2	6	8	4
2	6	8	4	2
6	8	4	2	6
8	4	2	6	8
4	2	6	8	4

41. a) Is the following mathematical system a group? Explain your answer.

b) Find an example showing that the operation is not associative for the given set of elements.

a	1	2	3	4
1	2	3	4	1
2	3	4	1	2
3	4	1	2	3
4	1	4	3	2

42. For the mathematical system

!	z	p	o	n
z	z	p	o	n
p	p	o	n	z
o	o	n	z	p
n	n	z	p	o

determine

a) the elements in the set.

b) the binary operation.

c) closure or nonclosure of the system.

d) $(z \,!\, o) \,!\, p$.

e) $p \,!\, (n \,!\, z)$.

f) the identity element.

g) the inverse of z.

h) the inverse of p.

In Exercises 43–48, for the mathematical system given, determine which of the five properties of a commutative group do not hold.

43.

3	○	△	□
○	○	△	□
△	△	○	□
□	□	△	○

44.

∧	w	x	y
w	w	y	x
x	y	x	w
y	x	a	y

45.

∅	~	*	?	L	P
~	L	P	~	*	*
*	P	~	*	~	L
?	~	*	?	L	P
L	*	~	L	P	?
P	*	L	P	?	~

46.

☺	a	b	π	0	△
a	△	a	b	π	0
b	a	b	π	0	△
π	b	π	0	△	a
0	π	0	△	a	b
△	0	△	π	b	a

47.

⊙	a	b	c	d	e
a	c	d	e	a	b
b	d	e	a	b	c
c	e	a	b	c	d
d	a	b	c	e	d
e	b	c	d	e	a

48.

#	0	1	2	3	4	5
0	0	0	0	0	0	0
1	0	1	2	3	4	5
2	0	2	4	0	2	4
3	0	3	0	3	0	3
4	0	4	2	0	4	2
5	0	5	4	3	2	1

49. a) Consider the set consisting of two elements $\{E, O\}$, where E stands for an even number and O stands for an odd number. For the operation of addition, complete the table.

+	E	O
E		
O		

b) Determine whether this mathematical system forms a commutative group under addition. Explain your answer.

50. a) Let E and O represent even numbers and odd numbers, respectively, as in Exercise 49. Complete the table for the operation of multiplication.

×	E	O
E		
O		

b) Determine whether this mathematical system forms a commutative group under the operation of multiplication. Explain your answer.

In Exercises 51 and 52, make up your own mathematical system that is a group. List the identity element and the inverses of each element. Do so with sets containing

51. three elements.

52. four elements.

53. The following table is an example of a noncommutative or nonabelian group.

∞	1	2	3	4	5	6
1	5	3	4	2	6	1
2	4	6	5	1	3	2
3	2	1	6	5	4	3
4	3	5	1	6	2	4
5	6	4	2	3	1	5
6	1	2	3	4	5	6

a) Show that these elements under the operation ∞ are a group. (It would be very time consuming to prove that the associative property holds, but you can give some examples to show that it appears to hold.)

b) Find a counterexample to show that the commutative property does not hold.

Problem Solving/Group Activities

54. If a mathematical system is defined by a four-element by four-element table, how many specific cases must be illustrated to prove the set of elements is associative under the given operation?

55. Repeat Exercise 54 for a five-element by five-element table.

56. Suppose that three books numbered 1, 2, and 3 are placed next to one another on a shelf. If we remove volume 3 and place it before volume 1, the new order of books is 3, 1, 2. Let's call this replacement R. We can write

$$R = \begin{pmatrix} 1 & 2 & 3 \\ 3 & 1 & 2 \end{pmatrix},$$

which indicates the books were switched in order from 1, 2, 3 to 3, 1, 2. Other possible replacements are S, T, U, V, and I, as indicated.

$$S = \begin{pmatrix} 1 & 2 & 3 \\ 2 & 1 & 3 \end{pmatrix} \quad T = \begin{pmatrix} 1 & 2 & 3 \\ 3 & 2 & 1 \end{pmatrix} \quad U = \begin{pmatrix} 1 & 2 & 3 \\ 1 & 3 & 2 \end{pmatrix},$$

$$V = \begin{pmatrix} 1 & 2 & 3 \\ 2 & 3 & 1 \end{pmatrix} \quad I = \begin{pmatrix} 1 & 2 & 3 \\ 1 & 2 & 3 \end{pmatrix}$$

Replacement set I indicates that the books were removed from the shelves and placed back in their original order. Consider the mathematical system with the set of elements R, S, T, U, V, I, with the operation *.

To evaluate $R * S$, write

$$R * S = \begin{pmatrix} 1 & 2 & 3 \\ 3 & 1 & 2 \end{pmatrix} * \begin{pmatrix} 1 & 2 & 3 \\ 2 & 1 & 3 \end{pmatrix}.$$

As shown in Fig. 10.3, R replaces 1 with 3, and S replaces 3 with 3 (no change), so $R * S$ replaces 1 with 3. R replaces 2 with 1, and S replaces 1 with 2, so $R * S$ replaces 2 with 2 (no change). R replaces 3 with 2, and S replaces 2 with 1, so $R * S$ replaces 3 with 1. $R * S$ replaces 1 with 3, 2 with 2, and 3 with 1.

$$\begin{array}{ccc} & R & S \\ 1 & \longrightarrow 3 & \longrightarrow 3 \\ 2 & \longrightarrow 1 & \longrightarrow 2 \\ 3 & \longrightarrow 2 & \longrightarrow 1 \\ & R & * \quad S \end{array}$$

$$R * S = \begin{pmatrix} 1 & 2 & 3 \\ 3 & 2 & 1 \end{pmatrix} = T$$

Figure 10.3

Since this result is the same as replacement set T, we write $R * S = T$.

a) Complete the table for the operation using the procedure outlined.

*	R	S	T	U	V	I
R		T				
S						
T						
U						
V						
I						

b) Is this mathematical system a group? Explain.
c) Is this mathematical system a commutative group? Explain.

Research Activities

57. In Section 7.3 we introduced matrices. Show that 2×2 matrices under the operation of addition form a commutative group.

58. Show that 2×2 matrices under the operation of multiplication do not form a commutative group.

10.3 MODULAR ARITHMETIC

Figure 10.4

Figure 10.5

Table 10.11

+	0	1	2	3	4	5	6
0	0	1	2	3	4	5	6
1	1	2	3	4	5	6	0
2	2	3	4	5	6	0	1
3	3	4	5	6	0	1	2
4	4	5	6	0	1	2	③
5	5	6	0	1	2	3	4
6	6	0	1	2	3	4	5

Figure 10.6

The clock arithmetic we discussed in the previous section is similar to modular arithmetic. The set of elements {0, 1, 2, 3, 4, 5, 6, 7, 8, 9, 10, 11} together with the operation of addition is called a modulo 12 or mod 12 system. There is one difference in notation between clock arithmetic and modulo 12 arithmetic. In the modulo 12 system the symbol 12 is replaced with the symbol 0.

A **modulo *m* system** consists of *m* elements, 0 through *m* − 1, and a binary operation. In this section we will discuss modular arithmetic systems and their properties.

If today is Sunday, what day of the week will it be in 23 days? The answer, Tuesday, is arrived at by dividing 23 by 7 and observing the remainder of 2. Twenty-three days represent three weeks plus two days. Since we are interested only in the day of the week that the twenty-third day will fall on, the three-week segment is unimportant to the answer. The remainder of 2 indicates that the answer will be two days later than Sunday, which is Tuesday.

If we place the days of the week on a clock face as shown in Fig. 10.4, in 23 days the hand would have made three complete revolutions and end on Tuesday. If we replace the days of the week with numbers, a modulo 7 arithmetic system will result. See Fig. 10.5: Sunday = 0, Monday = 1, Tuesday = 2, and so on. If we start at 0 and move the hand 23 places, we will end at 2. Table 10.11 shows a modulo 7 addition table.

If we start at 4 and add 6, we will end at 3 on the clock in Fig. 10.6. This number is circled in Table 10.11. The other numbers can be obtained in the same way.

A second method of determining the sum of 4 + 6 in modulo 7 arithmetic is to divide the sum, 10, by 7 and observe the remainder.

$$10 \div 7 = 1, \quad \text{remainder } 3$$

The remainder, 3, is the sum of 4 + 6 in a modulo 7 arithmetic system.

The concept of congruence is important in modular arithmetic.

> *a* is **congruent** to *b* modulo *m*, written, $a \equiv b \pmod{m}$, if *a* and *b* have the same remainder when divided by *m*.

We can show, for example, that $10 \equiv 3 \pmod 7$ by dividing both 10 and 3 by 7 and observing that we obtain the same remainder in each case.

$$10 \div 7 = 1, \quad \text{remainder } 3 \qquad \text{and} \qquad 3 \div 7 = 0, \quad \text{remainder } 3$$

Since the remainders are the same, 3 in each case, 10 is congruent to 3 modulo 7, and we may write $10 \equiv 3 \pmod 7$.

Now consider $37 \equiv 5 \pmod 8$. If we divide both 37 and 5 by 8, each will have the same remainder, 5.

In any modulo system we can develop a set of modulo classes by placing all numbers with the same remainder in the appropriate modulo class. In a modulo 7 system, every number must have a remainder of either 0, 1, 2, 3, 4, 5, or 6.

Table 10.12   Modulo 7 Classes

0	1	2	3	4	5	6
0	1	2	3	4	5	6
7	8	9	10	11	12	13
14	15	16	17	18	19	20
21	22	23	24	25	26	27
28	29	30	31	32	33	34
⋮	⋮	⋮	⋮	⋮	⋮	⋮

Thus a modulo 7 system has seven modulo classes. The seven classes are presented in Table 10.12.

Every number is congruent to a number from 0 to 6 in mod 7. For example, $24 \equiv 3 \pmod 7$ because 24 is in the same modulo class as 3.

The solution to a problem in modular arithmetic, if it exists, will always be a number from 0 through $m - 1$, where m is the modulus of the system. For example, in a modulo 7 system, because 7 is the modulus, the solution will be a number from 0 through 6.

EXAMPLE 1

Determine which number, from 0 through 6, the following numbers are congruent to in modulo 7.

a) 84   **b)** 65   **c)** 47

Solution: We could determine the answer by listing more entries in Table 10.12. Another method of finding the answer is to divide the given number by 7 and observe the remainder.

a) $84 \equiv ? \pmod 7$

To determine the value that 84 is congruent to in mod 7, divide 84 by 7.

$$84 \div 7 = 12, \quad \text{remainder } 0$$

Thus $84 \equiv 0 \pmod 7$.

b) $65 \equiv ? \pmod 7$

$$65 \div 7 = 9, \quad \text{remainder } 2$$

Thus $65 \equiv 2 \pmod 7$.

c) $47 \equiv ? \pmod 7$

$$47 \div 7 = 6, \quad \text{remainder } 5$$

Thus $47 \equiv 5 \pmod 7$.

EXAMPLE 2

Evaluate each of the following in mod 5.

a) $4 + 3$ **b)** $4 - 3$ **c)** $2 \cdot 4$

Solution:

a) $4 + 3 \equiv ? \pmod 5$
$7 \equiv ? \pmod 5$
Since $7 \div 5 = 1$, remainder 2, $4 + 3 \equiv 2 \pmod 5$.

b) $4 - 3 \equiv ? \pmod 5$
$1 \equiv ? \pmod 5$
$1 \equiv 1 \pmod 5$
Remember that we want to replace the question mark with a number between 0 and 4, inclusive. Thus $4 - 3 \equiv 1 \pmod 5$.

c) $2 \cdot 4 \equiv ? \pmod 5$
$8 \equiv ? \pmod 5$
Since $8 \div 5 = 1$, remainder 3, $8 \equiv 3 \pmod 5$. Thus $2 \cdot 4 \equiv 3 \pmod 5$.

*D**ID YOU KNOW***

MODULO 100,000

*Y*our car has an odometer that shows your mileage to be a little over 28,000 miles, but this is not the first time you've seen this reading. In fact, this is the third time around. The car's odometer uses a modulo 100,000 system. ∎

Note that in Table 10.12 every number in the same modulo class differs by a multiple of the modulo, in this case a multiple of 7. Adding (or subtracting) a multiple of the modulo number to (or from) a given number does not change the modulo class or congruence of the given number. For example, 3, 3 + 1(7), 3 + 2(7), 3 + 3(7), . . . , 3 + n(7) are all in the same modulo class, namely, 3. We use this fact in the solution to Example 3.

EXAMPLE 3

Find the replacement for the question mark that makes each of the following true.

a) $3 - 15 \equiv ? \pmod 7$ **b)** $2 - 4 \equiv ? \pmod 5$ **c)** $4 - ? \equiv 6 \pmod 8$

Solution:

a) In mod 7, adding 7 or a multiple of 7 to a number results in a sum that is in the same modulo class. Thus if we add 7, 14, 21, . . . to 3, the result will be a number in the same modulo class. We want to replace 3 with an equivalent mod 7 number that is greater than 15. Adding 14 to 3 yields a sum of 17, which is greater than 15.

$$3 - 15 \equiv ? \pmod 7$$
$$(3 + 14) - 15 \equiv ? \pmod 7$$
$$17 - 15 \equiv ? \pmod 7$$
$$2 \equiv 2 \pmod 7$$

Therefore ? = 2.

b) $2 - 4 \equiv ? \pmod 5$

In mod 5 adding 5 (or a multiple of 5) to a number will result in a number that is in the same modulo class. Thus we can add 5 to 2 so that $2 - 4 \equiv ? \pmod 5$ becomes $7 - 4 \equiv ? \pmod 5$.

$$7 - 4 \equiv ? \pmod 5$$
$$3 \equiv ? \pmod 5$$
$$3 \equiv 3 \pmod 5$$

Thus $2 - 4 \equiv 3 \pmod 5$. This outcome could be demonstrated with a modulo 5 clock. The minus sign indicates counting counterclockwise.

c) $4 - ? \equiv 6 \pmod 8$

In mod 8, adding 8 (or a multiple of 8) to a number results in a sum in the same modulo class. Thus we can add 8 to 4 so that the statement becomes

$$(8 + 4) - ? \equiv 6 \pmod 8$$
$$12 - ? \equiv 6 \pmod 8$$

We can see that $12 - 6 = 6$. Therefore ? = 6.   ▲

EXAMPLE 4

Find all replacements for the question mark that make the statements true.

a) $4 \cdot ? \equiv 3 \pmod 5$ **b)** $3 \cdot ? \equiv 0 \pmod 6$ **c)** $3 \cdot ? \equiv 2 \pmod 6$

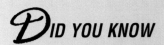

DID YOU KNOW

THE ENIGMA

A cipher, or secret code, actually has group properties: A well-defined operation turns plain text into a cipher and an inverse operation allows it to be deciphered. During World War II the Germans used an encrypting device based on a modulo 26 system (for the 26 letters of the alphabet) known as the enigma. It has four rotors, each with the 26 letters set out along the edge and an electrical contact for each letter. The rotors were wired to one another so that if an A were typed into the machine, the first rotor would contact a different letter on the second rotor, which contacted a different letter on the third rotor, and so on. The French secret service obtained the wiring instructions, and Polish and British cryptanalysts were able to crack the ciphers using group theory. This breakthrough helped to keep the Allies abreast of the deployment of the German navy. ■

Solution:

a) One method of determining the solution is to replace the question mark with the numbers 0–4 and then find the equivalent modulo class of the product. We use the numbers 0–4 because we are working in modulo 5.

$$4 \cdot ? \equiv 3 \pmod 5$$
$$4 \cdot 0 \equiv 0 \pmod 5$$
$$4 \cdot 1 \equiv 4 \pmod 5$$
$$4 \cdot 2 \equiv 3 \pmod 5$$
$$4 \cdot 3 \equiv 2 \pmod 5$$
$$4 \cdot 4 \equiv 1 \pmod 5$$

Therefore $? = 2$ since $4 \cdot 2 \equiv 3 \pmod 5$.

b) Replace the question mark with the numbers 0–5 and follow the procedure used in part (a).

$$3 \cdot ? \equiv 0 \pmod 6$$
$$3 \cdot 0 \equiv 0 \pmod 6$$
$$3 \cdot 1 \equiv 3 \pmod 6$$
$$3 \cdot 2 \equiv 0 \pmod 6$$
$$3 \cdot 3 \equiv 3 \pmod 6$$
$$3 \cdot 4 \equiv 0 \pmod 6$$
$$3 \cdot 5 \equiv 3 \pmod 6$$

Therefore replacing the question mark with 0, 2, or 4 results in true statements. The answers are 0, 2, and 4.

c) $3 \cdot ? \equiv 2 \pmod 6$

Examining the products in part (b) shows that there are no values that satisfy the statement. The answer is "no solution." ▲

Modular arithmetic systems under the operation of addition are commutative groups as illustrated in Example 5.

EXAMPLE 5

Construct a mod 5 addition table and show that the mathematical system is a commutative group. Assume that the associative property holds for the given operation.

Solution: The set of elements in modulo 5 arithmetic is {0, 1, 2, 3, 4}; the binary operation is +.

+	0	1	2	3	4
0	0	1	2	3	4
1	1	2	3	4	0
2	2	3	4	0	1
3	3	4	0	1	2
4	4	0	1	2	3

For this system to be a commutative group, it must satisfy the five properties of a commutative group.

1. *Closure:* Every entry in the table is a member of the set $\{0, 1, 2, 3, 4\}$, so the system is closed under addition (refer to the table on page 513).

2. *Identity element:* An easy way to determine whether there is an identity element is to look for a row in the table that is identical to the elements at the top of the table. Note that the row next to 0 is identical to the top of the table. This indicates that 0 *might be* the identity element. Now look at the column under the 0 at the top of the table. If this column is identical to the left-hand column, 0 is the identity element. Since the column under 0 is the same as the left-hand column, 0 is the additive identity element in modulo 5 arithmetic.

$$\text{Element} + \text{Identity} = \text{Element}$$
$$0 + 0 = 0$$
$$1 + 0 = 1$$
$$2 + 0 = 2$$
$$3 + 0 = 3$$
$$4 + 0 = 4$$

3. *Inverse elements:* Does every element have an inverse? Recall that an element plus its inverse must equal the identity element. In this example the identity element is 0. Therefore for each of the given elements 0, 1, 2, 3, and 4, we must find the element that when added to it results in a sum of zero. These elements will be the inverses.

$$\text{Element} + \text{Inverse} = \text{Identity}$$

0	+	?	=	0	Since $0 + 0 = 0$, 0 is its own inverse.
1	+	?	=	0	Since $1 + 4 = 0$, 4 is the inverse of 1.
2	+	?	=	0	Since $2 + 3 = 0$, 3 is the inverse of 2.
3	+	?	=	0	Since $3 + 2 = 0$, 2 is the inverse of 3.
4	+	?	=	0	Since $4 + 1 = 0$, 1 is the inverse of 4.

Note that each element has an inverse.

4. *Associative property:* It is given that the associative property holds. One example that illustrates the associative property is

$$(2 + 3) + 4 = 2 + (3 + 4)$$
$$0 + 4 = 2 + 2$$
$$4 = 4 \quad \text{True}$$

5. *Commutative property:* Is $a + b = b + a$ for *all* elements a and b of the given set? The table shows that the system is commutative because the

elements are symmetric about the main diagonal. We will give one example to illustrate the commutative property.

$$4 + 2 = 2 + 4$$
$$1 = 1 \quad \text{True}$$

All five properties are satisfied. Thus modulo 5 arithmetic under the operation of addition forms a commutative group.   ▲

Section 10.3 Exercises

1. What does a *modulo m* system consist of?

2. Explain the meaning of the statement, "*a* is congruent to *b* modulo *m*."

3. For a given modulo system how are modulo classes developed?

4. In a modulo 9 system, how many modulo classes will there be? Explain.

In Exercises 5–10, assume that today is Wednesday (day 3). Determine the day of the week it will be at the end of each period. (Assume no leap years.)

5. 17 days   **6.** 152 days   **7.** 366 days

8. 4 years   **9.** 3 years, 34 days   **10.** 463 days

In Exercises 11–16, consider 12 months to be a modulo 12 system. Determine the month it will be in the specified number of months. Use the current month as your reference point.

11. 15 months   **12.** 35 months

13. 2 years, 5 months   **14.** 4 years, 8 months

15. 8 years   **16.** 83 months

In Exercises 17–28, determine what number the sum, difference, or product is congruent to in mod 5.

17. $9 + 6$ **18.** $8 + 10$

19. $6 + 7 + 9$ **20.** $9 - 3$

21. $2 - 7$ **22.** $7 \cdot 4$

23. $5 \cdot 6$ **24.** $11 - 15$

25. $4 - 8$ **26.** $3 - 7$

27. $(15 \cdot 4) - 8$ **28.** $(4 - 9) \cdot 7$

In Exercises 29–40, find the modulo class to which each number belongs for the indicated modulo system.

29. 14, mod 8   **30.** 23, mod 7   **31.** 93, mod 15

32. 43, mod 6   **33.** 16, mod 8   **34.** 51, mod 9

35. 34, mod 7   **36.** 53, mod 4   **37.** −6, mod 7

38. −5, mod 4   **39.** −13, mod 11   **40.** −11, mod 13

In Exercises 41–54, find all replacements (less than the modulus) for the question mark that make the statement true.

41. $4 + 4 \equiv ? \pmod{5}$ **42.** $? + 9 \equiv 7 \pmod{8}$

43. $2 + ? \equiv 4 \pmod{5}$ **44.** $4 + ? \equiv 3 \pmod{6}$

45. $3 - ? \equiv 5 \pmod{6}$ **46.** $4 \cdot 5 \equiv ? \pmod{7}$

47. $5 \cdot ? \equiv 7 \pmod{9}$ **48.** $4 \cdot ? \equiv 5 \pmod{7}$

49. $2 \cdot ? \equiv 3 \pmod{4}$ **50.** $3 \cdot ? \equiv 3 \pmod{12}$

51. $3 \cdot ? \equiv 2 \pmod{8}$ **52.** $4 - 6 \equiv ? \pmod{8}$

53. $5 - 7 \equiv ? \pmod{9}$ **54.** $6 - ? \equiv 8 \pmod{9}$

55. Leap years are always presidential election years. Some leap years are . . . 1988, 1992, 1996, . . .
   a) List the next five presidential election years.
   b) What will be the first election year after the year 3000?
   c) List the election years between the years 2550 and 2575.

56. A pilot is scheduled to fly for five consecutive days and rest for three consecutive days. If today is the second day of her rest shift, determine whether she will be flying
   a) 60 days from today.
   b) 90 days from today.
   c) 240 days from today.
   d) Was she flying 6 days ago?
   e) Was she flying 20 days ago?

57. A nurse's work pattern at XYZ Hospital consists of working the 7 A.M.–3 P.M. shift for 3 weeks and then the 3 P.M.–11 P.M. shift for 2 weeks.
   a) If this is the third week of the pattern, what shift will the nurse be working 6 weeks from now?
   b) If this is the fourth week of the pattern, what shift will the nurse be working 7 weeks from now?
   c) If this is the first week of the pattern, what shift will the nurse be working 11 weeks from now?

58. A truck driver's routine is as follows: Drive 3 days from New York to Chicago; rest 1 day in Chicago; drive 3 days from Chicago to Los Angeles; rest 2 days in Los Angeles; drive 5 days to return to New York; rest 3 days in New York. Then the cycle begins again.

If the truck driver is starting his trip to Chicago today, what will he be doing

a) 30 days from today?

b) 70 days from today?

c) 2 years from today?

59. a) Construct a modulo 4 addition table.

b) Is the system closed? Explain.

c) Is there an identity element for the system? If so, what is it?

d) Does every element in the system have an inverse? If so, list the elements and their inverses.

e) The associative property holds for the system. Give an example.

f) Does the commutative property hold for the system? Give an example.

g) Is the system a commutative group?

h) Will every modulo system under the operation of addition be a commutative group? Explain.

60. Construct a modulo 8 addition table. Repeat parts (b)–(h) in Exercise 59.

61. a) Construct a modulo 4 multiplication table.

b) Is the system closed under the operation of multiplication?

c) Is there an identity element in the system? If so, what is it?

d) Does every element in the system have an inverse? Make a list showing the elements that have a multiplicative inverse and list the inverses.

e) The associative property holds for the system. Give an example.

f) Does the commutative property hold for the system? Give an example.

g) Is this mathematical system a commutative group? Explain.

62. Construct a modulo 7 multiplication table. Repeat parts (b)–(g) in Exercise 61.

Problem Solving/Group Activities

We have not discussed division in modular arithmetic. With what number or numbers, if any, can you replace the question marks to make the statement true?

63. $5 \div 7 \equiv ? \pmod 9$

64. $? \div 5 \equiv 5 \pmod 9$

65. $? \div ? \equiv 1 \pmod 4$

66. $1 \div 2 \equiv ? \pmod 5$

67. $2 \div ? \equiv 4 \pmod 6$

In Exercises 68–70, solve for x where k is any counting number.

68. $5k \equiv x \pmod 5$

69. $5k + 4 \equiv x \pmod 5$

70. $4k - 2 \equiv x \pmod 4$

71. One important use of modular arithmetic is in coding. One type of coding circle is given in Fig. 10.7. To use it, the person you are sending the message to must know the code key to decipher the code. The code key to this message is j. Can you decipher this code? (*Hint:* Subtract the code key from the code numbers.)

23 11 3 18 10 19 2 10 16 4 24

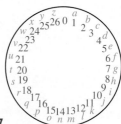

Figure 10.7

72. The procedure of *casting out nines* can be used to check arithmetic problems. The procedure is based on the fact that any whole number is congruent to the sum of its digits modulo 9. For example, the number 5783 and the sum of its digits, $5 + 7 + 8 + 3$, or 23, both have a remainder of 5 when divided by 9 and are therefore both congruent to 5 (mod 9): $5783 \equiv 5 \pmod 9$ and $23 \equiv 5 \pmod 9$.

a) Explain why the procedure works.

b) Perform the indicated operation and check your solution by casting out nines.

5321
+4786

73. Find the smallest positive number divisible by 5 to which 2 is congruent in modulo 6.

Research Activity

74. The concept and notation for modular systems were introduced by Carl Friedrich Gauss in 1801. Write a paper on Gauss's contribution to modular systems. References include *Historical Topics for the Classroom*, Thirty-First Yearbook, NCTM; Henrietta Midonick (ed.). *The Treasury of Mathematics.* New York: Philosophical Library, 1965; Ore, Oystein. *Number Theory and Its History.* New York: McGraw-Hill, 1948; Struik, Dirk. *A Source Book in Mathematics.* Cambridge, Mass.: Harvard University Press, 1969; and history of mathematics books.

CHAPTER 10 SUMMARY

Key Terms

10.1

associative property
binary operation
closure property
commutative (or abelian) group
commutative property
group

identity element
inverse
mathematical system

10.2

clock arithmetic
finite mathematical system

10.3

congruent
modulo *m* system
modulus
modulo classes

Important Facts

If the elements in a table are symmetric about the main diagonal, then the system is commutative.

	Addition	**Multiplication**
Commutative property	$a + b = b + a$	$a \cdot b = b \cdot a$
Associative property	$(a + b) + c$ $= a + (b + c)$	$(a \cdot b) \cdot c = a \cdot (b \cdot c)$

Properties of a Group and a Commutative Group

A mathematical system is a group if the first four conditions hold and a commutative group if all five conditions hold.

Commutative Group / Group

1. The set of elements is *closed* under the given operation.
2. An *identity element* exists for the set.
3. Every element in the set has an *inverse*.
4. The set of elements is *associative* under the given operation.
5. The set of elements is *commutative* under the given operation.

Chapter 10 Review Exercises

10.1–10.2

1. List the parts of a mathematical system.

Determine the value of the sum or difference in 12-hour clock arithmetic.

2. $7 + 11$ **3.** $5 + 8$ **4.** $6 - 10$

5. $5 + 7 + 9$ **6.** $7 - 10$ **7.** $7 - 4 + 6$

8. $2 - 5 - 7$

9. List the properties of a group, and explain what each property means.

10. What is an abelian group?

In Exercises 11–14, explain your answer.

11. Determine whether the set of integers under the operation of addition forms a group.

12. Determine whether the set of integers under the operation of multiplication forms a group.

13. Determine whether the set of rational numbers under the operation of addition forms a group.

14. Determine whether the set of rational numbers under the operation of multiplication forms a group.

In Exercises 15–17, for the mathematical system, determine which of the five properties of a commutative group do not hold.

15.

*	*a*	*c*	*b*
a	*a*	*b*	*c*
c	*b*	*c*	*a*
b	*c*	*a*	*b*

16.

:	*	?	△	*p*
*	?	△	*	*p*
?	△	*p*	?	*
△	*	?	△	*p*
p	*p*	*	*p*	△

17.

?	4	#	L	P
4	P	4	#	L
#	4	#	L	P
L	#	L	P	4
P	L	P	4	L

18. Consider the following mathematical system in which the operation is associative.

⊗	1	○	?	△
1	1	○	?	△
○	○	?	△	1
?	?	△	1	○
△	△	1	○	?

a) What are the elements of the set in this mathematical system?
b) What is the binary operation?
c) Is the system closed? Explain.
d) Is there an identity element for the system under the given operation?
e) Does every element in the system have an inverse? If so, give each element and its corresponding inverse.
f) Give an example to illustrate the associative property.
g) Is the system commutative? Give an example.
h) Is this mathematical system a commutative group? Explain.

10.3

In Exercises 19–27, find the modulo class to which the number belongs for the indicated modulo system.

19. 17, mod 3   **20.** 26, mod 5

21. 19, mod 7   **22.** 59, mod 8
23. 74, mod 13   **24.** 83, mod 4
25. 37, mod 6   **26.** 83, mod 14
27. 97, mod 11   **28.** 42, mod 11

In Exercises 29–38, find all replacements (less than the modulus) for the question mark that make the statement true.

29. $7 + 8 \equiv ?$ (mod 9)
30. $? - 3 \equiv 0$ (mod 4)
31. $5 \cdot ? \equiv 3$ (mod 6)
32. $3 - ? \equiv 5$ (mod 7)
33. $? \cdot 4 \equiv 0$ (mod 8)
34. $103 \equiv ?$ (mod 12)
35. $4 - 7 \equiv ?$ (mod 8)
36. $? \cdot 7 \equiv 3$ (mod 6)
37. $5 \cdot ? \equiv 3$ (mod 7)
38. $7 \cdot ? \equiv 2$ (mod 9)

39. Construct a modulo 6 addition table. Then determine whether the modulo 6 system forms a commutative group under the operation of addition.

40. Construct a modulo 4 multiplication table. Then determine whether the modulo 4 system forms a commutative group under the operation of multiplication.

41. Toni's work pattern at the fast food restaurant is as follows: She works 3 evenings, then has 2 evenings off, then she works 2 evenings, and then she has 3 evenings off; then the pattern repeats. If today is the first day of the work pattern,
a) will Toni be working 18 days from today?
b) will Toni have the evening off for a party that is being held in 38 days?

CHAPTER 10 TEST

1. What is a mathematical system?

2. List the requirements needed for a mathematical system to be a commutative group.

3. Is the set of whole numbers a commutative group under the operation of addition? Explain your answer completely.

4. Develop a four-hour clock arithmetic addition table.

5. Is four-hour clock arithmetic under the operation of addition a commutative group? Assume that the associative property holds. Explain your answer completely.

In Exercises 6 and 7, determine whether the mathematical system is a commutative group. Explain your answer completely.

6.

*	a	b	c
a	a	b	c
b	b	b	a
c	c	a	d

7.

?	1	2	3
1	3	1	2
2	1	2	3
3	2	3	1

Determine whether the mathematical system is a commutative group. Assume that the associative property holds.

8.

◯	@	$	&	%
@	%	@	$	&
$	@	$	&	%
&	$	&	%	@
%	&	%	@	$

In Exercises 9 and 10, determine the modulo class to which the number belongs for the indicated modulo system.

9. 73, mod 9   **10.** 58, mod 11

In Exercises 11–16, find all replacements for the question mark, less than the modulus, that make the statement true.

11. $6 + 9 \equiv ?$ (mod 7)   **12.** $? - 9 \equiv 4$ (mod 5)

13. $3 - ? \equiv 7$ (mod 9)   **14.** $4 \cdot 2 \equiv ?$ (mod 6)

15. $3 \cdot ? \equiv 2$ (mod 6)   **16.** $96 \equiv ?$ (mod 7)

17. a) Construct a modulo 5 multiplication table.
    b) Is this mathematical system a commutative group? Explain your answer completely.

GROUP PROJECTS

1. In arithmetic and algebra the statement, "If $a \cdot b = 0$, then $a = 0$ or $b = 0$" is true. That is, for the product of two numbers to be 0 at least one of the factors must be 0. Can the product of two nonzero numbers equal 0 in a specific modulo systems? If so, in what type of modulo systems can this result occur?
    a) Construct multiplication tables for mod. systems 3–9.
    b) Which, if any, of the multiplication tables in part (a) have products equal to 0 when neither factor is 0?
    c) Which, if any, of the multiplication tables in part (a) have products equal to 0 only when at least one factor is 0?
    d) Using the results in parts (b) and (c) can you write a conjecture as to which modulo systems have a product of 0 when neither factor is 0.

2. Are there certain modulo systems where all numbers have multiplicative inverses?
    a) If you have not worked Exercise 1, construct multiplication tables for modulo systems 3–9.
    b) Which of the multiplication tables in part (a) contain multiplicative inverses for all nonzero numbers?
    c) Which, if any, of the multiplication tables in part (a) do not contain multiplicative inverses for all nonzero numbers?
    d) Using the results in parts (b) and (c) can you write a conjecture as to which modulo systems contain multiplicative inverses for all nonzero numbers?

3. The square *ABCD* has a pin through it at point *P* so that the square can be rotated in a clockwise direction. The mathematical system consists of the set {*A, B, C, D*} and the operation ♣. The elements of the set represent different rotations of the square clockwise, as follows.

    A = rotate about the point *P* clockwise 90°.
    B = rotate about the point *P* clockwise 180°.
    C = rotate about the point *P* clockwise 270°.
    D = rotate about the point *P* clockwise 360°.

The operation ♣ means *followed by*. For example, *A* ♣ *B* means a clockwise rotation of 90° followed by a clockwise rotation of 180°, or a clockwise rotation of 270°. Since *C* represents a clockwise rotation of 270°, we write *A* ♣ *B* = *C*. This and two other results are shown in the following table.

♣	A	B	C	D
A		C		
B			A	
C	D			
D				

a) Complete the table and determine if the mathematical system is a commutative group. Explain.

Chapter 11 Consumer Mathematics

$\mathcal{M}$anaging the way you spend money takes a great deal of thought and planning. First, you need to know what your needs are on a daily basis. How much will you spend on transportation to get to school or work? How much on food? On a monthly and yearly basis, the questions are broader: What kind of house or apartment can you afford? Will you be able to save enough to buy a car, or go on a vacation? How much money do you need to keep on hand for unexpected emergencies? And then there are questions about your long-term goals. What would you like to accomplish financially 3, 5, and 10 years in the future? Will you be able to pay your debts? Do you want to own a home? Will you need to save money for raising children or for retirement?

The financial decisions you make daily all contribute to your longer range plans: As the saying goes, "Take care of the pennies and the dollars will take care of themselves." For example, say that you ride your bicycle every day instead of taking the bus. Over the course of a year, if the bus fare is $1.00, you may save as much as $500, which you could use to buy something immediately, or you could invest your savings and earn more money.

Many consumers shop carefully, using coupons and watching for sales. Sometimes, however, what is on "sale" isn't necessarily a bargain. For example, suppose that your local department store is having a sale on winter coats. You don't really need a new coat, but you see a big rack marked 25% off. Curious,

One variable studied by economists to determine the health of the national economy is a measure called the Index of Consumer Confidence. When consumers have confidence in their financial prospects, they purchase houses, cars, and retail goods, leading economists to predict economic growth for the nation.

Every year, an enormous number of magazines devoted to consumerism and financial planning are published. Most of these resources are available free of charge at your local library.

you try one on. Its original price reads $68, but the sale brings the price of the coat down to $51. If you buy the coat, it may seem that you are saving $17. But since you were not planning to buy a coat, you really are not saving $17, you're spending $51!

In order to make the best use of your money, you need to be able to make intelligent judgments. A solid understanding of consumer mathematics—including loans, interest rates, and mortgages—can help you set financial goals and plan how to achieve them. ■

Due to the use of telecommunications and computers, trading in international stock markets now continues 'round the clock. When the trading day ends in London, it begins again in Tokyo; when it ends there, it begins again in Chicago.

11.1   PERCENT

The study of mathematics is crucial to understanding how to make better financial decisions. A basic topic necessary for understanding the material in this chapter is percent. This section will give you a better understanding of the meaning of percent and its use in real-life situations.

The word *percent* comes from the Latin *per centum*, meaning "per hundred." A **percent** is simply a ratio of some number to 100. Thus $\frac{15}{100} = 15\%$, and $x/100 = x\%$.

Percents are useful in making comparisons. Consider Ross, who took two psychology tests. On the first test Ross answered 18 of the 20 questions correctly, and on the second test he answered 23 of 25 questions correctly. On which test did he make the higher score? One way to compare the results is to write a ratio of the correct answers to the number of questions on the test and then convert the ratios to percents. We find the grades in percent for each test by (a) writing a ratio of the number of correct answers to the total number of questions, (b) rewriting these ratios with a denominator of 100, and (c) expressing the ratios as percents.

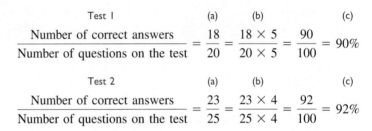

	Test 1	(a)	(b)	(c)
$\dfrac{\text{Number of correct answers}}{\text{Number of questions on the test}}$	$=$	$\dfrac{18}{20}$	$=\dfrac{18 \times 5}{20 \times 5}$	$=\dfrac{90}{100} = 90\%$

	Test 2	(a)	(b)	(c)
$\dfrac{\text{Number of correct answers}}{\text{Number of questions on the test}}$	$=$	$\dfrac{23}{25}$	$=\dfrac{23 \times 4}{25 \times 4}$	$=\dfrac{92}{100} = 92\%$

By changing the results of both tests to percents, we have a common standard for comparison. The results show that Ross scored 90% on the first test and 92% on the second test. He made a higher score on the second test.

Another procedure to change a fraction to a percent follows.

Procedure to Change a Fraction to a Percent
1. Divide the numerator by the denominator.
2. Multiply the quotient by 100 (which has the effect of moving the decimal point two places to the right).
3. Add a percent sign.

EXAMPLE 1

Change $\dfrac{3}{4}$ to percent.

Solution: Follow the steps listed in the procedure box.

1. $3 \div 4 = 0.75$ **2.** $0.75 \times 100 = 75$ **3.** 75%

Thus $\dfrac{3}{4} = 75\%$.

> **Procedure to Change a Decimal Number to a Percent**
> **1.** Multiply the decimal number by 100.
> **2.** Add a percent sign.

The procedure for changing a decimal number to a percent is equivalent to moving the decimal point two places to the right and adding a percent sign.

EXAMPLE 2

Change 0.235 to percent.

Solution: $0.235 = (0.235 \times 100)\% = 23.5\%$

> **Procedure to Change a Percent to a Decimal Number**
> **1.** Divide the number by 100.
> **2.** Remove the percent sign.

The procedure for changing a percent to a decimal number is equivalent to moving the decimal point two places to the left and removing the percent sign.

EXAMPLE 3

a) Change 35% to a decimal number.

b) Change $\frac{1}{2}\%$ to a decimal number.

Solution:

a) $35\% = \dfrac{35}{100} = 0.35$. Thus $35\% = 0.35$.

b) $\frac{1}{2}\% = 0.5\% = \dfrac{0.5}{100} = 0.005$. Thus $\frac{1}{2}\% = 0.005$.

EXAMPLE 4

Of 673 people surveyed, 418 said they favored the school budget. Find the percent in favor of the school budget.

Solution: The percent in favor is the ratio of the number in favor to the total number surveyed. To find the percent in favor, divide the number in favor by the total. Move the decimal point two places to the right and add a percent sign:

$$\text{Percent in favor} = \frac{418}{673} = 0.6211 = 62.11\%,$$

or   62.1% to the nearest tenth of a percent.

All answers in this section will be given to the nearest tenth of a percent. When rounding answers to the nearest tenth of a percent, we carry the division to four places after the decimal point (ten-thousandths) and then round to the nearest thousandths position.

EXAMPLE 5

At Metropolitan Community College the grades of students completing Mathematics 100 were as follows: 185 received an A, 348 received a B, 785 received a C, 216 received a D, and 96 received an F. Find the percent that received each grade.

Solution: The total number of students completing the course was

$$185 + 348 + 785 + 216 + 96 = 1630.$$

$$\text{Percent receiving an A} = \frac{185}{1630} = 0.1134 = \quad 11.3\%$$

$$\text{Percent receiving a B} = \frac{348}{1630} = 0.2134 = \quad 21.3\%$$

$$\text{Percent receiving a C} = \frac{785}{1630} = 0.4815 = \quad 48.2\%$$

$$\text{Percent receiving a D} = \frac{216}{1630} = 0.1325 = \quad 13.3\%$$

$$\text{Percent receiving an F} = \frac{96}{1630} = 0.0588 = \quad \underline{5.9\%}$$

$$\text{Sum of percents} = \quad 100.0\%$$

Note that every possible outcome was considered, so the sum of the percents is 100%. In some problems there may be a slight round-off error.   ▲

The percent increase or decrease, or percent change, over a period of time is found by the following formula.

$$\textbf{Percent change} = \frac{\left(\begin{array}{c}\text{Amount in}\\\text{latest period}\end{array}\right) - \left(\begin{array}{c}\text{Amount in}\\\text{previous period}\end{array}\right)}{\text{Amount in previous period}} \times 100$$

If the latest amount is greater than the previous amount, the answer will be positive and will indicate a percent increase. If the latest amount is smaller than the previous amount, the answer will be negative and will indicate a percent decrease.

EXAMPLE 6

In 1996 Yates County Dodge sold 937 new cars. In 1995 it sold 812 new cars. Find the percent increase or decrease in new car sales from 1995 to 1996.

Solution: The previous period is 1995, and the latest period is 1996.

$$\text{Percent change} = \frac{937 - 812}{812} \times 100$$

$$= \frac{125}{812} \times 100$$

$$= 0.1539 \times 100$$

$$= 15.4\% \text{ increase}$$

Cars sold

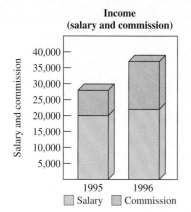

**Income
(salary and commission)**

Salary and commission

1995   1996

☐ Salary   ☐ Commission

EXAMPLE 7

In 1995 Ms. Erskine received a salary of $20,000 and a commission of $8000. In 1996 she received a salary of $22,000 and a commission of $15,000. Find the following:

a) The change in Ms. Erskine's income from 1995 to 1996.

b) The percent change in her income.

Solution:

a) Ms. Erskine's total income in 1995 was her salary plus commission, $20,000 + $8,000 = $28,000. Her total income in 1996 was $22,000 + $15,000 = $37,000. The change in her total income from 1995 to 1996 was $37,000 − $28,000 = $9,000.

b) Percent change $= \dfrac{\$37,000 - \$28,000}{\$28,000} \times 100$

$= \dfrac{\$9,000}{\$28,000} \times 100$

$= 0.321 \times 100$

$= 32.1\%$

Ms. Erskine's income rose 32.1%.   ▲

A similar formula is used to calculate percent markup or markdown on cost. A positive answer indicates a markup and a negative answer indicates a markdown.

$$\textbf{Percent markup on cost} = \frac{\text{Selling price} - \text{Dealer's cost}}{\text{Dealer's cost}} \times 100$$

EXAMPLE 8

Pickney Hardware stores pay $48.76 for glass fireplace screens. They regularly sell them for $79.88. At a sale they sell them for $69.99. Find the following.

a) The percent markup on the regular price

b) The percent markup on the sale price

c) The percent decrease of the sale price from the regular price

Solution:

a) We determine the percent markup on the regular price as follows.

$$\text{Percent markup} = \frac{\$79.88 - \$48.76}{\$48.76} \times 100$$

$$= 0.6382 \times 100$$

$$= 63.8\%$$

b) We determine the percent markup on the sale price as follows.

$$\text{Percent markup} = \frac{\$69.99 - \$48.76}{\$48.76} \times 100$$

$$= 0.4353 \times 100$$

$$= 43.5\%$$

c) Based on the regular price, we determine the percent decrease of the sale price.

$$\text{Percent decrease} = \frac{\$69.99 - \$79.88}{\$79.88} \times 100$$

$$= -0.1238 \times 100$$

$$= -12.4\%$$

The sale price is 12.4% lower than the regular price. ▲

In daily life we may need to know how to solve any one of the following three types of problems involving percent.

I. What is a 15% tip on a restaurant bill of $24.66? The problem can be stated as

15% of $24.66 is what number?

2. If Claryce made a sale of $500 and received a commission of $25, what percent of the sale is the commission? The problem can be stated as

What percent of $500 is $25?

3. If the price of a jacket was reduced by 25% or $12.50, what was the original price of the jacket? The problem can be stated as

25% of what number is $12.50?

To answer these questions we will write each problem as an equation. The word *is* means "is equal to," or =. In each problem we will represent the unknown quantity with the letter x. Therefore the preceding problems can be represented as

I. 15% of $24.66 = x **2.** x% of $500 = $25   **3.** 25% of x = $12.50

The word *of* in such problems indicates multiplication. To solve each problem change the percent to a decimal number, and express the problem as an equation; then solve the equation for the variable x. The solutions follow.

I. 15% of $24.66 = x

$\quad$ 0.15(24.66) = x (15% is written as 0.15 in decimal form.)

$\qquad\qquad$ 3.699 = x

Since 15% of $24.66 is $3.699, the tip would be $3.70.

2. x% of $500 = $25

$\quad$ (0.01x)500 = 25   (x% is written as 0.01x in decimal form.)

$\qquad\qquad$ 5x = 25

$\qquad\qquad \dfrac{5x}{5} = \dfrac{25}{5}$

$\qquad\qquad\quad x = 5$

Since 5% of $500 is $25, the commission is 5% of the sale.

3. 25% of x = $12.50

$$0.25(x) = 12.50$$

$$\frac{0.25x}{0.25} = \frac{12.50}{0.25}$$

$$x = 50$$

Since 25% of $50 is $12.50, the original price of the jacket was $50.00.

EXAMPLE 9

a) A sofa was purchased on an installment plan for $875. The down payment was 20% of the purchase price. How much was the down payment?

b) Nineteen, or 95%, of the students in a first-grade class have a pet. How many students are in the class?

c) Lalo scored 48 points on a 60-point test. What percent did he get correct?

Solution:

a) We want to find the amount of the down payment. Let x = the down payment. Then

$$x = 20\% \text{ of purchase price}$$
$$= 20\% \text{ of } \$875$$
$$= 0.20(875)$$
$$= 175.$$

The down payment is $175.

b) We want to find the number of students in the first-grade class. We know that 19 students represent 95% of the class. Let x = the number of students in the first-grade class. Then

$$95\% \text{ of } x = 19$$
$$0.95x = 19$$
$$\frac{0.95x}{0.95} = \frac{19}{0.95}$$
$$x = 20.$$

The number of students in the class is 20.

c) We want to determine the percent of answers that Lalo got correct. We need to determine what percent of 60 is 48. Let x = the percent that Lalo got correct. Then

$$x\% \text{ of } 60 = 48$$
$$0.01x(60) = 48$$
$$0.60x = 48$$
$$x = \frac{48}{0.60}, \quad \text{or} \quad 80.$$

Lalo scored 80% on the test.

Section 11.1 Exercises

1. What is a percent?

2. Explain in your own words how to change a fraction to a percent.

3. Explain in your own words how to change a decimal number to a percent.

4. Explain in your own words how to change a percent to a decimal number.

5. Explain in your own words how to determine percent change.

6. Explain in your own words how to determine percent markup on cost.

In Exercises 7–15, change the number to a percent. Express your answer to the nearest tenth of a percent.

7. $\frac{3}{4}$

8. $\frac{1}{8}$

9. $\frac{7}{8}$

10. $\frac{5}{8}$

11. 0.5688

12. 0.007654

13. 13.678

14. 3.78

15. 0.0036

In Exercises 16–24, change the percent to a decimal number.

16. 27%

17. 35.4%

18. 3.75%

19. 0.0005%

20. $\frac{1}{4}$%

21. $\frac{3}{8}$%

22. $\frac{1}{5}$%

23. 135.9%

24. 1%

25. A family estimates that $8000 of its total income of $42,500 is used to purchase food. What percent of the family's income is spent on food?

26. In 1994 approximately 58.407 million women were employed in the civilian labor force. This number is 57.9% of the eligible female population. Determine the total number of women that are eligible for employment.

27. In 1994 approximately 23.4 million CD-ROM (Compact Disk-Read Only Memory) disks were sold. Of those sold, 18.0 million were IBM-PC disks and 5.4 million were Macintosh disks. What percent of the disks sold were
    a) IBM-PC?   b) Macintosh?

In Exercises 28 and 29, use the circle graph to answer the questions. Three hundred people were surveyed about their favorite way to wash a pet. Each sector of the circle graph shows the number of people who favored a particular way or had no preference.

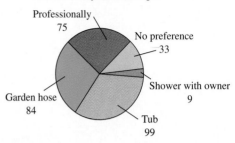

Favorite ways to wash a pet

Professionally 75

No preference — 33

Shower with owner 9

Tub 99

Garden hose 84

28. Find the percent of people who favor washing a pet in the tub.

29. Find the percent of people who favor washing a pet in the shower.

In Exercises 30 and 31, use the circle graph to answer the questions. The five top sports advertisers spent $146.8 million in the first five months of 1995.

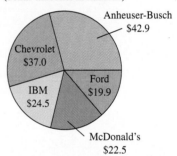

Top sports advertisers for the first three months of 1995 (dollar amounts in millions)

Anheuser-Busch — $42.9

Chevrolet $37.0

Ford $19.9

IBM $24.5

McDonald's $22.5

30. What percent of those sports advertising dollars were spent by Anheuser-Busch?

31. What percent of those sports advertising dollars were spent by Ford?

32. The population of the United States rose from approximately 226.5 million in 1980 to approximately 248.7 million in 1990.
    a) Find the percent increase from 1980 to 1990.
    b) What will the population be in the year 2000 if it increases at the same percent from 1990 to 2000 as it did from 1980 to 1990?

33. The consumer price index (CPI) in 1982–1984 was 100, in 1987 it was 113.6, and in 1993 it was 144.5. Find the
    a) percent increase in the CPI from 1982–1984 to 1987.
    b) percent increase from 1987 to 1993.

34. The graph shows the median family income in the United States for the years 1950 through 1993. Find
   a) the increase in median family income from 1950 to 1993.
   b) the percent increase in median family income for the same period.

Median family income in 1993 dollars

In Exercises 35–37, use the graph, which illustrates that major supermarkets are leaving the inner city in record numbers, to find

a) *the decrease in the number of supermarkets from 1970 to 1990.*
b) *the percent decrease in the number of supermarkets for the same period.*

Supermarkets are leaving inner cities

35. Manhattan   **36.** Chicago   **37.** Los Angeles

38. The maximum monthly social security benefits for 1985 and 1995 are shown in the graph. Find
   a) the dollar increase from 1985 to 1995.
   b) the percent increase in the maximum social security benefit for the same period.

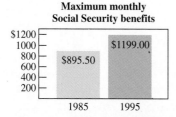

Maximum monthly Social Security benefits

In Exercises 39–44, write each statement as an equation and solve.

39. What percent of 75 is 15?
40. Fifty-four is 18% of what number?
41. What is 134% of 400?
42. What is 7% of 9?
43. Eleven is 20% of what number?
44. What percent of 346 is 86.5?

In Exercises 45–50, write the problem as an equation, then solve.

45. What is the sales tax on a dinner that cost $43.75 if the sales tax rate is 7%?

46. If the number of employees at the Golden Dragon restaurant increased by 25%, or ten employees, what was the original number of employees?

47. The Fastlock Company is hiring 57 new employees, which increases its staff by 30%. What was the original number of employees?

48. Eighteen students received an A on the third test, which is 150% of the students who received an A on the second test. How many students received an A on the second test?

49. In a survey of 300 people, 17% prefer Ranch dressing on their salad. How many people in the group prefer Ranch dressing?

50. Mr. Brown's present salary is $36,500. He is getting an increase of 7% in his salary next year. What will his new salary be?

51. In the United States in 1980 there were 29.3 million people classified as living in poverty. In 1992 there were 36.9 million classified as living in poverty. Find the percent increase or decrease in the number of people living in poverty from 1980 to 1992.

52. A Kirby vacuum cleaner dealership sold 430 units in 1994 and 407 units in 1995. Find the percent increase or decrease in units sold.

53. If the Kirby dealership in Exercise 52 pays $320 for each unit and sells it for $699, what is the percent markup?

54. The regular price of a Zenith color TV is $539.62. During a sale, Hill TV is selling it for $439. Find the percent decrease in the price of this TV.

55. The Dechs had eight great-grandchildren in 1994. In 1995 they had 12 great-grandchildren. Find the percent increase in the number of great-grandchildren from 1994 to 1995.

56. The number of deaths in the United States due to AIDS was 20,903 in 1988, 30,649 in 1990, and 29,084 in 1993. Find the

a) percent increase from 1988 to 1990.
b) percent decrease from 1990 to 1993.

57. The cost of a fish dinner to the owner of the Golden Wharf restaurant is $7.95. The fish dinner is sold for $11.95. Find the percent markup.

58. A football team played 12 games. They lost one game.
a) What percent of the games did they lose?
b) What percent did they win?

59. A furniture store advertised a table at a 15% discount. The original price was $115, and the sale price was $100. Was the sale price consistent with the ad? Explain.

60. Bonnie sold a truck and made a profit of $675. Her profit was 18% of the sale price. What was the sale price?

61. Quincy purchased a used car for $1000. He decided to sell the car for 10% above his purchase price. Quincy could not sell the car so he reduced his asking price by 10%. If he sells the car at the reduced price will he have a profit or loss or will he break even? Explain how you arrived at your answer.

Problem Solving/Group Activities

62. The Tie Shoppe paid $5901.79 for a shipment of 500 ties and wants to make a profit of 40% of the cost on the whole shipment. The store is having two special sales. At the first sale it plans on selling 100 ties for $9.00 each, and at the second special sale it plans on selling 150 ties for $12.50 each. What should be the selling price of the other 250 ties for the Tie Shoppe to make a 40% profit on the whole shipment?

63. a) A coat is marked down 10%, and the customer is given a second discount of 15%. Is this the same as a single discount of 25%? Explain.
b) The regular price of a chair is $189.99. Determine the sale price of the chair if the regular price is reduced by 10% and this price is then reduced another 15%.
c) Determine the sale price of the chair if the regular price of $189.99 is reduced by 25%.
d) Examine the answers obtained in parts (b) and (c). Does your answer to part (a) appear to be correct? Explain.

Research Activity

64. Find two circle graphs in newspapers or magazines whose data are not given in percents. Redraw the graphs and label them with percents.

11.2 PERSONAL LOANS AND SIMPLE INTEREST

Often consumers want to buy clothing, appliances, or furniture but do not have the cash to do so. They then have to determine whether the cost of borrowing the money is worth the convenience or pleasure of having the item today versus waiting several months or years and paying cash. If you are faced with this choice, you may decide that you must have the item today, and may choose to borrow the money from a bank or other lending institution. The money that a bank is willing to lend you is called the amount of **credit** extended or the **principal of the loan**.

The amount of credit and the interest rate that you may obtain depend on the assurance that you can give the lender that you will be able to repay the loan. Your credit is determined by your business reputation for honesty, by your earning power, and by what you can pledge as security to cover the loan. **Security** (or *collateral*) is anything of value pledged by the borrower that the lender may sell or keep if the borrower does not repay the loan. Acceptable security may be a business, a mortgage on a property, the title to an automobile, savings accounts, or stocks or bonds. The more marketable the security, the easier it is to obtain the loan, and in some cases marketability may help in getting a lower interest rate.

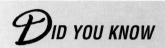

Bankers sometimes grant loans without security, but they require the signature of one or more other persons, called cosigners, who guarantee that the loan will be repaid. For either of the two types of loans, the secured loan or the cosigner loan, the borrower (and cosigner, if there is one) must sign an agreement called a personal note. This document states the terms and conditions of the loan.

The most common way for individuals to borrow money is through an installment loan or using a credit card. (Installment loans and credit cards will be discussed in Section 11.4.)

The concept of simple interest is essential to the understanding of notes and installment buying. Interest is the money the borrower pays for the use of the lender's money. One type of interest is called simple interest. Simple interest is based on the entire amount of the loan for the total period of the loan. The formula used to find simple interest follows.

Simple Interest Formula

$$\text{Interest} = \text{principal} \times \text{rate} \times \text{time}$$

$$i = prt$$

In the simple interest formula the principal, p, is the amount of money lent, the rate, r, is the rate of interest expressed as a percent, and the time, t, is the number of days, months, or years for which the money will be lent. Time is expressed in the same period as the rate. For example, if the rate is 2% per month, the time must be expressed in months. Typically rate means the annual rate unless otherwise stated. Principal and interest are expressed in dollars in the United States.

Ordinary Interest

The most common type of simple interest is called *ordinary interest*. For computing ordinary interest, each month has 30 days and a year has 12 months or 360 days. On the due date of a *simple interest note* the borrower must repay the principal plus the interest. (*Note:* Simple interest will mean ordinary interest unless stated otherwise.)

EXAMPLE 1

Find the simple interest for a three-month loan of $640 at 6% per year.

Solution: We use the formula $i = prt$. We know that $p = \$640$, $r = 6\%$ (or converted to a decimal, 0.06), and t in years $= \frac{3}{12} = \frac{1}{4} = 0.25$. We substitute the appropriate values in the formula.

$$i = p \times r \times t$$
$$= \$640 \times 0.06 \times 0.25$$
$$= \$9.60$$

The simple interest on $640 at 6% for three months is $9.60.   ▲

EXAMPLE 2

Bonnie borrowed $3000 from a bank for two months. The rate of interest charged by the bank is 0.0333%* per day.

a) Find the simple interest on the loan.

b) Find the amount (principal + interest) that Bonnie will pay the bank at the end of the two months.

Solution:

a) To find the interest on the loan, we use the formula $i = prt$. Since the rate is given in terms of days the time must be expressed in terms of days. Two months = 60 days. Substituting yields

$$i = p \times r \times t$$
$$= \$3000 \times 0.000333 \times 60$$
$$= \$60.00$$

b) The amount to be repaid is equal to the principal plus interest, or

$$A = p + i$$
$$= \$3000 + \$60.00$$
$$= \$3060.00 \qquad \blacktriangle$$

EXAMPLE 3

Jean lent her friend $300. Six months later her friend repaid the original $300 plus $30.00 interest. What annual rate of interest did Jean receive?

Solution: Using the formula $i = prt$, we get

$$\$30 = \$300 \times r \times 0.50$$
$$30 = 150r$$
$$\frac{30}{150} = r$$
$$0.2 = r$$

The annual rate of interest paid is 20%. $\qquad \blacktriangle$

In Examples 1, 2, and 3 we illustrated a simple interest note, for which the interest and principal are paid on the due date of the note. There is another type of loan, the **discount note**, for which the interest is paid at the time the borrower receives the loan. The interest charged in advance is called the **bank discount**. A Federal Reserve bill is a bank discount note issued by the U.S. government. Example 4 illustrates a discount note.

EXAMPLE 4

Heather borrowed $500 on a 10% discount note for a period of three months. Find the following:

a) The interest she must pay to the bank on the date she receives the loan.

*If you have forgotten the meaning of a bar above a digit or group of digits refer to page 210.

b) The net amount of money she receives from the bank.

c) The actual rate of interest for the loan.

Solution:

a) To find the interest, use the simple interest formula.

$$i = prt$$

$$= \$500 \times 0.10 \times \frac{3}{12}$$

$$= \$12.50$$

b) Since Heather must pay \$12.50 interest when she first receives the loan, the net amount she receives is \$500 − \$12.50 or \$487.50.

c) Calculate the actual rate of interest charged by substituting the net amount she received and the actual interest paid into the simple interest formula.

$$i = prt$$

$$\$12.50 = \$487.50 \times r \times \frac{3}{12}$$

$$12.50 = 121.875 \times r$$

$$\frac{12.50}{121.875} = r$$

$$0.1026 \approx r$$

Thus the actual rate of interest is about 10.3% rather than the quoted 10%.

▲

The United States Rule

A note has a date of maturity, at which time the principal and interest are due. It is possible to make payments on a note before the date of maturity. A Supreme Court decision specified the method by which these payments are credited. The procedure is called the **United States rule**.

The United States rule states that if a partial payment is made on the loan, interest is computed on the principal from the first day of the loan until the date of the partial payment. The partial payment is used to pay the interest first; then the rest of the payment is used to reduce the principal. The next time a partial payment is made, interest is calculated on the unpaid principal from the date of the previous date of payment. Again the payment goes first to pay the interest, with the rest of the payment used to reduce the principal. An individual can make as many partial payments as he or she wishes; the procedure is repeated for each payment. The balance due on the date of maturity is found by computing interest due since the last partial payment and adding this interest to the unpaid principal.

The **Banker's rule** is used to calculate simple interest when applying the United States rule. The Banker's rule considers a year to have 360 days, and any fractional part of a year is the exact number of days of the loan.

To determine the exact number of days in a period, we can use Table 11.1 on page 534. Example 5 illustrates how to use the table.

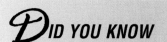

Table 11.1

Day of Month	Days in Each Month											
	31	**28**	**31**	**30**	**31**	**30**	**31**	**31**	**30**	**31**	**30**	**31**
	Jan	**Feb**	**Mar**	**Apr**	**May**	**Jun**	**Jul**	**Aug**	**Sep**	**Oct**	**Nov**	**Dec**
Day 1	1	32	60	91	121	152	182	213	244	274	305	335
Day 2	2	33	61	92	122	153	183	214	245	275	306	336
Day 3	3	34	62	93	123	154	184	215	246	276	307	337
Day 4	4	35	63	94	124	155	185	216	247	277	308	338
Day 5	5	36	64	95	125	156	186	217	248	278	309	339
Day 6	6	37	65	96	126	157	187	218	249	279	310	340
Day 7	7	38	66	97	127	158	188	219	250	280	311	341
Day 8	8	39	67	98	128	159	189	220	251	281	312	342
Day 9	9	40	68	99	129	160	190	221	252	282	313	343
Day 10	10	41	69	100	130	161	191	222	253	283	314	344
Day 11	11	42	70	101	131	162	192	223	(254)	284	315	345
Day 12	12	43	71	102	132	163	193	224	255	285	316	346
Day 13	13	44	72	103	133	164	194	225	256	286	317	347
Day 14	14	45	73	104	134	165	195	226	257	287	318	348
Day 15	15	46	(74)	105	135	166	196	227	258	288	319	349
Day 16	16	47	75	106	136	167	197	228	259	289	320	350
Day 17	17	48	76	107	137	168	198	229	260	290	321	351
Day 18	18	49	77	108	138	169	199	230	261	291	322	352
Day 19	19	50	78	109	139	170	200	231	262	292	323	353
Day 20	20	51	79	110	140	171	201	232	263	293	324	354
Day 21	21	52	80	111	141	172	202	233	264	294	325	355
Day 22	22	53	81	112	142	173	203	234	265	295	326	356
Day 23	23	54	82	113	143	174	204	235	266	296	327	357
Day 24	24	55	83	114	144	175	205	236	267	297	328	358
Day 25	25	56	84	115	145	176	206	237	268	298	329	359
Day 26	26	57	85	116	146	177	207	238	269	299	330	360
Day 27	27	58	86	117	147	178	208	239	270	300	331	361
Day 28	28	59	87	118	148	179	209	240	271	301	332	362
Day 29	29		88	119	149	180	210	241	272	302	333	363
Day 30	30		89	120	150	181	211	242	273	303	334	364
Day 31	31		90		151		212	243		304		365

Add 1 day for leap year if February 29 falls between the two dates.

EXAMPLE 5

Use Table 11.1 to find (a) the due date of a loan made on March 15 for 180 days and (b) the number of days from April 18 to July 31.

Solution:

a) To determine the due date of the note, do the following: In Table 11.1 find 15 in the left column (headed Day of Month), then move three columns to the right (heading at the top of the column is March) and find the number

74 (circled in black). Thus March 15 is the 74th day of the year. Add 74 to 180, since the note will be due 180 days after March 15:

$$74 + 180 = 254$$

Thus the due date of the note is the 254th day of the year. Find 254 (circled in blue) in Table 11.1 in the column headed September. The number in the same row as 254 in the left column (headed Day of Month) is day 11. Thus the due date of the note is September 11.

b) To determine the number of days from April 18 to July 31, use Table 11.1 to find that April 18 is the 108th day of the year and that July 31 is the 212th day of the year. Then find the difference: $212 - 108 = 104$. Thus the number of days from April 18 to July 31 is 104 days.   ▲

EXAMPLE 6

Find the simple interest on $300 at 12% for the period March 3 to May 3 using the Banker's rule.

Solution: The exact number of days from March 3 to May 3 is 61. The period of time in years is 61/360. Substituting in the simple interest formula gives

$$i = prt$$

$$= \$300 \times 0.12 \times \frac{61}{360}$$

$$= \$6.10$$

The interest due is $6.10.   ▲

The next example illustrates how a partial payment is credited under the United States rule. Making partial payments reduces the amount of interest paid and the cost of the loan.

EXAMPLE 7

On March 3, Benny took out a 90-day note for $475.00 at an interest rate of 15.5%. On April 1 he made a partial payment of $200.00, and on May 10 he made a partial payment of $150.00.

a) Find the due date of the note.

b) Find the interest and the amount credited to the principal on April 1.

c) Find the interest and the amount credited to the principal on May 10.

d) Find the amount that Benny must pay on the due date.

Solution:

a) We use Table 11.1 to determine the due date. March 3 is the 62nd day of the year. The sum of 62 and 90 is 152. The 152nd day of the year, June 1, is the due date.

b) The number of days from March 3 to April 1, the date of the first partial payment, is 29. Use the Banker's rule to compute the simple interest on $475 at 15.5% for 29 days. Substituting in the simple interest formula gives

$$i = \$475 \times 0.155 \times \frac{29}{360}$$

$$= \$5.93$$

The interest of $5.93 that is due on April 1 is deducted from the payment of $200.00. The remaining amount of the payment is then applied to the principal.

$200.00	Partial payment
− 5.93	Interest
$194.07	Amount to be applied to principal

$475.00	Original principal
− 194.07	Amount to be applied to principal
$280.93	New principal after April 1 payment

c) Use the Banker's rule to calculate the interest on the unpaid principal for the period from April 1 to May 10. The number of days from April 1 to May 10 is 39.

$$i = \$280.93 \times 0.155 \times \frac{39}{360}$$

$$= \$4.72$$

Now subtract the interest of $4.72 from the partial payment of $150.00 to obtain $145.28. Thus $145.28 is applied against the unpaid principal.

$280.93	Unpaid principal
− 145.28	Amount to be applied to principal
$135.65	Balance due on principal on date of maturity of loan

d) The date of maturity of the loan is June 1. The number of days from May 10 to June 1 is 22. The interest is computed on the remaining balance of $135.65 by using the simple interest formula.

$$i = \$135.65 \times 0.155 \times \frac{22}{360}$$

$$= \$1.28$$

Therefore the balance due on the maturity date of the loan is the sum of the principal and the interest, $135.65 plus $1.28, or $136.93. *Note:* The sum of the days in the three calculations, $29 + 39 + 22$, equals the total number of days of the note, 90.

Section 11.2 Exercises

1. What is credit?

2. What is security (or collateral)?

3. What is a cosigner?

4. What is a personal note?

5. What is interest?

6. In your own words, explain what each letter in the simple interest formula, $i = prt$, represents.

7. In your own words, explain the difference between ordinary interest and interest under the Banker's rule.

8. In your own words, explain the United States rule.

In Exercises 9–18, find the simple interest. (The rate is an annual rate unless otherwise noted. Assume 360 days in a year.)

9. $p = \$420$, $r = 9\%$, $t = 1$ year

10. $p = \$490$, $r = 6.5\%$, $t = 4$ years

11. $p = \$875$, $r = 12\%$, $t = 30$ days

12. $p = \$365.45$, $r = 11\frac{1}{2}\%$, $t = 8$ months

13. $p = \$587$, $r = 0.045\%$ per day, $t = 2$ months

14. $p = \$6742.75$, $r = 6.05\%$, $t = 90$ days

15. $p = \$4372.80$, $r = 5.25\%$, $t = 60$ days

16. $p = \$9752,\quad r = 1\frac{1}{4}\%$ per month, $t = 4$ months

17. $p = \$12{,}752,\quad r = 1\frac{1}{2}\%$ per month, $t = 7$ months

18. $p = \$12{,}752,\quad r = 0.055\%,\quad t = 90$ days

In Exercises 19–26, use the simple interest formula to find the missing value.

19. $p = ?,\quad r = 6\%,\quad t = 60$ days, $i = \$6.00$

20. $p = ?,\quad r = 8\%,\quad t = 3$ months, $i = \$12.00$

21. $p = \$800.00,\quad r = 6\%,\quad t = ?,\quad i = \$64.00$

22. $p = \$957.62,\quad r = 6.5\%,\quad t = ?,\quad i = \$124.49$

23. $p = \$1650.00,\quad r = ?,\quad t = 6.5$ years, $i = \$343.20$

24. $p = ?,\quad r = 10\%,\quad t = 4$ months, $i = \$9.50$

25. $p = \$675.00,\quad r = ?,\quad t = 4$ months, $i = \$29.25$

26. $p = \$1980,\quad r = ?,\quad t = 9$ months, $i = \$111.38$

27. Ward borrowed $1500 from his bank for 60 days at a simple interest rate of $10\frac{1}{2}\%$. He gave the bank stock certificates as security.
   a) How much did he pay for the use of the money?
   b) What is the amount he paid to the bank on the date of maturity?

28. Nita borrowed $3500 from the bank for 6 months. Her friend Ms. Harris was cosigner of Nita's personal note. The bank collected $12\frac{1}{2}\%$ simple interest on the date of maturity.
   a) How much did Nita pay for the use of the money?
   b) Find the amount she repaid to the bank on the due date of the note.

29. Kwame borrowed $2500 for 5 months from his bank, using U.S. government bonds as security. The bank discounted the loan at 13%.
   a) How much interest did Kwame pay the bank for the use of its money?
   b) How much did he receive from the bank?
   c) What was the actual rate of interest he paid?

30. Carol borrowed $3650 from the bank for 8 months. The bank discounted the loan at 12%.
   a) How much interest did Carol pay the bank for the use of its money?
   b) How much did she receive from the bank?
   c) What was the actual rate of interest she paid?

31. Enrico wants to borrow $350 for 6 months from his bank, using his savings account as security. The bank's policy is that the maximum amount that a person can borrow is 80% of the amount in the person's savings account. The interest rate is 2% higher than the interest rate being paid on the savings account. The current rate is $3\frac{1}{4}\%$ on the savings account.
   a) How much money must Enrico have in his account in order to borrow $350?
   b) What is the rate of interest the bank will charge for the loan?
   c) Find the amount Enrico must repay in six months.

In Exercises 32–37, find the exact time from the first date to the second date. Use Table 11.1.

32. April 4 to October 11

33. May 19 to September 17

34. June 19 to November 5

35. March 29 to November 10

36. February 15 to June 15

37. January 17 to May 30 (leap year)

In Exercises 38–42, determine the due date, using the exact time.

38. 90 days after April 18

39. 120 days after June 8

40. 120 days after April 15

41. 60 days after September 1

42. 180 days after August 15

In Exercises 43–49, a partial payment is made on the date indicated. Use the United States rule to determine the balance due on the note at the date of maturity. (The Effective Date is the date the note was written.)

		Effective Date	Maturity Date	Partial Payment(s)	
Principal	Rate			Amount	Date(s)
43. $2500	8%	July 1	Oct. 15	$300	Aug. 1
44. $2400	7%	April 10	July 7	$500	May 5
45. $8000	9%	May 1	Nov. 1	$2000	June 15
46. $7500	12%	April 15	Oct. 1	$1000	Aug. 1
47. $1000	12.5%	Jan. 1	Feb. 15	$300	Jan. 15
48. $1800	15%	Aug. 1	Nov. 1	$500	Sept. 1
				$500	Oct. 1
49. $5000	14%	Oct. 15	Jan. 1	$800	Nov. 15
				$800	Dec. 15

50. On March 1 the Zwick Balloon Company signed a $6500 note with simple interest of $10\frac{1}{2}\%$ for 180 days. The company made payments of $1750 on May 1 and $2350 on July 1. How much will the company owe on the date of maturity?

51. The Sweet Tooth Restaurant borrowed $3000 on a note dated May 15 with interest of 11%. The maturity date of the loan is September 1. They made partial payments of $875 on June 15 and $940 on August 1. Find the amount due on the maturity date of the loan.

52. The U.S. government borrows money by selling Treasury bills. A Treasury bill is a discounted note that has a time period of a year or less. Clint purchases a 26-week, $10,000 U.S. Treasury bill at an 8% discount.
   a) How much does Clint pay for the T-bill?
   b) How much will the bill be worth after 26 weeks?
   c) How much interest does Clint earn on the investment?

53. Mr. Harrison borrowed $600 for 3 months. The banker said he must repay the loan at the rate of $200 per month plus interest. The bank was charging a rate of 2% above the prime interest rate. The *prime interest rate* is the rate charged to preferred customers of the bank. The first month the prime rate was $8\frac{1}{2}$%, the second month 9%, and the third month 10%.
  - **a)** Find the amount he paid the bank at the end of the first month; at the end of the second month; at the end of the third month.
  - **b)** What was the total amount of interest he paid the bank?

Problem Solving/Group Activity

54. Treasury bills are discounted notes issued by the U.S. government. Mark sent the regional Federal Reserve bank nearest his home a check for $100,000 with a request to purchase a 52-week noncompetitive treasury bill (offered every 4 weeks). A short time later Mark received a notice from the regional Federal Reserve bank confirming his purchase of a 52-week Treasury bill and that the purchase price was $93.337 per $100 of par value (he paid $93.337 for each $100 value of the note).

  - **a)** What interest rate did Mark receive when he purchased the bill? Explain.
  - **b)** With the notice he received a check in his name. How much was the check made out for?
  - **c)** What was the actual interest rate he received?
  - **d)** On the day he received the check in part (b) he invested it in a 1-year certificate of deposit yielding 5% interest. What is the total amount of interest he will receive from both investments?

Research Activities

55. Three types of simple interest may be calculated with the simple interest formula. They are ordinary, Banker's rule, and exact time.
  - **a)** Visit a local bank to determine how exact time is used in the simple interest formula.
  - **b)** Compare the results for each method on a loan for $500 at 8% for the period January 2, 1996, to April 2, 1996.

56. Visit your local bank and investigate the procedure for obtaining a simple interest loan. Determine
  - **a)** what type of security the bank requires.
  - **b)** what rate of interest the bank will charge for the loan.
  - **c)** when the loan must be repaid.

11.3 COMPOUND INTEREST

Albert Einstein was once asked to name the greatest discovery of man. His reply was, "Compound interest."

In this section we discuss some ways to put your money to work for your benefit.

An **investment** is the use of money or capital for income or profit. We can divide investments into two classes: fixed investments and variable investments. In a **fixed investment** the amount invested as principal is guaranteed, and the interest is computed at a fixed rate. *Guaranteed* means that the exact amount invested will be paid back together with any accumulated interest. Examples of a fixed investment are savings accounts and certificates of deposit. Another fixed investment is a government savings bond. In a **variable investment** neither the principal nor the interest is guaranteed. Examples of variable investments are stocks, mutual funds and commercial bonds.

Simple interest, introduced earlier in the chapter, is calculated once for the period of a loan using the formula $i = prt$. The interest paid on savings accounts at most banks is compound interest. A bank computes the interest periodically (for example, daily or quarterly) and adds this interest to the original principal. The interest for the following period is computed by using the new principal (original principal plus interest). In effect the bank is computing interest on interest, which is called compound interest.

> Interest that is computed on the principal and any accumulated interest is called **compound interest**.

EXAMPLE 1

Find the amount, A, to which $1000 will grow at the end of 1 year if the interest is 12% compounded quarterly.

Solution: Compute the interest for the first quarter using the simple interest formula. Add this interest to the principal to find the amount at the end of the first quarter.

$$i = prt$$
$$= \$1000 \times 0.12 \times 0.25 = \$30.00$$
$$A = \$1000 + \$30 = \$1030$$

Find the amount at the end of the second quarter.

$$i = \$1030 \times 0.12 \times 0.25 = \$30.90$$
$$A = \$1030 + \$30.90 = \$1060.90$$

Find the amount at the end of the third quarter.

$$i = \$1060.90 \times 0.12 \times 0.25 = \$31.83$$
$$A = \$1060.90 + \$31.83 = \$1092.73$$

Find the amount at the end of the fourth quarter.

$$i = \$1092.73 \times 0.12 \times 0.25 = \$32.78$$
$$A = \$1092.73 + \$32.78 = \$1125.51$$

Hence the $1000 grows to a final value of $1125.51. ▲

This example shows the effect of earning interest on interest, or compounding interest. In 1 year the amount of $1000 has grown to $1125.51, compared to $1120 that would have been obtained with a simple interest rate of 12%. Thus in 1 year alone the gain was $5.51 more with compound interest than with simple interest.

A simpler and less time-consuming way to calculate compound interest is to use the compound interest formula and a calculator.

Compound Interest Formula

$$A = p\left(1 + \frac{r}{n}\right)^{nt}$$

In this formula, A is the amount, p is the principal, r is the annual rate of interest, t is the time in years, and n is the number of periods per year.

EXAMPLE 2

Calculate the amount, A, to which $1 will grow at 3% interest compounded annually for 6 years.

In 1626 Peter Du Minuit, representing the Dutch West India Company, traded beads and blankets to the native American inhabitants of Manhattan Island for the island. The trade was valued at 60 Dutch guilders, about $24. If at that time the $24 had been invested at 6% interest compounded annually, the investment would be worth $52,248,704,230 in 1995. As a comparison, the simple interest on $24 for 369 years at 6% is only $531.60. If the $24 had been compounded continuously, as is done in many savings accounts today, the investment would be worth approximately $98,784,750,000. These results dramatically illustrate the effect of compounding interest over time. ■

Solution: Since the interest is compounded annually, there is one period per year. Thus $n = 1$, $r = 0.03$, and $t = 6$. Substituting into the formula, we find the amount A.

$$A = p\left(1 + \frac{r}{n}\right)^{nt}$$
$$= 1\left(1 + \frac{0.03}{1}\right)^{(1)(6)}$$
$$= 1(1 + 0.03)^6$$
$$= 1(1.03)^6$$
$$= 1(1.1940523)$$
$$= \$1.19$$

In Example 2, to find the value of $(1.03)^6$ with a calculator you can use one of two methods, depending on whether your calculator is a four-function calculator or a scientific calculator with a $\boxed{y^x}$ key. On a four-function calculator multiply six factors of 1.03. The key strokes are

1.03 $\boxed{\times}$ 1.03 $\boxed{\times}$ 1.03 $\boxed{\times}$ 1.03 $\boxed{\times}$ 1.03 $\boxed{\times}$ 1.03 $\boxed{=}$ 1.1940523.

On a scientific calculator the key strokes are

1.03 $\boxed{y^x}$ 6 $\boxed{=}$ 1.1940523.

Note: Some scientific calculators have an $\boxed{x^y}$ or a $\boxed{\wedge}$ key in place of a $\boxed{y^x}$ key. In either case the procedure works the same: Enter the base then the key ($\boxed{y^x}$, $\boxed{x^y}$, or $\boxed{\wedge}$), the exponent, and then the $\boxed{=}$ key.

EXAMPLE 3

Calculate the interest on $650 at 8% compounded semiannually for 3 years, using the compound interest formula.

Solution: Since interest is compounded semiannually, there are two periods per year. Thus $n = 2$, $r = 0.08$, and $t = 3$. Substituting into the formula, we find the amount, A.

$$A = p\left(1 + \frac{r}{n}\right)^{nt}$$
$$= 650\left(1 + \frac{0.08}{2}\right)^{(2)(3)}$$
$$= 650(1.04)^6$$
$$= 650(1.2653190)$$
$$= \$822.46$$

Since the total amount is $822.46 and the original principal is $650, the interest must be $822.46 − $650, or $172.46.

In Example 3, the interest rate is stated as an annual rate of 8%, but the number of compounding periods per year is two. In applying the compound interest formula, the rate for one period is $r \div n$, which was 8% ÷ 2, or 4% per period.

Now we will calculate the amount, and interest, on $1 invested at 8% compounded semiannually for 1 year. The result is

$$A = 1\left(1 + \frac{0.08}{2}\right)^{(2)(1)} = 1.0816$$

Interest = Amount − Principal

$$i = 1.0816 - 1$$
$$= 0.0816$$

The interest for 1 year is 0.0816. This amount written as a percent, 8.16%, is called the **effective annual yield** (or **effective yield**). If $1 was invested at a simple interest rate of 8.16% and $1 was invested at 8% interest compounded semiannually (equivalent to an effective yield of 8.16%), the interest from both investments would be the same.

Many banks compound interest daily. When computing the effective annual yield they use 360 for the number of periods in a year. To find the effective annual yield for any interest rate, calculate the amount using the compound interest formula where p is $1. Then subtract $1 from that amount. The difference, written as a percent, is the effective annual yield, as illustrated in Example 4.

On the sign in the margin it shows that the annual rate is 5.25% and the effective annual yield is 5.39%. Show that an interest rate of 5.25% compounded daily has an effective yield of 5.39% now.

EXAMPLE 4

Determine the effective annual yield for $1 invested for 1 year at

a) 8% compounded daily.

b) 6% compounded quarterly.

$\mathcal{D}$ID YOU KNOW

THE PRINCIPAL/PRINCIPLE OF THE THING

$\mathcal{L}$ate in 1777 Jacob DeHaven lent General George Washington $50,000 in gold and $400,000 worth of supplies. Without such help, Washington was afraid that the army must either "disband or starve." The army went on to win the Revolutionary War. As repayment, DeHaven was offered Continental money, but he refused and asked to be repaid in gold. No gold was forthcoming, and this once-wealthy Pennsylvanian merchant died penniless.

In 1989 members of the DeHaven family decided to sue the U.S. Government over the 212-year-old debt. Allowing 6% interest compounded daily (the rate offered by the Continental Congress at the time), the $450,000 would have mushroomed into approximately $150.3 billion. (You can verify this yourself by using the compound interest formula.) As of this writing, the matter has yet to be resolved. ∎

Solution:

a) With daily compounding $n = 360$.

$$A = p\left(1 + \frac{r}{n}\right)^{nt}$$

$$= 1\left(1 + \frac{0.08}{360}\right)^{(1)(360)}$$

$$= 1.0832774 \approx 1.0833$$

$$i = A - 1$$

$$= 1.0833 - 1$$

$$= 0.0833$$

Thus when the interest is compounded daily, the effective annual yield is 8.33%.

b) With quarterly compounding, $n = 4$.

$$A = p\left(1 + \frac{r}{n}\right)^{nt}$$

$$= 1\left(1 + \frac{0.06}{4}\right)^{(1)(4)}$$

$$= 1.06136355 \approx 1.0614$$

$$i = 1.0614 - 1$$

$$= 0.0614$$

Thus when the interest is compounded quarterly, the effective annual yield is 6.14%. ▲

There are numerous types of savings accounts. Many savings institutions compound interest daily. Some pay interest from the day of deposit to the day of withdrawal, and others pay interest from the first of the month on all deposits made before the tenth of the month. In each of these accounts in which interest is compounded daily, interest is entered into the depositor's account only once each quarter. Some savings banks will not pay any interest on a day-to-day account if the balance falls below a set amount.

Present Value

You may wonder about what amount of money you must deposit in an account today to have a certain amount of money in the future. For example, how much must you deposit in an account today at a given rate of interest so that it will accumulate to $60,000 to pay your child's college costs in 15 years? The principal, p, which would have to be invested now is called the **present value**.

Present Value Formula

$$p = \frac{A}{\left(1 + \dfrac{r}{n}\right)^{nt}}$$

In the formula A represents the amount to be accumulated in n years. Note that the present value formula is a variation of the compound interest formula.

— **EXAMPLE 5**

Marshall is planning to spend $15,000 on a car in 5 years. How much principal, p, must be invested today at $8\frac{1}{2}\%$ compounded quarterly, to accumulate to $15,000 in 5 years?

Solution: To determine the principal we will use the present value formula. Since the interest is compounded quarterly, there are 4 periods per year. Thus $n = 4$, $r = 0.085$, $t = 5$, and $A = \$15,000$.

$$p = \frac{A}{\left(1 + \dfrac{r}{n}\right)^{nt}}$$

$$= \frac{15,000}{\left(1 + \dfrac{0.085}{4}\right)^{(4)(5)}}$$

$$= \frac{15,000}{(1.02125)^{20}}$$

$$= \frac{15,000}{1.5227948}$$

$$= 9850.309$$

Marshall needs to invest $9850.31 today in order to accumulate $15,000 in 5 years.   ▲

Section 11.3 Exercises

1. What is an investment?

2. What is a fixed investment?

3. What is a variable investment?

4. What is compound interest?

5. What is effective annual yield?

6. What is meant by present value in the Present Value Formula?

In Exercises 7–22, use the compound interest formula

$$A = p\left(1 + \frac{r}{n}\right)^{nt}$$

to compute

  a) *the total amount.*
  b) *the interest earned on each investment.*

7. $4000 for 2 years at 4% compounded annually

8. $3000 for 2 years at 4% compounded semiannually

9. $8000 for 3 years at 8% compounded quarterly

10. $7000 for 3 years at 12% compounded monthly

11. $2000 for 2 years at 6% compounded quarterly

12. $2000 for 2 years at 6% compounded monthly

13. $2000 for 2 years at 12% compounded quarterly

14. $2000 for 2 years at 12% compounded monthly

15. $1500 for 3 years at 8.5% compounded daily. (Use 1 year = 360 days.)

16. $4500 for 3 years at 12% compounded daily. (Use 1 year = 360 days.)

17. $7000 for 4 years at 12% compounded quarterly

18. $9500 for 4 years at 12% compounded annually

19. $7500 for 4 years at 8% compounded quarterly

20. $2500 for 1 year at 12% compounded monthly

21. $4500 for 2.5 years at 6% compounded semiannually

22. $2000 for 3.5 years at 8.5% compounded annually

23. When Steven was born, his father deposited $2000 in his name in a savings account. The account was paying 5% interest compounded semiannually.
   a) If the rate did not change, what was the value of the account after 15 years?
   b) If the money had been invested at 5% compounded quarterly, what would the value of the account have been after 15 years?

24. When Molly was born, her father deposited $2000 in her name at a savings and loan association. At the time that S&L was paying 6% interest compounded semiannually on savings. After 10 years the S&L changed to an interest rate of 6% compounded quarterly. How much had the $2000 amounted to after 18 years when the money was withdrawn for Molly to use to help pay her college expenses?

25. Michael borrowed $3000 from his brother Dave. He agreed to repay the money at the end of 2 years, giving Dave the same amount of interest that he would have received if the money had been invested at 8% compounded quarterly. How much money did Michael repay his brother?

26. Kathy borrowed $6000 from her daughter, Lynette. She repaid the $6000 at the end of 2 years. If Lynette had left the money in a bank account that paid $5\frac{1}{4}$% compounded monthly, how much interest would she have accumulated?

27. Joe invested $6000 at 8% compounded quarterly. Three years later he withdrew the full amount and used it for the down payment on a house. How much money did he put down on the house?

28. Compute the total amount and the interest paid on the principals (a)–(c) at 12% compounded monthly for 2 years.
   a) $100   **b)** $200   **c)** $400
   d) Is there a predictable outcome when the principal is doubled? Explain.

29. Determine the total amount and the interest paid on $1000 with interest compounded semiannually for 2 years at
   a) 2%.   **b)** 4%.   **c)** 8%.
   d) Is there a predictable outcome when the rate is doubled? Explain.

30. Compute the total amount and the interest paid on $1000 at 6% compounded semiannually for
   a) 2 years.   **b)** 4 years.   **c)** 8 years.
   d) Is there a predictable outcome when the time is doubled? Explain.

31. Show that the effective annual yield of a 6.77% interest rate compounded daily is 7.00%.

32. Determine the effective annual yield of $1 invested for 1 year at 6.5% compounded daily.

33. Determine the effective annual yield for $1 invested for 1 year at 7.5% compounded quarterly.

34. Determine the effective annual yield for $1 invested for 1 year at 6.5% compounded quarterly.

35. How many dollars should parents invest at the birth of their child in order to provide their child with $50,000 at age 18? Assume that the money earns interest at 8% compounded quarterly.

36. The Pearsons are planning on retiring in 20 years and feel that they will need $200,000 in addition to income from their retirement plan. How much must they invest today at 7.5% compounded quarterly to accomplish their goal?

37. How many dollars must you invest today to have $20,000 in 15 years? Assume that the money earns interest at 7% compounded quarterly.

38. How many dollars must you invest today to have $20,000 in 15 years? Assume that the money earns interest at 7% compounded monthly.

Problem Solving/Group Activities

39. For a total accumulated amount of $3586.58, a principal of $2000, and a time period of 5 years, use the compound interest formula to find r if interest is compounded monthly.

40. If the cost of a loaf of bread was $1.25 in 1988 and the annual average inflation rate is $3\frac{1}{2}$%, what will be the cost of a loaf of bread in 1998?

41. Dick borrowed $2000 from Diane. The terms of the loan are as follows: The period of the loan is 3 years, and the rate of interest is 8% compounded semiannually. What rate of simple interest would be equivalent to the rate Diane charged Dick?

42. A simple formula can help you estimate the number of years required to double your money. It's called the *Rule of 72*. You simply divide 72 by the interest rate (without the percent sign). For example, with an interest rate of 4%, your money would double in approximately 72 ÷ 4 or 18 years. In (a)–(d) determine the approximate number of years it will take for $1000 to double at the given interest rate.
   a) 3%   **b)** 6%   **c)** 8%   **d)** 12%
   e) If $120 doubles in approximately 22 years, estimate the rate of interest.

43. The question may be asked: If I deposit a fixed sum of money monthly, quarterly, semiannually, or annually at a fixed rate of interest, how much money will I have accumulated in x years? This is called an annuity. The formula for determining the value of the annuity is

$$S = \frac{R\left[\left(1 + \dfrac{r}{n}\right)^{nt} - 1\right]}{\dfrac{r}{n}}$$

where S is the value of the annuity in t years, R is the amount invested each period, r is the annual rate of interest, n is the number of payments per year, and t is the number of years. After Harriet's birth her parents invested $500 semiannually (every 6 months) at an annual rate of 5.5% compounded semiannually. What is the value of the annuity after 17 years?

44. a) To supplement her retirement income Chris is investing $50 each quarter at 8% compounded quarterly. How many dollars will she accumulate in 30 years? (See Exercise 43.)
   b) At the time of retirement, 30 years later, she invests the money accumulated in part (a) in a real-estate trust for 10 years. The investment agreement states that the company will pay 8% per year simple interest, and that the annual interest will be paid with 12 equal monthly checks. What will her monthly income be from this investment?

45. In Section 5.6 we discussed negative exponents. If you have not studied negative exponents or you have forgotten the rules, read Section 5.6 now. Rewrite the present value formula on page 542 using negative exponents.

Research Activities

46. Write a paper on the history of simple interest and compound interest. Answer the questions: When was interest first charged on loans? Historically, when and where were some of the highest rates of interest charged and what were the rates? (References include encyclopedias; and National Council of Teachers of Mathematics, *Historical Topics For the Classroom*, Thirty-First Yearbook.)

47. Visit the bank at which you have a checking account. Investigate and write a report on the various types of savings vehicles available through the bank (include savings accounts, money market accounts, CDs, annuities, and so on). In your report indicate the interest provided and whether the interest rate is fixed or varies over time. Indicate how the interest is determined for each vehicle. Also state the advantages and disadvantages of each of these savings vehicles.

11.4 INSTALLMENT BUYING

In Section 11.2 we discussed personal notes and discounted notes. When borrowing money by either of these methods, the borrower normally repays the loan as a single payment at the end of the specified time period. There may be circumstances under which it is more convenient for the borrower to repay the loan on a weekly or monthly basis or to use some other convenient time period. One method of doing so is to borrow money on the *installment plan*.

There are two types of installment loans, open end and fixed payment. An open-end installment loan is a loan on which you can make variable payments each month. Credit cards such as MasterCard, Visa, and Discover are actually open-end installment loans, used to purchase items such as clothing, textbooks, and meals. A fixed installment loan is one on which you pay a fixed amount of money for a set number of payments. Examples of items purchased with fixed-payment installment loans are college tuition loans, cars, boats, appliances, and furniture. These loans are generally repaid in 24, 36, 48, or 60 equal monthly payments.

Lenders give any individual wishing to borrow money or purchase goods or services on the installment plan a credit rating to determine if the borrower is likely to repay the loan. The lending institution determines whether the applicant is a good "credit risk" by examining the individual's income, assets, liabilities, and history of repaying debts.

The advantage of installment buying is that the buyer has the use of an article while paying for it. If the article is essential, installment buying may serve a real need. A disadvantage is that some people buy more on the installment plan than they can afford. Another disadvantage is the interest the borrower pays for the

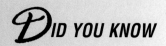

loan. The method of determining the interest charged on an installment plan may
vary with different lenders.

To provide the borrower with a way to compare interest charged, Congress
passed the Truth in Lending Act in 1969. The law requires that the lending institution tell the borrower two things: the annual percentage rate and the finance
charge. The annual percentage rate (APR) is the true rate of interest charged
for the loan. The APR is calculated by using a complex formula, so we will use
the tables provided by the government to determine the APR. The technique of
using the tables to find the APR will be illustrated in Example 1. The total finance
charge is the total amount of money the borrower must pay for its use. The finance
charge includes the interest plus any additional fees charged. The additional fees
may include service charges, credit investigation fees, mandatory insurance premiums, and so on.

The finance charge a consumer pays when purchasing goods or services on
the installment plan is the difference between the total installment price and the
cash price. The total installment price is the sum of all the monthly payments
and the down payment, if any.

Fixed Installment Loan

Example 1 shows how to determine the annual percentage rate on a loan that has
a schedule for repaying a fixed amount each period.

--- **EXAMPLE 1**

Jane is purchasing a new computer and accessories for $3600, including taxes.
She is making a $600 down payment and will finance the balance, $3000,
through a bank. The bank says that the payment will be $101 per month for
36 months.

a) Determine the finance charge.   **b)** Determine the APR.

Solution:

a) The total installment price is the down payment plus the total monthly
installment payments.

$$\text{Total installment price} = 600 + (36 \times 101)$$
$$= 600 + 3636 = \$4236$$

The finance charge is the total installment price minus the cash price.

$$\text{Finance charge} = 4236 - 3600 = \$636$$

b) To determine the annual percentage rate, use Table 11.2 as follows: First
divide the finance charge by the amount financed and multiply the quotient
by 100. The result is the finance charge per $100 of the amount financed.

$$\frac{\text{Finance charge}}{\text{Amount financed}} \times 100 = \frac{636}{3000} \times 100 = 0.212 \times 100 = 21.2$$

Thus the borrower pays $21.20 interest for each $100 being financed. To
use Table 11.2, look for 36 in the left column under the heading Number
of Payments. Then move across to the right until you find the value closest

Table 11.2   Annual Percentage Rate Table for Monthly Payment Plans

Number of Payments	Annual Percentage Rate												
	10.75%	11.00%	11.25%	11.50%	11.75%	12.00%	12.25%	12.50%	12.75%	13.00%	13.25%	13.50%	13.75%
	(Finance Charge per $100 of Amount Financed)												
6	3.16	3.23	3.31	3.38	3.45	3.53	3.60	3.68	3.75	3.83	3.90	3.97	4.05
12	5.92	6.06	6.20	6.34	6.48	6.62	6.76	6.90	7.04	7.18	7.32	7.46	7.60
18	8.73	8.93	9.14	9.35	9.56	9.77	9.98	10.19	10.40	10.61	10.82	11.03	11.24
24	11.58	11.86	12.14	12.42	12.70	12.98	13.26	13.54	13.82	14.10	14.38	14.66	14.95
30	14.48	14.83	15.19	15.54	15.89	16.24	16.60	16.95	17.31	17.66	18.02	18.38	18.74
36	17.43	17.86	18.29	18.71	19.14	19.57	20.00	20.43	20.87	(21.30)	21.73	22.17	22.60
48	23.48	24.06	24.64	25.23	25.81	26.40	26.99	27.58	28.18	28.77	29.37	29.97	30.57
60	29.71	30.45	31.20	31.96	32.71	33.47	34.23	34.99	35.75	36.52	37.29	38.06	38.33

to $21.20. The value closest to $21.20 is $21.30 (circled in blue). At the top of this column is the value 13.00%. Therefore the annual percentage rate is approximately 13.00%.

▲

Much more complete APR tables similar to Table 11.2 are available at your local bank.

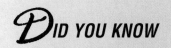

┌ **EXAMPLE 2**

Gary borrowed $9800 in 1996 to purchase a car. He does not recall the APR of the loan but remembers that there are 48 payments of $255.75. If he did not make a down payment on the car, determine the APR.

Solution: First determine the finance charge by subtracting the cash price from the total amount paid.

$$\text{Finance charge} = (255.75 \times 48) - 9800$$
$$= 12276 - 9800$$
$$= \$2476$$

Next divide the finance charge by the amount of the loan and multiply this quotient by 100.

$$\frac{2476}{9800} \times 100 = 25.27$$

Next find 48 payments in the left column of Table 11.2. Move to the right until you find the value that is closest to 25.27. The value closest to $25.27 is 25.23. At the top of the column is the APR of 11.50%.   ▲

In Example 2 Gary agreed to pay a finance charge of $2476. If he decides to repay the loan after making 30 payments, must he pay the total finance charge? The answer is no. Two methods are used to determine the finance charge when you repay an installment loan early: the **actuarial method** and the **rule of 78s**. The actuarial method uses the APR tables, whereas the more commonly used method, the rule of 78s, does not.

Actuarial Method	Rule of 78s
$$u = \frac{n \cdot P \cdot V}{100 + V}$$	$$u = \frac{f \cdot k(k + 1)}{n(n + 1)}$$
u = unearned interest	u = unearned interest
n = number of remaining monthly payments (excluding current payment)	f = original finance charge
	k = number of remaining monthly payments (excluding current payment)
P = monthly payment	
V = the value from the APR table that corresponds to the annual percentage rate for the number of remaining payments (excluding current payment)	n = original number of payments

These two formulas determine the unearned interest, u, which is the interest saved by paying off the loan early.

Example 3 illustrates the actuarial method and Example 4 the rule of 78s.

EXAMPLE 3

In Example 2 we determined the APR of Gary's loan to be 11.50%. Instead of making his 30th payment Gary wishes to pay his remaining balance and terminate the loan.

a) Use the actuarial method to determine how much interest Gary will save (the unearned interest, u) by repaying the loan early.

b) What is the total amount due on that day?

Solution:

a) After 30 payments have been made, 18 payments remain. Thus $n = 18$ and $P = \$255.75$. To determine V use the APR table (Table 11.2). In the Number of Payments column find the number of remaining payments, 18, and then look to the right until you reach the column headed by 11.50%, the APR. This row and column intersect at 9.35. Thus $V = 9.35$.

$$u = \frac{n \cdot P \cdot V}{100 + V}$$

$$= \frac{(18)(255.75)(9.35)}{100 + 9.35}$$

$$= 393.62$$

Gary will save $393.62 in interest by the actuarial method.

b) Since the remaining payments total 18($255.75) = $4603.50, Gary's remaining balance is

$4603.50	Total of remaining payments (which includes interest)
− 393.62	Interest saved (unearned interest)
$4209.88	Balance due

A payment of $4209.88 plus the 30th monthly payment of $255.75 will terminate Gary's installment loan. The total amount due is $4465.73. ▲

EXAMPLE 4

In Example 2 we determined the APR of Gary's loan to be 11.50%. Instead of making his 30th payment Gary decides to pay his remaining balance and terminate the loan.

a) Use the rule of 78s to determine how much interest Gary will save by repaying the loan early.

b) What is the total amount due on that day?

Solution:

a) After the 30 payments have been made, 18 remain. Thus $k = 18$, $n = 48$, and $f = \$2476$.

$$u = \frac{f \cdot k(k+1)}{n(n+1)}$$

$$= \frac{2476 \cdot 18(18+1)}{48(48+1)}$$

$$= \$360.03$$

Gary will save $360.03 in interest by the rule of 78s.

b) Gary's balance is computed in a manner similar to the method in Example 3.

$4603.50	Total of remaining payments
− 360.03	Interest saved
$4243.47	Balance due

A payment of $4243.47 plus a regular monthly payment of $255.75 will terminate Gary's installment loan. The total amount due is $4499.22. ▲

Most states use the rule of 78s. However, there is some discussion that the rule of 78s is outdated and is not fair to the borrower.

Open-End Installment Loan

A credit card is a popular way of making purchases or borrowing money. Use of a credit card is an example of an open-end installment loan. A typical charge account with a bank may have the terms in Table 11.3.

Typically, credit card monthly statements contain the following information: balance at the beginning of the period, balance at the end of the period (or new

Table 11.3   Credit Card Terms

Type of Charge	Periodic Rates*		Annual Percentage Rate*
	Monthly	Daily	
Purchases	1.30%		15.60%
Cash advances		0.04273%	15.60%

These rates vary with different charge accounts and localities.

balance), the transactions for the period, statement closing date (or billing date), payment due date, and the minimum payment due. For *purchases* there is no finance or interest charge if there is no previous balance due and you pay the entire new balance by the payment due date. However, if you borrow *money* (*cash advances*) through this account, a finance charge is applied from the date you borrowed the money until the date you repay the money. When you make purchases or borrow money, the minimum monthly payment is sometimes determined by dividing the balance due by 36 and rounding the answer up to the nearest whole dollar, thus ensuring repayment in 36 months. However, if the balance due for any month is less than $360, the minimum monthly payment is typically $10. These general guidelines may vary by bank and store.

EXAMPLE 5

While on vacation Sonya charged the following items to her MasterCard account.

Airplane ticket	$285
Hotel room	120
Meals	115
	$520

Her balance due on August 1 is $520, and she makes a payment of $100. The bank requires repayment within 36 months and charges an interest rate of 1.3% per month.

a) Determine the minimum payment due on August 1.

b) Find the balance due on September 1, assuming that there are no additional charges or cash advances.

Solution:

a) To determine the minimum payment we divide the balance due on August 1 by 36.

$$\frac{520}{36} = \$14.44$$

Rounding up to the nearest dollar, we determine that the minimum payment due on August 1 is $15.

b) With no additional purchases or cash advances and paying $100 on August 1 the balance on September 1 is $520 − $100 or $420. In addition she must pay interest on the outstanding balance of $420 for the month of August. The interest is

$$i = prt$$
$$= 420(0.013)(1)$$
$$= \$5.46$$

Therefore the balance due on September 1 is $420 + $5.46 or $425.46.

In Example 5 there were no additional transactions in the account for the period. When additional charges are made during the period, the finance

charges on open-end installment loans or credit cards are generally calculated in one of two ways: the *unpaid balance method* or the *average daily balance method*.

With the unpaid balance method, the borrower is charged interest or a finance charge on the unpaid balance from the previous charge period. Example 6 illustrates the unpaid balance method.

EXAMPLE 6

On November 5, the billing date, Ms. Storm had a balance due of $275 on her credit card. From November 5 through December 4 she charged articles totaling $320 and she made a payment of $145.

a) Find the finance charge due on December 5, using the unpaid balance method. Assume that the interest rate is 1.3% per month.

b) Find the new account balance on December 5.

Solution:

a) The finance charge is based on the $275 balance due on November 5. To find the finance charge due on December 5 we use the simple interest formula with a time of 1 month.

$$i = \$275 \times 0.013 \times 1 = \$3.58$$

The finance charge on December 5 is $3.58.

b) The balance due on December 5 is

$$\$275 + \$320 + \$3.58 - \$145 = \$453.58$$

The balance due on December 5 is $453.58. The finance charge on January 5 will be based on $453.58. ▲

Many lending institutions use the average daily balance method of calculating the finance charge because they believe that it is fairer to the customer.

With the average daily balance method, a balance is determined each day of the billing period for which there is a transaction in the account. The average daily balance method is illustrated in Example 7.

EXAMPLE 7

The balance on the Margolius's credit card account on July 1, their billing date, was $375.80. They had the following transactions during the month of July.

July 5	Payment	$150.00
July 10	Charge: Toy store	74.35
July 18	Charge: Garage	123.50
July 28	Charge: Restaurant	42.50.

a) Find the average daily balance for the billing period.

b) Find the finance charge to be paid on August 1. Assume that the interest rate is 1.3% per month.

c) Find the balance due on August 1.

Solution:

a) To determine the average daily balance we do the following: (i) Find the balance due for each transaction date.

July 1	$375.80
July 5	$375.80 − $150 = $225.80
July 10	$225.80 + $74.35 = $300.15
July 18	$300.15 + $123.50 = $423.65
July 28	$423.65 + $42.50 = $466.15

(ii) Find the number of days that the balance did not change between each transaction. Count the first day in the period but not the last day. Note that from July 28 through July 31 is 4 days. The next billing cycle starts on August 1. (iii) Multiply the balance due by the number of days the balance did not change. (iv) Find the sum of the products.

	(i)	(ii)	(iii)
Date	Balance Due	Number of Days Balance Did Not Change	(Balance)(Days)
July 1	$375.80	4	($375.80)(4) = $ 1503.20
July 5	$225.80	5	($225.80)(5) = $ 1129.00
July 10	$300.15	8	($300.15)(8) = $ 2401.20
July 18	$423.65	10	($423.65)(10) = $ 4236.50
July 28	$466.15	4	($466.15)(4) = $ 1864.60
		31	(iv) Sum = $11,134.50

(v) Divide this sum by the number of days in the billing cycle (in the month). The number of days may be found by summing the days in column ii.

$$\frac{\$11,134.50}{31} = \$359.18$$

Thus the average daily balance is $359.18.

b) The finance charge for the month is found using the simple interest formula with the average daily balance as the principal.

$$i = \$359.18 \times 0.013 \times 1 = \$4.67$$

c) Since the finance charge for the month is $4.67, the balance owed on August 1 is $466.15 + $4.67 or $470.82.   ▲

The calculations in Example 7 are tedious. However, these calculations are made almost instantaneously with computers.

Example 8 illustrates how a credit card is used to borrow money.

EXAMPLE 8

Don needed $250 to pay some bills. He borrowed the money from his charge account on September 1, and repaid it on September 21. How much interest did Don pay for the loan? Assume that the interest rate is 0.04273% per day.

Solution: The agreement says that the interest is calculated for the exact number of days of the loan, starting with the day the money is obtained. The time of the loan in this case is 20 days. Using the simple interest formula and

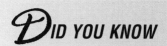
remembering that the rate and time are in terms of days, we find that the interest is $2.14.

$$i = prt$$
$$= \$250 \times 0.0004273 \times 20$$
$$= \$2.14$$

The amount due on September 21 was $250 + $2.14 = $252.14. ▲

Anyone purchasing a car or other costly items should consider a number of different sources for a loan. Example 9 illustrates one method of making a comparison.

EXAMPLE 9

Fran needed an additional $500 to buy a sailboat. She found that the Easy Loan Company charged 8.0% simple interest on the amount borrowed for the duration of the loan. It also required that the loan, including interest, be repaid in six equal monthly payments. The Friendly Loan Company offered loans of $500 to be repaid in 12 monthly payments of $44.50.

a) How much interest would be charged for the loan from the Easy Loan Company?

b) How much interest would be charged for the loan from the Friendly Loan Company?

c) What was the APR on the Easy Loan Company loan?

d) What was the APR on the Friendly Loan Company loan?

Solution:

a) To find the interest charged by the Easy Loan Company, we use the simple interest formula.

$$i = prt$$
$$= \$500 \times 0.08 \times \frac{6}{12}$$
$$= \$20.00$$

Thus the interest charged is $20.00 for the loan.

b) To find the interest charged by the Friendly Loan Company, we must subtract the amount borrowed from the total amount repaid.

$$\text{Interest} = \text{total amount repaid} - \text{amount borrowed}$$
$$= (\$44.50 \times 12) - \$500$$
$$= \$534 - \$500$$
$$= \$34.00$$

c) To find the APR on the loan from the Easy Loan Company, we divide the interest by the amount financed, multiply by 100, and determine the APR from Table 11.2.

$$\frac{\$20.00}{\$500} \times 100 = 4.00$$

The APR found in the table is 13.50%.

d) To find the APR on the loan from the Friendly Loan Company, we divide the interest by the amount financed, multiply by 100, and determine the APR from Table 11.2.

$$\frac{\$34}{\$500} \times 100 = 6.8$$

The APR found in the table is 12.25%.

Note that Fran would pay fewer dollars in interest on the loan from the Easy Loan Company ($20.00 versus $34). However, she would pay a higher rate of interest to the Easy Loan Company (13.50% versus 12.25%). The reason that the actual interest is less even though the APR is greater is the shorter time period for repayment to the Easy Loan Company than to the Friendly Loan Company.

Section 11.4 Exercises

1. Explain in your own words the difference between an open-end installment loan and a fixed installment loan.

2. What is an annual percentage rate (APR)?

3. What is a finance charge?

4. What is a total installment price?

5. Name the two methods used to determine the finance charge when a loan is repaid early.

6. Name the two methods used to determine the finance charge on an open-end installment loan.

7. Chris purchased a Macintosh computer on a monthly purchase plan. The computer sold for $2350. She paid $500 down and $88.50 a month for 24 months.
   a) What finance charge did Chris pay?
   b) What was the APR?

8. Jason can buy a knee-board (knee-boarding is a water sport) for a cash price of $675. The installment terms include a down payment of $175 and 12 monthly payments of $44.83.
   a) What finance charge will Jason pay?
   b) What is the APR?

9. Mr. and Mrs. Gerling want to buy furniture that has a cash price of $3450. On the installment plan they must pay one-third of the cash price as a down payment and make six monthly payments of $398.
   a) What finance charge will the Gerlings pay?
   b) What is the APR?

10. A diamond ring sells for $1750 cash. On the installment plan the ring may be purchased with a 25% down payment and the balance paid in 18 equal monthly installments. Compute the monthly payment if the total finance charge is $112.

11. Sara wants to purchase a new television set. The purchase price is $890. If she purchases the set today and pays cash, she must take money out of her savings account. Another option is to charge the TV on her credit card, take the set home today, and pay next month. Next month she will have cash and can pay her credit card balance without paying any interest. The simple interest rate on her savings account is $5\frac{1}{4}$%. How much is she saving by using the credit card instead of taking the money out of her savings account?

12. A certain automobile finance agency requires that the buyer make a down payment of one-third of the cash price. The agency subtracts the down payment from the cash price and then adds the cost of disability and life insurance.* The resulting sum is the unpaid balance. The total interest on the loan is computed on the unpaid balance at 6.9% simple interest for the duration of the loan. The Stearns family bought a second-hand car on these installment terms. The cash price of the car was $9600, the insurance was $150 per year, and the balance was to be paid in 18 monthly payments.
   a) What was the down payment?
   b) What was the interest charged?
   c) What was the APR?
   d) What was the monthly payment including interest and insurance?

_____

*Disability insurance will cover the monthly payments if the insured individual becomes disabled and cannot work. The life insurance will pay off the balance due on the car if the insured individual dies.

13. Penny needs $250. She finds that the ABC Loan Company charges 7.5% simple interest on the amount borrowed for the duration of the loan and requires the loan to be repaid in 6 equal monthly payments. The XYZ Loan Company offers loans of $250 to be repaid in 12 monthly payments of $22.39.
  a) How much interest is charged by the ABC Loan Company?
  b) How much interest is charged by the XYZ Loan Company?
  c) What is the APR on the ABC Loan Company loan?
  d) What is the APR on the XYZ Loan Company loan?

14. John borrowed $875 against his charge account on September 12 and repaid the loan on October 14 (32 days later). Assume that the interest rate is 0.04273% per day.
  a) How much interest did John pay on the loan?
  b) What amount did he pay the bank when he repaid the loan?

15. Amina bought a new Dodge Caravan for $24,850, including sales tax. She made a down payment of $8850 and financed the balance through her bank. She repaid the loan by paying $346.67 per month for 60 months. What was the APR?

16. Tina has a 48-month installment loan, with a fixed monthly payment of $193.75. The amount borrowed was $7500. Instead of making her 18th payment, Tina is paying the remaining balance on the loan.
  a) Determine the APR of the installment loan.
  b) How much interest will Tina save, computed by the actuarial method?
  c) What is the total amount due on that day?

17. The cash price for furniture for Melissa's apartment was $6520. She made a down payment of $3962 and financed the balance on a 24-month fixed payment installment loan. The monthly payments are $119.47. Instead of making her 12th payment, Melissa decides to pay the remaining balance.
  a) Determine the APR of the installment loan.
  b) How much interest will Melissa save, computed by the actuarial method?
  c) What is the total amount due on that day?

18. Tony Joseph is buying a $3000 sound system by making a down payment of $500 and 18 monthly payments of $151.39. Instead of making his 12th payment, Tony decides to pay the remaining balance.
  a) How much interest will Tony save, computed by the rule of 78s?
  b) What is the total amount due on that day?

19. Roger purchased woodworking tools for $2375. He made a down payment of $850 and financed the balance with a 12-month fixed payment installment loan. Instead of making his sixth monthly payment of $134.71 he decides to pay the remaining balance.

a) How much interest will Roger save, computed by the rule of 78s?
b) What is the total amount due on that day?

20. On September 5, the billing date, Verna had a balance due of $385.75 on her credit card. The transactions during the following month were:

September 8	Payment	$275.00
September 21	Charge: Airline ticket	330.00
September 27	Charge: Hotel bill	190.80
October 2	Charge: Clothing	84.75

  a) Find the finance charge on October 5, using the unpaid balance method. Assume that the interest rate is 1.40% per month.
  b) Find the new balance on October 5.

21. On April 15, the billing date, Lori had a balance due of $57.88 on her credit card. She is redecorating her apartment and has the following transactions.

April 16	Charge: Paint	$ 64.75
April 20	Payment	45.00
May 3	Charge: Curtains	72.85
May 10	Charge: Chair	135.50

  a) Find the finance charge on May 15, using the unpaid balance method. Assume that the interest rate is 1.35% per month.
  b) Find the new balance on May 15.

22. The balance on the Simpsons' credit card on May 12, their billing date, was $378.50. For the period ending June 12 they had the following transactions.

May 13	Charge: Toys	$129.79
May 15	Payment	50.00
June 1	Charge: Clothing	135.85
June 8	Charge: Housewares	37.63

  a) Find the average daily balance for the billing period.
  b) Find the finance charge to be paid on June 12. Assume an interest rate of 1.3% per month.
  c) Find the balance due on June 12.

23. The Calusines' credit card statement shows a balance due of $1578.25 on March 23, the billing date. For the period ending April 23 they had the following transactions.

March 26	Charge: Party supplies	$ 79.98
March 30	Charge: Restaurant meal	52.76
April 3	Payment	250.00
April 15	Charge: Clothing	190.52
April 22	Charge: Car repairs	190.85

  a) Find the average daily balance for the billing period.
  b) Find the finance charge to be paid on April 23. Assume an interest rate of 1.3% per month.
  c) Find the balance due on April 23.

24. The cash price of a motorboat and trailer is $12,500. The package may be purchased for a 10% down payment and 36 monthly payments. For convenience the dealer computes the finance charge at 7% simple interest on the amount financed.
  a) What is the finance charge on the boat and trailer?
  b) How much is the monthly payment?
  c) Determine the total cost of the package on the installment plan.
  d) What annual percentage rate must the dealer disclose to the buyer?

25. In Exercise 23, if the Calusines had made the payment on their account on March 24 rather than on April 3, how would it have affected
  a) the average daily balance?
  b) the finance charge?

Problem Solving/Group Activities

26. Ken bought a new car, but he cannot now remember the original purchase price. His payments are $379.50 per month for 36 months. He remembers that the salesperson said that the simple interest rate for the period of the loan was 6%. He also recalls that he was allowed $2500 on his old car. Find the original purchase price.

27. Suppose that the Gerlings in Exercise 9 use a credit card rather than the installment plan. Assume that they make the same down payment, there is no finance charge the first month, they make no additional purchases on their credit card and they pay $384 per month plus the finance charge starting with the second month. If the interest rate is 1.3% per month
  a) how many months will it take them to repay the loan?
  b) how much interest will they pay on the loan?
  c) which method of borrowing will cost the Gerlings the least amount of interest, the installment loan in Exercise 9 or the credit card?

28. Tina wants to buy a camera in time for a June 30 family gathering. She knows that she will not have the money to pay for the camera until August 5. The billing date on her credit card is the 25th of the month and she has a 20 day grace period from the billing date to pay the bill with no finance charge. Explain how she can buy the camera before June 30, pay for it after August 5, and pay no interest.

Research Activities

29. Write a brief report giving the advantages and disadvantages of leasing a car. Determine all the individual costs involved with leasing a car. Indicate why you would prefer to lease or purchase a car at the present time.

30. Assume that you are married and have a child. You don't own a washer and dryer and have no money to buy the appliances. Would it be cheaper to borrow money on an installment loan and buy the appliances or to continue going to the local coin-operated laundry for 5 years until you have saved enough to pay cash for a washer and dryer?

   With the aid of parents or friends, establish how many loads of laundry you would be doing each week. Then determine the cost of doing that number of loads at a coin-operated laundry. (Don't forget the cost of transportation to the laundry.) Shop around for a washer and dryer, and determine the total cost on an installment plan. Don't forget to include the cost of gas, electricity, and water. This information can be obtained from a local gas and electric company. With this information you should be able to make a decision about whether to buy now or wait for 5 years.

31. Do research on the features offered by MasterCard (Regular and Gold cards), Visa (Regular and Gold cards), Discover, American Express (Gold and Platinum cards), and AT&T Universal (Gold and Platinum cards). For example, features might include life insurance when traveling by a common carrier, miles toward air travel, discounts on automobiles, cash back at the end of the year, insurance on rental cars, and the like. Determine which card or cards have the most appropriate features for you and explain why you arrived at that conclusion.

32. By reading a financial newspaper or magazine, determine the five specific bank credit cards that offer the lowest interest rates. List each card and its interest rate and annual fee. Indicate where you obtained this information.

11.5   BUYING A HOUSE WITH A MORTGAGE

Buying a house is the largest purchase of a lifetime for most people. The purchaser will normally be committed to 10, 15, 20, 25, or 30 years of mortgage payments. Before selecting the "Dream House" the buyer should consider the following questions: "Can I afford it?" and "Does it suit my needs?"

The question "Can I afford the house?" must be answered carefully and accurately. If a family buys a house beyond its means, it will have a difficult time living within its income. When deciding whether to purchase a particular house, the purchaser must also consider crucial questions such as: "Do I have enough cash for the down payment?" "Can I afford the monthly payments with my current income?" These two items, down payment and mortgage payments over time, constitute the buyer's total cost of buying a house.

Buyers usually seek a *mortgage* from a bank or other lending institution. Before approving a mortgage, which is a long-term loan, the bank will require the buyer to have a specified minimum amount for the down payment. The **down payment** is the amount of cash the buyer must pay to the seller before the lending institution will grant the buyer a mortgage. If the buyer has the down payment and meets the other criteria for the mortgage, the lending institution prepares a written agreement called the mortgage, stating the terms of the loan. The loan specifies the repayment schedule, the duration of the loan, whether the loan can be assumed by another party, and the penalty if payments are late. The party borrowing the money accepts the terms of this agreement and gives the lending institution the title or deed to the property as security.

Homeowner's Mortgage

A long-term loan in which the property is pledged as security for payment of the difference between the down payment and the sale price.

The two most popular types of mortgage loans available today are the **adjustable-rate loan** and the **conventional loan**. The major difference between the two is that the interest rate for a conventional loan is fixed for the duration of the loan, whereas the interest rate for the variable-rate loan may change every period, as specified in the loan. We will first discuss the requirements that are the same for both types of loans.

The size of the down payment required depends on who is lending the money, how old the property is, and whether or not it is easy to borrow money at that particular time. The down payment required by the lending institution can vary from 5% to 50% of the purchase price. A larger down payment is required when money is "tight"—that is, when it is difficult to borrow money. Furthermore, most lending institutions tend to require larger down payments on older homes and smaller down payments on newer homes.

Most lending institutions may require the buyer to pay one or more **points** for their loan at the time of the **closing** (the final step in the sale process). **One point** amounts to 1% of the mortgage money (the amount being borrowed). By charging points, the bank reduces the rate of interest on the mortgage, thus reducing the size of the monthly payments and enabling more people to purchase houses. However, because they charge points, the rate of interest that banks state is not the APR (annual percentage rate) for the loan. The APR would be determined by adding the amount paid for points to the total interest paid and then using an APR table.

Conventional Loans

Example 1 illustrates purchasing a house with a conventional loan.

EXAMPLE 1

The Cohens want to purchase a house selling for $95,000. Their bank requires a 10% down payment and a payment of two points at the time of closing.

a) What is the required down payment?

b) With a 10% down payment, what is the mortgage on the property?

c) What is the cost of two points on the mortgage determined in part (b)?

Solution:

a) The required down payment is 10% of $95,000, or

$$0.10 \times \$95,000 = \$9500.$$

b) To find the mortgage on the property, subtract the down payment from the selling price of the house.

$$\$95,000 - \$9500 = \$85,500$$

c) To find the cost of two points on a mortgage of $85,500, find 2% of $85,500.

$$0.02 \times \$85,500 = \$1710$$

When the Cohens sign the mortgage, they must pay the bank $1710. The $9500 down payment is paid to the seller. ▲

Banks use a formula to determine the maximum monthly payment that they believe is within the purchaser's ability to pay. A mortgage loan officer will first determine the buyer's **adjusted monthly income** by subtracting from the gross monthly income (total income before any deductions) any fixed monthly payments with more than 10 months remaining (such as for a student loan, a car, furniture, or a television). The loan officer then multiplies the adjusted monthly income by 28%. (This percent, and the maximum number of payments remaining on other fixed loans, may vary in different locations.) In general, this product is the maximum monthly payment the lending institution believes the purchaser can afford to pay for a house. This estimated payment must cover principal, interest, property taxes, and insurance. Taxes and insurance are not necessarily paid to the bank; they may be paid directly to the tax collector and the insurance company. Example 2 shows how a bank uses the formula to determine whether a prospective buyer qualifies for a mortgage.

EXAMPLE 2

In Example 1, suppose that the Cohens' gross monthly income is $4100, and that they have 14 remaining payments of $225 per month on their car and 48 payments of $55 per month on a college loan. The taxes on the house they want to purchase are $150 per month, and insurance is $42 per month.

a) What maximum monthly payment does the bank's loan officer think the Cohens can afford?

b) The Cohens want a 20-year $85,500 mortgage. Do they qualify for the mortgage if the rate of interest is 8.5%?

Table 11.4   Monthly Payment per $1000 of Mortgage, Including Principal and Interest

Rate %	Number of Years				
	20	**25**	**30**	**35**	**40**
6.5	$7.46	$6.75	$6.32	$6.04	$5.85
7	7.75	7.70	6.65	6.39	6.21
7.5	8.06	7.39	6.99	6.74	6.58
8	8.36	7.72	7.34	7.10	6.95
8.5	(8.68)	8.05	7.69	7.47	7.33
9	9.00	8.40	8.05	7.84	7.72
9.5	9.33	8.74	8.41	8.22	8.11
10	9.66	9.09	8.70	8.60	8.50
10.5	9.98	9.44	9.15	8.98	8.89
11	10.32	9.80	9.52	9.37	9.28
11.5	10.66	10.16	9.90	9.76	9.68
12	11.01	10.53	10.29	10.16	10.08
12.5	11.36	10.90	10.67	10.55	10.49
13	11.72	11.28	11.06	10.95	10.90
13.5	12.07	11.66	11.45	11.35	11.30
14	12.44	12.04	11.85	11.76	11.71
14.5	12.80	12.42	12.25	12.16	12.12
15	13.17	12.81	12.64	12.57	12.53
15.5	13.54	13.20	13.05	12.98	12.94
16	13.91	13.59	13.45	13.38	13.36

Solution:

a) To find the maximum monthly payment that the bank's loan officer believes the Cohens can afford, first find their adjusted monthly income.

$$\begin{array}{ll} \$4100 & \text{Gross income} \\ -\ \ 280 & \text{Monthly payments (car, college loan)} \\ \hline \$3820 & \text{Adjusted monthly income} \end{array}$$

Next find 28% of the adjusted monthly income.

$$0.28 \times \$3820 = 1069.60$$

The loan officer determines that the Cohens can afford a maximum monthly payment of $1069.60 with their current income.

b) To determine whether the Cohens qualify for a 20-year conventional mortgage with their current income, calculate the monthly mortgage payment, which includes principal and interest. All lending institutions and most lawyers have a set of tables that contain a monthly mortgage payment per thousand dollars for a specific number of years at a specific interest rate. Table 11.4 shows selected monthly mortgage payments. With an interest rate of 8.5%, a 20-year loan would have a monthly payment of $8.68 per thousand dollars of mortgage (circled in blue in the table).

To determine the Cohens' payment, first divide the mortgage by $1000.

$$\frac{\$85,500}{\$1000} = 85.5$$

Then find the monthly mortgage payment by multiplying the number of thousands of dollars of mortgage, 85.5, by the value shown in Table 11.4.

$$\$8.68 \times 85.5 = \$742.14$$

To the $742.14 add the monthly cost of insurance, $42, and real estate taxes of $150, for a total cost of $934.14. Since $934.14 is less than the maximum monthly payment the loan officer has determined that the Cohens can afford, $1069.60, they will most likely be granted the loan.

▲

What is the effect on the monthly payments when only the period of time has been changed? The total monthly payments on the mortgage in Example 2 for different periods of time are $688.28 for 25 years, $657.50 for 30 years, $638.69 for 35 years, and $626.72 for 40 years. Increasing the length of time decreases the monthly payment but increases the total amount of interest paid because the borrower is paying for a longer period of time. The longer the term of a mortgage the more expensive is the total cost of the house.

EXAMPLE 3

The Cohen family of Examples 1 and 2 obtains a 20-year $85,500 conventional mortgage at 8.5% on a house selling for $95,000. Their monthly mortgage payment, including principal and interest, is $742.14. They paid two points at closing.

a) Determine the amount the Cohens will pay for their house over 20 years.

b) How much of the cost will be interest?

c) How much of the first payment on the mortgage is applied to the principal?

Solution:

a) To find the total amount the Cohens will pay for the house, compute as follows.

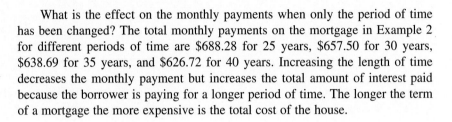

$	742.14	Mortgage payment for one month
×	12	
$	8905.68	Mortgage payments for one year
×	20	Number of years
	$178,113.60	Mortgage payments for 20 years
+	9,500.00	Down payment
+	1,710.00	Points
	$189,323.60	Total cost of the house

Note: The result might not be the exact cost of the house, since the final payment on the mortgage might be slightly more or less than the regular monthly payment.

b) To determine the amount of interest paid over 20 years, subtract the purchase price of the house and the cost of the points from the total cost.

	$189,323.60	Total cost
−	95,000.00	Purchase price
	$ 94,323.60	Total interest including points
−	1,710.00	Points
	$ 92,613.60	Total interest on mortgage payments

c) To find the amount of the first payment that is applied to the principal, subtract the amount of interest on the first payment from the monthly mortgage payment. Use the simple interest formula to find the interest on the first payment.

$$i = prt$$

$$= \$85,500 \times 0.085 \times \frac{1}{12}$$

$$= \$605.63$$

Now subtract the interest for the first month from the monthly mortgage payment. The difference will be the amount paid on the principal for the first month.

$742.14	Monthly mortgage payment
− 605.63	Interest paid for the first month
$136.51	Principal paid for the first month

Thus the first payment of $742.14 consists of $605.63 in interest and $136.51 in principal. The $136.51 is applied to reduce the loan. Thus the balance due after the first payment is $85,500 − $136.51, or $85,363.49.

By repeatedly using the simple interest formula month-to-month on the unpaid balance, you could calculate the principal and the interest for all the payments—a tedious task. However, with a computer a list containing the payment number, payment on the interest, payment on the principal, and balance of the loan can be easily prepared. Such a list is called a loan amortization schedule. Part of an amortization schedule for the Cohen's loan in Example 3 is given in Table 11.5. The schedule shows that with each payment the buyer pays a little less in interest and a little more in principal.

Table 11.5   Amortization Schedule

	Annual % rate: 8.5	Monthly payment: $742.14	
	Loan: $85,500	Term: Years 20, Months 0	
	Periods: 240		

Payment Number	Interest	Principal	Balance of Loan
1	$605.63	$136.51	$85,363.49
2	604.66	137.48	85,226.01
3	603.68	138.46	85,087.55
4	602.70	139.44	84,948.11
23	582.69	159.45	82,102.93
24	581.56	160.58	81,942.35
119	428.16	313.98	60,132.43
120	425.94	316.20	59,816.23
179	262.60	479.54	36,593.57
180	259.20	482.94	36,110.63
239	9.74	732.40	642.64
240	4.55	642.64	

Adjustable-Rate Mortgages

Now let's consider adjustable-rate mortgages, ARMs (also called *variable-rate mortgages*). The rules for ARMs vary from state to state and by local bank, so the material presented on ARMs may not apply in your state or at your local bank. Generally, with adjustable-rate mortgages, the monthly mortgage payment remains the same for a one-, two-, or five-year period even though the interest rate of the mortgage may change every three months, six months, or some other predetermined period. The interest rate in an adjustable-rate mortgage may be based on an index that is determined by the Federal Home Loan Bank Association. Or it may be based on the interest rate of a 3-month, 6-month, or 1-year Treasury bill. The interest rate of a 3-month Treasury bill may change every 3 months, a 6-month Treasury bill may change every 6 months, and so on. When the base is a Treasury bill, the actual interest rate charged for the mortgage is often determined by adding 3% to $3\frac{1}{2}$%, called the add on rate or margin to the rate of the Treasury bill. Thus if the rate of the Treasury bill is 6% and the add on rate is 3%, the interest rate charged is 9%.

EXAMPLE 4

The Jacksons purchased a house for $115,000 with a down payment of $23,100. They obtained a 30-year adjustable-rate mortgage with the following terms. The interest rate is based on a 6-month Treasury bill, the effective interest rate is 3% above the rate of the Treasury bill on the date of adjustment (3% is the add on rate), the interest rate is adjusted every 6 months, the interest rate will not change more than 1% (up or down) when the interest rate is adjusted, the maximum interest rate that can be charged for the duration of the loan is 16%, there is no lower limit on the interest rate, the initial mortgage interest rate is 8%, and the monthly payment (including principal and interest) is adjusted every 5 years.

a) Determine the initial monthly payment.

b) Determine the adjusted interest rate in 6 months if the interest rate on the Treasury bill at that time is 4.5%.

Solution:

a) To determine the initial monthly payment of interest and principal, divide the amount of the loan, $115,000 − $23,000 = $91,900, by $1000. The result is 91.9. Now multiply the number of thousands of dollars of mortgage, 91.9, by the value found in Table 11.4 with $r = 8\%$ for 30 years.

$$\$7.34 \times 91.9 = \$674.55$$

Thus the initial monthly payment for principal and interest is $674.55. This amount will not change for the first 5 years of the mortgage.

b) The adjusted interest rate in 6 months will be the Treasury bill rate plus the add on rate.

$$4.5\% + 3\% = 7.5\%$$

▲

In Example 4(b), note that the rate after 6 months is lower than the initial rate of 8%. Since the monthly payment remains the same, the additional money paid the bank is applied to reduce the principal. The monthly interest and principal payment of $674.55 would pay off the loan in 30 years if the interest remained constant at 8%. What happens if the interest rate drops and stays lower than the initial 8% for the length of the loan? In this case, at the end of each 5-year period the bank reduces the monthly payment so that the loan will be paid off in 30 years. What happens if the interest rates increase above the initial 8% rate? In this case part or, if necessary, all of the principal part of the monthly payment would be used to meet the interest obligation. At the end of the 5-year period the bank will increase the monthly payment so that the loan can be repaid by the end of the 30-year period. Or the bank may increase the time period of the loan beyond 30 years so that the monthly payment is affordable.

To prevent rapid increases in interest rates some banks have a rate cap. A rate cap limits the maximum amount that the interest rate may change. A periodic rate cap limits the amount that the interest rate may increase in any one period. For example, your mortgage could provide that, even if the index increases by 2% in 1 year, your rate can only go up 1% per year. An aggregate rate cap limits the interest rate increase and decrease over the entire life of the loan. If the initial interest rate is 9% and the aggregate rate cap is 5%, the interest rate could go no higher than 14% and no lower than 4% over the life of the mortgage. A payment cap limits the amount that the monthly payment may change but does not limit changes in interest rates. If interest rates increase rapidly on a loan with a payment cap, the monthly payment may not be large enough to pay the monthly principal and interest on the loan. If that happens the borrower could end up paying interest on interest.

Other Types of Mortgages

Conventional mortgages and variable-rate mortgages are not the only methods of financing the purchase of a house. We next briefly describe four other methods. Two of these, FHA and VA mortgages, are backed by the government.

FHA Mortgage   A house can be purchased with a smaller down payment than with a conventional mortgage if the individual qualifies for a Federal Housing Administration (FHA) loan. The loan application is made through a local bank. The bank's loan officer determines the maximum monthly payment that a loan applicant can afford by taking 29% of the adjusted monthly income. The bank provides the money, but the FHA insures the loan. The down payment for an FHA loan is as low as 2.5% of the purchase price, rather than the standard 5 to 50%. Another advantage is that FHA loans can be assumed at the original rate of interest on the loan. For example, if you purchase a home today that already has an 8% FHA mortgage, you, the new buyer, will be able to assume that 8% mortgage regardless of the current interest rates. However, to be able to assume a mortgage, the purchaser must be able to make a down payment equal to the difference between the purchase price and the balance due on the original mortgage. The government sets the maximum interest rate that the lender may charge.

One drawback to FHA loans is that the borrower must pay an FHA insurance premium as part of the monthly mortgage payment. The insurance premium is calculated at a rate of one-half percent (0.5%) of the unpaid balance of the loan on the anniversary date of the loan. Thus even though the insurance premium decreases each year, it adds to the monthly payments.

VA Mortgage   A veteran certified by the Department of Veterans Affairs (VA) who wants to purchase a house applies for a mortgage with a bank or lending institution. The individual must meet the requirements set by the bank or lending institution for a mortgage. The VA guarantees a certain percentage of the loan. For example, if the loan is less than $45,000 the VA guarantees 50% of the loan.

For loans from $45,000 to $144,000 the VA guarantees the lesser of 40% of the loan or $36,000. This guarantee takes the place of the down payment. A veteran who can make the monthly mortgage payments may therefore obtain a certain mortgage without a down payment.

The government sets the maximum interest rate that the lender may charge. A VA loan is always assumable. With a VA loan the seller may be asked to pay points, but there is no monthly insurance premium.

Graduated Payment Mortgage (GPM)   A GPM mortgage is designed so that for the first 5 to 10 years the size of the mortgage payment is smaller than the payments for the remaining time of the mortgage. After the 5- to 10-year period the mortgage payments remain constant for the duration of the mortgage. Depending on the size of the mortgage, the lender might find that the monthly payments made for the first few years may actually be less than the interest owed on the loan for those few years. The interest not paid during the first few years of the mortgage is then added to the original loan. This type of loan is strictly for those who are confident that their annual incomes will increase as rapidly as the mortgage payments do.

Balloon-Payment Mortgage (BPM)   This type of loan could be for the person who needs time to find a permanent loan. It may work this way: The individual pays the interest for 3 to 8 years, the period of the loan. At the end of the 3- to 8-year period, the buyer must repay the entire principal unless the lender agrees to a loan extension. Balloon-payment loans generally offer lower rates than conventional mortgages. This type of loan may be advantageous if the buyer plans to sell the house before the maturity date of the balloon-payment mortgage.

For further information on any of these mortgages, consult your local banker.

Additional information on buying a house may be obtained from the following pamphlets. A common source for most of these and other pamphlets is the U.S. Government Printing Office in Pueblo, CO 81009.

1. *Consumer Handbook on Adjustable Rate Mortgages.* Federal Reserve Board Office of Thrift Supervision.

2. *A Quick Guide for Homebuyers and Real Estate Professionals.* VA Pamphlet 26-91-1. Department of Veterans Affairs, 810 Vermont Avenue, N.W., Washington, DC 20420.

3. *The Mortgage Money Guide.* Federal Trade Commission, 6th and Pennsylvania Avenue, N.W., Washington, DC 20580.

Section 11.5 Exercises

1. What is a mortgage?

2. What is a down payment?

3. Explain in your own words the difference between a variable-rate mortgage and a conventional mortgage.

4. **a)** What are points in a mortgage agreement?
   b) Explain in your own words how to determine the cost of *x* points.

5. Explain in your own words how a banks loan officer determines a buyer's adjusted monthly income.

6. What is an amortization schedule?

7. What is an add on rate or margin?

8. Who insures an FHA loan? Who provides the money for an FHA loan?

9. Eric is buying a house selling for $55,000. The bank is requiring a minimum down payment of 25%. The current mortgage rate is 8%.
   a) Determine the amount of the required down payment.

b) Determine the monthly mortgage payment for a 30-year loan with a minimum down payment.

10. Ann and Ted Thomas are buying a house selling for $48,000. The bank is requiring a minimum down payment of 10%. The current mortgage rate is 8.5%.
  a) Determine the amount of the required down payment.
  b) Determine the monthly mortgage payment for a 25-year loan with a minimum down payment.

11. Barbara is buying a condominium selling for $65,000. The bank is requiring a minimum down payment of 15%. To obtain a 30-year mortgage at 9%, she must pay two points at the time of closing.
  a) What is the required down payment?
  b) With the 15% down payment, what is the amount of the mortgage on the property?
  c) What is the cost of two points on the mortgage determined in part (b)?

12. The Imans are buying a house that sells for $52,000. The bank is requiring a minimum down payment of 15%. To obtain a 30-year mortgage at 9.5%, they must pay three points at the time of closing.
  a) What is the required down payment?
  b) With the 15% down payment, what is the amount of the mortgage on the property?
  c) What is the cost of three points on the mortgage determined in part (b)?

13. Marlon's gross monthly income is $2500. He has 12 remaining payments of $95 on furniture and appliances. The taxes and insurance on the house are $105 per month.
  a) What maximum monthly payment does the bank's loan officer feel that Marlon can afford?
  b) Marlon would like to get a 30-year, $52,200 mortgage. Does he qualify for this mortgage with a 9% interest rate?

14. The Dechs' gross monthly income is $7200. They have 15 remaining payments of $290 on a new car. The taxes on the house are $350 per month and insurance is $50 per month.
  a) What maximum monthly payment does the bank's loan officer feel that the Dechs can afford?
  b) The Dechs want a 30-year, $140,000 mortgage. Do they qualify for this mortgage with a 9% interest rate?

15. Andrea obtains a 30-year, $63,750 conventional mortgage at 8.5% on a house selling for $75,000. Her monthly payment, including principal and interest, is $490.24.
  a) Determine the total amount that Andrea will pay for her house.
  b) How much of the cost will be interest?
  c) How much of the first payment on the mortgage is applied to the principal?

16. Mr. and Mrs. Georgi obtain a 25-year, $110,000 conventional mortgage at 10.5% on a house selling for $160,000. Their monthly mortgage payment, including principal and interest, is $1038.40.
  a) Determine the amount the Georgis will pay for their house.
  b) How much of the cost will be interest?
  c) How much of the first payment on the mortgage is applied to the principal?

17. The Schollers purchased a home selling for $58,000 with a 20% down payment and were required to pay two points at the time of closing. The period of the conventional mortgage is 35 years, and the rate of interest is 9.5%.
  a) Determine the amount of the down payment.
  b) Determine the cost of the two points.
  c) Determine the monthly mortgage payment.
  d) Determine the total cost of the house (including points).
  e) Determine the total interest paid.
  f) Determine how much of the first payment on the loan is applied to the principal.

18. The Washingtons found a house that is selling for $113,500. The taxes on the house are $1200 per year, and insurance is $320 per year. They are requesting a conventional loan from the local bank. The bank is currently requiring a 28% down payment and three points, and the interest rate is 10%. The Washingtons' monthly income is $4750. They have more than 10 monthly payments remaining on a car, a boat, and furniture. The total monthly payments for these items is $420.
  a) Determine the required down payment.
  b) Determine the cost of the three points.
  c) Determine their adjusted monthly income.
  d) Determine the maximum monthly payment the bank's loan officer believes that they can afford.
  e) Determine the monthly payments of principal and interest for a 20-year loan.
  f) Determine their total monthly payment, including insurance and taxes.
  g) Determine whether the Washingtons qualify for the 20-year loan.
  h) Determine how much of the first payment on the loan is applied to the principal.

19. Kathy wants to buy a condominium selling for $95,000. The taxes on the property are $1500 per year, and insurance is $336 per year. Kathy's gross monthly income is $4000. She has 15 monthly payments of $135 remaining on her van. She has $20,000 in her savings account. The bank is requiring 20% down and is charging 9.5% interest.
  a) Determine the required down payment.
  b) Determine the maximum monthly payment that the bank's loan officer believes Kathy can afford.
  c) Determine the monthly payment of principal and interest for a 35-year loan.

d) Determine her total monthly payments, including insurance and taxes.
e) Does Kathy qualify for the loan?
f) Determine how much of the first payment on the mortgage is applied to the principal.
g) Determine the total amount she pays for the condominium with a 35-year conventional loan. (Do not include taxes or insurance.)
h) Determine the total interest paid for the 35-year loan.

20. The Christophers are negotiating with two banks for a mortgage to buy a house selling for $105,000. The terms at bank A are a 10% down payment, an interest rate of 10%, a 30-year conventional mortgage, and three points to be paid at the time of closing. The terms at bank B are a 20% down payment, an interest rate of 11.5%, a 25-year conventional mortgage, and no points. Which loan should the Christophers select in order for the total cost of the house to be less?

21. The Kellys purchased a house for $95,000 with a down payment of $13,000. They obtained a 30-year adjustable-rate mortgage. The terms of the mortgage are as follows: The interest rate is based on a 3-month Treasury bill, the effective interest rate is 3.25% above the rate of the Treasury bill on the date of adjustment, the interest rate is adjusted every 3 months, the interest rate will not change more than 1% (up or down) when the interest rate is adjusted, the maximum interest rate that can be charged for the duration of the loan is 16%, there is no lower limit on the interest rate, the initial mortgage interest rate is 8.5%, and the monthly payment of interest and principal is adjusted annually.
a) Determine the initial monthly payment for interest and principal.
b) Determine the interest rate in 3 months if the interest rate on the Treasury bill at the time is 5.65%.
c) Determine the interest rate in 6 months if the interest rate on the Treasury bill at the time is 4.85%.

Problem Solving/Group Activities

22. The Wongs can afford to make $950 a month mortgage payments. If the bank will give them a 25-year conventional mortgage at 9% and requires a 25% down payment, what is
a) the maximum mortgage the bank will grant the Wongs?
b) the highest-priced house they can afford?

23. The Kalecks purchased a house for $105,000 with a down payment of $5000. They obtained a 30-year adjustable-rate mortgage. The terms of the mortgage are as follows: The interest rate is based on a 3-month Treasury bill, the effective interest rate is 3.25% above the rate of the Treasury bill on the date of adjustment, the interest rate is adjusted every 3 months, the interest rate will not change more than 1% (up or down) when the interest rate is adjusted, the maximum interest rate

that can be charged for the duration of the loan is 16%, there is no lower limit on the interest rate, the initial mortgage interest rate is 9%, and the monthly payment of interest and principal is adjusted semiannually.
a) Determine the initial monthly payment for interest and principal.
b) Determine an amortization schedule for months 1–3.
c) Determine the interest rate for months 4–6 if the interest rate on the Treasury bill at the time is 6.13%.
d) Determine an amortization schedule for months 4–6.
e) Determine the interest rate for months 7–9 if the interest rate on the Treasury bill at the time is 6.21%.

24. The MacPhersons are applying for a $90,000 mortgage. They can choose between a conventional mortgage and a variable-rate mortgage. The interest rate on a 30-year conventional mortgage is 9.5%. The terms of the variable-rate mortgage are 6.5% the first year, an annual cap of 1%, and an aggregate cap of 6%. The interest rates and the mortgage payments are adjusted annually. Assume that the interest rates for the variable-rate mortgage increase by the maximum amount each year. Then the monthly mortgage payments for the variable-rate mortgage for years 1–6 are $568.86, $628.05, $688.29, $749.35, $811.02, and $873.11, respectively.
a) Knowing that they will be in the house for only 6 years, which mortgage will be the least expensive for that period?
b) How much will they save by choosing the less expensive mortgage?

Research Activity

25. An important part of buying a house is the closing. The exact procedures for the closing differ with individual cases and in different parts of the country. In any closing, however, both the buyer and the seller have certain expenses. To determine what is involved in the closing of a property in your community, contact a lawyer, a real estate agent, or a banker. Explain that you are a student and that your objective is to understand the procedure for closing a real estate purchase and the costs to both buyer and seller. Select a specific piece of property that is for sale. Use the asking price to determine the total closing costs to both buyer and seller. The following is a partial list of the most common costs. Consider them in your research.
a) Fee for title search and title insurance
b) Credit report on buyer
c) Fees to the lender for services in granting the loan
d) Fee for property survey
e) Fee for recording of the deed
f) Appraisal fee
g) Lawyer's fee
h) Escrow accounts (taxes, insurance)
i) Mortgage assumption fee

CHAPTER 11 SUMMARY

Key Terms

11.1

percent
percent change
percent markup on cost

11.2

bank discount
Banker's rule
cosigners
credit
discount note
interest
ordinary interest
personal note
principal
rate
security

simple interest
time
United States rule

11.3

compound interest
effective annual yield
fixed investment
investment
present value
variable investment

11.4

annual percentage rate (APR)
average daily balance method
finance charge
fixed installment loan

open-end installment loan
total installment price
unpaid balance method

11.5

adjustable-rate mortgages
adjusted monthly income
aggregate rate cap
amortization schedule
closing
conventional loan
down payment
homeowner's mortgage
payment cap
periodic rate cap
points
rate cap
variable-rate loan

Important Facts

Ordinary Interest

When computing ordinary interest, each month is considered to have 30 days and a year is considered to have 360 days.

Banker's Rule

When computing interest with the Banker's rule a year is considered to have 360 days and any fractional part of a year is the exact number of days.

Simple Interest Formula

Interest = principal × rate × time
or $i = prt$

Percent Change $= \dfrac{\text{amount in latest period} - \text{amount in previous period}}{\text{amount in previous period}} \times 100$

Percent Markup on Cost $= \dfrac{\text{selling price} - \text{dealer's cost}}{\text{dealer's cost}} \times 100$

Compound Interest Formula $A = p\left(1 + \dfrac{r}{n}\right)^{nt}$

Present Value Formula $p = \dfrac{A}{\left(1 + \dfrac{r}{n}\right)^{nt}}$

Actuarial Method $u = \dfrac{n \cdot P \cdot V}{100 + V}$

Rule of 78s $u = \dfrac{f \cdot k(k + 1)}{n(n + 1)}$

Chapter 11 Review Exercises

11.1

Change the following to percents. Express your answer to the nearest tenth of a percent.

1. $\dfrac{1}{8}$ **2.** $\dfrac{1}{3}$ **3.** $\dfrac{1}{2}$

4. 0.045   **5.** 0.0075   **6.** 1.256

Change the following percents to decimal numbers.

7. 37%   **8.** 13.8%   **9.** 0.237%

10. $\dfrac{2}{5}$%   **11.** $\dfrac{5}{6}$%   **12.** 0.00045%

13. In 1991 Corinne purchased an automobile for $18,000. The 1996 model of the same automobile cost $25,500. What is the percent increase in the price of the automobile?

14. Hugo the happy salesman earned a commission of $47,800 in 1993 and $42,870 in 1995. What is the percent change in his commission from 1993 to 1995?

In Exercises 15–19, write the statement as an equation and solve.

15. What percent of 80 is 30?

16. Forty-four is 20% of what number?

17. What is 17% of 670?

18. If your restaurant bill is $48.93 and the suggested tip is 15%, what is the tip on the meal?

19. If the number of people in your card club increased by 20% or 8, what was the original number of people in the club?

20. An organization had 75 members and increased the number of members to 95. What is the percent increase in the number of members?

11.2

In Exercises 21–24, find the missing quantity by using the simple interest formula.

21. $p = \$1800$, $r = 7\%$, $t = 30$ days, $i = ?$

22. $p = \$1575$, $r = ?$, $t = 100$ days, $i = \$41.56$

23. $p = ?$, $r = 8\frac{1}{2}\%$, $t = 3$ years, $i = \$114.75$

24. $p = \$5500$, $r = 11\frac{1}{2}\%$, $t = ?$, $i = \$316.25$

25. Bobby borrowed $4500 from the bank for 12 months. His mother cosigned the personal note. The rate of interest on the note was $11\frac{1}{2}\%$. How much did Bobby pay the bank on the date of maturity?

26. Sue borrowed $3500 from her bank for 180 days at a simple interest rate of 12%.
   a) How much interest did she pay for the use of the money?
   b) How much did she pay the bank on the date of maturity?

27. The Manbecks borrowed $4000 for 18 months from the bank, using stock as security. The bank discounted the loan at $11\frac{1}{2}\%$.
   a) How much interest did the Manbecks pay the bank for the use of the money?

b) How much did they receive from the bank?
   c) What was the actual rate of interest?

28. Charlotte borrowed $800 for 6 months from her bank, using her savings account as security. A bank rule limits the amount that can be borrowed in this manner to 85% of the amount in the borrower's savings account. The rate of interest is 2% higher than the interest rate being paid on the savings account. The current rate on the savings account is $5\frac{1}{2}\%$.
   a) What rate of interest will the bank charge for the loan?
   b) Find the amount that Charlotte must repay in 6 months.
   c) How much money must she have in her account in order to borrow $800?

29. How many dollars must you invest today to have $40,000 in 20 years? Assume that the money earns 7% interest compounded quarterly.

11.3

30. Determine the amount and the interest when $1500 is invested for 3 years at 5%
   a) compounded annually.
   b) compounded semiannually.
   c) compounded quarterly.

31. Donald deposited $7500 in a savings account that pays $5\frac{1}{4}\%$ interest compounded quarterly. What will be the total amount of money in the account 10 years from the day of deposit?

32. Determine the effective annual yield of an investment if the interest is compounded daily at an annual rate of 5.6%.

11.4

33. Wanda has a 48-month installment loan, with a fixed monthly payment of $193.75. The amount borrowed was $7500. Instead of making her 24th payment, Wanda is paying the remaining balance on the note.
   a) Determine the APR of the installment loan.
   b) How much interest will Wanda save, computed by the actuarial method?
   c) What is the total amount due on that day?

34. Reggie is buying a book collection that costs $4000. He is making a down payment of $500 and 24 monthly payments of $163.33. Instead of making his 12th payment, Reggie decides to pay the total remaining balance and terminate the loan.
   a) How much interest will Reggie save, computed by the rule of 78s?
   b) What is the total amount due on that day?

35. Jessica paid $3420 for a new wardrobe. She made a down payment of $860 and financed the balance on a 24-month fixed payment installment loan. The monthly payments are $119.47. Instead of making her 12th payment, Jessica decides to pay the total remaining balance and terminate the loan.

a) Determine the APR of the installment loan.

b) How much interest will Jessica save, computed by the actuarial method?

c) What is the total amount due on that day?

36. On November 1, the billing date, Lydia had a balance due of $485.75 on her credit card. The transactions during the month of November are:

November 4	Payment	$375.00
November 8	Charge: Airline ticket	370.00
November 21	Charge: Hotel bill	175.80
November 28	Charge: Clothing	184.75

a) Find the finance charge on December 1 by using the unpaid balance method. Assume that the interest rate is 1.3% per month.

b) Find the new account balance on December 1.

37. On March 5, the billing date, Janine had a balance due of $185.72 on her credit card. She is redecorating her apartment and had the following transactions.

March 8	Charge: Clothing	$ 85.75
March 10	Payment	75.00
March 15	Charge: Draperies	72.85
March 21	Charge: Chair	275.00

a) Find the finance charge on April 5 by using the average daily balance method. Assume that the interest rate is 1.4% per month.

b) Find the new account balance on April 5.

38. Patrick bought a new Dodge Spirit for $12,500. He was required to make a 25% down payment. He financed the balance with the dealer on a 48-month payment plan. The salesperson told him that the interest rate was 7.5% simple interest on the principal for the duration of the loan.

a) Find the amount of the down payment.

b) Find the amount to be financed.

c) Find the total interest paid.

d) Find the APR.

39. Joyce can buy a cross-country skiing outfit for $135. The store is offering the following terms: $35 down and 12 monthly payments of $8.95.

a) Find the interest paid.

b) Find the APR.

11.5

40. The Clars have decided to build a new house. The contractor quoted them a price of $135,700. The taxes on the house will be $3450 per year, and insurance will be $350 per year. They have applied for a conventional loan from a local bank. The bank is requiring a 25% down payment, and the interest rate on the loan is 9.5%. The Clars' annual income is $64,000. They have more than 10 monthly payments remaining on each of the following: $218 on a car, $120 on new furniture, and $190 on a camper. Determine

a) the required down payment.

b) their adjusted monthly income.

c) the maximum monthly payment the bank's loan officer believes that they can afford.

d) the monthly payment of principal and interest for a 40-year loan.

e) their total monthly payment, including insurance and taxes.

f) Do the Clars qualify for the mortgage?

41. The Bakunins purchased a home selling for $89,900 with a 15% down payment. The period of the mortgage is 30 years and the interest rate is 11.5%. Determine the

a) amount of the down payment.

b) monthly mortgage payment.

c) amount of the first payment that is applied to the principal.

d) total cost of the house.

e) total interest paid.

42. The Browns purchased a house for $105,000 with a down payment of $26,250. They obtained a 30-year adjustable-rate mortgage. The terms of the mortgage are as follows: The interest rate is based on the 6-month Treasury bill, the effective interest rate is 3.00% above the rate of the Treasury bill on the date of adjustment, the interest rate is adjusted every 6 months, the interest rate will not change more than 1% (up or down) when the interest rate is adjusted, the maximum interest rate that can be charged for the duration of the loan is 16%, there is no lower limit on the interest rate, the initial mortgage interest rate is 7.5%, and the monthly payment of interest and principal is adjusted annually. Determine the

a) initial monthly payment for interest and principal.

b) interest rate in 6 months if the interest rate on the Treasury bill at the time is 5.00%.

c) interest rate in 6 months if the interest rate on the Treasury bill at the time is 4.75%.

CHAPTER 11 TEST

In Exercises 1 and 2, find the missing quantity by using the simple interest formula.

1. $i = ?$, $p = \$700$, $r = 8\%$, $t = 8$ months

2. $i = \$144$, $p = \$600$, $r = 8\%$, $t = ?$

In Exercises 3 and 4, Greg borrowed \$5000 from a bank for 18 months. The rate of interest charged is 11.4%.

3. How much interest did he pay for the use of the money?

4. What is the amount he repaid to the bank on the due date of the loan?

In Exercises 5–7, a new computer sells for \$2350. To finance the computer through a bank, the bank will require a down payment of 15% and monthly payments of \$94.50 for 24 months.

5. How much money will the purchaser borrow from the bank?

6. What finance charge will the individual pay the bank?

7. What is the APR?

8. Jacques purchased a fishing boat for \$6750. He made a down payment of \$1550 and financed the balance with a 12-month fixed-payment installment note. Instead of making the sixth monthly payment of \$465.85 he decides to pay the remaining balance.
   a) How much interest will Jacques save, computed by the rule of 78s?
   b) What is the total amount due on that day?

9. The balance on the Bishops's credit card on May 8, their billing date, was \$378.50. For the period ending June 8 they had the following transactions.

May 13	Charge: Gifts	\$276.79
May 15	Payment	250.00
May 18	Charge: Books	144.85
May 29	Charge: Restaurant bill	55.63

   a) Find the finance charge on June 8 by the average daily balance method. Assume the interest rate is 1.3% per month.
   b) Find the new balance on June 8.

10. The O'Grady's credit card statement shows a balance due of \$878.25 on March 23, the billing date. For the period ending on April 23 they had the following transactions.

March 26	Charge: Auto supplies	\$ 95.89
March 30	Charge: Restaurant bill	68.76
April 3	Payment	450.00
April 15	Charge: Clothing	90.52
April 22	Charge: Lumber	450.85

   a) Find the finance charge due March 23 by using the unpaid balance method. Assume that the interest rate is 1.4% per month.
   b) Find the new account balance on April 23.

In Exercises 11 and 12, Brian signed a \$5400 note with interest at 12.5% for 90 days on August 1. Brian made a payment of \$3000 on September 15.

11. How much did he owe the bank on the date of maturity?

12. What total amount or interest did he pay on the loan?

In Exercises 13 and 14, compute the amount and the compound interest.

	Principal	Time	Rate	Compounded
13.	\$8600	4 years	12%	Quarterly
14.	\$4800	2 years	16%	Semiannually

In Exercises 15–22, the Klines decided to build a new house. The contractor quoted them a price of \$144,500, including the lot. The taxes on the house would be \$3200 per year, and insurance would cost \$450 per year. They have applied for a conventional loan from a bank. The bank is requiring a 15% down payment, and the interest rate is $10\frac{1}{2}\%$. The Klines' annual income is \$86,500. They have more than 10 monthly payments remaining on each of the following: \$220 for a car, \$175 for new furniture, and \$210 on a college education loan.

15. What is the required down payment?

16. Determine their adjusted monthly income.

17. What is the maximum monthly payment the bank's loan officer believes that the Klines can afford?

18. Determine the monthly payments of principal and interest for a 20-year loan.

19. Determine their total monthly payments including insurance and taxes.

20. Does the bank's loan officer believe that the Klines meet the requirements for the mortgage?

21. a) Find the total cost of the house after 20 years.
b) How much of the total cost is interest?

GROUP PROJECTS

Mortgage Loan

1. The Young family is purchasing a $130,000 house with a VA mortgage. The bank is offering them a 25-year mortgage with an interest rate of 9.5%. They have $20,000 invested that could be used for a down payment. Since they do not need a down payment, Mr. Young wants to keep the money invested. Mrs. Young feels that they should make a down payment of $20,000.
a) Determine the total cost of the house with no down payment.
b) Determine the total cost of the house if they make a down payment of $20,000.
c) Mr. Young feels that the $20,000 investment will have an annual rate of return of 10% compounded quarterly. Assuming that Mr. Young is right, calculate the value of the investment in 25 years. (See Section 11.3.)
d) If the Youngs use the $20,000 as a down payment, their monthly payments will decrease. Determine the difference of the monthly payments in parts (a) and (b).
e) Assume that the difference in monthly payments, part (d), is invested each month at a rate of 6% compounded monthly for 25 years. Determine the value of the investment in 25 years. (See Exercise 43 in Section 11.3.)
f) Use the information from parts (a)–(e) to analyze the problem. Would you recommend that the Youngs make the down payment of $20,000 and invest the difference in their monthly payments as in part (e), or that they do not make the down payment and keep the $20,000 invested as in part (c). Explain.

Credit Card Terms

2. With each credit card comes a credit agreement (or security agreement) that the cardholder must sign. Select two members of your group who have a major credit card (MasterCard, Visa, Discover, or American Express). If possible one of the credit cards should be a gold card. Obtain a copy of the credit agreement signed by each cardholder and answer questions (a)–(p).

a) What are the cardholder's responsibilities?
b) What is the cardholder's maximum line of credit?
c) What restrictions apply to the use of the credit card?
d) How many days after the billing date does the cardholder have to make a payment without being charged interest?
e) What is the minimum monthly payment required, and how is it determined?
f) What is the interest charged on purchases?
g) How does the bank determine when to start charging interest on purchases?
h) Is there an annual fee for the credit card? If so, what is it?
i) What late charge applies if payments are not made on time?
j) What information is given on the monthly statement?
k) How is the finance charge computed?
l) What other fees, if any, may the bank charge you?
m) If the card is lost, what is the responsibility of the cardholder?
n) If the card is lost, what is the liability of the cardholder?
o) What are the advantages of a gold card over a regular card?
p) In your opinion which of the two cards is more desirable? Explain.

Retirement Plans

3. Working men and women often have the benefit of contributing to various retirement plans. Such plans include IRAs, 403(b) plans, 401(k) plans, IRA–SEP plans, and Keogh Plans. Research each of these plans and answer questions (a)–(d).
a) What type of employees are eligible to use each type of plan?
b) What is the maximum annual contribution that can be made to each plan?
c) What type of investments can be made through each of these plans?
d) What are the tax advantages of each type of plan?

Chapter 12 Probability

"With his elbows propped, his head between his hands, he seemed to lose himself in the study of an abstruse problem in mathematics." So sat the character George Massy in the Joseph Conrad story *The End of the Tether* (1902). Massy had won the second-greatest prize in the Manila lottery. So obsessed was he with winning again that he pored over lists of winning numbers. "There was a hint there of a definite rule! He was afraid of missing some recondite principle in the overwhelming wealth of his material. What could it be?" It could be, and indeed is, a principle belonging to an area of mathematics known as probability.

The fascination that the fictional character George Massy displayed in a game of chance is shared in some degree by the millions of Americans each year who "gamble" on state lotteries or play bingo. If you play the lottery, your hope is to beat the odds and be the person with the winning numbers.

It is now legal to place some kind of bet in all but 2 of the 50 United States (Utah and Hawaii are the exceptions).

The laws of probability have applications in the science of genetics. The larger and more diverse the population, the greater is the probability that the species will have the characteristics necessary to adapt to changes in the environment. The cheetah, the world's fastest-running mammal, faces extinction because it lacks the genetic diversity necessary to survive disease. Once found worldwide, the species now lives wild in only a few areas of Africa.

Governments and private industry apply the counting principle of probability when they set the patterns of numbers and letters used to identify individuals. This application includes social security numbers, telephone numbers, and license plates.

Mathematicians of the sixteenth, seventeenth, and eighteenth centuries, not satisfied with leaving things to chance, invented the study of probability to use mathematics to determine the likelihood of an event (winning numbers, for example) occurring.

Although the rules of probability were first applied to gaming, they have many other applications. The quality of the food you eat, the pedigree of your cat or dog, the cost of your car insurance—all have to do with probability. In the business of insurance underwriting, the likelihood of an event such as an automobile accident, given the age, sex, and location of the driver, are facts used in determining the cost of the insurance. ■

Insurers use the laws of probability to set rates and define risk. One of the best-known insurers is Lloyd's of London. Lloyd's has the reputation of insuring just about anything, from the diamonds of actress Elizabeth Taylor to the riches of the Tut-Ankhamen exhibition.

12.1   THE NATURE OF PROBABILITY

A die (one of a pair of dice) contains six surfaces. Each surface contains a unique number of dots, from 1 to 6. The sum of the dots on opposite surfaces is 7.

History

Probability is presently used in many areas, including public finance, medicine, insurance, elections, manufacturing, educational tests and measurements, genet-ics, weather forecasting, investments, opinion polls, the natural sciences, games of chance, and a multitude of others. The study of probability originated from the study of games of chance. Archaeologists have found artifacts used in games of chance in Egypt dating from about 3000 B.C.

Mathematical problems relating to games of chance were studied by a number of mathematicians of the Renaissance. Italy's Girolamo Cardano (1501–1576) in his *Liber de Ludo Aleae* (book on the games of chance) presents one of the first systematic computations of probabilities. Although it is basically a gambler's manual, many consider it the first book ever written on probability. A short time later, two French mathematicians, Blaise Pascal (1623–1662) and Pierre de Fer-mat (1601–1665), worked together studying "the geometry of the die." In 1657 the Dutch mathematician Christian Huygens (1629–1695) published *De Ratio-ciniis in Luno Aleae* (on ratiocination in dice games), which contained the first documented reference to the concept of mathematical expectation (see Section 12.4). The Swiss mathematician Jacob Bernoulli (1654–1705), whom many con-sider the founder of probability theory, is said to have fused pure mathematics with the empirical methods used in statistical experiments. The works of Pierre-Simon de Laplace (1749–1827) dominated probability throughout the nineteenth century.

The Nature of Probability

Before we discuss the meaning of the word *probability* and learn how to calculate probabilities, we must introduce a few definitions.

An **experiment** is a controlled operation that yields a set of results.

The process by which medical researchers administer experimental drugs to pa-tients to determine their reaction is one type of experiment.

The possible results of an experiment are called its **outcomes**.

For example, the possible outcomes from administering an experimental drug may be a favorable reaction, no reaction, or an adverse reaction.

An **event** is a subcollection of the outcomes of an experiment.

For example, when a die is rolled, the event of rolling a number greater than 2 can be satisfied by any one of four outcomes—3, 4, 5, or 6. The event of rolling a 5 can be satisfied by only one outcome, the 5 itself. The event of rolling an even number can be satisfied by any of three outcomes—2, 4, or 6.

Probability is classified as either *empirical* (experimental) or *theoretical* (mathematical). Empirical probability is the relative frequency of occurrence of an event and is determined by actual observations of an experiment. Theoretical probability is determined through a study of the possible *outcomes* that can occur for the given experiment. We will indicate the probability of an event E by $P(E)$.

In this section we will briefly discuss empirical probability. The emphasis in the remaining sections is on theoretical probability. Following is the formula for computing empirical probability, or relative frequency.

Empirical Probability (Relative Frequency)

$$P(E) = \frac{\text{number of times event } E \text{ has occurred}}{\text{total number of times the experiment has been performed}}$$

The probability of an event, whether empirical or theoretical, will always be a number between 0 and 1, inclusive, and may be expressed as a decimal, a fraction, or a percent. An empirical probability of 0 indicates that the event has never occurred. An empirical probability of 1 indicates that the event has always occurred.

EXAMPLE 1

In 100 tosses of a fair coin, 58 landed heads up. Find the empirical probability of the coin landing heads up.

Solution: Let E be the event that the coin lands heads up. Then

$$P(E) = \frac{58}{100} = 0.58.$$

EXAMPLE 2

A pharmaceutical company is testing a new drug that is supposed to reduce cholesterol. The drug is given to 500 individuals with the following outcomes.

Cholesterol Reduced	Cholesterol Unchanged	Cholesterol Increased
307	120	73

If this drug is given to an individual, find the empirical probability that the cholesterol level is (a) reduced, (b) not affected, (c) increased.

Solution:
a) Let E be the event that the cholesterol level is reduced.

$$P(E) = \frac{307}{500} = 0.614$$

b) Let E be the event that the cholesterol level is not affected.

$$P(E) = \frac{120}{500} = 0.240$$

c) Let E be the event that the cholesterol level is increased.

$$P(E) = \frac{73}{500} = 0.146$$

Empirical probability is used when probabilities cannot be theoretically calculated. For example, life insurance companies use empirical probabilities to determine the chance of an individual in a certain profession, with certain risk factors, living to age 65.

Empirical Probability in Genetics

Using empirical probability, Gregor Mendel (1822–1884) developed the laws of heredity by crossbreeding different types of ''pure'' pea plants and observing the relative frequencies of the resulting offspring. These laws became the foundation for the study of genetics. For example, when he crossbred a pure yellow pea plant and a pure green pea plant, the resulting offspring (the first generation) were always yellow (see Fig. 12.1a). When he crossbred a pure round-seeded pea plant and a pure wrinkled-seeded pea plant, the resulting offspring (the first generation) were always round (Fig. 12.1b).

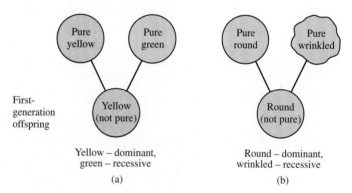

Figure 12.1

Traits such as yellow color and round seeds Mendel called **dominant** because they overcame or ''dominated'' the other trait. The green and wrinkled traits he labeled **recessive**.

Mendel then crossbred the offspring of the first generation. The resulting second generation had both the dominant and the recessive traits of their grandparents (see Fig. 12.2a and b). What's more, these traits always appeared in approximately a 3 to 1 ratio of dominant to recessive.

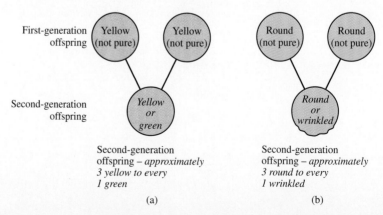

Figure 12.2

Table 12.1   Second Generation Offspring

Dominant Trait	Number with Dominant Trait	Recessive Trait	Number with Recessive Trait	Ratio of Dominant to Recessive	P (dominant trait)
Yellow seeds	6022	Green seeds	2001	3.01 to 1	$\frac{6022}{8023} = 0.75$
Round seeds	5474	Wrinkled seeds	1850	2.96 to 1	$\frac{5474}{7324} = 0.75$

Table 12.1 lists some of the results of Mendel's experiments with pea plants. Note that the ratio of dominant trait to recessive trait in the second generation offspring is about 3 to 1 for each of the experiments. The empirical probability of the dominant trait has also been calculated. How would you find the empirical probability of the recessive trait?

From his work Mendel concluded that the sex cells (now called gametes) of the pure yellow (dominant) pea plant carried some factor that caused the offspring

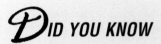

$\mathcal{D}$ID YOU KNOW

THE ROYAL DISEASE

$\mathcal{T}$he effect of genetic inheritance is dramatically demonstrated by the occurrence of hemophilia in some of the royal families of Europe. Great Britain's Queen Victoria (1819–1901) was the initial carrier. The disease was subsequently introduced into the royal lines of Prussia, Russia, and Spain through the marriages of her children.

Hemophilia is a disease that keeps blood from clotting. As a result, even a minor bruise or cut can be dangerous. It is also a recessive sex-linked disease. Females

Queen Victoria had 9 children, 26 grandchildren, and 34 great-grandchildren. Among them 1 son and 9 grandsons were hemophiliacs, and 2 daughters and 4 granddaughters were carriers of the gene for hemophilia. The genetic line of the present-day royal family is free of the disease.

have a second gene that enables the blood to clot, which blocks the effects of the recessive carrier gene; males do not. So even though both males and females are carriers, the disease afflicts only males. ∎

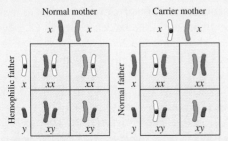

Chances of carrier daughter (xx): 100%   Chances of carrier daughter (xx): 50%
Chances of hemophilic son (xy): 0%   Chances of hemophilic son (xy): 50%

In humans, genes are located on 23 pairs of *chromosomes*. Each parent contributes one member of each pair to a child. The gene that affects blood clotting is carried on the X chromosome. Females have two X chromosomes; males have one X chromosome and one Y chromosome.

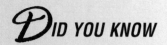

*I*f Ken Griffey, Jr., gets three hits in his first three at bats of the season, he is batting a thousand (1.000). But over the course of the 162 games of the season (with three or four at bats per game), his batting average will fall closer to .323 (his 1994 average). In 1994 out of 433 at bats, Griffey had 140 hits, an average above most players' but much less than 1.000. His batting average is a relative frequency (or empirical probability) of hits to at bats. It is only the long-term average that we take seriously because it is based on the law of large numbers. ■

"The laws of probability, so true in general, so fallacious in particular."
Edward Gibbon, 1796

to be yellow, and the gametes of the green variety had a variant factor that "induced the development of green plants." In 1909 the Danish geneticist W. Johannsen called these factors "genes." Mendel's work led to the understanding that each pea plant contains two genes for color, one that comes from the mother and the other from the father. If the two genes are alike, for instance both for yellow plants or both for green plants, the plant will be that color. If the genes for color are different the plant will grow the color of the dominant gene. Thus if one parent contributes a gene for the plant to be yellow (dominant) and the other parent contributes a gene for the plant to be green (recessive), the plant will be yellow.

The Law of Large Numbers

Most of us accept the fact that if a "fair coin" is tossed many, many times, it will land heads up approximately half of the time. Intuitively, we can guess that the probability that a fair coin will land heads up is $\frac{1}{2}$. Does this mean that if a coin is tossed twice, it will land heads up exactly once? If a fair coin is tossed 10 times, will there necessarily be five heads? The answer is clearly no. What then does it mean when we state that the probability that a fair coin will land heads up is $\frac{1}{2}$? To answer this question, let's examine Table 12.2, which shows what may occur when a fair coin is tossed a given number of times.

Table 12.2

Number of Tosses	Expected Number of Heads	Actual Number of Heads Observed	Relative Frequency of Heads
10	5	4	$\frac{4}{10} = 0.4$
100	50	43	$\frac{43}{100} = 0.43$
1,000	500	540	$\frac{540}{1,000} = 0.54$
10,000	5,000	4,852	$\frac{4,852}{10,000} = 0.4852$
100,000	50,000	49,770	$\frac{49,770}{100,000} = 0.49770$

The last column of Table 12.2, the relative frequency of heads, is a ratio of the number of heads observed to the total number of tosses of the coin. The relative frequency is the empirical probability, as defined earlier. Note that as the number of tosses increases, the relative frequency of heads gets closer and closer to $\frac{1}{2}$, or 0.5, which is what we expect.

The nature of probability is summarized by the law of large numbers.

> The **law of large numbers** states that probability statements apply in practice to a large number of trials—not to a single trial. It is the relative frequency over the long run that is accurately predictable, not individual events or precise totals.

What does it mean to say that the probability of rolling a 2 on a die is $\frac{1}{6}$? It means that over the long run, on the average, one of every six rolls will result in a 2.

Section 12.1 Exercises

1. What is an experiment?

2. **a)** What are outcomes of an experiment?
   b) What is an event?

3. What is empirical probability and how is empirical probability determined?

4. What are theoretical probabilities based on?

5. Explain in your own words the law of large numbers.

6. Explain in your own words why empirical probabilities are used in determining premiums for life insurance policies.

7. A probability statement is often interpreted differently by different people. Consider the statement ''The National Weather Service predicts that the probability of rain today in the Denver, Colorado, area is 70%.'' Below are five possible interpretations of that statement. Which, if any, do you think is its actual meaning?
   a) In 70% of the Denver area there will be rain.
   b) There is a 70% chance that at least any one point in the Denver area will receive rain.
   c) It has rained on 70% of the days with similar weather conditions in the Denver area.
   d) There is a 70% chance that a specific location in the Denver area will receive rain.
   e) There is a 70% chance that the entire Denver area will receive rain.

8. The probability of rolling a 4 on a die is $\frac{1}{6}$. Does this probability mean that, if a die is rolled six times, one 4 will appear? If not, what does it mean?

9. In order to determine premiums, life insurance companies must compute the probable date of death. On the basis of a great deal of research Mr. Bennett, age 35, is expected to live another 42.34 years. Does this determination mean that Mr. Bennett will live until he is 77.34 years old? If not, what does it mean?

10. **a)** Explain how you would find the empirical probability of rolling a 5 on a die.
    b) What do you believe is the empirical probability of rolling a 5?
    c) Find the empirical probability of rolling a 5 by rolling a die 50 times.

11. **a)** Explain how you would find the empirical probability of rolling an even number on a die.
    b) What do you believe is the empirical probability of rolling an even number?
    c) Find the empirical probability of rolling an even number by rolling a die 50 times.

12. Roll a die 50 times and record the results. Find the empirical probability of rolling (a) a 1 and (b) a 6. (c) Does the probability of rolling a 1 appear to be the same as the probability of rolling a 6?

13. Roll a pair of dice 60 times and record the sums. Compute the empirical probability of rolling a sum of (a) 2 and (b) 7. (c) Does the probability of rolling a sum of 2 appear to be the same as the probability of rolling a sum of 7?

14. Toss two coins 50 times and record the number of times exactly one head was obtained. Compute the empirical probability of tossing exactly one head.

15. The last 20 birds that fed at the Zwicks' bird feeder were 10 finches, 7 cardinals, and 3 blue jays. Use this information to determine the empirical probability that the next bird to feed from the feeder is a
    a) finch.
    b) cardinal.
    c) blue jay.

16. In a sample of 50,000 births, 6 were found to have Prader-Willi syndrome. Find the empirical probability that Mrs. Gardina's first child will be born with this syndrome.

17. Of 100 Harris carrot seeds planted, 92 germinated and produced carrots.
    a) Find the empirical probability of germination of Harris carrot seeds.
    b) How many of these seeds should you plant if you want to grow 368 carrots?

18. A certain community recorded 3000 births last year, of which 1450 were females. What is the empirical probability that the next child born in the community will be
    a) female?
    b) male?

19. The following chart shows the gain made by the Dow-Jones Industrial Average (DJIA) in each year that ends in a 5 since records have been kept.

YEARS ENDING IN 5	
Year	**DJIA return**
1885	27.7%
1895	2.3%
1905	38.2%
1915	86.5%
1925	30.0%
1935	38.5%
1945	26.6%
1955	20.8%
1965	10.9%
1975	38.3%
1985	27.7%
1995	36.8%

a) What is the empirical probability that the DJIA will increase in a year ending in 5?

b) Is it possible that the DJIA could have a loss in a year ending in 5? Explain.

20. The following circle graphs were obtained in a CNN/Gallup poll in April 1995. What is the empirical probability that a person selected at random chose ''protecting the environment'' in

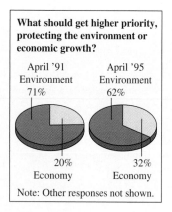

What should get higher priority, protecting the environment or economic growth?

April '91
Environment
71%

April '95
Environment
62%

20%
Economy

32%
Economy

Note: Other responses not shown.

a) 1991?

b) 1995?

c) If the 1995 survey contained the responses of 1007 adults, estimate the number who selected ''protecting the environment'' in 1995.

21. The pattern of hits shown on the target resulted from a marksman firing 20 rounds. For a single shot

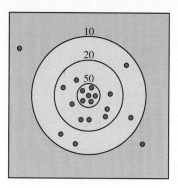

a) find the empirical probability that the marksman hits the 50-point bulls-eye (the center of the target).

b) find the empirical probability that the marksman does not hit the bulls-eye.

c) find the empirical probability that the marksman scores at least 20 points.

d) find the empirical probability that the marksman does not score any points (the area outside the large circle).

22. The following chart was taken from the April 30, 1995, *Sun-Sentinel* (Orlando, Florida) newspaper.

LOTTERY ANALYSIS
Lotto

Top 10 Lotto numbers for the past 52 weeks, not including Saturday night's drawing.

	No.	Freq.		No.	Freq.
1.	34	25%	6.	29	17%
2.	18	19%	7.	02	17%
3.	09	19%	8.	48	15%
4.	47	17%	9.	39	15%
5.	32	17%	10.	38	15%

a) Based on the lotto drawings over the past 52 weeks, find the empirical probability that the number 34 will be selected in next week's lotto drawing.

b) How many times was the number 34 drawn in the past 52 weeks?

23. A faulty candy machine next to the classroom accepts dollar bills. A student obtains the information to form the following chart by observing fellow students' experience with the machine.

Outcome	Frequency of Occurrence
Gives the candy and returns the proper change	25
Gives the candy but does not return the proper change	8
Does not give the candy and keeps the money	7
	40

a) If the machine is tried a 41st time, what is the empirical probability that it will work properly?

b) What is the empirical probability of obtaining candy from the machine?

24. In a survey, 1600 people were asked to rate the president's performance as good, fair, or poor. The results were as follows.

Number of People	Rating
380	Good
602	Fair
490	Poor
128	No opinion

If this sample is representative of the general population, find the empirical probability that the next person surveyed

a) rates the president's performance as good.

b) rates the president's performance as poor.

c) has no opinion.

25. An experimental serum was injected into 500 guinea pigs. Initially, 150 of the guinea pigs had circular cells, 250 had elliptical cells, and 100 had irregularly shaped cells. After the serum was injected, none of the guinea pigs with circular cells were affected, 50 with elliptical cells were affected, and all of those with irregular cells were affected. Find the empirical probability that a guinea pig with (a) circular cells, (b) elliptical cells, and (c) irregular cells will be affected by injection of the serum.

26. Jim finds an irregularly shaped five-sided rock. He labels each side and tosses the rock 100 times. The results of his tosses are shown in the table. Find the empirical probability that the rock will land on side 4 if tossed again.

Side	1	2	3	4	5
Frequency	32	18	15	13	22

27. In one of Mendel's experiments, he crossbred nonpure purple flower pea plants. These purple pea plants had two traits for flowers, purple (dominant) and white (recessive). The result of this crossbreeding was 705

second-generation plants with purple flowers and 224 second-generation plants with white flowers. Find the empirical probability of a second-generation plant having

a) white flowers.

b) purple flowers.

28. In another experiment Mendel crossbred nonpure tall pea plants. As a result the second-generation offspring were 787 tall plants and 277 short plants. Find the empirical probability of a second-generation plant being (a) tall and (b) short.

29. In the United States more male babies are born than female. In 1991, 2,101,518 males were born and 2,009,389 females were born. Find the empirical probability of an individual being born (a) male and (b) female.

Problem Solving/Group Activities

30. a) Do you believe that the word *a* or the word *the* is used more frequently?

b) Design an experiment to determine the empirical probabilities (or relative frequencies) of the words *a* and *the* appearing in a book or newspaper article.

c) Perform the experiment in part (b) and determine the empirical probabilities.

d) Which word appears to occur more frequently?

31. Can people selected at random distinguish Coke from Pepsi? Which do they prefer?

a) Design an experiment to determine the empirical probability that a person selected at random can select Coke when given samples of both Coke and Pepsi.

b) Perform the experiment in part (a) and determine the probability.

c) Determine the empirical probability that a person selected at random will prefer Coke over Pepsi.

Research Activities

32. Write a paper on how insurance companies use empirical probabilities in determining insurance premiums. An insurance agent may be able to direct you to a source of information.

33. Write a paper on how Gregor Mendel's use of empirical probability led to the development of the science of genetics. You may want to check with a biology professor to determine references to use.

12.2 THEORETICAL PROBABILITY

To be able to do the problems in this section and the remainder of the chapter, you must have a thorough understanding of fractions. If you have forgotten how to use fractions, we strongly suggest that you review Section 5.3 before beginning this section.

Should you spend the 32 cents for a stamp to return the *Reader's Digest*

Sweepstakes ticket? What are your chances of winning a lottery? If you go to a carnival or bazaar, which games provide the greatest chance of winning? These and similar questions can be answered once you have an understanding of theoretical probability. *In the remainder of this chapter, the word* probability *will refer to theoretical probability.*

Recall from Section 12.1 that the results of an experiment are called outcomes. When you roll a die and observe the number of points that face up, the possible outcomes are 1, 2, 3, 4, 5, and 6. It is equally likely that you will roll any one of the possible numbers.

> If each outcome of an experiment has the same chance of occurring as any other outcome, they are said to be **equally likely outcomes**.

Can you think of a second set of equally likely outcomes when a die is rolled? An odd number is as likely to be rolled as an even number. Therefore odd and even are another set of equally likely outcomes.

If an event has *equally likely outcomes*, the probability of event E, symbolized by $P(E)$, may be calculated with the following formula.

"It is remarkable that a science which began with the consideration of games of chance should have become the most important object of human knowledge."
Marquis de Laplace (1749–1827)

Probability

$$P(E) = \frac{\text{number of outcomes favorable to } E}{\text{total number of possible outcomes}}$$

Example 1 illustrates how to use this formula.

EXAMPLE 1

A die is rolled. Find the probability of rolling

a) a 5.   **b)** an odd number.   **c)** a number greater than 4.
d) a 7.   **e)** a number less than 7.

Solution:

a) There are six possible equally likely outcomes: 1, 2, 3, 4, 5, and 6. The event of rolling a 5 can occur in only one way.

$$P(5) = \frac{\text{number of outcomes that will result in a 5}}{\text{total number of possible outcomes}} = \frac{1}{6}$$

b) The event of rolling an odd number can occur in three ways (1, 3, 5).

$$P(\text{odd number}) = \frac{\text{number of outcomes that result in an odd number}}{\text{total number of possible outcomes}}$$

$$= \frac{3}{6} = \frac{1}{2}$$

c) Two numbers are greater than 4, namely, 5 and 6.

$$P(\text{number greater than 4}) = \frac{2}{6} = \frac{1}{3}$$

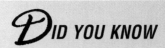

d) No outcomes will result in a 7. Thus the event cannot occur and the probability is 0.

$$P(7) = \frac{0}{6} = 0$$

e) All the outcomes 1 through 6 are less than 7. Thus the event must occur and the probability is 1.

$$P(\text{number less than } 7) = \frac{6}{6} = 1$$

Four important facts about probability follow.

Important Facts

1. The probability of an event that cannot occur is 0.

2. The probability of an event that must occur is 1.

3. Every probability will be a number between 0 and 1 inclusive; that is, $0 \le P(E) \le 1$.

4. The sum of the probabilities of all possible outcomes of an experiment is 1.

EXAMPLE 2

The names of 15 birds are listed in Table 12.3. Each name is listed on a slip of paper, and the 15 slips are deposited in a bag. One slip is to be selected at random from the bag. Find the probability that the slip containing

a) a sparrow is selected.

b) a bird that has a high attractiveness to peanut kernels is selected.

Table 12.3

Bird	Peanut Kernels	Cracked Corn	Black Striped Sunflower Seeds
American goldfinch	L	L	H
Blue jay	H	M	H
Chickadee	M	L	H
Common grackle	M	H	H
Evening grosbeak	L	L	H
House finch	M	L	H
House sparrow	L	M	M
Mourning dove	L	M	M
Northern cardinal	L	L	H
Purple finch	L	L	H
Scrub jay	H	L	H
Song sparrow	L	L	M
Tufted titmouse	H	L	H
White-crowned sparrow	H	M	H
White-throated sparrow	H	H	H

H = high attractiveness; M = medium attractiveness; L = low attractiveness.
Source: How to Attract Birds *(Ortho Books)*

c) a bird that has a low attractiveness to peanut kernels, a low attractiveness to cracked corn, *and* a high attractiveness to black striped sunflower seeds is selected.

d) a bird that has a high attractiveness to either peanut kernels *or* cracked corn (or both) is selected.

Solution:

a) Four of the 15 birds listed are sparrows.

$$P(\text{sparrow}) = \frac{4}{15}$$

b) Five of the 15 birds listed have a high attractiveness to peanut kernels.

$$P(\text{high attractiveness to peanut kernels}) = \frac{5}{15} = \frac{1}{3}$$

c) Reading across the rows reveals that 4 birds have a low attractiveness to peanut kernels and cracked corn and a high attractiveness to black striped sunflower seeds (American goldfinch, evening grosbeak, northern cardinal, and purple finch).

$$P(\text{low to peanuts, low to corn, and high to black sunflower seeds}) = \frac{4}{15}$$

d) The birds that have a high attractiveness to either peanut kernels or to cracked corn (or both) are the bluejay, common grackle, jay scrub, tufted titmouse, white-crowned sparrow, and white-throated sparrow.

$$P(\text{high to peanut kernels or cracked corn}) = \frac{6}{15} = \frac{2}{5}$$

In any experiment an event must either occur or not occur. *The sum of the probability that an event will occur and the probability that it will not occur is 1.* Thus for any event A we conclude that

$$P(A) + P(\text{not } A) = 1$$
$$\text{or} \quad P(\text{not } A) = 1 - P(A).$$

For example, if the probability that event A will occur is $\frac{5}{12}$, the probability that event A will not occur is $1 - \frac{5}{12}$, or $\frac{7}{12}$. Similarly, if the probability that event A will not occur is 0.3, the probability that event A will occur is $1 - 0.3 = 0.7$ or $\frac{7}{10}$. We make use of this concept in Example 3.

EXAMPLE 3

A deck of 52 playing cards is shown. It consists of four suits: spades, diamonds, hearts, and clubs. Each suit has 13 cards, including numbered cards ace (1) through 10 and three picture (or face) cards the jack, the queen and the king. Hearts and diamonds are red; clubs and spades are black. There are 12 picture cards, consisting of 4 jacks, 4 queens, and 4 kings. One card is to be selected

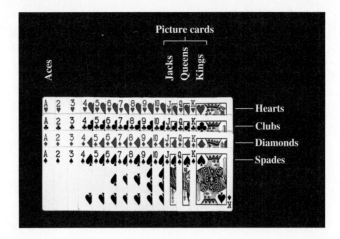

at random from the deck of cards. Find the probability that the card selected is

a) a 3.   **b)** not a 3.

c) a spade.   **d)** a jack *or* queen *or* king (a picture card).

e) a heart *and* a club. **f)** a card greater than 6 *and* less than 9.

Solution:

a) There are four 3s in a deck of 52 cards.

$$P(3) = \frac{4}{52} = \frac{1}{13}$$

b) $P(\text{not a } 3) = 1 - P(3) = 1 - \frac{1}{13} = \frac{12}{13}$

This probability could also have been found by noting that there are 48 cards that are not 3s in a deck of 52 cards.

$$P(\text{not a } 3) = \frac{48}{52} = \frac{12}{13}$$

c) There are 13 spades in the deck.

$$P(\text{spade}) = \frac{13}{52} = \frac{1}{4}$$

d) There are 4 jacks, 4 queens, and 4 kings, or a total of 12 picture cards.

$$P(\text{jack } or \text{ queen } or \text{ king}) = \frac{12}{52} = \frac{3}{13}$$

e) The word *and* means that *both* events must occur. Since it is not possible to select one card that is both a heart and a club, the probability is 0.

$$P(\text{heart and club}) = \frac{0}{52} = 0$$

f) The cards that are both greater than 6 and less than 9 are 7s and 8s. There are four 7s and four 8s, or a total of eight cards.

$$P(\text{greater than 6 } and \text{ less than 9}) = \frac{8}{52} = \frac{2}{13}$$

Section 12.2 Exercises

1. What are equally likely outcomes?

2. Explain in your own words how to find the theoretical probability of an event.

3. State the relationship that exists for $P(A)$ and $P(\text{not } A)$.

4. How many of each of the following are there in a standard deck of cards?
   a) Total cards
   b) Hearts
   c) Red cards
   d) Fives
   e) Black cards
   f) Picture cards
   g) Aces
   h) Queens

5. a) Using the definition of probability, explain in your own words why the probability of an event that cannot occur is 0.
   b) Using the definition of probability, explain in your own words why the probability of an event that must occur is 1.

6. a) Between what two numbers (inclusively) will all probabilities lie?
   b) What is the sum of all the probabilities of all possible outcomes of an experiment?

7. A multiple choice test has five possible answers for each question.
   a) If you guess at an answer, what is the probability that you select the correct answer for one particular question?
   b) If you eliminate one of the five possible answers and guess from the remaining possibilities, what is the probability that you select the correct answer to that question?

8. A TV remote has keys for channels 0 through 9. If you select one key at random,
   a) what is the probability that you press channel 3?
   b) what is the probability that you press a key for an even number?
   c) what is the probability that you press a key for a number less than 7?

In Exercises 9–17, one card is selected at random from a deck of cards. Find the probability that the card selected is

9. a 9.

10. a 9 or a 10.

11. not a 9.

12. the ace of spades.

13. a red card.

14. a red card or a black card.

15. a red card and a black card.

16. a card greater than 6 and less than 10.

17. a king and a diamond.

In Exercises 18–22, assume that the spinner cannot land on a line. Find the probability that the spinner lands on (a) red, (b) blue, (c) green.

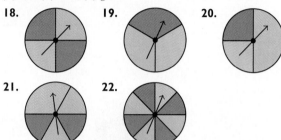

18.   19.   20.

21.   22.

In Exercises 23–26, a bin contains 100 batteries (all size D). Forty are Eveready, 35 are Duracell, 20 are Kodak, and 5 are Rayovac. One battery is selected at random from the bin. Find the probability that the battery selected is a

23. Duracell.

24. Duracell or Eveready.

25. Duracell, Eveready, or Kodak.

26. Fuji.

In Exercises 27–30, use the small replica of the Wheel of Fortune.

If the wheel is spun at random, find the probability of the sector indicated stopping under the pointer.

27. $200

28. Lose a turn or Bankrupt

29. A number greater than $500

30. $2500 or Surprise

In Exercises 31–34, 15 CDs including five country CDs, four rhythm and blues CDs, two oldies CDs, two classical CDs and two jazz CDs are each wrapped in plain brown paper. Ms. Dunn selects one CD at random. Find the probability that it is

31. a country CD.

32. not a country CD.

33. a jazz or classical CD.

34. an oldies or a rhythm and blues or a country CD.

In Exercises 35–38, a traffic light is red for 30 sec, yellow for 5 sec, and green for 40 sec. What is the probability that, when you reach the light,

35. the light is red.

36. the light is yellow.

37. the light is not red.

38. the light is not red or yellow.

In Exercises 39–44, each individual letter of the word ''pfeffernuesse'' is placed on a piece of paper, and all 13 pieces of paper are placed in a hat. If one letter is selected at random from the hat find the probability that

39. the letter *f* is selected.

40. the letter *f* is not selected.

41. a vowel is selected.

42. the letter *f* or *e* is selected.

43. the letter *e* is not selected.

44. the letter *g* is selected.

In Exercises 45–48, 12 pencils are in a cup. Five are sharpened and have erasers, four are sharpened but do not have erasers, and three are not sharpened but have erasers. If one pencil is selected at random from the cup, find the probability that the pencil

45. is sharpened.

46. is not sharpened.

47. has an eraser.

48. does not have an eraser.

In Exercises 49–52, the minimum age required to obtain a regular driver's license in 1993 (without restrictions) in the 50 states and the District of Columbia is summarized as follows.

Minimum Age for License*	Number of Locations
15	1
16	21
17	5
18	23
21	1

*In some states a license may be obtained at a younger age if an applicant completes a driver education course.

If one of these locations is selected at random, find the probability that the minimum age needed for a regular driver's license is

49. 15

50. less than 17.

51. 16 or 17.

52. greater than or equal to 16.

In Exercises 53–56, a dart is thrown randomly and sticks on the circular dart board shown.

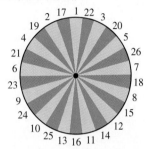

Assuming that the dart cannot land on the black area or on a border between colors, find the probability that the dart lands on

53. the area marked 25.

54. a green area.

55. an area marked with a number that is greater than or equal to 24.

56. an area marked with a number greater than 6 and less than or equal to 9.

In Exercises 57–62, refer to the following table, which contains information about a psychology class.

	Freshmen	Sophomores	Total
Females	8	6	14
Males	12	9	21
Total	20	15	35

If one individual is selected at random from this class, find the probability that the individual selected is a

57. female.　　　　　　**58.** male.

59. freshman.　　　　　**60.** sophomore.

61. female sophomore.　**62.** male freshman.

In Exercises 63–66, refer to the table on page 588, which summarizes the travel of clients at the World Travel Agency.

	Domestic Travel	Foreign Travel	Total
Cost under $2000	20	12	32
Cost $2000 or more	8	10	18
Total	28	22	50

If one client is selected at random, find the probability that the

63. client traveled within the United States.

64. client's cost was under $2000.

65. client's cost was under $2000 on foreign travel.

66. client's cost was $2000 or more on domestic travel.

In Exercises 67–71, a bean bag is randomly thrown onto the square table and does not touch a line.

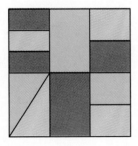

Find the probability that the bean bag lands on

67. a red area.

68. a green area.

69. a blue area.

70. a red or green area.

71. a green or blue area.

Before working Exercises 72–73, reread the material on genetics in Section 12.1.

72. Cystic fibrosis is an inherited disease that occurs in about 1 in every 2000 Caucasian births in North America and about 1 of every 250,000 non-Caucasian births. Let's denote the cystic fibrosis gene as *c* and a disease-free gene as *C*. Since the disease-free gene is dominant, only a person with *cc* genes will have the disease. A person who has *Cc* genes is a carrier of cystic fibrosis but does not actually have the disease. If one parent has *CC* genes and the other parent has *cc* genes, find the probability that

  a) an offspring will inherit cystic fibrosis, that is, *cc* genes.

  b) an offspring will be a carrier of cystic fibrosis but not contract the disease.

73. Sickle cell anemia is an inherited disease that occurs in about 1 in every 500 black baby births and about 1 in

every 160,000 non-black baby births. Unlike cystic fibrosis, in which the cystic fibrosis gene is recessive, sickle cell anemia is *codominant*. In other words, a person inheriting two sickle cell genes will have sickle cell anemia, whereas a person inheriting only one of the sickle cell genes will have a mild version of sickle cell anemia, called *sickle cell trait*. Let's call the disease-free gene s_1 and the sickle cell gene s_2. If both parents have $s_1 s_2$ genes, determine the probability that

  a) an offspring will have sickle cell anemia.

  b) an offspring will have the sickle cell trait.

  c) an offspring will have neither sickle cell anemia nor the sickle cell trait.

Problem Solving/Group Activities

In Exercises 74–76, the solutions involve material that we will discuss in later sections of the chapter. Try to solve them before reading ahead.

74. Consider the figure accompanying Exercises 53–56. If one dart is thrown, find the probability that

  a) the dart sticks on a red area.

  b) the dart sticks on an even number.

  c) the dart sticks on a red and even numbered area.

  d) the dart sticks on a red area or an even numbered area. (To be discussed in Section 12.6)

  e) Using the probabilities found in parts (a)–(d), can you discover a formula for finding *P*(dart lands on red *or* an even numbered area)?

75. A bottle contains two red and two green marbles, and a second bottle also contains two red and two green marbles. If you select one marble at random from each bottle, find the probability (to be discussed in Section 12.6) that you obtain

  a) two red marbles.

  b) two green marbles.

  c) a red marble from the first bottle and a green marble from the second bottle.

76. Consider Table 12.3. Suppose you are told that one bird's name was selected from the birds listed and that the bird selected has a low attractiveness to peanut kernels. Find the probability (to be discussed in Section 12.7) that

  a) the bird is a sparrow.

  b) the bird has a high attractiveness to cracked corn.

  c) the bird has a high attractiveness to black striped sunflower seeds.

Research Activity

77. On page 574 we briefly discuss Jacob Bernoulli. The Bernoulli family produced several prominent mathematicians, including Jacob I, Johann I and Daniel. Write a paper on the Bernoulli family, indicating some of the accomplishments of each of the three Bernoullis named and their relationship. Indicate which Bernoulli the Bernoulli numbers are named after, which Bernoulli the Bernoulli theorem in statistics is named after, and which Bernoulli the Bernoulli theorem of fluid dynamics is named after. References include encyclopedias and history of mathematics books.

12.3   ODDS

The odds against winning a lottery are 89,000 to 1; the odds against being audited by the IRS this year are 50 to 1. We see the word *odds* daily in newspapers and magazines and often use it ourselves. Yet there is a widespread misunderstanding of its meaning. In this section we will explain the meaning of odds.

The odds given at horse races, at craps, and at all gambling games in Las Vegas and other casinos throughout the world are always *odds against* unless they are otherwise specified. The **odds against** an event is a ratio of the probability that the event will fail to occur (failure) to the probability that the event will occur (success). Thus *in order to find odds you must first know or determine the probability of success and the probability of failure.*

> "We figured the odds as best we could, and then we rolled the dice."
> Jimmy Carter, on his decision to run for president

$$\text{Odds against event} = \frac{P(\text{event fails to occur})}{P(\text{event occurs})} = \frac{P(\text{failure})}{P(\text{success})}$$

EXAMPLE 1

Find the odds against rolling a 3 on one roll of a die.

Solution: Before we can determine the odds, we must first determine the probability of rolling a 3 (success) and the probability of not rolling a 3 (failure). When a die is rolled there are six possible outcomes: 1, 2, 3, 4, 5, and 6.

$$P(\text{rolls a 3}) = \frac{1}{6} \qquad P(\text{fails to roll a 3}) = \frac{5}{6}$$

Now that we know the probabilities of success and failure, we can determine the odds against rolling a 3.

$$\text{Odds against rolling a 3} = \frac{P(\text{fails to roll a 3})}{P(\text{rolls a 3})}$$

$$= \frac{5/6}{1/6} = \frac{5}{\cancel{6}} \cdot \frac{\cancel{6}}{1} = \frac{5}{1}$$

The ratio 5/1 is commonly written as 5 : 1 and is read "5 to 1." Thus the odds against rolling a 3 are 5 to 1.

Note: The denominators of the probabilities in an odds problem will always divide out.

In Example 1 consider the possible outcomes of the die—1, 2, 3, 4, 5, 6. Over the long run, one of every six rolls will result in a 3, and five of every six rolls will result in a number other than 3. Therefore, for each dollar bet in favor of the rolling of a 3, five dollars should be bet against the rolling of a 3 if it is to be a fair game. The person betting in favor of the rolling of a 3 will either lose one dollar (if a number other than a 3 is rolled) or win five dollars (if a 3 is rolled). The person betting against the rolling of a 3 will either win one dollar (if a number other than a 3 is rolled) or lose five dollars (if a 3 is rolled). If this game is played for a long enough period, each player theoretically will break even.

EXAMPLE 2

An estimated 3 of every 10 new businesses in a specific location go bankrupt within their first year of operation. If you open a new business there, what are the odds against its going bankrupt?

Solution: The probability that your business will go bankrupt is $\frac{3}{10}$. Therefore the probability that your business will not go bankrupt is $1 - \frac{3}{10}$, or $\frac{7}{10}$.

$$\frac{\text{Odds against your}}{\text{business going bankrupt}} = \frac{P(\text{the business fails to go bankrupt})}{P(\text{the business goes bankrupt})}$$

$$= \frac{7/10}{3/10} = \frac{7}{10} \cdot \frac{10}{3} = \frac{7}{3} \text{ or } 7:3$$

Although odds are generally given against an event, at times they may be given in favor of an event. The **odds in favor of** an event are expressed as a ratio of the probability that the event will occur to the probability that the event will fail to occur.

$$\text{Odds in favor of event} = \frac{P(\text{event occurs})}{P(\text{event fails to occur})} = \frac{P(\text{success})}{P(\text{failure})}$$

If the odds *against* an event are $a:b$, the odds *in favor of* the event will be $b:a$.

EXAMPLE 3

Two percent of the U.S. population is born gifted (IQ of 130 or above). Find

a) the odds against an individual's being born gifted.

b) the odds in favor of an individual's being born gifted.

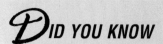

𝐵efore you buy your next lottery ticket, consider the odds. Compare the odds of being struck by lightning (approximately 700,000 to 1) with your chances of winning a lottery. In Colorado, for example, where you must select 6 of 42 numbers, the odds against winning the lottery are about 5.2 million to 1. Thus you are about 7.5 times more likely to be struck by lightning than to win the lottery. Given those odds, consider the case of a Colorado man, Don Whittman, 29, who won half of a $4 million jackpot on December 23, 1989, and won a $2.2 million jackpot again on October 22, 1991! ∎

Solution:

a) The probability of being born gifted is 0.02 or 2/100. The probability of not being born gifted is therefore $1 - (2/100) = 98/100$.

$$\text{Odds against being born gifted} = \frac{P(\text{not born gifted})}{P(\text{born gifted})}$$

$$= \frac{98/100}{2/100}$$

$$= \frac{\overset{49}{\cancel{98}}}{\underset{1}{\cancel{100}}} \cdot \frac{\overset{1}{\cancel{100}}}{\underset{1}{\cancel{2}}} = \frac{49}{1} \text{ or } 49:1$$

b) The odds in favor of being born gifted are $1:49$. ▲

Finding Probabilities from Odds

When odds are given, either in favor of or against a particular event, it is possible to determine the probabilities of that event. The denominators of the probabilities are found by adding the numbers in the odds statement. The numerators of the probabilities are the numbers given in the odds statements.

EXAMPLE 4

The odds against Jennifer being admitted to the college of her choice are $9:2$. Find the probability that (a) Jennifer is admitted and (b) Jennifer is not admitted.

Solution:

a) We have been given odds against and have been asked to find probabilities.

$$\text{Odds against being admitted} = \frac{P(\text{fails to be admitted})}{P(\text{is admitted})}$$

Since the odds statement is $9:2$, the denominators of both the probability of success and failure must be $9 + 2$ or 11. To get the odds ratio of $9:2$ the probabilities must be $\frac{9}{11}$ and $\frac{2}{11}$. Since odds against is a ratio of failure to success, the $\frac{9}{11}$ and $\frac{2}{11}$ represent the probabilities of failure and success, respectively. Thus the probability that Jennifer is admitted (success) is $\frac{2}{11}$.

b) The probability that Jennifer is not admitted (failure) is $\frac{9}{11}$. ▲

Odds and probability statements are sometimes stated incorrectly. For example, consider the statement, ''The odds of being selected to represent the district are 1 in 5.'' Odds are given using the word *to*, not *in*. Thus there is a mistake in this statement. The correct statement might be, ''The odds of being selected to represent the district are 1 to 5'' or ''The probability of being selected to represent the district is 1 in 5.'' Without additional information it is not possible to tell which is the correct interpretation.

Section 12.3 Exercises

1. Explain in your own words how to determine the odds against an event.

2. Explain in your own words how to determine the odds in favor of an event.

3. Which odds are generally quoted, odds against or odds in favor?

4. Explain how to determine probabilities when you are given an odds statement.

5. The odds in favor of winning the door prize are 2 to 15. Find the odds against winning the door prize.

6. The odds against Speedo winning the horse race are 5:2. Find the odds in favor of Speedo winning.

7. Mr. Egan's closet contains 18 shirts, of which 8 are white. Mr. Egan selects one shirt at random. Find
   a) the probability that it is white.
   b) the probability that it is not white.
   c) the odds against it being white.
   d) the odds in favor of it being white.

8. There are four collies, six German shepherds, and three huskies in the animal shelter. If Lalo selects one dog at random, find
   a) the probability that it is a German shepherd.
   b) the probability that it is not a German shepherd.
   c) the odds in favor of it being a German shepherd.
   d) the odds against it being a German shepherd.

In Exercises 9–12, a die is tossed. Find the odds against rolling

9. a 2.

10. an odd number.

11. a number less than 2.

12. a number greater than 3.

In Exercises 13–16, a card is picked from a deck of cards. Find the odds against and the odds in favor of selecting

13. a 9.   14. a picture card.

15. a club.   16. a card greater than 7 (ace is low).

In Exercises 17–20, assume that the spinner cannot land on a line. Find the odds against the spinner landing on the color red.

17.   18.

19.   20.

21. One million tickets are sold for a lottery.
   a) If you purchase a ticket, find your odds against winning.
   b) If you purchase 10 tickets, find your odds against winning.

22. One person is selected at random from a class of 18 males and 11 females. Find the odds against selecting
   a) a female.
   b) a male.

In Exercises 23–27, use the rack of 15 billiard balls shown.

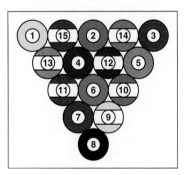

23. If one ball is selected at random, find the odds against it containing a stripe. (Balls numbered 1 through 7 contain stripes.)

24. If one ball is selected at random, find the odds in favor of it being an even-numbered ball.

25. If one ball is selected at random, find the odds in favor of it being a ball other than the 8 ball.

26. If one ball is selected at random, find the odds against it containing a number greater than or equal to 6.

27. If one ball is selected at random, find the odds against it containing any yellow coloring (solid or striped).

28. The odds in favor of an event are 8:3. Find the probability that the event
   a) occurs.
   b) does not occur.

29. The odds in favor of Becky winning a racquetball tournament are 1:8. Find the probability that Becky will
   a) win the tournament.
   b) not win the tournament.

30. The odds against Vic getting a teaching position next fall are 2:9. Find the probability that Vic gets a teaching position next fall.

31. The odds against Man of Peace winning the fifth race are 5:2. Find the probability that
a) Man of Peace wins.
b) Man of Peace loses.

32. Suppose that the probability that you sell your car this week is 0.4. Find the odds against selling your car this week.

33. Suppose that the probability that you are asked to work overtime this week is 3/8. Find the odds in favor of your being asked to work overtime.

34. Suppose that the probability that the mechanic fixes your car right the first time is 0.8. Find the odds against your car being repaired right on the first attempt.

35. Gout constitutes about 5% of all systemic arthritis, and it is uncommon in women. The male to female ratio of gout is estimated to be 20 to 1.
a) If J. Douglas has gout, what are the odds against J. Douglas's being female?
b) If J. Douglas has gout, what is the probability that J. Douglas is a male?

36. One in 40 individuals whose salaries range between $10,000–$40,000 will be randomly selected to have their income tax returns audited for this year. Mr. Frank is in this income tax range. Find
a) the probability that Mr. Frank will be audited.
b) the odds against Mr. Frank being audited.

Problem Solving/Group Activities

37. Find the odds against an even number or a number greater than 3 being rolled on a die.

38. Racetracks quote the approximate odds against each horse winning on a large board called a *tote board*. The odds quoted on a tote board for a race with five horses is as follows.

Horse Number	Odds
1	7:2
2	2:1
3	15:1
4	7:5
5	1:1

Find the probability of each horse winning the race. (Do not be concerned that the sum of the probabilities is not 1.)

39. Turn to the roulette wheel illustrated on page 600. If the wheel is spun, find
a) the probability that the ball lands on black.
b) the odds against the ball landing on black.
c) the probability that the ball lands on 0 or 00.
d) the odds in favor of the ball landing on 0 or 00.

Research Activities

40. Determine whether your state has a lottery. If so, do research and write a paper indicating
a) the probability of winning the grand prize.
b) the odds against winning the grand prize.
c) Explain, using real objects such as pennies or ping pong balls, what these odds really mean.

41. There are many types of games of chance to choose from at casinos. The house has the advantage in each game, but the advantages differ according to the game.
a) List the games available at a typical casino.
b) List those for which the house has the smallest advantage.
c) List those for which the house has the greatest advantage of winning.

Your local library has many books available on this topic.

12.4 EXPECTED VALUE (EXPECTATION)

Expected value is often used to determine the expected results of an experiment or business venture *over the long run*. Expectation is used to make important decisions in many different areas. In business, for example, expectation is used to predict future profits of a new product. In the insurance industry, expectation is used to determine how much each insurance policy should cost for the company to make an overall profit. Expectation is also used to predict the expected gain or loss in games of chance such as the lottery, roulette, craps, and slot machines.

Consider the following: Tim tells Barbara that he will give her $1 if she can roll an even number on a single die. If she fails to roll an even number, she must give Tim $1. Who would win money in the long run if this game were played many times? We would expect in the long run that half the time Tim would win

$1 and half the time he would lose $1, therefore breaking even. Mathematically, we could find Tim's expected gain or loss by the following procedure:

$$\text{Tim's expected gain or loss} = P\left(\begin{array}{c}\text{Tim}\\\text{wins}\end{array}\right)\left(\begin{array}{c}\text{amount}\\\text{Tim wins}\end{array}\right) + P\left(\begin{array}{c}\text{Tim}\\\text{loses}\end{array}\right)\left(\begin{array}{c}\text{amount}\\\text{Tim loses}\end{array}\right)$$

$$= \frac{1}{2}\,(\$1) + \frac{1}{2}\,(-\$1) = \$0.$$

Note that the loss is written as a negative number. This procedure indicates that Tim has an expected gain or loss (or expected value) of $0. The expected value of zero indicates that he would indeed break even, as we had anticipated. Thus the game is a *fair game*. If his expected value were positive, it would indicate a gain; if negative, a loss.

The expected value can be used to determine the expected gain or loss of an experiment or business venture *over the long run*. The expected value, E, is calculated by multiplying the probability of an event occurring by the *net* amount that will be gained or lost if the event occurs. If there are a number of different events and amounts to be considered, we use the following formula.

Expected Value

$$E = P_1 A_1 + P_2 A_2 + P_3 A_3 + \cdots + P_n A_n$$

The symbol P_1 represents the probability that the first event will occur, and A_1 represents the net amount won or lost if the first event occurs. P_2 is the probability of the second event, and A_2 is the net amount won or lost if the second event occurs, and so on. The sum of these products of the probabilities and their respective amounts is the expected value. The expected value is the average (or mean) result that would be obtained if the experiment were performed a great many times.

EXAMPLE 1

Smitty's Construction Company, after considerable research, is planning to bid on a building contract. From past experience, Smitty estimates that if the company's bid is accepted, there is a 50% chance of making a $500,000 profit, a 10% chance of breaking even, and a 40% chance of losing $200,000, depending on weather conditions, inflation, and a possible strike. How much can Smitty "expect" to make or lose on this contract if his bid is accepted?

Solution: Three amounts are to be considered: a gain of $500,000, breaking even at $0, and a loss of $200,000. The probability of gaining $500,000 is 0.5, the probability of breaking even is 0.1, and the probability of losing $200,000 is 0.4.

$$\text{Smitty's expectation} = \overbrace{P_1 A_1}^{\text{Gain}} + \overbrace{P_2 A_2}^{\substack{\text{Break}\\\text{even}}} + \overbrace{P_3 A_3}^{\text{Loss}}$$

$$= (0.5)(\$500,000) + (0.1)(\$0) + (0.4)(-\$200,000)$$

$$= \$250,000 + \$0 - \$80,000$$

$$= \$170,000$$

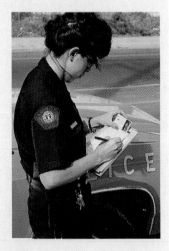

ᗪID YOU KNOW

TOUGH DECISION

*T*he concept of expected value can be used to help evaluate the consequences of many decisions. You use this concept when you consider whether to double-park your car for a few minutes. The probability of being caught may be low, but the penalty, a parking ticket, may be high. You weigh these factors when you decide whether to spend time looking for a legal parking spot. ∎

In the long run, Smitty would have an average gain of $170,000 on each bid of this type. However, there is still a 40% chance that he would lose $200,000 on this *particular* bid.   ▲

EXAMPLE 2

Maria is taking a multiple choice exam in which there are five possible answers for each question. The instructions indicate that she will be awarded 2 points for each correct response, that she will lose $\frac{1}{2}$ point for each incorrect response, and that no points will be added or subtracted for answers left blank.

a) If Maria does not know the correct answer to a question is it to her advantage or disadvantage to guess at an answer?

b) If she can eliminate one of the possible choices, is it to her advantage or disadvantage to guess at the answer?

Solution:

a) Let's determine the expected value if Maria guesses at an answer. Only one of five possible answers is correct.

$$P(\text{guesses correctly}) = \frac{1}{5} \qquad P(\text{guesses incorrectly}) = \frac{4}{5}$$

$$\text{Maria's expectation} = \overbrace{P_1 A_1}^{\substack{\text{Guesses} \\ \text{correctly}}} + \overbrace{P_2 A_2}^{\substack{\text{Guesses} \\ \text{incorrectly}}}$$

$$= \frac{1}{5}\,(2) \quad + \frac{4}{5}\left(-\frac{1}{2}\right)$$

$$= \frac{2}{5} - \frac{2}{5} = 0$$

Thus Maria's expectation is zero when she guesses. Thus over the long run she will neither gain nor lose points by guessing.

b) If Maria can eliminate one possible choice, one of four answers will be correct.

$$P(\text{guesses correctly}) = \frac{1}{4} \qquad P(\text{guesses incorrectly}) = \frac{3}{4}$$

$$\text{Maria's expectation} = \overbrace{P_1 A_1}^{\substack{\text{Guesses} \\ \text{correctly}}} + \overbrace{P_2 A_2}^{\substack{\text{Guesses} \\ \text{incorrectly}}}$$

$$= \frac{1}{4}\,(2) \quad + \frac{3}{4}\left(-\frac{1}{2}\right)$$

$$= \frac{2}{4} - \frac{3}{8} = \frac{4}{8} - \frac{3}{8} = \frac{1}{8}$$

Since the expectation is a positive $\frac{1}{8}$, Maria will, on average, gain $\frac{1}{8}$ point each time she guesses when she can eliminate one possible choice.   ▲

The amounts in the expectation formula are **net amounts**, or the actual amounts won or lost. Consider Example 3, which illustrates the use of net amounts.

EXAMPLE 3

One hundred lottery tickets are sold for $2 each. One prize of $50 will be awarded. Nick purchases one ticket. Find his expectation.

Solution: Two amounts are to be considered: the net amount that Nick may win and the cost of the ticket. If Nick wins, his net or actual winnings are $48 (the $50 prize minus his $2 cost of the ticket). If he does not win, he loses the $2 paid for the ticket. Since only one prize will be awarded, Nick's probability of winning is $\frac{1}{100}$. His probability of losing is therefore $\frac{99}{100}$.

$$E = P(\text{Nick wins})(\text{amount won}) + P(\text{Nick loses})(\text{amount lost})$$

$$= \frac{1}{100}(48) + \frac{99}{100}(-2)$$

$$= \frac{48}{100} - \frac{198}{100} = \frac{-150}{100} = -\$1.50$$

Nick's expectation is $-\$1.50$ per ticket.   ▲

The **fair price** of a game of chance is the amount that should be charged for the game to be fair and result in an expectation of 0. If the fair price and cost to play a game are known, the expectation may be found by the formula

Expectation = fair price − cost to play.

If any two of the three items in the formula are known, the third may be found. For instance, in Example 3, Nick's expectation is $-\$1.50$ and the cost of a ticket is $2.00; thus the fair price, f, may be calculated as follows.

$$\text{Expectation} = \text{fair price} - \text{cost to play}$$
$$-1.50 = f - 2.00$$
$$2.00 - 1.50 = f - 2.00 + 2.00$$
$$0.50 = f$$

Thus the fair price of a ticket is $0.50, or 50¢. This price makes sense because, if the price of a ticket were reduced by $1.50 to 50¢, Nick's expectation would be $0.00.

Expectation problems requiring money to be paid in advance, as in Example 3, can be solved by an alternative procedure. First, determine the fair price to play by using the formula

Fair price $= P_1G_1 + P_2G_2 + \cdots + P_nG_n,$

where each P represents the probability of winning and each corresponding G represents the **gross amount** won. The gross amounts do not include the cost to play the game. After determining the fair price, use the formula Expectation = fair price − cost to play to determine the expectation.

The expectation of the lottery ticket in Example 3 can be found as follows.

$$\text{Fair price} = P_1 G_1 = \frac{1}{100} (\$50) = \frac{50}{100} = \$0.50$$

$$\text{Expectation} = \text{fair price} - \text{cost to play}$$
$$= \$0.50 - \$2.00 = -\$1.50$$

Using either procedure, we see that Nick's expectation is $-\$1.50$.

To obtain the fair price, using gross amounts when more than one amount is awarded, find the sum of the products of the probabilities and their respective amounts. If, for example, two amounts are awarded, we would use the formula

$$\text{Fair price} = P\begin{pmatrix} \text{winning} \\ \text{amount 1} \end{pmatrix}\begin{pmatrix} \text{gross} \\ \text{amount 1} \end{pmatrix} + P\begin{pmatrix} \text{winning} \\ \text{amount 2} \end{pmatrix}\begin{pmatrix} \text{gross} \\ \text{amount 2} \end{pmatrix}.$$

EXAMPLE 4

One thousand lottery tickets are sold for $1 each. One grand prize of $500 and two consolation prizes of $100 will be awarded. Using *net amounts*, find

a) Carol's expectation if she purchases one ticket.

b) Carol's expectation if she purchases five tickets.

c) What is the fair price of a ticket?

Solution:

a) Three amounts are to be considered: the net gain in winning the grand prize, the net gain in winning the consolation prize, and the loss of the cost of the ticket. If Carol wins the grand prize, her net gain is $499 ($500 minus $1 spent for the ticket). If Carol wins the consolation prize, her net gain is $99 ($100 minus $1). We assume that the winners' names are replaced in the pool after being selected. The probability that Carol wins the grand prize is $\frac{1}{1000}$. Since two consolation prizes will be awarded, the probability that she wins a consolation prize is $\frac{2}{1000}$. The probability that she does not win either prize is $1 - \frac{3}{1000} = \frac{997}{1000}$.

$$E = P_1 A_1 + P_2 A_2 + P_3 A_3$$

$$= \frac{1}{1000}(\$499) + \frac{2}{1000}(\$99) + \frac{997}{1000}(-\$1)$$

$$= \frac{499}{1000} + \frac{198}{1000} - \frac{997}{1000} = \frac{-300}{1000} = -\$0.30 \text{ or } -30¢$$

b) On average Carol loses 30¢ on each ticket purchased. On five tickets her expectation is $(-\$0.30)(5)$, or $-\$1.50$.

c) We learned in part (a) that Carol's expectation on one ticket is $-\$0.30$.

$$\text{Expectation} = \text{fair price} - \text{cost}$$
$$-0.30 = f - 1.00$$
$$0.70 = f$$

The fair price of a ticket is 70¢.

The fair price and expectation of the ticket in Example 4 may also be found using *gross amounts* as follows.

$$\text{Fair price} = \frac{1}{1000}\,(\$500) + \frac{2}{1000}\,(\$100) = \frac{500}{1000} + \frac{200}{1000} = \frac{700}{1000} = \$0.70$$

$$\text{Expectation} = \text{fair price} - \text{cost to play}$$
$$= \$0.70 - \$1.00 = -\$0.30$$

EXAMPLE 5

A highway crew repairs 30 potholes a day in dry weather and 12 potholes a day in wet weather. If the weather in this region is wet 40% of the time, find the expected (average) number of potholes that can be repaired per day.

Solution: The amounts in this problem are the number of potholes repaired. Since the weather is wet 40% of the time, it will be dry 60% (100% − 40%) of the time:

$$E = P(\text{dry})(\text{amount repaired}) + P(\text{wet})(\text{amount repaired})$$
$$= 0.60(30) + 0.40(12) = 18.0 + 4.8 = 22.8$$

Thus the average, or expected, number of potholes repaired per day is 22.8.

Section 12.4 Exercises

1. What does the expected value of an experiment or business venture represent?

2. What does an expected value of 0 mean?

3. What is the fair price of a game of chance?

4. Write the formula that would be used to find the expected value of an experiment with three possible amounts.

5. On a $2 bet Mary's expected value is −$0.40.
   a) What is Mary's expected value on a $10 bet?
   b) How much can Mary expect to win or lose if she places a $10 bet?

6. If on a $1 bet Pam's expected value is $0.20,
   a) what is Pam's expected value on a $5 bet?
   b) How much can Pam expect to win or lose if she places a $5 bet? Explain.

7. An investment club is considering purchasing a certain stock option. After considerable research the club members determine that there is a 40% chance of making $7000, a 10% chance of breaking even, and a 50% chance of losing $4500. Find the expectation of this purchase.

8. An investment counselor is advising her client on a particular investment. She estimates that, if the tax law does not change, the client will make $40,000 but if the tax law changes, the client will lose $12,000. Find the client's expected value if there is a 60% chance that the tax law will change.

9. In July in Seattle, the grass grows $\frac{1}{2}$ in. a day on a sunny day and $\frac{1}{4}$ in. a day on a cloudy day. In Seattle in July 75% of the days are sunny and 25% are cloudy.
   a) Find the expected amount of grass growth on a typical day in July in Seattle.
   b) Find the expected total grass growth in the month of July in Seattle.

10. Cal and Phil play the following game: Cal picks a card from a deck of cards. If he selects a club, Phil gives him $10. If not, he gives Phil $4.
    a) Find Cal's expectation.
    b) Find Phil's expectation.

11. A multiple choice exam has four possible answers for each question. For each correct answer you are awarded 5 points. For each incorrect answer, 2 points are subtracted from your score. For answers left blank, no points are added or subtracted.
    a) If you do not know the correct answer to a particular question, is it to your advantage to guess?
    b) If you do not know the correct answer but can eliminate one possible choice, is it to your advantage to guess?

12. At a Chinese restaurant, a slip in the fortune cookies indicates a dollar amount that will be subtracted from your total bill. A bag of 10 fortune cookies (each individually wrapped) is given to you from which you will select 1. If 7 fortune cookies contain "$1 off," 2 contain "$2 off," and 1 contains "$5 off," find the expectation of a selection.

13. One thousand raffle tickets are sold for $1 each. One prize of $500 is to be awarded.
   a) Olga purchases one ticket. Find her expected value.
   b) Find the fair price of a ticket.
   c) If the vendor sells all 1000 tickets, how much profit will he make?

14. Ten thousand raffle tickets are sold for $5 each. Four prizes will be awarded—one for $10,000, one for $5000, and two for $1000. Sidhardt purchases one of these tickets.
   a) Find his expected value.
   b) Find the fair price of a ticket.

In Exercises 15–18, assume that you spin the pointer and are awarded the amount indicated by the pointer.
   a) *Find the fair price to play the game.*
   b) *If it costs $2 to play the game, find the expectation of a person who plays the game.*

15.   **16.**

17.   **18.**

19. According to a mortality table, the probability that a 20-year-old woman will survive one year is 0.994, and the probability that she will die within one year is 0.006. If she buys a $10,000 one-year term policy for $100, what is the company's expected gain or loss?

20. It will cost an oil drilling company $30,000 to sink a test well. If it hits oil, the company will make a net profit of $500,000. If it hits natural gas, the net profit will be $100,000. If it hits nothing, it will lose its $30,000. If the probability of hitting oil is 0.08 and the probability of hitting gas is 0.20, what is the expectation of the oil drilling company? Should it sink the test well? Explain.

21. An instrumentation contractor is considering making a bid on a water pollution control project. She determines that if her bid is accepted and she completes the project on schedule her net profit would be $380,000. If she completes the project between 0 and 3 months after the contract deadline date her net profit drops to $150,000.

If she completes the project more than 3 months late her net loss is $450,000. The probability that she completes the job on schedule is 0.5, the probability that she completes the job between 0 and 3 months late is 0.3, and the probability that she completes the job more than 3 months late is 0.2. Find her expected gain or loss for this bid.

22. A die is rolled many times, and the points facing up are recorded. Find the expected (average) number of points facing up over the long run.

23. United Airlines is planning its staffing needs for next year. On January 1 the Civil Aeronautics Board will inform United Airlines whether it will be granted the new routes it has requested. If the new routes are approved, United will hire 850 new employees. If the new routes are not granted, United will hire only 150 new employees. If the probability that the Board will grant United Airlines' request is 0.25, what is the expected number of new employees to be hired by United Airlines?

24. The Gecko Manufacturing Company is negotiating a contract with its employees. The probability that the union will go on strike is 5%. If the union goes on strike, the company estimates that it will lose $345,000 for the year. If the union does not go on strike, the company estimates that it will make $1.2 million. Find the company's expected gain or loss for the year.

25. On a clear day in Reading the AAA makes an average of 125 service calls for motorist assistance; on a rainy day it makes an average of 190 service calls; and on a snowy day it makes an average of 280 service calls. If the weather in Reading is clear 200 days of the year, rainy 100 days of the year, and snowy 65 days of the year, find the expected number of service calls made by the AAA in a given day.

26. The expenses for Dale, a realtor, to list, advertise, and attempt to sell a house are $1000. If Dale succeeds in selling the house, he will receive 6% of the sales price. If a realtor with a different company sells the house, Dale will still receive 3% of the sales price. If the house is unsold after 3 months, Dale loses the listing and receives nothing. Suppose that the probability Dale sells a $100,000 house is 0.2, the probability that another realtor sells the house is 0.5, and the probability that the house is unsold after 3 months is 0.3. Find Dale's expectation if he accepts this house for listing. Should Dale list the house? Explain.

27. An insurance company has written life insurance policies on people of age 25. There are 200 policies of $10,000, 400 policies of $5000, and 1000 policies of $1000. Experience shows that the probability of an individual dying at age 25 is 0.002. Determine the amount the insurance company can expect to pay out on these policies this year.

In Exercises 28 and 29, assume that you are blindfolded and throw a dart at the dart board shown. Assuming your dart sticks in the dart board, determine

    a) *the probabilities that the dart lands on $1, $10, $20, and $100, respectively.*

    b) *If you win the amount of money indicated by the section of the board where the dart lands, find your expectation when you throw the dart.*

    c) *If the game is to be fair, how much should you pay to play?*

28.

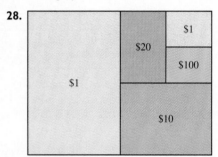

29.

30. During those years when financial responsibilities are greatest, such as when children are in school or there is a mortgage on the house, a person may choose to buy a term life insurance policy. The insurance company will pay the face value of the policy if the insured person dies during the term of the policy. For how much should an insurance company sell a 10-year term policy with a face value of $40,000 to a 30-year-old man in order for the company to make a profit? The probability of a 30-year-old man living to age 40 is 0.97. Explain your answer.

Problem Solving/Group Activities

In Exercises 31 and 32, use the roulette wheel illustrated. A roulette wheel typically contains slots with numbers 1–36 and slots marked 0 and 00. A ball is spun on the wheel and comes to rest in one of the 38 slots. Eighteen numbers are colored red, and 18 numbers are colored black. The 0 and 00 are colored green. If you bet on one particular number and the ball lands on that number the house pays off odds of 35 to 1. If you bet on a red number or black number and win, the house pays 1 to 1 (even money).

31. Find the expected value of betting $1 on a particular number.

32. Find the expected value of betting $1 on red.

33. The following is a miniature version of the Wheel of Fortune. When Mr. Dudley spins the wheel he is awarded the amount on the wheel indicated by the pointer. If the wheel points to Bankrupt, he loses the total amount he has accumulated and also loses his turn. Assume that the wheel stops on a position at random and that each position is equally likely to occur.

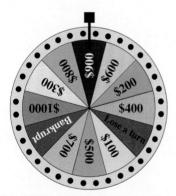

    a) Find Mr. Dudley's expectation when he spins the wheel at the start of the game (he has no money to lose if he lands on Bankrupt).

    b) If Mr. Dudley presently has a balance of $1800, find his expectation when he spins the wheel.

34. The dealer shuffles five black cards and five red cards and spreads them out on the table face down. You choose two at random. If both cards are red or both cards are black, you win a dollar. Otherwise, you lose a dollar. Determine whether the game favors you, is fair, or favors the dealer. Explain your answer.

35. A dealer has three cards. One is red on both sides, another is black on both sides, and the third is red on one side and black on the other side. The dealer shuffles the cards in a hat, draws one at random, and places it flat on the table. The side showing is red. If the other side is red, you win a dollar. If the other side is black, you lose a dollar. Determine whether the game favors the player, is fair, or favors the dealer.

36. Do not throw away the next sweepstakes ticket that you get as part of an advertisement.

a) Use the information provided to determine the expectation of the ticket.

b) If you consider the current cost of a stamp, is it to your benefit to mail in the ticket? Explain.

37. Is it possible to determine your expectation when you purchase a lottery ticket? Explain.

12.5   TREE DIAGRAMS

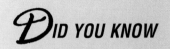

We stated earlier that the possible results of an experiment are called its outcomes. In order to solve more difficult probability problems we must first be able to determine all the possible outcomes of the experiment. The counting principle can be used to determine the number of outcomes of an experiment and is helpful in constructing tree diagrams.

> **Counting Principle**
>
> If a first experiment can be performed in M distinct ways and a second experiment can be performed in N distinct ways, then the two experiments in that specific order can be performed in $M \cdot N$ distinct ways.

If we wanted to find the number of possible outcomes when a coin is tossed and a die is rolled, we could reason: The coin has two possible outcomes—heads and tails. The die has six possible outcomes—1, 2, 3, 4, 5, and 6. Thus the two experiments together have $2 \cdot 6$, or 12, possible outcomes.

A list of all the possible outcomes of an experiment is called a **sample space**. Each individual outcome in the sample space is called a **sample point**. **Tree diagrams** are helpful in determining sample spaces.

A tree diagram illustrating all the possible outcomes when a coin is tossed and a die is rolled (see Fig. 12.3) will have two initial branches, one for each of the possible outcomes of the coin. Each of these branches will have six branches

Possible Outcomes from Coin	Possible Outcomes from Die	Sample Space
H	1	H1
	2	H2
	3	H3
	4	H4
	5	H5
	6	H6
T	1	T1
	2	T2
	3	T3
	4	T4
	5	T5
	6	T6

Figure 12.3

emerging from them, one for each of the possible outcomes of the die. This will give a total of 12 branches, the same number of possible outcomes found by using the counting principle. We can obtain the sample space by listing all the possible combinations of branches. Note that this sample space consists of 12 sample points.

Example 1 uses the phrase ''without replacement.'' This phrase tells us that once an item is selected it cannot be selected again, making it impossible to select the same item twice.

EXAMPLE I

Two balls are to be selected *without replacement* from a bag that contains one red, one blue, one green, and one orange ball (see Fig. 12.4).

a) Use the counting principle to determine the number of points in the sample space.

b) Construct a tree diagram and list the sample space.

c) Find the probability that one red ball is selected.

d) Find the probability that a green ball followed by a red ball is selected.

Solution:

a) The first selection may be any one of the four balls. Once the first ball is selected, only three balls remain for the second selection. Thus there are $4 \cdot 3$, or 12, sample points in the sample space.

b) The first ball selected can be red, blue, green, or orange. Since this experiment is done without replacement, the same colored ball cannot be selected twice. For example, if the first ball selected is red, the second ball selected must be either blue, green, or orange. The tree diagram and sample space are shown in Fig. 12.5. The sample space contains 12 points. That result checks with the answer obtained with the counting principle.

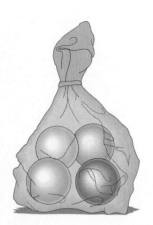

Figure 12.4

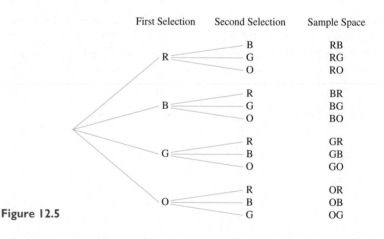

Figure 12.5

c) If we know the sample space, we can compute probabilities using the formula

$$P(E) = \frac{\text{number of outcomes favorable to } E}{\text{total number of outcomes}}.$$

The total number of outcomes will be the number of points in the sample space. From Fig. 12.5 we determine that there are 12 possible outcomes. Six outcomes have one red ball: RB, RG, RO, BR, GR, and OR.

$$P(\text{one red ball is selected}) = \frac{6}{12} = \frac{1}{2}$$

d) One possible outcome meets the criteria of the problem: GR.

$$P(\text{green followed by red}) = \frac{1}{12}$$

▲

The counting principle can be extended to any number of experiments, as illustrated in Example 2.

EXAMPLE 2

There are two highways from New York to Cleveland, three highways from Cleveland to Chicago, and two highways from Chicago to San Francisco, as illustrated in Fig. 12.6.

Figure 12.6

a) Use the counting principle to determine the number of different routes from New York to San Francisco.

b) Use a tree diagram to determine the routes.

c) If a route from New York to San Francisco is selected at random and all routes are considered equally likely, find the probability that both routes *a* and *g* are used.

d) Find the probability that neither routes *d* nor *f* are used.

Solution:

a) Using the counting principle, we can determine that there are $2 \cdot 3 \cdot 2$, or 12, routes from New York to San Francisco.

b) The tree diagram illustrating the 12 possibilities is given in Fig. 12.7.

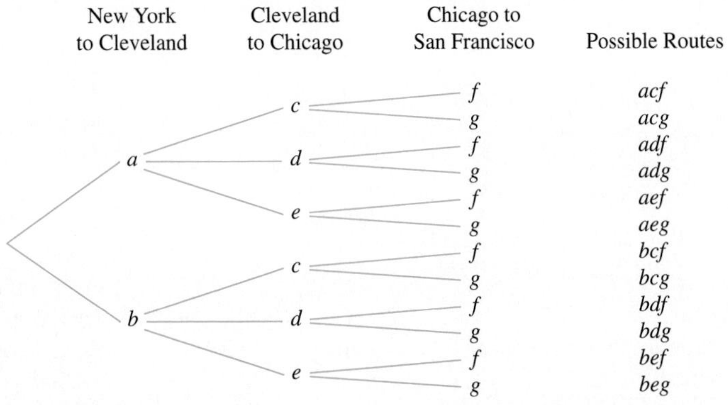

New York to Cleveland	Cleveland to Chicago	Chicago to San Francisco	Possible Routes
		f	acf
	c	g	acg
		f	adf
a	d	g	adg
		f	aef
	e	g	aeg
		f	bcf
	c	g	bcg
		f	bdf
b	d	g	bdg
		f	bef
	e	g	beg

Figure 12.7

c) Of the 12 possible routes, 3 use both a and g (acg, adg, aeg).

$$P(\text{routes } a \text{ and } g \text{ are both used}) = \frac{3}{12} = \frac{1}{4}$$

d) Of the 12 possible routes, 4 use neither d nor f (acg, aeg, bcg, beg).

$$P(\text{neither } d \text{ nor } f \text{ is used}) = \frac{4}{12} = \frac{1}{3}$$

Section 12.5 Exercises

1. Explain the counting principle.

2. a) What is a sample space?
   b) What is a sample point?

3. A jar contains five different colored marbles of the same size. If two marbles are to be selected at random *without replacement*, how many sample points will be in the sample space? Explain.

4. Repeat Exercise 3 for selection *with replacement*.

5. A bag contains five different light bulbs. The bulbs are the same size, but each has a different wattage. If you select three bulbs at random from the bag, *with replacement*, how many sample points will be in the sample space? Explain.

6. Repeat Exercise 5 for selection *without replacement*.

In Exercises 7–20, use the counting principle to determine the number of points in the sample space before constructing the tree diagram. Assume that each event is equally likely to occur.

7. Two coins are tossed.
   a) Construct a tree diagram and list the sample space.
   Find the probability that
   b) no heads are tossed.
   c) exactly one head is tossed.
   d) two heads are tossed.

8. A bag contains three cards: a jack, a queen, and a king.

Two cards are to be selected at random with replacement.
a) Construct a tree diagram and determine the sample space.

Find the probability that
b) two jacks are selected.
c) a jack and then a queen are selected.
d) at least one king is selected.

9. Repeat Exercise 8 without replacement.

10. A hat contains four marbles: one yellow, one red, one blue, and one green. Two marbles are to be selected at random without replacement from the hat.

a) Construct a tree diagram and list the sample space.
Find the probability of selecting
b) exactly one red marble.
c) at least one marble that is not red.
d) no green marbles.

11. A couple plan to have two children.
   a) Construct a tree diagram and list the sample space of the possible arrangements of boys and girls.
   Find the probability that the family has
   b) two girls.
   c) at least one girl.

12. Cecelia owns three cats including a Siamese, Angora, and Persian, and four dogs including a collie, poodle, beagle, and husky. She selects one cat and one dog at random to take with her to the park.
   a) Construct a tree diagram and determine the sample space.
   Find the probability that she selects
   b) the Persian cat.
   c) a cat other than the Persian.
   d) the Siamese cat and either the collie or poodle.
   e) the poodle or the husky.

13. Mr. and Mrs. Jones are vacationing in the Orlando, Florida, area. They have listed what they consider the major attractions in the area in two groups: Disney attractions and non-Disney attractions. The Disney attractions include the Magic Kingdom, Epcot Center, MGM Studios, and Pleasure Island. The non-Disney attractions include Sea World, Universal Studios, and Busch Gardens (Tampa). Because of time limitations they decided to first visit one Disney and then one non-Disney attraction. They will select the name of one Disney attraction from one hat and the name of one non-Disney attraction from another hat.

a) Construct a tree diagram and determine the sample space.

Find the probability that they select

b) the Magic Kingdom or Epcot Center.

c) MGM Studios or Universal Studios.

d) MGM Studios and Universal Studios.

e) the Magic Kingdom and either Sea World or Busch Gardens.

14. Gustave Ruckert is starting his own publishing company. He is considering six different printers, including Abe's, Bonnie's, Charlie's, Do It Right, Economy, and First Class; and three different binders, including Goldman's, Hunt's, and Ivanhoe's. He needs to select a printer and a binder.

a) Construct a tree diagram of the possible combinations of printers and binders and list the sample space.

Find the probability that

b) Goldman's is selected.

c) Abe's or Economy is selected.

d) Charlie's and Ivanhoe's are not both selected.

15. Two dice are rolled.

a) Construct a tree diagram and list the sample space.

Find the probability that

b) a double is rolled.

c) a sum of 7 is rolled.

d) a sum of 2 is rolled.

e) Are you as likely to roll a sum of 2 as you are of rolling a sum of 7? Explain your answer.

16. Three coins are tossed.

a) Construct a tree diagram and list the sample space.

Find the probability that

b) exactly three heads are tossed.

c) exactly one head is tossed.

d) at least one head is tossed.

17. A couple plans to have three children.

a) Construct a tree diagram and list the sample space.

Find the probability that the family has

b) no boys.

c) at least one girl.

d) either exactly two boys or two girls.

18. Two coins are tossed and a die is rolled.

a) Construct a tree diagram and list the sample space.

Find the probability of

b) tossing two heads and rolling the number 3.

c) tossing exactly one head.

d) tossing at least one head and rolling an even number.

19. A bag contains three chips: one red, one blue, and one green. Three chips are to be selected at random with replacement.

a) Construct a tree diagram and list the sample space. Find the probability of selecting

b) three red chips.

c) a red chip, then a blue chip, and then a green chip.

d) at least one chip that is not blue.

20. Repeat Exercise 19 without replacement.

21. An individual can be classified as male or female with red, brown, black, or blonde hair and with brown, blue, or green eyes.

a) How many different classifications are possible?

b) Construct a tree diagram to determine the sample space (for example, red-headed, blue-eyed, and male).

c) If each outcome is equally likely, find the probability that the individual will be a male with black hair and blue eyes.

d) Find the probability that the individual will be a female with blonde hair.

22. A pea plant must have exactly one of each of the following pairs of traits: short (*s*) or tall (*t*); round (*r*) or wrinkled (*w*) seeds; yellow (*y*) or green (*g*) peas; and white (*wh*) or purple (*p*) flowers (for example, short, wrinkled, green pea with white flowers).

a) How many different classifications of pea plants are possible?

b) Use a tree diagram to determine all the classifications possible.

c) If each characteristic is equally likely, find the probability that the pea plant will have round peas.

d) Find the probability that the pea plant will be short, be wrinkled, have yellow seeds, and have purple flowers.

Problem Solving/Group Activity

23. Use the tree diagram to answer the following questions.

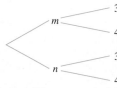

a) What are the possible outcomes of the first experiment?

b) What are the possible outcomes of the second experiment?

c) List the sample space indicated by this tree diagram.

d) From the tree diagram, is it possible to determine the probability that the outcome of the first experiment is *m*? Explain.

e) From the tree diagram, is it possible to determine whether the probability of getting *m* followed by 3 is the same as the probability of *n* followed by 4? Explain.

f) Do your responses to parts (d) and (e) change if you are told that the probabilities of experiment 1 are equally likely and that the probabilities of experiment 2 are also equally likely? Explain.

12.6 *OR* AND *AND* PROBLEMS

In Section 12.5 we showed how to work probability problems by constructing sample spaces. Often it is inconvenient or too time-consuming to solve a problem by first constructing a sample space. For example, if an experiment consists of selecting two cards with replacement from a deck of 52 cards, there would be 52 · 52 or 2704 points in the sample space. Trying to list all these sample points could take hours. In this section we will learn how to solve compound probability problems that contain the words *and* or *or* without constructing a sample space.

Or Problems

The *or probability problem* requires obtaining a "successful" outcome for *at least one* of the given events. For example suppose that we roll one die and we are interested in finding the probability of rolling an even number *or* a number greater than 4. For this situation rolling either a 2, 4, or 6 (an even number) or a 5 or 6 (a number greater than 4) would be considered successful. Note that the number 6 satisfies both criteria. A formula for finding the probability of event *A* or event *B*, symbolized *P*(*A* or *B*), follows.

$$P(A \text{ or } B) = P(A) + P(B) - P(A \text{ and } B).$$

Since we add (and subtract) probabilities to find *P*(*A* or *B*) this formula is sometimes referred to as the *addition formula*. We explain the use of the *or* formula in Example 1.

─ **EXAMPLE 1**

Each of the numbers 1, 2, 3, 4, 5, 6, 7, 8, 9, and 10 is written on a separate piece of paper. The 10 pieces of paper are then placed in a hat, and one piece is randomly selected. Find the probability that the piece of paper selected contains an even number or a number greater than 6.

Solution: We are asked to find the probability that the number selected *is even* or *greater than 6*. Let's use set *A* to represent the statement, "the number is even," and set *B* to represent the statement, "the number is greater than 6." Figure 12.8 is a Venn diagram, as introduced in Chapter 2, with sets *A* (even) and *B* (greater than 6). There are a total of 10 numbers, of which five are even (2, 4, 6, 8, and 10). Thus the probability of selecting an even number is $\frac{5}{10}$. Four numbers are greater than 6: the 7, 8, 9, and 10. Thus the probability of selecting a number greater than 6 is $\frac{4}{10}$. Two numbers are both even and greater than 6: the 8 and 10. Thus the probability of selecting a number that is both even and greater than 6 is $\frac{2}{10}$.

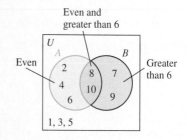

Figure 12.8

If we substitute the appropriate statements for A and B in the formula we obtain

$$P(A \text{ or } B) = P(A) + P(B) - P(A \text{ and } B);$$

$$P\left(\begin{matrix} \text{even or} \\ \text{greater than 6} \end{matrix}\right) = P(\text{even}) + P\left(\begin{matrix} \text{greater} \\ \text{than 6} \end{matrix}\right) - P\left(\begin{matrix} \text{even and} \\ \text{greater than 6} \end{matrix}\right)$$

$$= \frac{5}{10} + \frac{4}{10} - \frac{2}{10}$$

$$= \frac{7}{10}$$

Example 1 illustrates that when finding the probability of A or B, we sum the probabilities of events A and B and then subtract the probability of both events occurring simultaneously.

EXAMPLE 2

Consider the same sample space, the numbers 1 through 10, as in Example 1. If one piece of paper is selected, find the probability that it contains a number less than 4 or a number greater than 6.

Solution: Let A represent the statement, ''the number is less than 4,'' and let B represent the statement, ''the number is greater than 6.'' A Venn diagram illustrating these statements is shown in Fig. 12.9.

$$P(\text{selecting a number less than 4}) = \frac{3}{10}$$

$$P(\text{selecting a number greater than 6}) = \frac{4}{10}$$

Since there are no numbers that are both less than 4 and greater than 6, $P(\text{selecting a number less than 4 and greater than 6}) = 0$. Therefore

$$P\left(\begin{matrix} \text{less than 4 or} \\ \text{greater than 6} \end{matrix}\right) = P(\text{less than 4}) + P\left(\begin{matrix} \text{greater} \\ \text{than 6} \end{matrix}\right) - P\left(\begin{matrix} \text{less than 4 and} \\ \text{greater than 6} \end{matrix}\right)$$

$$= \frac{3}{10} + \frac{4}{10} - 0 = \frac{7}{10}$$

In Example 2 it is impossible to select a number that is both less than 4 and greater than 6 when only one number is to be selected. Events such as these are said to be mutually exclusive.

Two events A and B are **mutually exclusive** if it is impossible for both events to occur simultaneously.

If events A and B are mutually exclusive, then $P(A \text{ and } B) = 0$, and the addition formula simplifies to $P(A \text{ or } B) = P(A) + P(B)$.

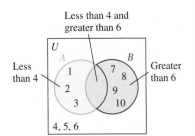

Less than 4 and greater than 6

Less than 4

Greater than 6

Figure 12.9

EXAMPLE 3

One card is selected from a deck of cards. Determine whether the following pairs of events are mutually exclusive, and find $P(A \text{ or } B)$.

a) A = an ace, B = a king

b) A = an ace, B = a spade

c) A = a red card, B = a black card

d) A = a picture card, B = a red card

Solution:

a) It is impossible to select both an ace and a king when only one card is selected. Therefore these events are mutually exclusive.

$$P(\text{ace or king}) = P(\text{ace}) + P(\text{king}) = \frac{4}{52} + \frac{4}{52} = \frac{8}{52} = \frac{2}{13}$$

b) The ace of spades is both an ace and a spade; therefore these events are not mutually exclusive.

$$P(\text{ace}) = \frac{4}{52} \qquad P(\text{spade}) = \frac{13}{52} \qquad P(\text{ace and spade}) = \frac{1}{52}$$

$$P(\text{ace or spade}) = P(\text{ace}) + P(\text{spade}) - P(\text{ace and spade})$$

$$= \frac{4}{52} + \frac{13}{52} - \frac{1}{52}$$

$$= \frac{16}{52} = \frac{4}{13}$$

c) It is impossible to select one card that is both a red card and a black card. Therefore the events are mutually exclusive.

$$P(\text{red or black}) = P(\text{red}) + P(\text{black})$$

$$= \frac{26}{52} + \frac{26}{52} = \frac{52}{52} = 1$$

Therefore a red card or a black card must be selected.

d) Six picture cards are red: jack, queen, and king of hearts and jack, queen, and king of diamonds. Thus these events are not mutually exclusive.

$$P\binom{\text{picture card}}{\text{or red card}} = P\binom{\text{picture}}{\text{card}} + P\binom{\text{red}}{\text{card}} - P\binom{\text{picture card}}{\text{and red card}}$$

$$= \frac{12}{52} + \frac{26}{52} - \frac{6}{52}$$

$$= \frac{32}{52} = \frac{8}{13}$$

And Problems

A second type of problem is the *and* **probability problem** which requires obtaining a favorable outcome in *each* of the given events. For example suppose that *two* cards are to be selected from a deck of cards and that we are interested in the probability of selecting two aces (one ace *and* then a second ace). Only if *both* cards selected are aces would this experiment be considered successful. A formula for finding the probability of events *A* and *B*, symbolized *P*(*A* and *B*), follows.

$$P(A \text{ and } B) = P(A) \cdot P(B), \text{ assuming that event } A \text{ has occurred*}$$

Since we multiply to find *P*(*A* and *B*), this formula is sometimes referred to as the *multiplication formula*. When using the multiplication formula, we will not always write the words "assuming that event *A* has occurred." However, since this type of problem required obtaining a favorable outcome in *both* of the given events, **we must always assume that event *A* has occurred before calculating the probability of event *B*** (even when "assuming that event *A* has occurred" has not been written).

┌─ **EXAMPLE 4**

Two cards are to be selected with replacement from a deck of cards. Find the probability that two aces will be selected.

Solution: Since the deck of 52 cards contains four aces, the probability of selecting an ace on the first draw is $\frac{4}{52}$. The card selected is then returned to the deck. Therefore the probability of selecting an ace on the second draw remains $\frac{4}{52}$.

If we let *A* represent the selection of the first ace and *B* represent the selection of the second ace, the formula may be written

$$P(A \text{ and } B) = P(A) \cdot P(B);$$

$$P(2 \text{ aces}) = P(\text{ace } and \text{ ace}) = P(\text{ace } 1) \cdot P(\text{ace } 2)$$

$$= \frac{4}{52} \cdot \frac{4}{52}$$

$$= \frac{1}{13} \cdot \frac{1}{13} = \frac{1}{169} \quad \blacktriangle$$

┌─ **EXAMPLE 5**

Repeat Example 4 without replacement.

Solution: The probability of selecting an ace on the first draw is $\frac{4}{52}$. When calculating the probability of selecting the second ace, we must assume that the first ace has been selected. Once this first ace has been selected only 51 cards, including 3 aces, remain in the deck. The probability of selecting an ace on the second draw becomes $\frac{3}{51}$. The probability of selecting two aces without replacement is

$$P(2 \text{ aces}) = P(\text{ace } 1) \cdot P(\text{ace } 2)$$

$$= \frac{4}{52} \cdot \frac{3}{51}$$

$$= \frac{1}{13} \cdot \frac{1}{17} = \frac{1}{221} \quad \blacktriangle$$

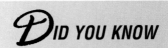

*T*he odds against a family having eight girls in a row is 255 to 1. Thus, for every 256 families with eight children, on average, one family will have all girls and 255 will not. However, should another child be born to that one family, the probability of the ninth child being a girl is still $\frac{1}{2}$! ∎

---

*P(B), assuming that event A has occurred, may be denoted $P(B|A)$, which is read "the probability of B, given A." We will discuss this type of probability (conditional probability) further in Section 12.7.

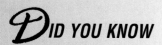

DID YOU KNOW

THE BIRTHDAY PROBLEM

*A*mong 24 people chosen at random, what would you guess is the probability that at least 2 of them will have the same birthday? It might surprise you to learn that it is greater than 0.5. There are 365 days on which the first person selected can have a birthday. That person has a 365/365 chance of having a birthday on one of those days. The probability that the second person's birthday is on any other day is 364/365. The probability that the third person's birthday is on a day different from the first two is 363/365, and so on. The probability that the 24th person has a birthday on any other day than the first 23 people is 342/365. Thus the probability, *P*, that of 24 people, no 2 have the same birthday is (365/365) × (364/365) × (363/365) × · · · × (342/365) = 0.462. Then the probability of at least 2 people of 24 having the same birthday is 1 − *P* = 1 − 0.462 = 0.538, or slightly larger than $\frac{1}{2}$. ∎

Event *A* and event *B* are **independent events** if the occurrence of either event in no way affects the probability of occurrence of the other.

Rolling dice and tossing coins are examples of independent events. In Example 4 the events are independent since the first card was returned to the deck. The probability of selecting an ace on the second draw was not affected by the first selection. The events in Example 5 are not independent since the probability of the selection of the second ace was affected by removing the first ace selected from the deck. Such events are called dependent events. *Experiments done with replacement will result in independent events, and those done without replacement will result in dependent events.*

EXAMPLE 6

A bowl of jelly beans contains 20 green and 15 orange jelly beans. Samantha, a mathematician, is going to pick and eat 3 jelly beans at random one after the other. She is interested in determining the probability that she picks 3 orange jelly beans. Are the events of picking and eating the jelly beans independent or dependent events in this experiment? Explain your answer.

Solution: The events are dependent events since each time she picks 1 orange jelly bean it changes the probability that the next jelly bean selected will be an orange jelly bean.

The multiplication formula may be extended to more than two events, as illustrated in Example 7.

EXAMPLE 7

A package of 25 zinnia seeds contain 8 seeds for red flowers, 12 seeds for white flowers, and 5 seeds for yellow flowers. Three seeds are randomly selected and planted. Find the probability of each of the following.

a) All three seeds will produce red flowers.

b) The first seed selected will produce a red flower, the second seed will produce a white flower, and the third seed will produce a red flower.

c) None of the seeds will produce red flowers.

Solution: Each time a seed is selected and planted, the number of seeds remaining decreases by one.

a) The probability that the first seed selected produces a red flower is $\frac{8}{25}$. If the first seed selected is red, only 7 red seeds in 24 are left. The probability of selecting a second red seed is $\frac{7}{24}$. If the second seed selected is red, only 6 red seeds in 23 are left. The probability of selecting a third red seed is $\frac{6}{23}$.

$$P(3 \text{ red seeds}) = P(\text{red seed 1}) \cdot P(\text{red seed 2}) \cdot P(\text{red seed 3})$$

$$= \frac{8}{25} \cdot \frac{7}{24} \cdot \frac{6}{23} = \frac{14}{575}$$

b) The probability that the first seed selected produces a red flower is $\frac{8}{25}$. Once a seed for a red flower is selected, only 24 seeds are left. Twelve of the remaining 24 will produce white flowers. Thus the probability that the second seed selected will produce a white flower is $\frac{12}{24}$. After the second seed has been selected, there are 23 seeds left, 7 of which will produce red flowers. The probability that the third seed produces a red flower is therefore $\frac{7}{23}$.

$$P\left(\begin{array}{l}\text{first red and second}\\\text{white and third red}\end{array}\right) = P(\text{first red}) \cdot P(\text{second white}) \cdot P(\text{third red})$$

$$= \frac{8}{25} \cdot \frac{12}{24} \cdot \frac{7}{23} = \frac{28}{575}$$

c) If no flowers are to be red, they must either be white or yellow. Seventeen seeds will not produce red flowers (12 for white and 5 for yellow). The probability that the first seed does not produce a red flower is $\frac{17}{25}$. After the first seed has been selected, 16 of the remaining 24 seeds will not produce red flowers. After the second seed has been selected, 15 of the remaining 23 seeds will not produce red flowers.

$$P(\text{none red}) = P(\text{first not red}) \cdot P(\text{second not red}) \cdot P(\text{third not red})$$

$$= \frac{17}{25} \cdot \frac{16}{24} \cdot \frac{15}{23} = \frac{34}{115}$$

Which Formula to Use

It is sometimes difficult to determine when to use the *or formula* and when to use the *and formula*. The following information may be helpful in deciding which formula to use.

Or Formula

Or problems will almost always contain the word *or* in the statement of the problem. For example, find the probability of selecting a heart *or* a 6. *Or* problems in this book generally involve only *one* selection. For example, ''one card is selected'' or ''one die is rolled.''

And Formula

And problems will often *not* use the word *and* in the statement of the problem. For example, ''find the probability that both cards selected are red,'' or ''find the probability that none of those selected is a banana'' are both *and*-type problems. *And* problems in this book will generally involve *more than 1* selection. For example, the problem may read ''two cards are selected'' or ''three coins are flipped.''

Section 12.6 Exercises

1. a) Give the formula for $P(A \text{ or } B)$.
   b) In your own words, explain how to find $P(A \text{ or } B)$ with the formula.

2. a) What events are mutually exclusive? Give an example.

   b) How do you calculate $P(A \text{ or } B)$ when A and B are mutually exclusive?

3. a) Give the formula for $P(A \text{ and } B)$.
   b) In your own words, explain how to find $P(A \text{ and } B)$ with the formula.

4. What are independent events? Give an example.

5. When finding $P(B)$ using the formula $P(A$ and $B)$, what do we always assume?

6. What are dependent events? Give an example.

7. **a)** Write a problem that you would use the *or formula* to solve. Solve the problem and give the answer.
   b) Write a problem that you would use the *and formula* to solve. Solve the problem and give the answer.

8. A family is selected at random. Let event A be that the mother has a college degree. Let event B be that the daughters will get a college degree.
   a) Are events A and B mutually exclusive? Explain.
   b) Are they independent events? Explain.

9. An individual is selected at random. Let event A be that the individual is happy. Let event B be that the individual is healthy.
   a) Are events A and B mutually exclusive? Explain.
   b) Are they independent events? Explain.

10. A family is selected at random. Let event A be that the father smokes. Let event B be that the mother smokes.
    a) Are events A and B mutually exclusive? Explain.
    b) Are they independent events? Explain.

11. If events A and B are mutually exclusive, explain why the formula $P(A$ or $B) = P(A) + P(B) - P(A$ and $B)$ can be simplified to $P(A$ or $B) = P(A) + P(B)$.

In Exercises 12–14, find the indicated probability.

12. If $P(A) = 0.4$, $P(B) = 0.5$, and $P(A$ and $B) = 0.2$, find $P(A$ or $B)$.

13. If $P(A$ or $B) = 0.9$, $P(A) = 0.5$, and $P(B) = 0.6$, find $P(A$ and $B)$.

14. If $P(A$ or $B) = 0.8$, $P(A) = 0.4$, and $P(A$ and $B) = 0.1$, find $P(B)$.

In Exercises 15–18, a single die is rolled. Find the probability of rolling

15. a 2 or 3.

16. an odd number or a number greater than 2.

17. a number greater than 5 or less than 3.

18. a number greater than 3 or less than 5.

In Exercises 19–24, one card is selected from a deck of cards. Find the probability of selecting

19. an ace or a king.

20. a king or a heart.

21. a picture card or a black card.

22. a club or a red card.

23. a card less than 8 or a diamond. (*Note:* The ace is considered a low card.)

24. a card greater than 9 or a black card.

In Exercises 25–32, a board game uses the deck of 20 cards shown.

Two cards are selected at random from this deck. Find the probability of the following

    a) *with replacement.*
    b) *without replacement.*

25. They both show birds.

26. They both show the number 1.

27. The first shows a monkey, and the second shows a lion.

28. The first shows a 1, and the second shows a 5.

29. The first shows a frog, and the second shows a yellow bird.

30. They both show odd numbers.

31. Neither shows an odd number.

32. The first shows a monkey, and the second shows a red bird.

In the deck of cards used in Exercises 25–32, if one card is drawn, find the probability that the card shows

33. a monkey or an odd number.

34. a yellow bird or a number greater than 4.

35. a monkey or a 5.

36. a lion or an even number.

In Exercises 37–46, assume that the pointer cannot land on the line and assume that each spin is independent.

Figure 12.10

If the pointer in Fig. 12.10 is spun twice, find the probability that the pointer lands on

37. red on both spins.   38. red and then yellow.

If the pointer in Fig. 12.11 is spun twice, find the probability that the pointer lands on

39. red and then green.　　**40.** green on both spins.

Figure 12.11

If the pointer in Fig. 12.12 is spun twice. Find the probability that the pointer lands on

41. yellow on both spins.

42. a color other than red on both spins.

Figure 12.12

In Exercises 43–46, assume that the pointer in Fig. 12.10 is spun and then the pointer in Fig. 12.11 is spun. Find the probability of the pointers landing on

43. red on both spins.

44. red on the first spin and yellow on the second spin.

45. a color other than yellow on both spins.

46. yellow on the first spin and a color other than yellow on the second spin.

47. The Haefners plan to have five children. Find the probability that all their children will be boys. (Assume that $P(\text{boy}) = \frac{1}{2}$, and assume independence.)

48. The McDonalds presently have five boys. Mrs. McDonald is expecting another child. Find the probability that it will be a boy.

In Exercises 49–52, a family has three children. Assuming independence, find the probability that

49. all three are girls.

50. all three are boys.

51. the youngest child is a boy and the older children are girls.

52. the youngest child is a girl, the middle child is a boy, and the oldest child is a girl.

In Exercises 53–56, a box contains four marbles: two red, one blue, and one green. Two marbles will be selected at random. Find the probability of selecting each of the following

　a) *with replacement.*
　b) *without replacement.*

53. A red marble and then a blue marble

54. Two blue marbles

55. No red marbles

56. No green marbles

In Exercises 57–60, each individual letter of the word "pfeffernuesse" is placed on a piece of paper, and all 13 pieces of paper are placed in a hat. Three letters are selected at random from the hat. Find the probability of selecting each of the following

　a) *with replacement.*
　b) *without replacement.*

57. Three *f*s

58. No *f*s

59. Three vowels

60. The first letter selected is not a vowel and the second and third letters selected are vowels.

In Exercises 61–64, a sample of 150 people indicates that 50 favor candidate A, 65 favor candidate B, 20 favor candidate C, and 15 have no opinion. Three people from the sample are randomly selected. Find the probability of each of the following.

61. The first favors candidate A, the second favors candidate B, and the third favors candidate C.

62. The first two have no opinion, and the third favors candidate C.

63. All three favor candidate A.

64. None favors candidate C.

In Exercises 65–68, an experimental drug was given to a sample of 100 hospital patients with an unknown sickness. Of the total, 70 patients reacted favorably, 10 reacted unfavorably, and 20 were unaffected by the drug. Assume that this sample is representative of the entire population. If this drug is given to Mr. and Mrs. Connors and their son Jimmy, what is the probability of each of the following? (Assume independence.)

65. Mrs. Connors reacts favorably.

66. Mr. and Mrs. Connors react favorably, and Jimmy is unaffected.

67. All three react favorably.

68. None reacts favorably.

In Exercises 69–74, each question of a five-question multiple choice exam has four possible answers. Gurshawn picks an answer at random for each question. Find the probability that he selects the correct answer on

69. any one question.

70. only the first question.

71. only the third and fourth questions.

72. all five questions.

73. none of the questions.

74. at least one of the questions.

In Exercises 75–78, consider the slot machine shown.

Most people who play slot machines end up losing money because the machines are designed to favor the casino (the house). A slot machine generally has three or four reels that spin independently of each other. When you pull the handle the reels spin and the pictures on the reels come to rest in a random order. The following is a list of the number of symbols of each type on each of the three reels (for one particular slot machine).

Pictures on Reels	Reels		
	1	**2**	**3**
Cherries 🍒	2	4	4
Oranges ⬤	5	4	5
Plums ⬤	6	3	4
Bells 🔔	2	4	3
Melons ⬤	2	2	2
Bars ▭	2	2	1
7s 𝟟	1	1	1

For this slot machine, find the probability of obtaining

75. a cherry on the first reel.

76. oranges on all three reels.

77. no 7s.

78. three 7s.

In Exercises 79–82, the following double wheel is spun. The outer wheel is spun clockwise and the inner wheel is spun counterclockwise.

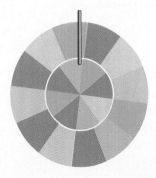

Assuming that the wheels are independent and each outcome is equally likely, find the probability that

79. red on both wheels stop under the pointer.

80. red on the outer wheel and blue on the inner wheel stop under the pointer.

81. red on neither wheel stops under the pointer.

82. red on at least one wheel stops under the pointer.

In Exercises 83–86, the probability that a heat-seeking torpedo will hit its target is 0.4. If the first torpedo hits its target, the probability that the second torpedo will hit the target increases to 0.9 because of the extra heat generated by the first explosion. If two heat-seeking torpedos are fired at a target, find the probability that

83. neither hits the target.

84. the first hits the target and the second misses the target.

85. both hit the target.

86. the first misses the target and the second hits the target.

87. Certain birth defects and syndromes are *polygenetic* in nature. Typically, the chance that an offspring will be born with a polygenetic affliction is small. However, once an offspring is born with the affliction, the probability that future offspring of the same parents will be born with the same affliction increases. Let's assume that the probability of a child's being born with affliction *A* is 0.001. If a child is born with this affliction, the probability of a future child's being born with the same affliction becomes 0.04.

a) Are the events of the births of two children in the same family with affliction *A* independent? Explain.

b) A couple plans to have one child. Find the probability that the child will be born with this affliction.

c) A couple plans to have two children. Find the probability that
   i) both children will be born with the affliction.
   ii) the first has the affliction and the second does not.
   iii) neither has the affliction.

In Exercises 88–91, use the fact that the Internal Revenue Service claims that 24 in every 1000 people in the $10,000–$50,000 income bracket are audited yearly. Assuming that the returns to be audited are selected at random and that each year's selections are independent of the previous year's selections, find the probability that a person in this income bracket will be audited

88. this year.

89. the next two years in succession.

90. this year but not next year.

91. neither this year nor next year.

92. Ms. Runningdeer has a lottery ticket with a three-digit number in the range 000 to 999. Three balls are to be selected at random with replacement from a bin. An equal number of balls are marked with the digits 0, 1, 2, . . . , 9. Find the probability that Ms. Runningdeer's number is selected.

A die has 1 dot on one side, 2 dots on two sides, and 3 dots on three sides. If the die is rolled once, find the probability of rolling

93. a 2.

94. a 3.

95. an even number or a number less than 3.

96. an odd number or a number greater than 1.

Problem Solving/Group Activities

97. A bag contains five red chips, three blue chips, and two yellow chips. Two chips are selected from the bag without replacement. Find the probability that two chips of the same color are selected.

98. Ron has ten coins from Japan: three 1-yen coins, one 10-yen coin, two 20-yen coins, one 50-yen coin, and three 100-yen coins. He selects two coins at random without replacement. Assuming that each coin is

equally likely to be selected, find the probability that Ron selects at least one 1-yen coin.

99. An investment advisor is considering the purchase of stock of 5 drug companies and 10 computer companies. Knowing that each company had the same record of gain over the last year, the investment advisor decides to select 3 stocks at random from the 15 stocks. What is the probability that 2 stocks are computer companies and that 1 is a drug company?

100. Two playing cards are dealt to you from a well-shuffled deck of 52 cards. If either card is a diamond, or both are, you win; otherwise, you lose. Determine whether this game favors you, is fair, or favors the dealer. Explain your answer.

101. You have three cards: an ace, a king, and a queen. A friend shuffles the cards, selects two of them at random, and discards the third. You ask your friend to show you a picture card, and she turns over the king. What is the probability that she also has the queen?

Research Activity

102. Girolamo Cardano (1501–1576) wrote *Liber de Ludo Aleae*, which is considered to be the first book on probability. Cardano had a number of different vocations. Do research and write a paper on the life and accomplishments of Girolamo Cardano. References include encyclopedias and history of mathematics books.

12.7 CONDITIONAL PROBABILITY

In Section 12.6 we indicated that events are independent when the outcome of either event has no effect on the outcome of the other event. For example, selecting two cards from a deck of cards *with replacement* represents independent events. However, not all events are independent events. Consider this problem: Find the probability of selecting two aces from a deck of cards *without replacement*. The probability of selecting the first ace is $\frac{4}{52}$. The probability that the second ace is selected becomes $\frac{3}{51}$, since we assume that an ace was removed from the deck with the first selection. Since the probability of the second ace being selected is affected by the first ace being selected, these two events are *dependent*. Probability problems involving dependent events can be solved by using conditional probability.

> **Conditional Probability**
>
> In general, the probability of event E_2 occurring, given that an event E_1 has happened (or will happen—the time relationship does not matter) is called a **conditional probability** and is written $P(E_2 | E_1)$.

The symbol $P(E_2 | E_1)$, read "the probability of E_2, given E_1," represents the probability of E_2 occurring, assuming that E_1 has already occurred (or will occur).

EXAMPLE 1

A single card is selected from a deck of cards. Find the probability that it is a club, given that it is black.

Solution: We are told that it is black. Thus only 26 cards are possible, of which 13 are clubs. Therefore

$$P(\text{Club} | \text{Black}) \text{ or } P(C | B) = \frac{13}{26} = \frac{1}{2}.$$

EXAMPLE 2

Given a family with two children, and assuming that boys and girls are equally likely, find the probability that the family has

a) two girls.

b) two girls if you know at least one of the children is a girl.

c) two girls given that the older child is a girl.

Solution:

a) The sample space of two children, BB, BG, GB, GG, can be determined by a tree diagram (see Fig. 12.13).

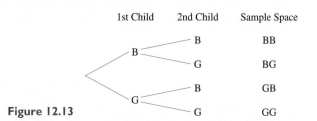

Figure 12.13

Since there are four possibilities, of which only one has two girls,

$$P(2 \text{ girls}) = \frac{1}{4}.$$

b) We are told that at least one of the children is a girl. Therefore for this problem the sample space is BG, GB, GG. Since there are three possibilities, of which only one has two girls,

$$P(\text{both girls} | \text{at least one is a girl}) = \frac{1}{3}.$$

c) If the older child is a girl, the sample space reduces to GB, GG. Thus

$$P(\text{both girls} | \text{older child is a girl}) = \frac{1}{2}.$$

There are a number of formulas that can be used to find conditional probabilities. The one we will use in this book follows.

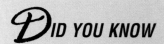

Conditional Probability

For any two events E_1 and E_2,

$$P(E_2|E_1) = \frac{n(E_1 \text{ and } E_2)}{n(E_1)}$$

In the formula $n(E_1 \text{ and } E_2)$ represents the number of sample points common to both event 1 and event 2, and $n(E_1)$ is the number of sample points in event E_1, the given event. Since the intersection of E_1 and E_2, $E_1 \cap E_2$, represents the sample points common to both E_1 and E_2, the formula can also be expressed as

$$P(E_2|E_1) = \frac{n(E_1 \cap E_2)}{n(E_1)}$$

Figure 12.14 is helpful in explaining conditional probability.

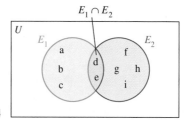

Figure 12.14

Here, the number of elements in E_1 is five, the number of elements in E_2 is six, and the number of elements in both E_1 and E_2, or $E_1 \cap E_2$, is two.

$$P(E_2|E_1) = \frac{n(E_1 \text{ and } E_2)}{n(E_1)} = \frac{2}{5}$$

Thus for this situation the probability of selecting an element from E_2, given that the element is in E_1, is $\frac{2}{5}$.

EXAMPLE 3

Each person in a sample of 200 residents in Washington County was asked whether he or she favored having a single countywide police department. The county consists of one large city and a number of small townships. The response of those sampled is given in the following table.

Residence	Favor	Oppose	Total
Live in city	80	50	130
Live outside city	60	10	70
Total	140	60	200

If one person from the sample is selected at random, find the probability that the person

a) favors a single combined police force.

b) favors a single combined police force if the person lives in the city.

c) opposes a single countywide police force if the person lives outside the city.

Solution:

a) The total number of respondents is 200, of which 140 favor a single police force.

$$P(\text{favors a single police force}) = \frac{140}{200} = \frac{7}{10}$$

b) Let E_1 be the given information "the person lives in the city." Let E_2 be "the person favors a single countywide police force." We are being asked to find $P(E_2 | E_1)$. The number of people who live in the city, $n(E_1)$, is 130. The number of people who live in the city and favor a single countywide police force, $n(E_1 \text{ and } E_2)$, is 80. Thus

$$P(E_2 | E_1) = \frac{n(E_1 \text{ and } E_2)}{n(E_1)} = \frac{80}{130} = \frac{8}{13}.$$

c) Let E_1 be the given information "the person lives outside the city." Let E_2 be "the person opposes a single countywide police force." We are asked to find $P(E_2 | E_1)$. The number of people who live outside the city, $n(E_1)$, is 70. The number of people who live outside the city and oppose the countywide police force, $n(E_1 \text{ and } E_2)$, is 10. Thus

$$P(E_2 | E_1) = \frac{n(E_1 \text{ and } E_2)}{n(E_1)} = \frac{10}{70} = \frac{1}{7}.$$

Section 12.7 Exercises

1. What does the notation $P(E_2 | E_1)$ mean?

2. Give the formula for $P(E_2 | E_1)$.

In Exercises 3–8, consider the circles shown.

Assume that one circle is selected at random and that each circle is equally likely to be selected. Find the probability of selecting

3. a 5, given that the circle is orange.

4. a 5, given that the circle is yellow.

5. an even number, given that the number is greater than 2.

6. a number less than 2, given that the number is less than 5.

7. a green number, given that the circle is orange.

8. a number greater than 3, given that the circle is yellow.

In Exercises 9–16, consider the following wheel.

If the wheel is spun and each section is equally likely to stop under the pointer, find the probability that the pointer lands on

9. a nine, given that the color is purple.

10. an even number, given that red stops under the pointer.

11. purple, given that the number is odd.

12. a number greater than 4, given that red stops under the pointer.

13. a number greater than 4, given that purple stops under the pointer.

14. an even number, given that red or purple stops under the pointer.

15. purple, given that a number greater than 5 stops under the pointer.

16. yellow, given that a number greater than 10 stops under the pointer.

In Exercises 17–20, assume that a hat contains four bills: a $1, a $5, a $10, and a $20 bill. Two bills are to be selected at random with replacement. Construct a sample space as was done in Example 2 and find the probability that

17. both bills are $5 bills.

18. both bills are $5 if the first selected is a $5 bill.

19. both bills are $5 if at least one of the bills is a $5 bill.

20. both bills are greater than $5 if the second bill is a $10 bill.

In Exercises 21–24, assume that a couple has three children and that boys and girls are equally likely. Construct a sample space and find the probability that

21. they have three boys.

22. they have at least two boys if the first born is a boy.

23. they have three girls if the first and second born are girls.

24. they have no girls if the first is a boy.

In Exercises 25–30, two dice are rolled. Construct a sample space and find the probability that the sum of the points on the dice total

25. 7.

26. 7 if the first die is a 1.

27. 7 if the first die is a 3.

28. an even number if the second die is a 2.

29. a number greater than 7 if the second die is a 5.

30. a 7 or 11 if the first die is a 5.

The results of a survey of the service at a local restaurant are summarized as follows.

Customer	Service Good	Service Poor	Total
Male	40	10	50
Female	55	20	75
Total	95	30	125

In Exercises 31–34, use the survey to find the probability that the service was rated

31. good.

32. good, given that the customer was a male.

33. poor, given that the customer was a female.

34. poor, given that the customer was a male.

In Exercises 35–40, a quality control inspector is checking a sample of light bulbs for defects. The following table summarizes her findings.

Wattage	Good	Defective	Total
20	80	15	95
50	100	5	105
100	120	10	130
Total	300	30	330

If one of the light bulbs is selected at random, find the probability that the light bulb is

35. good.

36. good, given that it is 50 watts.

37. defective, given that it is 20 watts.

38. good, given that it is 100 watts.

39. good, given that it is 50 or 100 watts.

40. defective, given that it is not 50 watts.

In Exercises 41–45, 210 individuals are asked which evening news they watch most often. The results are summarized as follows.

Viewers	ABC	NBC	CBS	Other	Total
Men	30	20	40	25	115
Women	50	10	20	15	95
Total	80	30	60	40	210

If one of these individuals is selected at random find the probability that the person watches

41. ABC or NBC.

42. ABC, given that the individual is a woman.

43. ABC or NBC, given that the individual is a man.

44. a station other than CBS, given that the individual is a woman.

45. ABC, NBC, or CBS, given that the individual is a man.

Problem Solving/Group Activities

In Exercises 46–52, suppose that each circle is equally likely to be selected. One circle is selected at random.

Find the probability indicated.

46. P(green circle | + obtained)

47. P(+ | orange circle obtained)

48. P(yellow circle | − obtained)

49. P(green + | + obtained)

50. P(green or orange circle | green + obtained)

51. P(orange circle with green + | + was obtained)

52. P(green + or green − | orange circle was obtained)

In Exercises 53–58, a pair of dice are rolled. Find the probability indicated.

53. P(sum = 7 | at least one die is a 3)

54. P(sum = 12 | at least one die is a 5)

55. P(sum ≠ 4 | at least one die is a 2)

56. P(sum = 7 | sum is odd)

57. P(double rolled | sum is 8)

58. P(sum = 6 | double is rolled)

59. Consider the Venn diagram that follows. The numbers in the regions of the circle indicate the number of items that belong to that region. For example, 60 items are in set A but not in set B.

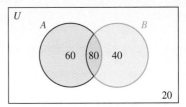

Find **a)** $n(A)$ **b)** $n(B)$ **c)** $p(A)$ **d)** $p(B)$

Use the formula on page 617 to find:

e) $p(A | B)$ **f)** $p(B | A)$

g) Explain why $p(A | B) \neq p(A) \cdot p(B)$.

60. A formula we gave for conditional probability is

$$p(E_2 | E_1) = \frac{n(E_1 \text{ and } E_2)}{n(E_1)}$$

This formula may be derived from the formula

$$p(E_2 | E_1) = \frac{p(E_1 \text{ and } E_2)}{p(E_1)}$$

Can you explain why? (*Hint:* Consider what happens to the denominators of $p(E_1 \text{ and } E_2)$ and $p(E_1)$ when they are expressed as fractions and the fractions are divided out.)

61. Given that $p(A) = 0.3$, $p(B) = 0.4$, and $p(A \text{ and } B) = 0.12$, use the formula

$$p(E_2 | E_1) = \frac{p(E_1 \text{ and } E_2)}{p(E_1)}$$

to find

a) $p(A | B)$ **b)** $p(B | A)$

c) Are A and B independent? Explain.

12.8 THE COUNTING PRINCIPLE AND PERMUTATIONS

In Section 12.5 we introduced the counting principle, which is repeated here for your convenience.

> **Counting Principle**
>
> If a first experiment can be performed in M distinct ways and a second experiment can be performed in N distinct ways, then the two experiments in that specific order can be performed in $M \cdot N$ distinct ways.

The counting principle is illustrated in Examples 1 and 2.

EXAMPLE 1

A license plate is to consist of two letters followed by three digits. Determine how many different license plates are possible if

a) repetition of letters and digits is permitted.

b) repetition of letters and digits is not permitted.

c) the first letter must be a vowel (*a, e, i, o, u*) and the first digit cannot be a 0, and repetition of letters and digits is not permitted.

Solution: There are 26 letters and 10 digits (0–9). We have five positions to fill, as indicated.

$$\underline{\text{L}} \ \underline{\text{L}} \ \underline{\text{D}} \ \underline{\text{D}} \ \underline{\text{D}}$$

a) Since repetition is permitted, there are 26 possible choices for both the first and second positions. There are 10 possible choices for the third, fourth, and fifth positions.

$$\frac{26}{\text{L}} \ \frac{26}{\text{L}} \ \frac{10}{\text{D}} \ \frac{10}{\text{D}} \ \frac{10}{\text{D}}$$

Since $26 \cdot 26 \cdot 10 \cdot 10 \cdot 10 = 676{,}000$, there are 676,000 different possible arrangements.

b) There are 26 possibilities for the first position. Since repetition of letters is not permitted, there are only 25 possibilities for the second position. The same reasoning is used when determining the number of digits for positions 3 through 5.

$$\frac{26}{\text{L}} \ \frac{25}{\text{L}} \ \frac{10}{\text{D}} \ \frac{9}{\text{D}} \ \frac{8}{\text{D}}$$

Since $26 \cdot 25 \cdot 10 \cdot 9 \cdot 8 = 468{,}000$, there are 468,000 different possible arrangements.

c) Since the first letter must be an *a, e, i, o,* or *u,* there are five possible choices for the first position. The second position can be filled by any of the letters except for the vowel selected for the first position. Therefore there are 25 possibilities for the second position.

Since the first digit cannot be a 0, there are nine possibilities for the third position. The fourth position can be filled by any digit except the one selected for the third position. Thus there are nine possibilities for the fourth position. Since the last position cannot be filled by any of the two digits previously used, there are eight possibilities for the last position:

$$\frac{5}{\text{L}} \ \frac{25}{\text{L}} \ \frac{9}{\text{D}} \ \frac{9}{\text{D}} \ \frac{8}{\text{D}}$$

Since $5 \cdot 25 \cdot 9 \cdot 9 \cdot 8 = 81{,}000$, there are 81,000 different arrangements that meet the conditions specified. ▲

EXAMPLE 2

At a baseball card show, Kristen DiMarco wishes to display five different Mickey Mantle cards.

a) In how many different ways can she place the five cards in a straight row?

b) If she wants to place Mickey Mantle's rookie card in the middle, in how many ways can she arrange the cards?

Solution:

a) There are five positions to fill, using the five cards. In the first position, on the left, she can use any one of the five cards. In the second position she can use any of the four remaining cards. In the third position she can use any of the three remaining cards, and so on. The number of distinct possible arrangements is

$$\underline{5} \cdot \underline{4} \cdot \underline{3} \cdot \underline{2} \cdot \underline{1} = 120.$$

b) We begin by satisfying the specified requirements stated. In this case the rookie card must be placed in the middle. Therefore there is only one possibility for the middle position.

$$\underline{} \quad \underline{} \quad \underline{1} \quad \underline{} \quad \underline{}$$

For the first position there are now four possibilities. For the second position, there will be three possibilities. For the fourth position there will be two possibilities. Finally, in the last position, there is only one possibility.

$$\underline{4} \cdot \underline{3} \cdot \underline{1} \cdot \underline{2} \cdot \underline{1} = 24$$

Thus under the condition stated there are 24 different possible arrangements. ▲

Permutations

Now we introduce the definition of a permutation.

> A **permutation** is any *ordered arrangement* of a given set of objects.

Peter, Paul, Mary and Mary, Paul, Peter represent two different permutations of the same three names. In Example 2(a) there are 120 different ordered arrangements, or permutations, of the five Mickey Mantle cards. In Example 2(b) there are 24 different arrangements, or permutations, possible if Mickey Mantle's rookie card must be displayed in the middle.

When determining the number of permutations possible we assume that repetition of an item is not permitted. To help you understand and visualize permutations, we illustrate the various permutations possible when a triangle, rectangle, and circle are to be placed in a line.

Six Permutations

Note that for this set of three shapes six different arrangements, or six permutations, are possible. We can obtain the number of permutations by using the counting principle. For the first position there are three choices. There are then two choices for the second position, and only one choice is left for the third position.

$$\text{Number of permutations} = 3 \cdot 2 \cdot 1 = 6$$

The product $3 \cdot 2 \cdot 1$ is referred to as 3 factorial, and written 3!. Thus

$$3! = 3 \cdot 2 \cdot 1 = 6.$$

Number of Permutations

The number of permutations of n distinct items is n factorial symbolized $n!$, where

$$n! = n(n - 1)(n - 2) \cdots (3)(2)(1).$$

It is important to note that 0! is defined to be 1. Many calculators have a factorial key, generally symbolized as $\boxed{x!}$ or $\boxed{n!}$. To find 5! on your calculator you would press

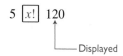

EXAMPLE 3

In how many ways can eight different books be arranged on a shelf?

Solution: Since there are eight different books, the number of permutations is 8!

$$8! = 8 \cdot 7 \cdot 6 \cdot 5 \cdot 4 \cdot 3 \cdot 2 \cdot 1 = 40,320$$

▲

Example 4 illustrates how to use the counting principle to determine the number of permutations possible when only a part of the total number of items is to be selected and arranged.

EXAMPLE 4

Consider the five letters p, q, r, s, t. In how many distinct ways can three letters be selected and arranged if repetition is not allowed?

Solution: We are asked to select and arrange only three of the five possible letters. Using the counting principle, we find that there are five possible letters for the first choice, four possible letters for the second choice, and three possible letters for the third choice:

$$5 \cdot 4 \cdot 3 = 60.$$

Thus there are 60 different possible ordered arrangements, or permutations.

▲

In Example 4 we determined the number of different ways in which we could select and arrange three of the five items. We can indicate this by using the notation $_5P_3$. The notation $_5P_3$ is read ''the number of permutations of five items taken three at a time.'' The notation $_nP_r$, is read ''the number of permutations of n items taken r at a time.''

We use the counting principle in the following problems to evaluate permutations of the form $_nP_r$.

One more than 7 − 4

$$_7P_4 = 7 \cdot 6 \cdot 5 \cdot 4$$

One more than 9 − 3

$$_9P_3 = 9 \cdot 8 \cdot 7$$

One more than 10 − 5

$$_{10}P_5 = 10 \cdot 9 \cdot 8 \cdot 7 \cdot 6$$

Notice to evaluate $_nP_r$ we begin with n and form a product of r consecutive descending factors. For example, to evaluate $_{10}P_5$, we start with 10 and form a product of 5 consecutive descending factors (see the preceding illustration).

In general, the number of permutations of n items taken r at a time, $_nP_r$, may be found by the formula

One more than $n − r$

$$_nP_r = n(n - 1)(n - 2) \cdots (n - r + 1)$$

Therefore when evaluating $_{20}P_{15}$ we would find the product of consecutive decreasing integers from 20 to $(20 - 15 + 1)$ or 6, which is written as $20 \cdot 19 \cdot 18 \cdot 17 \cdot \cdots \cdot 6$.

Now let's develop an alternative formula that we can use to find the number of permutations possible when r objects are selected from n objects:

$$_nP_r = n(n - 1)(n - 2) \cdots (n - r + 1),$$

or

$$_nP_r = n(n - 1)(n - 2) \cdots (n - r + 1) \times \frac{(n - r)!}{(n - r)!}.$$

For example,

$$_{10}P_5 = 10 \cdot 9 \cdot \cdots \cdot 6 \times \frac{5!}{5!},$$

or

$$_{10}P_5 = \frac{10 \cdot 9 \cdot \cdots \cdot 6 \times 5!}{5!}.$$

The expression for $_nP_r$ can be rewritten as

$$_nP_r = \frac{n(n - 1)(n - 2) \cdots (n - r + 1)\overbrace{(n - r)(n - r - 1) \cdots (3)(2)(1)}^{(n - r)!}}{(n - r)!}.$$

Since the numerator of this expression is $n!$, we can write

$$_nP_r = \frac{n!}{(n - r)!}.$$

For example,

$$_{10}P_5 = \frac{10!}{(10 - 5)!}.$$

The number of permutations possible when r objects are selected from n objects is found by the **permutation formula**

$$_nP_r = \frac{n!}{(n - r)!}.$$

In Example 4 we found that when selecting three of five letters there were 60 permutations. We can obtain the same result using the permutation formula:

$$_5P_3 = \frac{5!}{(5 - 3)!} = \frac{5!}{2!} = \frac{5 \cdot 4 \cdot 3 \cdot \cancel{2 \cdot 1}}{\cancel{2 \cdot 1}} = 60.$$

EXAMPLE 5

You are among eight people forming a bicycle club. Collectively you decide to put everyone's name in a hat and to randomly select a president, a vice-president, and a secretary. How many different arrangements or permutations of officers are possible?

Solution: Since there are eight people, $n = 8$, of which three are to be selected; thus $r = 3$.

$$_8P_3 = \frac{8!}{(8 - 3)!} = \frac{8!}{5!} = \frac{8 \cdot 7 \cdot 6 \cdot \cancel{5 \cdot 4 \cdot 3 \cdot 2 \cdot 1}}{\cancel{5 \cdot 4 \cdot 3 \cdot 2 \cdot 1}} = 336$$

Thus with eight people there can be 336 different arrangements for president, vice-president, and secretary. ▲

In Example 5, the fraction

$$\frac{8 \cdot 7 \cdot 6 \cdot \cancel{5 \cdot 4 \cdot 3 \cdot 2 \cdot 1}}{\cancel{5 \cdot 4 \cdot 3 \cdot 2 \cdot 1}} = 336$$

can be also expressed as

$$\frac{8 \cdot 7 \cdot 6 \cdot \cancel{5!}}{\cancel{5!}} = 336.$$

The solution to Example 5, like other permutation problems, can also be obtained with the counting principle.

EXAMPLE 6

The Stereo Shop's warehouse has 10 different receivers in stock. The owner of the chain calls the warehouse requesting that a different receiver be sent to each of its 6 stores. How many ways can the distribution of the receivers to the stores be made?

Solution: There are 10 receivers from which 6 are to be selected and distributed. Thus $n = 10$ and $r = 6$.

$$_{10}P_6 = \frac{10!}{(10-6)!} = \frac{10!}{4!} = \frac{10 \cdot 9 \cdot 8 \cdot 7 \cdot 6 \cdot 5 \cdot \cancel{4!}}{\cancel{4!}}$$

$$= 151{,}200$$

There are 151,200 different permutations of receivers possible.   ▲

We have worked permutation problems (selecting and arranging, without replacement, *r* items out of *n distinct* items) by using the counting principle and using the permutation formula. When you are given a permutation problem, unless specified by your instructor, you may use either technique to determine its solution.

Permutations of Duplicate Items

So far, all the examples we have discussed in this section have involved arrangements with distinct items. Now we will consider permutation problems in which some of the items to be arranged are duplicates. For example, the word BOB contains three letters, of which the two Bs are duplicates. How many permutations of the word BOB are possible? If the two Bs were distinguishable (one red and the other blue), there would be six permutations.

BOB   BBO   OBB

BOB   BBO   OBB

However if the Bs are not distinguishable (replacing all colored Bs with black print), we see that there are only three permutations

BOB   BBO   OBB

The number of permutations of BOB can be computed as

$$\frac{3!}{2!} = \frac{3 \cdot \cancel{2 \cdot 1}}{\cancel{2 \cdot 1}} = 3,$$

where 3! represents the number of permutations of three letters, assuming that none are duplicates, and 2! represents the number of ways the two items that are duplicates can be arranged (BB or BB). In general, we have the following rule.

> **Permutations of Duplicate Objects**
>
> The number of distinct permutations of *n* objects where n_1 of the objects are identical, n_2 of the objects are identical, ..., n_r of the objects are identical is found by the formula
>
> $$\frac{n!}{n_1! n_2! \cdots n_r!}.$$

EXAMPLE 7

In how many different ways can the letters of the word *statistics* be arranged?

Solution: Of the 10 letters, three are *t*'s, three are *s*'s, and two are *i*'s. The number of possible arrangements is

$$\frac{10!}{3!3!2!} = \frac{10 \cdot 9 \cdot 8 \cdot 7 \cdot 6 \cdot 5 \cdot 4 \cdot 3 \cdot 2 \cdot 1}{3 \cdot 2 \cdot 1 \cdot 3 \cdot 2 \cdot 1 \cdot 2 \cdot 1} = 50,400.$$

There are 50,400 different possible arrangements of the letters in the word ''statistics.''

Section 12.8 Exercises

1. In your own words, state the meaning of the counting principle.

2. In your own words, describe a permutation.

3. In your own words, explain how to find $n!$ for any whole number n.

4. Give the formula for the number of permutations of n distinct items.

5. How do you read $_nP_r$? When you evaluate $_nP_r$, what does the outcome represent?

6. Give the formula for the number of permutations when r objects are selected from n objects.

7. Give the formula for the number of permutations when $n_1, n_2, \ldots, n_r$, of the objects are identical.

8. Does $_1P_1 = {}_1P_0$? Explain.

In Exercises 9–20, evaluate the expression.

9. $6!$ 10. $8!$ 11. $7!$ 12. $_5P_2$

13. $0!$ 14. $_6P_4$ 15. $_9P_0$ 16. $_5P_0$

17. $_8P_7$ 18. $_4P_4$ 19. $_8P_8$ 20. $_9P_6$

21. To use an automated teller machine, you generally must enter a four-digit code, using the digits 0–9. How many four-digit codes are possible if repetition of digits is permitted?

22. The daily double at most race tracks consists of selecting the winning horse in both the first and second races. If the first race has seven entries and the second race has eight entries, how many daily double tickets must you purchase to guarantee a win?

23. Some doors on cars can now be opened by pressing the correct sequence of buttons. A display of the five buttons by the door handle of a car follows.

```
  0-1    2-3    4-5    6-7    8-9
```

The correct sequence of five buttons must be pressed to unlock the door. If the same button may be pressed consecutively,

a) how many possible ways can the five buttons be pressed (repetition is permitted)?

b) If five buttons are pressed at random, find the probability that a sequence that unlocks the door will be entered.

24. Secured doors at airports and other locations are opened by pressing the correct sequence of numbers on a control panel. If the control panel contains the digits 0–9 and a three-digit code must be entered (repetition is permitted),

a) how many different codes are possible?

b) If one code is entered at random, find the probability the correct code is entered.

25. A social security number consists of nine digits. How many different social security numbers are possible if repetition of digits is permitted?

26. A college identification number consists of a letter followed by four digits. How many different identification numbers are possible if repetition is not permitted?

27. The operator of the Audio Warehouse is planning a grand opening. He wishes to advertise that the store has many different sound systems available. They stock 10 different tape decks, 12 different receivers, and 9 different speakers. Assuming that a sound system will consist of one of each, and that all pieces are compatible, how many different sound systems can they advertise?

28. The trifecta at most race tracks consists of selecting the first-, second-, and third-place finishers in a particular race in their proper order. If there are six entries in the trifecta race, how many tickets must you purchase to guarantee a win?

29. Consider the five figures shown.

In how many different ways can the figures be arranged
a) from left to right?
b) from top to bottom if placed one under the other?
c) from left to right if the triangle is to be placed on the far right?
d) from left to right if the circle is to be placed on the far left and the triangle is to be placed on the far right?

30. The six pictures shown are to be placed side by side along a wall.

In how many ways can they be arranged from left to right if
a) they can be arranged in any order?
b) the bird must be on the far left?
c) the bird must be on the far left and the giraffe must be next to the bird?
d) a four-legged animal must be on the far right?

31. If a club consists of eight members, how many different arrangements of president and vice-president are possible?

32. Three door prizes—a baseball cap, a mug, and a magazine subscription—are to be awarded (gag gifts) at a get-together of seven people. If three people are to be selected at random and awarded the gifts, how many different ways can the gifts be awarded?

33. Each book registered in the Library of Congress must have an ISBN code number. For an ISBN number of the form D-DD-DDDDDD-D, where D represents a digit from 0 through 9, how many different ISBN numbers are possible if repetition of digits is allowed? (See research activity Exercise 65.)

34. At the reception line of a wedding, the bride, the groom, the best man, the maid of honor, the four ushers, and the four bridesmaids must line up to receive the guests.
a) If these individuals can line up in any order, how many arrangements are possible?
b) If the groom must be the last in line and the bride must be next to the groom, and the others can line up in any order, how many arrangements are possible?
c) If the groom is to be last in line, the bride next to the groom, and males and females are to alternate, how many arrangements are possible?

In Exercises 35–38, an identification code is to consist of two letters followed by two digits. How many different codes are possible if

35. repetition is permitted?

36. repetition is not permitted?

37. the first letter must be a W, X, Y, or Z and repetition is permitted?

38. the first two entries must both be the same letter and repetition of the digits is not permitted?

In Exercises 39–42, a license plate is to consist of four digits followed by two letters. Determine the number of different license plates possible if

39. repetition of numbers and letters is permitted.

40. repetition of numbers and letters is not permitted.

41. the first and second digits must be odd, and repetition is not permitted.

42. the first digit cannot be zero, and repetition is not permitted.

43. A telephone number consists of seven digits with the restriction that the first digit cannot be 0 or 1.
a) How many distinct telephone numbers are possible?
b) How many distinct telephone numbers are possible with three-digit area codes preceding the seven-digit number, where the first digit of the area code is not 0 or 1?
c) With the increasing use of cellular phones and paging systems our society is beginning to run out of usable phone numbers. Various phone companies are developing phone numbers that use 11 digits instead of 7. How many distinct phone numbers can be made with 11 digits, assuming that the area code remains three digits and the first digit of the area code and the phone number cannot be 0 or 1?

44. Find the number of permutations of the letters in the word PEANUT.

45. A teacher decides to give six different prizes to 6 of the 30 students in her class. In how many ways can she do so?

46. A bookcase contains 20 different books. Three books will be selected at random and given away as prizes to three individuals. In how many ways can the selection be made?

47. A night guard visits 10 different offices every hour. The pattern is varied each night so that the guard will not follow a specific routine. In how many different ways can this pattern be varied?

48. Find the number of permutations of the colors in the spectrum that are shown on page 629.

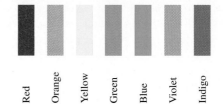

Red Orange Yellow Green Blue Violet Indigo

49. A bank has three drive-through stations. Assuming that each is equally likely to be selected by customers, how many different ways can the next six drivers select a station?

50. A man has 8 pairs of pants, 12 shirts, 15 ties, and 6 sport coats. Assuming he must wear one of each, how many different outfits can he wear?

51. a) To open a combination lock, you must know the lock's three-number sequence in its proper order. Repetition of numbers is permitted. Why is this lock more like a permutation lock than a combination lock? Why is it not a true permutation problem?
   b) Assuming that a combination lock has 40 numbers, determine how many different three-number arrangements are possible if repetition of numbers is allowed.
   c) Answer the question in part (b) if repetition is not allowed.

52. In how many ways can the letters in the word WINNING be arranged?

53. In how many ways can the letters in the word KISSIMMEE be arranged?

54. In how many ways can the letters in the word SASSAFRAS be arranged?

55. In how many ways can the digits of the number 1,242,332 be arranged?

56. In one question of a history test the student is asked to match ten dates with 10 events; each date can only be matched with 1 event. In how many different ways can this question be answered?

57. Five different colored flags will be placed on a pole, one beneath another. The arrangement of the colors indicates the message. How many messages are possible if five flags are to be selected from eight different colored flags?

58. In how many ways can the manager of a National League baseball team arrange his batting order of nine players if the pitcher must bat last?

59. Five Rembrandts are to be displayed in a museum.
   a) In how many different ways can they be arranged if they must be next to one another?
   b) In how many different ways can they be displayed if a specific one is to be in the middle?

Problem Solving/Group Activities

60. Door keys for a certain automobile are made from a blank key on which five cuts are made. Each cut may be one of five different depths.
   a) How many different keys can be made?
   b) If 400,000 of these automobiles are made and the same number of each type key is made, how many cars can be opened by a specific key?
   c) If one of these cars is selected at random, what is the probability that the key selected at random will open the car?

61. Patricia Burgess, who is playing Scrabble with Tonyo Poweiga, has seven different letters. She decides to test each five-letter permutation before her next move. If each permutation takes 5 sec, how long will Tonyo need to wait before Patricia's next move?

62. In Exercise 61, assume that, of Patricia's seven letters, three are identical and two are identical. How long will Tonyo need to wait if Patricia plans to try all different permutations of her seven letters?

63. On a ballot, each committee member is asked to rank three of seven candidates for recommendation for promotion, giving their first, second, and third choices (no ties). What is the minimum number of ballots that must be cast in order to guarantee that at least two ballots are the same?

64. Does $_nP_r = {}_nP_{(n-r)}$ for all whole numbers, where $n \geq r$? Explain.

Research Activity

65. When a book is published it is assigned a ten-digit code number called the International Standard Book Number (ISBN). Do research and write a report on how this coding system works. Ask a school librarian for help or check *The UMAP Journal*, 1985, 6(1):41–54, for an article entitled *International Standard Code Number* by P. Tuchinsky.

12.9   COMBINATIONS

When the order of the selection of the items is important to the final outcome, the problem is a permutation problem. When the order of the selection of the items is unimportant to the final outcome, the problem is a combination problem.

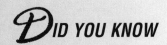

Recall from Section 12.8 that permutations are *ordered* arrangements. Thus, for example, *a*, *b*, *c* and *b*, *c*, *a* are two different permutations because the ordering of the three letters is different. The letters *a*, *b*, *c* and *b*, *c*, *a* represent the same combination of letters because the *same letters* are used in each set. However, the letters *a*, *b*, *c* and *a*, *b*, *d* represent two different combinations of letters because the letters contained in each set are different.

> A **combination** is a distinct group (or set) of objects without regard to their arrangement.

EXAMPLE I

Determine whether the situation represents a permutation or combination problem.

a) A group of five friends, Arline, Inez, Judy, Dan, and Eunice, are forming a club. The group will elect a president and a treasurer. In how many different ways can the president and treasurer be selected?

b) Of the five individuals named two will be attending a meeting together. How many different ways can they do so?

Solution:

a) Since the president's position is different from the treasurer's position, this problem is a permutation problem. Judy as president with Dan as treasurer is different from Dan as president with Judy as treasurer.

b) Since the order in which the two individuals selected to attend the meeting is not important, this is a combination problem. There is no difference if Judy is selected and then Dan is selected, or if Dan is selected and then Judy is selected. ▲

In Section 12.8 you learned that $_nP_r$ represents the number of permutations when r items are selected from n distinct items. *Similarly $_nC_r$ represents the number of combinations when r items are selected from n distinct items.*

Consider the set of elements $\{a, b, c, d, e\}$. The number of permutations of two letters from the set is represented as $_5P_2$ and the number of combinations of two letters from the set is represented as $_5C_2$. Twenty permutations of two letters and 10 combinations of two letters are possible from these five letters. Thus $_5P_2 = 20$ and $_5C_2 = 10$, as shown.

Permutations	**Combinations**
$\left.\begin{array}{l} ab,\ ba,\ ac,\ ca,\ ad,\ da,\ ae,\ ea,\ bc,\ cb, \\ bd,\ db,\ be,\ eb,\ cd,\ dc,\ ce,\ ec,\ de,\ ed \end{array}\right\} 20$	$\left.\begin{array}{l} ab,\ ac,\ ad,\ ae,\ bc, \\ bd,\ be,\ cd,\ ce,\ de \end{array}\right\} 10$

When discussing both combination and permutation problems, we always assume that the experiment is performed without replacement. That is why duplicate letters such as *aa* or *bb* are not included in the preceding example.

Note that from one combination of two letters two permutations can be formed. For example, the combination *ab* gives the permutations *ab* and *ba*, or twice as many permutations as combinations. Thus for this example we may write

$$_5P_2 = 2(_5C_2).$$

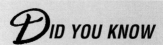

Since $2 = 2!$ we may write

$$_5P_2 = 2!(_5C_2).$$

If we repeated this same process for comparing the number of permutations in $_nP_r$ with the number of combinations in $_nC_r$, we would find that

$$_nP_r = r!(_nC_r).$$

Dividing both sides of the equation by $r!$ gives

$$_nC_r = \frac{_nP_r}{r!}.$$

Since $_nP_r = n!/(n - r)!$, the combination formula may be expressed as

$$_nC_r = \frac{n!/(n - r)!}{r!} = \frac{n!}{(n - r)!r!}.$$

> The number of combinations possible when r objects are selected from n objects is found by the **combination formula**
>
> $$_nC_r = \frac{n!}{(n - r)!r!}.$$

EXAMPLE 2

Eight colleagues staying at the same hotel want to go to a restaurant that is not within walking distance. Only one car that seats five is available. Assuming that any of the eight can drive the car, determine the number of different groups of people that can ride in the car.

Solution: This problem is a combination problem because the order in which the five people are selected is unimportant. There are a total of eight people, so $n = 8$. Five are to be selected, so $r = 5$.

$$_8C_5 = \frac{8!}{(8 - 5)!5!} = \frac{8!}{3!5!} = \frac{8 \cdot 7 \cdot \cancel{6} \cdot \cancel{5 \cdot 4 \cdot 3 \cdot 2 \cdot 1}}{\cancel{8} \cdot \cancel{2} \cdot 1 \cdot \cancel{5 \cdot 4 \cdot 3 \cdot 2 \cdot 1}} = 56$$

Thus 56 different combinations are possible when five from a group of eight are to be selected. ▲

EXAMPLE 3

An exam consists of six questions. Any four may be selected for answering. In how many ways can this selection be made?

Solution: This is a combination problem because the order in which the four questions are answered is immaterial.

$$_6C_4 = \frac{6!}{(6 - 4)!4!} = \frac{6!}{2!4!} = \frac{\overset{3}{\cancel{6}} \cdot 5 \cdot \cancel{4 \cdot 3 \cdot 2 \cdot 1}}{\cancel{2} \cdot 1 \cdot \cancel{4 \cdot 3 \cdot 2 \cdot 1}} = 15$$

There are 15 different ways that four of the six questions can be selected. List them now. ▲

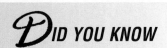
EXAMPLE 4

At Su Wong's Chinese Restaurant, dinner for eight consists of three items from column A, four items from column B, and three items from column C. If columns A, B, and C have five, seven, and six items, respectively, how many different dinner combinations are possible?

Solution: For column A, three of five items must be selected. This can be represented as $_5C_3$. For column B, four of seven items must be selected. This can be represented as $_7C_4$. For column C, three of six items must be selected, or $_6C_3$.

$$_5C_3 = 10, \qquad _7C_4 = 35, \quad \text{and} \quad _6C_3 = 20$$

Using the counting principle, we can determine the total number of dinner combinations by multiplying the number of choices from columns A, B, and C:

$$\text{Total number of dinner choices} = {}_5C_3 \cdot {}_7C_4 \cdot {}_6C_3$$
$$= 10 \cdot 35 \cdot 20 = 7000$$

Therefore 7000 different combinations are possible under these conditions. ▲

We have presented various counting methods, including the counting principle, permutations, and combinations. You will often need to decide which method to use to solve a problem. Table 12.4 may help you in selecting the procedure to use.

Table 12.4

Counting Principle: If a first experiment can be performed in M distinct ways and a second experiment can be performed in N distinct ways, then the two experiments in that specific order can be performed in $M \cdot N$ distinct ways. May be used with or without repetition. Use when determining the number of different ways that two or more experiments can occur. Also use when there are specific placement requirements, such as the first digit must be a 0 or 1.	Determining the Number of Ways of Selecting r Items from n Items. Repetition Not Permitted.	
	Permutations	**Combinations**
	Use when order is important. For example, a, b, c and b, c, a are two different permutations of the same three letters. $$_nP_r = \frac{n!}{(n-r)!}$$ Problems solved with the permutation formula may also be solved by using the counting principle.	Use when order is not important. For example, a, b, c and b, c, a are the same combination of three letters. But $a, b, c,$ and a, b, d are two different combinations of three letters. $$_nC_r = \frac{n!}{r!(n-r)!}$$

Section 12.9 Exercises

1. In your own words, explain what is meant by a combination.

2. What does $_nC_r$ mean?

3. Give the formula for finding $_nC_r$.

4. What is the relationship between $_nC_r$ and $_nP_r$?

5. In your own words, explain the difference between a permutation and a combination.

6. Assume that you have five different objects and that you are going to select three without replacement. Will

there be more combinations or more permutations of the three items? Explain.

In Exercises 7–18, evaluate the expression.

7. $_5C_4$ **8.** $_6C_2$

9. $_7C_3$ **10.** $_8C_0$

11. $_9C_5$ **12.** $_{12}C_8$

13. $_{10}C_3$ **14.** $_7C_7$

15. $\dfrac{_5C_3}{_5P_3}$ **16.** $\dfrac{_6C_2}{_6P_2}$

17. $\dfrac{_8C_5}{_8C_2}$ **18.** $\dfrac{_6C_6}{_8C_0}$

19. An ice cream parlor has 20 different flavors. Cynthia orders a banana split and has to select 3 different flavors. How many different selections are possible?

20. Mr. DeVito, a teacher, decides to give identical prizes to 8 of the 20 students in his class. In how many ways can he do so?

21. A student must select and answer four of five essay questions on a test. In how many ways can she do so?

22. A textbook search committee is considering eight books for possible adoption. The committee has decided to select three of the eight for further consideration. In how many ways can it do so?

23. Six of a sample of 15 television sets will be selected and tested for defects. In how many ways can the quality control engineer do so?

24. Three of eight finalists will be selected and awarded $5000 scholarships. In how many ways can this selection be made?

25. A disc jockey has 24 records that she wants to play during the next hour, but she has time to play only 14 of them. How many different ways can she select the 14?

26. At the beginning of the semester your instructor informs the class that she marks on a curve and that exactly 5 of the 26 students in the class will receive a final grade of A. In how many ways can this result occur?

In Exercises 27–30, a double wheel shown is to be spun. The outer wheel is to be spun clockwise and the inner wheel counterclockwise.

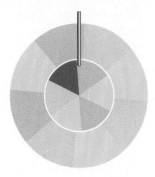

Assuming that each outcome is equally likely, find

27. the number of ways that green on both wheels can stop under the pointer.

28. the number of ways that the outer wheel can stop at orange and the inner wheel can stop at green under the pointer.

29. the number of ways that the outer wheel can stop at green and the inner wheel can stop at orange under the pointer.

30. the number of different color combinations possible when the two wheels stop under the pointer. Treat, for example, yellow on the outer wheel and orange on the inner wheel as the same combination as orange on the outer wheel and yellow on the inner wheel. Orange on both wheels counts as one combination. List the possible combinations.

31. A quinella bet consists of selecting the first- and second-place winners, in any order, in a particular event. For example, suppose you select a 2–5 quinella. If 2 wins and 5 finishes second, or if 5 wins and 2 finishes second, you win. In the game of jai alai, 8 two-man teams play against one another. How many quinella tickets are necessary to guarantee a win?

32. On an English test, Tito must write an essay for three of the five questions in Part 1 and four of the six questions in Part 2. How many different combinations of questions can he answer?

33. The singing group Alabama is planning to make a compact disc consisting of 6 fast songs and 4 slow songs. If they have 10 fast songs and 7 slow songs to choose from, how many different possible combinations do they have?

34. A new-car dealership has six different cars. A business is to select and purchase three of the cars. How many different selections are possible?

35. In the New York State lottery you must select the 6 winning numbers (in any order without replacement) out of 54 possible numbers to win the grand prize.
  a) How many combinations of the 6 winning numbers can be selected?

b) How many combinations of 6 numbers out of the 54 are possible?

36. An editor has eight manuscripts for mathematics books and five manuscripts for computer science books. If he is to select five mathematics and three computer science books for publication, how many different choices does he have?

37. Michael is sent to the store to get 5 different bottles of regular soda and 3 different bottles of diet soda. If there are 10 different regular sodas and 7 different diet sodas to choose from, how many different choices does Michael have?

38. How many different committees can be formed from 6 teachers and 50 students if the committee is to consist of 2 teachers and 3 students?

39. A teacher is constructing a mathematics test consisting of 10 questions. She has a pool of 28 questions, which are classified by level of difficulty as follows: 6 difficult questions, 10 average questions, and 12 easy questions. How many different 10-question tests can she construct from the pool of 28 questions if her test is to have 3 difficult, 4 average, and 3 easy questions?

40. General Mills is testing six oat cereals, five wheat cereals, and four rice cereals. If it plans to market three of the oat cereals, two of the wheat cereals, and two of the rice cereals, how many different combinations are possible?

41. A catering service is making up trays of hors d'oeuvres. The hors d'oeuvres are categorized as inexpensive, average, and expensive. If the client must select three of the seven inexpensive, five of the eight average, and two of the four expensive hors d'oeuvres, how many different choices are possible?

Problem Solving/Group Activities

42. Subsets were discussed on page 50. Consider set $A = \{a, b, c, d, e, f, g, h\}$. Determine the number of different subsets of set A that contain
   a) one element.
   b) three elements.
   c) six elements.
   d) How are subsets and combinations similar? Explain.

43. Consider a 10-question test in which each question can be answered either right or wrong.
   a) How many different ways are there to answer the questions so that eight are right and two are wrong?
   b) How many ways are there to answer the questions so that at least eight are right?

44. The notation $_nC_r$ may be written $\binom{n}{r}$.
   a) Use this notation to evaluate each of the combinations in the following array. Form a triangle of the results, similar to the one given, by placing the answer to each combination in the same relative position in the triangle.

$$\binom{0}{0}$$

$$\binom{1}{0} \quad \binom{1}{1}$$

$$\binom{2}{0} \quad \binom{2}{1} \quad \binom{2}{2}$$

$$\binom{3}{0} \quad \binom{3}{1} \quad \binom{3}{2} \quad \binom{3}{3}$$

$$\binom{4}{0} \quad \binom{4}{1} \quad \binom{4}{2} \quad \binom{4}{3} \quad \binom{4}{4}$$

   b) Using the number pattern in part (a), find the next row of numbers of the triangle (known as Pascal's triangle).

45. a) Four people at dinner make a toast. If each person is to tap glasses with each other person, how many taps will take place?
   b) Repeat part (a) with five people.
   c) How many taps will there be if there are n people at the dinner table?

46. Determine the number of combinations possible in a state lottery where you must select
   a) 6 of 46 numbers.
   b) 6 of 47 numbers.
   c) 6 of 48 numbers.
   d) 6 of 49 numbers.
   e) Does the number of combinations increase by the same amount going from part (a) to part (b) as from part (b) to part (c)?
   f) Guess the number of combinations possible for 6 of 50 numbers.
   g) Determine the number of combinations possible for 6 of 50 numbers. Compare your answer to your answer in part (f).

47. Prove that $_nC_r = {_nC_{(n-r)}}$.

48. a) How many distinct ways can four people be seated in a row?
   b) How many distinct ways can four people be seated at a circular table?

12.10 SOLVING PROBABILITY PROBLEMS BY USING COMBINATIONS

In Section 12.9 we discussed combination problems. Now we will use combinations to solve probability problems.

Suppose that we want to find the probability of selecting two picture cards (jacks, queens, or kings) when two cards are selected, without replacement, from a deck of 52 cards. Using the *and* probability formula discussed in Section 12.6, we could reason as follows.

$$P(2 \text{ picture cards}) = P(\text{1st picture card}) \cdot P(\text{2nd picture card})$$

$$= \frac{12}{52} \cdot \frac{11}{51} = \frac{132}{2652}, \quad \text{or} \quad \frac{11}{221}$$

Since the order of the two picture cards that are selected is not important to the final answer, this problem can be considered a combination probability problem. We can find the probability of selecting two picture cards, using combinations, by finding the number of possible successful outcomes (selecting two picture cards) and dividing that answer by the total number of possible outcomes (selecting any two cards).

The number of ways in which two picture cards can be selected from the 12 picture cards in a deck is $_{12}C_2$, or

$$_{12}C_2 = \frac{12!}{(12-2)!2!} = \frac{\overset{6}{\cancel{12}} \cdot 11 \cdot \cancel{10!}}{\cancel{10!} \cdot \cancel{2} \cdot 1} = \frac{66}{1} = 66.$$

The number of ways in which two cards can be selected from a deck of 52 cards is $_{52}C_2$, or

$$_{52}C_2 = \frac{52!}{(52-2)!2!} = \frac{\overset{26}{\cancel{52}} \cdot 51 \cdot \cancel{50!}}{\cancel{50!} \cdot \cancel{2} \cdot 1} = 1326.$$

Thus

$$P(\text{selecting 2 picture cards}) = \frac{_{12}C_2}{_{52}C_2} = \frac{66}{1326} = \frac{11}{221}.$$

Note that the same answer is obtained with either method. To give you more exposure to counting techniques, we will work the problems in this section using combinations.

┌─ **EXAMPLE 1**

A club consists of four men and five women. Three members are to be selected at random to form a committee. What is the probability that the committee will consist of three women?

Solution:

$$P\binom{\text{committee consists}}{\text{of 3 women}} = \frac{\text{number of possible committees with 3 women}}{\text{total number of possible 3-member committees}}$$

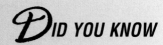

The number of possible committees with three women is $_5C_3 = 10$.
The total number of possible three-member committees is $_9C_3 = 84$.

$$P(\text{committee consists of 3 women}) = \frac{10}{84} = \frac{5}{42}$$

The probability of randomly selecting a committee with three women is $\frac{5}{42}$. ▲

EXAMPLE 2

An airplane has 10 empty seats left: 6 aisle seats and 4 window seats. If three people about to board the plane are given seat numbers at random, what is the probability that they are all given aisle seats?

Solution:

$$P\left(\begin{array}{c}\text{all 3 given}\\\text{aisle seats}\end{array}\right) = \frac{\begin{array}{c}\text{number of possible combinations}\\\text{of aisle seats for 3 people}\end{array}}{\begin{array}{c}\text{total number of possible combinations}\\\text{of seats for 3 people}\end{array}}$$

The number of possible combinations of seating the three people in the 6 aisle seats is $_6C_3 = 20$. The total number of possible combinations of seating the three people in the 10 available seats is $_{10}C_3 = 120$.

$$P(\text{all 3 given aisle seats}) = \frac{_6C_3}{_{10}C_3} = \frac{20}{120} = \frac{1}{6}$$

Thus the probability of all three passengers getting aisle seats is $\frac{1}{6}$. ▲

EXAMPLE 3

You are dealt a flush in the game of poker when you are dealt five cards of the same suit. If you are dealt a five-card hand, find the probability that you will be dealt a heart flush.

Solution:

$$P(\text{heart flush}) = \frac{\text{number of possible 5-card heart flushes}}{\text{total number of possible 5-card hands}}$$

The order in which the five cards are received is immaterial. Thus the number of possible hands can be found by the combination formula. Since there are 13 hearts in a deck of cards, the number of possible five-card flush hands is $_{13}C_5 = 1287$. The total number of possible five-card hands in a deck of 52 cards is $_{52}C_5 = 2,598,960$.

$$P(\text{heart flush}) = \frac{_{13}C_5}{_{52}C_5} = \frac{1287}{2,598,960} = \frac{33}{66,640}$$

The probability of being dealt a heart flush is $\frac{33}{66,640}$. ▲

EXAMPLE 4

The Kellogg Company is testing 12 new cereals for possible production. They are testing 3 oat cereals, 4 wheat cereals, and 5 rice cereals. If we assume that each of the 12 cereals has the same chance of being selected and 4 new cereals will be produced, find the probability that

a) no wheat cereals are selected,

b) at least 1 wheat cereal is selected,

c) 2 wheat cereals and 2 rice cereals are selected.

Solution:

a) If no wheat cereals are to be selected, then only oat and rice cereals must be selected. A total of 8 cereals are oat or rice. Thus the number of ways that 4 oat or rice cereals may be selected from the 8 possible oat or rice cereals is $_8C_4$. The total number of possible selections is $_{12}C_4$.

$$P(\text{no wheat cereals}) = \frac{_8C_4}{_{12}C_4} = \frac{70}{495} = \frac{14}{99}$$

b) When 4 cereals are selected, the choice must contain either no wheat cereal or at least 1 wheat cereal. Since one of these outcomes must occur, the sum of the probabilities must be 1, or

$$P(\text{no wheat cereal}) + P(\text{at least 1 wheat cereal}) = 1.$$

Therefore

$$P(\text{at least 1 wheat cereal}) = 1 - P(\text{no wheat cereal})$$

$$= 1 - \frac{14}{99} = \frac{99}{99} - \frac{14}{99} = \frac{85}{99}$$

Note that the probability of selecting no wheat cereals was found in part (a).

c) The number of ways of selecting 2 wheat cereals out of 4 wheat cereals is $_4C_2$, which equals 6. The number of ways of selecting 2 rice cereals out of 5 rice cereals is $_5C_2$, which equals 10. The total number of possible selections when 4 cereals are selected from the 12 choices is $_{12}C_4$. Since both the 2 wheat *and* the 2 rice cereals must be selected, the probability is calculated as follows.

$$P(\text{2 wheat and 2 rice}) = \frac{_4C_2 \cdot {_5C_2}}{_{12}C_4} = \frac{6 \cdot 10}{495} = \frac{60}{495} = \frac{4}{33}$$

Section 12.10 Exercises

In Exercises 1–10, the problems are to be done without replacement. Use combinations to determine probabilities.

1. Each of the numbers 1–6 is written on a piece of paper, and the six pieces of paper are placed in a hat. If two numbers are selected at random, find the probability that both numbers selected are even.

2. An urn contains five red balls and four blue balls. You plan to draw three balls at random. Find the probability of selecting three red balls.

3. A hat contains four 1-dollar bills, three 5-dollar bills, and one 10-dollar bill. If you select two bills at random from the hat, find the probability of selecting two 1-dollar bills.

4. A box contains four good and four defective batteries. If you select three at random, find the probability that you select three good batteries.

5. Each of the digits 0, 1, 2, 3, 4, 5, 6, 7, 8, and 9 is written on a slip of paper, and the slips are placed in a hat. If three slips of paper are selected at random, find the probability that the three numbers selected are greater than 4.

6. The 10 finalists in a scholarship contest consist of 6 women and 4 men. If two equal prizes are to be awarded, find the probability that both prizes are awarded to women. Assume that each applicant has an equal chance of being selected.

7. A three-person committee is to be selected at random from five Democrats and three Republicans. Find the probability that all three selected are Democrats.

8. A committee of four is to be randomly selected from a group of seven teachers and eight students. Find the probability that the committee will consist of four students.

9. A lottery consists of 46 numbers. You select 6 numbers and if they match the 6 numbers selected by the lottery commission, you win the grand prize. Find the probability of winning the grand prize.

10. You are dealt 5 cards from a deck of 52 cards. Find the probability that you are dealt 5 red cards.

In Exercises 11–13, a television game show has five doors, of which the contestant must pick two. Behind two of the doors are expensive cars, and behind the other three doors are consolation prizes. The contestant gets to keep the items behind the two doors she selects. Find the probability that the contestant wins

11. both cars.

12. no cars.

13. at least one car.

In Exercises 14–16, your friend agrees to give you a Spider Man comic book in exchange for 3 of your record albums he selects at random. You have a total of 10 albums: 3 Beatles, 4 Boston Symphony, and 3 Reba McIntire. (You like a wide variety of music.) Assuming each album has the same chance of being selected, find the probability that

14. all 3 of your Beatle albums will be selected.

15. two Boston Symphony albums and 1 Reba McIntire album will be selected.

16. no Reba McIntire album will be selected.

In Exercises 17–20, a jury pool has 17 men and 22 women, from which 12 will be selected. Assuming that each person is equally likely to be selected and that the jury is selected at random, find the probability that the jury consists of

17. all women.

18. 8 women and 4 men.

19. 6 men and 6 women.

20. at least 1 man.

In Exercises 21–24, drug A is given to 5 patients. Drug B is given to 4 patients, and drug C is given to 6 patients. If 4 of these 15 patients are selected at random, find the probability that

21. 2 were given drug A and 2 were given drug C.

22. 3 were given drug C and 1 was given drug A.

23. at least 1 patient was given drug C.

24. 1 was given drug A, 2 were given drug B, and 1 was given drug C.

In Exercises 25–28, five men and six women are going to be assigned to a specific row of seats in the theater. If the 11 tickets for the numbered seats are given out at random, find the probability that

25. five women are given the first five seats next to the center aisle.

26. at least one woman is in one of the first five seats.

27. exactly one woman is in one of the first five seats.

28. three women are seated in the first three seats and two men are seated in the next two seats.

29. Among 24 employees who work at McDonald's three are brothers. If six of the 24 are to be selected at random to work a late shift, find the probability that the three brothers are selected.

30. A full house in poker consists of getting three of one card and two of another card in a five-card hand. For example, if a hand contains three kings and two 5s, it is a full house. If 5 cards are dealt at random from a deck of 52 cards, without replacement, find the probability of getting three kings and two 5s.

31. A royal flush consists of the ace, king, queen, jack, and 10 all in the same suit. If 7 cards are dealt at random from a deck of 52 cards, find the probability of getting a
   a) royal flush in spades.
   b) royal flush in any suit.

Problem Solving/Group Activities

In Exercises 32–35, the first reel of a three-reel slot machine consists of three oranges, four cherries, two lemons, six plums, two bells, and one bar. The second reel consists of two oranges, five cherries, three lemons, five plums, two bells, and one bar. The third reel consists of four oranges, four cherries, three lemons, three plums, two bells, and two bars. When you pull the handle of the slot machine, find the probability of obtaining

32. cherries on all three reels.

33. cherry, cherry, bell.

34. bars on all three reels.

35. a cherry on at least one reel.

36. If three men and three women are to be assigned at random to six seats in a row at a theater, find the probability that they will alternate by sex.

37. A club consists of 15 people including Ali, Kendra, Ted, Alice, Marie, Dan, Linda, and Frank. From the 15 members a president, vice-president, and treasurer will be selected at random, and an advisory committee of 5 other individuals will also be selected at random.
   a) Find the probability that Ali is selected president, Kendra is selected vice-president, Ted is selected treasurer, and the other 5 individuals named form the advisory committee.
   b) Find the probability that 3 of the 8 individuals named are selected for the three officers' positions and the other 5 are selected for the advisory board.

38. A pair of aces and a pair of 8s is often referred to as the "dead man's hand." (See the Did You Know on page 636.)
   a) Find the probability of being dealt the dead man's hand when dealt 5 cards, without replacement, from a deck of 52 cards.
   b) The actual cards Hickok was holding when shot were the aces of spades and clubs, the 8s of spades and clubs, and the 9 of diamonds. If you are dealt five cards without replacement, find the probability of being dealt this exact hand.

39. A number is written with a magic marker on each card of a deck of 52 cards. The number 1 is put on the first card, 2 on the second, and so on. The cards are then shuffled and cut. What is the probability that the top 4 cards will be in ascending order? (For example, the top card is 12, the second 22, the third 41, and the fourth 51.)

12.11 BINOMIAL PROBABILITY FORMULA

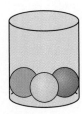

Figure 12.15

Suppose that a basket contains 3 identical balls, except for their color. One is red, one is blue, and one is yellow (Fig. 12.15). Suppose further that we are going to select 3 balls *with replacement* from the basket. We can determine specific probabilities by examining the tree diagram shown in Fig. 12.16. Note that 27 different selections are possible, as indicated in the sample space.

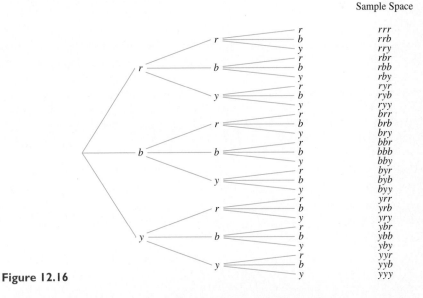

Figure 12.16

Our three selections may yield 0, 1, 2, or 3 red balls. We can determine the probability of selecting exactly 0, 1, 2, or 3 red balls by using the sample space. To determine the probability of selecting 0 red balls we count those outcomes that do not contain a red ball. There are 8 of them (*bbb*, *bby*, *byb*, *byy*, *ybb*, *yby*, *yyb*, *yyy*). Thus the probability of obtaining exactly 0 red balls is 8/27. We determine the probability of selecting exactly 1 red ball by counting the sample points that contain exactly 1 red ball. There are 12 of them. Thus the probability is 12/27, or 4/9.

We can determine the probability of selecting exactly 2 red balls and exactly 3 red balls in a similar manner. The probabilities of selecting exactly 0, 1, 2, and 3 red balls is illustrated in Table 12.5.

Table 12.5

Number of Red Balls (x)	Probability of Selecting the Number of Red Balls, $P(x)$
0	$\dfrac{8}{27}$
1	$\dfrac{12}{27}$
2	$\dfrac{6}{27}$
3	$\dfrac{1}{27}$
	Sum $= \dfrac{27}{27} = 1$

Note that the sum of the probabilities is 1. This table is an example of **probability distribution**, which shows the probabilities associated with each specific outcome of an experiment. In a probability distribution every possible outcome must be listed, and the sum of the probabilities must be 1.

Let us specifically consider the probability of selecting 1 red ball in 3 selections. We see from Table 12.5 that this probability is $\frac{12}{27}$ or $\frac{4}{9}$. Can we determine this probability without developing a tree diagram? The answer is yes.

Suppose we consider selecting a red ball success, S, and a nonred ball failure, F. Furthermore, suppose we let p represent the probability of success and q the probability of failure on any trial. Then $p = \frac{1}{3}$ and $q = \frac{2}{3}$. We can obtain 1 success in three selections in the following ways:

$$\text{SFF} \qquad \text{FSF} \qquad \text{FFS}$$

We can compute the probabilities of each of these outcomes using the multiplication formula because each of the selections is independent.

$$P(\text{SFF}) = P(\text{S}) \cdot P(\text{F}) \cdot P(\text{F}) = p \cdot q \cdot q = pq^2 = \frac{1}{3}\left(\frac{2}{3}\right)^2 = \frac{4}{27}$$

$$P(\text{FSF}) = P(\text{F}) \cdot P(\text{S}) \cdot P(\text{F}) = q \cdot p \cdot q = pq^2 = \frac{1}{3}\left(\frac{2}{3}\right)^2 = \frac{4}{27}$$

$$P(\text{FFS}) = P(\text{F}) \cdot P(\text{F}) \cdot P(\text{S}) = q \cdot q \cdot p = pq^2 = \frac{1}{3}\left(\frac{2}{3}\right)^2 = \frac{4}{27}$$

$$\text{Sum} = \frac{12}{27} = \frac{4}{9}$$

We obtained an answer of 4/9, the same answer that was obtained using the tree diagram. Note that each of the 3 sets of outcomes above has one success and two failures. Rather than listing all the possibilities containing 1 success and 2 failures, we can use the combination formula to determine the number of possible combinations of 1 success in 3 trials. To do so evaluate $_3C_1$.

$$_3C_1 = \frac{3!}{(3-1)!1!} = \frac{3 \cdot 2 \cdot 1}{2 \cdot 1 \cdot 1} = 3$$

Number of trials   Number of successes

Thus we see that there are 3 ways that the 1 success could occur in 3 trials. To compute the probability of 1 success in 3 trials we can multiply the probability of success in any one trial, $p \cdot q^2$, by the number of ways the 1 success can be arranged among the 3 trials, $_3C_1$. Thus the probability of selecting 1 red ball, $P(1)$, in 3 trials may be found as follows,

$$P(1) = (_3C_1)p^1q^2 = 3\left(\frac{1}{3}\right)\left(\frac{2}{3}\right)^2 = \frac{12}{27} = \frac{4}{9}$$

The binomial probability formula, which we introduce shortly, explains how to obtain expressions like $P(1) = (_3C_1)p^1q^2$, and is very useful in finding certain types of probabilities.

To use the binomial probability formula the following two conditions must hold.

To Use the Binomial Probability Formula

1. Each trial has two possible outcomes, *success* and *failure*.
2. There are n repeated independent trials.

Before going further let's discuss why we can use the binomial probability formula to find the probability of selecting a specific number of red balls when balls are selected with replacement. First, we may consider selecting a red ball as success and selecting any ball of another color as failure. Second, since we are doing each trial *with replacement*, the trials are independent of each other. Now let's discuss the binomial probability formula.

Binomial Probability Formula

The probability of obtaining exactly x successes, $P(x)$, in n independent trials is given by

$$P(x) = (_nC_x)p^xq^{n-x},$$

where p is the probability of success on a single trial and q is the probability of failure on a single trial.

In the formula, p will be a number from 0 through 1, inclusive, and $q = 1 - p$. Therefore, if $p = 0.2$, then $q = 1 - 0.2 = 0.8$. If $p = 3/5$, then $q = 1 - 3/5 = 2/5$. Note that $p + q = 1$ and that the values of p and q remain the same for each independent trial. The combination $_nC_x$ is called the *binomial coefficient*.

In Example 1 we use the binomial probability formula to solve the same problem that we recently solved by using a tree diagram.

EXAMPLE 1

A basket contains 3 balls: 1 red, 1 blue, and 1 yellow. Three balls are going to be selected with replacement from the basket. Find the probability that

a) no red balls are selected.

b) exactly 1 red ball is selected.

c) exactly 2 red balls are selected.

d) exactly 3 red balls are selected.

Solution:

a) We will consider selecting a red ball a success and selecting a ball of any other color a failure. Since only 1 of the 3 balls is red, the probability of success on any single trial, p, is $1/3$. The probability of failure on any single trial, q, is $1 - 1/3 = 2/3$. We do not want any successes (red balls), so $x = 0$. There are 3 independent selections (or trials), so $n = 3$. We determine the probability of 0 successes, or $P(0)$, as follows.

$$P(x) = (_nC_x)p^x q^{n-x}$$

$$P(0) = (_3C_0)\left(\frac{1}{3}\right)^0\left(\frac{2}{3}\right)^{3-0}$$

$$= (1)(1)\left(\frac{2}{3}\right)^3$$

$$= \left(\frac{2}{3}\right)^3 = \frac{8}{27}$$

b) We are finding the probability of obtaining exactly 1 red ball or exactly 1 success in 3 independent selections. Thus $x = 1$ and $n = 3$. We find the probability of exactly 1 success, or $P(1)$, as follows.

$$P(x) = (_nC_x)p^x q^{n-x}$$

$$P(1) = (_3C_1)\left(\frac{1}{3}\right)^1\left(\frac{2}{3}\right)^{3-1}$$

$$= 3\left(\frac{1}{3}\right)\left(\frac{2}{3}\right)^2$$

$$= 3\left(\frac{1}{3}\right)\left(\frac{4}{9}\right) = \frac{4}{9}$$

c) We are finding the probability of selecting exactly 2 red balls in 3 independent trials. Thus $x = 2$ and $n = 3$. We find $P(2)$ as follows.

$$P(x) = (_nC_x)p^xq^{n-x}$$

$$P(2) = {}_3C_2\left(\frac{1}{3}\right)^2\left(\frac{2}{3}\right)^{3-2}$$

$$= 3\left(\frac{1}{3}\right)^2\left(\frac{2}{3}\right)^1$$

$$= 3\left(\frac{1}{9}\right)\left(\frac{2}{3}\right) = \frac{2}{9}$$

d) We are finding the probability of selecting exactly 3 red balls in 3 independent trials. Thus $x = 3$ and $n = 3$. We find $P(3)$ as follows.

$$P(x) = (_nC_x)p^xq^{n-x}$$

$$P(3) = (_3C_3)\left(\frac{1}{3}\right)^3\left(\frac{2}{3}\right)^{3-3}$$

$$= 1\left(\frac{1}{3}\right)^3\left(\frac{2}{3}\right)^0$$

$$= 1\left(\frac{1}{27}\right)(1) = \frac{1}{27}$$

All the probabilities obtained in Example 1 agree with the answers obtained by using the tree diagram. Whenever you obtain a value for $P(x)$, you should obtain a value between 0 and 1, inclusive. If you obtain a value greater than 1, you have made a mistake.

EXAMPLE 2

A manufacturer of matches knows that 0.1% of the matches produced by the company are defective.

a) Write the binomial probability formula that would be used to determine the probability that exactly x matches in a package of n matches are defective.

b) Write the binomial probability formula that would be used to find the probability that exactly 3 matches in a 50-match package will be defective.

Solution:

a) Since we want to find the probability of selecting exactly x defective matches, selecting a defective match is considered success. The probability that an individual match is defective is 0.1%, or 0.001 in decimal form. The probability that a match is not defective, q, is $1 - 0.001$, or 0.999. The general formula for finding the probability of selecting exactly x defective matches from n matches is

$$P(x) = (_nC_x)p^xq^{n-x}.$$

Substituting 0.001 for p and 0.999 for q, we obtain the formula

$$P(x) = (_nC_x)(0.001)^x(0.999)^{n-x}$$

b) We want to determine the probability of selecting exactly 3 defective matches from 50. Thus $x = 3$ and $n = 50$. Substituting these values into the formula in part (a) gives

$$P(3) = (_{50}C_3)(0.001)^3(0.999)^{50-3}$$
$$= (_{50}C_3)(0.001)^3(0.999)^{47}.$$

The answer to part (b) of Example 2 may be obtained from a table in a statistics book or by using a statistical calculator.

EXAMPLE 3

The local weather person has been accurate in her "within 2 degrees" temperature forecast 80% of the time. Find the probability that she is accurate

a) exactly 3 of the next 5 days.

b) exactly 4 of the next 4 days.

Solution:

a) We want to find the probability that she is successful (within 2°) exactly 3 of the next 5 days. Thus $x = 3$ and $n = 5$. The probability of success on any one day, p, is 80%, or 0.8. The probability of failure, q, is $1 - 0.8 = 0.2$. Substituting these values into the binomial probability formula yields

$$P(x) = (_nC_x)p^xq^{n-x}$$
$$P(3) = (_5C_3)(0.8)^3(0.2)^{5-3}$$
$$= 10(0.512)(0.04)$$
$$= 0.2048.$$

b) We want to find $P(4)$.

$$P(x) = (_nC_x)p^xq^{n-x}$$
$$P(4) = (_4C_4)(0.8)^4(0.2)^{4-4}$$
$$= 1(0.8)^4(0.2)^0$$
$$= 1(0.4096)(1)$$
$$= 0.4096$$

EXAMPLE 4

The probability that an individual selected at random has blue eyes is 0.4. Find the probability that

a) none of four people selected at random has blue eyes.

b) at least one of four people selected at random has blue eyes.

Solution: Success is selecting a person with blue eyes. Thus $p = 0.4$ and $q = 1 - 0.4 = 0.6$. We want to find the probability of 0 successes in 4 trials.

Thus $x = 0$ and $n = 4$. We find the probability of 0 successes, or $P(0)$, as follows.

$$P(x) = (_nC_x)p^x q^{n-x}$$

$$P(0) = (_4C_0)(0.4)^0(0.6)^{4-0}$$

$$= 1(1)(0.6)^4$$

$$= 1(1)(0.1296)$$

$$= 0.1296$$

b) The probability that at least one person of the four has blue eyes can be found by subtracting from 1 the probability that none of the people has blue eyes. We worked problems of this type in earlier sections of the chapter.

In part (a) we determined that the probability that none of the people has blue eyes is 0.1296. Thus

$$P(\text{at least 1 has blue eyes}) = 1 - P(\text{none has blue eyes})$$

$$= 1 - 0.1296$$

$$= 0.8704.$$

Section 12.11 Exercises

1. What is a probability distribution?

2. What are the two requirements for use of the binomial probability formula?

3. Write the binomial probability formula.

4. In the binomial probability formula what do p and q represent?

In Exercises 5–10, assume that each of the n trials is independent and that p is the probability of success on a given trial. Use the binomial probability formula to find P(x).

5. $n = 4, x = 1, p = 0.1$ **6.** $n = 3, x = 2, p = 0.6$

7. $n = 5, x = 4, p = 0.3$ **8.** $n = 3, x = 3, p = 0.9$

9. $n = 4, x = 0, p = 0.5$ **10.** $n = 5, x = 3, p = 0.4$

11. An egg distributor determines that the probability that any individual egg has a crack is 0.15.
   a) Write the binomial probability formula to determine the probability that exactly x of n eggs are cracked.
   b) Write the binomial probability formula to determine the probability that exactly 2 in a one dozen–egg carton are cracked.

12. The probability that a family selected at random will be audited by the Internal Revenue Service (IRS) this year is 0.0237.
   a) Write the binomial probability formula to determine the probability that exactly x out of n families selected at random will be audited by the IRS this year.

   b) Write the binomial probability formula to determine the probability that exactly 5 of 20 families selected at random will be audited by the IRS this year.

In Exercises 13–21, use the binomial probability formula to answer the question.

13. A family plans on having 6 children. Find the probability the family has exactly 2 girls. Assume that births are independent and that the probability of a girl being born is $1/2$.

14. The probability of rolling a sum of 7 when a pair of dice is rolled is $1/6$. Find the probability that, if a pair of dice is rolled 6 times, exactly one sum of 7 is rolled.

15. Janet Geterin makes 70% of her free throws in a basketball game. Find the probability that she makes exactly 4 of the next 6 free throws.

16. Ninety-six percent of Dr. William's radical keratotomy patients end up with 20–30, or better, vision. Find the probability that exactly 2 of her next 3 patients end up with 20–30, or better, vision.

17. A quality control engineer at a Panorama film processing plant finds that 1% of its rolls of film are defective. Find the probability that exactly 2 of the next 4 rolls of film made are defective.

18. Twenty percent of North Shore Airlines flights are overbooked in the month of December. Find the probability that, of 6 North Shore flights selected at random in December, exactly 4 are overbooked.

19. The probability of rolling a number greater than 1 on a die is 5/6. Find the probability that on the next 4 rolls of a die exactly 3 show a number greater than 1.

20. At a Circuit City store 1/7 of those purchasing color televisions purchase a large-screen TV. Find the probability that
  a) none of the next four people who purchase a color television at Circuit City purchases a large-screen TV.
  b) at least one of the next four people who purchase a color television at Circuit City purchases a large-screen TV.

21. Sixty percent of the eligible voting residents of a certain community support the incumbent candidate. If five of the residents are selected at random, find the probability that
  a) none supports the candidate.
  b) at least one supports the candidate.

Problem Solving/Group Activities

22. In a random sample of working mothers in Duluth, Minnesota, the following data indicating how they get to work were obtained.

Mode of Transportation	Number of Mothers
Car	40
Bus	20
Bike	16
Other	4

If this sample is representative of all working mothers in Duluth, find the probability that exactly 3 of 5 working mothers selected at random
  a) take a car to work.
  b) take a bus to work.

23. Six cards are selected from a deck of playing cards with replacement. Find the probability that
  a) exactly 3 picture cards are obtained.
  b) exactly 2 spades are obtained.

24. The probability that a person visiting Dr. Dent's office is over 60 years old is 0.7. Find the probability that
  a) exactly 3 of the next 5 people visiting the office are over 60 years old.
  b) at least 3 of the next 5 people visiting the office are over 60 years old.

CHAPTER 12 SUMMARY

Key Terms

12.1
empirical (experimental) probability
event
experiment
outcomes
relative frequency
theoretical (mathematical) probability

12.2
equally likely outcomes

12.3
odds against
odds in favor

12.4
expected value (expectation)
fair price

12.5
counting principle
sample point
sample space
tree diagrams

12.6
and probability problem
compound probability
dependent events
independent events
mutually exclusive events
or probability problem

12.7
conditional probability

12.8
permutation

12.9
combination

12.11
probability distribution

Important Facts

Empirical Probability

$$P(E) = \frac{\text{number of times event } E \text{ has occurred}}{\text{total number of times the experiment has been performed}}$$

The Law of Large Numbers

Probability statements apply in practice to a large number of trials, not to a single trial. It is the relative frequency over the long run that is accurately predictable, not individual events or precise totals.

Theoretical Probability

$$P(E) = \frac{\text{number of outcomes favorable to } E}{\text{total number of possible outcomes}}$$

The probability of an event that cannot occur is 0. The probability of an event that must occur is 1. Every probability must be a number between 0 and 1 inclusively; that is,

$$0 \leq P(E) \leq 1.$$

The sum of the probabilities of all possible outcomes of an event is 1.

$$P(A) + P(\text{not } A) = 1$$

Odds Against an Event

$$\text{Odds against} = \frac{P(\text{event fails to occur})}{P(\text{event occurs})} = \frac{P(\text{failure})}{P(\text{success})}$$

Odds in Favor of an Event

$$\text{Odds in favor} = \frac{P(\text{event occurs})}{P(\text{event fails to occur})} = \frac{P(\text{success})}{P(\text{failure})}$$

Expected Value

$$E = P_1A_1 + P_2A_2 + P_3A_3 + \cdots + P_nA_n$$

Expected value = Fair price − Cost to play

Counting Principle

If a first experiment can be performed in M distinct ways and a second experiment can be performed in N distinct ways, then the two experiments in that specific order can be performed in $M \cdot N$ distinct ways.

Or and And Problems

$$P(A \text{ or } B) = P(A) + P(B) - P(A \text{ and } B)$$

$$P(A \text{ and } B) = P(A) \cdot P(B)$$

Conditional Probability

$$P(E_2 | E_1) = \frac{n(E_1 \text{ and } E_2)}{n(E_1)}$$

The **number of permutations** of n items is $n!$.

$$n! = n(n-1)(n-2) \cdots (3)(2)(1)$$

Permutation Formula

$$_nP_r = \frac{n!}{(n-r)!}$$

The number of different permutations of n objects where $n_1, n_2, \ldots, n_r$ of the objects are identical is

$$\frac{n!}{n_1!n_2! \cdots n_r!}$$

Combination Formula

$$_nC_r = \frac{n!}{(n-r)!r!}$$

Binomial Probability Formula: $P(x) = (_nC_x)p^xq^{n-x}$

Chapter 12 Review Exercises

12.1–12.10

1. In your own words, explain the law of large numbers.

2. Explain how empirical probability can be used to determine whether a die is ''loaded'' (not a fair die).

3. Of 80 tosses of a trick coin, 50 resulted in heads. Find the empirical probability of the coin landing heads up.

4. Select a card from a deck of cards 50 times with replacement and compute the empirical probability of selecting a spade.

5. In 100 births at County Hospital 58 were males. Find the empirical probability that the next baby born at the hospital will be a male.

In Exercises 6–9, each of the digits 0, 1, 2, 3, 4, 5, 6, 7, 8, 9 is written on a piece of paper, and all the pieces of paper are placed in a hat. One number is selected at random. Find the probability that the number selected is

6. odd.

7. even or greater than 4.

8. greater than 4 or less than 6.

9. even and greater than 4.

In Exercises 10–13, assume that the ages of a sample of children in a junior high school resulted in the following data.

Age	Number
11	13
12	39
13	50
14	55
15	43

If one individual is selected at random from this sample, find the probability that the individual selected is

10. 14 years old.

11. younger than 14.

12. 12 or 13 years old.

13. older than 13.

14. The probability that the Mellincamps win the doubles tennis tournament is $\frac{1}{5}$. Find the odds
   a) against their winning.
   b) in favor of their winning.

15. Mikey, a mischievous little boy, removed labels on the eight cans of vegetables in the cabinet. Mikey's father knows that there are three cans of corn, three cans of beans, and two cans of carrots. If the father selects and opens one can at random, find the odds against his selecting a can of corn.

16. Assuming that the odds against winning a game of chess are 2:3, find the probability of winning the game.

17. The probability that Dorothy and Toto get lost on their way to Auntie Em's house is 0.3. Find the odds in favor of their getting lost.

18. A thousand raffle tickets are sold at $2 each. Three prizes of $200 and two prizes of $100 will be awarded.
   a) Find the expectation of a person who purchases a ticket.
   b) Find the expectation of a person who purchases three tickets.
   c) Find the fair price to pay for a ticket.

19. If Cameron selects a picture card from a deck of cards, Lindsey will give him $9. If Cameron does not select a picture card, he must give Lindsey $3.
   a) Find Cameron's expectation.
   b) Find Lindsey's expectation.
   c) If Cameron plays this game 100 times, how much can he expect to lose or gain?

20. If the day is sunny, 1000 people will attend the baseball game. If the day is cloudy, only 500 people will attend. If it rains, only 100 people will attend. The local meteorologist states that the probability of a sunny day is 0.4, of a cloudy day is 0.5, and of a rainy day is 0.1. Find the number of people that are expected to attend.

21. Tina, Jake, Gina, and Carla form a club. They plan to select a president and a vice-president.
   a) Construct a tree diagram showing all the possible outcomes.
   b) List the sample space.
   c) Find the probability that Gina is selected president and Jake is selected vice-president.

22. A coin is flipped and then a marble is to be selected from a bag. The bag contains four marbles: one red, one blue, one green, and one purple.
   a) Construct a tree diagram showing all the possible outcomes.
   b) List the sample space.
   c) Find the probability that a head is flipped and either a red or purple marble is selected.

In Exercises 23–29, the following double wheel is spun. The outer wheel is spun clockwise and the inner wheel is spun counterclockwise.

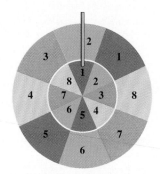

Assuming equally likely outcomes, find the probability that

23. even numbers on both wheels stop under the pointer.

24. numbers greater than 5 on both wheels stop under the pointer.

25. an even number on the outer wheel and a number less than 4 on the inner wheel stop under the pointer.

26. an even number or a number less than 6 on the outer wheel stops under the pointer.

27. an even number or a color other than blue on the inner wheel stops under the pointer.

28. blue on the outer wheel and a color other than blue on the inner wheel stop under the pointer.

29. blue on at least one wheel stops under the pointer.

A dog pound has eight dogs to give away—three rottweilers, three yellow labs, and two golden retrievers. A family decides to select at random two of these dogs. Find the probability that

30. both are yellow labs.

31. both are golden retrievers.

32. the first selected is a yellow lab and the second is a rottweiler.

In Exercises 33–36, assume that the spinner cannot land on a line.

If spun once, find

33. the probability that the spinner lands on red.

34. the odds against and in favor of the spinner landing on red.

35. If you are awarded $5 if the spinner lands on red, find a fair price to play the game.

36. If the spinner is spun twice, find the probability that it lands on red and then gray (assume independence).

In Exercises 37–40, assume that the spinner cannot land on a line.

If spun once, find

37. the probability that the spinner does not land on gray.

38. the odds in favor of and against the spinner landing on gray.

39. A person wins $10 if the spinner lands on gray, wins $5 if the spinner lands on red, and loses $20 if the spinner lands on yellow. Find the expectation of a person who plays this game.

40. If the spinner is spun three times, find the probability that at least one spin lands on red.

In Exercises 41–44, a sample of 180 new cars were checked for defects. The following table shows the results of the survey.

Car	Fewer than Six Defects	Six or More Defects	Total
American built	89	17	106
Foreign built	55	19	74
Total	144	36	180

Find the probability that if one car is selected from this sample the car has

41. fewer than six defects if it is American built.

42. fewer than six defects if it is foreign built.

43. six or more defects if it is foreign built.

44. six or more defects if it is American built.

In Exercises 45–48, a sample of voters is made to determine which of three candidates, Mr. Spock, Captain Kirk, or Dr. McCoy, should run against the incumbent governor. The results are shown in the following table.

Gender	Mr. Spock	Captain Kirk	Dr. McCoy	Total
Male	60	40	70	170
Female	20	100	50	170
Total	80	140	120	340

If one voter is selected from this sample, find the probability that the voter favors

45. Mr. Spock if the voter is male.

46. Captain Kirk if the voter is female.

47. a candidate other than Dr. McCoy if the voter is male.

48. Mr. Spock or Captain Kirk if the voter is female.

49. Four finalists remain in a lottery drawing. The prizes to be awarded among the finalists are $100, $500, $1000, and $10,000.
  a) In how many different ways can the winners be selected?
  b) What is the expectation of a finalist?

50. Five finalists remain in a talent contest. Two will receive $50 each, two will receive $100 each, and one will receive $500. How many different arrangements of prizes are possible?

51. Each of Mr. Vargas' three children is planning to select his or her own pet rabbit from a litter of eight rabbits. In how many different ways can they do so?

52. Three of nine astronauts must be selected for a mission. One will be the captain, one will be the navigator, and one will explore the surface of the moon. In how many ways can a three-person crew be selected so that each person has a different assignment?

53. Dr. Fox has three doses of serum for influenza type A. Six patients in the office require the serum. In how many different ways could Dr. Fox dispense the serum?

54. a) Ten of 15 huskies are to be selected to pull a dogsled. In how many ways can this selection be made?
  b) How many different arrangements of the 10 huskies on a dogsled are possible?

55. In the New York State lottery game called Lotto, you must pick the correct 6 from a possible 54 numbers to win. Then, if all 6 numbers you picked are selected, you win the grand prize. Find the probability of winning the grand prize at Lotto.

56. A club consists of six males and eight females. Three males and three females are to be selected to represent the club at a dinner. How many different combinations are possible?

57. In a psychology research laboratory, one room contains eight men, numbered 1–8, and another room contains five women, numbered 1–5. Three men and two women are to be selected at random to be given a psychological test. How many different combinations of these people are possible?

58. Two cards are selected at random, without replacement, from a deck of 52 cards. Find the probability that two aces are selected (use combinations).

In Exercises 59–62, a bag contains five red chips, three white chips, and two blue chips. Three chips are to be selected at random, without replacement. Find the probability that

59. all are red.

60. the first two are red and the third is blue.

61. the first is red, the second is white, and the third is blue.

62. at least one is red.

In Exercises 63–66, Agway Lawn and Garden Center carries 5 pine, 6 maple, and 3 birch trees. Gene Rodenberry plans to select 3 trees at random from the 14 mentioned.

Assuming that each tree has the same chance of being selected, find the probability that

63. 3 maple trees are selected.

64. 2 pine trees and 1 maple tree are selected.

65. no pine trees are selected.

66. at least one pine tree is selected.

67. In the community of Lockwood, 60% of the homes that are purchased cost more than $80,000.
   a) Write the binomial probability formula to determine the probability that exactly x of the next n homes that are purchased in Lockwood cost more than $80,000.
   b) Write the binomial probability formula to determine the probability that exactly 75 of the next 100 home purchases cost more than $80,000.

68. At the Pretty Pink Flower Shop, 1/5 of those ordering flowers select long-stemmed roses. Find the probability that exactly 3 of the next 5 customers ordering flowers select long-stemmed roses.

69. During any semester at City College, 60% of the students are taking a mathematics class. Find the probability that of four students selected at random
   a) none is taking a mathematics class this semester.
   b) at least one is taking a mathematics course this semester.

CHAPTER 12 TEST

1. In a certain community, of the last 90 traffic tickets, 25 were for speeding. Find the empirical probability that the next ticket given will be for speeding.

In Exercises 2–5, each of the numbers 1–9 is written on a sheet of paper, and the nine sheets of paper are placed in a hat. If one sheet of paper is selected at random from the hat, find the probability that the number selected is

2. greater than 6.

3. even.

4. even or greater than 4.

5. odd and greater than 4.

In Exercises 6–9, if 2 of the same 9 sheets of paper are selected, without replacement, from the hat, find the probability that

6. both numbers are greater than 4.

7. both numbers are even.

8. the first number is odd and the second number is even.

9. neither of the numbers is greater than 6.

10. One card is selected at random from a deck of cards. Find the probability that the card selected is a red card or a picture card.

In Exercises 11–15, one die is rolled and one letter—a, b, or c—is selected at random.

11. Use the counting principle to determine the number of sample points in the sample space.

12. Construct a tree diagram illustrating all the possible outcomes, and list the sample space.

In Exercises 13–15, by observing the sample space of the same die and three letters, determine the probability of obtaining

13. the number 6 and the letter *a*.

14. the number 6 or the letter *a*.

15. an even number or the letter *c*.

16. A code is to consist of three digits followed by three letters. Find the number of possible codes if the first digit cannot be a 0 or 1 and replacement is not permitted.

17. A class consists of 16 females and 9 males. If one person is selected at random from the class, find the odds against the person's being female.

18. The odds against Tracy winning the racquetball tournament are 5:3. Find the probability that Tracy wins the tournament.

19. You get to select one card at random from a deck of cards. If you pick a club, you win $8. If you pick a heart, you win $4. If you pick any other suit, you lose $6. Find your expectation for this game.

20. A biologist is studying the habits of squirrels at two national parks. The following table indicates the number and type of squirrels that were studied at the two parks.

National Park	Gray Squirrels	Other Types of Squirrel	Total
Great Smoky Mountain	82	33	115
Yosemite	60	45	105
Total	142	78	220

If one of the squirrels being studied is spotted at Yosemite, find the probability that it is a gray squirrel.

21. Three of six people are to be selected and given small prizes. One will be given a book, one will be given a calculator, and one will be given a $10 bill. In how many different ways can these prizes be awarded?

In Exercises 22 and 23, a bin contains a total of 20 batteries of which 8 are defective. If you select 2 at random, without replacement, find the probability that

22. none is good.

23. at least one is good.

24. Five green apples and seven red apples are in a bucket. Five apples are to be selected at random, without replacement. Find the probability that three red apples and two green apples are selected.

25. At the Fernwood Association for the Prevention of Cruelty to Animals (FAPCA), 3/5 of the people who adopt pets adopt dogs. Find the probability that exactly 3 of the next 4 people who adopt pets from FAPCA adopt a dog.

GROUP PROJECTS

The Probability of an Exact Measured Value

1. Your car's speedometer indicates that you are traveling at 65 mph. What is the probability that you are traveling at *exactly* 65 mph? Explain your answer.

Probability and Genetics

2. Exercise 87 in the Section 12.6 Exercises, dealt with polygenetic birth defects. If you have not worked that exercise, do so now and check your answers.

One polygenetic birth defect is hydrocephalus. The probability that an individual will be born with hydrocephalus is about 1 in 15,000. If the first child is born with hydrocephalus the probability that another child in the family is born with the same affliction becomes 1 in 100. If two children in the family are born with the affliction the probability that a future child is born with the affliction increases to about 1 in 10.

A family plans to have 3 children. Find

a) P(2nd child is born with hydrocephalus | 1st child is not born with hydrocephalus).

b) P(2nd child is born with hydrocephalus | 1st child is born with hydrocephalus).

c) P(3rd child is born with hydrocephalus | 1st child is born with hydrocephalus and 2nd child is not born with hydrocephalus).

d) P(3rd child is born with hydrocephalus | 1st child is born with hydrocephalus and 2nd child is born with hydrocephalus).

e) P(none of the 3 children is born with hydrocephalus).

f) P(all 3 children are born with hydrocephalus).

g) P(the 1st child is born with hydrocephalus, the 2nd child is not born with hydrocephalus, and the 3rd child is born with hydrocephalus).

h) P(the first two children are not born with hydrocephalus but the third child is born with hydrocephalus).

i) P(the first two children are born with hydrocephalus, but the third child is not born with hydrocephalus).

Taking a Test

3. A 10-question multiple choice exam is given, and each question has five possible answers. Pascal takes this exam and guesses at every question. Use the binomial probability formula to find the probability (to 5 decimal places) that

a) he gets exactly 2 questions correct.

b) he gets no questions correct.

c) he gets at least 1 question correct (use the information from part (b) to answer this part).

d) he gets at least 9 questions correct.

e) Without using the binomial probability formula, determine the probability that he gets exactly 2 questions correct.

f) Compare your answers to parts (a) and (f). If they are not the same explain why.

Keyless Entry

4. Many cars now have keyless entry. To open the lock you may press a 3-digit code on a set of buttons like that illustrated. The code may include repeated digits like 114 or 555.

a) How many different 3-digit codes can be made using the 10 digits if repetition is permitted?

b) How many different ways are there of pressing 3 buttons if repetition is allowed?

c) A burglar is going to press 3 buttons at random, with repetition allowed. Find the probability that the burglar hits the sequence to open the door.

d) Suppose that each button had only one number associated with it as illustrated on page 653. How many different 3-digit codes can be made with the 5 digits if repetition is permitted?

e) Using the buttons labeled 1–5, how many different ways are there to press three buttons if repetition is allowed?

f) A burglar is going to press 3 buttons of those labeled 1–5 at random with repetition allowed. Find the probability that the burglar hits the sequence to open the door.

g) Is a burglar more likely, less likely, or does he or she have the same likelihood of pressing 3 buttons and opening the car door if the buttons are labeled as in the first illustration or as in the second illustration? Explain your answer.

h) Can you see any advantages in labeling the buttons as in the first illustration? Explain.

Chapter 13 Statistics

Benjamin Disraeli (1804–1881), once Prime Minister of Britain, said that there are three kinds of lies: "Lies, damned lies, and statistics." Do numbers lie? Numbers are, after all, the foundation of all statistical information. The "lie" comes in when, either intentionally or carelessly, a number is used in such a way as to lead us to a conclusion that is unjustified or incorrect.

The desire to record data about the human population and its activities has been with us for thousands of years. The earliest censuses were undertaken either to provide a basis for taxation or to register individuals for military service. The first large-scale survey was commissioned in 1086 by William the Conqueror of England. Called the Domesday Book (meaning "day of judgment"), the survey was intended to establish the Crown's holdings and to list the resources of the country for taxation purposes. Completed in only 1 year, the survey details the property holdings of more than 13,000 medieval towns and villages. The first modern census was taken in the United States in August 1790. It was the first census taken for use as a basis for government representation. A census has been taken in the United States every 10 years since then, and is now used for many purposes.

Census taking represents an attempt to gather data household by household, a practice that is time-consuming and costly. Today there are methods by which statistics obtained from a small sample are used to represent a much larger population. In fact, a sample as small as 1600 may be used to

Now in its 111th edition and weighing over 2 pounds, the *Statistical Abstract of the United States* is a compilation of facts and figures taken from the U.S. Census. In it you'll find many items, including what Americans owe (in 1989, 956.9 million credit cards were used to charge $430.4 billion worth of goods) and what kinds of pets Americans have (38.2% of households had dogs, 30.5% had cats, and 5.7% had birds).

The results of statistical surveys are closely related to the way the questions are asked. A classic example is a 1940 survey by Elmer Roper in which people were asked essentially the same question in two different ways: "Do you think the United States should forbid public speeches against democracy?" (about half the interviewees responded "yes"); "Do you think the United States should allow public speeches against democracy?" (here three-quarters of the interviewees responded "no").

predict the outcome of a national election. Such information gathering is conducted by a variety of people, including medical researchers, scientists, advertisers, and political pollsters. You are likely to find examples of statistics every day on the front page of your newspaper.

When evaluating statistical information, remember that you need to judge the sampling method as well as the numbers given. Ask yourself who conducted the study and whether they have a bias, how large the sample was and whether it was representative, and where the study appeared and whether there was an opposing side. Numbers may not lie, but they can be manipulated and misinterpreted. ■

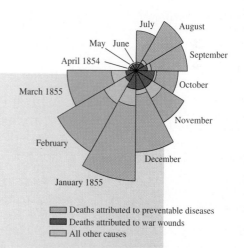

Deaths attributed to preventable diseases
Deaths attributed to war wounds
All other causes

Florence Nightingale (1820–1910), the British nurse considered to be the founder of modern medical nursing, used statistical data and graphs as evidence in her fight to improve sanitary conditions in hospitals. She used this polar-area diagram (her own invention) to show dramatically that British soldiers hospitalized in the Crimean War were far more likely to die from infection and disease than from war wounds. Using these graphs, she lobbied the British government and public to make sanitary reforms, which saved many lives.

13.1   SAMPLING TECHNIQUES

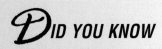

Statistics is the art and science of gathering, analyzing, and making inferences (predictions) from numerical information obtained in an experiment. This numerical information is referred to as data. The use of statistics, originally associated with numbers gathered for governments, has grown significantly and is now applied in all walks of life.

Governments use statistics to estimate the amount of unemployment and the cost of living. Thus statistics has become an indispensable tool in attempts to regulate the economy. In psychology and education the statistical theory of tests and measurements has been developed to compare achievements of individuals from diverse places and backgrounds. Another use of statistics with which we are all familiar is the public opinion poll. Newspapers and magazines carry the results of different Harris and Gallup polls on topics ranging from the president's popularity to the number of cans of beer consumed. In recent years these polls have attained a high degree of accuracy. The A. C. Nielsen rating is a public opinion poll that determines the country's most and least watched TV shows. Statistics is used in scores of other professions; in fact, it is difficult to find one that does not depend on some aspect of statistics.

Statistics is divided into two main branches: descriptive and inferential. Descriptive statistics is concerned with the collection, organization, and analysis of data. Inferential statistics is concerned with making generalizations or predictions from the data collected.

Probability and statistics are closely related. Someone in the field of probability is interested in computing the chance of occurrence of a particular event when all the possible outcomes are known. A statistician's interest lies in drawing conclusions about possible outcomes through observations of only a few particular events.

If a probability expert and a statistician find identical boxes, the probability expert might open the box, observe the contents, replace the cover, and proceed to compute the probability of randomly selecting a specific object from the box. The statistician might select a few items from the box without looking at the contents and make a prediction as to the total contents of the box. The few items the statistician randomly selected from the box constitute a *sample*. The entire contents of the box constitute the population. The statistician often uses a subset of the population, the sample, to make predictions concerning the population.

When a statistician draws a conclusion from a sample, there is always the possibility that the conclusion is incorrect. For example, suppose that a jar contains 90 blue marbles and 10 red marbles, as shown in Fig. 13.1. If the statistician selects a random sample of five marbles from the jar and all are blue, he or she may wrongly conclude that the jar contains all blue marbles. If the statistician takes a larger sample, say, 15 marbles, he or she is likely to select some red marbles. At that point the statistician may make a prediction about the contents of the jar based on the sample selected. Of course, the most accurate result would occur if every object in the jar, the entire population, were observed. However, in most statistical experiments, that is not practical.

Statisticians use samples instead of the entire population for two reasons: (a) it is often impossible to obtain an entire population; and (b) sampling is less expensive because collecting the data takes less time and effort. For example, suppose that you wanted to determine the number of each species of all the fish

Figure 13.1

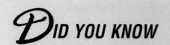

THE BIRTH OF INFERENTIAL STATISTICS

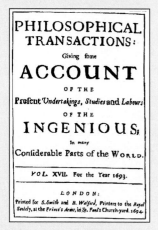

*T*he London merchant John Gaunt is credited with first making statistical predictions, or inferences, from a set of data rather than basing them simply on the laws of chance. He studied the vital statistics (births, deaths, marriages) contained in the Bills of Mortality published during the years of the Great Plague. He observed that more males were born than females and that women lived longer than men. From these observations, he made predictions about life expectancies. The keeping of mortality statistics was stimulated considerably by the growth of the insurance industry. ■

in a lake. To do so would be almost impossible without using a sample. If you did try to obtain this information from the entire population, the cost would be astronomical.

Later in this chapter we will discuss statistical measures such as the mean and the standard deviation. When statisticians calculate the mean and the standard deviation of the entire population they use different symbols and formulas than when they calculate the mean and standard deviation of a sample. The following chart shows the symbols used to represent the mean and standard deviation of a sample and of a population. Note that the mean and standard deviation of a population are symbolized by Greek letters.

Measure	Sample	Population
Mean	$\bar{x}$ (read "ex bar")	μ (mu)
Standard deviation	s	σ (sigma)

In this book we will always assume we are working with a sample and so we will use $\bar{x}$ and s. If you take a course in statistics you will use all four symbols and different formulas for a sample and for a population.

Consider the task of determining the political strength of a certain candidate running in a national election. It is not possible for pollsters to ask each of the approximately 185 million eligible voters his or her preference for a candidate. Thus pollsters must select and use a sample of the population to obtain their information. How large a sample do you think they use to make predictions about an upcoming national election? You might be surprised to learn that pollsters use only about 1600 registered voters in their national sample. How can a pollster using such a small percentage of the population make an accurate prediction?

The answer lies in the fact that, when pollsters select a sample, they use sophisticated statistical techniques to obtain a sample that is unbiased. An **unbiased sample** is one that is a small replica of the entire population with regard to income, education, sex, race, religion, political affiliation, age, and so on. The procedures that statisticians use to obtain unbiased samples are quite complex. The following sampling techniques will give you a brief idea of how statisticians obtain unbiased samples.

Random Sampling

If a sample is drawn in such a way that each time an item is selected, each item in the population has an equal chance of being drawn, the sample is said to be a **random sample**. Under these conditions, one combination of a specified number of items will have the same probability of being selected as any other combination. When all the items in the population are similar with regard to the specific characteristic we are interested in, a random sample can be expected to produce satisfactory results. For example, consider a large container holding 300 tennis balls that are identical except for color. One-third of the balls are yellow, one-third are white, and one-third are green. If the balls can be thoroughly mixed between each draw of a tennis ball so that each ball has an equally likely chance of being selected, randomness is not difficult to achieve. However, if the objects or items are not all the same size, shape, or texture, it might be impossible to obtain a random sample by reaching into a container and selecting an object.

The best procedure for selecting a random sample is to use a random number generator or a table of random numbers. A random number generator is a device,

usually a calculator or computer program, that produces a list of random numbers. A random number table is a collection of random digits in which each digit has an equal chance of appearing. To select a random sample, first assign a number to each element in the population. Numbers are usually assigned in order. Then select the number of random numbers needed, which is determined by the sample size. Each numbered element from the population that corresponds to a selected random number becomes part of the sample.

Systematic Sampling

When a sample is obtained by drawing every *n*th item on a list or production line, the sample is a systematic sample. The first item should be determined by using a random number.

It is important that the list from which a systematic sample is chosen include the entire population being studied. See the Did You Know on the *Literary Digest*. Another problem that must be avoided when this method of sampling is used is the constantly recurring characteristic. For example, on an assembly line, every tenth item could be the work of robot X. If only every tenth item is checked, the work of other robots doing the same job may not be checked and may be defective.

Cluster Sampling

The cluster sample is sometimes referred to as an *area sample* because it is frequently applied on a geographical basis. Essentially, the sampling consists of a random selection of groups of units. For example, geographically we might select blocks of a city or some other geographical subdivision to use as a sample unit. Another example would be to select *x* boxes of screws from a whole order, count the number of defective screws in the *x* boxes, and use this number to determine the expected number of defective screws in the whole order.

Stratified Sampling

When a population is divided into parts, called strata, for the purpose of drawing a sample, the procedure is known as stratified sampling. Stratified sampling involves dividing the population by characteristics called *stratifying factors*, such as sex, race, religion, or income. When a population has varied characteristics, it is desirable to separate the population into classes with similar characteristics and then take a random sample from each stratum (or class).

The use of stratified sampling requires some knowledge of the population. For example, to obtain a cross section of voters in a city, we must know where various groups are located and the approximate numbers in each location.

Section 13.1 Exercises

1. Explain the difference between descriptive and inferential statistics.

2. Define *statistics* in your own words.

3. When you hear the word *statistics*, what specific words or ideas come to mind?

4. Name five areas other than those mentioned in this section in which statistics is used.

5. Attempt to list at least two professions in which no aspect of statistics is used.

6. Explain the difference between probability and statistics.

7. **a)** What is a population?
   b) What is a sample?

8. **a)** What is a random sample?
   b) How might a random sample be selected?

9. **a)** What is a systematic sample?
   b) How might a systematic sample be selected?

10. **a)** What is a cluster sample?
    b) How might a cluster sample be selected?

11. **a)** What is a stratified sample?
    b) How might a stratified sample be selected?

12. The principal of an elementary school wishes to determine the "average" family size of the children who attend the school. To obtain a sample, the principal visits each room and selects the four students closest to each corner of the room. The principal asks each of these students how many people are in his or her family.
    a) Will this technique result in an unbiased sample? Explain your answer.
    b) If the sample is biased, will the average be greater than or less than the true family size? Explain.

In Exercises 13–20, identify the sampling technique used to obtain a sample. Explain your answer.

13. A door prize is given away at a conference. Tickets are placed in a bin and the tickets are mixed up. Then three tickets are selected by a blindfolded person.

14. Every 20th tree in an orchard is checked for the presence of fruit flies.

15. A city has numbered streets going in one direction (1st Street, 2nd Street, and so on) and numbered avenues going in the other direction (1st Avenue, 2nd Avenue, and so on). Individuals from the four blocks around the intersection of streets and avenues ending in 0 (such as 10th Street and 10th Avenue) are selected as the sample.

16. A group of people are classified according to age and then random samples of people from each group are taken.

17. Every 25th roll of film coming off the assembly line is selected and checked for defects.

18. At Disneyworld, everyone whose birthday is October 13 will be used as a sample.

19. BINGO balls in a bin are shaken and then balls are selected from the bin.

20. The Food and Drug Administration randomly selects five stores from each of four randomly selected sections of a large city and checks food items for freshness. These stores are used as a representative sample of the entire city.

21. **a)** Select a topic and population of interest to which a random sampling technique can be applied to obtain data.
    b) Explain how you would obtain a random sample for your population interest.
    c) Actually obtain the sample by the procedure stated in part (b).

Problem Solving/Group Activity

22. Some subscribers of *Consumers Reports* respond to an annual questionnaire regarding their satisfaction with new appliances, cars, and other items. The information obtained from these questionnaires is then used as a sample from which frequency of repairs and other ratings are made by the magazine. Are the data obtained from these returned questionnaires representative of the entire population or are they biased? Explain your answer.

Research Activity

23. We have briefly introduced sampling techniques. There are many statistics books that discuss this topic in more depth. Select one type of sampling technique (it may be one that we have not discussed in this section) and write a report on how statisticians obtain that type of sample. Also indicate when that type of sampling technique may be preferred. List two examples of when the sampling technique may be used.

13.2 THE MISUSES OF STATISTICS

Statistics, when used properly, is a valuable tool to society. However, many individuals, businesses, and advertising firms misuse statistics to their own advantage. You should examine statistical statements very carefully before accepting them as fact. Two questions you should ask yourself are: Was the sample used to gather the statistical data unbiased and of sufficient size? Is the statistical statement ambiguous; that is, can it be interpreted in more than one way?

Let's examine two advertisements. "Three out of five dentists recommend sugarless gum for their patients who chew gum." In this advertisement, we do

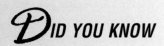

not know the sample size and the number of times the experiment was performed to obtain the desired results. The advertisement does not mention that possibly only 1 of 100 dentists recommended gum at all.

In a golf ball commercial, a "type A" ball is hit, and a second ball is hit in the same manner. The type A ball travels farther. We are supposed to conclude that the type A is the better ball. The advertisement does not mention the number of times the experiment was previously performed or the results of the earlier experiments. Possible sources of bias include (1) wind speed and direction, (2) the fact that no two swings are identical, and (3) the fact that the ball may land on a rough or smooth surface.

Vague or ambiguous words also lead to statistical misuses or misinterpretations. The word *average* is one such culprit. There are at least four different "averages," some of which are discussed in Section 13.5. Each is calculated differently, and each may have a different value for the same sample. During contract negotiations it is not uncommon for an employer to state publicly that the average salary of its employees is $30,000, while the employees' union states that the average is $25,000. Who is lying? Actually, both may be telling the truth. Each will use the average that best suits its needs to arrive at a figure. Advertisers also use the average that most enhances their products. Consumers often misinterpret this average as the one with which they are most familiar.

Another vague word is *largest*. For example, ABC claims that it is the largest department store in the United States. Does that mean largest profit, largest sales, largest building, largest staff, largest acreage, or largest number of outlets?

Still another deceptive technique used in advertising is to state a claim from which the public may draw irrelevant conclusions. For example, a disinfectant manufacturer claims that its product killed 40,760 germs in a laboratory in 5 seconds. "To prevent colds, use disinfectant A." It may well be that the germs killed in the laboratory were not related to any type of cold germ. In another example, company C claims that its paper towels are heavier than its competition's towels. Therefore they will hold more water. Is weight a measure of absorbency? A rock is heavier than a sponge, yet a sponge is more absorbent.

An insurance advertisement claims that in Duluth, Minnesota, 212 people switched to insurance company Z. One may conclude that this company is offering something special to attract these people. What may have been omitted from the advertisement is that 415 people in Duluth, Minnesota, dropped insurance company Z during the same period.

A foreign car manufacturer claims that 9 of every 10 of a popular model that it sold in the United States during the previous 10 years were still on the road. From this statement the public is to conclude that this foreign car is well manufactured and would last for many years. The commercial neglects to state that this model has been selling in the United States for only a few years. The manufacturer could just as well have stated that 9 of every 10 of these cars sold in the United States in the previous 100 years were still on the road.

Charts and graphs can also be misleading or deceptive. In Fig. 13.2 two graphs show performance of two stocks. Which stock would you purchase? Actually, the two graphs present identical information; the only difference is that the vertical scale of the graph for stock B has been exaggerated.

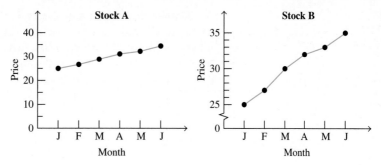

Figure 13.2

The two graphs in Fig. 13.3 show the same change. However, the graph in part (a) appears to show a greater increase than the graph in part (b), again because of a different scale.

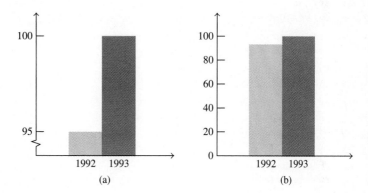

Figure 13.3

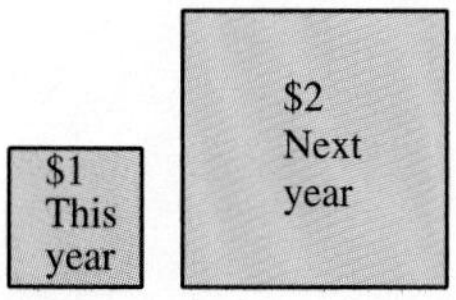

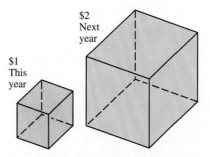

Figure 13.4

Consider a claim that, if you invest $1, by next year you will have $2. This type of claim is sometimes misrepresented as in Fig. 13.4. Actually your investment has only doubled, but the area of the square on the right is four times that of the square on the left. By expressing the amounts as cubes (Fig. 13.5), you increase the volume eightfold.

Figure 13.5

Despite the examples presented in this section, you should not be left with the impression that statistics is used solely for the purpose of misleading or cheating the consumer. As stated earlier, there are many important and necessary uses of statistics. Most statistical reports are accurate and useful. You should realize, however, the importance of being an aware consumer.

Section 13.2 Exercises

1. Find five advertisements or commercials that may be statistically misleading. Explain why each may be misleading.

In Exercises 2–14, discuss the statement and tell what possible misuses or misinterpretations may exist.

2. Most accidents occur on Saturday night. This means that people do not drive carefully on Saturday night.

3. Morgan's is the largest department store in New York. So shop at Morgan's and save money.

4. A majority of U.S. voters favor a three-party system. Therefore, if a third party candidate runs for president, that person will be elected.

5. At West High School, half the students are below average in mathematics. Therefore the school should receive more federal aid to raise student scores.

6. Eighty percent of all automobile accidents occur within 10 miles of the driver's home. Therefore it is safer to take long trips.

7. Arizona has the highest death rate for asthma in the country. Therefore it is unsafe to go to Arizona if you have asthma.

8. Florida has the highest death rate in the country. Therefore Florida should be condemned as unsafe.

9. More men than women are involved in automobile accidents. Therefore women are better drivers.

10. Five of every six dentists surveyed preferred Fresh and Clean toothpaste. Therefore it is the best toothpaste.

11. The average depth of the pond is only 3 ft, so it is safe to go wading.

12. Brand A has 70 percent fewer calories than the best-selling brand. So eat brand A and stay healthy.

13. Females have a higher average score than males on the English part of the Scholastic Aptitude Test. Therefore on this test a particular female selected at random will outperform a particular male selected at random.

14. A recent survey showed that 65% of those surveyed preferred Red Dog beer to Ice House beer. Therefore more people buy Red Dog than Ice House.

15. The following table shows the average 30-year mortgage rate (to the nearest tenth percent) from May 1994–May 1995.

Month	Rate
May	8.7%
June	8.5
July	8.7
August	8.6
September	8.7
October	9.0
November	9.3
December	9.4
January	9.2
February	9.0
March	8.6
April	8.5
May	8.1

a) Draw a line graph that makes the rates appear stable.

b) Draw a line graph that makes the rates appear to have a large fluctuation.

16. In 1995 the U.S. Census Bureau reported that the average lifetime earnings of a person who holds a high school diploma* is $821,000 and that the average lifetime earnings of a person who holds an associate's degree is $1 million. Draw a bar graph that appears to show

a) a small increase.

b) a large increase.

---

*Lifetime earnings for bachelor's degree holders average $1.5 million; for master's degree holders, $1.7 million; for doctorate holders, $2.1 million; and for professional degree holders, $3 million.

17. The average distance of a home run in the major leagues was 382 feet in 1993 and 387 feet in 1994. Draw a bar graph that appears to show
a) a small increase.
b) a large increase.

18. More and more women are fighting sex discrimination at work. The following data, provided by the U.S. Equal Employment Opportunity Commission, show the number of sex discrimination charges filed from 1990 through 1994.

Year	Charges Filed
1990	30,040
1991	31,270
1992	36,290
1993	38,130
1994	39,990

Draw a line graph that makes the increase in the number of charges filed appear to be
a) small.
b) large.

19. The following graph shows the average interest rate charged on credit cards in July 1994 and July 1995.

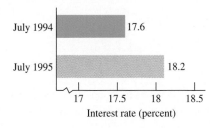

a) Draw a graph that shows the entire scale from 0 to 20%.
b) Does the new graph give a different impression? Explain.

Problem Solving/Group Activity

20. Consider the following graph which shows the 1995 U.S. population and the projected U.S. population in 2050.

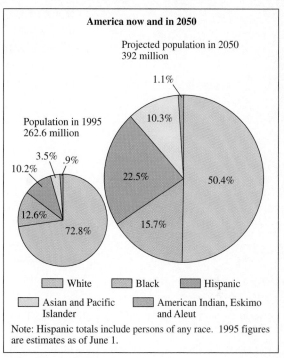

America now and in 2050

Projected population in 2050
392 million

Population in 1995
262.6 million

White   Black   Hispanic
Asian and Pacific Islander   American Indian, Eskimo and Aleut

Note: Hispanic totals include persons of any race. 1995 figures are estimates as of June 1.

Source: U.S. Census Bureau

a) Compute the projected percent increase in population from 1995 to 2050 by using the formula given on page 524.
b) Measure the radius and then compute the area of the circle representing 1995.
c) Repeat part (b) for the circle representing 2050.
d) Compute the percent increase in the size of the area of the circle from 1995 to 2050.
e) Are the circle graphs misleading? Explain your answer.

Research Activity

21. Read the book *How to Lie with Statistics* by Darrell Huff and write a book report on it. Select three illustrations from the book that show how people manipulate statistics.

13.3 FREQUENCY DISTRIBUTIONS

It is not uncommon for statisticians and others to have to analyze thousands of pieces of data. A *piece of data* is a single response to an experiment. When the amount of data is large, it is usually advantageous to construct a frequency distribution. A *frequency distribution* is a listing of the observed values and the corresponding frequency of occurrence of each value.

EXAMPLE 1

The number of children per family is recorded for 64 families surveyed. Construct a frequency distribution of the following data:

```
0  1  1  2  2  3  4  5
0  1  1  2  2  3  4  5
0  1  1  2  2  3  4  6
0  1  2  2  2  3  4  6
0  1  2  2  2  3  4  7
0  1  2  2  3  3  4  8
0  1  2  2  3  3  5  8
0  1  2  2  3  3  5  9
```

Solution: Listing the number of children (observed values) and the number of families (frequency) gives the following frequency distribution.

Number of Children (Observed Values)	Number of Families (Frequency)
0	8
1	11
2	18
3	11
4	6
5	4
6	2
7	1
8	2
9	1
	64

Eight families had no children, 11 families had one child, 18 families had two children, and so on. Note that the sum of the frequencies is equal to the original number of pieces of data, 64. ▲

Often data are grouped in classes to provide information about the distribution that would be difficult to observe if the data were ungrouped. Graphs called *histograms* and *frequency polygons* can be made of grouped data as will be explained in Section 13.4. These graphs also provide a great deal of useful information.

When data are grouped in classes, certain rules should be followed.

Rules for Data Grouped by Classes

1. The classes should be of the same "width."
2. The classes should not overlap.
3. Each piece of data should belong to only one class.

In addition it is often suggested that a frequency distribution should be constructed with 5 to 12 classes. If there are too few or too many classes, the distribution may become difficult to interpret.

To understand these rules, let's consider a set of observed values that go from a low of 0 to a high of 26. Let's assume that the first class is arbitrarily selected to go from 0 through 4. Thus any of the data with values of 0, 1, 2, 3, 4 would belong in this class. We say that the class width is 5, since there are five integral values that belong to the class. This first class ended with 4, so the second class must start with 5. If this class is to have a width of 5, at what value must it end? The answer is 9 (5, 6, 7, 8, 9). The second class is 5–9. Continuing in the same manner we obtain the following set of classes.

Classes

$$\text{Lower class limits} \left\{ \begin{array}{l} 0–4 \\ 5–9 \\ 10–14 \\ 15–19 \\ 20–24 \\ 25–29 \end{array} \right\} \text{Upper class limits}$$

We need not go beyond the 25–29 class because the largest value we are considering is 26. The classes meet our three criteria: They have the same width, there is no overlap among the classes, and each of the values from a low of 0 to a high of 26 belongs to one and only one class.

The choice of the first class, 0–4, was arbitrary. If we wanted to have more classes or fewer, we would make the class widths smaller or larger, respectively.

The numbers 0, 5, 10, 15, 20, 25 are called the lower class limits, and the numbers 4, 9, 14, 19, 24, 29 are called the upper class limits. Each class has a width of 5. Note that the class width, 5, can be obtained by subtracting the first lower class limit from the second lower class limit: $5 - 0 = 5$. The difference between any two consecutive lower class or upper class limits is also 5.

EXAMPLE 2

The following table shows the 50 largest U.S. corporations in terms of annual sales, rounded to the nearest $ billion, in 1993.

Corporation	Sales	Corporation	Sales
General Motors	134	Xerox	15
Ford Motor	109	Sara Lee	15
Exxon	98	McDonnell Douglas	14
Intl. Business Machines	63	Digital Equipment	14
General Electric	61	Johnson & Johnson	14
Mobil	57	Minnesota Mining & Mfg.	14
Philip Morris	51	Coca-Cola	14
Chrysler	44	International Paper	14
Texaco	34	Tenneco	13
E.I. du Pont de Nemours	33	Lockheed	13
Chevron	32	Georgia-Pacific	12
Procter & Gamble	30	Phillips Petroleum	12
Amoco	25	Allied Signal	12
Boeing	25	IBP	12
PepsiCo	25	Goodyear Tire	12
Conagra	22	Caterpillar	12
Shell Oil	21	Westinghouse Electric	12
United Technologies	21	Anheuser-Busch	12
Hewlett-Packard	20	Bristol-Myers Squibb	11
Eastman Kodak	20	Rockwell International	11
Dow Chemical	18	Merck	10
Atlantic Richfield	17	Coastal	10
Motorola	17	Archer Daniels Midland	10
USX	17	Ashland Oil	10
RJR Nabisco Holdings	15	Weyerhaeuser	10

Construct a frequency distribution of the data, letting the first class be $10 billion–$24 billion.

Solution: Fifty pieces of data are given in *descending order* from highest to lowest. The first class is 10–24. The second class must therefore start at 25. We subtract 10, the lower class limit of the first class, from 25, the lower class limit of the second class, to obtain a class width of 15 units. The upper class limit of the second class is found by adding the class width, 15, to the upper class limit of the first class, 24. The upper class limit of the second class is 24 + 15, or 39. Thus

$$10\text{–}24 = \text{first class}$$

$$25\text{–}39 = \text{second class.}$$

The remaining classes will be 40–54, 55–69, 70–84, 85–99, 100–114, 115–129, and 130–144. Since the highest value observed is 134, there is no need to go any farther. Note that each two consecutive lower class limits differ by 15, as do each two consecutive upper class limits. There are 35 pieces of data in the 10–24 class (10, 10, 10, 10, 10, 11, 11, 12, 12, 12, 12, 12, 12, 12, 12, 13, 13, 14, 14, 14, 14, 14, 14, 15, 15, 15, 17, 17, 17, 18, 20, 20, 21, 21, 22). There are 7 pieces of data in the 25–39 class, 2 in the 40–54 class, 3 in the 55–69 class, 0 in the 70–84 class, 1 in the 85–99 class, 1 in the 100–114 class, 0 in the 115–129 class, and 1 in the 130–144 class. The complete frequency distribution of nine classes follows. The number of companies totals 50, so we have included each piece of data.

Sales ($ billions)	Number of Corporations
10–24	35
25–39	7
40–54	2
55–69	3
70–84	0
85–99	1
100–114	1
115–129	0
130–144	1
	50

The modal class of a frequency distribution is the class with the greatest frequency. In Example 2 the modal class is 10–24. The midpoint of a class, also called the class mark, is found by adding the lower and upper class limits and dividing the sum by 2. The midpoint of the first class in Example 2 is

$$\frac{10 + 24}{2} = \frac{34}{2} = 17.$$

Note the difference between successive class marks is the class width.

EXAMPLE 3

The following set of data represents the median family income (in thousands of dollars, rounded to the nearest hundred) of 15 randomly selected families.

31.5   16.8   30.8   29.7   25.9

50.2   37.4   17.6   38.7   33.8

20.5   25.3   24.8   41.3   35.7

Construct a frequency distribution with a first class of 16.5–22.6.

Solution: First rearrange the data from lowest to highest (or highest to lowest) so that the data will be easier to categorize.

$$16.8 \quad 24.8 \quad 29.7 \quad 33.8 \quad 38.7$$
$$17.6 \quad 25.3 \quad 30.8 \quad 35.7 \quad 41.3$$
$$20.5 \quad 25.9 \quad 31.5 \quad 37.4 \quad 50.2$$

The first class goes from 16.5 to 22.6. Since the data are in tenths, the class limits will also be given in tenths. The first class ends with 22.6; therefore the second class must start with 22.7. The class width of the first class is 22.7–16.5, or 6.2. The upper class limit of the second class must therefore be 22.6 + 6.2, or 28.8. The frequency distribution is as follows.

Income ($1000)	Number of Families
16.5–22.6	3
22.7–28.8	3
28.9–35.0	4
35.1–41.2	3
41.3–47.4	1
47.5–53.6	1
	15

Note in Example 3 that the class width is 6.2, that the modal class is 28.9–35.0, and that the class mark of the first class is (16.5 + 22.6)/2, or 19.55.

Section 13.3 Exercises

1. What is a frequency distribution?

2. How can a class width be determined using class limits?

3. Suppose that the first class of a frequency distribution is 9–15.
   a) What is the width of this class?
   b) What is the second class?
   c) What is the lower class limit of the second class?
   d) What is the upper class limit of the second class?

4. Repeat Exercise 3 for a frequency distribution whose first class is 12–20.

5. What is the modal class of a frequency distribution?

6. What is another name for the midpoint of a class? How is the midpoint of a class determined?

In Exercises 7 and 8, use the given frequency distribution to determine

   a) *the total number of observations.*
   b) *the width of each class.*
   c) *the midpoint of the second class.*
   d) *the modal class (or classes).*
   e) *the class limits of the next class if an additional class had to be added.*

7.

Class	Frequency
7–13	4
14–20	2
21–27	3
28–34	0
35–41	4
42–48	6

8.

Class	Frequency
20–29	8
30–39	6
40–49	4
50–59	3
60–69	8
70–79	2

9. The results of a quiz given to a statistics class are shown. Construct a frequency distribution, letting each class have a width of 1 (as in Example 1).

```
0  1  2  3  5  6  7  8  8  9
0  1  2  3  5  6  7  8  8  9
0  1  2  4  5  6  7  8  8  9
0  1  3  4  6  6  8  8  8  10
1  1  3  4  6  6  8  8  8  10
```

10. The heights of a class of third-grade children, rounded to the nearest inch, are shown. Construct a frequency distribution, letting each class have a width of 1.

```
35  38  39  41  42
37  39  40  41  42
37  39  40  42  43
38  39  40  42  43
38  39  40  42  43
```

Note: No one had a height of 36 inches. However, it is customary to include a missing value as an observed value and assign to it a frequency of 0.

In Exercises 11–14, use the following data, which show the result of 50 sixth-grade I.Q. scores.

```
80  89  92  95   97  100  102  106  110  120
81  89  93  95   98  100  103  108  113  120
87  90  94  97   99  100  103  108  114  122
88  91  94  97  100  100  103  108  114  128
89  92  94  97  100  101  104  109  119  135
```

Construct a frequency distribution with a first class of

11. 78–86.   **12.** 80–88.
13. 80–90.   **14.** 80–92.

In Exercises 15–18, use the following data, which represent the ages of the U.S. presidents at their first inauguration.

```
57  57  49  52  50  51  51  56  46
61  61  64  56  47  56  60  61
57  54  50  46  55  55  62  52
57  68  48  54  54  51  43  69
58  51  65  49  42  54  55  64
```

Construct a frequency distribution with a first class of

15. 40–45.   **16.** 42–47.
17. 42–46.   **18.** 40–44.

In Exercises 19–22, use the data in Example 2 on page 665 to construct a frequency distribution with a first class of

19. 10–30.   **20.** 5–25.
21. 10–25.   **22.** 0–20.

In Exercises 23–26, use the following data, which represent the 1995 population of the world's 35 largest cities in millions (rounded to the nearest 100,000).

```
28.4  14.1  11.7  10.4   8.8   7.2   5.9
23.9  13.5  11.3  10.1   7.9   6.8   5.8
21.5  12.9  11.2   9.8   7.6   6.7   5.8
19.1  12.8  11.2   9.3   7.5   6.6   5.7
14.6  12.2  10.7   8.8   7.4   6.5   5.6
```

Construct a frequency distribution with a first class of

23. 5.0–8.0.   **24.** 5.5–8.4.
25. 5.0–7.7.   **26.** 5.1–7.5.

In Exercises 27–30, use the data in the following table.

Density of Population by State*
(Per square mile, land area only)

State	1990	State	1990	State	1990
AL	79.6	LA	96.9	OH	264.9
AK	1.0	MA	767.6	OK	45.8
AZ	32.3	MD	489.2	OR	29.6
AR	45.1	ME	39.8	PA	265.1
CA	190.8	MI	163.6	RI	960.3
CO	31.8	MN	55.0	SC	115.8
CT	678.4	MO	74.3	SD	9.2
DE	340.8	MS	54.9	TN	118.3
FL	239.6	MT	5.5	TX	64.9
GA	111.9	NB	20.5	UT	21.0
HI	172.5	NC	136.1	VA	156.3
IA	49.7	ND	9.3	VT	60.8
ID	12.2	NH	123.7	WA	73.1
IL	205.6	NJ	1042.0	WI	90.1
IN	154.6	NM	12.5	WV	74.5
KS	30.3	NV	10.9	WY	4.7
KY	92.8	NY	381.0		

Construct a frequency distribution with a first class of

27. 0.5–100.5.   **28.** 0.5–90.5.
29. 0.5–150.5.   **30.** 0.5–250.5.

*Washington, D.C., has a population density of 9882.8 per square mile. The U.S. average is 70.3 people per square mile.

13.4   STATISTICAL GRAPHS

Now we will consider three types of graphs: the circle graph, the histogram, and the frequency polygon. Each can be used to represent statistical data.

Circle graphs (also known as pie charts) are often used to compare parts of one or more components of the whole to the whole. The circle graph in Fig. 13.6

shows the percentage of the worldwide PC software market as of June 1995. Since the total circle represents 100%, the sum of the percents of the sectors should be 100%, and it is.

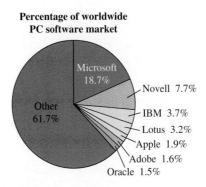

Percentage of worldwide PC software market

Figure 13.6

In order to construct circle graphs we must be able to find central angles. A *central angle* of a circle is an angle whose vertex is at the center of the circle. Since every circle contains 360°, we find the central angles of each class (sector) by multiplying the percent of the total of each class by 360°.

┌─ **EXAMPLE 1**

The following table indicates the number of passenger cars, in hundreds of thousands, manufactured in the United States by various manufacturers.

Company	Cars Produced (100,000s)
Chrysler Corporation	4.9
Ford Motor Company	14.9
General Motors	25.4
Honda	4.0
Mazda	2.2
Nissan	2.9
Toyota	3.6
Diamond Star	1.4
Subaru	0.5
	59.8

Use this information to construct a circle graph illustrating the percent produced by Chrysler, Ford, General Motors, and others.

Solution: First determine the total for each category. All the companies except Chrysler, Ford, and General Motors will be grouped under Others. The sum of the cars produced by others is 4.0 + 2.2 + 2.9 + 3.6 + 1.4 + 0.5 = 14.6. Then determine the measure of the corresponding central angle, as illustrated in the following table.

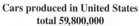
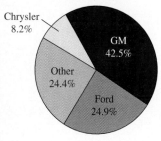

Cars produced in United States
total 59,800,000

Chrysler 8.2%

GM 42.5%

Other 24.4%

Ford 24.9%

Figure 13.7

Company	Cars Produced (100,000s)	Percent of Total (to the Nearest Tenth of a Percent)	Measure of Central Angle
Chrysler	4.9	$\frac{4.9}{59.8} \times 100 = 8.2\%$	$0.082 \times 360° = 29.5°$
Ford	14.9	$\frac{14.9}{59.8} \times 100 = 24.9\%$	$0.249 \times 360° = 89.6°$
GM	25.4	$\frac{25.4}{59.8} \times 100 = 42.5\%$	$0.425 \times 360° = 153.0°$
Other	14.6	$\frac{14.6}{59.8} \times 100 = 24.4\%$	$0.244 \times 360° = 87.8°$
Total	59.8	100%	359.9°*

Now use a protractor to construct a circle graph and label it properly, as illustrated in Fig. 13.7. The measure of the central angle for Chrysler is about 29.5°, for Ford it is about 89.6°, and so on. ▲

Histograms and frequency polygons are statistical graphs used to illustrate frequency distributions. A **histogram** is a graph with observed values on its horizontal scale and frequencies on its vertical scale. A bar is constructed above each observed value (or class when classes are used), indicating the frequency of that value. The horizontal scale need not start at zero, and the calibrations on the horizontal and vertical scales do not have to be the same. The vertical scale must start at zero. To accommodate large frequencies on the vertical scale, it may be necessary to break the scale. Because histograms and other bar graphs are easy to interpret visually, they are used a great deal in newspapers and magazines.

EXAMPLE 2

The frequency distribution developed in Example 1, Section 13.3, is repeated here. Construct a histogram of this frequency distribution.

Number of Children (Observed Values)	Number of Families (Frequency)
0	8
1	11
2	18
3	11
4	6
5	4
6	2
7	1
8	2
9	1

Solution: The vertical scale must extend at least to the number 18, since that is the greatest recorded frequency (see Fig. 13.8). The horizontal scale must include the numbers 0–9, the number of children observed. Eight families have

*Due to a rounding error, we get 359.9°, not exactly 360°. If the measure of the central angle were rounded to hundredths, the sum would be exactly 360°.

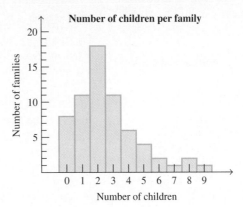

Figure 13.8

no children. We indicate this by constructing a bar above the number 0 on the horizontal scale extended up to 8 on the vertical scale. Eleven families have one child, so we construct a bar extending to 11 above the number 1 on the horizontal scale. We continue this procedure for each observed value. Both the horizontal and vertical scales should be labeled, the bars should be the same width, and the histogram should have a title. In a histogram the bars should always touch.   ▲

Frequency polygons are line graphs with scales the same as those of the histogram; that is, the horizontal scale indicates observed values and the vertical scale indicates frequency. To construct a frequency polygon, place a dot at the corresponding frequency above each of the observed values. Then connect the dots with straight-line segments. When constructing frequency polygons, always put in two additional class marks, one at the lower end and one at the upper end on the horizontal scale (values for these added class marks are not needed on the frequency polygon). Since the frequency at these added class marks is 0, the endpoints of the frequency polygon will always be on the horizontal scale.

EXAMPLE 3

Construct a frequency polygon of the frequency distribution in Example 2.

Solution: Since eight families have no children, place a mark above the 0 at 8 on the vertical scale, as shown in Fig. 13.9. Because there are 11 families with one child, place a mark above the 1 at the 11 on the vertical scale, and so on. Connect the dots by straight-line segments, and bring the endpoints of the graph down to the horizontal scale, as shown.

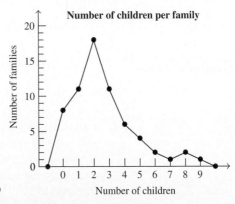

Figure 13.9

Table 13.1

Mileage (mpg)	Number of Cars
10–14	4
15–19	32
20–24	36
25–29	8
30–34	5
35–39	4
40–44	1

EXAMPLE 4

The frequency distribution of average gas mileage for 1996 automobiles is listed in Table 13.1. Construct a histogram and then construct a frequency polygon on the histogram.

Solution: The histogram can be constructed with either class limits or class marks on the horizontal scale. Frequency polygons are constructed with class marks on the horizontal scale. Since we will construct a frequency polygon on the histogram, we will use class marks. Recall that class marks, or class midpoints, are found by adding the lower class limit and upper class limit and dividing the sum by 2. For the first class, the class mark is $(10 + 14)/2$, or 12. Since the class widths are 5 units, the class marks will also differ by 5 units (see Fig. 13.10).

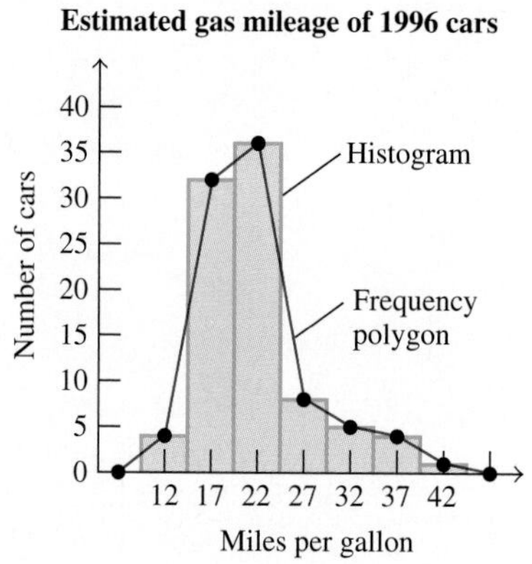

Estimated gas mileage of 1996 cars

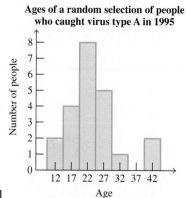

Figure 13.10

EXAMPLE 5

Construct the frequency distribution from the histogram shown in Fig. 13.11.

Ages of a random selection of people who caught virus type A in 1995

Figure 13.11

Table 13.2

Age	Number of People
10–14	2
15–19	4
20–24	8
25–29	5
30–34	1
35–39	0
40–44	2

Solution: There are five units between class midpoints, so each class width must also be five units. Since 12 is the midpoint of the first class, there must be two units below and two above it. The first class must be 10–14. The second class must therefore be 15–19, and so on. The frequency distribution is given in Table 13.2.

Section 13.4 Exercises

1. In your own words, explain how to construct a circle graph from a table of values.

2. **a)** What is listed on the horizontal axis of a histogram and frequency polygon?
   b) What is listed on the vertical axis of a histogram and frequency polygon?

3. In your own words, explain how to construct a histogram from a set of data.

4. In your own words, explain how to construct a frequency polygon from a set of data.

5. **a)** In your own words, explain how to construct a frequency polygon from a histogram.
   b) Construct a frequency polygon from the histogram in Fig. 13.12.

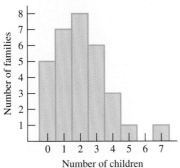

Children in selected families

Figure 13.12   Number of children

6. **a)** In your own words, explain how to construct a histogram from a frequency polygon.
   b) Construct a histogram from the frequency polygon in Fig. 13.13.

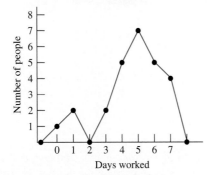

Figure 13.13

7. In 1992, 148 million computers were in use worldwide. The following circle graph shows, by percent, where these computers were being used. Estimate the number of computers in use in 1992 in each of the five categories.

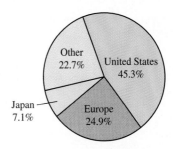

8. The circle graph shows the percent of female military personnel on active duty in 1993 by military branch. If the total number of female military personnel on active duty in 1993 was 203,500, estimate the number of female military personnel on active duty in each branch.

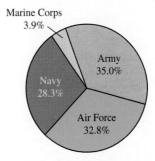

9. The table shows the percent of the recording industry's income according to products sold in 1993. Construct a circle graph to illustrate this information.

Product	Percent
CDs	64.8
Cassettes	29.0
Cassette singles	3.0
Music videos	2.1
Other	1.1

10. Listed are the number of gold medals and total medals won by selected countries in the 1994 winter Olympics in Lillehammer, Norway.

Country	Gold	Total
Norway	10	26
Germany	9	24
Russia	11	23
Italy	7	20
United States	6	13
Canada	3	13
Others	15	64

a) Construct a circle graph indicating the percents of gold medals won by the countries indicated.

b) Repeat part (a) for total medals won.

11. The frequency distribution shown indicates the number of years each of the 38 employees have worked for the Wheatsworth Company.

Years	Number of Employees
0	6
1	4
2	2
3	1
4	6
5	8
6	3
7	3
8	5

a) Construct a histogram of the frequency distribution.

b) Construct a frequency polygon of the frequency distribution.

12. The frequency distribution shown indicates the ages of a group of 40 people attending a party.

Age	Number of People
20	6
21	3
22	0
23	4
24	6
25	3
26	8
27	10

a) Construct a histogram of the frequency distribution.

b) Construct a frequency polygon of the frequency distribution.

13. The frequency distribution shown represents the annual salaries in thousands of dollars of the people in management positions at the Bradley Milton Corporation.

Salary (in $1000)	Number of People
20–25	4
26–31	6
32–37	8
38–43	9
44–49	8
50–55	5
56–61	3

a) Construct a histogram of the frequency distribution.

b) Construct a frequency polygon of the frequency distribution.

14. The frequency distribution shown indicates the gross weekly salaries of families whose children receive free lunch at Villdale Elementary School.

Salary	Number of Families
133–141	8
142–150	6
151–159	8
160–168	4
169–177	3
178–186	1

a) Construct a histogram of the frequency distribution.

b) Construct a frequency polygon of the frequency distribution.

15. Use the histogram shown in Fig. 13.14 to answer the following questions.

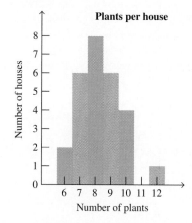

Figure 13.14

a) How many houses were surveyed?

b) In how many houses were nine plants observed?

c) What is the modal class?

d) How many plants were observed?

e) Construct a frequency distribution from this histogram.

16. Use the histogram shown in Fig. 13.15 to answer the following questions.

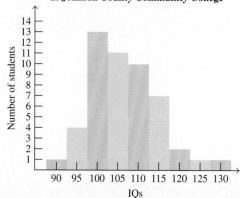

Figure 13.15

a) What are the upper and lower class limits of the first and second classes?
b) How many students had IQs in the class with a class mark of 100?
c) How many students were surveyed?
d) Construct a frequency distribution.

17. Use the frequency polygon in Fig. 13.16 to answer the following questions.

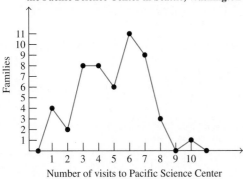

Number of visits selected families have made to the Pacific Science Center in Seattle, Washington

Figure 13.16

a) How many families visited the Pacific Science Center four times?
b) How many families visited the Pacific Science Center at least six times?
c) How many families were surveyed?
d) Construct a frequency distribution from the frequency polygon.
e) Construct a histogram from the frequency distribution in part (d).

18. The frequency polygon shown indicates the sales price (rounded to thousands) of houses sold in a specific eight-block area of a city.

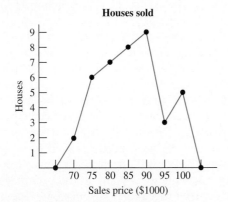

Houses sold

a) How many houses were sold during this period?
b) Construct a frequency distribution from the frequency polygon.

c) Construct a histogram from the frequency distribution in part (b).
d) Is it possible to tell from the frequency polygon how many houses sold for between $75,000 and $80,000? Explain.

19. Construct a histogram and a frequency polygon from the frequency distribution given in Exercise 7 for Section 13.3. See page 667.

20. Construct a histogram and a frequency polygon from the frequency distribution given in Exercise 8 for Section 13.3. See page 667.

21. Starting salaries (rounded to the nearest thousand) for chemical engineers with B.S. degrees and no experience are shown for 25 different companies.

25	26	27	29	31
26	26	27	29	31
26	26	28	30	31
26	27	28	30	32
26	27	28	30	32

a) Construct a frequency distribution.
b) Construct a histogram.
c) Construct a frequency polygon.

22. The ages of a random sample of U.S. ambassadors are

40	43	45	50	52	55	59	64
41	43	46	50	53	55	60	65
41	44	47	50	54	55	60	65
42	44	48	51	54	57	60	66
43	45	48	51	54	58	62	67

a) Construct a frequency distribution with the first class 40–44.
b) Construct a histogram.
c) Construct a frequency polygon.

23. The weights of a random sample of forty-five twelfth-grade girls follows.

96	100	108	112	113	115	125	132	137
98	101	109	112	113	117	125	132	143
99	104	109	112	114	119	127	135	157
99	106	110	113	114	122	129	135	160
100	107	112	113	115	124	132	137	162

a) Construct a frequency distribution with the first class 96–102.
b) Construct a histogram.
c) Construct a frequency polygon.

24. The following table lists the circulation of the 30 best-selling U.S. magazines as of December 31, 1993.
a) Construct a frequency distribution with the first class 1.7–4.9 million.
b) Construct a histogram.
c) Construct a frequency polygon.

Magazine	Circulation
1. *Reader's Digest*	16,261,968
2. *TV Guide*	14,122,915
3. *National Geographic*	9,390,787
4. *Better Homes and Gardens*	7,600,960
5. *Good Housekeeping*	5,162,597
6. *Ladies' Home Journal*	5,153,565
7. *Family Circle*	5,114,030
8. *Woman's Day*	4,858,625
9. *McCall's*	4,605,441
10. *Time*	4,103,722
11. *People*	3,446,569
12. *National Enquirer*	3,403,330
13. *Playboy*	3,402,619
14. *Sports Illustrated*	3,356,729
15. *Redbook*	3,345,451
16. *Newsweek*	3,220,763
17. *Prevention*	3,156,192
18. *Star*	2,957,915
19. *Cosmopolitan*	2,627,491
20. *Southern Living*	2,368,678
21. *Glamour*	2,304,769
22. *U.S. News and World Report*	2,281,369
23. *Smithsonian*	2,212,418
24. *Money*	2,100,039
25. *Field & Stream*	2,007,901
26. *Discover*	1,977,214
27. *Seventeen*	1,940,601
28. *Ebony*	1,939,500
29. *Popular Science*	1,815,819
30. *Parent's Magazine*	1,776,470

Problem Solving/Group Activities

25. a) What do you believe a histogram of the months in which the students in your class were born (January is month 1 and December is month 12) would look like? Explain.

b) By asking, determine the month in which the students of your class were born (include yourself).

c) Construct a frequency distribution containing 12 classes.

d) Construct a histogram from the frequency distribution in part (c).

e) Construct a frequency polygon of the frequency distribution in part (c).

26. Repeat Exercise 25 for the last digit of the students' social security numbers. Include classes for the digits 0–9.

Research Activity

27. Over the years many changes have been made in the U.S. Social Security System.

a) Do research at your local library and determine the number of people receiving social security benefits for the years 1945, 1950, 1955, 1960, . . . , 1995. Then construct a frequency distribution and histogram of the data.

b) Determine the maximum amount that self-employed individuals had to pay into social security (the FICA tax) for the years 1945, 1950, 1955, 1960, . . . , 1995. Then construct a frequency distribution and a histogram of the data.

13.5 MEASURES OF CENTRAL TENDENCY

Most people have an intuitive idea of what is meant by an "average." The term is used daily in many familiar ways: "This car averages 19 miles per gallon." "The average test grade was 78." "The average height of adult males is 5 feet 8 inches."

An average is a number that is representative of a group of data. There are at least four different averages: the mean, the median, the mode, and the midrange. Each is calculated differently and may yield different results for the same set of data. Each will result in a number near the center of the data; for this reason, averages are commonly referred to as measures of central tendency.

The arithmetic mean, or simply the mean, is symbolized either by $\bar{x}$ (read "x bar") or by the Greek letter mu, μ. The symbol $\bar{x}$ is used when the mean of a *sample* of the population is calculated. The symbol μ is used when the mean of the *entire population* is calculated. We will assume that the data featured in this book represent samples, and therefore we will always use $\bar{x}$ for the mean.

The Greek letter sigma, Σ, is used to indicate "summation." The notation Σx, read "the sum of x," is used to indicate the sum of all the data. For example, if there are five pieces of data, 4, 6, 1, 0, 5, then $\Sigma x = 4 + 6 + 1 + 0 + 5 = 16$.

Now we can discuss the procedure for finding the mean of a set of data.

> The mean, $\bar{x}$, is the sum of the data divided by the number of pieces of data. The formula for calculating the mean is
>
> $$\bar{x} = \frac{\Sigma x}{n},$$
>
> where Σx represents the sum of all the data and n represents the number of pieces of data.

The most common use of the word *average* is the mean.

EXAMPLE 1

Find Tom's mean grade if he received grades of 49, 75, 40, 78, and 78 on his five statistics exams.

Solution:

$$\bar{x} = \frac{\Sigma x}{n} = \frac{49 + 75 + 40 + 78 + 78}{5} = \frac{320}{5} = 64$$

Therefore the mean, $\bar{x}$, is 64.   ▲

The mean represents "the balancing point" of a set of scores. For example, if a seesaw were pivoted at the mean and uniform weights were placed at points corresponding to the test scores, the seesaw would balance. Figure 13.17 shows the five test scores given in Example 1 and the calculated mean.

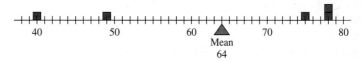

Figure 13.17

A second average is the *median*. To find the median of a set of data, *rank the data* from smallest to largest, or largest to smallest, and determine the value in the middle of the set of *ranked data*. This value will be the median.

> The **median** is the value in the middle of a set of *ranked data*.

EXAMPLE 2

Find the median of Tom's grades in Example 1.

Solution: Ranking the data from smallest to largest gives 40, 49, 75, 78, 78. Since 75 is the value in the middle of this set of ranked data (two pieces of data above it and two pieces below it), 75 is the median.   ▲

If there are an even number of pieces of data, the median will be halfway between the two middle pieces. To find the median, add the two middle pieces and divide this sum by 2.

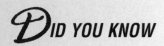

EXAMPLE 3

Find the median of the following sets of data.

a) 1, 4, 6, 8, 9, 8, 3, 3

b) 2, 3, 3, 3, 4, 5

Solution:

a) Ranking the data gives 1, 3, 3, 4, 6, 8, 8, 9. There are eight pieces of data. Therefore the median will lie halfway between the two middle pieces, the 4 and the 6. The median is (4 + 6)/2, or 10/2, or 5.

b) There are six pieces of data and they are already ranked. Therefore the median will lie halfway between the two middle pieces. Both middle pieces are 3s. The median is (3 + 3)/2, or 6/2, or 3. ▲

A third average is the mode.

The **mode** is the piece of data that occurs most frequently.

EXAMPLE 4

Find the mode of Tom's grades in Example 1.

Solution: Tom's grades were 49, 75, 40, 78, 78. The grade 78 is the mode because it occurs twice and the other values occur only once. ▲

If no one piece of data occurs more frequently than every other piece of data, the set of data has no mode. For example, neither of the following two sets of data has a mode.

$$1, 2, 3, 4, 5 \qquad \text{no mode}$$
$$1, 1, 2, 3, 3, 4, 5 \qquad \text{no mode*}$$

The last average that we will discuss is the midrange. The midrange is the value halfway between the lowest (L) and highest (H) values in a set of data. It is found by adding the lowest and highest values and dividing the sum by 2. A formula for finding the midrange follows.

$$\textbf{Midrange} = \frac{\text{lowest value } + \text{ highest value}}{2}$$

EXAMPLE 5

Find the midrange of Tom's grades given in Example 1: 49, 75, 40, 78, 78.

Solution: Tom's lowest grade is 40, and his highest grade is 78.

$$\text{Midrange} = \frac{\text{lowest } + \text{ highest}}{2}$$
$$= \frac{40 + 78}{2} = \frac{118}{2} = 59$$

▲

---

*Some textbooks refer to sets of data of this type as bimodal.

Tom's "average" grade for the exam scores of 49, 75, 40, 78, 78 can be considered any one of the following values: 64 (mean), 75 (median), 78 (mode), or 59 (midrange). Which average do you feel is most representative of his grades? We will discuss this question later in this section.

EXAMPLE 6

The salaries of eight selected chemical engineers rounded to the nearest thousand dollars are 40, 25, 28, 35, 42, 60, 60, and 73. For this set of data find the (a) mean, (b) median, (c) mode, and (d) midrange, and then (e) rank the measures of central tendency from lowest to highest.

Solution:

a) $\bar{x} = \dfrac{40 + 25 + 28 + 35 + 42 + 60 + 60 + 73}{8} = \dfrac{363}{8} = 45.375$

b) Listing the data from the smallest to largest gives

$$25, 28, 35, 40, 42, 60, 60, 73.$$

Since there are an even number of pieces of data, the median is halfway between 40 and 42. The median $= (40 + 42)/2 = 82/2 = 41$.

c) The mode is the piece of data that occurs most frequently. The mode is 60.

d) The midrange $= (L + H)/2 = (25 + 73)/2 = 98/2 = 49$.

e) The averages from lowest to highest are the median, mean, midrange, and mode. Their values are 41, 45.375, 49, and 60, respectively. ▲

$\mathcal{D}$ID YOU KNOW

MEASURES OF LOCATION

$\mathcal{W}$hen you took the Scholastic Aptitude Test (SAT) before applying to college, your score was described as a measure of location rather than a measure of central tendency. Measures of location are often used to make comparisons, such as comparing the scores of individuals from different populations, and are generally used when the amount of data is large.

Two measures of location are percentiles and quartiles. Percentiles divide the set of data into 100 equal parts (see Fig. a). For example, suppose that you scored 490 on the math half of the SAT, and that the score of 490 was reported to be in the 78th percentile of high school students. This *does not* mean that 78% of your answers were correct; it *does* mean that you outperformed about 78% of all those taking the exam. In general, a score in the *n*th percentile means you outperformed about *n*% of the population who took the test, and that $(100 - n)$% of the people taking the test performed better than you did.

Measures of location

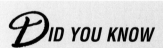

(a) Percentiles

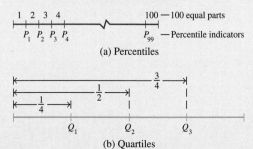

(b) Quartiles

Quartiles divide data into four equal parts: The first quartile is a value higher than 1/4 or 25% of the population. It is the same as the 25th percentile. The second quartile is higher than 1/2 the population and is the same as the 50th percentile, or the median. The third quartile is higher than 3/4 of the population and is the same as the 75th percentile (see Fig. b). ■

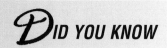

At this point you should be able to calculate the four measures of central tendency. Now let's examine the circumstances in which each is used.

The mean is used when each piece of data is to be considered and "weighed" equally. It is the most commonly used average. It is the only average that can be affected by *any* change in the set of data; for this reason it is the most sensitive of all the measures of central tendency (see Exercise 22).

Occasionally, one or more pieces of data may be much greater or much smaller than the rest of the data. When this occurs, these "extreme" values have the effect of increasing or decreasing the mean significantly so that the mean will not be representative of the set of data. Under these circumstances the median should be used instead of the mean. The median is often used in describing average family incomes because a relatively small number of families have extremely large incomes. These few families would inflate the mean income, making it nonrepresentative of the millions of families in the population.

Consider a set of exam scores from a mathematics class: 0, 16, 19, 65, 65, 65, 68, 69, 70, 72, 73, 73, 75, 78, 80, 85, 88, 92. Which average would best represent these grades? The mean is 64.06. The median is 71. Since only 3 of the 18 scores fall below the mean, the mean would not be considered a good representative score. The median of 71 probably would be the better average to use.

The mode is the piece of data, if any, that occurs most frequently. Builders planning houses are interested in the most common family size. Retailers ordering shirts are interested in the most common shirt size. An individual purchasing a thermometer might select one with the most common reading. These examples illustrate how the mode may be used.

The midrange is sometimes used as the average when the item being studied is constantly fluctuating. Average daily temperature, used to compare temperatures in different areas, is calculated by adding the lowest and highest temperatures for the day and dividing the sum by 2. The midrange is actually the mean of the high value and the low value of a set of data. Occasionally, the midrange is used to estimate the mean, since it is much easier to calculate.

Sometimes an average itself is of little value, and care must be taken in interpreting its meaning. For example Jim is told that the average depth of Willow Pond is only 3 ft. He is not a good swimmer but decides that it is safe to go out a short distance in this shallow pond. After he is rescued, he exclaims, "I thought this pond was only three feet deep." Jim didn't realize that an average does not indicate extreme values or the spread of the values. The spread of data will be discussed in Section 13.6.

Section 13.5 Exercises

1. Describe the mean and explain how to find it.

2. What is *ranked data*?

3. Describe the median and explain how to find it.

4. Describe the mode and explain how to find it.

5. Describe the midrange and explain how to find it.

6. When might the median be the preferred average to the mean? Give an example.

7. When might the mode be the preferred average to use? Give an example.

8. When might the midrange be the preferred average to use? Give an example.

In Exercises 9–21, find the mean, median, mode, and midrange of the set of data. Round your answer to the nearest tenth.

9. 1, 1, 4, 6, 8, 9, 9, 9, 34

10. 4, 5, 10, 12, 10, 9, 8, 372, 40, 37 ✓

11. 60, 72, 80, 84, 86, 45, 96

12. 5, 3, 6, 6, 6, 9, 11 ✓

13. 1, 3, 5, 7, 9, 11, 13, 15

14. 1, 7, 11, 27, 36, 14, 12, 9, 1

15. 40, 50, 30, 60, 90, 100, 140

16. 1, 1, 1, 1, 4, 4, 4, 4, 6, 8, 10, 12, 15, 21 ✓

17. 5, 7, 11, 12, 10, 9, 12, 14, 16

18. 2, 10, 2, 10, 2, 10, 2, 10 ✓

19. 237, 463, 812, 487, 387

20. 2432, 16203, 962, 72, 3

21. 78, 60, 72, 83, 92, 96

22. The mean is the "most sensitive" average because it is affected by any change in the data.
    a) Determine the mean, median, mode, and midrange for 1, 2, 3, 5, 5, 7, 11.
    b) Change the 7 to a 10 in part (a). Find the mean, median, mode, and midrange.
    c) Which averages were affected by changing the 7 to a 10?
    d) Which averages will be affected by changing the 11 to a 10 in part (a)?

23. In 1994 the National Center for Health Statistics indicated a new record "average life expectancy" of 76.3 years for the total U.S. population. The average life expectancy for men was 73.0 years, and for women it was 79.6 years. Which "average" do you think the National Center for Health is using? Explain your answer.

24. To get a grade of B, a student must have a mean average of 80. Jim has a mean average of 79 for 10 quizzes. He approaches his teachers and asks for a B, reasoning that he missed a B by only one point. What is wrong with Jim's reasoning?

25. The salaries of 10 employees of a small company follow.

$22,000	$58,000
19,000	18,000
25,000	21,000
20,000	75,000
20,000	23,000

    Calculate the
    a) mean.
    b) median.
    c) mode.
    d) midrange.
    e) Which is the best measure of central tendency for this set of data? Explain your answer.

26. Ted tosses a frisbee as far as he can 15 times. The following distances, in feet, were recorded.

190	102	188	270	105
201	95	196	284	200
164	191	110	169	212

    Calculate the
    a) mean.
    b) median.
    c) mode.
    d) midrange.
    e) Which average is the best measure of central tendency for this set of data? Explain.

27. The Webers' monthly telephone charges for a 1-year period are shown.

$45.11	$85.53	$128.94
97.15	67.78	72.13
85.29	74.33	84.19
73.48	74.68	87.15

    Find the
    a) mean.
    b) median.
    c) mode.
    d) midrange.

28. The nine National Parks (NPs) or National Recreation Areas (NRAs) where consumers spent the most money in 1995 are listed.

Area	Amount Spent ($ millions)
Lake Mead NRA	938.7
Yosemite NP	850.9
Great Smokey Mountains NP	750.2
Grand Canyon NP	495.9
Glen Canyon NRA	405.5
Sequoia NP	398.6
Kings Canyon NP	320.3
Grand Tetons NP	227.2
Hawaii Volcano NP	215.7

    Determine the
    a) mean.
    b) median.
    c) mode.
    d) midrange.

29. Susan's mean average on six exams is 87. Find the sum of her scores.

30. Mary's mean average on five exams is 72. Find the sum of her scores.

31. In the 50 United States, in 1993, approximately 4 million motorcycles were registered. Wyoming had the fewest, 15,000, and California had the most, 600,000. Considering all the motorcycles registered

in the United States, determine whether it is possible to find

a) the mean number of motorcycles registered for all 50 states.

b) the median number of motorcycles registered.

c) the mode.

d) the midrange.

e) Find all the measures of central tendency that can be found with the information given and explain why the others cannot be found.

32. Construct a set of five pieces of data in which the mode has a lower value than the median and the median has a lower value than the mean.

33. Construct a set of five pieces of data with a mean of 75 and no two pieces of data the same.

34. Construct a set of six pieces of data with a mean, median, and midrange of 75 and no two pieces of data the same.

35. A mean average of 80 for five exams is needed for a final grade of B. Jorge's first four exam grades are 68, 78, 83, and 80. What grade does Jorge need on the fifth exam to get a B in the course?

36. A mean average of 60 on seven exams is needed to pass a course. On her first six exams, Sheryl received grades of 49, 72, 80, 60, 57, and 69.

a) What grade must she receive on her last exam to pass the course?

b) An average of 70 is needed to get a C in the course. Is it possible for Sheryl to get a C? If so, what grade must she receive on the seventh exam?

c) If her lowest grade is to be dropped and only her six best exams are counted, what grade must she receive on her last exam to pass the course?

d) If her lowest grade is to be dropped and only her six best exams are counted, what grade must she receive on her last exam to get a C in the course?

37. Which of the measures of central tendency *must* be an actual piece of data in the distribution? Explain.

38. Construct a set of six pieces of data such that if only one piece of data is changed, the mean, median, and mode will all change.

39. Consider the set of data 1, 1, 1, 2, 2, 2. If one 2 is changed to a 3, which of the following will change: mean, median, mode, midrange? Explain.

40. Is it possible to construct a set of six different pieces of data such that by changing only one piece of data you cause the mean, median, mode, and midrange to change? Explain.

In Exercises 41–47, refer to the Did You Know on Measures of Location on page 679, and then answer the following questions.

41. For any set of data, what must be done to the data before percentiles can be determined?

42. Liz scored in the 87th percentile on the verbal part of her College Board Test. What does that mean?

43. When a national sample of the heights of kindergarten children was taken, Kevin was told that he was in the 48th percentile. Explain what that means.

44. A union leader is told that, when all workers' salaries are considered, the first quartile is $17,750. Explain what that means.

45. The third quartile for the time required for students to travel one way to Central Valley College is 47.6 minutes. Explain what that means.

46. Give the names of two other statistics that have the same value as the 50th percentile.

47. Jonathan took an admission test for the University of California and scored in the 85th percentile. The following year Jonathan's sister Kendra took a similar admission test for the University of California and scored in the 90th percentile.

a) Is it possible to determine which of the two answered the higher percent of questions correctly on their respective exams? Explain your answer.

b) Is it possible to determine which of the two was in a better relative position with regard to their respective populations? Explain.

48. The following statistics represent weekly salaries at the Midtown Construction Company:

Mean	$490	First quartile	$450
Median	$480	Third quartile	$515
Mode	$470	83rd percentile	$555

a) What is the most common salary?

b) What salary did half the employees surpass?

c) About what percent of employees surpassed $515?

d) About what percent of employees' salaries were below $450?

e) About what percent of employees surpassed $555?

f) If the company has 100 employees, what is the total weekly salary of all employees?

Problem Solving/Group Activities

49. Consider the following five sets of values.

i)   5   6   7   7   8   9   14
ii)   3   6   8   9
iii) 1   1   1   2   5
iv)   6   8   9   12   15
v)   50   51   55   60   80   100

a) Compute the mean of each of the 5 sets of data.

b) Compute the mean of the five means in part (a).

c) Find the mean of the 27 pieces of data.

d) Compare your answer in part (b) to your answer in part (c). Are the values the same? Does your answer make sense? Explain.

50. Consider the following table, which appeared in the May 2, 1995, issue of many publications.

Average Lifetime Earnings

Professional degree	$3-million
Doctorate	$2.1-million
Master's degree	$1.7-million
Bachelor's degree	$1.5-million
Associate's degree	$1-million
High school diploma	$821,000

Source: U.S. Census and National Science Foundation

a) What average do you believe is being presented? Explain your answer.

b) Compute the average hourly salary for each category, assuming that the average person works for 40 years, for 40 hours per week, for 48 weeks a year (disregard paid holidays and vacations).

51. The following tables compare the batting performances for selected years for two well-known former baseball players, Babe Ruth and Mickey Mantle.

BABE RUTH
Boston Red Sox 1914–1919
New York Yankees 1920–1934

Year	At Bats	Hits	Pct.
1925	359	104	
1930	518	186	
1933	459	138	
1916	136	37	
1922	406	128	
Total	1878	593	

MICKEY MANTLE
New York Yankees 1951–1968

Year	At Bats	Hits	Pct.
1954	543	163	
1957	474	173	
1958	519	158	
1960	527	145	
1962	377	121	
Total	2440	760	

a) For each player, compute the batting average percent (pct.) for each year by dividing the hits by the at bats. Round off to thousandths. Place the answers in the pct. column.

b) Going across each of the five horizontal lines (for example Ruth, 1925, vs. Mantle, 1954), compare the percents (pct.) and determine which is greater in each case.

c) For each player, compute the mean batting average percent for the 5 given years by determining the total at bats and hits over the 5 years and dividing the total hits by the total at bats. Which is greater, Ruth's or Mantle's?

d) Based on your answer in part (b), does your answer in part (c) make sense? Explain.

e) Find the mean percent for each player by adding the five pcts. and dividing by 5. Which is greater, Ruth's or Mantle's?

f) Why do the answers obtained in parts (c) and (e) differ? Explain.

g) Who would you say has the better batting average percent for the 5 years selected? Explain.

Research Activity

52. Two other measures of location that we did not mention in the Did You Know on page 679 are *stanines* and *deciles*. Use statistics books and books on educational testing and measurements to determine what stanines and deciles are and when percentiles, quartiles, stanines, and deciles are used.

13.6   MEASURES OF DISPERSION

The measures of central tendency by themselves do not always give sufficient information to analyze a situation and make decisions. As an example, two manufacturers of airplane engines are being considered for a contract. Manufacturer A's engines have an average (mean) life of 1000 hours of flying time before they must be rebuilt. Manufacturer B's engines have an average life of 950 hours of flying time before they must be rebuilt. If you assume that both cost the same,

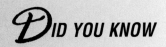

which engines should be purchased? The average engine life may not be the most important factor. The fact that manufacturer A's engines have an average life of 1000 hours could mean that half will last about 500 hours and the other half will last about 1500 hours. If in fact all of manufacturer B's engines have a life span of between 900 and 1000 hours, then B's engines are more consistent and reliable. If A's engines were purchased, they would all have to be rebuilt every 300 hours or so because it would be impossible to determine which ones would fail first. If B's engines were purchased, they could go much longer before having to be rebuilt. This example is of course an exaggeration used to illustrate the importance of knowing something about the *spread*, or *variability*, of the data.

Measures of dispersion are used to indicate the spread of the data. The range and standard deviation* are the measures of dispersion that will be discussed in this book.

The range is the difference between the highest and lowest values; it indicates the total spread of the data.

Range = highest value − lowest value

EXAMPLE I

Find the range of the following weights: 140, 195, 203, 175, 85, 197, 220, 212, 415.

Solution: Range = highest value − lowest value = 415 − 85 = 330. The range of this data is 330 pounds. ▲

The second measure of dispersion, the standard deviation, measures how much the data *differs from the mean*. It is symbolized either by the letter *s* or by the Greek letter sigma, σ.[†] The *s* is used when the standard deviation of a *sample* is calculated. The σ is used when the standard deviation of the entire *population* is calculated. Since we are assuming that all data presented here are for samples, we will use *s* to represent the standard deviation in this book (note, however, that on the doctors' charts in Fig. 13.19, σ is used). The larger the spread of the data about the mean, the larger is the standard deviation. Consider the following two sets of data.

$$5, 8, 9, 10, 12, 13 \qquad 8, 9, 9, 10, 10, 11$$

Both have a mean of 9.5. Which set of values on the whole do you feel differs less from the mean of 9.5? Figure 13.18 may make the answer more apparent. The scores in the second set of data are closer to the mean and have a smaller standard deviation. You will soon be able to verify such relationships yourself.

Sometimes only a very small standard deviation is desirable or acceptable. Consider a cereal box that is to contain 8 oz of cereal. If the amount of cereal put into the box varies too much—sometimes underfilling, sometimes overfilling—

Variance, another measure of dispersion, is the square of the standard deviation.

[†]Our alphabet uses both uppercase and lowercase letters—for example, *A* and *a*. The Greek alphabet also uses both uppercase and lowercase letters. The symbol Σ is the capital Greek letter sigma, and σ is the lowercase Greek letter sigma.

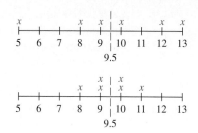

Figure 13.18

the manufacturer will soon be in trouble with consumer groups and government agencies.

At other times a larger spread of data is desirable or expected. For example, intelligence quotients (IQs) are expected to exhibit a considerable spread about the mean, since everyone is different.

To Find the Standard Deviation of a Set of Data:

1. Find the mean of the set of data.
2. Make a chart having three columns:

$$\text{Data} \qquad \text{Data} - \text{Mean} \qquad (\text{Data} - \text{Mean})^2$$

3. List the data vertically under the column marked Data.
4. Complete the Data − Mean column for each piece of data.
5. Square the values obtained in the Data − Mean column and record these values in the (Data − Mean)^2 column.
6. Find the sum of the values in the (Data − Mean)^2 column.
7. Divide the sum obtained in step 6 by $n - 1$, where n is the number of pieces of data.*
8. Find the square root of the number obtained in step 7. This number is the standard deviation for the set of data.

Example 2 illustrates the procedure to follow to find the standard deviation of a set of data.

EXAMPLE 2

Find the standard deviation of

$$7, 9, 11, 15, 18.$$

Solution: First, determine the mean:

$$\bar{x} = \frac{\Sigma x}{n} = \frac{7 + 9 + 11 + 15 + 18}{5} = \frac{60}{5} = 12.$$

---

*To find the standard deviation of a sample, divide the sum of (Data − Mean)^2 column by $n - 1$. To find the standard deviation of a population, divide the sum by n. In this book we assume that the set of data represents a sample and divide by $n - 1$. The quotient obtained in step 7 represents a measure of dispersion called the *variance*.

Next, construct a table with three columns, as illustrated in Table 13.3 and list the data in the first column (it is often helpful to list the data in ascending or descending order). Complete the second column by subtracting the mean, 12 in this case, from each piece of data in the first column.

Table 13.3

Data	Data − Mean	(Data − Mean)^2
7	$7 - 12 = -5$	
9	$9 - 12 = -3$	
11	$11 - 12 = -1$	
15	$15 - 12 = 3$	
18	$18 - 12 = 6$	
	0	

The sum of the values in the Data − Mean column should always be zero; if not, you have made an error. (If the mean is a decimal, there may be a slight round-off error.)

Next square the values in the second column and place the squares in the third column (Table 13.4).

Table 13.4

Data	Data − Mean	(Data − Mean)^2
7	-5	$(-5)^2 = (-5)(-5) = 25$
9	-3	$(-3)^2 = (-3)(-3) = 9$
11	-1	$(-1)^2 = (-1)(-1) = 1$
15	3	$(3)^2 = (3)(3) = 9$
18	6	$(6)^2 = (6)(6) = 36$
	0	80

Add the squares in the third column. In this case the sum is 80. Divide this sum by one less than the number of pieces of data ($n - 1$). In this case the number of pieces of data is 5. Therefore we divide by 4 and get

$$\frac{80}{4} = 20.*$$

Finally, take the square root of this number. The square root of 20 can be found by using a calculator. Since $\sqrt{20} \approx 4.5$, the standard deviation, symbolized s, is 4.5 (to the nearest tenth). ▲

Now we will develop a formula for finding the standard deviation of a set of data. If we call the individual data x and the mean $\bar{x}$, we could write the three column heads Data, Data − Mean, and (Data − Mean)^2 in Table 13.4 as

$$x \qquad x - \bar{x} \qquad (x - \bar{x})^2$$

Let's follow the procedure we used to obtain the standard deviation in Example 2. We found the sum of the (Data − Mean)^2 column, which is the same as the

---

*20 is the variance, symbolized s^2, of this set of data.

sum of the $(x - \bar{x})^2$ column. We can represent the sum of the $(x - \bar{x})^2$ column by using the summation notation, $\Sigma(x - \bar{x})^2$. Thus in Table 13.4, $\Sigma(x - \bar{x})^2 = 80$. We then divided this number by 1 less than the number of pieces of data, $n - 1$. Thus we have

$$\frac{\Sigma(x - \bar{x})^2}{n - 1}.$$

Finally, we took the square root of this value to obtain the standard deviation.

Standard Deviation

$$s = \sqrt{\frac{\Sigma(x - \bar{x})^2}{n - 1}}$$

EXAMPLE 3

Find the standard deviation of

$$20, 30, 32, 35, 40, 40, 41, 42, 43, 47.$$

Solution: The mean, $\bar{x}$, is

$$\bar{x} = \frac{\Sigma x}{n}$$

$$= \frac{20 + 30 + 32 + 35 + 40 + 40 + 41 + 42 + 43 + 47}{10} = \frac{370}{10}$$

$$= 37.$$

Now find $\Sigma(x - \bar{x})^2$. Table 13.5 shows that $\Sigma(x - \bar{x})^2 = 562$. For the 10 pieces of data, $n - 1$ is $10 - 1$, or 9.

$$s = \sqrt{\frac{\Sigma(x - \bar{x})^2}{n - 1}} = \sqrt{\frac{562}{9}} = \sqrt{62.4} \approx 7.9$$

The standard deviation is 7.9.   ▲

Table 13.5

x	$x - \bar{x}$	$(x - \bar{x})^2$
20	-17	289
30	-7	49
32	-5	25
35	-2	4
40	3	9
40	3	9
41	4	16
42	5	25
43	6	36
47	10	100
	0	$\Sigma(x - \bar{x})^2 = 562$

Standard deviation will be used in Section 13.7 to find the percent of data between any two values in a normal curve. Standard deviations are also often used in determining norms for a population (see Exercise 27).

Section 13.6 Exercises

1. Explain how to find the range of a set of data.

2. What does the standard deviation of a set of data measure?

3. Explain how to find the standard deviation of a set of data.

4. What is the standard deviation of a set of data in which all the data is the same? Explain.

5. Why is measuring variation (dispersion) in observed data important?

6. Can you think of any situations in which a large standard deviation may be desirable? Explain.

7. Can you think of any situations in which a small standard deviation may be desirable? Explain.

8. Without actually doing the calculations, decide which of the following two sets of data will have the greater standard deviation: 13, 16, 17, 18, 20, 24, or 16, 17, 17, 18, 18, 19. Explain why.

9. Of the following two sets of data, which would you expect to have the larger standard deviation? Explain.

$$2, 4, 6, 8, 10 \quad \text{or} \quad 102, 104, 106, 108, 110$$

10. By studying the standard deviation formula, explain why the standard deviation of a set of data will always be greater than or equal to 0.

In Exercises 11–24, find the range and standard deviation of the set of data.

11. 8, 2, 0, 6, 9

12. 6, 6, 10, 12, 4, 4

13. 150, 151, 152, 153, 154, 155, 156

14. 4, 0, 3, 6, 9, 12, 2, 3, 4, 7

15. 4, 8, 9

16. 9, 9, 9, 9, 9, 9

17. 10, 9, 14, 15, 17

18. 5, 6, 5, 4, 4, 5, 6

19. 42, 40, 44, 49, 30, 33, 54, 52

20. 9, 4, 9, 4, 9, 4, 7, 2

21. 5, 7, 12, 20, 17, 8, 13, 10, 10, 8

22. 60, 72, 84, 88, 76

23. 3, 4, 5, 9, 3, 7, 4, 4, 9, 2

24. 103, 106, 109, 112, 115, 118, 121

25. a) Pick any five numbers. Compute the mean and the standard deviation of this distribution.
    b) Add 20 to each of the numbers in your original distribution and compute the mean and the standard deviation of this new distribution.
    c) Subtract 5 from each number in your original distribution and compute the mean and standard deviation of this new distribution.
    d) What conclusions can you draw about changes in the mean and the standard deviation when the same number is added to or subtracted from each piece of data in a distribution?
    e) How will the mean and standard deviation of the numbers 6, 7, 8, 9, 10, 11, 12 differ from the mean and standard deviation of the numbers 596, 597, 598, 599, 600, 601, 602? Find the mean and standard deviation of both sets of numbers.

26. a) Pick any five numbers. Compute the mean and the standard deviation of this distribution.
    b) Multiply each number in your distribution by 4 and compute the mean and the standard deviation of this new distribution.

c) Multiply each number in your original distribution by 9 and compute the mean and the standard deviation of this new distribution.
d) What conclusions can you draw about changes in the mean and the standard deviation when each value in a distribution is multiplied by the same number?
e) The mean and standard deviation of the distribution 1, 3, 4, 4, 5, 7 are 4 and 2, respectively. Use the conclusion drawn in part (d) to determine the mean and standard deviation of the distribution 5, 15, 20, 20, 25, 35.

27. The chart shown in Fig. 13.19 uses the symbol σ to represent the standard deviation. Note that 2σ represents the value that is two standard deviations above the mean; -2σ represents the value that is two standard deviations below the mean. The unshaded areas, from two standard deviations below the mean to two standard deviations above the mean, are considered the normal range. For example, the average (mean) 8-year-old boy has a height of about 50 inches. But any heights between approximately 46 inches and 55 inches are considered normal for 8-year-old boys. Refer to Fig. 13.19 to answer the following questions.

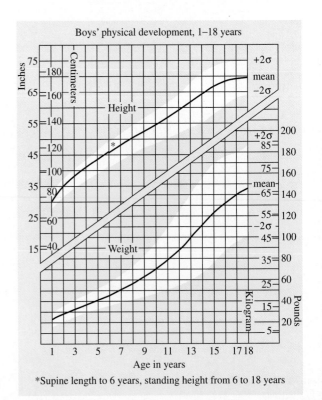

*Supine length to 6 years, standing height from 6 to 18 years

Figure 13.19

a) What happens to the standard deviation for weights of boys as the age of boys increases? What is the significance of this fact?

b) At age 16, what is the mean weight, in pounds, of boys?

c) What is the approximate standard deviation of boys' weights at age 16?

d) Find the mean weight and normal range for boys at age 13.

e) Find the mean height and normal range for boys at age 13.

f) Assuming that this chart was constructed so that approximately 95% of all boys are always in the normal range, determine what percentage of boys will not be in the normal range.

Problem Solving/Group Activities

28. Consider the following two bank-customer waiting systems.

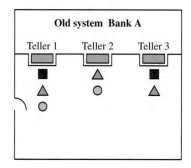

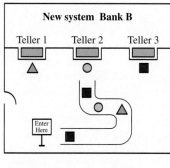

a) How would you expect the mean waiting time in Bank A to compare to the mean waiting time in Bank B? Explain your answer.

b) How would you expect the standard deviation of waiting times in Bank A to compare to the standard deviation of waiting times in Bank B? Explain your answer.

29. The following chart appeared in the May 10, 1994, issue of *USA Today*.

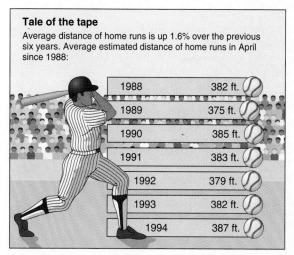

Source: *The Physics of Baseball*, STATS, Inc., USA TODAY research.

a) Which average do you feel is being used? Explain.

b) The chart states that the average distance of home runs in 1994 (387 ft) is 1.6% greater than the average for the preceding 6 years. Show how this 1.6% figure was obtained.

c) Find the average distance of home runs for the 7 years by adding the 7 distances and dividing the sum by 7.

d) If the individual distances of all the home runs from 1988 through 1994 were added and this sum was divided by the total number of home runs for this period, would the average obtained be the same as the average obtained in part (c)? Explain your answer.

e) Compute the standard deviation of the seven average distances shown in the chart. This value (the standard deviation of a set of averages) is sometimes called the *standard error of the means*. Round the mean in part (c) to tenths to find the standard deviation.

30. Is it possible for the standard deviation of a set of data to be greater than the mean of the same set of data? Explain your answer.

31. The following display appeared in the August 14, 1995, issue of *Newsweek*.

a) Without doing any calculations, do you believe that the mean of the platform temperatures or the mean of the outside temperatures is greater? Explain.

b) Without doing any calculations, do you believe that the standard deviation of the platform temperatures

---

Métro or Microwave

It's August, and subway travel worldwide is 1 percent transporation, 99 percent perspiration. Here are subway versus outside temps at eight stations last week:

Temperatures at midday on Aug. 3

CITY	STATION	TEMPERATURE PLATFORM	OUTSIDE
Paris	Franklin Roosevelt	93*	73°
New York City	Columbus Circle	92*	81°
Washington, D.C.	Farragut West	86°	89°
London	Marble Arch	85°	68°
Chicago	Randolph Street	84°	75°
Tokyo	Ginza	79°	91°
Moscow	Kievskaya	75°	77°
San Francisco	Embarcadero	75°	65°

*FAHRENHEIT. SOURCE: OUTDOOR TEMPERATURES PROVIDED BY THE NAT. METEOROLOGICAL CENTER.

Timing is everything

For stock market investors, knowing when to sell is just as important as knowing when and what to buy. This chart shows the monthly trading range (from high to low prices) for stock in Toys "Я" Us Inc., the toy retailing chain based in Paramus, N.J.

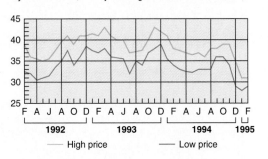

— High price   — Low price

An investor who bought the stock three years ago for $33 a share had many opportunities to sell at $40 per share or more. However, holding on too long turned out to be painful for their portfolios. The stock recently has been selling for $30 a share. The stock trades on the New York Stock Exchange under the symbol TOY.

Source: St. Petersburg Times

Month	Low	High	Difference
July	33	36	
August	33	38	
September	36	38	
October	36	39	
November	34	39	
December	29	35	

---

or the standard deviation of the outside temperatures is greater? Explain.

c) Compute the mean of the platform temperatures and the mean of the outside temperatures and determine whether your answer to part (a) is correct.

d) Compute the standard deviation of the platform temperatures and the standard deviation of the outside temperatures and determine whether your answer to part (b) is correct. Round each mean to the nearest tenth to determine standard deviations.

32. Consider the following graph and text in the box.

a) If the difference between the high and low values is calculated for each month (for example, for February 1995, it would be approximately $31 - 29.5$) and the standard deviation of these differences is calculated, would you expect it to be large or small? Explain.

The table below the box gives approximate high and low values of the stock (Toys "Я" Us, Inc.) for the months July–December 1994.

Find the standard deviation of the

b) lows.

c) highs.

d) Compute the difference of the high and low values for each month and find the standard deviation of the differences.

e) Compare the standard deviations in parts (b), (c), and (d). Does the standard deviation of the highs minus the standard deviation of the lows equal the standard deviation of the differences? Explain why or why not.

f) If you were to purchase $1000 worth of stock at the low price of the month and sell it at the high price of the month for each month July–December, what would your profit be for the 6-month period? (Disregard commissions.)

13.7 THE NORMAL CURVE

Certain shapes of distributions of data are more common than others. In this section we will illustrate and discuss a few of the more common ones. In each case the vertical scale is the frequency and the horizontal scale is the observed values.

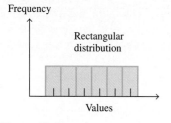

Figure 13.20

In a **rectangular distribution** (Fig. 13.20), all the observed values occur with the same frequency. If a die is rolled many times, we would expect the numbers 1–6 to occur with about the same frequency. The distribution representing the outcomes of the die is rectangular.

In **J-shaped distributions** the frequency is either constantly increasing (Fig. 13.21a) or constantly decreasing (Fig. 13.21b). The number of hours studied per week by students may have a distribution like that in Fig. 13.21(b). The bars might represent (from left to right) 0–5, 6–10, 11–15 hours, and so on.

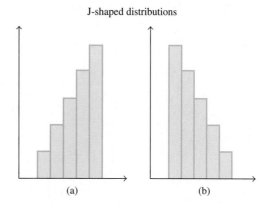

Figure 13.21

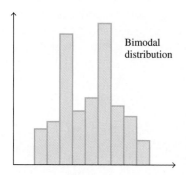

Figure 13.22

A **bimodal distribution** (Fig. 13.22) is one in which two nonadjacent values occur more frequently than any other values in a set of data. For example, if an equal number of men and women were weighed, the distribution of their weights would probably be bimodal, with one mode for the women's weights and the second for the men's weights.

The life expectancy of light bulbs has a bimodal distribution: a small peak very near 0 hours of life, resulting from the bulbs that burned out very quickly because of a manufacturing defect, and a much broader peak representing the nondefective bulbs. A bimodal frequency distribution generally means that you are dealing with two distinct populations, in this case, defective and nondefective bulbs.

Another distribution, called a **skewed distribution**, has more of a ''tail'' on one side than the other. A skewed distribution with a tail on the right (Fig. 13.23a) is said to be skewed to the right. If the tail is on the left (Fig. 13.23b), the distribution is referred to as skewed to the left.

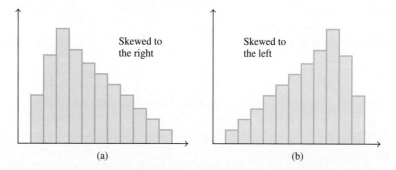

Figure 13.23

The number of children per family might be a distribution that is skewed to the right. Some families have no children, more families may have one child, the greatest percentage may have two children, fewer three children, still fewer four children, and so on.

Since few families have high incomes, distributions of family incomes might be skewed to the right.

Smoothing the histograms of the skewed distributions shown in Fig. 13.23 to form curves gives the curves illustrated in Fig. 13.24.

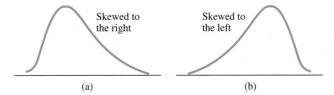

Figure 13.24

In Fig. 13.24(a) the greatest frequency appears on the left side of the curve, and the frequency decreases from left to right. Since the mode is the value with the greatest frequency, the mode would appear on the left side of the curve.

Every value in the set of data is considered in determining the mean. The values on the far right side of the curve in Fig. 13.24(a) would tend to increase the value of the mean. Thus the value of the mean would be farther to the right than the mode. The median would be between the mode and the mean. The relationship between the mean, median, and mode for curves that are skewed to the right and left is given in Fig. 13.25.

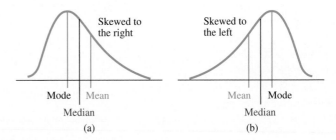

Figure 13.25

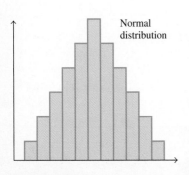

Figure 13.26

Each of these distributions is useful in describing sets of data. However, the most important distribution is the **normal** or **Gaussian distribution**, named for Carl Friedrich Gauss. The histogram of a normal distribution is illustrated in Fig. 13.26.

The normal distribution is important because many sets of data are normally distributed, or they closely resemble a normal distribution. Such distributions include intelligence quotients, heights and weights of males, heights and weights of females, lengths of full-grown boa constrictors, weights of watermelons, wear-out mileage of automobile brakes, and life spans of refrigerators—to name a few.

The normal distribution is symmetric about the mean. If you were to fold the histogram down the middle, the left side would fit the right side exactly. **In a normal distribution the mean, median, and mode all have the same value.**

When the histogram of a normal distribution is smoothed to form a curve, the curve is bell-shaped. The bell may be high and narrow or short and wide. Each of the three curves in Fig. 13.27 represents a normal curve. Curve 13.27(a) has the smallest standard deviation (spread from the mean); curve 13.27(c) has the largest.

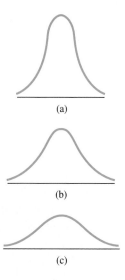

(a)

(b)

(c)

Figure 13.27

Since the curve is symmetric, 50% of the data always falls above (to the right of) the mean and 50% of the data below (to the left of) the mean. In addition, every normal distribution has approximately 68% of the data between the value that is one standard deviation below the mean and the value that is one standard deviation above the mean (Fig. 13.28). Approximately 95% of the data falls between the value that is two standard deviations below the mean and the value that is two standard deviations above the mean.

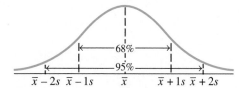

Figure 13.28

Thus if a normal distribution has a mean of 100 and a standard deviation of 10, then approximately 68% of all the data will fall between $100 - 10$ and $100 + 10$, or between 90 and 110. Approximately 95% of the data will fall between $100 - 20$ and $100 + 20$, or between 80 and 120. In fact, given any normal distribution with a known standard deviation and mean, it is possible through the use of Table 13.6 on page 695 (the z-table) to determine the percent of data between any two given values.

We use *z-scores* (or *standard scores*) to determine how far, in terms of standard deviations, a given score is from the mean of the distribution. For example, a score that has a z-value of 1.5 means that the score is 1.5 standard deviations above the mean. The standard or z-score is calculated as follows.

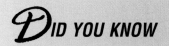

WHAT CONCLUSIONS CAN YOU DRAW?

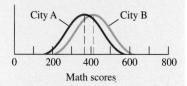

Math scores

*B*ased on the figure, which shows the distribution of scores on the mathematics part of the SAT test for two different cities, can we say that any given person selected at random from city B has outperformed any given person selected at random from city A? Both distributions appear normal, and the mean of city A is slightly smaller than the mean of city B. But consider two randomly selected students who took this test: Sally, from city A, and Kendra, from city B. The graph shows that many students in city A outperformed students from city B, so we cannot conclude that Sally scored higher than Kendra or that Kendra scored higher than Sally. ▪

$$z = \frac{\text{value of the piece of data} - \text{mean}}{\text{standard deviation}}$$

Letting x represent the value of the given piece of data, $\bar{x}$ the mean, and s the standard deviation, we can symbolize the formula as follows.

$$z = \frac{x - \bar{x}}{s}$$

In this book the notation z_x represents the z-score, or standard score, of the value x. For example, if a normal distribution has a mean of 86 with a standard deviation of 12, a score of 110 has a standard or z-score of

$$z_{110} = \frac{110 - 86}{12} = \frac{24}{12} = 2.$$

Therefore a value of 110 in this distribution has a z-score of 2. The score of 110 is two standard deviations above the mean.

Data below the mean will always have negative z-scores; data above the mean will always have positive z-scores. The mean will always have a z-score of 0.

EXAMPLE I

A normal distribution has a mean of 100 and a standard deviation of 10. Find z-scores for the following values.

a) 110   **b)** 115   **c)** 100   **d)** 88

Solution:

a)

$$z = \frac{\text{value} - \text{mean}}{\text{standard deviation}}$$

$$z_{110} = \frac{110 - 100}{10} = \frac{10}{10} = 1$$

A score of 110 is one standard deviation above the mean.

b)

$$z_{115} = \frac{115 - 100}{10} = \frac{15}{10} = 1.5$$

A score of 115 is 1.5 standard deviations above the mean.

c)

$$z_{100} = \frac{100 - 100}{10} = \frac{0}{10} = 0$$

The mean always has a z-score of 0.

d)

$$z_{88} = \frac{88 - 100}{10} = \frac{-12}{10} = -1.2$$

A score of 88 is 1.2 standard deviations below the mean.   ▲

Let's now consider finding areas under the normal curve. The total area under the normal curve is 1.00. Table 13.6 will be used to determine the area under the normal curve between any two given points (the values in the table have been rounded). **Table 13.6 gives the area under the normal curve from the mean (a z-value of 0) to a z-value to the right of the mean.**

For example, between the mean and $z = 2.00$ the table shows a value of 0.477. Thus there is 0.477 of the total area under the curve between the mean

Table 13.6   Areas under the Standard Normal Curve (the z-table)

Area found in table

The column under A gives the area under the entire curve that is between $z = 0$ and a positive value of z.

z	A	z	A	z	A	z	A	z	A	z	A	z	A	z	A	z	A
.00	.000	.37	.144	.74	.270	1.11	.367	1.48	.431	1.85	.468	2.22	.487	2.59	.495	2.96	.499
.01	.004	.38	.148	.75	.273	1.12	.369	1.49	.432	1.86	.469	2.23	.487	2.60	.495	2.97	.499
.02	.008	.39	.152	.76	.276	1.13	.371	1.50	.433	1.87	.469	2.24	.488	2.61	.496	2.98	.499
.03	.012	.40	.155	.77	.279	1.14	.373	1.51	.435	1.88	.470	2.25	.488	2.62	.496	2.99	.499
.04	.016	.41	.159	.78	.282	1.15	.375	1.52	.436	1.89	.471	2.26	.488	2.63	.496	3.00	.499
.05	.020	.42	.163	.79	.285	1.16	.377	1.53	.437	1.90	.471	2.27	.488	2.64	.496	3.01	.499
.06	.024	.43	.166	.80	.288	1.17	.379	1.54	.438	1.91	.472	2.28	.489	2.65	.496	3.02	.499
.07	.028	.44	.170	.81	.291	1.18	.381	1.55	.439	1.92	.473	2.29	.489	2.66	.496	3.03	.499
.08	.032	.45	.174	.82	.294	1.19	.383	1.56	.441	1.93	.473	2.30	.489	2.67	.496	3.04	.499
.09	.036	.46	.177	.83	.297	1.20	.385	1.57	.442	1.94	.474	2.31	.490	2.68	.496	3.05	.499
.10	.040	.47	.181	.84	.300	1.21	.387	1.58	.443	1.95	.474	2.32	.490	2.69	.496	3.06	.499
.11	.044	.48	.184	.85	.302	1.22	.389	1.59	.444	1.96	.475	2.33	.490	2.70	.497	3.07	.499
.12	.048	.49	.188	.86	.305	1.23	.391	1.60	.445	1.97	.476	2.34	.490	2.71	.497	3.08	.499
.13	.052	.50	.192	.87	.308	1.24	.393	1.61	.446	1.98	.476	2.35	.491	2.72	.497	3.09	.499
.14	.056	.51	.195	.88	.311	1.25	.394	1.62	.447	1.99	.477	2.36	.491	2.73	.497	3.10	.499
.15	.060	.52	.199	.89	.313	1.26	.396	1.63	.449	2.00	.477	2.37	.491	2.74	.497	3.11	.499
.16	.064	.53	.202	.90	.316	1.27	.398	1.64	.450	2.01	.478	2.38	.491	2.75	.497	3.12	.499
.17	.068	.54	.205	.91	.319	1.28	.400	1.65	.451	2.02	.478	2.39	.492	2.76	.497	3.13	.499
.18	.071	.55	.209	.92	.321	1.29	.402	1.66	.452	2.03	.479	2.40	.492	2.77	.497	3.14	.499
.19	.075	.56	.212	.93	.324	1.30	.403	1.67	.453	2.04	.479	2.41	.492	2.78	.497	3.15	.499
.20	.079	.57	.216	.94	.326	1.31	.405	1.68	.454	2.05	.480	2.42	.492	2.79	.497	3.16	.499
.21	.083	.58	.219	.95	.329	1.32	.407	1.69	.455	2.06	.480	2.43	.493	2.80	.497	3.17	.499
.22	.087	.59	.222	.96	.332	1.33	.408	1.70	.455	2.07	.481	2.44	.493	2.81	.498	3.18	.499
.23	.091	.60	.226	.97	.334	1.34	.410	1.71	.456	2.08	.481	2.45	.493	2.82	.498	3.19	.499
.24	.095	.61	.229	.98	.337	1.35	.412	1.72	.457	2.09	.482	2.46	.493	2.83	.498	3.20	.499
.25	.099	.62	.232	.99	.339	1.36	.413	1.73	.458	2.10	.482	2.47	.493	2.84	.498	3.21	.499
.26	.103	.63	.236	1.00	.341	1.37	.415	1.74	.459	2.11	.483	2.48	.493	2.85	.498	3.22	.499
.27	.106	.64	.239	1.01	.344	1.38	.416	1.75	.460	2.12	.483	2.49	.494	2.86	.498	3.23	.499
.28	.110	.65	.242	1.02	.346	1.39	.418	1.76	.461	2.13	.483	2.50	.494	2.87	.498	3.24	.499
.29	.114	.66	.245	1.03	.349	1.40	.419	1.77	.462	2.14	.484	2.51	.494	2.88	.498	3.25	.499
.30	.118	.67	.249	1.04	.351	1.41	.421	1.78	.463	2.15	.484	2.52	.494	2.89	.498	3.26	.499
.31	.122	.68	.252	1.05	.353	1.42	.422	1.79	.463	2.16	.485	2.53	.494	2.90	.498	3.27	.500
.32	.126	.69	.255	1.06	.355	1.43	.424	1.80	.464	2.17	.485	2.54	.495	2.91	.498	3.28	.500
.33	.129	.70	.258	1.07	.358	1.44	.425	1.81	.465	2.18	.485	2.55	.495	2.92	.498	3.29	.500
.34	.133	.71	.261	1.08	.360	1.45	.427	1.82	.466	2.19	.486	2.56	.495	2.93	.498	3.30	.500
.35	.137	.72	.264	1.09	.362	1.46	.428	1.83	.466	2.20	.486	2.57	.495	2.94	.498	3.31	.500
.36	.141	.73	.267	1.10	.364	1.47	.429	1.84	.467	2.21	.487	2.58	.495	2.95	.498	3.32	.500

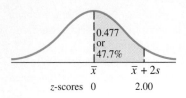

Figure 13.29

and $z = 2.00$ (Fig. 13.29). To change this area of 0.477 to a percent, simply multiply by 100%: $0.477 \times 100\%$ is 47.7%. Thus 47.7% of all scores will be between the mean and the score that is two standard deviations above the mean.

When you are finding the area under the normal curve, it is often helpful to draw a picture such as the one in Fig. 13.29, indicating the area or percent to be found.

The normal curve is symmetric about the mean. Thus the same percent of data is between the mean and a positive z-score as between the mean and the corresponding negative z-score. For example, there is the same area under the normal curve between a z of 1.60 and the mean as between a z of -1.60 and the mean. Both have an area of 0.445 (Fig. 13.30).

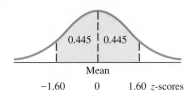

Figure 13.30

You now have the necessary knowledge to find the percent of data between any two values in a normal distribution.

To Find the Percent of Data Between any Two Values:
1. Use the formula $z = (x - \bar{x})/s$ to convert the given values to z-scores.
2. Look up the percent that corresponds to each z-score in Table 13.6.
3. a) When finding the percent of data between two z-scores on the opposite side of the mean (when one z-score is positive and the other is negative) you find the sum of the individual percents.
  b) When finding the percent of data between two z-scores on the same side of the mean (when both z-scores are positive or both are negative) subtract the smaller percent from the larger percent.

EXAMPLE 2

Intelligence quotients are normally distributed with a mean of 100 and a standard deviation of 15. Find the percent of individuals with IQs in the following ranges.

a) Between 100 and 115   **b)** Between 70 and 100

c) Between 70 and 115   **d)** Between 115 and 130

e) Below 130   **f)** Above 122.5

Solution:
a) We want to find the area under the normal curve between the values of 100 and 115, as illustrated in Fig. 13.31(a). Converting 100 to a z-score yields a z-score of 0.

$$z_{100} = \frac{100 - 100}{15} = \frac{0}{15} = 0$$

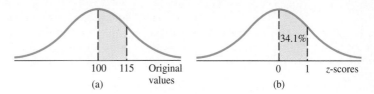

Figure 13.31

Converting 115 to a z-score yields a z-score of 1.00.

$$z_{115} = \frac{115 - 100}{15} = \frac{15}{15} = 1.00$$

The percent of individuals with IQs between 100 and 115 is the same as the percent of data between z-scores of 0 and 1 (Fig. 13.31b).

From Table 13.6 we determine that 0.341 of the area, or 34.1% of all the data, is between z-scores of 0 and 1.00. Therefore 34.1% of individuals have IQs between 100 and 115.

b)
$$z_{70} = \frac{70 - 100}{15} = \frac{-30}{15} = -2.00$$

$z_{100} = 0$ (from part a)

The percent of data between scores of 70 and 100 is the same as the percent between $z = -2$ and $z = 0$ (Fig. 13.32). The percent of data between the mean and two standard deviations below the mean is the same as between the mean and two standard deviations above the mean. From Table 13.6 we determine that 47.7% of the data is between $z = 0$ and $z = 2$. Thus 47.7% of the data is also between $z = -2$ and $z = 0$. Therefore 47.7% of all individuals have IQs between 70 and 100.

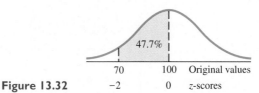

Figure 13.32

c) In parts (a) and (b) we determined that $z_{115} = 1.00$ and $z_{70} = -2.00$. Since the values are on opposite sides of the mean, the percent of data between the two values is found by adding the individual percents: 34.1% + 47.7% = 81.8% (Fig. 13.33). Thus 81.8% of the IQs are between 70 and 115.

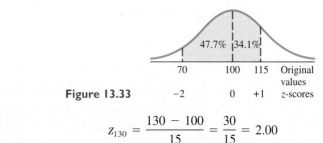

Figure 13.33

d)
$$z_{130} = \frac{130 - 100}{15} = \frac{30}{15} = 2.00$$

$z_{115} = 1.00$ (from part a)

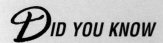

Since both values are on the same side of the mean (Fig. 13.34), the smaller percent must be subtracted from the larger percent to obtain the percent of data in the shaded area: 47.7% − 34.1% is 13.6%. Thus 13.6% of all the individuals have IQs between 115 and 130.

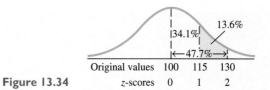

Figure 13.34

e) The percent of IQs below 130 is the same as the percent of data below a z-score of 2. Between $z = 2$ and the mean is 47.7% of the data (Fig. 13.35). To this 47.7% we add the 50% of the data below the mean to give 97.7%. Thus 50% + 47.7%, or 97.7%, of all IQs are below 130.

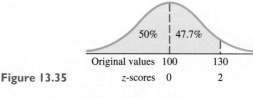

Figure 13.35

f)
$$Z_{122.5} = \frac{122.5 - 100}{15} = \frac{22.5}{15} = 1.50$$

The percent of IQs above 122.5 is the same as the percent of data above $z = 1.5$ (Fig. 13.36). Fifty percent of the data is to the right of the mean. Since 43.3% of the data is between the mean and $z = 1.5$, 50% − 43.3%, or 6.7%, of the data is greater than $z = 1.5$. Thus 6.7% of all IQs are greater than 122.5.

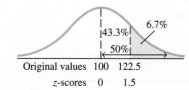

Figure 13.36

EXAMPLE 3

The wearout mileage of a certain tire is normally distributed with a mean of 35,000 miles and standard deviation of 2500 miles.

a) Find the percent of tires that will last at least 35,000 miles.

b) Find the percent of tires that will last between 30,000 and 37,500 miles.

c) Find the percent of tires that will last at least 39,000 miles.

d) If the manufacturer guarantees the tires to last at least 30,000 miles, what percent of tires will fail to live up to the guarantee?

e) If 200,000 tires are produced, how many will last at least 39,000 miles?

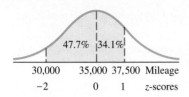

Figure 13.37

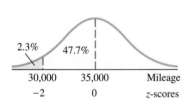

Figure 13.38

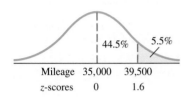

Figure 13.39

Solution:

a) In a normal distribution, half the data are always above the mean. Since 35,000 miles is the mean, half or 50% of the tires will last at least 35,000 miles.

b) Convert 30,000 miles and 37,500 miles to *z*-scores.

$$z_{30,000} = \frac{30,000 - 35,000}{2500} = -2.00$$

$$z_{37,500} = \frac{37,500 - 35,000}{2500} = 1.00$$

Now look up the areas in Table 13.6. The percent of tires that will last between 30,000 and 37,500 miles is 47.7% + 34.1%, or 81.8% (Fig. 13.37).

c)
$$z_{39,000} = \frac{39,000 - 35,000}{2500} = 1.60$$

Figure 13.38 shows that 44.5% of the data is between the mean and $z = 1.60$. Therefore the percent of data above $z = 1.60$ is 50% − 44.5% = 5.5%. Thus 5.5% of the tires will last at least 39,000 miles.

d) To solve this problem, we must find the percent of tires that last less than 30,000 miles: $z_{30,000} = -2.00$. From Table 13.6 we determine that 47.7% of the data is between $z = -2.00$ and $z = 0$ (Fig. 13.39). The percent to the left of $z = -2.00$ is found by subtracting 47.7% from 50% to obtain 2.3%. Thus 2.3% of all the tires will last less than 30,000 miles and fail to live up to the guarantee.

e) In part (c) we determined that 5.5% of all tires will last at least 39,000 miles. We now multiply 0.055 times 200,000 to determine the number of tires that will last at least 39,000 miles: $0.055 \times 200,000 = 11,000$ tires.

▲

Section 13.7 Exercises

In Exercises 1–6, describe

1. a rectangular distribution.

2. a J-shaped distribution.

3. a bimodal distribution.

4. a distribution that is skewed to the right.

5. a distribution that is skewed to the left.

6. a normal distribution.

In Exercises 7–10, give an example of the type of distribution.

7. Rectangular   **8.** J-shaped

9. Skewed   **10.** Bimodal

11. In a distribution that is skewed to the right, which has the greatest value—the mean, median, or mode? Which has the smallest value? Explain.

12. In a distribution that is skewed to the left, which has the greatest value—the mean, median, or mode? Which has the smallest value? Explain.

13. List three populations other than those given in the text that may be normally distributed.

14. List three populations other than those given in the text that may not be normally distributed.

15. In a normal distribution, what is the relationship between the mean, median, and mode?

16. What does the z, or standard score, measure?

17. When will a z-score be negative?

18. Explain in your own words how to find the z-score.

19. What is the value of the z-score of the mean of a set of data?

20. In a normal distribution, approximately what percent of the data is between
   a) one standard deviation below the mean to one standard deviation above the mean?
   b) two standard deviations below the mean to two standard deviations above the mean?

In Exercises 21–32, use Table 13.6 to find the specified area.

21. Above the mean

22. Below the mean

23. Between two standard deviations below the mean and one standard deviation above the mean

24. Between 1.10 and 1.80 standard deviations above the mean

25. To the right of $z = 1.47$

26. To the left of $z = 1.32$

27. To the left of $z = -1.78$

28. To the right of $z = -1.78$

29. To the right of $z = 2.08$

30. To the left of $z = 1.96$

31. To the left of $z = -1.21$

32. To the left of $z = -0.20$

In Exercises 33–42, use Table 13.6 to find the percent of data specified.

33. Between $z = 0$ and $z = 0.96$

34. Between $z = -0.20$ and $z = -0.77$

35. Between $z = -1.34$ and $z = 2.24$

36. Less than $z = -1.90$

37. Greater than $z = -1.90$

38. Greater than $z = 2.66$

39. Less than $z = 1.96$

40. Between $z = 0.74$ and $z = 1.65$

41. Between $z = -1.75$ and $z = -0.92$

42. Between $z = -2.15$ and $z = 3.31$

In Exercises 43–46, assume that the heights of first-grade children are normally distributed with a mean of 42 in. and a standard deviation of 6 in.

43. What percent of first-grade children are between 42 in. and 48 in. tall?

44. What percent of first-grade children are less than 48 in. tall?

45. What percent of first-grade children are between 36 in. and 48 in. tall?

46. If 1000 first-grade children are selected at random, how many will be less than 45 in. tall?

In Exercises 47–52, assume that the results of an English examination are normally distributed with a mean of 72 and a standard deviation of 8. Find the percent of individuals that scored

47. between 56 and 80.

48. less than 76.

49. higher than 76.

50. between 76 and 84.

51. between 58 and 70.

52. higher than 68.

In Exercises 53–56, the lengths of long distance phone calls from a particular city are normally distributed with a mean of 7.2 min and a standard deviation of 3.1 min.

53. What percent of the phone calls are longer than 9.3 min?

54. What percent of phone calls are between 6.0 and 8.0 min?

55. What percent of phone calls are longer than 2.1 min?

56. If 500 phone calls are selected at random, how many will be longer than 9.2 min?

In Exercises 57–62, the weights of chicken eggs are normally distributed with a mean of 1.6 oz and a standard deviation of 0.4 oz.

57. What percent of chicken eggs are heavier than 1.6 oz?

58. What percent of chicken eggs are between 1.2 and 2.6 oz?

59. What percent of chicken eggs are lighter than 1.8 oz?

60. What percent of chicken eggs are heavier than 2.0 oz?

61. If 150 chicken eggs are selected at random, how many will weigh more than 2.0 oz?

62. A chicken egg is considered extra large if it weighs more than 2.2 oz. What percent of chicken eggs are extra large?

In Exercises 63–66, a vending machine is designed to dispense a mean of 6.7 oz of coffee into a 7-oz cup. If the standard deviation of the amount of coffee dispensed is 0.4 oz and the amount of coffee dispensed is normally distributed, find

63. the percent of times the machine will dispense from 6.5 to 6.8 oz.

64. the percent of times it will dispense less than 6.0 oz.

65. the percent of times it will dispense less than 6.8 oz.

66. the percent of times the 7-oz cup will overflow.

In Exercises 67–71, the life expectancy of nondefective GE light bulbs is normally distributed, with a mean life of 1500 hr and a standard deviation of 100 hr.

67. Find the percent of bulbs that will last more than 1450 hr.

68. Find the percent of bulbs that will last between 1400 hr and 1550 hr.

69. Find the percent of bulbs that will last less than 1480 hr.

70. If 80,000 of these bulbs are produced, how many will last 1500 hr or more?

71. If 80,000 of these bulbs are produced, how many will last between 1400 hr and 1600 hr?

72. A weight-loss clinic guarantees that its new customers will lose at least 5 lb by the end of their first month of participation or their money will be refunded. If the loss of weight of customers at the end of their first month is normally distributed, with a mean of 6.7 lb and a standard deviation of 0.81 lb, find the percent of customers that will be able to claim a refund.

73. The warranty on the motor of a dishwasher is 8 yr. If the breakdown times of this motor are normally distributed, with a mean of 10.2 yr and a standard deviation of 1.8 yr, find the percent of motors that can be expected to require repair or replacement under warranty.

74. A vending machine that dispenses coffee does not appear to be working correctly. The machine rarely gives the proper amount of coffee. Some of the time the cup is underfilled, and some of the time the cup overflows. Does this variation indicate that the mean number of ounces dispensed has to be adjusted or that the standard deviation of the amount of coffee dispensed by the machine is too large? Explain your answer.

75. Mr. Brittain marks his class on a normal curve. Those with z-scores above 1.8 will receive an A, those between 1.8 and 1.1 will receive a B, those between 1.1 and −1.2 will receive a C, those between −1.2 and −1.9 will receive a D, and those under −1.9 will receive an F. Find the percent of grades that will be A, B, C, D, and F.

76. Professor Hart marks his classes on the normal curve. His statistics class has a mean of 72, with a standard deviation of 8. He has decided that 10% of the class will get an A, 20% will get a B, 40% will get a C, 20% will get a D, and 10% will get an F. Find
  a) the minimum grade needed to get an A.
  b) the minimum grade needed to pass the course (D or better).
  c) the range of grades that will result in a C.

Problem Solving/Group Activities

77. The breaking strength of a particular type of rope is normally distributed with a mean of 150 lb, with a standard deviation of 22 lb. What is the probability that one of these ropes selected at random will break when subjected to a force of up to 185 lb?

78. How can you determine whether a distribution is approximately normal? A statistical theorem called **Chebyshev's theorem** states that the minimum percent of data between plus and minus K standard deviations from the mean ($K > 1$) in *any distribution* can be found by the formula

$$\text{Minimum percent} = 1 - \frac{1}{K^2}$$

Thus, for example, between ±2 standard deviations from the mean, there will always be a minimum of 75% of data. This minimum percent is true for any distribution. For $K = 2$,

$$\text{Minimum percent} = 1 - \frac{1}{2^2}$$

$$= 1 - \frac{1}{4} = \frac{3}{4}, \quad \text{or} \quad 75\%.$$

Likewise, between ±3 standard deviations from the mean, there will always be a minimum of 89% of the data. For $K = 3$,

$$\text{Minimum percent} = 1 - \frac{1}{3^2}$$

$$= 1 - \frac{1}{9} = \frac{8}{9}, \quad \text{or} \quad 89\%.$$

The following table lists the minimum percent of data in *any distribution* and the actual percent of data in *the normal distribution* between ±1.1, ±1.5, ±2.0, and ±2.5 standard deviations from the mean. The minimum percents of data in any distribution were calculated by using Chebyshev's theorem. The actual percents of data for the normal distribution were calculated by using the area given in the standard normal, or z, table.

	$K = 1.1$	$K = 1.5$	$K = 2$	$K = 2.5$
Minimum (for any distribution)	17.4%	55.6%	75%	84%
Normal distribution	72.8%	86.6%	95.4%	99.8%
Given distribution				

The third row of the chart has been left blank for you to fill in the percents when you reach part (e).

Consider the following 30 pieces of data obtained from a quiz.

1, 1, 1, 1, 2, 2, 2, 2, 3, 3, 4, 4, 4, 5, 6,
6, 6, 7, 7, 7, 8, 8, 8, 8, 9, 9, 9, 10, 10

a) Find the mean of the set of scores.
b) Find the standard deviation of the set of scores.

c) Determine the values that correspond to 1.1, 1.5, 2, and 2.5 standard deviations above the mean. (For example, the value that corresponds to 1.5 standard deviations above the mean is $\bar{x} + 1.5s$.)

Then determine the values that correspond to 1.1, 1.5, 2, and 2.5 standard deviations below the mean. (For example, the value that corresponds to 1.5 standard deviations below the mean is $\bar{x} - 1.5s$.)

d) By observing the 30 pieces of data, determine the actual percent of quiz scores between

± 1.1 standard deviations from the mean.
± 1.5 standard deviations from the mean.
± 2 standard deviations from the mean.
± 2.5 standard deviations from the mean.

e) Place the percents found in part (d) in the third row of the chart.

f) Compare the percents in the third row of the chart with the minimum percents in the first row and the normal percents in the second row, then make a judgment as to whether this set of 30 scores is approximately normally distributed. Explain your answer.

79. Obtain a set of test scores from your teacher.
a) Find the mean, median, mode, and midrange of the test scores.
b) Find the range and standard deviation of the set of scores. (You may round the mean to the nearest tenth when finding the standard deviation.)
c) Construct a frequency distribution of the set of scores. Select your first class so that there will be between 5 and 12 classes.

d) Construct a histogram and frequency polygon of the frequency distribution in part (c).
e) Does the histogram in part (d) appear to represent a normal distribution? Explain.
f) Use the procedure explained in Exercise 78 to determine whether the set of scores represented is a normal distribution. Explain.

Research Activity

80. In this project you actually become the statistician.
a) Select a project of interest to you in which data must be collected.
b) Write a proposal and submit it to your instructor for approval. In the proposal, discuss the aims of your project and how you plan to gather the data to make your sample unbiased.
c) After your proposal has been approved, gather 50 pieces of data by the method you proposed.
d) Rank the data from smallest to largest.
e) Compute the mean, median, mode, and midrange.
f) Determine the range and standard deviation of the data. You may round the mean to the nearest tenth when computing the standard deviation.
g) Construct a frequency distribution, histogram, and frequency polygon of your data. Select your first class so that there will be between 5 and 12 classes. Be sure to label your histogram and frequency polygon.
h) Does your distribution appear to be normal? Explain your answer. Does it appear to be another type of distribution discussed? Explain.
i) Determine whether your distribution is approximately normal by using the technique discussed in Exercise 78.

13.8 LINEAR CORRELATION AND REGRESSION

In this section we discuss two important statistical topics: correlation and regression. Correlation is used to determine whether there is a relationship between two quantities and, if so, how strong that relationship is. Regression is used to determine the equation that relates the two quantities. Although there are other types of correlation and regression, in this section we discuss only linear correlation and linear regression. We begin by discussing linear correlation.

Do you believe that there is a relationship between

a) the time a person studied for an exam and the exam grade received?

b) the age of a car and the value of the car?

c) the height and weight of adult males?

d) a person's IQ and income?

Correlation is used to answer questions of this type. The linear correlation coefficient, r, is a unitless measure that describes the strength of the linear relationship between two variables. A positive value of r, or a positive correlation, means that as one variable increases, the other variable also increases. A negative value of r, or a negative correlation, means that as one variable increases the other variable decreases. The correlation coefficient, r, will always be a value between -1 and 1 inclusive. A value of 1 indicates the strongest possible positive correlation, the value of -1 indicates the strongest negative correlation, and a value of 0 indicates no correlation (Fig. 13.40).

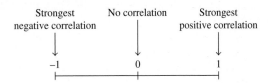

Figure 13.40

A visual aid used with correlation is the scatter diagram, a plot of data points. To explain how to construct a scatter diagram consider 10 students who provided the following information about the number of hours they studied for a test and their test grade.

Student	**1**	**2**	**3**	**4**	**5**	**6**	**7**	**8**	**9**	**10**
Time studied (minutes)	20	80	52	60	90	50	100	40	40	55
Grade received (percent)	40	92	80	76	89	70	95	45	60	64

For each of the 10 students there are two pieces of data, time studied and test grade. We call the set of data bivariate data. When we have a set of bivariate data, we generally denote the quantity that can be controlled, the independent variable, x. The other variable, the dependent variable, is denoted y. In this problem, the time can be controlled, so we call time studied x and grade received y. If we plot the 10 pieces of bivariate data in the Cartesian coordinate system, we get a scatter diagram, as shown in Fig. 13.41.

Figure 13.41 shows that, generally, the longer a person studied the higher was the grade received.

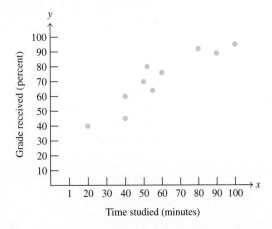

Figure 13.41

In Figure 13.42 we show some scatter diagrams and indicate the corresponding strength of correlation between the quantities on the horizontal and vertical axes.

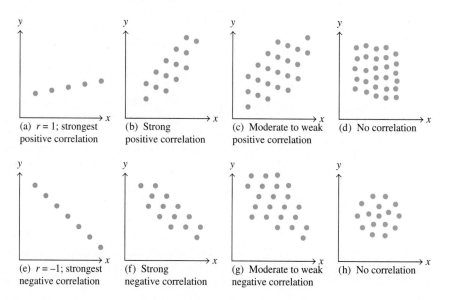

(a) $r = 1$; strongest positive correlation

(b) Strong positive correlation

(c) Moderate to weak positive correlation

(d) No correlation

(e) $r = -1$; strongest negative correlation

(f) Strong negative correlation

(g) Moderate to weak negative correlation

(h) No correlation

Figure 13.42

Earlier we mentioned that r will always be a value between -1 and 1 inclusive. A value of $r = 1$ will be obtained only when every point of the bivariate data on a scatter diagram lies in a straight line and the line is increasing from left to right (see Fig. 13.42a). In other words, the line has a positive slope, as discussed in Section 6.6.

A value of $r = -1$ will be obtained only when every point of the bivariate data on a scatter diagram lies in a straight line and the line is decreasing from left to right (see Fig. 13.42e). In other words, the line has a negative slope.

The value of r is a measure of how far a set of points varies from a straight line. The greater the spread, the weaker is the correlation and the closer the value of r is to 0. Figure 13.42 shows that the more the dots diverge from a straight line the weaker the correlation becomes.

The following formula is used to calculate r.

$$r = \frac{n(\Sigma xy) - (\Sigma x)(\Sigma y)}{\sqrt{n(\Sigma x^2) - (\Sigma x)^2}\sqrt{n(\Sigma y^2) - (\Sigma y)^2}}.$$

To determine the correlation coefficient, r, and the equation of the line of best fit (to be discussed shortly) a statistical calculator, which costs under \$20, may be used. On the calculator you simply enter the ordered pairs, (x, y), and press the appropriate keys.

In Example 1 we show how to determine r for a set of bivariate data without the use of a statistical calculator. We will use the same set of bivariate data given on page 703.

EXAMPLE 1

Ten students provided the following data about the lengths of time they studied for an exam and the grades they received on the exam.

Determine the correlation coefficient between length of time studied and grade received.

Student	1	2	3	4	5	6	7	8	9	10
Time studied (minutes)	20	80	52	60	90	50	100	40	40	55
Grade received (percent)	40	92	80	76	89	70	95	45	60	64

Solution: We plotted this set of data on the scatter diagram in Fig. 13.41. Since time studied may be controlled, we will call it x. We will call the grade received y. We list the values of x and y and calculate the necessary sums: Σx, Σy, Σxy, Σx^2, and Σy^2. We determine the values in the column labeled x^2 by squaring the x's (multiplying the x's by themselves). We determine the values in the column labeled y^2 by squaring the y's. We determine the values in the column labeled xy by multiplying each x value by its corresponding y value.

(Time) x	(Grade) y	x^2	y^2	xy
20	40	400	1600	800
80	92	6400	8464	7360
52	80	2704	6400	4160
60	76	3600	5776	4560
90	89	8100	7921	8010
50	70	2500	4900	3500
100	95	10,000	9025	9500
40	45	1600	2025	1800
40	60	1600	3600	2400
55	64	3025	4096	3520
587	711	39,929	53,807	45,610

Thus $\Sigma x = 587$, $\Sigma y = 711$, $\Sigma x^2 = 39{,}929$, $\Sigma y^2 = 53{,}807$, and $\Sigma xy = 45{,}610$. In the formula we use both $(\Sigma x)^2$ and Σx^2. Note that $(\Sigma x)^2 = (587)^2 = 344{,}569$ and that $\Sigma x^2 = 39{,}929$. Similarly, $(\Sigma y)^2 = (711)^2 = 505{,}521$ and $\Sigma y^2 = 53{,}807$.

The n in the formula represents the number of pieces of bivariate data. Here, $n = 10$. Now let's determine r.

$$r = \frac{n(\Sigma xy) - (\Sigma x)(\Sigma y)}{\sqrt{n(\Sigma x^2) - (\Sigma x)^2}\sqrt{n(\Sigma y^2) - (\Sigma y)^2}}$$

$$= \frac{10(45{,}610) - (587)(711)}{\sqrt{10(39{,}929) - (587)^2}\sqrt{10(53{,}807) - (711)^2}}$$

$$= \frac{456{,}100 - 417{,}357}{\sqrt{10(39{,}929) - 344{,}569}\sqrt{10(53{,}807) - 505{,}521}}$$

$$= \frac{38{,}743}{\sqrt{399{,}290 - 344{,}569}\sqrt{538{,}070 - 505{,}521}}$$

$$= \frac{38{,}743}{\sqrt{54{,}721}\sqrt{32{,}549}} \approx 0.92$$

Since the maximum possible value for r is 1.00, a correlation coefficient of 0.92 is a strong positive correlation. This result implies that, generally, the more time a person studied, the higher was the grade received. ▲

Table 13.7 Correlation Coefficient, *r*

n	α = 0.05	α = 0.01
4	0.950	0.990
5	0.878	0.959
6	0.811	0.917
7	0.754	0.875
8	0.707	0.834
9	0.666	0.798
10	0.632	0.765
11	0.602	0.735
12	0.576	0.708
13	0.553	0.684
14	0.532	0.661
15	0.514	0.641
16	0.497	0.623
17	0.482	0.606
18	0.468	0.590
19	0.456	0.575
20	0.444	0.561
22	0.423	0.537
27	0.381	0.487
32	0.349	0.449
37	0.325	0.418
42	0.304	0.393
47	0.288	0.372
52	0.273	0.354
62	0.250	0.325
72	0.232	0.302
82	0.217	0.283
92	0.205	0.267
102	0.195	0.254

The derivation of this table is beyond the scope of this course. It shows the critical values of the Pearson correlation coefficient.

In Example 1, had we found r to be a value greater than 1 or less than -1 it would have indicated that we had made an error. Also, from the scatter diagram we should realize that r should be a positive value and not negative.

In Example 1 there appears to be a cause–effect relationship. That is, the greater the time studied the higher was the test grade. *However, a correlation does not necessarily indicate a cause–effect relationship.* For example, there is a positive correlation between police officers' salaries and drug use over the past 10 years (both have increased), but that does not mean that the increase in police officers' salaries caused the increase in drug use.

Suppose in Example 1 that r had been 0.53. Would this value have indicated a correlation? What is the minimum value of r needed to assume that a correlation exists between the variables? To answer this question we introduce the term *level of significance*. The level of significance, denoted α (alpha), is used to identify the cutoff between results attributed to chance and results attributed to an actual relationship between the two variables. Table 13.7 gives critical values* (or cutoff values) that are sometimes used for determining whether two variables are related. The table indicates two different levels of significance: $\alpha = 0.05$ and $\alpha = 0.01$. A level of significance of 5%, written $\alpha = 0.05$, means that there is a 5% chance that, when you say the variables are related, they actually are *not* related. Similarly, a level of significance of 1%, or $\alpha = 0.01$, means that there is a 1% chance that, when you say the variables are related, they actually are *not* related. More complete critical value tables are available in statistics books.

To explain the use of the table we use absolute value, symbolized $|\quad|$. The absolute value of a nonzero number is the positive value of the number and the absolute value of 0 is 0. Therefore

$$|3| = 3, \quad |-3| = 3, \quad |5| = 5, \quad |-5| = 5, \quad \text{and} \quad |0| = 0.$$

If the absolute value of r, written $|r|$, is *greater than* the value given in the table under the specified α and appropriate sample size n, we assume that a correlation does exist between the variables. If $|r|$ is less than the table value, we assume that no correlation exists.

Returning to Example 1, if we want to determine whether there is a correlation at a 5% level of significance, we find the critical value (or cutoff value) to the right of $n = 10$ (there are 10 pieces of bivariate data) and under the $\alpha = 0.05$ column. Here, the critical value is 0.632. From the formula we had obtained $r = 0.92$. Since $|0.92| > 0.632$, or $0.92 > 0.632$, we assume that a correlation between the variables exists.

Note in Table 13.7 that the larger the sample size, the smaller is the value of r needed for a significant correlation.

┌─ **EXAMPLE 2**
│
│ To test the length of time that an infection-fighting drug will stay in a person's
│ bloodstream, a doctor gives 300 milligrams of the drug to eight patients. Once
│ each hour, for 8 hours, one of the eight patients is selected at random and that
│ person's blood is tested to determine the amount of the drug remaining in the
│ bloodstream. The results are as follows.

*This table of values may be used only under certain conditions. If you take a statistics course you will learn more about which critical values to use to determine whether a linear correlation exists.

Patient	1	2	3	4	5	6	7	8
Time (hr)	1	2	3	4	5	6	7	8
Drug remaining (mg)	250	230	200	210	140	120	210	100

Determine at a level of significance of 5% whether a correlation exists between the time elapsed and the amount of drug remaining.

Solution: Let time be represented by x and the amount of drug remaining by y. We first draw a scatter diagram (Fig. 13.43).

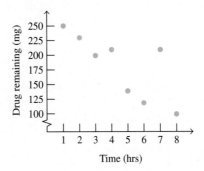

Figure 13.43

The scatter diagram indicates that, if a correlation exists, it will be negative. We now construct a table of values and calculate r.

x	y	x^2	y^2	xy
1	250	1	62,500	280
2	230	4	52,900	460
3	200	9	40,000	600
4	210	16	44,100	840
5	140	25	19,600	700
6	120	36	14,400	720
7	210	49	44,100	1470
8	100	64	10,000	800
36	1460	204	287,600	5840

$$r = \frac{n(\Sigma xy) - (\Sigma x)(\Sigma y)}{\sqrt{n(\Sigma x^2) - (\Sigma x)^2}\sqrt{n(\Sigma y^2) - (\Sigma y)^2}}$$

$$= \frac{8(5840) - (36)(1460)}{\sqrt{8(204) - (36)^2}\sqrt{8(287,600) - (1460)^2}}$$

$$= \frac{-5840}{\sqrt{336}\sqrt{169,200}} \approx \frac{-5840}{7539.97}$$

$$\approx -0.775$$

From Table 13.7, for $n = 8$ and $\alpha = 0.05$ we get 0.707. Since $|-0.775| = 0.775$ and $0.775 > 0.707$, a correlation exists. The correlation is negative, which indicates that the longer the time period the smaller is the amount of drug remaining.   ▲

In Example 2, if we had used a 1% level of significance, the critical value would be 0.834. We would then assume that there is no correlation between time and drug remaining because $|-0.775| < 0.834$.

Let's now turn to regression. **Linear regression** is the process of determining the linear relationship between two variables. Recall from Section 6.6 that the slope–intercept form of a straight line is $y = mx + b$, where m is the slope and b is the y intercept.

Using the set of bivariate data, we will determine the equation of **the line of best fit**. The line of best fit is also called **the regression line**, or **the least squares line**. The *line of best fit* is the line such that the sum of the vertical distances from the line to the data points (on the scatter diagram) will be a minimum, as shown in Fig. 13.44.

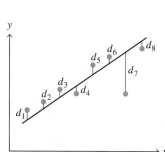

Figure 13.44

In Fig. 13.44, the line of best fit will minimize the sum of d_1 through d_8. To determine the equation of the line of best fit, $y = mx + b$*, we must find m and then b. The formulas for finding m and b are as follows.

The equation of the line of best fit is $y = mx + b$, where

$$m = \frac{n(\Sigma xy) - (\Sigma x)(\Sigma y)}{n(\Sigma x^2) - (\Sigma x)^2}, \quad \text{and} \quad b = \frac{\Sigma y - m(\Sigma x)}{n}.$$

Note that the numerator of the fraction used to find m is identical to the numerator used to find r. Therefore, if you have previously found r, you do not need to repeat the calculation. Also, the denominator of the fraction used to find m is identical to the radicand of the first square root in the denominator of the fraction used to find r.

EXAMPLE 3

a) Use the data in Example 1 to find the equation of the line of best fit that relates time studied and test grade.

b) Graph the equation of the line of best fit on a scatter diagram that illustrates the set of bivariate points.

---

*Some statistics books use $y = ax + b$, $y = b_0 + b_1x$, or something similar for the equation of the line of best fit. In any case, the letter next to the variable x represents the slope of the line of best fit, and the other letter represents the y intercept of the graph.

Solution:

a) In Example 1 we found $n(\Sigma xy) - (\Sigma x)(\Sigma y) = 38{,}743$ and $n(\Sigma x^2) - (\Sigma x)^2 = 54{,}721$. Thus

$$m = \frac{n(\Sigma xy) - (\Sigma x)(\Sigma y)}{n(\Sigma x^2) - (\Sigma x)^2} = \frac{38{,}743}{54{,}721} \approx 0.71.$$

Now we find the y intercept, b. In Example 1 we found $n = 10$, $\Sigma x = 587$, and $\Sigma y = 711$.

$$b = \frac{\Sigma y - m(\Sigma x)}{n}$$

$$= \frac{711 - 0.71(587)}{10} = \frac{294.23}{10}$$

$$\approx 29.4$$

Therefore the equation of the line of best fit is

$$y = mx + b$$

$$y = 0.71x + 29.4,$$

where x represents the number of minutes studied and y represents the predicted grade.

b) To graph $y = 0.71x + 29.4$, we need to plot at least two points. We will plot three points and then draw the graph.

	y = 0.71x + 29.4	**x**	**y**
$x = 20$	$y = 0.71(20) + 29.4 = 43.6$	20	43.6
$x = 50$	$y = 0.71(50) + 29.4 = 64.9$	50	64.9
$x = 80$	$y = 0.71(80) + 29.4 = 86.2$	80	86.2

These three calculations indicate that the average student who studies 20 minutes will obtain a grade of about 44, the average student who studies 50 minutes will receive a grade of about 65, and the average student who studies 80 minutes will receive a grade of about 86. The scatter diagram and graph are plotted in Fig. 13.45.

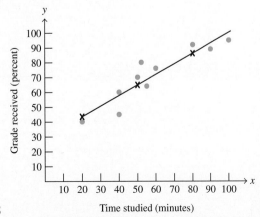

Figure 13.45

In Example 3, had we extended the line of best fit to the y-axis, the graph would intersect the y-axis at 29.4, the value we determined for b in part (a).

EXAMPLE 4

a) Determine the equation of the line of best fit between time elapsed and amount of drug remaining in a person's bloodstream from Example 2.

b) If the average person is given 300 mg of the drug, how much will remain in the person's bloodstream after 5 hr?

Solution:

a) From the scatter diagram on page 707 we see that the slope of the line of best fit, m, will be negative. In Example 2, we found that $n(\Sigma xy) - (\Sigma x)(\Sigma y) = -5840$ and that $n(\Sigma x^2) - (\Sigma x)^2 = 336$. Thus

$$m = \frac{n(\Sigma xy) - (\Sigma x)(\Sigma y)}{n(\Sigma x^2) - (\Sigma x)^2}$$

$$= \frac{-5840}{336}$$

$$\approx -17.4$$

From Example 2, $n = 8$, $\Sigma x = 36$, and $\Sigma y = 1460$.

$$b = \frac{\Sigma y - m(\Sigma x)}{n}$$

$$= \frac{1460 - (-17.4)(36)}{8}$$

$$= 260.8$$

Thus the equation of the line of best fit is

$$y = mx + b$$

$$y = -17.4x + 260.8$$

where x is the elapsed time and y is the amount of drug remaining.

b) We evaluate $y = -17.4x + 260.8$ at $x = 5$.

$$y = -17.4x + 260.8$$

$$y = -17.4(5) + 260.8 = 173.8$$

Thus, after 5 hr about 174 mg of the drug remain in the average person's bloodstream. ▲

Section 13.8 Exercises

1. What does the correlation coefficient measure?

2. What is the purpose of regression?

3. What value of r represents the maximum positive correlation?

4. What value of r represents the maximum negative correlation?

5. What value of r represents no correlation between the variables?

6. What does a positive correlation between two variables indicate?

7. What does a negative correlation between two variables indicate?

8. What does the line of best fit represent?

In Exercises 9–16, assume that a sample of bivariate data yields the correlation coefficient, r, indicated. Use Table 13.7 for the specified sample size and level of significance to determine whether a linear correlation exists.

9. $r = 0.62$ when $n = 10$ at $\alpha = 0.05$

10. $r = 0.54$ when $n = 8$ at $\alpha = 0.01$

11. $r = -0.72$ when $n = 15$ at $\alpha = 0.05$

12. $r = 0.6$ when $n = 42$ at $\alpha = 0.05$

13. $r = -0.23$ when $n = 102$ at $\alpha = 0.01$

14. $r = -0.49$ when $n = 18$ at $\alpha = 0.01$

15. $r = 0.82$ when $n = 6$ at $\alpha = 0.01$

16. $r = 0.96$ when $n = 5$ at $\alpha = 0.01$

In Exercises 17–24, (a) draw a scatter diagram, (b) find the value of r, rounded to the nearest thousandth, (c) determine whether a correlation exists at $\alpha = 0.05$, and (d) determine whether a correlation exists at $\alpha = 0.01$.

17.

x	y
2	5
3	8
4	10
5	10
8	12

18.

x	y
4	8
6	7
9	5
12	6
15	4

19.

x	y
18	25
30	32
27	22
38	15
45	35

20.

x	y
1.4	5.6
1.8	7.3
2.2	7.4
3.8	12.6
4.2	12.0

21.

x	y
5.3	10.3
4.7	9.6
8.4	12.5
12.7	16.2
4.9	9.8

22.

x	y
12	15
16	19
13	45
24	30
100	60
50	28

23.

x	y
100	2
80	3
60	5
60	6
40	6
20	8

24.

x	y
90	90
70	70
65	65
60	60
50	50
40	40
15	15

In Exercises 25–32, find the equation of the line of best fit from the data in the exercise indicated. Round both the slope and the y-intercept to the nearest tenth.

25. Exercise 17

26. Exercise 18

27. Exercise 19

28. Exercise 20

29. Exercise 21

30. Exercise 22

31. Exercise 23

32. Exercise 24

33. The consumer price index (CPI) levels from 1990 to 1995 (measured on June 30 each year) were:

Year	90	91	92	93	94	95
CPI	130	136	140	144	148	153

a) Determine the correlation coefficient between year and CPI.

b) Determine whether a correlation exists at $\alpha = 0.05$.

c) Find the equation of the line of best fit for year and CPI.

34. The average 1-year bank CD interest rates from 1990 to 1995 (measured on June 30 of each year) were:

Year	90	91	92	93	94	95
Rate (%)	8.0	6.1	4.0	3.2	3.9	5.3

a) Determine the correlation coefficient between year and interest rate.

b) Determine whether a correlation exists at $\alpha = 0.01$.

c) Find the equation of the line of best fit for year and interest rate.

35. The blue book value of a Ford Mustang, based on its age from 1 to 8 years, is as follows.

Age (yr)	1	2	3	4	5	6	7	8
Value ($ thousands)	17.6	10.9	9.8	8.6	7.0	5.6	2.6	3.2

a) Determine the correlation coefficient for age and value.

b) Determine whether a correlation exists at $\alpha = 0.05$.

c) Find the equation of the line of best fit for age and value.

d) Use the equation in part (c) to estimate the blue book value of the Mustang when it is 5.5 years old.

36. The height and weight of a sample of eight boys in a first-grade class are as follows.

Height (in)	30	40	37	32	42	40	30
Weight (lb)	32	40	30	46	50	50	46

a) Determine the correlation coefficient for height and weight.

b) Determine whether a correlation exists at $\alpha = 0.05$.

c) Find the equation of the line of best fit for height and weight.

d) Use the equation in part (c) to estimate the average weight of a first-grade boy who is 35 inches tall.

37. In a certain section of a city muggings have been a problem. The number of police officers patrolling that section of the city has varied. The following chart shows the number of police officers and the number of muggings for 10 successive days.

Police Officers	20	12	18	15	22	10	20	12
Muggings	8	10	12	9	6	15	7	18

a) Determine the correlation for number of police officers and number of muggings.
b) Determine whether a correlation exists at $\alpha = 0.05$.
c) Find the equation of the line of best fit for number of police officers and number of muggings.
d) Use the equation in part (c) to estimate the average number of muggings when 14 police officers are patrolling that section of the city.

38. The cumulative weight loss of a customer after several weeks on a diet plan is shown.

Number of Weeks	1	2	3	4	5	6	7
Cumulative Weight Loss (lb)	3	5	5	6	5	7	9

a) Determine the correlation coefficient for number of weeks on a diet and cumulative weight loss.
b) Determine whether a correlation exists at $\alpha = 0.05$.
c) Find the equation of best fit for number of weeks and cumulative weight loss.
d) Use the equation in part (c) to estimate the average cumulative weight loss after 5 weeks.

39. The following table shows the amount of time required for unprotected skin to burn at various temperatures at a specific location.

Temperature (°)	80	70	85	90	75	95	87
Time to Burn (min)	30.2	40.8	36.0	20.0	36.0	18.0	24.0

a) Determine the correlation coefficient between temperature and time to burn.
b) Determine whether a correlation exists at $\alpha = 0.01$.
c) Determine the equation of the line of best fit for temperature and time to burn.
d) Use the equation in part (c) to estimate the average time to burn at a temperature of 90°.

40. A popular newspaper rates movies from 1 to 4 stars (four stars is the highest rating). The following table shows the ratings of 10 movies selected at random and the gross earnings of each movie.

Rating (stars)	4	4	3	2	1	3	4	2	4	1
Earnings ($ millions)	100	67	80	120	40	90	60	60	90	100

a) Determine the correlation coefficient between number of stars and movies' earnings.
b) Determine whether a correlation exists at $\alpha = 0.05$.
c) Determine the equation of the line of best fit for number of stars and movies' earnings.
d) Use the equation in part (c) to estimate the average gross earnings of a movie having a rating of 3 stars.

41. A gallon of chlorine is put into a swimming pool. Each hour later for the following 6 hr the percent of chlorine that remains in the pool is measured. The following information is obtained.

Time	1	2	3	4	5	6
Chlorine remaining (%)	80.0	76.2	68.7	50.1	30.2	20.8

a) Determine the correlation coefficient for time and percent of chlorine remaining.
b) Determine whether a correlation exists at $\alpha = 0.01$.
c) Determine the equation of the line of best fit for time and amount of chlorine remaining.
d) Use the equation in part (c) to estimate the average amount of chlorine remaining after 4.5 hr.

42. a) Match the first 9 digits in your phone number (including area code) with the 9 digits in your social security number. Match the first digit in your phone number with the first digit in your social security number to get one ordered pair. Match the second digits to get a second ordered pair. Continue this process until you get a total of nine ordered pairs.
b) Do you believe that this set of bivariate data will have a positive correlation, a negative correlation, or no correlation? Explain your answer.
c) Construct a scatter diagram for the nine ordered pairs.
d) Calculate the correlation coefficient, r.
e) Is there a correlation at $\alpha = 0.05$? Explain.
f) Calculate the equation of the line of best fit.
g) Using the equation in part (f) estimate the digit in a social security number that corresponds with a 7 in a telephone number.

43. a) Do you believe that there is a positive correlation, a negative correlation, or no correlation between speed of a car and stopping distance when the brakes are applied? Explain.
b) Do you believe that there is a stronger correlation between speed of a car and stopping distance on wet or dry roads? Explain.
c) Use Fig. 13.46 to construct two scatter diagrams, one for dry pavement and the other for wet pavement. Place the speed of the car on the horizontal axis.

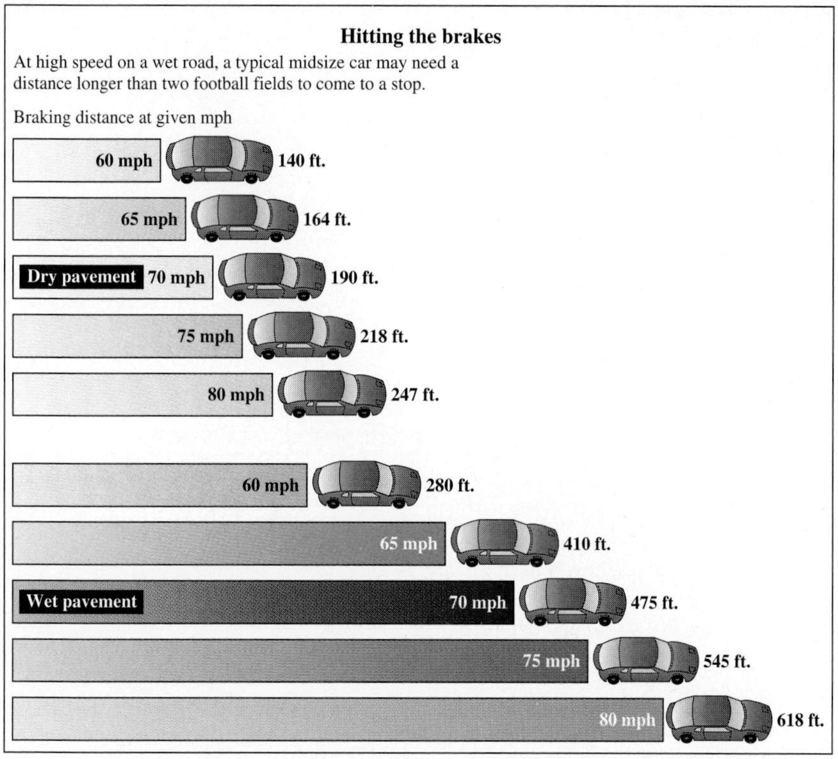

Hitting the brakes

At high speed on a wet road, a typical midsize car may need a distance longer than two football fields to come to a stop.

Braking distance at given mph

Dry pavement	
60 mph	140 ft.
65 mph	164 ft.
70 mph	190 ft.
75 mph	218 ft.
80 mph	247 ft.

Wet pavement	
60 mph	280 ft.
65 mph	410 ft.
70 mph	475 ft.
75 mph	545 ft.
80 mph	618 ft.

Source: *Car and Driver*, American Automobile Association

Figure 13.46

d) Compute the correlation coefficient for speed of the car and stopping distance for dry pavement.

e) Repeat part (d) for wet pavement.

f) Were your answers to parts (a) and (b) correct? Explain.

g) Determine the equation of the line of best fit for dry pavement.

h) Repeat part (g) for wet pavement.

i) Use the equations in parts (g) and (h) to estimate the stopping distance of a car going 77 mph on both dry and wet pavements.

Problem Solving/Group Activities

44. a) Assume that a set of bivariate data yields a specific correlation coefficient. If the x and y values are interchanged and the correlation coefficient is recalculated, will the correlation coefficient change? Explain.

b) Make up a table of five pieces of bivariate data and determine r using the data. Then switch the values of the x's and y's and recompute the correlation coefficient. Has the value of r changed?

45. a) Do you believe that a correlation exists between a person's height and the length of a person's forearm? Explain.

b) Select 10 people from your class and measure (in inches) their heights and the lengths of their forearms.

c) Plot the 10 ordered pairs on a scatter diagram.

d) Calculate the correlation coefficient, r.

e) Determine the equation of the line of best fit.

f) Estimate the length of the forearm of a person who is 58 in. tall.

46. a) Have your group select a category of bivariate data that it thinks has a strong positive correlation. Designate the independent variable and the dependent variable. Explain why your group believes the bivariate data have a strong positive correlation.

b) Collect at least 10 pieces of bivariate data that can be used to determine the correlation coefficient. Explain how your group chose these data.

c) Plot a scatter diagram.

d) Calculate the correlation coefficient.

e) Does there appear to be a strong positive correlation? Explain your answer.
f) Calculate the equation of the line of best fit.
g) Explain how the equation in part (f) may be used.

47. In Exercise 33 we presented the following table.

Year	90	91	92	93	94	95
CPI	130	136	140	144	148	153

a) If you have not previously done so, calculate r.
b) If 90 is subtracted from each year, the table obtained becomes:

Year	0	1	2	3	4	5
CPI	130	136	140	144	148	153

If r is calculated from these values, how will it compare with the r determined in part (a)? Explain.
c) Calculate r from the values in part (b) and compare the results with the value of r found in part (a). Are they the same? If not, explain why.

48. a) There are equivalent formulas that can be used to find the correlation coefficient and the equation of the line of best fit. A formula used in some statistics books to find the correlation coefficient is

$$r = \frac{SS(xy)}{\sqrt{SS(x)SS(y)}},$$

where

$$SS(x) = \Sigma x^2 - \frac{(\Sigma x)^2}{n},$$

$$SS(y) = \Sigma y^2 - \frac{(\Sigma y)^2}{n},$$

and $$SS(xy) = \Sigma xy - \frac{(\Sigma x)(\Sigma y)}{n}.$$

Use this formula to find the correlation coefficient of the set of bivariate data given in Example 1 on page 705.
b) Compare your answer with the answer obtained in Example 1.

Research Activities

49. a) Obtain a set of bivariate data from a newspaper or magazine.
b) Plot the information on a scatter diagram.
c) Indicate whether you believe that the data show a positive correlation, a negative correlation, or no correlation. Explain your answer.
d) Calculate r and determine whether your answer to part (b) was correct.
e) Determine the equation of the line of best fit for the bivariate data.

50. Find a scatter diagram in a newspaper or magazine and write a paper on what the diagram indicates. Indicate whether you believe that the bivariate data show a positive correlation, a negative correlation, or no correlation and explain why.

CHAPTER 13 SUMMARY

Key Terms

13.1

cluster sample
data
descriptive statistics
inferential statistics
population
random sample
sample
stratified sample
systematic sample
unbiased sample

13.3

class mark
class width
classes

frequency distribution
lower class limits
midpoint of a class
modal class
piece of data
upper class limits

13.4

central angle
circle graph
frequency polygon
histogram

13.5

mean
median
midrange

measures of central tendency
measures of location
mode
percentiles
quartiles
ranked data

13.6

measures of dispersion
range
standard deviation

13.7

bimodal distribution
Gaussian distribution
J-shaped distribution

normal distribution	**13.8**	line of best fit
rectangular distribution		linear correlation
skewed distribution	bivariate data	linear correlation coefficient
standard scores	correlation	linear regression
z-scores	critical values	regression
	level of significance	

Important Facts

Rules for Data Grouped by Classes

1. The classes should be the same width.
2. The classes should not overlap.
3. Each piece of data should belong to only one class.

Measures of Central Tendency

The **mean** is the sum of the data divided by the number of pieces of data: $\bar{x} = \dfrac{\Sigma x}{n}$.

The **median** is the value in the middle of a set of ranked data.

The **mode** is the piece of data that occurs most frequently (if there is one).

The **midrange** is the value halfway between the lowest and highest values: midrange $= \dfrac{L + H}{2}$.

Measures of Dispersion

The **range** is the difference between the highest value and lowest value in a set of data.

The **standard deviation**, s, is a measure of the spread of a set of data about the mean: $s = \sqrt{\dfrac{\Sigma(x - \bar{x})^2}{n - 1}}$.

z-scores

$$z = \frac{x - \bar{x}}{s}$$

Chebyshev's Theorem

Minimum percentage $= 1 - \dfrac{1}{k^2}, \, k > 1$

Linear Correlation and Regression

Linear correlation coefficient, r, is

$$r = \frac{n(\Sigma xy) - (\Sigma x)(\Sigma y)}{\sqrt{n(\Sigma x^2) - (\Sigma x)^2}\sqrt{n(\Sigma y^2) - (\Sigma y)^2}}$$

Equation of the Line of the Best Fit is

$y = mx + b$, where

$$m = \frac{n(\Sigma xy) - (\Sigma x)(\Sigma y)}{n(\Sigma x^2) - (\Sigma x)^2}, \text{ and}$$

$$b = \frac{\Sigma y - m(\Sigma x)}{n}$$

Chapter 13 Review Exercises

13.1

1. a) What is a population?
   b) What is a sample?

2. What is a random sample?

13.2

In Exercises 3 and 4, tell what possible misuses or misinterpretations may exist in the statements.

3. The Stay Healthy Candy Bar indicates on its label that it has no cholesterol. Therefore it is safe to eat as many of these candy bars as you want.

4. More copies of *Time* magazine are sold than are copies of *Money* magazine. Therefore *Time* is a more profitable magazine than *Money*.

5. The value of a 1-oz gold coin on December 31, 1996, was $432. The value of the same coin on December 31, 1995, was $420. Draw a graph that appears to show (a) a small decrease in value, and (b) a large decrease in value.

13.3–13.4

6. Consider the following set of data.

35	37	38	41	43
36	37	38	41	43
36	37	39	41	43
36	37	39	41	44
37	37	39	42	45

a) Construct a frequency distribution.
b) Construct a histogram.
c) Construct a frequency polygon.

7. Consider the following set of data.

30	40	52	58	63	72	78	84
32	44	53	60	67	72	78	86
36	46	54	60	68	74	78	88
38	48	55	60	70	76	80	92
39	50	57	62	72	78	82	98

a) Construct a frequency distribution with the first class 30–38.

b) Construct a histogram.

c) Construct a frequency polygon.

13.5–13.6

In Exercises 8–13, for the set of data 58, 67, 73, 82, 85, 97, find the

8. mean.

9. median.

10. mode.

11. midrange.

12. range.

13. standard deviation.

In Exercises 14–19, for the set of data 4, 5, 12, 14, 19, 7, 12, 23, 7, 17, 15, 21, find the

14. mean.

15. median.

16. mode.

17. midrange.

18. range.

19. standard deviation.

13.7

In Exercises 20–24, assume that anthropologists have determined that a certain type of primitive animal had a mean head circumference of 42 cm with a standard deviation of 5 cm. Given that head sizes were normally distributed, determine the percent of heads that are

20. between 37 and 47 cm.

21. between 32 and 52 cm.

22. less than 50 cm.

23. greater than 50 cm.

24. greater than 39 cm.

In Exercises 25–28, assume that the useful life of an electric motor is normally distributed with a mean of 7.6 years and a standard deviation of 0.5 year. Find the percent of motors with a useful life

25. between 7.6 years and 8.1 years.

26. less than 7.4 years.

27. between 7.8 years and 8.8 years.

28. If the manufacturer warranties its motors for 7.3 years, what percent will need to be replaced under warranty?

13.8

29. The following chart shows the average amounts of time the FDA took to approve or reject new drug applications for the years 1987–1992.

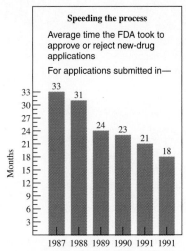

Source: *USN&WR* General Accounting Office

a) Construct a scatter diagram with the year on the horizontal axis.

b) Use the scatter diagram in part (a) to determine whether you believe that a correlation exists between year and average time. If so, is it a positive or negative correlation? Explain.

c) Calculate the correlation coefficient (use only the last two numbers in each year in your calculations; for example, use 87 for 1987).

d) Determine whether a correlation exists at $\alpha = 0.05$. Explain how you arrived at your answer.

30. Use the data in Exercise 29 to determine the equation of the line of best fit between year and average time (use only the last two numbers in each year in your calculations).

31. The following table shows the yield (interest), to the nearest tenth percent, for specific Treasury bill maturities in November 1995.

Why go long?	
The current narrow yield spread in Treasuries makes long bonds relatively unattractive.	
3-month T-bill	**5.1%**
1-year T-bill	**5.5%**
5-year T-note	**5.8%**
10-year T-note	**6.1%**
30-year T-bond	**6.4%**
USN&WR—Basic data: Technical Data Inc.	

a) Determine the correlation coefficient. Use 0.25 yr instead of 3 mo in your calculations.

b) Is there a correlation between length of time and yield at $\alpha = 0.05$? If so, is it a positive or negative correlation?

c) Determine the equation of the line of best fit between length of time and yield.

d) Use the equation in part (c) to estimate the yield on a 7-yr T-note (if there were such a maturity).

32. The following table shows the 10 highest yielding stocks in the Dow-Jones Industrial Average, along with their current price and price to earnings ratio (PE ratio) on January 22, 1996.

BEAT-THE-DOW STOCKS
10 highest-yielding Dow industrials

	RECENT PRICE	CURRENT YIELD	P/E RATIO*
Philip Morris	$94	4.3%	12
J.P. Morgan	80	4.1	11
Texaco	79	4.1	17
Exxon	81	3.7	16
Chevron	54	3.7	16
DuPont	72	2.9	11
3M	63	3.0	16
International Paper	38	2.6	12
General Electric	74	2.5	15
General Motors	47	2.5	7

*based on consensus 1996 earnings estimates provided by First Call Corp. Data to January 22

a) Construct a scatter diagram with price on the horizontal axis and current yield on the vertical axis.

b) By looking at the diagram in part (a) do you believe there is a linear correlation? If so, is it positive or negative? Explain.

c) Compute the correlation coefficient.

d) At $\alpha = 0.05$ does there appear to be a correlation? Explain.

e) Determine the equation of the line of best fit between price and yield.

33. Repeat Exercise 32 for stock price (place on the horizontal axis) and P/E ratio (place on the vertical axis).

13.5–13.7

In Exercises 34–41, use the following data obtained from a study of the weights of adult men.

Mean	187 lb	First quartile	173 lb
Median	180 lb	Third quartile	227 lb
Mode	175 lb	86th percentile	234 lb
Standard deviation	23 lb		

34. What is the most common weight?

35. What weight did half of those surveyed exceed?

36. About what percent of those surveyed weighed more than 227 lb?

37. About what percent of those surveyed weighed less than 173 lb?

38. About what percent of those surveyed weighed more than 234 lb?

39. If 100 men were surveyed, what is the total weight of all the men?

40. What weight represents two standard deviations above the mean?

41. What weight represents 1.8 standard deviations below the mean?

13.2–13.7

The following list shows the names of the 41 U.S. presidents and the number of children in their families.

Washington	0	Cleveland	5
J. Adams	5	B. Harrison	3
Jefferson	6	McKinley	2
Madison	0	T. Roosevelt	6
Monroe	2	Taft	3
J. Q. Adams	4	Wilson	3
Jackson	0	Harding	0
Van Buren	4	Coolidge	2
W. H. Harrison	10	Hoover	2
Tyler	14	F. D. Roosevelt	6
Polk	0	Truman	1
Taylor	6	Eisenhower	2
Fillmore	2	Kennedy	3
Pierce	3	L. B. Johnson	2
Buchanan	0	Nixon	2
Lincoln	4	Ford	4
A. Johnson	5	Carter	4
Grant	4	Reagan	4
Hayes	8	Bush	6
Garfield	7	Clinton	1
Arthur	3		

In Exercises 42–53, use the data above to determine the following.

42. Mean number of children

43. Mode   **44.** Median

45. Midrange   **46.** Range

47. Standard deviation (round the mean to the nearest tenth).

48. Construct a frequency distribution; let the first class be 0–1.

49. Construct a histogram.

50. Construct a frequency polygon.

51. Does this distribution appear to be normal? Explain.

52. On the basis of this sample, do you think that the number of children per family in the United States would have a normal distribution? Explain.

53. Do you believe that this sample is representative of the population? Explain.

CHAPTER 13 TEST

In Exercises 1–6, for the set of data 10, 21, 25, 25, 19 find the

1. mean.

2. median.

3. mode.

4. midrange.

5. range.

6. standard deviation.

In Exercises 7–9, use the set of data

26	28	35	46	49	56
26	30	36	46	49	58
26	32	40	47	50	58
26	32	44	47	52	62
27	35	46	47	54	66

to construct

7. a frequency distribution; let the first class be 25–30.

8. a histogram of the frequency distribution.

9. a frequency polygon of the frequency distribution.

In Exercises 10–16, use the following data on weekly salaries at the Hoperchang Publishing Company.

Mean	$500	First quartile	$450
Median	$470	Third quartile	$505
Mode	$495	79th percentile	$512
Standard deviation	$40		

10. What is the most common salary?

11. What salary did half the employees exceed?

12. About what percent of employees' salaries exceeded $450?

13. About what percent of employees' salaries was less than $512?

14. If the company has 100 employees, what is the total weekly salary of all employees?

15. What salary represents one standard deviation above the mean?

16. What salary represents 1.5 standard deviations below the mean?

In Exercises 17–20, the mileage of 5-year-old cars is normally distributed with a mean of 75,000 and a standard deviation of 12,000 miles.

17. What percent of 5-year-old cars have mileage between 50,000 and 70,000 miles?

18. What percent of 5-year-old cars have mileage greater than 60,000 miles?

19. What percent of 5-year-old cars have mileage greater than 90,000 miles?

20. If a random sample of 300 five-year-old cars is selected, how many would have mileage between 60,000 and 70,000 miles?

21. The following chart shows the percent of Americans that had "a great deal of confidence" in American national leaders and institutions for the years 1966, 1975, 1985, and 1994.

Trashing Our Leaders

Feelings of disappointment have devastated faith in national leaders and institutions. In 1966, 42% of Americans had a "great deal" of confidence in Congress. In 1994, only 8% did.

	1966	1975	1985	1994
Congress	42%	13%	16%	8%
Executive branch	41	13	15	12
The press	41	13	15	12
Major companies	41	13	15	12
Colleges, universities	41	13	15	12
Medicine	41	13	15	12

SOURCE: "THE GOOD LIFE AND ITS DISCONTENTS"

a) Construct a scatter diagram placing percents in 1966 on the horizontal axis and percents in 1994 on the vertical axis.

b) Do you believe that a correlation exists between 1966 and 1994 percents? If so is it positive or negative? Explain.

c) Calculate the correlation coefficient.

d) At $\alpha = 0.05$ does a correlation exist?

e) Calculate the equation of the line of best fit using the 1966 and 1994 percents.

f) In 1966, if 37% of Americans had indicated that they had a great deal of confidence in a particular institution, what percent would feel that way in 1994?

GROUP PROJECTS

Watching TV

1. Do you think that men or women, aged 17–20, watch more hours of TV weekly, or do you think that they watch the same number of hours?

   a) Write a procedure to use to determine the answer to that question. In your procedure use a sample of 30 men and 30 women. State how you will obtain an unbiased sample.

   b) Collect 30 pieces of data from men aged 17–20 and 30 pieces of data from women aged 17–20. Round answers to the nearest 0.5 hr. Follow the procedure developed in part (a) to obtain your unbiased sample.

   c) Compute the mean for your two groups of data to the nearest tenth.

   d) Using the means obtained in part (c) answer the question asked at the beginning of the problem.

   e) Is it possible that your conclusion in part (d) is wrong? Explain.

   f) Compute the standard deviation of both groups of data. Round each mean to the nearest tenth. How do the standard deviations compare?

   g) Do you believe that the distribution of data from either or both groups resembles a normal distribution? Explain.

   h) Add the two groups of data to get one group of 60 pieces of data. If these 60 pieces of data are added and divided by 60, will you obtain the same mean as when you add the two means from part (c) and divide the sum by 2? Explain.

   i) Compute the mean of the 60 pieces of data by using both methods in part (h). Are they the same? If so, why? If not, why not?

   j) Do you believe that this group of 60 pieces of data represents a normal distribution? Explain.

Binomial Probability Experiment

2. **a)** Have your group select a category of bivariate data that it thinks has a strong negative correlation. Indicate the variable that you will designate as the independent variable and the variable that you will designate as the dependent variable. Explain why your group believes the bivariate data have a strong negative correlation.

   b) Collect at least 10 pieces of bivariate data that can be used to determine the correlation coefficient. Explain how your group chose these data.

   c) Plot a scatter diagram.

   d) Calculate the correlation coefficient.

   e) Does there appear to be a negative correlation at $\alpha = 0.05$? Explain your answer.

   f) Calculate the equation of the line of best fit.

   g) Explain how the equation in part (f) may be used.

Statistics and Insurance

3. Insurance companies rely heavily on probability and statistics. It is essential for insurance companies to know how to measure the risk of certain things happening. In order to set premiums, a fire insurance company must predict how many fires will occur. An automobile insurance company must be able to predict the number of accidents involving injury, loss of life, and property damage. A life insurance company must predict the number of deaths by age in a given group of policyholders.

Let's explore life insurance. Life insurance companies base premiums on mortality tables, which are based on prior results (empirical probability). Insurers construct a mortality table that lists deaths per 1000 at each age from 0 through at least 120 years. Figure 13.47 shows the death rate per 1000 for ages 0 through 70, based on a mortality table, and Figure 13.48 shows the number of deaths expected at each age based on 10,000,000 people starting at age 0.

a) If there were 9,664,994 20-year-olds living at the start of 1995 and in 1995 17,300 20-year-olds died, determine, to two decimal places, the number of deaths per 1000 for 20-year-olds.

Use Figures 13.47 and 13.48 to answer the following questions.

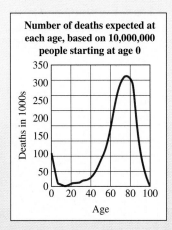

Figure 13.48

b) What is the death rate per 1000 at age 10?

c) At what age does the death rate reach 20 per 1000?

d) Based on 10,000,000 people, how many deaths are expected at age 30?

e) Based on 10,000,000 people, how many deaths are expected at age 70?

f) At what age is the number of deaths the greatest?

g) At what age is the number of deaths the least?

h) If the death rate for 16-year-olds in 1995 was 1.54 per 1000 and an insurance company insured 50,000 16-year-olds, how many did the insurance company expect to die in 1995?

i) Suppose that you were going to start a life insurance company. You have a group of 1000 sixteen-year-olds that desire insurance, and you agree to insure each of them for $1000 for 1 year. How much in premiums will you need to collect from each of them, based on the death rate in part (h), for your company to break even?

j) If your company wants to increase the premiums in part (i) by 12% to cover costs and make a small profit, determine the premium that should be charged.

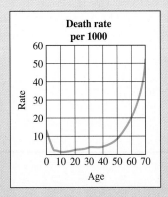

Figure 13.47

Answers

CHAPTER I

Section 1.1, Page 7

1. a) 1, 2, 3, 4, 5, . . . **b)** Counting numbers

3. A conjecture is a belief based on specific observations that has not been proven or disproven.

5. Deductive reasoning is the process of reasoning to a specific conclusion from a general statement.

7. Inductive reasoning, because a generalization was made from specific cases

9. $1234 \times 10 + 5 = 12{,}345$ **11.** $11 \times 14 = 154$

13. △ **15.**

17. 13, 16, 19 **19.** $-5, -7, -9$ **21.** 162, -486, 1458

23. 36, 49, 64 **25.** 80, 99, 120 **27.** 89,991

29. 25 or 5^2
36 or 6^2

31. a) 28, 36
b) To obtain the nth triangular number (after the first), add n to the previous triangular number.
c) No

33. a) 19 **b)** 100

35. Blue: 1, 5, 7, 10, 12 Red: 2, 4, 6, 9, 11 Yellow: 3, 8

37. a) You should obtain twice the original number.
b) You should obtain twice the original number.
c) The result is always twice the original number.
d) $n, 6n, 6n + 3, \dfrac{6n + 3}{3} = 2n + 1, 2n + 1 - 1 = 2n$

39. a) You should obtain the original number.
b) You should obtain the original number.
c) The result is always the original number.
d) $n, n + 5, \dfrac{n + 5}{5} = \dfrac{n}{5} + 1, \dfrac{n}{5} + 1 - 1 = \dfrac{n}{5},$

$\dfrac{n}{5} \cdot 5 = n$

41. 1×3 is not divisible by 2.

43. $3 \times 3 = 9$, and 9 is not even.

45. $11 + 13 + 15 = 39$

47. $(3 + 2)/2$ is not an even number.

49. a) 360°
c) The sum of the interior angles of a quadrilateral is 360°.

51. b) $t(t + 1)$ followed by 25, where t is the tens digit
c) 3025, 5625, 9025, 11025

53. 1881, 8008, 8118

Section 1.2, Page 16

(Answers in this section will vary depending on how you round your numbers. All answers are approximate.)

1. 860 **3.** 800,000,000 **5.** 5000 **7.** 100

9. 3,738,000,000 **11.** 1,260,000,000 **13.** $6.00

15. 20 lb **17.** $280 **19.** $780 **21.** 24 mpg **23.** $7.00

25. $6 **27.** 143 min, or 2 hr 23 min. **29.** 4600

31. a) ≈ 3.5 mi **b)** ≈ 6 km

33. 2.4 mi

35. a) ≈ 6 billion **b)** ≈ 65 years **c)** ≈ 40 years
d) ≈ 2.5 billion **e)** ≈ 3 billion
f) the projected growth from 1990 to 2020

37. a) $\approx$ $0.8 billion **b)** $\approx$ $0.10 billion
c) $\approx$ $1 billion

39. 18 **41.** 200 **43.** 120° **45.** 10% **47.** 9

49. ≈ 325 **51.** 38 in.

61. a) $\approx$ $350 **b)** $\approx$ $424

Section 1.3, Page 29

1. 8.39 ft **3.** $6\dfrac{5}{6}$ in. or ≈ 6.83 in. **5.** $5.40

7. 7.5 oz **9.** ≈ 667 times faster **11.** $12.50 per week

13. $60 **15.** 64 **17.** 1 out of 10,000, or $\dfrac{1}{10{,}000}$

19. $240 **21.** ≈ 18.7 mpg **23.** $82.08

25. $188,598.61

27. a) 4106.25 gal **b)** $17.25

29. a) ≈ 34.733 gal **b)** $\approx$ $41.68
c) $\approx 4{,}863{,}000{,}000$ gal

31. Less than the original price

33. a) 112, which can be obtained in two ways: 5 boxes of 20 and 1 box of 12 = 112 tapes; or 2 boxes of 20 and 6 boxes of 12 = 112 tapes
b) $270

35. a) water/milk: 3 cups; salt: 3/8 tsp; cream: 9 tbsp (or 9/16 cup)

b) water/milk: 2 7/8 cups; salt: 3/8 tsp; cream: 5/8 cup (or 10 tbsp)

c) water/milk: 2 3/4 cups; salt: 3/8 tsp; cream: 9/16 cup (or 9 tbsp)

d) Differences exist in water/milk because the amount for 4 servings is not twice that for 2 servings. Differences also exist in Cream of Wheat because 1/2 cup is not twice 3 tbsp.

37. 144 **39.** The area is 4 times as large.

41. $7 + 7 - (7 \div 7) = 13$ **43.** 36 squares

45. a) Place 5g, 1g, and 3g on one side and 9g on the other side.

b) Place 16g, 9g, and 3g on one side and 27g and 1g on the other side.

47. i) Place slice 1 and slice 2 in pan and cook for $\frac{1}{2}$ min.

ii) Turn slice 1 over and replace slice 2 with slice 3, cook $\frac{1}{2}$ min.

iii) Remove slice 1 and replace with slice 2 and turn slice 3 over and cook for $\frac{1}{2}$ min.

49.

8	6	16
18	10	2
4	14	12

51. Multiply the middle number by 3.

53. 216 cm³ **55.** 4

57.

59. 5 switch changes

61.

	7	
3	1	4
5	8	6
	2	

Other answers are possible, but 1 and 8 must appear in the center.

63. a) $244 **b)** $\approx 80\%$ **c)** $\approx 20\%$

65. 16 tsp. **67.** It will take longer if the wind blows.

69. 714 sq units

71. $2500 + 2501 + \cdots + 2550 = 2551 + \cdots + 2600$

Review Exercises, Page 34

1. 11, 13, 15 **2.** 25, 36, 49 **3.** 64, -128, 256

4. 25, 32, 40 **5.** 10, 4, -3 **6.** 21, 34, 55

7. **8.**

9. a) The original number and the final number are the same.

b) The original number and the final number are the same.

c) The final number is the same as the original number.

d) $n, 2n, 2n + 10, \dfrac{2n + 10}{2} = n + 5, n + 5 - 5 = n$

10. This process will always result in an answer of 3.

11. $6^2 - 4^2 = 36 - 16 = 20.$

The answers to Exercises 12–20 will vary, depending upon how you round the numbers. All answers are approximate.

12. 400,000,000 **13.** 30 **14.** 1570 **15.** 200

17. $8.00 **18.** $12.00 **19.** 3 mph **20.** $14.00

21. 2 mi **22.** $2300, $4700 **23.** $\approx$ $5400

24. 12 sq units **25.** Length = 22 ft; height = 8 ft

26. $3.95 **27.** $1.70 **28.** $1.20

29. Johnson's is cheaper by $20.00. **30.** $8.45 **31.** $311

32. 7.05 mg **33.** $665.00 **34.** 6 hr 45 min

35. July 26, 11:00 A.M.

36. a) 6.45 cm² **b)** 16.39 cm³ **c)** 1 cm = 0.39 in.

37. The one 1-in. hose **38.** 201 **39.** 6 **40.** 6

41. a) 6, 10, 15, 55 **b)** $\dfrac{n(n + 1)}{2}$ **c)** 210 **d)** 1275

42.

21	7	8	18
10	16	15	13
14	12	11	17
9	19	20	6

43.

23	25	15
13	21	29
27	17	19

44.

120	140	40
20	100	180
160	60	80

45. 59 min 59 sec

46.
$25	Room
$ 3	Men
$ 2	Clerk
$30	

47. 6 **48.** 140 lb **49.** 40 posts

50.

A	B	C	D	E
E	A	B	C	D
D	E	A	B	C
C	D	E	A	B
B	C	D	E	A

51. Yes; 3 quarters and 4 dimes, or 1 half dollar, 1 quarter and 4 dimes, or 1 quarter and 9 dimes. Other answers are possible.

52. Place six coins in each pan with one coin off to the side. If it balances, the heavier coin is the one on the side. If the pan does not balance, take the six coins on the heavier side and split them into two groups of three. Select the three heavier coins and weigh two coins. If the pan balances, it is the third coin. If the pan does not balance, you can identify the heavier coin.

53. 125,250

54. There is the same amount of wine in the red cup as water in the blue cup.

55. 16 blue **56.** 90

57. The fifth figure will be an octagon with equal sides. Inside the octagon will be a 7-sided figure with each side of equal length. The figure will have one antenna. The nth figure will have $n + 3$ sides of equal length. Inside the figure will be a figure of $n + 2$ sides of equal length. If n is an odd number, it will have one antenna. If n is an even number, it will have 2 antennas.

58. Some possible answers are shown. Others are possible.

```
        4                          4
3   12      10   1         1   12      11   2
    8        5                 7        5
2   6      11   7          3   8      9   6
        9                          10
                    1
              3   11    8   4
                12        7
              2    5    9   10
                    6
```

59. If each number is written out in words, the number of letters in the words form a magic square, each row, column, and diagonal having a sum of 27.

60. a) 2 **b)** 6 **c)** 24 **d)** 120
  e) $n(n - 1)(n - 2) \cdots 1$, (or $n!$), where $n =$ the number of people in line

61.   749
        $\underline{+853}$
          1602
  Other answers are possible.

Chapter Test, Page 38

1. 16, 20, 24 **2.** $\frac{1}{81}, \frac{1}{243}, \frac{1}{729}$

3. a) The result is the original number plus 1.
  b) The result is the original number plus 1.
  c) The result will always be the original number plus 1.
  d) n, $5n$, $5n + 10$, $\dfrac{5n + 10}{5} = n + 2$,
  $n + 2 - 1 = n + 1$

The answers for Exercises 4–6 are approximate.

4. 360 **5.** 30,000,000 **6.** 4

7. a) $\approx 970, \approx 900, \approx 270$ **b)** ≈ 330 **c)** Yes

8. 113 therms **9.** 25 cans **10.** $7\frac{1}{2}$ min

11. $\approx$ 30 in. by 22 in. (The actual dimensions are 77 cm by 53 cm.)

12. $50.00

13.

40	15	20
5	25	45
30	35	10

14. Less time if she had driven at 45 mph for the entire trip

15. $2 \cdot 6 \cdot 8 \cdot 9 \cdot 13$; 11 does not divide 11,232.

16. 243 beans

17. a) $11.97 **b)** $11.81
  c) Save 16 cents by using the 25% off coupon.

18. 8

CHAPTER 2

Section 2.1, Page 48

1. A set is a collection of objects.

3. Description, roster form, set-builder notation; The set of even counting numbers less than 7. $\{2, 4, 6\}$; $\{x \mid x \in N \text{ and } x < 7\}$

5. A set that is not finite

7. The cardinal number of set A, symbolized by $n(A)$, is the number of elements in set A.

9. $N = \{1, 2, 3, 4, 5, \ldots\}$

11. An empty set is a set that contains no elements.

13. Not well defined **15.** Well defined **17.** Well defined

19. Infinite **21.** Finite **23.** Infinite

25. {Africa, Antarctica, Asia, Australia, Europe, North America, South America}

27. $\{16, 17, 18, 19, \ldots, 274\}$ **29.** $C = \{8\}$ **31.** $\{ \ \}$

33. $E = \{5, 6, 7, 8, \ldots, 79\}$

35. $A = \{x \mid x \in N \text{ and } x < 10\}$, or $A = \{x \mid x \in N \text{ and } x \leq 9\}$

37. $C = \{x \mid x \in N \text{ and } x \text{ is even}\}$

39. $E = \{x \mid x \in N \text{ and } x \text{ is odd}\}$

41. $G = \{x \mid x \text{ is one of the three largest manufacturers of pickup trucks in the United States}\}$

43. A is the set of natural numbers less than 10.

45. L is the set of Great Lakes in the United States.

47. S is the set of the seven dwarfs from the movie *Snow White and the Seven Dwarfs*.

49. E is the set that contains no elements.

51. False; $\{a\}$ is a subset of the set, not an element of the set.

53. True **55.** True

57. False; Peter Pan is a brand of peanut butter.

59. 4 **61.** 0 **63.** Both **65.** Neither **67.** Equivalent

69. a) A is the set of natural numbers greater than 2.
  B is the set of all numbers greater than 2.

b) Set A contains only natural numbers. Set B contains other types of numbers, including fractions and decimal numbers.

c) $A = \{3, 4, 5, 6, \ldots\}$

d) No; set B cannot be written in roster form, since we cannot list all the elements in set B.

71. Cardinal   **73.** Ordinal

Section 2.2, Page 54

1. Set A is a subset of set B, symbolized $A \subseteq B$, if and only if all the elements of set A are also elements of set B.

3. If $A \subseteq B$, then every element of set A is an element of set B.
If $A \subset B$, then every element of set A is an element of set B and set $A \neq$ set B.

5. The number of proper subsets is determined with the formula $2^n - 1$, where n is the number of elements in the set.

7. False; Bach is an element of the set, not a subset.

9. True   **11.** True

13. False; the set $\{\varnothing\}$ contains the element $\varnothing$.

15. True

17. False; $\{5\}$ is a subset of the set, not an element of the set.

19. False; no set is a proper subset of itself.

21. True   **23.** $B \subseteq A, B \subset A$ **25.** $A = B, A \subseteq B, B \subseteq A$

27. None   **29.** $A = B, A \subseteq B, B \subseteq A$ **31.** { }

33. { }, {sheep}, {goat}, {sheep, goat}

35. a) { }, $\{a\}$, $\{b\}$, $\{c\}$, $\{d\}$, $\{a, b\}$, $\{a, c\}$, $\{a, d\}$, $\{b, c\}$, $\{b, d\}$, $\{c, d\}$, $\{a, b, c\}$, $\{a, b, d\}$, $\{a, c, d\}$, $\{b, c, d\}$, $\{a, b, c, d\}$

b) $\{a, b, c, d\}$

37. False   **39.** False   **41.** True   **43.** True   **45.** False

47. True   **49.** $E = F$ **51.** 32   **53.** 64

55. a) Yes   **b)** No   **c)** Yes

57. The set of drivers of the cars

Section 2.3, Page 61

1.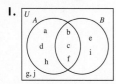

3. A' is the set of all the students attending a college or university in Florida during the academic years 1994–1996, who attended at least one social function at the school they were attending during that period.

5. A' is the set of students who will not pass the course.

7. The set of students in your college registered for the same mathematics course as yourself

9. The set of nurses employed at Community Hospital who have experience working in the emergency room and the delivery room

11. The set of nurses employed at Community Hospital who have experience working in the delivery room and no experience in the emergency room

13. The set of cities in the state of North Carolina with a population over 50,000 or with a philharmonic orchestra

15. The set of cities in the state of North Carolina without a professional sports team and with a philharmonic orchestra

17. The set of cities in the state of North Carolina that do not have a population over 50,000 or do not have a professional sports team

19. $\{2, 3, 4, 6\}$ **21.** $\{2, 3, 4, 5, 6, 7, 8, 9, 10\}$

23. $\{2, 3\}$ **25.** $\{7, 8\}$

27. {MI, IL, CA, NY, NJ, PA}

29. {MI, IL, CA, NY, NJ, PA, CT, AK, DE, MD, MA, OR, RI, WI}

31. {NY, NJ, PA}   **33.** {MD, MA, OR, RI, WI}

35. $\{1, 2, 3, 4, 5, 6, 8\}$ **37.** $\{1, 5, 7, 8\}$ **39.** $\{7\}$

41. { }   **43.** $\{7\}$ **45.** $\{b, e, h, j, k\}$ **47.** $\{a, f, i\}$

49. $\{a, b, c, d, f, g, i\}$ **51.** $\{a, c, d, e, f, g, h, i, j, k\}$

53. $\{b\}$ **55.** $\{1, 2, 3, 4, 5, 6, 7, 8, 9\}$, or U

57. $\{2, 4, 6, 8\}$, or B **59.** $\{1, 3, 5, 7, 9\}$, or A

61. $\{1, 3, 5, 6, 7, 8, 9\}$ **63.** $\{1, 2, 3, 4, 5, 7, 9\}$

65. $\{2, 4, 6, 8\}$, or B **67.** { }

69. A set and its complement will always be disjoint for all sets. For example, if $U = \{1, 2, 3\}$, $A = \{1, 2\}$, and $A' = \{3\}$, then $A \cap A' = \{\ \}$.

73. A **75.** B **77.** C

79. $\{2, 6, 10, 14, 18, \ldots\}$ **81.** $\{2, 6, 10, 14, 18, \ldots\}$

83. U **85.** A **87.** $\varnothing$ **89.** A **91.** $B \subseteq A$

93. A and B are disjoint sets.   **95.** $A \subseteq B$

97. **99.**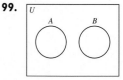

101. $\{a, i\}$ **103.** $\{b, c, g\}$ **105.** $\{1, 6, 8\}$

107. $\{2, 4, 5, 7, 9, 11\}$ **109.** $\{13\}$

111. $\{(1, a), (1, b), (1, c), (2, a), (2, b), (2, c)\}$ **113.** 6

115. Yes   **117.** $m \times n$

Section 2.4, Page 68

1.
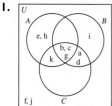

3.
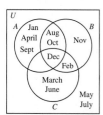

5.
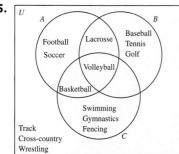

7. VI   **9.** VI   **11.** VII   **13.** II   **15.** II   **17.** III
19. VIII   **21.** VIII   **23.** III   **25.** III   **27.** VI   **29.** V
31. II   **33.** VIII   **35.** {1, 2, 3, 4, 5, 6}
37. {4, 5, 6, 7, 8, 10}   **39.** {3, 4, 5}
41. {1, 2, 3, 6, 9, 10, 11, 12}
43. {1, 2, 3, 4, 5, 6, 7, 8, 9, 12}   **45.** {9, 11, 12}
47. {7, 8, 9, 10, 11, 12}

49.
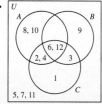

51. No   **53.** No   **55.** No   **57.** No   **59.** Yes   **61.** No
63. Yes   **65.** Yes   **67.** No

69. a) Yes   **c)** No

71.
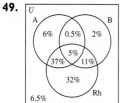

73. a)
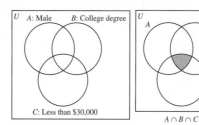

b)
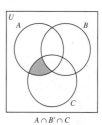

$A \cap B \cap C$

c)
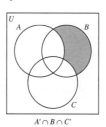

$A' \cap B \cap C'$

d)
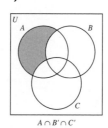

$A \cap B' \cap C$

e)
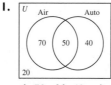

$A \cap B' \cap C'$

f)
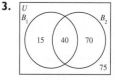

$A' \cap B' \cap C$

75. $n(A \cup B \cup C) = n(A) + n(B) + n(C)$
$$- 2n(A \cap B \cap C)$$
$$- n(A \cap B \cap C')$$
$$- n(A \cap B' \cap C)$$
$$- n(A' \cap B \cap C)$$

Section 2.5, Page 74

1.
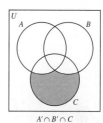

a) 70   **b)** 40   **c)** 20

3.

a) 15   **b)** 75   **c)** Yes

5.

a) 247   **b)** 49   **c)** 153   **d)** 96   **e)** 19

7.

a) 185   **b)** 10   **c)** 25   **d)** 401

9.

a) 93   **b)** 58   **c)** 4   **d)** 126   **e)** 142

11.

No; only 99 of 100 interviewed are accounted for.

13.

a) 6   **b)** 15   **c)** 24   **d)** 15

15.

a) 410   **b)** 35   **c)** 90   **d)** 50

Section 2.6, Page 80

1. $\{\,3\,,\ 4\,,\ 5\,,\ 6\,,\dots,\ n+2\,,\dots\}$
$\quad\downarrow\ \ \downarrow\ \ \downarrow\ \ \downarrow\qquad\quad\downarrow$
$\{\,4\,,\ 5\,,\ 6\,,\ 7\,,\dots,\ n+3\,,\dots\}$

3. $\{\,2\,,\ 3\,,\ 4\,,\ 5\,,\dots,\ n+1\,,\dots\}$
$\quad\downarrow\ \ \downarrow\ \ \downarrow\ \ \downarrow\qquad\quad\downarrow$
$\{\,3\,,\ 4\,,\ 5\,,\ 6\,,\dots,\ n+2\,,\dots\}$

5. $\{\,4\,,\ 7\,,\ 10\,,\ 13\,,\dots,\ 3n+1\,,\dots\}$
$\quad\downarrow\ \ \downarrow\ \ \downarrow\ \ \downarrow\qquad\qquad\downarrow$
$\{\,7\,,\ 10\,,\ 13\,,\ 16\,,\dots,\ 3n+4\,,\dots\}$

7. $\{\,8\,,\ 10\,,\ 12\,,\ 14\,,\dots,\ 2n+6\,,\dots\}$
$\quad\downarrow\ \ \ \downarrow\ \ \ \downarrow\ \ \ \downarrow\qquad\qquad\downarrow$
$\{10,\ 12,\ 14,\ 16,\dots,\ 2n+8\,,\dots\}$

9. $\left\{\,1\,,\ \tfrac13\,,\ \tfrac15\,,\ \tfrac17\,,\dots,\ \dfrac{1}{2n-1}\,,\cdots\right\}$
$\quad\ \downarrow\ \ \downarrow\ \ \downarrow\ \ \downarrow\qquad\qquad\downarrow$
$\left\{\,\tfrac13\,,\ \tfrac15\,,\ \tfrac17\,,\ \tfrac19\,,\dots,\ \dfrac{1}{2n+1}\,,\cdots\right\}$

11. $\{\,1\,,\ 2\,,\ 3\,,\ 4\,,\dots,\ n\,,\dots\}$
$\quad\ \downarrow\ \ \downarrow\ \ \downarrow\ \ \downarrow\qquad\quad\downarrow$
$\{\,3\,,\ 6\,,\ 9\,,\ 12\,,\dots,\ 3n\,,\dots\}$

13. $\{\,1\,,\ 2\,,\ 3\,,\ 4\,,\dots,\quad n\quad,\dots\}$
$\quad\ \downarrow\ \ \downarrow\ \ \downarrow\ \ \downarrow\qquad\qquad\downarrow$
$\{\,4\,,\ 6\,,\ 8\,,\ 10\,,\dots,\ 2n+2\,,\dots\}$

15. $\{\,1\,,\ 2\,,\ 3\,,\ 4\,,\dots,\quad n\quad,\dots\}$
$\quad\ \downarrow\ \ \downarrow\ \ \downarrow\ \ \downarrow\qquad\qquad\downarrow$
$\{\,2\,,\ 5\,,\ 8\,,\ 11\,,\dots,\ 3n-1\,,\dots\}$

17. $\{\,1\,,\ 2\,,\ 3\,,\ 4\,,\dots,\quad n\quad,\dots\}$
$\quad\ \downarrow\ \ \downarrow\ \ \downarrow\ \ \downarrow\qquad\qquad\downarrow$
$\{\,5\,,\ 8\,,\ 11\,,\ 14\,,\dots,\ 3n+2\,,\dots\}$

19. $\{\,1\,,\ 2\,,\ 3\,,\ 4\,,\dots,\quad n\quad,\dots\}$
$\quad\ \downarrow\ \ \downarrow\ \ \downarrow\ \ \downarrow\qquad\qquad\downarrow$
$\left\{\,\tfrac13\,,\ \tfrac14\,,\ \tfrac15\,,\ \tfrac16\,,\dots,\ \dfrac{1}{n+2}\,,\cdots\right\}$

21. $\{\,1\,,\ 2\,,\ 3\,,\ 4\,,\dots,\ n\,,\dots\}$
$\quad\ \downarrow\ \ \downarrow\ \ \downarrow\ \ \downarrow\qquad\quad\downarrow$
$\{\,1\,,\ 4\,,\ 9\,,\ 16\,,\dots,\ n^2\,,\dots\}$

23. $\{\,1\,,\ 2\,,\ 3\,,\ 4\,,\dots,\ n\,,\dots\}$
$\quad\ \downarrow\ \ \downarrow\ \ \downarrow\ \ \downarrow\qquad\quad\downarrow$
$\{\,3\,,\ 9\,,\ 27,\ 81,\dots,\ 3^n\,,\dots\}$

Review Exercises, Page 81

1. True

2. False; the words ''too much'' make the statement not well defined.

3. False; crow is not a member of the set.

4. False; no set is a proper subset of itself.

5. False; the elements 6, 12, 18, 24, . . . are members of both sets.

6. True

7. False; both sets do not contain exactly the same elements.

8. True   **9.** True   **10.** True   **11.** True   **12.** True

13. True   **14.** True   **15.** $A = \{5, 7, 9, \dots, 17\}$

16. $B = \{\text{TX, LA, MS, AL, FL}\}$

17. $C = \{1, 2, 3, 4, \dots, 453\}$

18. $D = \{4, 5, 6, 7, \dots, 35\}$

19. $A = \{x \mid x \in N \text{ and } 25 < x < 150\}$

20. $B = \{x \mid x \in N \text{ and } x > 79\}$

21. $C = \{x \mid x \in N \text{ and } x < 3\}$

22. $D = \{x \mid x \in N \text{ and } 23 \le x \le 41\}$

23. A is the set of letters in the English alphabet from E through M, inclusive.

24. B is the set of U.S. coins with a value of less than a dollar.

25. C is the set of the three commercial television networks in the United States with the largest number of viewers.

26. D is the set of English teachers who own a red car and do not live in Nebraska.

27. $A \cap B = \{5, 7\}$

28. $A \cup B' = \{1, 3, 5, 7, 11, 15\}$

29. $A' \cap B = \{9, 13\}$

30. $(A \cup B)' \cup C = \{1, 7, 11, 13, 15\}$

31. 16 **32.** 15

33.

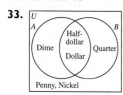

34. {a, c, d, e, f, g, i, k} **35.** {i, d}

36. {a, b, c, d, e, f, g, i, k} **37.** {e} **38.** {a, d, e}

39. {a, b, d, e, f, g} **40.** True **41.** True **42.** $450

43.

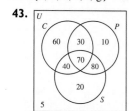

a) 315 **b)** 10 **c)** 30 **d)** 110

44.

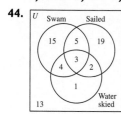

a) 13 **b)** 11 **c)** 35 **d)** 4 **e)** 1

45. $\{2, 4, 6, 8, \ldots, \quad 2n \quad, \ldots\}$
$\quad\downarrow \;\downarrow \;\downarrow \;\downarrow \qquad\quad \downarrow$
$\{4, 6, 8, 10, \ldots, 2n + 2, \ldots\}$

46. $\{3, 5, 7, 9, \ldots, 2n + 1, \ldots\}$
$\quad\downarrow \;\downarrow \;\downarrow \;\downarrow \qquad\quad \downarrow$
$\{5, 7, 9, 11, \ldots, 2n + 3, \ldots\}$

47. $\{1, 2, 3, 4, \ldots, \quad n \quad, \ldots\}$
$\quad\downarrow \;\downarrow \;\downarrow \;\downarrow \qquad\quad \downarrow$
$\{5, 8, 11, 14, \ldots, 3n + 2, \ldots\}$

48. $\{1, 2, 3, 4, \ldots, \quad n \quad, \ldots\}$
$\quad\downarrow \;\downarrow \;\downarrow \;\downarrow \qquad\quad \downarrow$
$\{4, 9, 14, 19, \ldots, 5n - 1, \ldots\}$

Chapter Test, Page 83

1. True

2. False; the sets do not contain exactly the same elements.

3. True **4.** True

5. False; the second set has no subset that contains the element 7.

6. True

7. False; the set has 2^3, or 8 subsets. **8.** True

9. False; for any set A, $A \cup A' = U$, not { }.

10. True

11. $A = \{1, 2, 3, 4, 5\}$

12. Set A is the set of natural numbers less than 6.

13. {3, 5, 7, 9, 11, 13} **14.** {5, 7, 9}

15. {3, 5, 7, 9}, or A **16.** 2

17.

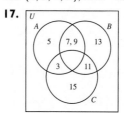

18. Equal

19. a)

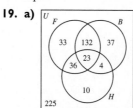

b) 80 **c)** 225 **d)** 172 **e)** 195 **f)** 79 **g)** 36

20. $\{7, 8, 9, 10, \ldots, n + 6, \ldots\}$
$\quad\downarrow \;\downarrow \;\downarrow \;\downarrow \qquad\quad \downarrow$
$\{8, 9, 10, 11, \ldots, n + 7, \ldots\}$

21. $\{1, 2, 3, 4, \ldots, \quad n \quad, \ldots\}$
$\quad\downarrow \;\downarrow \;\downarrow \;\downarrow \qquad\quad \downarrow$
$\{1, 3, 5, 7, \ldots, 2n - 1, \ldots\}$

CHAPTER 3

Section 3.1, Page 97

1. A statement that conveys only one idea

3. a) Some are **b)** All are **c)** Some are not **d)** None are

5. $\sim p$. (p: The ink *is* purple.)

7. When two simple statements are on the same side of the comma, they are placed together in parentheses when translated into symbolic form.

9. Compound; conjunction, $\wedge$

11. Compound; biconditional, $\leftrightarrow$

13. Compound; disjunction, $\vee$ 15. Simple statement

17. Compound; negation, $\sim$

19. Compound; conjunction, $\wedge$

21. Compound; conditional, $\rightarrow$

23. No toys cost more than $15.

25. All dogs have fleas.

27. Some plants are not living things.

29. No orchestras are having financial difficulty.

31. Someone likes a bully.

33. All rain forests are being destroyed.

35. Some nurses do not have a degree.   37. $\sim q$

39. $\sim q \vee \sim p$ 41. $\sim p \rightarrow \sim q$ 43. $p \rightarrow q$

45. $\sim p \wedge \sim q$ 47. The horse is not five years old.

49. The horse is five years old and the horse won the race.

51. If the horse is not five years old, then the horse won the race.

53. It is false that the horse won the race or the horse is five years old.

55. The horse is not five years old or the horse has not won the race.

57. $(p \wedge q) \vee r$ 59. $p \rightarrow (q \vee \sim r)$

61. $(\sim p \leftrightarrow \sim q) \vee \sim r$ 63. $(r \leftrightarrow q) \wedge p$

65. $(r \vee \sim q) \leftrightarrow p$

67. The water is 70° or the sun is shining, and we go swimming.

69. If the sun is shining then the water is 70°, or we go swimming.

71. If we do not go swimming, then the sun is shining and the water is 70°.

73. If we go swimming then the sun is shining, and the water is 70°.

75. The sun is shining if and only if the water is 70°, and we go swimming.

77. Not permissible   79. Not permissible

81. a) $(\sim r) \rightarrow p$ b) Conditional

83. a) $q \wedge (\sim r)$ b) Conjunction

85. a) $(p \vee q) \rightarrow r$ b) Conditional

87. a) $r \rightarrow (p \vee q)$ b) Conditional

89. a) $(\sim p) \leftrightarrow (\sim q \rightarrow r)$ b) Biconditional

91. a) $(r \wedge \sim q) \rightarrow (q \wedge \sim p)$ b) Conditional

93. a) $\sim[(p \wedge q) \leftrightarrow (p \vee r)]$ b) Negation

95. a) $\sim f \vee s$ b) Disjunction

97. a) $(\sim s \vee e) \rightarrow m$ b) Conditional

99. a) $\sim(d \rightarrow \sim r)$ b) Negation

101. a) $c \leftrightarrow (s \vee a)$ b) Biconditional

103. a) $(c \leftrightarrow s) \vee a$ b) Disjunction

105. $p \wedge q$ (p: Each language has a beauty of its own; q: Each language has forms of expression which are duplicated nowhere else.)

107. $(p \wedge q) \wedge r$ (p: The Constitution requires that the president be at least 35 years old; q: The Constitution requires that the president be a natural-born citizen; r: The Constitution requires that the president be a resident of the country for 14 years.)

109. $p \rightarrow q$ (p: you want to kill an idea in the world today; q: get a committee working on it.)

111. $\sim p \wedge q$ (p: I know the dignity of their birth; q: I know the glory of their death.)

113. $[(\sim q) \rightarrow (r \vee p)] \leftrightarrow [(\sim r) \wedge q]$; biconditional

115. a) The conjunction and disjunction have the same dominance.

Section 3.2, Page 107

1. a) 4   b)

p	q
T	T
T	F
F	T
F	F

3. a)

p	q	$p \wedge q$
T	T	T
T	F	F
F	T	F
F	F	F

3. b) Only when both simple statements are true

5. F   7. T   9. F   11. T   13. T   15. T
F   F   F   F   T   F
          T   T   T   F   T
          T   F   T   T   T
                                                     T
                                                     F
                                                     T
                                                     F

17. F   19. T   21. $p \wedge q$ 23. $p \wedge \sim q$
F   F   T   F
T   T   F   T
F   T   F   F
F   T   F   F
F   F
T   T
T   F

25. $(p \vee q) \wedge \sim r$

F
T
F
T
F
T
F
F

27. $p \wedge (q \wedge r)$

T
F
F
F
F
F
F
F

29. $(p \wedge q) \vee \sim q$

T
T
F
T

31. False **33.** False **35.** True **37.** True **39.** False
41. True **43.** False **45.** True **47.** True **49.** True

51. $\sim p \wedge q$

F
F
T
F

True in case 3 when p is false and q is true.

53. $p \vee \sim q$

T
T
F
T

True in cases 1, 2, and 4, when p is true, or when p and q are both false.

55. $(p \wedge q) \vee r$

T
T
T
F
T
F
T
F

True in cases 1, 2, 3, 5, and 7. True except when p, q, r have truth values of TFF or FTF or FFF.

57. $q \vee (p \wedge \sim r)$

T
T
F
T
T
T
F
F

True in cases 1, 2, 4, 5, and 6. True except when p, q, r have truth values TFT, FFT, or FFF.

59. a) Ms. Bell and Mrs. Spatz qualify.
  b) Mr. Vapmek does not qualify, since he makes less than $40,000.

61. a) Mr. James qualifies; the other four do not.
  b) Ms. Vela is returning on April 2. Mr. Davis is returning on a Monday. Ms. Chang is not staying over on a Saturday. Ms. Ghandi is returning on a Monday.

63. T
T
T
T
F
T
F
T

65. Yes

Section 3.3, Page 115

1.

p	q	$p \rightarrow q$
T	T	T
T	F	F
F	T	T
F	F	T

3. a) Substitute the truth values for the simple statements. Then evaluate the compound statement, using the assigned truth values.
  b) False

5. A tautology is a compound statement that is always true.

7. F	**9.** F	**11.** F	**13.** T	**15.** F
T	T	T	F	T
T	T	T	T	T
T	F	F	T	T

17. T	**19.** T	**21.** F	**23.** T	**25.** T
T	F	F	T	T
T	F	T	T	T
F	T	F	F	F
T	F	T	T	T
T	T	T	F	T
T	F	F	T	F
T	T	T	F	T

27. $p \rightarrow (\sim q \wedge r)$

F
F
T
F
T
T
T
T

29. $(p \leftrightarrow q) \vee \sim r$

T
T
F
T
F
T
T
T

31. $(\sim p \rightarrow q) \vee \sim r$

T
T
T
T
T
T
F
T

33. Neither **35.** Self-contradiction **37.** Tautology
39. Not an implication **41.** Implication **43.** Implication
45. True **47.** True **49.** False **51.** True **53.** True
55. True **57.** False **59.** False **61.** True **63.** True
65. True **67.** True **69.** False **71.** No

73. T
F
F
T
T
F
F
F

77. Helen Hurry—chemist; Sam Swift—historian; Nick Nickelson—mathematician; Peter Perkins—biologist.

Section 3.4, Page 127

1. Equivalent statements are statements that have exactly the same truth values in the answer columns of the truth tables.

3. The two statements are equivalent.

5. a) $q \rightarrow p$ **b)** $\sim p \rightarrow \sim q$ **c)** $\sim q \rightarrow \sim p$

7. Equivalent **9.** Not equivalent **11.** Equivalent

13. Not equivalent **15.** Equivalent **17.** Equivalent

19. Equivalent **21.** Equivalent **23.** Not equivalent

25. Not equivalent

27. Both $(p \rightarrow q) \wedge (q \rightarrow p)$ and $p \leftrightarrow q$ have the same truth values, T, F, F, T. Therefore they are equivalent.

29. The grass does not need watering or the trees do not need fertilizing.

31. It is false that Jerry does not subscribe to *National Geographic* and Jerry knows geography.

33. It is not true that the number is odd or the number is even.

35. We will sell the house, or it is not true that we will repair the roof or we will replace the water heater.

37. You do not do your workout or your metabolism will increase.

39. If Riva does not watch the CNN news, then she will watch the CNBC channel.

41. If today is not Monday, then I will not go to my piano lesson.

43. A number is even if and only if it is divisible by 2.

45. If you need to pay taxes then you receive income, and if you receive income then you need to pay taxes.

47. Converse: If we name him Cameron, then the baby is a boy.
Inverse: If the baby is not a boy, then we will not name him Cameron.
Contrapositive: If we do not name him Cameron, then the baby is not a boy.

49. Converse: If the calculator can illustrate graphics, then the calculator is a TI-82.
Inverse: If the calculator is not a TI-82, then the calculator cannot illustrate graphics.

Contrapositive: If the calculator cannot illustrate graphics, then the calculator is not a TI-82.

51. Converse: If you have not eaten, then you are hungry.
Inverse: If you are not hungry, then you have eaten.
Contrapositive: If you have eaten, then you are not hungry.

53. Converse: If our society is in trouble, then we did not stop all the crime.
Inverse: If we stop all the crime, then our society will not be in trouble.
Contrapositive: If our society is not in trouble, then we have stopped all the crime.

55. Converse: If the opposite sides are parallel and equal, then the figure is a parallelogram.
Inverse: If the figure is not a parallelogram, then the opposite sides are not parallel or the opposite sides are not equal.
Contrapositive: If the opposite sides are not parallel or the opposite sides are not equal, then the figure is not a parallelogram.

57. If the sum of the digits is divisible by 3, then the number is divisible by 3. True

59. If 2 divides the units digit of the counting number, then 2 divides the counting number. True

61. If two lines are not parallel, then the two lines intersect in at least one point. True

63. If the polygon is a quadrilateral, then the sum of the interior angles of the polygon measure 360°. True

65. a), b), and c) are equivalent.

67. a) and c) are equivalent.

69. b) and c) are equivalent.

71. b) and c) are equivalent.

73. b) and c) are equivalent.

75. None are equivalent.

77. a) and c) are equivalent.

79. a) and b) are equivalent.

81. Yes **83.** Yes

Section 3.5, Page 135

1. The conclusion necessarily follows from the premises.

3. If $(p_1 \wedge p_2) \rightarrow c$ is a tautology, then the argument is valid.

5. $p \rightarrow q$ **7.** $p \rightarrow q$ **9.** $p \rightarrow q$
$\underline{p}$ $\underline{\sim q}$ $\underline{q}$
$\therefore q$ $\therefore \sim p$ $\therefore p$

11. Invalid **13.** Valid **15.** Valid **17.** Invalid

19. Invalid **21.** Valid **23.** Valid **25.** Invalid

27. Invalid **29.** Valid

31. $j \rightarrow s$　**33.** $s \rightarrow l$　**35.** $r \vee p$
　$\underline{j}$　　　$\underline{l}$　　　$\underline{p \rightarrow \sim r}$
　$\therefore s$　　　$\therefore s$　　　$\therefore \sim p$
　Valid　　Invalid　　Invalid

37. $\sim t$　　**39.** $a \vee b$　**41.** $f \rightarrow t$
　$\underline{t \rightarrow m}$　　$\underline{\sim b \rightarrow a}$　$\underline{t \rightarrow e}$
　$\therefore m$　　$\therefore b \vee a$　$\therefore f \rightarrow e$
　Invalid　　Valid　　Valid

43. $\sim b \vee f$　**45.** $b \rightarrow p$　**47.** $h \wedge z$
　$\underline{b}$　　$\underline{\sim p}$　$\underline{\sim z \vee h}$
　$\therefore f$　　$\therefore \sim b$　$\therefore \sim h$
　Valid　　Valid　　Invalid

49. $f \rightarrow d$　**51.** $l \rightarrow p$
　$\underline{d \rightarrow \sim s}$　　$\underline{u \rightarrow p}$
　$\therefore f \rightarrow s$　$\therefore l \rightarrow u$
　Invalid　　Invalid

53. Therefore, you will be healthy.

55. Therefore, John fixes the car.

57. Therefore, the dog is not brown.

59. Therefore, if you overdraw your checking account, then you will have to pay a fee.

61. Therefore, if the logs are not dry, then you will have to dry the logs. (or, Therefore, if you do not have to dry the logs, then the logs are dry.)

Section 3.6, Page 141

1. The conclusion necessarily follows from the premises.

3. Yes　**5.** Valid　**7.** Valid　**9.** Fallacy　**11.** Valid

13. Fallacy　**15.** Fallacy　**17.** Fallacy　**19.** Valid

21. Fallacy　**23.** Fallacy　**25.** Valid　**27.** Valid

Review Exercises, Page 144

1. Some cameras do not use film.

2. Some dogs have fleas.　**3.** All apples are red.

4. No locks are keyless.　**5.** All people wear glasses.

6. Some rabbits wear glasses.

7. I opened a can of soda and I spilled some soda.

8. I did not open a can of soda or the soda is red.

9. If I opened a can of soda, then I spilled some soda and the soda is not red.

10. I opened a can of soda if and only if I spilled some soda.

11. I opened a can of soda or I did not spill some soda, and the soda is not red.

12. I did not open a can of soda, if and only if the soda is red and I did not spill some soda.

13. $q \vee \sim r$　**14.** $p \wedge r$　**15.** $r \rightarrow (\sim p \vee q)$

16. $(q \leftrightarrow p) \wedge \sim r$　**17.** $(r \wedge p) \vee q$　**18.** $\sim (q \vee r)$

19.	**20.**	**21.**	**22.**	**23.**	**24.**	**25.**	**26.**
T	F	T	F	T	F	T	F
T	F	T	T	F	T	T	T
F	F	T	T	T	F	T	T
F	T	F	F	T	F	T	T
				F	T	T	T
				F	T	F	T
				F	T	F	T
				F	T	T	T

27. True　**28.** True　**29.** True　**30.** True　**31.** True

32. True　**33.** True　**34.** False　**35.** False　**36.** True

37. Equivalent　**38.** Not equivalent　**39.** Equivalent

40. Not equivalent

41. If the stapler is not empty, then the stapler is jammed.

42. It is not true that the Kings do not score goals or the Kings do not win games.

43. New Orleans is not in California and Chicago is a city.

44. There is water in the vase or the flowers will wilt.

45. It is not true that I went to the party or I finished my special report.

46. If you do not have to stop, then the railroad crossing light is not flashing red.

47. If John's eyes do not have to be checked, then John is not having difficulty seeing.

48. If I am not at work, then today is a holiday.

49. If it is not Sam's computer, then it does not have a color monitor or it does not have an extended keyboard.

50. Converse: If I get a passing grade, then I studied.
Inverse: If I do not study, then I will not get a passing grade.
Contrapositive: If I do not get a passing grade, then I did not study.

51. a), b), and c) are equivalent.　**52.** None are equivalent.

53. a) and c) are equivalent.　**54.** b) and c) are equivalent.

55. Invalid　**56.** Valid　**57.** Invalid　**58.** Valid

59. Invalid　**60.** Invalid

Chapter Test, Page 147

1. $(r \vee q) \wedge p$　**2.** $(p \wedge \sim r) \rightarrow \sim q$　**3.** $\sim (r \leftrightarrow q)$

4. It is false that if Raul is a boy then Raul does not live in New Jersey.

5. Raul is a boy, if and only if Raul is 10 years old and Raul lives in New Jersey.

6.	**7.**
F	T
T	T
F	T
F	T
F	F
F	T
F	T
F	F

8. True   **9.** False   **10.** True   **11.** True

12. Equivalent   **13.** a) and b) are equivalent.

14. a) and b) are equivalent.

15. $s \rightarrow f$ **16.** Valid
$f \rightarrow p$
$\therefore s \rightarrow p$
Valid

17. Some birds are not black.   **18.** No people are funny.

19. Inverse: If the apple is not red, then it is not a delicious apple.
Converse: If it is a delicious apple, then the apple is red.
Contrapositive: If it is not a delicious apple, then the apple is not red.

20. Yes

CHAPTER 4

Section 4.1, Page 159

1. A number is a quantity, and it answers the question, "How many?" A numeral is a symbol used to represent the number.

3. A system of numeration consists of a set of numerals and a scheme or rule for combining the numerals to represent numbers.

5. In a multiplicative system there are numerals for each number less than the base and for powers of the base. Each numeral less than the base is multiplied by a numeral for the power of the base and these products are added to obtain the number.

7. 213   **9.** 1533   **11.** 334,214   **13.** 999∩∩ı

15. ⌡999∩∩∩∩ıııııı

17. ∝ııııııı ⌡⌡⌡999999999∩∩∩ıııııı

19. 14   **21.** 147   **23.** 1492   **25.** 1945   **27.** 13,675

29. 9464   **31.** XLII   **33.** CXXXIX   **35.** MCMXCVII

37. MMMMDCCXCIII   **39.** $\overline{\text{IX}}$CMXCIX

41. $\overline{\text{XX}}$DCXLIV   **43.** 54   **45.** 3081   **47.** 2650

49. 三
十
二

51. 四
百
六
十
五

53. 三
千
五
百
七
十

55. 364   **57.** 42,505   **59.** 9007

61. $\lambda\beta$ **63.** $\upsilon\zeta\epsilon$ **65.** $\lambda'\epsilon'\psi\delta$

67. Advantages: Can write some numbers more compactly.
Disadvantages: There are more numerals to memorize.

69. Advantages: Can write some numbers more compactly.
Disadvantages: There are more numerals to memorize.

71. 1936, ⌡999999999∩∩∩ıııııı, $\alpha'\pi\lambda\zeta$, 一
千
九
百
三
十
六

73. 422, 9999∩∩ıı, CDXXII, 四
百
二
十
二

75. $\pi'Q'\theta'\pi Q \theta$

Section 4.2, Page 165

1. A base ten place-value system

3. Write each numeral times its corresponding positional value.

5. a) The lack of a symbol for zero led to a great deal of ambiguity and confusion.
b) ≪ ‹ııı for both numbers

7. $(3 \times 10) + (4 \times 1)$

9. $(2 \times 100) + (3 \times 10) + (7 \times 1)$

11. $(8 \times 100) + (9 \times 10) + (7 \times 1)$

13. $(4 \times 1000) + (6 \times 100) + (4 \times 10) + (8 \times 1)$

15. $(5 \times 10,000) + (6 \times 1000) + (8 \times 100) + (3 \times 10) + (2 \times 1)$

17. $(3 \times 100,000) + (4 \times 10,000) + (6 \times 1000) + (8 \times 100) + (6 \times 10) + (1 \times 1)$

19. 32   **21.** 724   **23.** 5528   **25.** ı ≪   **27.** ıı ı

29. ı ıııı ≪≪≪‹ııı   **31.** 70   **33.** 6841   **35.** 4000

37. ═   **39.** ═   **41.** ⦁⦁⦁

43. Advantages: In general, they are more compact; large and small numbers can be written more easily; there are fewer symbols to memorize.
Disadvantages: If many of the symbols in the numeral represent zero, the place value system may be less compact.

45. 33, ⦁⦁⦁

47. a) No largest number
b) ıııı ≪≪≪‹ııı ≪≪≪ ıııııı ≪≪≪‹ıı

49. ıııı ‹‹ıııı   **51.** ⦁⦁⦁

53. 612, 630, 711, 810

Section 4.3, Page 170

3. 6   **5.** 15   **7.** 11   **9.** 90   **11.** 393   **13.** 1367

15. 193   **17.** 83   **19.** 6597   **21.** 1000_2 **23.** 11001_2

25. 663_8 **27.** $E93_{12}$ **29.** 1022_6 **31.** $17TE_{12}$

33. 1111110011_2 **35.** 4550_8 **37.** 1844   **39.** 447,415

41. 133_{16} **43.** 1566_{16} **45.** 11111001100_2 **47.** 30441_5

49. 3714_8

51. The numeral is written incorrectly; there is no 6 in the set of numerals for base 5.

53. Written correctly

55. b) 10213_5 **c)** 1373_8

59. $2^7 = 128$

Section 4.4, Page 178

1. a) b^0, b^1, b^2, b^3, b^4 **b)** $6^0, 6^1, 6^2, 6^3, 6^4$

5. 131_4 **7.** 11301_5 **9.** $7T5_{12}$ **11.** 2100_3

13. 24001_7 **15.** 10001_2 **17.** 203_4 **19.** 1134_5

21. 325_{12} **23.** 11_2 **25.** 3616_7 **27.** 1011_3 **29.** 134_5

31. 3051_7 **33.** 20212_4 **35.** 6072_9 **37.** 100011_2

39. 25_6 **41.** 54_8 R 2 **43.** 86_{12} R 1 **45.** 13_5

47. a) 21252_8 **b)** 306 and 29 **c)** 8874 **d)** 8874
e) Yes

49. 2302_5, 327

Section 4.5, Page 182

1. Duplation and mediation, the galley method, and Napier rods

3. 405 **5.** 1458 **7.** 9503 **9.** 7225 **11.** 2555

13. 1113 **15.** 900 **17.** 78,324 **19.** 111 **21.** 600

23. 625 **25.** 60,678 **27.** ꝯꝯ∩∩∩∩∩∩∩∩∩|||||||

29. |||||| ≪≪|| **31.** 12331_5 **33.** 16001_7

Review Exercises, Page 183

1. 1101 **2.** 1112 **3.** 1212 **4.** 2114 **5.** 3214

6. 2312 **7.** bbbbaaa **8.** cbbbbbbaaaaaaa

9. cccbbbbbbbbbaaaaaaaaaa

10. dccccccccccbbbbbbbbbbaaaaaa

11. ddddddddccccccbbbbaa **12.** ddddddccccbaaaa **13.** 93

14. 27 **15.** 648 **16.** 4068 **17.** 5648 **18.** 4809

19. exg **20.** bygxf **21.** fyixd **22.** bzbx **23.** fzd

24. bza **25.** 84 **26.** 503 **27.** 246 **28.** 46,883

29. 70,082 **30.** 60,529 **31.** mb **32.** snb **33.** vrc

34. BArg **35.** LDxqe **36.** QFvrf **37.** ⨍ꝯꝯꝯꝯ∩∩∩∩∩||

38. MCDLXII **39.** 一千四百六十二 **40.** $\alpha'\upsilon\xi\beta$ **41.** ≪≪|||| ≪≪||

42. ⋮⋮ **43.** 122,025 **44.** 8254 **45.** 585 **46.** 1991

47. 1277 **48.** 1971 **49.** 34 **50.** 5 **51.** 117

52. 1510 **53.** 1451 **54.** 186 **55.** 1231_7 **56.** 122011_3

57. 111001111_2 **58.** 13033_4 **59.** 327_{12} **60.** 717_8

61. 121_8 **62.** 101111_2 **63.** 176_{12} **64.** 1023_7

65. 13101_5 **66.** 10423_8 **67.** 3411_7 **68.** 100_2

69. $3E4_{12}$ **70.** 3324_5 **71.** 630_8 **72.** 1102_3 **73.** 124_5

74. 1203_4 **75.** 5656_{12} **76.** 12221_3 **77.** 110111_2

78. 13632_8 **79.** 1011_2 **80.** 132_4 **81.** 30_5 **82.** 433_6

83. 305_6 R 2 **84.** 664_8 R 2 **85.** 3408 **86.** 3408

87. 3408

Chapter Test, Page 185

1. A number is a quantity and it answers the question "How many?" A numeral is a symbol used to represent the number.

2. 2647 **3.** 1275 **4.** 8090 **5.** 1302 **6.** 12,342

7. 9999 **8.** ꝯꝯꝯ∩∩∩||||| **9.** $\beta'\upsilon o \zeta$

10. ⋯⋮⋮⋯

11. ≪≪|||||| ≪≪≪||||||

12. MMCCCLXXVIII

13. In an additive system the number represented by a particular set of numerals is the sum of the values of the numerals.

14. In a multiplicative system there are numerals for each number less than the base and for powers of the base. Each numeral less than the base is multiplied by a numeral for the power of the base and these products are added to obtain the numbers.

15. In a ciphered system the number represented by a particular set of numerals is the sum of the values of the numerals. There are numerals for each number up to and including the base and multiples of the base.

16. In a place-value system each number is multiplied by a power of the base. The position of the numeral indicates the power of the base by which it is multiplied.

17. 14 **18.** 22 **19.** 45 **20.** 305 **21.** 100100_2

22. 2221_3 **23.** 1444_{12} **24.** 125246_7 **25.** 1113_5

26. 336_7 **27.** 2003_6 **28.** 220_5 **29.** 392 **30.** 8428

CHAPTER 5

Section 5.1, Page 199

1. Number theory is the study of numbers and their properties.

3. a) *a divides b* means that the quotient of *b* divided by *a* has a remainder of 0.
b) *a is divisible by b* means that the quotient of *a* divided by *b* has a remainder of 0.

5. A composite number is a natural number that is divisible by a number other than itself and 1.

7. a) The greatest common divisor of a set of natural numbers is the largest natural number that divides (without remainder) every number in the set.
  c) 8

9. Prime numbers of the form $2^n - 1$ are Mersenne primes.

11. A conjecture is a supposition (or hypothesis) that has not been proved or disproved.

13. True **15.** True **17.** True **19.** False **21.** True

23. 2, 3, 4, 6 **25.** 3, 5 **27.** 2, 3, 4, 5, 6, 8, 10

29. 60 (Other answers are possible.) **31.** $2^2 \cdot 13$

33. $3 \cdot 19$ **35.** $2^3 \cdot 3 \cdot 13$ **37.** $2 \cdot 3 \cdot 5 \cdot 17$ **39.** $2^2 \cdot 449$

41. $2^2 \cdot 3^3 \cdot 19$ **43.** 15 **45.** 14 **47.** 30 **49.** 4 **51.** 8

53. 180 **55.** 168 **57.** 1800 **59.** 5088 **61.** 384

63. True **65.** 11 and 13, 17 and 19

71. 72 chips in each stack **73.** 120 days **75.** 3:20 A.M.

77. a) They all do.
  b) Every prime number greater than 3 differs by 1 from a multiple of the number 6.

79. 5 **81.** 2 **83.** 30 **85.** Not a perfect number

87. Perfect number

89. a) 8 **b)** 1, 2, 3, 4, 6, 8, 12, 24

91. The product of three consecutive numbers may be written $(n + 1)(n + 2)(n + 3)$. Both 2 and 3 are factors of this product, so the product will be divisible by 6.

Section 5.2, Page 207

1. The product of two numbers with like signs is a positive number. The product of two numbers with unlike signs is a negative number.

3. 0 **5.** 3 **7.** 8 **9.** -4 **11.** 1 **13.** -20

15. -3 **17.** -13 **19.** -2 **21.** -6 **23.** -2 **25.** -20

27. 64 **29.** 135 **31.** -105 **33.** -720 **35.** 4

37. -1 **39.** -9 **41.** -12 **43.** -20 **45.** False

47. False **49.** True **51.** True **53.** False **55.** 6

57. -17 **59.** -12 **61.** -5 **63.** -6

65. $-5, -3, -1, 0, 4, 5$ **67.** $-6, -5, -4, -3, -2, -1$

69. 30% **71.** 1769 ft **73.** Increase of $19

75. a) 9 hours **b)** 2 hours

77. -1

79. $0 + 1 - 2 + 3 + 4 - 5 + 6 - 7 - 8 + 9 = 1$ (Other answers are possible.)

Section 5.3, Page 218

1. The set of rational numbers is the set of numbers of the form p/q, where p and q are integers and $q \neq 0$.

3. A set of numbers is dense if, between any two distinct members of the set, there exists a third distinct member of the set.

5. b) $\dfrac{14}{45}$ **7. b)** $\dfrac{32}{19}$ **9. b)** $\dfrac{31}{36}$

11. Terminating **13.** Repeating **15.** Repeating

17. Repeating **19.** Terminating **21.** 0.75

23. $0.\overline{714285}$ **25.** 0.625 **27.** $5.\overline{6}$ **29.** $5.\overline{3}$

31. $\dfrac{4}{10} = \dfrac{2}{5}$ **33.** $\dfrac{76}{1000} = \dfrac{19}{250}$ **35.** $\dfrac{6001}{1000}$

37. $\dfrac{1246}{1000} = \dfrac{623}{500}$ **39.** $\dfrac{20{,}678}{10{,}000} = \dfrac{10{,}339}{5{,}000}$ **41.** $\dfrac{6}{9} = \dfrac{2}{3}$

43. $\dfrac{2}{1}$ **45.** $\dfrac{19}{11}$ **47.** $\dfrac{1085}{333}$ **49.** $\dfrac{109}{90}$ **51.** $\dfrac{6}{35}$

53. $\dfrac{2}{5}$ **55.** $\dfrac{25}{36}$ **57.** $\dfrac{36}{35}$ **59.** $\dfrac{5}{14}$ **61.** $\dfrac{2}{3}$ **63.** $\dfrac{12}{23}$

65. $\dfrac{13}{27}$ **67.** $\dfrac{24}{53}$ **69.** $\dfrac{4}{11}$ **71.** $\dfrac{15}{56}$ **73.** $\dfrac{23}{130}$

75. $\dfrac{23}{54}$ **77.** $\dfrac{17}{144}$ **79.** $\dfrac{-109}{600}$ **81.** $\dfrac{17}{12}$ **83.** $\dfrac{41}{28}$

85. $\dfrac{76}{96} = \dfrac{19}{24}$ **87.** $\dfrac{23}{60}$ **89.** $\dfrac{4}{11}$ **91.** $\dfrac{23}{42}$

In Exercises 93–108, other answers are possible.

93. 0.255 **95.** -2.1755 **97.** 3.1234505

99. 4.8725 **101.** $\dfrac{7}{10}$ **103.** $\dfrac{3}{40}$ **105.** $\dfrac{9}{40}$ **107.** $\dfrac{11}{200}$

109. Yes, between every two rational numbers there exists another rational number.

113. $\dfrac{5}{20} = \dfrac{1}{4}$ **115.** 129 in., or 10 ft 9 in. **117.** $1\frac{1}{16}$ ft

119. $1\frac{2}{3}$ tsp **121.** $58\frac{7}{8}$ in.

123. a) $7\frac{11}{16}$ in. from either end **b)** $7\frac{5}{8}$ in.

125. $-\dfrac{13}{9}$ **127.** $\dfrac{5}{6}$

129. $26\frac{1}{16}$ in.

131. a) 37 ft 10 in., or $37.8\overline{3}$ ft **b)** 88 ft^2 **c)** $806.\overline{6}$ ft^3

Section 5.4, Page 225

1. A rational number when expressed as a decimal number will be a terminating or a repeating decimal number. An irrational number will not.

3. Any number that is a square of a natural number is a perfect square.

5. b) $8\sqrt{3}$

7. To rationalize a denominator means to remove the radical expression from the denominator.

9. Rational **11.** Rational **13.** Irrational **15.** Rational

17. Irrational **19.** 9 **21.** 7 **23.** -13 **25.** -15

27. -10

29. Natural number, integer, rational number

31. Natural number, integer, rational number

33. Irrational number

35. Rational number

37. Rational number **39.** 8 **41.** $3\sqrt{5}$ **43.** $5\sqrt{5}$

45. $4\sqrt{3}$ **47.** $3\sqrt{6}$ **49.** $5\sqrt{5}$ **51.** $-4\sqrt{5}$

53. $-13\sqrt{3}$ **55.** $4\sqrt{3}$ **57.** $23\sqrt{2}$ **59.** $\sqrt{10}$

61. $2\sqrt{10}$ **63.** $6\sqrt{2}$ **65.** $\sqrt{2}$ **67.** 3 **69.** $\dfrac{2\sqrt{3}}{3}$

71. $\dfrac{\sqrt{14}}{2}$ **73.** 2 **75.** $\dfrac{3\sqrt{2}}{2}$ **77.** $\dfrac{\sqrt{15}}{3}$ **79.** True

81. False **83.** False **85.** $\sqrt{3} + 2\sqrt{3} = 3\sqrt{3}$

87. $\sqrt{2} \cdot \sqrt{3} = \sqrt{6}$

89. No, $\sqrt{3}$ is an irrational number and 1.732 is a rational number ($\sqrt{3} \approx 1.732$).

91. $\sqrt{4 + 9} \neq \sqrt{4} + \sqrt{9}, (\sqrt{13} \neq 5)$

93. $\sqrt{54.5} \approx 7.4$ ft **95.** ≈ 1.2 sec

97. No, the sum of two irrational numbers may not be an irrational number. For example, $\sqrt{3} + (-\sqrt{3}) = 0$.

Section 5.5, Page 231

1. The set of real numbers is formed by the union of the rational numbers and the irrational numbers.

3. If an operation is performed on any two elements of a set and the result is an element of the set, the set is closed under the given operation.

5. The order in which two numbers are multiplied is immaterial. One example is $2 \cdot 3 = 3 \cdot 2$.

7. When multiplying three real numbers, you may place parentheses around any two adjacent numbers. One example is $(2 \cdot 3) \cdot 4 = 2 \cdot (3 \cdot 4)$.

9. Closed **11.** Closed **13.** Closed **15.** Not closed

17. Closed **19.** Closed **21.** Not closed **23.** Not closed

25. Closed **27.** Not closed **29.** Commutative property

31. $(-3) + (-4) = (-4) + (-3)$

33. No, $3 - 2 \neq 2 - 3$

35. No, $3 \div 2 \neq 2 \div 3$

37. $[(-3) + (-4)] + (-5) = (-3) + [(-4) + (-5)]$

39. No, $(1 - 2) - 3 \neq 1 - (2 - 3)$

41. No, $(5 - 3) - 2 \neq 5 - (3 - 2)$

43. No, $3 + (5 \cdot 6) \neq (3 + 5) \cdot (3 + 6)$

45. Commutative property of multiplication

47. Distributive property of multiplication over addition

49. Commutative property of addition

51. Commutative property of addition

53. Distributive property of multiplication over addition

55. Commutative property of addition

57. Commutative property of multiplication

59. $4x + 20$ **61.** $2\sqrt{3} + 3\sqrt{2}$ **63.** $x\sqrt{3} + 3$

65. Distributive property of multiplication over addition

67. Distributive property of multiplication over addition

69. Commutative property of addition

71. Distributive property of multiplication over addition

73. Associative property of addition

75. No **77.** Yes **79.** No **81.** No **83.** No **85.** Yes

87. No, $0 \div a = 0$ and $a \div 0$ is undefined.

Section 5.6, Page 240

1. The 4 is the base and the 6 is the exponent.

3. b) $3^3 \cdot 3^5 = 3^8$

5. b) $3^{-5} = \dfrac{1}{3^5}$

7. b) $(4^4)^3 = 4^{12}$

9. b) 4.26×10^{-4}

11. a) The number is greater than or equal to 10.
    b) The number is greater than or equal to 1 but less than 10.
    c) The number is less than 1.

13. 25 **15.** 9 **17** -8 **19.** $\dfrac{25}{36}$ **21.** 576

23. 27 **25.** $\dfrac{1}{125}$ **27.** 1 **29.** 16 **31.** $\dfrac{1}{4}$

33. $3^6 = 729$ **35.** 25 **37.** $\dfrac{1}{125}$ **39.** 81 **41.** 3

43. a) $25, 25$ **b)** $-25, -25$ **c)** $125, -125$
    d) $-125, 125$

45. 3.7×10^4 **47.** 6.0×10^2 **49.** 5.3×10^{-2}

51. 1.9×10^4 **53.** 1.86×10^{-6} **55.** 4.23×10^{-6}

57. 7.11×10^2 **59.** 1.53×10^{-1} **61.** 4200

63. 60,000,000 **65.** 0.0000213 **67.** 0.312

69. 9,000,000 **71.** 231 **73.** 35,000 **75.** 10,000

77. 120,000,000 **79.** 0.0153 **81.** 320 **83.** 0.0021

85. 20 **87.** 4.2×10^{12} **89.** 4.5×10^{-7}

91. 2.0×10^3 **93.** 2.0×10^{-7} **95.** 3.0×10^8

97. $8.3 \times 10^{-4}, 3.2 \times 10^{-1}, 4.6, 5.8 \times 10^5$

99. $8.3 \times 10^{-5}, 0.00079, 4.1 \times 10^3, 40,000$

101. 3.72×10^9 **103.** 18,000 hr

105. 2,500,000 sec or ≈ 29 days **107.** 1.8×10^4

109. a) $15,817.12
  b) $2,923,000,000,000, or 2.923×10^{12}
  c) ≈ 3.56 times greater

111. a) 5.321×10^{10}, or $53,210,000,000$
  b) 1.806×10^{12}, or $1,806,000,000,000$
  c) 5.321×10^{10}
  d) 1.806×10^{10}
  e) more $1 bills, 3.515×10^{10} more $1 bills than
    $100 bills

113. 1000

115. a) 1.0×10^6, 1.0×10^9, 1.0×10^{12}
  b) 1000 days, or about 2.74 yr
  c) 1,000,000 days, or about 2740 yr
  d) 1,000,000,000 days or about 2,740,000 yr
  e) 1000

117. a) $\approx 5.87 \times 10^{12}$ mi
  b) 500 sec, or 8 min 20 sec

Section 5.7, Page 248

1. A sequence is a list of numbers that are related to each other by a rule. One example is 1, 2, 3, 4, . . .

3. A sequence in which each term after the first differs from the preceding term by a constant amount is called an arithmetic sequence. One example is 1, 4, 7, 10, . . .

5. A geometric sequence is one in which the ratio of any term to the term that directly precedes it is a constant. One example is 1, 3, 9, 27, . . .

7. 4, 7, 10, 13, 16   **9.** −4, 1, 6, 11, 16

11. 6, 4, 2, 0, −2   **13.** 24, 16, 8, 0, −8   **15.** $\frac{1}{2}, 1, \frac{3}{2}, 2, \frac{5}{2}$

17. $a_5 = 36$ **19.** $a_8 = 52$ **21.** $a_{10} = -138$

23. $a_{15} = \frac{139}{2}$ **25.** $a_n = 4n$ **27.** $a_n = 10n - 2$

29. $a_n = n - \frac{7}{2}$ **31.** $a_n = \frac{5}{2}n - \frac{15}{2}$

33. $a_n = -\frac{1}{2}n - \frac{5}{2}$ **35.** 78   **37.** 110   **39.** −36

41. −85.5   **43.** −35   **45.** 3, 6, 12, 24, 48
47. 5, −10, 20, −40, 80   **49.** −2, 2, −2, 2, −2

51. −2, −1, $-\frac{1}{2}, -\frac{1}{4}, -\frac{1}{8}$ **53.** 5, 3, $\frac{9}{5}, \frac{27}{25}, \frac{81}{125}$

55. $a_6 = 486$ **57.** $a_5 = 48$ **59.** $a_4 = 216$ **61.** $a_3 = \frac{3}{4}$

63. $a_5 = 8$ **65.** $a_n = 2(2)^{n-1}$ or $a_n = 2^n$

67. $a_n = 1(3)^{n-1}$ or $a_n = 3^{n-1}$ **69.** $a_n = (-8)(1/2)^{n-1}$

71. $a_n = (-6)(-3)^{n-1}$ **73.** $a_n = (-4)(2/3)^{n-1}$

75. $s_4 = 30$ **77.** $s_5 = 2343$ **79.** $s_5 = 484$

81. $s_5 = -33$ **83.** $s_{50} = 2550$ **85.** $s_{20} = 630$

87. a) $a_{12} = 63$ in.   **b)** $s_{12} = 954$ in.

89. $s_{31} = \$496$

91. $a_{15} = \$45,218.00$ (or $45,218.08, depending on the capacity of the calculator)

93. $a_5 = 12.288$ ft   **95.** $s_6 = 161.4375$ **97.** 267

99. 191.3568 ft

Section 5.8, Page 255

3. 0.618

7. a) 1.618   **b)** 0.618   **c)** 1.00

9. Alternate increasing then decreasing.

15. a) 3 in. × 5 in. and 5 in. × 8 in.

19. Yes, 29, 47   **21.** No   **23.** Yes, 195, 315

29. a) 1, 3, 4, 7, 11, 18, 29, 47
  b) $8 + 21 = 29$
    $13 + 34 = 47$
  c) It is the Fibonacci sequence

Review Exercises, Page 258

1. 2, 3, 4, 6, 8   **2.** 2, 3, 4, 6, 9   **3.** $2^3 \cdot 37$

4. $2^4 \cdot 3 \cdot 5$ **5.** $2^3 \cdot 3^2 \cdot 5$ **6.** $2^2 \cdot 3^2 \cdot 5 \cdot 7$

7. $2^6 \cdot 3 \cdot 5$ **8.** GCD = 16, LCM = 48

9. GCD = 5, LCM = 600

10. GCD = 4, LCM = 7992

11. GCD = 40, LCM = 6720

12. GCD = 4, LCM = 480

13. GCD = 36, LCM = 432   **14.** −4   **15.** 2

16. −5   **17.** −5   **18.** −7   **19.** 1   **20.** 0   **21.** 4

22. −6   **23.** 24   **24.** −30   **25.** −24   **26.** 5

27. −4   **28.** 6   **29.** 6   **30.** 3   **31.** 0.6   **32.** 0.8

33. $0.\overline{6}$ **34.** 3.75   **35.** $0.\overline{285714}$ **36.** $0.41\overline{6}$

37. 0.375   **38.** 0.875   **39.** $0.\overline{428571}$

40. $\frac{582}{1000} = \frac{291}{500}$ **41.** $\frac{2}{3}$ **42.** $\frac{243}{100}$ **43.** $\frac{170}{99}$

44. $\frac{12,083}{1000}$ **45.** $\frac{21}{5000}$ **46.** $\frac{97}{45}$ **47.** $\frac{232}{99}$

48. $\frac{2506}{495}$

In Exercises 49–54, other answers are possible.

49. 0.00615   **50.** $\frac{27}{40}$ **51.** 3.5095   **52.** $\frac{19}{24}$

53. 0.505   **54.** 0.25   **55.** $\frac{9}{20}$ **56.** $\frac{1}{10}$ **57.** $\frac{37}{36}$

58. $\frac{28}{45}$ **59.** $\frac{35}{54}$ **60.** $\frac{53}{28}$ **61.** $\frac{1}{6}$ **62.** $\frac{13}{40}$ **63.** $\frac{8}{15}$

64. $2\sqrt{3}$ **65.** $6\sqrt{2}$ **66.** $4\sqrt{2}$ **67.** $-3\sqrt{3}$

68. $8\sqrt{2}$ **69.** $-20\sqrt{3}$ **70.** $8\sqrt{3}$ **71.** $3\sqrt{2}$

72. $4\sqrt{3}$ **73.** 3   **74.** $2\sqrt{7}$ **75.** $\dfrac{3\sqrt{2}}{2}$ **76.** $\dfrac{\sqrt{15}}{5}$

77. $15 + 5\sqrt{5}$ **78.** $4\sqrt{3} + 3\sqrt{2}$

79. Commutative property of addition

80. Commutative property of multiplication

81. Associative property of addition

82. Distributive property of multiplication over addition

83. Commutative property of addition

84. Commutative property of addition

85. Associative property of multiplication

86. Commutative property of multiplication

87. Distributive property of multiplication over addition

88. Commutative property of multiplication

89. Closed   **90.** Closed   **91.** Not closed

92. Closed   **93.** Not closed   **94.** Not closed   **95.** 27

96. $\dfrac{1}{9}$ **97.** 5   **98.** 216   **99.** 1   **100.** $\dfrac{1}{125}$ **101.** 64

102. 81   **103.** 6.31×10^3 **104.** 4.23×10^{-5}

105. 2.75×10^{-3} **106.** 4.95×10^6 **107.** 870

108. 0.0000387   **109.** 0.000175   **110.** 100,000

111. 6.4×10^2 **112.** 1.38×10^5 **113.** 2.1×10^1

114. 3.0×10^0 **115.** 33,600,000,000   **116.** 22.5

117. 3200   **118.** 5.0   **119.** 25   **120.** Arithmetic, 23, 28

121. Geometric, -324, 972   **122.** Arithmetic, -16, -20

123. Geometric, $\dfrac{1}{16}$, $\dfrac{1}{32}$ **124.** Arithmetic, 16, 19

125. Geometric, -2, 2   **126.** 4, 9, 14, 19, 24

127. $-6, -9, -12, -15, -18$

128. $-\dfrac{1}{2}, -\dfrac{5}{2}, -\dfrac{9}{2}, -\dfrac{13}{2}, -\dfrac{17}{2}$ **129.** 4, 8, 16, 32, 64

130. 16, 8, 4, 2, 1   **131.** $\dfrac{1}{2}, -\dfrac{1}{4}, \dfrac{1}{8}, -\dfrac{1}{16}, \dfrac{1}{32}$ **132.** 14

133. -36 **134.** 92   **135.** 216   **136.** $\dfrac{1}{4}$ **137.** -48

138. -84 **139.** -25 **140.** 632   **141.** 52   **142.** 170

143. 33   **144.** Arithmetic, $a_n = -3n + 10$

145. Arithmetic, $a_n = 5n - 5$

146. Arithmetic, $a_n = -\dfrac{3}{2}n + \dfrac{11}{2}$

147. Geometric, $a_n = 3(2)^{n-1}$

148. Geometric, $a_n = 4(-1)^{n-1}$

149. Geometric, $a_n = 5\left(\dfrac{1}{3}\right)^{n-1}$ **150.** Yes, 21, 34

151. No   **152.** No   **153.** Yes, 26, 42

Chapter Test, Page 261

1. 2, 3, 4, 6, 8, 9   **2.** $2^3 \cdot 5 \cdot 7$ **3.** -7 **4.** -24

5. -175 **6.** 0.8735   **7.** $\dfrac{17}{144}$ **8.** 0.625

9. $\dfrac{645}{100} = \dfrac{129}{20}$ **10.** $\dfrac{121}{240}$ **11.** $\dfrac{19}{30}$ **12.** $8\sqrt{2}$ **13.** $\dfrac{\sqrt{30}}{6}$

14. Closed

15. Associative property of addition

16. Distributive property of multiplication over addition

17. 36   **18.** $4^5 = 1024$ **19.** $\dfrac{1}{81}$ **20.** 8.0×10^6

21. $a_n = -4n$ **22.** -187 **23.** 243   **24.** 1023

25. $3(2)^{n-1}$ **26.** 1, 1, 2, 3, 5, 8, 13, 21

CHAPTER 6

Section 6.1, Page 269

1. Letters of the alphabet used to represent numbers are called *variables*.

3. An *algebraic expression* is a collection of variables, numbers, parentheses, and operation symbols. An example is $5x^2y - 11$.

5. a) The 4 is the base and the 5 is the exponent.

7. 36   **9.** -64 **11.** 686   **13.** 21   **15.** 28   **17.** 15

19. $-\dfrac{83}{18}$ **21.** 7   **23.** 1   **25.** 0   **27.** No   **29.** Yes

31. No   **33.** Yes   **35.** Yes

37. a) $34.50   **b)** $80.50

39. 16,000,000   **41.** 780 baskets of oranges   **43.** 1.71 in.

45. $1^n = 1$ for all natural numbers, since $1 \cdot 1 \cdot 1 \cdot \;\cdots\; \cdot 1 = 1$ for n factors of 1.

Section 6.2, Page 279

1. The parts that are added or subtracted in an algebraic expression are called *terms*. In $3x - 2y$, the $3x$ and $-2y$ are terms.

3. The numerical part of a term is called its *numerical coefficient*. For the term $3x$, 3 is the numerical coefficient.

5. A *linear equation* is one in which the exponent on the variables is 1. An example of a linear equation is $3x + 4 = 2$.

7. If $a = b$, then $a - c = b - c$ for all real numbers a, b, and c. An example of the subtraction property of equality is: If $2x + 3 = 5$, then $2x + 3 - 3 = 5 - 3$.

9. If $a = b$, then $a/c = b/c$ for all real numbers a, b, and c, $c \neq 0$. An example of the division property of equality is: If $4x = 8$, then $\dfrac{4x}{4} = \dfrac{8}{4}$.

11. A ratio is a quotient of two quantities.

13. $12x$ **15.** $5x - 7$ **17.** $3x + 11y$ **19.** $-8x + 2$

21. $-5x + 3$ **23.** $2.6x + 6.4$ **25.** $\dfrac{1}{12}x - 2$

27. $10x - 5y + 4$ **29.** $10r + 6$ **31.** $-2.2x + 5$

33. $-\dfrac{7}{20}x + \dfrac{17}{15}$ **35.** $4.52x - 13.5$ **37.** 5 **39.** 1

41. $\dfrac{44}{9}$ **43.** $\dfrac{2}{3}$ **45.** 3 **47.** 3 **49.** 21 **51.** -15

53. No solution **55.** All real numbers

57. $-\dfrac{0.73}{0.34} \approx -2.1$ **59.** 3 **61.** 12 **63.** $61.78

65. a) 336 lb **b)** 12 bags

67. 384 mi **69.** 3.75 mm **71.** 0.3 cc

73. b) $x = -1$

75. a) An equation that has no solution.
b) When solving an equation, if you obtain a false statement such as $0 = 3$, the equation is inconsistent.

77. a) $2:5$ **b)** $m:(m + n)$

Section 6.3, Page 287

1. A formula is an equation that typically has a real-life application.

3. Subscripts are numbers (or letters) placed below and to the right of variables. They are used to help clarify a formula.

5. 330 **7.** 32 **9.** 4 **11.** 83 **13.** 37.1 **15.** 2

17. 2000 **19.** 5.57 **21.** 10 **23.** 2 **25.** 500 **27.** 233

29. $-\dfrac{3}{2}$ **31.** 4 **33.** 0.5 **35.** 14

37. $y = \dfrac{5x - 7}{3}$, or $y = \dfrac{5}{3}x - \dfrac{7}{3}$

39. $y = \dfrac{-4x + 14}{7}$, or $y = -\dfrac{4}{7}x + 2$

41. $y = \dfrac{2x + 6}{3}$, or $y = \dfrac{2}{3}x + 2$

43. $y = \dfrac{-5x - 14}{7}$, or $y = -\dfrac{5}{7}x - 2$

45. $y = \dfrac{9x + 4z - 7}{8}$, or $\dfrac{9}{8}x + \dfrac{1}{2}z - \dfrac{7}{8}$

47. $t = d/r$ **49.** $c = p - a - b$ **51.** $h = \dfrac{V}{lw}$

53. $d = \dfrac{C}{\pi}$ **55.** $b = y - mx$ **57.** $w = \dfrac{P - 2l}{2}$

59. $c = 3A - a - b$ **61.** $h = \dfrac{3V}{B}$ **63.** $C = \dfrac{5}{9}(F - 32)$

65. $x_1 = \dfrac{y_1 - y_2 + mx_2}{m}$ **67.** $2600 **69.** 6.54 in^3

71. 18.8 mg **73.** $\approx \$171{,}920{,}595{,}200{,}000$

75. 1051.5 in^3

Section 6.4, Page 292

1. A *mathematical expression* is a collection of variables, numbers, parentheses, and operation symbols. An *equation* is two algebraic expressions joined by an equals sign.

3. $4x + 9$ **5.** $5r - 13$ **7.** $11 - 2p$ **9.** $x + 7$

11. $\dfrac{7 + r}{9}$ **13.** $(5y - 6) + 3$ **15.** $x - 5 = 22, 27$

17. $5x = 95, 19$ **19.** $13 - 3x = 40, -9$

21. $3x = x - 12, -6$ **23.** $x + 11 = 3x + 1, 5$

25. $x + 3 = 5(x + 7), -8$ **27.** $x + 4x = 240, 48, 192$

29. $6200 + 500y = 12{,}200, 12$ yr

31. $x + 0.07x = 3.00, \$2.80$

33. $x + 3x = 28$, 7 tons of cardboard, 21 tons of newspaper

35. $0.20x + 60 = 100, 200$ mi

37. a) $x + x + 3x = 45{,}000, 9000, 9000, 27{,}000$
b) Yes

39. $x + \dfrac{1}{6}x = 203, 174$ lb

41. $2(w + 9) + 2w = 186, w = 42$ ft, $l = 51$ ft

43. $70x = 760, \approx 11$ months

45. $\dfrac{r}{2} + 0.07r = 227, \$426.13$

47. Deduct $720 from Mr. Neuland's income and $2920 from Mrs. Neuland's income.

49. $x + (x + 1) + (x + 2) = 3(x + 2) - 3$
$$3x + 3 = 3x + 6 - 3$$
$$3x + 3 = 3x + 3$$

Section 6.5, Page 300

1. In direct variation, as one quantity increases the other quantity increases, and vice versa.

3. In joint variation one quantity varies directly as the product of two or more other quantities.

5. Direct **7.** Direct **9.** Inverse **11.** Direct

13. Direct **15.** Inverse **17.** Inverse **19.** Inverse

23. a) $x = ky$ **b)** 72

25. a) $y = kR$ **b)** 306

27. a) $R = \dfrac{k}{W}$ **b)** $\dfrac{1}{20}$

29. a) $A = \dfrac{kB}{C}$ **b)** 9

31. a) $T = \dfrac{kD^2}{F}$ b) 51.2

33. a) $Z = kWY$ b) 420

35. a) $S = \dfrac{k}{G}$ b) 0.96

37. a) $x = \dfrac{k}{P^2}$ b) 0.9

39. a) $F = \dfrac{kq_1q_2}{d^2}$ b) 672

41. a) $V = \dfrac{k}{P}$ b) 6400 cc

43. a) $I = \dfrac{k}{d^2}$ b) 31.25 footcandles

45. a) $R = \dfrac{kA}{P}$ b) 4600 tapes

47. a) $W = kI^2R$ b) 40 watts

49. a) $N = \dfrac{kp_1p_2}{d}$ b) 121,528 calls

51. a) Inversely b) stays 0.3

53. $W = \dfrac{kTA\sqrt{F}}{R}$, $124.92

Section 6.6, Page 306

1. When both sides of an inequality are being multiplied or divided by a negative number, the order of the inequality must be reversed.

3. Yes

29. (number line with filled points at $-16, -15, -14, -13$)

31. (number line with filled points at $-21, -20, -19, -18$)

33. (number line with filled points at $-3, -2, -1, 0, 1, 2, 3$)

35. (number line with filled points at $-1, 0, 1, 2$)

37. (number line with filled points at $-5, -4, -3, -2$)

39. (number line with filled points at $0, 1, 2, 3, 4$)

41. (number line with filled points at $6, 7$)

43. $190 < 110 + 0.20x,\ x > 400$ miles

45. a) $180 + 60x \le 1200$ b) 17 boxes

47. $2x > 300 + 1.5x,\ x > 600$

49. $80 \le \dfrac{78 + 64 + 88 + 76 + x}{5} < 90;\ 94 \le x \le 100$, assuming 100 is the highest grade possible

51. $18{,}000 \le 8000 + 175x \le 24{,}000$, minimum 58, maximum 91

53. $81.6 \le x \le 100$, assuming that 100 is the highest grade possible

Section 6.7, Page 317

1. A *graph* is an illustration of all the points whose coordinates satisfy the equation.

3. To find the *y*-intercept, set $x = 0$ and solve the equation for *y*.

5. Plotting points, using intercepts, and using the slope and the *y* intercept

For Exercises 7, 9, 11, 13 see the following figure

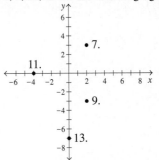

For Exercises 15, 17, 19, 21 see the following figure

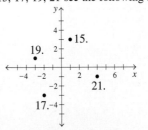

23. $(4, 3)$ **25.** $(-2, 4)$ **27.** $(-3, 1)$ **29.** $(-5, -3)$

31. $(2, -3)$

33. a) $(-3, -2)$ **b)** 42 square units

35. $(7, 2)$ or $(-1, 2)$ **37.** $b = -4$ **39.** $b = 3$

41. $(2, 4)$ **43.** $(0, 5), (-\frac{5}{2}, 0)$ **45.** $(0, \frac{4}{3}), (8, 0)$

47. $(-1, 1), (4, 3)$

49. 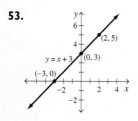   **51.**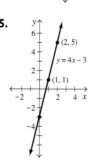

53. 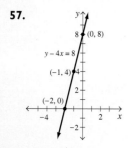   **55.**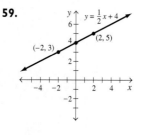

57. 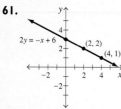   **59.**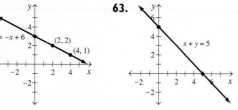

61. 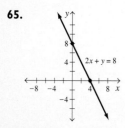   **63.**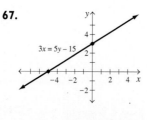

65. 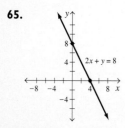   **67.**

69. 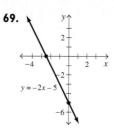   **71.**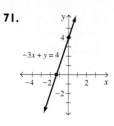

73. -1 **75.** $\dfrac{6}{5}$ **77.** 0 **79.** Undefined   **81.** $-\dfrac{4}{3}$

83. 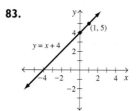   **85.**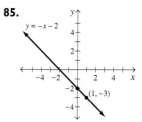

87. 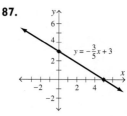   **89.**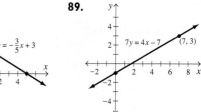

91.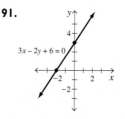

93. $y = \dfrac{1}{2}x - 1$ **95.** $y = -\dfrac{3}{4}x + 3$

97. a)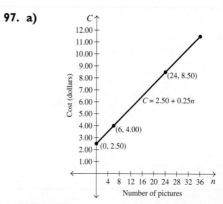

b) $7.50

c) 36 prints

99. a)

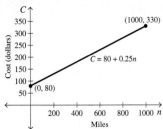

b) ≈$230.00 **c)** ≈400 mi

101. a) Solve the equations for y to put them in slope–intercept form. Then compare the slopes and y intercepts. If the slopes are equal but the y intercepts are different, then the lines are parallel.
  b) The lines are parallel.

Section 6.8, Page 321

1.

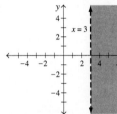

3.

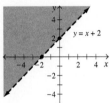

5.

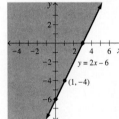

7.

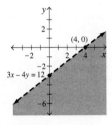

9.

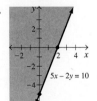

11.

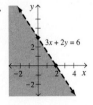

13.

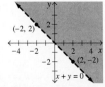

15.

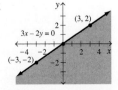

17.

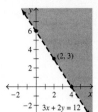

19.

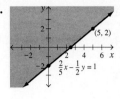

21.

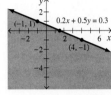

23. a) $2l + 2w \le 30,\ 0 \le l \le 15,\ 0 \le w \le 15$
  b)

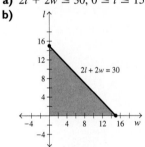

25. a) No **b)**

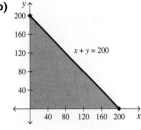

Section 6.9, Page 330

1. A *trinomial* is an expression containing three terms in which each exponent that appears on a variable is a whole number.

5. $(x + 2)(x + 6)$ **7.** $(x - 1)(x + 7)$

9. $(x + 6)(x - 4)$ **11.** $(x - 3)(x + 1)$

13. $(x - 3)(x - 7)$ **15.** $(x - 7)(x + 7)$

17. $(x - 4)(x + 7)$ **19.** $(x - 7)(x + 9)$

21. $(2x - 1)(x + 2)$ **23.** $(x - 5)(3x + 2)$

25. $(5x + 2)(x + 2)$ **27.** $(5x + 3)(x - 2)$

29. $(x - 2)(5x - 3)$ **31.** $(3x + 4)(x - 6)$ **33.** $6, -3$

35. $-\dfrac{5}{3}, \dfrac{3}{4}$ **37.** $-2, -4$ **39.** $3, 5$ **41.** $-3, 5$ **43.** $1, 3$

45. $-9, 9$ **47.** $-9, 4$ **49.** $-1, \dfrac{2}{3}$ **51.** $-\dfrac{1}{5}, -2$

53. $\frac{1}{3}$, 1   **55.** $\frac{1}{4}$, 2   **57.** 4, 5   **59.** $-2, 5$ **61.** $-1, 9$

63. No real solution   **65.** $\frac{5 \pm \sqrt{41}}{2}$ **67.** $\frac{13 \pm \sqrt{313}}{6}$

69. $\frac{1 \pm \sqrt{17}}{8}$ **71.** $-\frac{5}{2}, -1$ **73.** $1, \frac{7}{3}$

75. No real solution   **77.** 193

79. a) Two   **b)** One   **c)** None

Section 6.10, Page 339

1. Any set of ordered pairs

3. The set of values that can be used for the independent variable

5. If a vertical line touches more than one point on the graph, then for each value of x there is not a unique value for y, and the graph is not a function.

7. Function, domain: $x = -2, -1, 1, 2, 3$; range: $-1, 1, 2, 3$

9. Function, domain: all real numbers, $\mathbb{R}$; range: all real numbers, $\mathbb{R}$

11. Function, domain: $\mathbb{R}$; range: $y = 2$

13. Function, domain: $\mathbb{R}$; range: $y \geq -4$

15. Function, domain: $0 \leq x \leq 8$; range: $-1 \leq y \leq 1$

17. Function, domain: $0 \leq x < 12$; range: $y = 1, 2, 3$

19. Not a function   **21.** Function, domain: $\mathbb{R}$; range: $y > 0$

23. Yes   **25.** No   **27.** Yes   **29.** 3   **31.** 18   **33.** 15

35. 52   **37.** 4   **39.** -38 **41.** -48

43. a) Upward   **b)** $x = 0$ **c)** $(0, -9)$ **d)** $(0, -9)$
e) $(-3, 0), (3, 0)$
f)

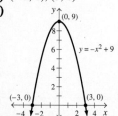

g) Domain: $\mathbb{R}$; range: $y \geq -9$

45. a) Downward   **b)** $x = 0$ **c)** $(0, 9)$ **d)** $(0, 9)$
e) $(-3, 0), (3, 0)$
f)

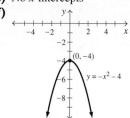

g) Domain: $\mathbb{R}$; range: $y \leq 9$

47. a) Downward   **b)** $x = 0$ **c)** $(0, -4)$ **d)** $(0, -4)$

e) No x-intercepts
f)

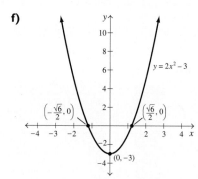

g) Domain: $\mathbb{R}$; range: $y \leq -4$

49. a) Upward   **b)** $x = 0$ **c)** $(0, -3)$ **d)** $(0, -3)$
e) $\left(\frac{-\sqrt{6}}{2}, 0\right), \left(\frac{\sqrt{6}}{2}, 0\right)$

f)

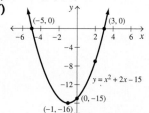

g) Domain: $\mathbb{R}$; range: $y \geq -3$

51. a) Upward   **b)** $x = -1$ **c)** $(-1, -16)$ **d)** $(0, -15)$
e) $(-5, 0), (3, 0)$
f)

g) Domain: $\mathbb{R}$; range: $y \geq -16$

53. a) Upward   **b)** $x = -\frac{5}{2}$ **c)** $(-2.5, -0.25)$
d) $(0, 6)$ **e)** $(-3, 0), (-2, 0)$
f)

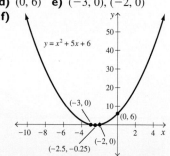

g) Domain: $\mathbb{R}$; range: $y \geq -0.25$

55. a) Downward **b)** $x = 2$ **c)** $(2, -2)$ **d)** $(0, -6)$
 e) No x intercepts
 f)

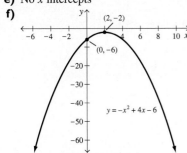

 g) Domain: $\mathbb{R}$; range: $y \leq -2$

57. a) Downward **b)** $x = \dfrac{7}{3}$ **c)** $\left(\dfrac{7}{3}, \dfrac{25}{3}\right)$ **d)** $(0, -8)$

 e) $\left(\dfrac{2}{3}, 0\right)$, $(4, 0)$
 f)

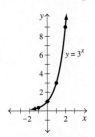

 g) Domain: $\mathbb{R}$; range: $y \leq \dfrac{25}{3}$

59. Domain: $\mathbb{R}$; range: $y > 0$

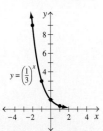

61. Domain: $\mathbb{R}$; range: $y > 0$

63. Domain: $\mathbb{R}$; range: $y > 1$

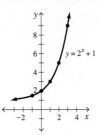

65. Domain: $\mathbb{R}$; range: $y > 1$

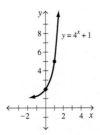

67. Domain: $\mathbb{R}$; range: $y > 0$

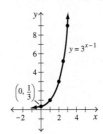

69. Domain: $\mathbb{R}$; range: $y > 0$

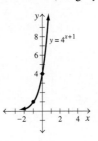

71. a) 6.4 **b)** 9.6 **c)** 12

73. a) 93.28 m **b)** 64.8 m

75. a) 1,073,741,824 **b)** \$10,737,418.24

77. a) $\approx$ \$116,328; the value of the house after 8 years is about \$116,328.
 b) 15 years

79. a) 241,800 mi **b)** $d = 11,160,000t$
 c) 92,628,000 mi

Review Exercises, Page 343

1. 20   **2.** -1 **3.** 69   **4** $\frac{1}{4}$ **5.** -65

6. 23   **7.** $2x - 2$ **8.** $11x - 15$ **9.** $3x - \frac{13}{2}$

10. 5   **11.** 7   **12.** 11   **13.** -31 **14.** $\frac{54}{5}$

15. $\frac{1}{2}$ cup   **16.** 375 min, or 6 hr 15 min   **17.** 36

18. 173.1   **19.** 101.5   **20.** 6

21. $y = \dfrac{3x - 6}{2}$, or $y = \dfrac{3}{2}x - 3$

22. $y = \dfrac{-3x + 8}{4}$, or $y = -\dfrac{3}{4}x + 2$

23. $y = \dfrac{2x + 22}{3}$, or $y = \dfrac{2}{3}x + \dfrac{22}{3}$

24. $y = \dfrac{-3x + 5z - 4}{4}$, or $y = -\dfrac{3}{4}x + \dfrac{5}{4}z - 1$

25. $l = \dfrac{A}{w}$ **26.** $w = \dfrac{P - 2l}{2}$

27. $l = \dfrac{L - 2wh}{2h}$, or $l = \dfrac{L}{2h} - w$ **28.** $d = \dfrac{a_n - a_1}{n - 1}$

29. $8 - 7x$ **30.** $5x - 3$ **31.** $6r + 7$ **32.** $\dfrac{8}{q} - 11$

33. $12 - 3x = 21; x = -3$

34. $3x + 8 = x - 6, x = -7$

35. $5(x - 4) = 45, x = 13$

36. $10x + 14 = 8(x + 12); x = 41$

37. $x + \dfrac{1}{3}x = 48{,}000$; Wesley's income is $12,000.

38. $10{,}000 + 8.50x = 85{,}000$, 8823 units

39. $x + (x + 40) = 200$; systolic = 120, diastolic = 80

40. $30x = 480$, 16 hr   **41.** 20   **42.** 1   **43.** 20

44. 426.7   **45.** 5 in.   **46.** $119.88   **47.** 400 ft

48. 200.96

49.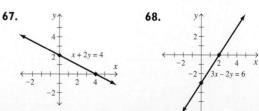

50.

51.

52.

53.

54.

55.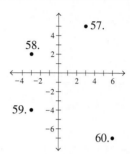

56.

For Exercises 57, 58, 59, 60 see the following figure

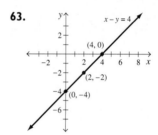

61. $D(-3, -1)$, Area = 20 square units

62. $D(4, 1)$, Area = 21 square units

63.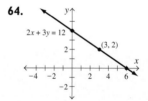

64.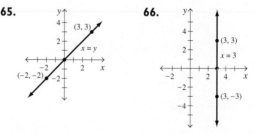

65.   **66.**

67.   **68.**

69.

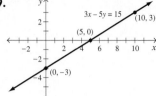

70.

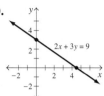

71. $-\dfrac{1}{4}$ **72.** $-\dfrac{5}{2}$ **73.** $\dfrac{7}{6}$ **74.** $\dfrac{1}{3}$

75.

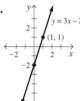

76.

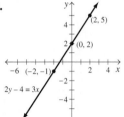

77.

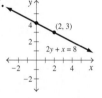

78.

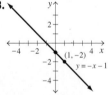

79. $y = 2x + 4$ **80.** $y = -x + 1$

81. a)

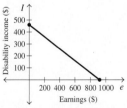

b) About \$160   **c)** About \$160

82. a)

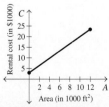

b) About \$6400   **c)** About 4120 ft²

83.

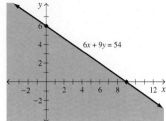

84.

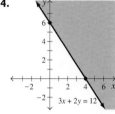

85.

   86.

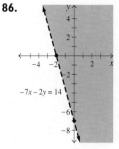

87. $(x + 3)(x + 7)$ **88.** $(x + 3)(x - 2)$

89. $(x - 6)(x - 4)$ **90.** $(x - 5)(x - 4)$

91. $(x - 3)(2x + 7)$ **92.** $(3x - 1)(x + 2)$

93. $-1, -2$ **94.** $1, 5$ **95.** $\dfrac{2}{3}, 5$ **96.** $-2, -\dfrac{1}{3}$

97. $\dfrac{5 \pm \sqrt{21}}{2}$ **98.** $1, 2$ **99.** No real solution

100. $-1, \dfrac{3}{2}$

101. Function, domain: $x = -2, -1, 2, 3$; range: $y = -1, 0, 2$

102. Not a function   **103.** Not a function

104. Function, domain: $\mathbb{R}$; range: $\mathbb{R}$ **105.** 3   **106.** 17

107. 39   **108.** -27

109. a) Downward **b)** $x = -2$ **c)** $(-2, 25)$
d) $(0, 21)$ **e)** $(-7, 0), (3, 0)$
f)

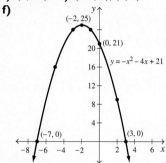

g) Domain: $\mathbb{R}$; range: $y \leq 25$

110. a) Upward **b)** $x = 4$ **c)** $(4, -78)$ **d)** $(0, -30)$
e) $(4 - \sqrt{26}, 0), (4 + \sqrt{26}, 0)$
f)

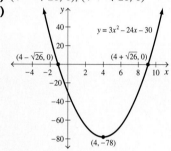

g) Domain: $\mathbb{R}$; range: $y \geq -78$

111. Domain: $\mathbb{R}$; range: $y > 0$

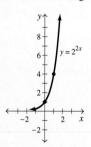

112. Domain: $\mathbb{R}$; range: $y > 0$

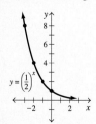

113. 22.8 mpg

114. a) 4208 **b)** 4250

115. 68.7%

Chapter Test, Page 347

1. 9 **2.** $\dfrac{19}{5}$ **3.** 18 **4.** $5x - 4 = 26, 6$

5. $7.75x - 4.35x = 60, \approx 18$ units **6.** 84

7. $y = \dfrac{5x - 17}{8}$, or $y = \dfrac{5}{8}x - \dfrac{17}{8}$ **8.** 1.125

9. 2.38 min

10.

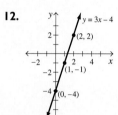

11. $-\dfrac{22}{15}$

12.

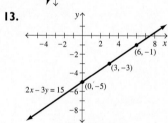

13.

14.

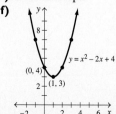

15. $-7, 5$ **16.** $\dfrac{4}{3}, -2$ **17.** It is a function. **18.** 11

19. a) Upward **b)** $x = 1$ **c)** $(1, 3)$ **d)** $(0, 4)$
e) No x-intercepts
f)

g) Domain: $\mathbb{R}$; range: $y \geq 3$

CHAPTER 7

Section 7.1, Page 357

1. Two or more linear equations form a system of linear equations.

3. A system of equations that has a solution

5. A *dependent system* of equations is one that has an infinite number of solutions.

7.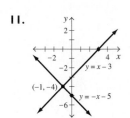

9.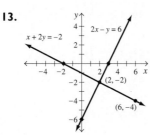

11.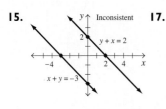

13.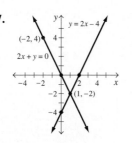

15.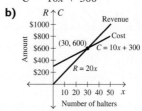

17.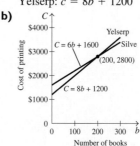

19.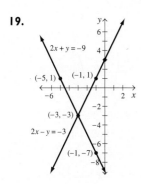

21.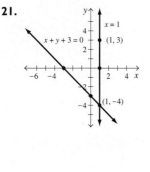

23.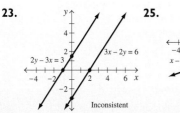

25.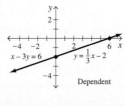

27. a) One unique solution; the lines intersect at one and only one point.
b) No solution; the lines do not intersect.
c) Infinitely many solutions; the lines coincide.

29. No solution   **31.** One solution

33. An infinite number of solutions   **35.** No solution

37. One solution   **39.** One solution   **41.** Perpendicular

43. Not perpendicular

45. a) Sivle: $c = 6b + 1600$
Yelserp: $c = 8b + 1200$

b)

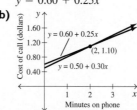

c) 200 books   **d)** Yelserp is less expensive.

47. a) $R = 20x$
$C = 10x + 300$

b)

c) 30 halters   **d)** Loss of $200   **e)** 130 halters

49. Solve each equation for y to determine the slope and y-intercept for each. If the slopes are not equal, the equations are consistent. If the slopes and the y-intercepts are equal, the equations are dependent. If the slopes are equal and the y-intercepts are not equal, the equations are inconsistent.

51. a) $y = 0.50 + 0.30x$
$y = 0.60 + 0.25x$

b)

c) 2 min

Section 7.2, Page 367

3. The system is dependent if the same value is obtained on both sides of the equals sign.

5. $(-4, -1)$ **7.** $(0, -3)$

9. No solution; inconsistent system   **11.** $(3, 3)$

13. An infinite number of solutions; dependent system

15. $(-5, 2)$

17. $\left(\dfrac{11}{5}, -\dfrac{13}{5}\right)$ **19.** $\left(\dfrac{2}{5}, -\dfrac{1}{5}\right)$

21. No solution; inconsistent system   **23.** $(1, -1)$

25. $\left(\dfrac{2}{3}, 2\right)$ **27.** $(-2, 0)$ **29.** $(-1, 8)$ **31.** $(-1, -2)$

33. $(1, -2)$ **35.** No solution; inconsistent system

37. $(5, -3)$

39. $p + w = 280$
$p - w = 240$
Speed of the plane in still air = 260 mph; speed of the wind = 20 mph.

41. $x + y = 45$
$2x + 3y = 101$
Thirty-four 2-point baskets and eleven 3-point baskets

43. $175 = x + 15y$
$200 = x + 20y$
The fixed cost is \$100, and the charge per mile is \$5.

45. $x + y = 30$
$6x + 8y = 200$
Mix 20 lb of the \$6 coffee with 10 lb of the \$8 coffee.

47. $x + y = 100$
$0.80x + 0.50y = 0.75(100)$
Mix $83\frac{1}{3}$ liters of 80% acid solution and $16\frac{2}{3}$ liters of 50% acid solution.

49. $0.10A + 0.20B = 20$; 80 g of Mix A, 60 g of Mix B
$0.06A + 0.02B = 6$

51. $c = 0.10x + 6$,   **a)** 40 checks   **b)** Bank B
$c = 0.20x + 2$

53. \$66.67

Section 7.3, Page 375

1. A *matrix* is a rectangular array of elements.

3. A *square matrix* contains the same number of rows and columns.

5. b) $\begin{bmatrix} -1 & 2 & 7 \\ -1 & 1 & -2 \end{bmatrix}$ **7. b)** $\begin{bmatrix} 11 & -14 \\ 10 & -15 \end{bmatrix}$

9. $\begin{bmatrix} 5 & 3 \\ 9 & 4 \end{bmatrix}$ **11.** $\begin{bmatrix} 2 & 4 \\ -6 & 9 \\ 4 & 8 \end{bmatrix}$ **13.** $\begin{bmatrix} 4 & 8 \\ 7 & -7 \end{bmatrix}$

15. $\begin{bmatrix} 1 & 0 & -7 \\ 9 & 8 & -7 \\ 6 & 1 & -9 \end{bmatrix}$ **17.** $\begin{bmatrix} 6 & 4 \\ 10 & 0 \end{bmatrix}$ **19.** $\begin{bmatrix} 0 & 13 \\ 22 & 0 \end{bmatrix}$

21. $\begin{bmatrix} 13 & 0 \\ 7 & 0 \end{bmatrix}$ **23.** $\begin{bmatrix} 12 & 21 \\ 8 & 22 \end{bmatrix}$ **25.** $\begin{bmatrix} 15 \\ 22 \end{bmatrix}$

27. $\begin{bmatrix} -5 & -8 \\ -4 & -5 \end{bmatrix}$

29. $\begin{bmatrix} 1 & 5 & 4 \\ 4 & 2 & 8 \end{bmatrix}$; cannot be multiplied.

31. Cannot be added, $A \times B = \begin{bmatrix} 26 & 38 \\ 24 & 24 \end{bmatrix}$

33. Cannot be added, $\begin{bmatrix} 1 \\ -1 \end{bmatrix}$

35. $A + B = B + A = \begin{bmatrix} 12 & 7 \\ 3 & 2 \end{bmatrix}$

37. $A + B = B + A = \begin{bmatrix} 8 & 0 \\ 6 & -8 \end{bmatrix}$

39. $(A + B) + C = A + (B + C) = \begin{bmatrix} 6 & -4 \\ 11 & 7 \end{bmatrix}$

41. $(A + B) + C = A + (B + C) = \begin{bmatrix} 5 & 5 \\ 7 & -37 \end{bmatrix}$

43. No   **45.** No   **47.** Yes

49. $(A \times B) \times C = A \times (B \times C) = \begin{bmatrix} 35 & 17 \\ 44 & 20 \end{bmatrix}$

51. $(A \times B) \times C = A \times (B \times C) = \begin{bmatrix} 16 & -10 \\ -24 & 2 \end{bmatrix}$

53. $(A \times B) \times C = A \times (B \times C) = \begin{bmatrix} 17 & 0 \\ -7 & 0 \end{bmatrix}$

55.   Small   Large
$\begin{bmatrix} 38 & 50 \\ 56 & 72 \\ 17 & 26 \\ 10 & 14 \end{bmatrix}$

57. [\$36.04   \$47.52]   **59.** No

61. Yes, $A \times B = B \times A = \begin{bmatrix} 1 & 0 \\ 0 & 1 \end{bmatrix}$ **63.** False

65. a) \$28.70   **b)** \$60.10   **c)** $\begin{bmatrix} 28.7 & 24.6 \\ 41.3 & 35.7 \\ 69.3 & 60.1 \end{bmatrix}$

Section 7.4, Page 383

1. An *augmented matrix* is a matrix formed with the coefficients of the variables and the constants. The coefficients of the variables are separated from the constants by a vertical bar.

3. $(2, -1)$ **5.** $(-2, -1)$ **7.** Dependent system

9. $(-2, 1)$ **11.** $(2, 1)$

13. No solution; inconsistent system   **15.** $(3, 5)$

17. Cherries: \$4 per pound; mints: \$5 per pound

19. Pizza: $15; beer: $4

21. Nonrefillable pencils: 125; refillable pencils: 75

Section 7.5, Page 386

1. The *solution set* of a system of linear inequalities is the set of points that satisfy all inequalities in the system.

3.

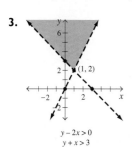

$$y - 2x > 0$$
$$y + x > 3$$

5.

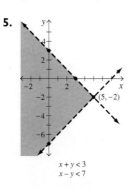

$$x + y < 3$$
$$x - y < 7$$

7.

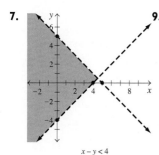

$$x - y < 4$$
$$x + y < 5$$

9.

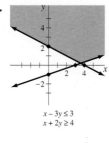

$$x - 3y \leq 3$$
$$x + 2y \geq 4$$

11.

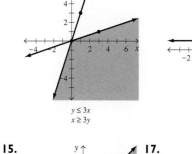

$$y \leq 3x$$
$$x \geq 3y$$

13.

$$x \geq 1$$
$$y \leq 1$$

15.

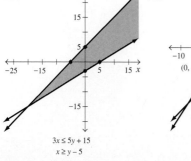

$$3x \leq 5y + 15$$
$$x \geq y - 5$$

17.

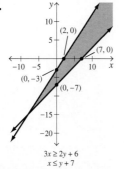

$$3x \geq 2y + 6$$
$$x \leq y + 7$$

19. a) $x + y < 500$, $x \geq 150$, $y \geq 150$

b)

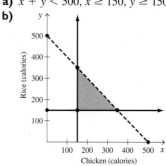

c) One example is (220, 220); approximately 3.7 oz of chicken, 8.8 oz of rice

21. a) No   **b)** One example is $x + y > 4$
$$x + y < 1$$

Section 7.6, Page 390

1. *Constraints* are restrictions that are represented as a linear inequality.

3. Vertices

5. a)

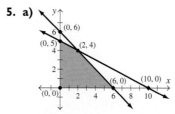

b) Max (2, 4) = 22
Min (0, 0) = 0

7. a)

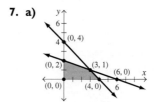

b) Max (3, 1) = 25
Min (0, 0) = 0

9. a)

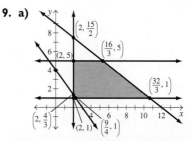

b) Max $\left(\dfrac{32}{3}, 1\right) = 25.12$

Min $\left(\dfrac{9}{4}, 1\right)$ or $\left(2, \dfrac{4}{3}\right) = 6.60$

11. a) $x + y \le 18$
$x \ge 2$
$x \le 5$
$y \ge 2$

b) $P = 20x + 25y$

c)

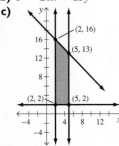

d) $(2, 2)$, $(2, 16)$, $(5, 13)$, $(5, 2)$
e) Type A: 2; Type B: 16
f) $440

13. a) $60x + 50y \ge 300$, $8x + 20y \ge 80$, $6x + 30y \ge 90$,
$x \ge 0$, $y \ge 0$
b) $C = 0.25x + 0.32y$
c)

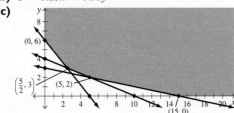

d) $(0, 6)$, $\left(\dfrac{5}{2}, 3\right)$, $(5, 2)$, $(15, 0)$

e) Trimfit: 2.5 cups; Usave: 3 cups
f) $1.59

15. Two 4-cylinder and seven 6-cylinder, $2050

Review Exercises, Page 392

1.

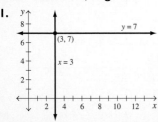

2.

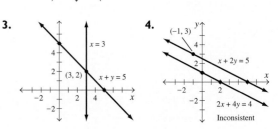

3.   **4.**

5. An infinite number of solutions   **6.** No solution

7. One solution   **8.** One solution   **9.** $(5, 1)$

10. $(0, 2)$ **11.** $(-2, -8)$ **12.** No solution; inconsistent

13. $(-9, 3)$ **14.** $(-7, 16)$ **15.** $(-1, -1)$ **16.** $(2, 0)$

17. $(30, -15)$

18. An infinite number of solutions; dependent

19. $\begin{bmatrix} -1 & -3 \\ -3 & 3 \end{bmatrix}$ **20.** $\begin{bmatrix} -3 & 9 \\ 11 & -13 \end{bmatrix}$ **21.** $\begin{bmatrix} -6 & 9 \\ 12 & -15 \end{bmatrix}$

22. $\begin{bmatrix} -7 & 24 \\ 29 & -34 \end{bmatrix}$ **23.** $\begin{bmatrix} -23 & 36 \\ 39 & -64 \end{bmatrix}$

24. $\begin{bmatrix} -26 & 33 \\ 46 & -61 \end{bmatrix}$ **25.** $(0, 2)$ **26.** $(-2, 2)$ **27.** $(1, 2)$

28. $(1, 0)$ **29.** $\left(\dfrac{12}{11}, \dfrac{7}{11}\right)$ **30.** $(3, 1)$

31. a) 100 months, or 8 yr 4 months   **b)** Moneywell

32. $\approx$ 25 gal of water to produce 1 lb of wheat, $\approx$ 2500 gal of water to produce 1 lb of meat

33. $500 salary, 4% commission rate

34. nickels: 5; dimes: 55

35. a) 32.5 months   **b)** Model 6070B

36.   **37.**

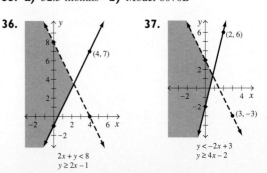

38.

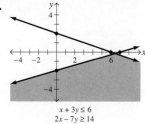

$$x + 3y \le 6$$
$$2x - 7y \ge 14$$

39.

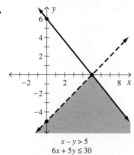

$$x - y > 5$$
$$6x + 5y \le 30$$

40. Max (9, 0) = 54

Chapter Test, Page 394

1. If the lines do not intersect (or are parallel) the system of equations is inconsistent. The system of equations is consistent if the lines intersect only once. If both equations represent the same line, then the system of equations is dependent.

2.

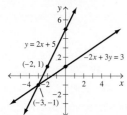

3. One solution   **4.** (−4, 3)   **5.** (6, 30)   **6.** (3, −1)

7. (−1, 3)   **8.** $\left(\frac{1}{2}, 1\right)$ **9.** (−2, 2)   **10.** $\begin{bmatrix} 7 & -7 \\ 7 & -3 \end{bmatrix}$

11. $\begin{bmatrix} 17 & -1 \\ 5 & -17 \end{bmatrix}$ **12.** $\begin{bmatrix} -2 & -34 \\ -17 & -25 \end{bmatrix}$

13. $y \le -2x + 4,\ 2x - y > 8$

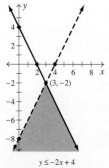

$$y \le -2x + 4$$
$$2x - y > 8$$

14. 8 lb of $5.40 per lb and 2 lb of $8.40 per lb

15. 12 one bedroom, 8 two bedroom

16. a)

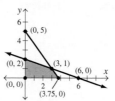

b) Max (3, 1) = 9
Min (0, 0) = 0

CHAPTER 8

Section 8.1, Page 402

1. The metric system

3. It is the standard of measurement accepted worldwide. There is only one basic unit of measurement for each quantity. It is based on the number 10, which makes many calculations easier.

5. b) 0.002 09 km   **c)** 92 400 dm

9. Yard   **11.** 5   **13.** (e)   **15.** (d)   **17.** (b)   **19.** Liter

21. Meter   **23.** Gram   **25.** Kilometer

27. Degree Celsius

29. a) 100   **b)** 0.001   **c)** 1000   **d)** 0.01   **e)** 10   **f)** 0.1

31. a) mg, 0.001 g

33. dg, 0.1 g   **35.** dag, 10 g   **37.** kg, 1000 g   **39.** 30

41. 2.4   **43.** 0.001 34   **45.** 14 270   **47.** 0.000 076

49. 0.013 hm   **51.** 1.049 m   **53.** 34.72 dℓ   **55.** 14 ℓ

57. 67 cm, 680 mm, 64 dm   **59.** 420 cℓ, 4.3 ℓ, 0.045 kℓ

61. 0.045 kℓ, 460 dℓ, 48 000 cℓ   **63.** Cindy

65. 1 hm in 10 min

67. a) 1.3 m   **b)** 130 cm

69. 310 cm   **71.** 1 hr 23.3 min   **73.** 3.2 km

75. a) 2160 mℓ   **b)** 2.16 ℓ   **c)** $1.13/ℓ

77. 1000

79. $1 \times 10^{24} = 1{,}000{,}000{,}000{,}000{,}000{,}000{,}000{,}000$

81. ≈ 30 eggs   **83.** ≈ 3.9 cups   **85.** 5 m   **87.** 4 dm

89. 5 hm

Section 8.2, Page 410

1. Meters or centimeters   **3.** Kilometers

5. Meters   **7.** Centimeters

9. Centimeters or meters   **11.** Millimeters

13. Kilometers   **17.** (a)   **19.** (a)   **21.** (a)   **23.** (a)

35. Kilometer   **37.** Meter   **39.** Kilometer

41. Square meters

43. Square millimeters or square centimeters

45. Square meters **47.** Square centimeters

49. Square centimeters **51.** (a) **53.** (b) **55.** (a)

57. (a) **59.** 1 000 000 mm^2 **61.** 100 hm^2

63. 0.0001 m^2

71. $r = 1.2$ cm, $C = 2.4\pi \approx 7.54$ cm,
$A = 1.44\pi \approx 4.52$ cm^2

73. a) 2881 m^2 **b)** 321 m^2

75. 0.022 11 ha **77.** Liters **79.** Milliliters

81. Liters **83.** Liters **85.** Milliliters **87.** (a)

89. (a) **91.** (c) **93.** (a)

95. b) 0.75 m^3

97. b) 0.25 m^3

101. 620 **103.** 76 **105.** 54 ℓ **107.** $304 **109.** 2.7 ℓ

111. 100 times larger **113.** 1000 times larger

115. 140,000,000 **117.** 279,936 in^3

Section 8.3, Page 418

1. Kilograms **3.** Grams **5.** Grams **7.** Tonnes

9. Grams **11.** (a) **13.** (c) **15.** (b) **23.** 281 g

25. a) 2304 m^3 **b)** 2304 kℓ **c)** 2304 t

27. 0.0012 t **29.** 42 600 000 **31.** (c) **33.** (b) **35.** (b)

37. (c) **39.** 86 **41.** 22.2 **43.** 176.7 **45.** 98.6

47. -10.6 **49.** 68 **51.** 45 **53.** Take an aspirin.

55. -40

Section 8.4, Page 425

1. 36 kg **3.** 1.26 m **5.** 12 m^2 **7.** 333.3 lb

9. 1687.5 acres **11.** 55.1 ℓ **13.** 1.52 fl oz **15.** 54 kg

17. 200 km **19.** 5280 km **21.** 37.5 mi/hr **23.** 400 mi

25. 360 m^3

27. a) ≈ 1.44 T **b)** ≈ 2888.9 lb

29. ≈ 9078.95 gal **31.** An ounce, a pound **33.** Ounce

35. 5 foot 2 **37.** 10 yd **39.** 0.025 g

41. a) 900 cm^2 **b)** 27 000 cm^3

43. 25.2 mg **45.** 6840 mg, or 6.84 g

47. 2250 mg, or 2.25 g **49.** $16.75

51. a) 25 mg **b)** 900 mg

53. a) 200 mg **b)** 0.35 mℓ **c)** 4800 mg

55. 0.12 ℓ graham crackers, 336 g nuts, 224 g chocolate, 0.32 ℓ coconut, 0.32 ℓ milk; use 22.86-cm $\times$ 33.02-cm baking pan; bake at 176.7°C; makes about 2 doz 3.81-cm $\times$ 7.62-cm bars.

57. b) 12.5 oz **c)** 375 mℓ

59. 1.0 cc

61. a) 2200 cm^3 **b)** 134.3 in^3

Review Exercises, Page 428

1. $\frac{1}{100}$ of base unit **2.** 1000 $\times$ base unit

3. $\frac{1}{1000}$ of base unit **4.** 100 $\times$ base unit

5. 10 times base unit **6.** $\frac{1}{10}$ of base unit **7.** 0.072 g

8. 2420 cℓ **9.** 9.97 mm **10.** 1 kg **11.** 4620 ℓ

12. 19 260 dg **13.** 3000 mℓ, 14 630 cℓ, 2.67 kℓ

14. 0.047 km, 47 000 cm, 4700 m **15.** Centimeters

16. Kilograms or grams **17.** Degrees Celsius

18. Millimeters **19.** Square meters

20. Milliliters or cubic centimeters **21.** Millimeters

22. Kilograms **23.** Kilometers **24.** Liters **27.** (c)

28. (b) **29.** (c) **30.** (a) **31.** (a) **32.** (b) **33.** 1.64 t

34. 6 300 000 g **35.** 82.4°F **36.** 20°C **37.** -21.1°C

38. 102.2°F

39. $l = 4$ cm, $w = 1.6$ cm, $P = 11.2$ cm, $A = 6.4$ cm^2

40. $b = 3.2$ cm, $h = 2.5$ cm, hypotenuse = 4.1 cm, $P = 9.8$ cm, $A = 4$ cm^2

41. a) 64 m^3 **b)** 64 000 kg

42. a) 660 m^2 **b)** 0.000 66 km^2

43. a) 96 000 cm^3 **b)** 0.096 m^3 **c)** 96 000 mℓ
d) 0.096 kℓ

44. 10,000 times larger **45.** 38.1 cm **46.** ≈ 233.3 lb

47. 198 m **48.** ≈ 111.1 yd **49.** 88 km/hr **50.** 81 kg

51. 57 ℓ **52.** ≈ 52.6 yd^3 **53.** 2.25 kg **54.** 3.8 ℓ

55. 11.4 m^3 **56.** 240 km **57.** 0.9 ft **58.** 82.55 mm

59. a) 1050 kg **b)** ≈ 2333.3 lb

60. 512 km **61.** 32.4 m^2

62. a) 1154 km **b)** 721.25 mi

63. a) 56 km/hr **b)** 56 000 m/hr

64. a) 252 ℓ **b)** 252 kg

Chapter Test, Page 430

1. 0.004 06 hm **2.** 14 800 000 mg

3. 100 times greater **4.** 1.8 km **5.** (b) **6.** (a)

7. (c) **8.** (c) **9.** (b) **10.** 10,000 times greater

11. 1,000,000,000 times greater **12.** ≈ 339.4 in.

13. 104 km/hr **14.** ≈ 16.7°C **15.** 86°F

16. ≈ 360 cm or 365.8 cm, depending on which conversion factor you used

17. a) 3200 m^3 **b)** 3 200 000 ℓ (or 3200 kℓ)
c) 3 200 000 kg

18. $245

19. a) A unit fraction is a fraction in which the numerator and denominator contain different units and the value of the fraction is 1.
c) 307.775 keruts **d)** ≈ 0.486

CHAPTER 9

Section 9.1, Page 441

1. Axioms are statements that are accepted as being true on the basis of their "obviousness" and their relation to the physical world. Theorems are statements that have been proved by using undefined terms, definitions, and axioms.

3. A half-line doesn't include the endpoint. A ray includes the endpoint.

5. Two lines that do not lie in the same plane and do not intersect are called skewed lines.

7. Two angles the sum of whose measures is 90° are called complementary angles.

9. An angle whose measure is 90° is a right angle.

11. An angle whose measure is between 90° and 180° is an obtuse angle.

13. Line segment, $\overline{AB}$ **15.** Open line segment, $\overset{\circ\!\!\!-\!\!\!\circ}{AB}$

17. Half-line, $\overset{\circ\!\!\!\rightarrow}{BA}$ **19.** $\angle BFC$ **21.** $\overrightarrow{BD}$ **23.** $\{C\}$

25. $\overline{BC}$ **27.** $\overrightarrow{AD}$ **29.** { } **31.** { } **33.** $\overleftrightarrow{BC}$

35. $\angle ABE$ **37.** $\angle CBE$ **39.** $\overleftrightarrow{AC}$ **41.** $\overset{\circ\!\!\!-\!\!\!\circ}{BE}$

43. Straight **45.** Obtuse **47.** 47° **49.** $22\frac{1}{2}°$

51. 89° **53.** 86° **55.** 65° **57.** $90\frac{3}{4}°$ **59.** d

61. b **63.** c **65.** 33°, 57° **67.** 20°, 160°

69. $\angle 1 = 48°$, $\angle 2 = 48°$, $\angle 3 = 132°$, $\angle 4 = 132°$, $\angle 5 = 48°$, $\angle 6 = 48°$, $\angle 7 = 132°$

71. $\angle 1 = 22°$, $\angle 2 = 68°$ **73.** $\angle 1 = 78.5°$, $\angle 2 = 11.5°$

75. $\angle 1 = 135.5°$, $\angle 2 = 44.5°$

77. $\angle 1 = 26°$, $\angle 2 = 154°$ **79.** A line **81.** A point

83. An angle, $\angle BAC$ **85.** HGB, GDC, and IDG

87. $\overline{GB}$ and GFE **89.** $\overleftrightarrow{AB}$ and $\overrightarrow{GD}$

91. $GHB \cap GDC \cap GED \cap GID$

93. It laid the foundation for Euclidean (or plane) geometry.

95. Sometimes true **97.** Sometimes true

99. Never true

101.

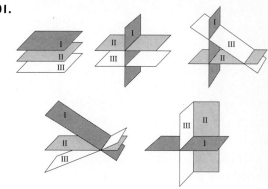

Section 9.2, Page 448

1. Two polygons are similar if their corresponding angles have the same measure and their corresponding sides are in proportion.

3. Hexagon **5.** Pentagon **7.** Quadrilateral

9. Scalene **11.** Scalene **13.** Acute **15.** Obtuse

17. Right **19.** Parallelogram **21.** Rhombus **23.** 17°

25. 150°

27. a)

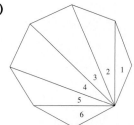

b) 180° × 6 = 1080° **c)** (8 − 2)180° = 1080°

29. 1260° **31.** 540° **33.** 60°; 120° **35.** 128.6°; 51.4°

37. 108°; 72° **39.** 55° **41.** 35° **43.** 80° **45.** 100°

47. $\overline{EF} = \frac{16}{5}$, or 3.2, $\overline{DF} = 4$ **49.** $\overline{BC} = 6$, $\overline{DF} = 3.2$

51. 6 **53.** $\frac{25}{12}$ **55.** $\frac{12}{5}$ **57.** 120° **59.** 60° **61.** 7

63. 14 **65.** 28° **67.** 4 **69.** 8 **71.** 70°

73. a) $\angle CED = \angle CBA$
   $\angle EDC = \angle BAC$
   $\angle ECD = \angle BCA$
b) 96 ft

75. a) Corresponding angles are equal, $m\angle F = m\angle B$, $m\angle HMF = m\angle TMB$, and $m\angle H = m\angle T$
b) 44 ft

Section 9.3, Page 458

1. 27 in^2 **3.** 2.5 yd^2 **5.** 32 in^2, 26 in.

7. 36 ft^2, 24 ft

9. 6000 cm^2 or 0.6 m^2, 6.54 m or 654 cm

In Exercises 11–14, we used the π key on the calculator to determine the answer. If you use 3.14 for π, your answer may vary slightly.

11. ≈ 50.27 in^2, ≈ 25.13 in.   **13.** 38.48 ft^2, ≈ 21.99 ft

15. 25 ft   **17.** $\sqrt{99}$ ≈ 9.95 m   **19.** 21.46 m^2 **21.** 8 in^2

23. 15.27 m^2 **25.** 6 mm^2 **27.** 1.26 yd^2 **29.** 164.7 ft^2

31. 156,000 cm^2 **33.** 0.0608 m^2 **35.** $100,200

37. $837.20   **39.** 53.3 gal

41. a) ≈ 17,518 ft^2 **b)** 3.5 bags
   c) $119.80, assuming that they purchase four bags

43. It becomes nine times as large.

45. Its area is doubled.   **47.** The area remains the same.

49. The area remains the same.

51. a) 177.1 m^2 **b)** 0.01771 ha

53. The round burger from Burger King is about 0.62 in^2 larger.

55. a) 2410 ft^2 **b)** 9.64 gal
   c) $249.50 plus tax, assuming that 10 gal are purchased

57. 94 ft^2 **59.** $lw - 2s^2$ **61.** 24 cm^2

Section 9.4, Page 467

3. A prism is a special type of polyhedron whose bases are congruent polygons and whose sides are parallelograms. A pyramid has only one base and its sides are triangles.

5. 4 ft^3

7. 75.36 in^3 (75.40 in^3, using the calculator value of π)

9. 2400 in^3

11. 292.5 ft^3 **13.** 900 in^3

15. 30.51 ft^3 (30.49 ft^3, using the calculator value of π)

17. 31.4 m^3 **19.** 24 ft^3 **21.** 108 ft^3 **23.** 7.85 yd^3

25. 1,500,000 cm^3 **27.** 4 m^3

29. a) 28,750 in^3 **b)** 16.64 ft^3

31. ≈ 2.5 qt

33. a) 120,000 cm^3 **b)** 120,000 mℓ   **c)** 120 ℓ

35. a) 87.5 ft^3 **b)** 3.24 yd^3

37. 82,944,000 ft^3 **39.** ≈ 0.904 mm^3

41. a) Area of round pan, 63.6 in^2; area of rectangular pan, 63 in^2
   b) Volume of round pan, 127.2 in^3; volume of rectangular pan, 126 in^3
   c) The round pan has the largest volume.

43. 6 cuts   **45.** 4 faces   **47.** 6 vertices   **49.** 14 edges

51. Increases the volume by eight-fold

53. Increase the volume by eight-fold

55. a) 36.75 ft^3 **b)** 2296.9 lb   **c)** 276.7 gal

57. 10 rolls

59. The grapefruit with the 7-cm radius is the better buy; it costs less per cubic cm.

61. 25 920 mℓ, or 25.92 ℓ

Section 9.5, Page 473

1. A vertex is a designated point.

3. To determine whether a vertex is odd or even, count the number of arcs attached to the vertex. If the number of arcs is odd, the vertex is odd. If the number of arcs is even, the vertex is even.

5. 5 vertices, 7 arcs   **7.** 7 vertices, 11 arcs

9. Each graph has the same number of arcs from the corresponding vertices.

11. Yes; start at C and end at D, or start at D and end at C.

13. Yes; start at any point and end where you started.

15. No; 4 odd vertices.

17. Yes; start at A and end at C, or start at C and end at A.

19. a) 0 odd, 5 even   **b)** Yes
   c) Start in any room and end where you began.

21. a) 2 odd, 4 even   **b)** Yes
   c) Start at B and end at F, or start at F and end at B.

23. a) 4 odd, 1 even   **b)** No

25. a) 3 odd, 2 even   **b)** No

27. The door must be placed in room D.

29. Yes; two odd vertices. Begin at either the island on the left or on the right and end at the other island.

31. No, it is not possible, assuming that your starting and ending points are considered vertices.

33. a) Kentucky, Virginia, North Carolina, Georgia, Alabama, Mississippi, Arkansas, Missouri
   b) Illinois, Arkansas, Tennessee

35. a) 4   **b)** 4   **c)** 11

37.

39. Arcs = Vertices + Regions − 2

Section 9.6, Page 478

1. A Mobius strip is a one-sided, one-edged surface.

3. An object that resembles a bottle but has only one side

5. Two figures are topologically equivalent if one figure can be elastically twisted, bent, shrunk, or straightened into the other.

7. One edge   **9.** One strip

11. a) No   **b)** 2   **c)** 2
   d) Two strips, one inside the other

13. No, it does not.

15. a) **b)**

17. a) 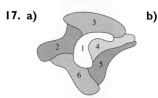 **b)**

19. a)  **b)**

21. Inside of curve

23. Two **25.** One **27.** Three **29.** Three or more (five)

Section 9.7, Page 485

7. a) The sum of the measures of the angles of a triangle is 180°.
   b) The sum of the measures of a triangle is less than 180°.
   c) The sum of the measures of a triangle is greater than 180°.

9. A sphere

11. Each type of geometry can be used in its own frame of reference.

13. Positive curvature: Riemannian geometry
   Zero curvature: Euclidean geometry
   Negative curvature: hyperbolic geometry

15. a)

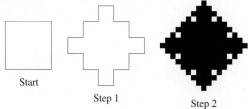

Start   Step 1   Step 2

   b) Infinite **c)** Finite

17. a) ℓ_2 is parallel to ℓ_3
   b) The sum of the measures of the angles is less than 180°.

Review Exercises, Page 488

1. $\{B\}$ **2.** $\overline{BC}$ **3.** $\triangle BFC$ **4.** $\{F\}$ **5.** $\overleftrightarrow{BH}$

6. $\{\ \}$ **7.** 2 in. **8.** 10.2 in. **9.** 92° **10.** 58°

11. $\angle 1 = 65°$, $\angle 2 = 65°$, $\angle 3 = 115°$, $\angle 4 = 115°$, $\angle 5 = 65°$, $\angle 6 = 65°$

12. 30 cm² **13.** 49 in² **14.** 13 in² **15.** 84 in²

16. 113.10 cm² **17.** 216 cm³ **18.** 402.12 in²

19. 66.09 mm² **20.** 28 ft³ **21.** 540 m³ **22.** 3140 mm³

23. 960 ft³ **24.** $483.47

25. a) 67.9 ft³
   b) Total weight: 4618.8 lb; Yes, it will support the trough filled with water.
   c) 511.3 gal

26. Yes, start at vertex C or E.

27. No; more than two odd vertices

28. Yes, start at any vertex.

29. Yes, start at vertex C or E.

30. No; more than two odd vertices

32. Euclidean: Given a line and a point not on the line, one and only one line can be drawn parallel to the given line through the given point.
   Elliptical: Given a line and a point not on the line, no line can be drawn through the given point parallel to the given line.
   Hyperbolic: Given a line and a point not on the line, two or more lines can be drawn through the given point parallel to the given line.

Chapter Test, Page 490

1. $\overset{\circ}{\overline{EF}}$ **2.** $\triangle BCD$ **3.** $\{D\}$ **4.** $\overleftrightarrow{AC}$ **5.** 36.8°

6. 100.2° **7.** 37° **8.** 720° **9.** ≈ 5.38 cm

10. a) 20 in. **b)** 60 in. **c)** 150 in²

11. ≈ 1767.15 cm³ **12.** 4550.8 ft³ **13.** 96 ft³

14. Yes, start at any vertex. **15.** No

CHAPTER 10

Section 10.1, Page 499

1. A binary operation is an operation, or rule, that can be performed on two and only two elements of a set. The result is a single element.

3. a) When we add two numbers, the sum is one number: $4 + 5 = 9$.
   b) When we subtract two numbers, the difference is one number: $5 - 4 = 1$.
   c) When we multiply two numbers, the product is one number: $5 \times 4 = 20$.
   d) When we divide two numbers, the quotient is one number: $20 \div 5 = 4$.

5. An identity element is an element in a set such that when a binary operation is performed on it and any given element in the set, the result is the given element. The additive identity element is 0, and the multiplicative identity element is 1.

7. $(a + b) + c = a + (b + c)$, for any real numbers a, b, and c; $(3 + 4) + 5 = 3 + (4 + 5)$.

9. $a + b = b + a$, for any real numbers a and b; $3 + 4 = 4 + 3$.

11. A specific example illustrating that a specific property is not true is called a counterexample.

13. $12 \div 4 \neq 4 \div 12$

$$3 \neq \frac{1}{3}$$

15. $(18 \div 6) \div 3 \neq 18 \div (6 \div 3)$

$$3 \div 3 \neq 18 \div 2$$

$$1 \neq 9$$

17. A mathematical system is a commutative group if all five conditions hold.
  1. The set of elements is closed under the given operation.
  2. An identity element exists for the set.
  3. Every element in the set has an inverse.
  4. The set of elements is associative under the given operation.
  5. The set of elements is commutative under the given operation.

19. No; no identity element   **21.** Yes

23. No; not associative   **25.** No; not closed   **27.** Yes

Section 10.2, Page 507

3. 12   **5.** 9   **7.** 11   **9.** 7   **11.** 4   **13.** 3   **15.** 4

17. 9   **19.** 6

21. If the first number is greater than the second, subtract to obtain the answer. Otherwise add 12 to the first number and then subtract to obtain the answer. For example, $6 - 9 = (6 + 12) - 9 = 18 - 9 = 9.$

23. 1   **25.** 7   **27.** 4   **29.** 4   **31.** 4   **33.** 6

35. Yes; satisfies the five required properties

37. a) $\{0, 1, 2, 3\}$ **b)** $*$ **c)** Yes   **d)** Yes; 0
  e) Yes: 0–0, 1–3, 2–2, 3–1
  f) $(2 * 3) * 1 = 2 * (3 * 1)$ **g)** Yes; $2 * 3 = 3 * 2$
  h) Yes

39. a) $\{2, 3, 7, 11\}$ **b)** ∞ **c)** Yes   **d)** Yes; 7
  e) Yes; 2–2, 3–11, 7–7, 11–3
  f) $(2 \infty 3) \infty 11 = 2 \infty (3 \infty 11)$
  g) Yes; $2 \infty 3 = 3 \infty 2$ **h)** Yes

41. a) No; no identity element
  b) $(1 @ 3) @ 4 \neq 1 @ (3 @ 4)$

43. Not associative: $(\triangle \ 3 \ \square) \ 3 \ \triangle \neq \triangle \ 3 \ (\square \ 3 \ \triangle)$, not commutative: $\square \ 3 \ \triangle \neq \triangle \ 3 \ \square$

45. No inverse for $\sim$ or for $*$, not associative

47. No identity element, no inverses, not associative, not commutative

49. a)

+	E	O
E	E	O
O	O	E

  b) Yes; satisfies the five properties

53. a) Is closed; identity element is 6; inverses: 1–5, 2–2, 3–3, 4–4, 5–1, 6–6; is associative—for example, $(2 \infty 5) \infty 3 = 2 \infty (5 \infty 3)$

$$3 \infty 3 = 2 \infty 2$$

$$6 = 6$$

  b) $3 \infty 1 \neq 1 \infty 3$

$$2 \neq 4$$

55. $5^3 = 125$

Section 10.3, Page 515

1. A modulo m system consists of m elements, 0 through $m - 1$, and a binary operation.

3. In any modulo system, modulo classes are developed by placing all numbers with the same remainder in the same modulo class.

5. Saturday   **7.** Friday   **9.** Friday   **17.** 0   **19.** 2

21. 0   **23.** 0   **25.** 1   **27.** 2   **29.** 6   **31.** 3   **33.** 0

35. 6   **37.** 1   **39.** 9   **41.** 3   **43.** 2   **45.** 4   **47.** 5

49. { }   **51.** 6   **53.** 7

55. a) 2000, 2004, 2008, 2012, 2016   **b)** 3004
  c) 2552, 2556, 2560, 2564, 2568, 2572

57. a) 3 P.M.–11 P.M.   **b)** 7 A.M.–3 P.M.
  c) 7 A.M.–3 P.M.

59. a)

+	0	1	2	3
0	0	1	2	3
1	1	2	3	0
2	2	3	0	1
3	3	0	1	2

  b) Yes   **c)** Yes, 0   **d)** Yes, 0–0, 1–3, 2–2, 3–1
  e) $(1 + 2) + 3 = 1 + (2 + 3)$
  f) Yes, $2 + 3 = 3 + 2$ **g)** Yes   **h)** Yes

61. a)

×	0	1	2	3
0	0	0	0	0
1	0	1	2	3
2	0	2	0	2
3	0	3	2	1

  b) Yes   **c)** Yes, 1
  d) No; no inverse for 0 or for 2, inverse of 1 is 1, inverse of 3 is 3
  e) $(1 \times 2) \times 3 = 1 \times (2 \times 3)$
  f) Yes, $2 \times 3 = 3 \times 2$ **g)** No

63. 2   **65.** 1, 2, 3   **67.** 5   **69.** 4   **71.** math is fun.

73. 20

Review Exercises, Page 517

1. A set of elements and a binary operation.   **2.** 6

3. 1   **4.** 8   **5.** 9   **6.** 9   **7.** 9   **8.** 2

9. Closure, identity element, inverses, associative property

10. A commutative group   **11.** Yes

12. No; no inverse for any integer except 1   **13.** Yes

14. No; no inverse for 0   **15.** No identity element

16. Not associative. For example, $(* : P) : ? \neq * : (p : ?)$

17. Not every element has an inverse; not associative. For example, $(P \, ? \, P) \, ? \, 4 \neq P \, ? \, (P \, ? \, 4)$

18. a) $\{1, 0, ?, \triangle\}$ **b)** $\otimes$ **c)** Yes **d)** Yes, 1
  e) Yes; 1–1, $\bigcirc$–$\triangle$, ?–?, $\triangle$–$\bigcirc$
  f) $(\bigcirc \otimes ?) \otimes \triangle = \bigcirc \otimes (? \otimes \triangle)$
  g) Yes; $\bigcirc \otimes ? = ? \otimes \bigcirc$ **h)** Yes

19. 2 **20.** 1 **21.** 5 **22.** 3 **23.** 9 **24.** 3 **25.** 1

26. 13 **27.** 9 **28.** 9 **29.** 6 **30.** 3 **31.** 3 **32.** 5

33. 0, 2, 4, 6 **34.** 7 **35.** 5 **36.** 3 **37.** 2 **38.** 8

39.

+	0	1	2	3	4	5
0	0	1	2	3	4	5
1	1	2	3	4	5	0
2	2	3	4	5	0	1
3	3	4	5	0	1	2
4	4	5	0	1	2	3
5	5	0	1	2	3	4

Yes, a commutative group

40.

×	0	1	2	3
0	0	0	0	0
1	0	1	2	3
2	0	2	0	2
3	0	3	2	1

No; no inverse for 0

41. a) No, she will be off.
  b) Yes, she will have the evening off.

Chapter Test, Page 518

1. A set of elements and a binary operation

2. Closure, identity element, inverses, associative property, commutative property

3. No, not all elements have inverses.

4.

+	1	2	3	4
1	2	3	4	1
2	3	4	1	2
3	4	1	2	3
4	1	2	3	4

5. Yes, it is a commutative group.   **6.** No; not closed

7. Yes, it is a commutative group.

8. Yes, it is a commutative group. **9.** 1 **10.** 3 **11.** 1

12. 3 **13.** 5 **14.** 2 **15.** { } **16.** 5

17. a)

×	0	1	2	3	4
0	0	0	0	0	0
1	0	1	2	3	4
2	0	2	4	1	3
3	0	3	1	4	2
4	0	4	3	2	1

b) No; no inverse for 0

CHAPTER 11

Section 11.1, Page 528

1. A percent is a ratio of some number to 100.

3. Multiply the decimal number by 100 and add a percent sign.

5.
$$\text{Percent change} = \frac{\left(\begin{array}{c}\text{Amount in} \\ \text{latest period}\end{array}\right) - \left(\begin{array}{c}\text{Amount in} \\ \text{previous period}\end{array}\right)}{\text{Amount in previous period}}$$

7. 75.0% **9.** 87.5% **11.** 56.9% **13.** 1367.8%

15. 0.4% **17.** 0.354 **19.** 0.000005 **21.** 0.00375

23. 1.359 **25.** 18.8%

27. a) 76.9% **b)** 23.1%

29. 3% **31.** 13.6%

33. a) 13.6% **b)** 27.2%

35. a) 125 **b)** 41.4% decrease

37. a) 374 **b)** 35.0% decrease

39. $0.01x(75) = 15$, 20% **41.** $1.34(400) = x$, 536

43. $0.20x = 11$, 55 **45.** $0.07(43.75) = x$, $3.06

47. $0.3x = 57$, 190 **49.** $0.17(300) = x$, 51

51. 25.9% increase **53.** 118.4% markup

55. 50% increase **57.** 50.3% markup

59. The sale price is not consistent with the ad. The sale price was $2.25 higher.

61. He will have a loss of $10.

63. a) No, the 25% discount is greater. **b)** $145.34
  c) $142.49 **d)** Yes

Section 11.2, Page 536

1. The amount of money that a bank is willing to lend to a person is called credit.

3. A cosigner is a person, other than the person who received the loan, who guarantees that a loan will be repaid.

5. Interest is the money the borrower pays for the use of the lender's money.

7. The difference between ordinary interest and interest calculated using the Banker's rule is the way in which time is used in the simple interest formula. Ordinary interest: A month is 30 days, and a year is 360 days. Banker's rule: Any fractional part of a year is the exact number of days, and a year is 360 days.

9. $37.80 **11.** $8.75 **13.** $15.85 **15.** $38.26

17. $1338.96 **19.** $600 **21.** $1\frac{1}{3}$ yr **23.** 3.2% **25.** 13%

27. a) $26.25 **b)** $1526.25

29. a) $135.42 **b)** $2364.58 **c)** 13.7%

31. a) $437.50 **b)** $5\frac{1}{4}$% **c)** $359.19

33. 121 days **35.** 226 days **37.** 134 days

39. October 6 **41.** October 31 **43.** $2254.17

45. $6301.63 **47.** $712.45 **49.** $3533.18 **51.** $1256.14

53. a) $205.25 first month, $203.67 second month, $202.00 third month
b) $10.92

Section 11.3, Page 543

1. An investment is the use of money nor capital for income or profit.

3. With a variable investment neither the principal nor the interest is guaranteed.

5. The effective annual yield on an investment is the simple interest rate that gives the same amount of interest as a compound rate over the same period of time.

7. a) $4326.40 **b)** $326.40

9. a) $10,145.93 **b)** $2145.93

11. a) $2252.99 **b)** $252.99

13. a) $2533.54 **b)** $533.54

15. a) $1935.63 **b)** $435.63

17. a) $11,232.95 **b)** $4232.95

19. a) $10,295.89 **b)** $2795.89

21. a) $5216.73 **b)** $716.73

23. a) $4195.14 **b)** $4214.36

25. $3514.98 **27.** $7609.45

29. a) $1040.60, $40.60 **b)** $1082.43, $82.43
c) $1169.86, $169.86 **d)** no

33. 7.714% **35.** $12,015.94 **37.** $7062.61 **39.** 11.7%

41. 8.84% simple rate **43.** $27,550.11

45. $p = A\left(1 + \dfrac{r}{n}\right)^{-nt}$

Section 11.4, Page 554

3. The finance charge is the total amount of money the borrower must pay for the use of money.

5. The Actuarial method and the rule of 78's

7. a) $274.00 **b)** 13.75%

9. a) $88 **b)** 13.00%

11. $3.89

13. a) $9.38 **b)** $18.68 **c)** 12.75% **d)** 13.50%

15. 10.75%

17. a) 11.25% **b)** $83.70 **c)** $1469.41

19. a) $24.64 **b)** $918.33

21. a) $0.78 **b)** $286.76

23. a) $1585.37 **b)** $20.61 **c)** $1862.97

25. a) The average daily balance is $1504.72, or $80.65 less.
b) The finance charge is $19.56, or $1.05 less.

27. a) 6 months **b)** $74.62
c) They pay the least amount of interest using the credit card. They save $13.38.

Section 11.5, Page 564

1. A mortgage is a long-term loan for which the property is pledged as security for payment of the difference between the down payment and the sale price.

7. An add-on rate, or margin, is the percent added to the interest rate on which the adjustable rate mortgage is based.

9. a) $13,750.00 **b)** $302.78

11. a) $9750.00 **b)** $55,250.00 **c)** $1105.00

13. a) $673.40 **b)** Yes

15. a) $187,736.40 **b)** $112,736.40 **c)** $38.68

17. a) $11,600.00 **b)** $928.00 **c)** $381.41
d) $172,720.20 **e)** $113,792.20 **f)** $14.08

19. a) $19,000.00 **b)** $1082.20 **c)** $624.72
d) $777.72 **e)** Yes **f)** $23.05 **g)** $281,382.40
h) $186,382.40

21. a) $630.58 **b)** 8.9% **c)** 8.1%

23. a) $805.00
b)

Payment Number	Interest	Principal	Balance of Loan
1	$750.00	$55.00	$99,945.00
2	$749.59	$55.41	$99,889.59
3	$749.17	$55.83	$99,833.76

c) 9.38%
d)

Payment Number	Interest	Principal	Balance of Loan
4	$780.37	$24.63	$99,809.13
5	$780.17	$24.83	$99,784.30
6	$779.98	$25.02	$99,759.28

e) 9.46%

Review Exercises, Page 567

1. 12.5% **2.** 33.3% **3.** 50.0% **4.** 4.5% **5.** 0.8%

6. 125.6% **7.** 0.37 **8.** 0.138 **9.** 0.00237

10. 0.004 **11.** 0.008$\overline{3}$ **12.** 0.0000045

13. 41.7% increase **14.** 10.31% decrease

15. $0.01x(80) = 30$, 37.5% **16.** $0.20x = 44$, 220

17. $0.17(670) = x$, 113.9 **18.** $0.15(48.93) = x$, $7.34

19. $0.20x = 8$, 40 **20.** 26.7% **21.** $10.50 **22.** 9.5%

23. $450.00 **24.** 6 months **25.** $5017.50

26. a) $210.00 **b)** $3710.00

27. a) $690.00 **b)** $3310.00 **c)** 13.9%

28. a) $7\frac{1}{2}$% **b)** $830.00 **c)** $941.18

29. $9984.05

30. a) $1736.44, $236.44 **b)** $1739.54, $239.54
c) $1741.13, $241.13

31. $12,635.21 **32.** 5.76%

33. a) 11.00% **b)** $493.02 **c)** $4350.73

34. a) $109.18 **b)** $2014.11

35. a) 11.25% **b)** $83.70 **c)** $1469.41

36. a) $6.31 **b)** $847.61

37. a) $5.36 **b)** $549.68

38. a) $3125.00 **b)** $9375 **c)** $2812.50 **d)** 13.50%

39. a) $7.40 **b)** 13.5%

40. a) $33,925.00 **b)** $4805.33 **c)** $1345.49
d) $825.40 **e)** $1142.06 **f)** Yes

41. a) $13,485.00 **b)** $756.51 **c)** $24.20
d) $285,828.60 **e)** $195,928.60

42. a) $550.46 **b)** 8.00% **c)** 7.75%

Chapter Test, Page 570

1. $37.33 **2.** 3 yr **3.** $855.00 **4.** $5855.00

5. $1997.50 **6.** $270.50 **7.** 12.50%

8. a) $105.05 **b)** $3155.90

9. a) $6.93 **b)** $612.70

10. a) $12.30 **b)** $1146.57

11. $2523.20 **12.** $123.20 **13.** $13,800.48, $5200.48

14. $6530.35, $1730.35 **15.** $21,675.00 **16.** $6603.33

17. $1848.93 **18.** $1225.79 **19.** $1529.96 **20.** Yes

21. $315,864.60 **22.** $171,364.60

CHAPTER 12

Section 12.1, Page 579

1. An experiment is a controlled operation that yields a set of results.

3. Empirical probability is the relative frequency of occurrence of an event. It is determined by actual observation of an experiment.

$$P(E) = \frac{\text{Number of times event has occurred}}{\text{Number of times experiment was performed}}$$

7. The official definition by the Weather Service is (**d**). The specific location is the place where the rain gauge is. The Weather Service uses sophisticated mathematical equations to calculate these probabilities.

9. No, it means that the average person with traits similar to Mr. Bennett's will live another 42.34 years.

11. a) Roll a die many times and find the relative frequency of even numbers to the total number of rolls.

15. a) $\frac{1}{2}$ **b)** $\frac{7}{20}$ **c)** $\frac{3}{20}$

17. a) 0.92 **b)** 400

19. a) 1 **b)** Yes

21. a) $\frac{6}{20} = \frac{3}{10}$ **b)** $\frac{14}{20} = \frac{7}{10}$ **c)** $\frac{14}{20} = \frac{7}{10}$ **d)** $\frac{2}{20} = \frac{1}{10}$

23. a) $\frac{25}{40}$ or 0.625 **b)** $\frac{33}{40}$ or 0.825

25. a) 0 **b)** $\frac{50}{250} = 0.2$ **c)** 1

27. a) 0.24 **b)** 0.76

29. a) 0.51 **b)** 0.49

Section 12.2, Page 586

1. If each outcome of an experiment has the same chance of occurring as any other outcome, they are said to be equally likely outcomes.

3. $P(A) + P(\text{not } A) = 1$

7. a) $\frac{1}{5}$ **b)** $\frac{1}{4}$

9. $\frac{1}{13}$ **11.** $\frac{12}{13}$ **13.** $\frac{1}{2}$ **15.** 0 **17.** $\frac{1}{52}$

19. a) $\frac{1}{3}$ **b)** $\frac{1}{3}$ **c)** $\frac{1}{3}$

21. a) $\frac{1}{3}$ **b)** $\frac{1}{3}$ **c)** $\frac{1}{3}$

23. $\frac{35}{100} = \frac{7}{20}$ **25.** $\frac{95}{100} = \frac{19}{20}$ **27.** $\frac{1}{12}$ **29.** $\frac{1}{3}$ **31.** $\frac{1}{3}$

33. $\frac{4}{15}$ **35.** $\frac{2}{5}$ **37.** $\frac{3}{5}$ **39.** $\frac{3}{13}$ **41.** $\frac{5}{13}$ **43.** $\frac{9}{13}$ **45.** $\frac{3}{4}$

47. $\frac{2}{3}$ **49.** $\frac{1}{51}$ **51.** $\frac{26}{51}$ **53.** $\frac{1}{26}$ **55.** $\frac{3}{26}$ **57.** $\frac{2}{5}$ **59.** $\frac{4}{7}$

61. $\frac{6}{35}$ **63.** $\frac{14}{25}$ **65.** $\frac{6}{25}$ **67.** $\frac{13}{36}$ **69.** $\frac{1}{3}$ **71.** $\frac{23}{36}$

73. a) $\frac{1}{4}$ **b)** $\frac{1}{2}$ **c)** $\frac{1}{4}$

75. a) $\frac{1}{4}$ **b)** $\frac{1}{4}$ **c)** $\frac{1}{4}$

Section 12.3, Page 592

3. Odds against **5.** 15 to 2

7. a) $\frac{8}{18} = \frac{4}{9}$ **b)** $\frac{10}{18} = \frac{5}{9}$ **c)** 5:4 **d)** 4:5

9. 5:1 **11.** 5:1 **13.** 12:1, 1:12 **15.** 3:1, 1:3

17. 1:1 **19.** 3:1

21. a) 999,999:1 **b)** 999,990:10 = 99,999:1

23. 8:7 **25.** 14:1 **27.** 13:2

29. a) $\frac{1}{9}$ **b)** $\frac{8}{9}$

31. a) $\frac{2}{7}$ **b)** $\frac{5}{7}$

33. 3:5

35. a) 20:1 **b)** $\frac{20}{21}$

37. 1:2

39. a) $\frac{18}{38} = \frac{9}{19}$ **b)** 10:9 **c)** $\frac{2}{38} = \frac{1}{19}$ **d)** 1:18

Section 12.4, Page 598

1. The expected gain or loss of an experiment over the long run

3. The fair price is the amount that should be charged for the game to be fair and result in an expectation of 0.

5. a) −$2 **b)** Lose $2

7. $550

9. a) $\frac{7}{16}$ in. or 0.4375 in. **b)** $\frac{217}{16}$ or 13.56 in.

11. a) No, a disadvantage **b)** Yes, it is an advantage.

13. a) −$0.50 **b)** $0.50 **c)** $500

15. a) $3 **b)** $1

17. a) $4.25 **b)** $2.25 **19.** $40.00 gain

21. $145,000 **23.** 325 **25.** 170.41 calls **27.** $10,000

29. a) $\frac{5}{12}, \frac{1}{6}, \frac{7}{24}, \frac{1}{8}$ **b)** $20.42 **c)** $20.42

31. −5.3¢

33. a) $458.33 **b)** $308.33

35. Favors the player

Section 12.5, Page 604

1. If a first experiment can be performed in *M* distinct ways and a second experiment can be performed in *N* distinct ways, then the two experiments in that specific order can be performed in $M \cdot N$ distinct ways.

3. 20 **5.** 125

7. a)

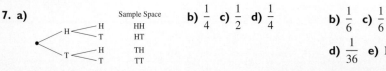

Sample Space

b) $\frac{1}{4}$ **c)** $\frac{1}{2}$ **d)** $\frac{1}{4}$

9. a)

Sample Space

b) 0 **c)** $\frac{1}{6}$ **d)** $\frac{2}{3}$

11. a)

Sample Space

b) $\frac{1}{4}$ **c)** $\frac{3}{4}$

13. a)

Sample Space

b) $\frac{1}{2}$ **c)** $\frac{1}{2}$

d) $\frac{1}{12}$ **e)** $\frac{1}{6}$

15. a)

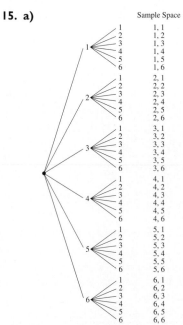

Sample Space

b) $\frac{1}{6}$ **c)** $\frac{1}{6}$

d) $\frac{1}{36}$ **e)** No

17. a)

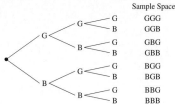

	Sample Space
	GGG
	GGB
	GBG
	GBB
	BGG
	BGB
	BBG
	BBB

b) $\dfrac{1}{8}$ **c)** $\dfrac{7}{8}$ **d)** $\dfrac{3}{4}$

19. a)

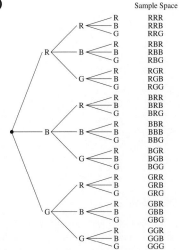

	Sample Space
	RRR
	RRB
	RRG
	RBR
	RBB
	RBG
	RGR
	RGB
	RGG
	BRR
	BRB
	BRG
	BBR
	BBB
	BBG
	BGR
	BGB
	BGG
	GRR
	GRB
	GRG
	GBR
	GBB
	GBG
	GGR
	GGB
	GGG

b) $\dfrac{1}{27}$ **c)** $\dfrac{1}{27}$ **d)** $\dfrac{26}{27}$

21. a) 24
b)

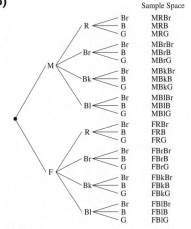

	Sample Space
	MRBr
	MRB
	MRG
	MBrBr
	MBrB
	MBrG
	MBkBr
	MBkB
	MBkG
	MBlBr
	MBlB
	MBlG
	FRBr
	FRB
	FRG
	FBrBr
	FBrB
	FBrG
	FBkBr
	FBkB
	FBkG
	FBlBr
	FBlB
	FBlG

c) $\dfrac{1}{24}$ **d)** $\dfrac{1}{8}$

23. a) m or n **b)** 3 or 4 **c)** $m3, m4, n3, n4$
d) No, not unless we know that all the outcomes are equally likely

e) No, not unless we know that all the outcomes are equally likely
f) Yes

Section 12.6, Page 611

1. a) $P(A \text{ or } B) = P(A) + P(B) - P(A \text{ and } B)$

3. a) $P(A \text{ and } B) = P(A) \cdot P(B)$, assuming that event A has occurred

5. We assume that event A has already occurred.

9. a) No, both events can occur at the same time.
b) No, being healthy can affect your happiness.

11. a) Events A and B cannot occur at the same time. Therefore, $P(A \text{ and } B) = 0$.

13. 0.2 **15.** $\dfrac{1}{3}$ **17.** $\dfrac{1}{2}$ **19.** $\dfrac{2}{13}$ **21.** $\dfrac{8}{13}$ **23.** $\dfrac{17}{26}$

25. a) $\dfrac{1}{16}$ **b)** $\dfrac{5}{76}$

27. a) $\dfrac{1}{16}$ **b)** $\dfrac{5}{76}$

29. a) $\dfrac{1}{40}$ **b)** $\dfrac{1}{38}$

31. a) $\dfrac{4}{25}$ **b)** $\dfrac{14}{95}$

33. $\dfrac{7}{10}$ **35.** $\dfrac{2}{5}$ **37.** $\dfrac{1}{4}$ **39.** $\dfrac{1}{8}$ **41.** $\dfrac{9}{64}$ **43.** $\dfrac{1}{8}$ **45.** $\dfrac{3}{8}$

47. $\dfrac{1}{32}$ **49.** $\dfrac{1}{8}$ **51.** $\dfrac{1}{8}$

53. a) $\dfrac{1}{8}$ **b)** $\dfrac{1}{6}$

55. a) $\dfrac{1}{4}$ **b)** $\dfrac{1}{6}$

57. a) $\dfrac{27}{2197}$ **b)** $\dfrac{1}{286}$

59. a) $\dfrac{125}{2197}$ **b)** $\dfrac{5}{143}$

61. $\dfrac{325}{16,539}$ **63.** $\dfrac{196}{5513}$ **65.** 0.7 **67.** 0.343 **69.** $\dfrac{1}{4}$

71. $\dfrac{27}{1024}$ **73.** $\dfrac{243}{1024}$ **75.** $\dfrac{1}{10}$ **77.** $\dfrac{6859}{8000}$ **79.** $\dfrac{1}{8}$ **81.** $\dfrac{5}{12}$

83. 0.36 **85.** 0.36

87. a) No **b)** 0.001 **c1)** 0.00004 **c2)** 0.00096
c3) 0.998001

89. $\dfrac{9}{15,625}$, or 0.000576 **91.** $\dfrac{14,884}{15,625}$, or 0.952576

93. $\dfrac{1}{3}$ **95.** $\dfrac{1}{2}$ **97.** $\dfrac{14}{45}$ **99.** $\dfrac{45}{91}$ **101.** $\dfrac{1}{2}$

Section 12.7, Page 618

1. The probability of E_2 given that E_1 has occurred

3. $\frac{1}{3}$ **5.** $\frac{1}{2}$ **7.** $\frac{2}{3}$ **9.** $\frac{1}{5}$ **11.** $\frac{1}{3}$ **13.** $\frac{3}{5}$ **15.** $\frac{3}{7}$

17. $\frac{1}{16}$ **19.** $\frac{1}{7}$ **21.** $\frac{1}{8}$ **23.** $\frac{1}{2}$ **25.** $\frac{1}{6}$ **27.** $\frac{1}{6}$ **29.** $\frac{2}{3}$

31. $\frac{19}{25}$ **33.** $\frac{4}{15}$ **35.** $\frac{10}{11}$ **37.** $\frac{3}{19}$ **39.** $\frac{44}{47}$ **41.** $\frac{11}{21}$

43. $\frac{10}{23}$ **45.** $\frac{18}{23}$ **47.** $\frac{2}{3}$ **49.** $\frac{1}{3}$ **51.** $\frac{1}{3}$ **53.** $\frac{2}{11}$

55. $\frac{10}{11}$ **57.** $\frac{1}{5}$

59. a) 140 **b)** 120 **c)** $\frac{7}{10}$ **d)** $\frac{3}{5}$ **e)** $\frac{2}{3}$ **f)** $\frac{4}{7}$

g) Because A and B are not independent events

61. a) 0.3 **b)** 0.4 **c)** Yes; $P(A \mid B) = P(A) \cdot P(B)$.

Section 12.8, Page 627

5. The number of permutations of n items taken r at a time.

7. $\frac{n!}{n_1! \, n_2! \, \ldots \, n_r!}$ **9.** 720 **11.** 5040 **13.** 1 **15.** 1

17. 40,320 **19.** 40,320 **21.** 10,000

23. a) $5^5 = 3125$ **b)** $\frac{1}{3125} = 0.00032$

25. 1,000,000,000 **27.** 1080 systems

29. a) 120 **b)** 120 **c)** 24 **d)** 6

31. 56 **33.** 10,000,000,000 **35.** 67,600 **37.** 10,400

39. 6,760,000 **41.** 728,000

43. a) 8,000,000 **b)** 6,400,000,000
c) 64,000,000,000,000

45. 427,518,000 **47.** 3,628,800 **49.** 729

51. a) The order is important. Since the numbers may be repeated, it is not a true permutation lock.
b) 64,000 **c)** 59,280

53. 22,680 **55.** 420 **57.** 6720

59. a) 120 **b)** 24

61. 12,600 sec, or 3.5 hr **63.** 211

Section 12.9, Page 632

3. $_nC_r = \frac{n!}{(n-r)!r!}$ **7.** 5 **9.** 35 **11.** 126 **13.** 120

15. $\frac{1}{6}$ **17.** 2 **19.** 1140 **21.** 5 **23.** 5005

25. 1,961,256 **27.** 6 **29.** 6 **31.** 28 **33.** 7350

35. a) 1 **b)** 25,827,165

37. 8820 **39.** 924,000 **41.** 11,760

43. a) 45 **b)** 56

45. a) 6 **b)** 10 **c)** $_nC_2$

Section 12.10, Page 637

1. $\frac{1}{5}$ **3.** $\frac{3}{14}$ **5.** $\frac{1}{12}$ **7.** $\frac{5}{28}$ **9.** $\frac{1}{9,366,819}$ **11.** $\frac{1}{10}$

13. $\frac{7}{10}$ **15.** $\frac{3}{20}$ **17.** $\frac{646,646}{3,910,797,436} \approx 0.0001653$

19. $\frac{923,410,488}{3,910,797,436} \approx 0.236$ **21.** $\frac{10}{91}$ **23.** $\frac{59}{65}$ **25.** $\frac{1}{77}$

27. $\frac{5}{77}$ **29.** $\frac{5}{506} \approx 0.010$

31. a) $\frac{1}{123,760}$ **b)** $\frac{1}{30,940}$

33. $\frac{5}{729}$ **35.** $\frac{821}{1458}$

37. a) $\frac{1}{2,162,160}$ **b)** $\frac{1}{6435}$

39. $\frac{1}{24}$

Section 12.11, Page 645

1. A probability distribution shows the probability associated with each specific outcome of an experiment. In a probability distribution every possible outcome must be listed and the sum of all the probabilities must be 1.

3. $P(x) = (_nC_x)p^x q^{n-x}$ **5.** 0.2916 **7.** 0.02835

9. 0.0625

11. a) $P(x) = (_nC_x)(0.15)^x(0.85)^{n-x}$
b) $P(2) = (_{12}C_2)(0.15)^2(0.85)^{10}$

13. 0.234375 **15.** 0.324135 **17.** 0.0005881

19. $\frac{125}{324} \approx 0.38580$

21. a) 0.01024 **b)** 0.98976

23. a) ≈ 0.1119 **b)** ≈ 0.2966

Review Exercises, Page 647

3. 0.625 **5.** 0.58 **6.** $\frac{1}{2}$ **7.** $\frac{4}{5}$ **8.** 1 **9.** $\frac{1}{5}$

10. $\frac{11}{40}$ **11.** $\frac{51}{100}$ **12.** $\frac{89}{200}$ **13.** $\frac{49}{100}$

14. a) 4:1 **b)** 1:4

15. 5:3 **16.** $\frac{3}{5}$ **17.** 3:7

18. a) −$1.20 **b)** −$3.60 **c)** $0.80

19. a) −$0.23 **b)** $0.23 **c)** Lose $23.08

20. 660 people

21. a)
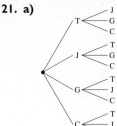

b) Sample Space

TJ
TG
TC
JT
JG
JC
GT
GJ
GC
CT
CJ
CG

c) $\dfrac{1}{12}$

22. a)

b) Sample Space

HR
HB
HG
HP
TR
TB
TG
TP

c) $\dfrac{1}{4}$

23. $\dfrac{1}{4}$ **24.** $\dfrac{9}{64}$ **25.** $\dfrac{3}{16}$ **26.** $\dfrac{7}{8}$ **27.** 1 **28.** $\dfrac{3}{16}$ **29.** $\dfrac{7}{16}$

30. $\dfrac{3}{28}$ **31.** $\dfrac{1}{28}$ **32.** $\dfrac{9}{56}$ **33.** $\dfrac{1}{4}$

34. Against, 3:1; in favor, 1:3 **35.** $1.25 **36.** $\dfrac{1}{8}$ **37.** $\dfrac{5}{8}$

38. In favor, 3:5; against, 5:3 **39.** $3.75 **40.** $\dfrac{7}{8}$

41. $\dfrac{89}{106}$ **42.** $\dfrac{55}{74}$ **43.** $\dfrac{19}{74}$ **44.** $\dfrac{17}{106}$ **45.** $\dfrac{6}{17}$ **46.** $\dfrac{10}{17}$

47. $\dfrac{10}{17}$ **48.** $\dfrac{12}{17}$

49. a) 24 **b)** $2900

50. 30 **51.** 336 **52.** 504 **53.** 20

54. a) 3003 **b)** 3,628,800

55. $\dfrac{1}{25,827,165}$ **56.** 1120 **57.** 560 **58.** $\dfrac{1}{221}$ **59.** $\dfrac{1}{12}$

60. $\dfrac{1}{18}$ **61.** $\dfrac{1}{24}$ **62.** $\dfrac{11}{12}$ **63.** $\dfrac{5}{91}$ **64.** $\dfrac{15}{91}$ **65.** $\dfrac{3}{13}$

66. $\dfrac{10}{13}$

67. a) $P(x) = (_nC_x)(0.6)^x(0.4)^{n-x}$
b) $P(75) = (_{100}C_{75})(0.6)^{75}(0.4)^{25}$

68. 0.0512

69. a) 0.0256 **b)** 0.9744

Chapter Test, Page 651

1. $\dfrac{5}{18}$ **2.** $\dfrac{1}{3}$ **3.** $\dfrac{4}{9}$ **4.** $\dfrac{7}{9}$ **5.** $\dfrac{1}{3}$ **6.** $\dfrac{5}{18}$ **7.** $\dfrac{1}{6}$

8. $\dfrac{5}{18}$ **9.** $\dfrac{5}{12}$ **10.** $\dfrac{8}{13}$ **11.** 18

12.
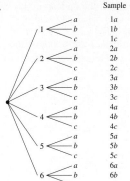

Sample

1a
1b
1c
2a
2b
2c
3a
3b
3c
4a
4b
4c
5a
5b
5c
6a
6b
6c

13. $\dfrac{1}{18}$ **14.** $\dfrac{4}{9}$ **15.** $\dfrac{2}{3}$ **16.** 8,985,600 **17.** 9:16 **18.** $\dfrac{3}{8}$

19. $0 **20.** $\dfrac{4}{7}$ **21.** 120 **22.** $\dfrac{14}{95}$ **23.** $\dfrac{81}{95}$ **24.** $\dfrac{175}{396}$

25. $\dfrac{216}{625} = 0.3456$

CHAPTER 13

Section 13.1, Page 658

1. Descriptive statistics is concerned with the collection, organization, and analysis of data. Inferential statistics is concerned with making generalizations or predictions from the data collected.

7. a) A population is the entire group of items being studied
b) A sample is a subset of the population

9. a) A systematic sample is a sample obtained by selecting every nth item on a list or production line
b) Use the random number table to select the first item, then select every nth item after that.

11. a) and **b)** Divide the population into parts (strata), then draw random samples from each stratum.

13. Random sample **15.** Cluster sample

17. Systematic sample **19.** Random sample

Section 13.2, Page 662

3. Being the largest department store does not imply that it is inexpensive.

5. Half the students in a population are expected to be below average.

7. People with asthma may move to Arizona in their later years because of its climate. Therefore more people with asthma may live in Arizona.

9. The males may drive more miles than females, and males may drive in worse driving conditions (like snow).

11. There may be deep sections in the pond, so it may not be safe to go wading.

13. Averages apply to a set of data. Thus, although the female average score may be greater, some males have higher scores than some females.

15. a)

b)

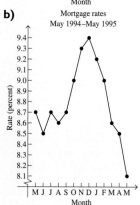

17. a)

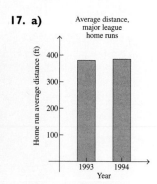

b)

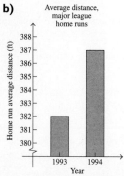

19.

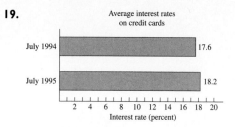

Section 13.3, Page 667

1. A frequency distribution is a listing of observed values and the corresponding frequency of occurrence of each value.

3. a) 7 **b)** 16−22 **c)** 16 **d)** 22

5. The modal class is the class with the greatest frequency.

7. a) 19 **b)** 7 **c)** 17 **d)** 42–48 **e)** 49–55

9.

Grade	Number of Students
0	4
1	6
2	3
3	4
4	3
5	3
6	7
7	3
8	12
9	3
10	2

11.

IQ	Number of Students
78–86	2
87–95	15
96–104	18
105–113	7
114–122	6
123–131	1
132–140	1

13.

IQ	Number of Students
80–90	8
91–101	22
102–112	11
113–123	7
124–134	1
135–145	1

15.

Age	Number of Presidents
40–45	2
46–51	12
52–57	16
58–63	6
64–69	5

17.

Age	Number of Presidents
42–46	4
47–51	10
52–56	12
57–61	9
62–66	4
67–71	2

19.

Sales ($ billions)	Number of Corporations
10–30	39
31–51	5
52–72	3
73–93	0
94–114	2
115–135	1

21.

Sales ($ billions)	Number of Corporations
10–25	38
26–41	4
42–57	3
58–73	2
74–89	0
90–105	1
106–121	1
122–137	1

23.

Population (to nearest 100,000)	Number of Countries
5.0–8.0	14
8.1–11.1	7
11.2–14.2	9
14.3–17.3	1
17.4–20.4	1
20.5–23.5	1
23.6–26.6	1
26.7–29.7	1

25.

Population (to nearest 100,000)	Number of Countries
5.0–7.7	13
7.8–10.5	7
10.6–13.3	8
13.4–16.1	3
16.2–18.9	0
19.0–21.7	2
21.8–24.5	1
24.6–27.3	0
27.4–30.1	1

27.

Population Density per Square Mile	Number of States
0.5– 100.5	29
100.6– 200.6	10
200.7– 300.7	4
300.8– 400.8	2
400.9– 500.9	1
501.0– 601.0	0
601.1– 701.1	1
701.2– 801.2	1
801.3– 901.3	0
901.4–1001.4	1
1001.5–1101.5	1

29.

Population Density per Square Mile	Number of States
0.5– 150.5	34
150.6– 300.6	9
300.7– 450.7	2
450.8– 600.8	1
600.9– 750.9	1
751.0– 901.0	1
901.1–1051.1	2

Did You Know on page 664: There are 6 *F*'s in the statement.

Section 13.4, Page 673

5. b)

Children in selected families

7. United States ≈ 67.04 million; Europe ≈ 36.85 million; Japan ≈ 10.51 million; other ≈ 33.60 million

9.

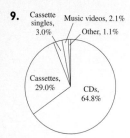

Cassette singles, 3.0%
Music videos, 2.1%
Other, 1.1%
Cassettes, 29.0%
CDs, 64.8%

11. a) and **b)**

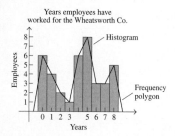

Years employees have worked for the Wheatsworth Co.

Histogram
Frequency polygon

13. a) and **b)**

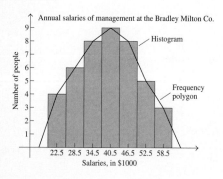

Annual salaries of management at the Bradley Milton Co.

Histogram
Frequency polygon

15. a) 27 **b)** 6 **c)** 8 **d)** 224

e)

Number of Plants	Number of Houses
6	2
7	6
8	8
9	6
10	4
11	0
12	1

17. a) 8 **b)** 24 **c)** 52

d)

Number of Visits	Number of Families
1	4
2	2
3	8
4	8
5	6
6	11
7	9
8	3
9	0
10	1

e)

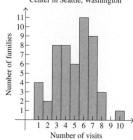

Number of visits selected families have made to the Pacific Science Center in Seattle, Washington

19.

21. a)

Salaries ($1000)	Number of Companies
25	1
26	7
27	4
28	3
29	2
30	3
31	3
32	2

b) and **c)**

Starting salaries of employees at 25 different companies

23. a)

Weight (lb)	Number of Twelfth Graders
96–102	7
103–109	6
110–116	13
117–123	3
124–130	5
131–137	7
138–144	1
145–151	0
152–158	1
159–165	2

b), c)

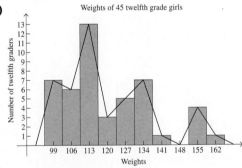

Weights of 45 twelfth grade girls

Section 13.5, Page 680

1. The mean is the balancing point of a set of data. It is the sum of the data divided by the number of pieces of data.

3. The median is the value in the middle of a set of ranked data. To find the median, rank the data and select the value in the middle.

5. The midrange is halfway between the lowest and highest values. To find the midrange, add low and high values and divide the sum by 2.

7. The mode may be used when you are primarily interested in the most popular value, or the one that occurs the most often; for example, when buying clothing for a store.

9. 9, 8, 9, 17.5   **11.** 74.7, 80, none, 70.5

13. 8, 8, none, 8   **15.** 72.9, 60, none, 85

17. 10.7, 11, 12, 10.5   **19.** 477.2, 463, none, 524.5

21. 80.2, 80.5, none, 78

25. a) $30,100   **b)** $21,500   **c)** $20,000   **d)** $46,500

27. a) $81.31   **b)** $79.44   **c)** No mode   **d)** $87.03

29. 522

31. a) Yes   **b)** No   **c)** No   **d)** Yes
e) Mean = 80,000; midrange = 307,500

33. One example is 55, 65, 75, 85, 95.   **35.** 91

37. Mode

39. The mean, mode, and midrange will all change.

41. The data must be ranked.

43. Kevin was taller than approximately 48% of all kindergarten children.

45. About 75% of the students travel no more than 47.6 minutes one way to Central Valley College.

47. a) No
b) Yes, Kendra was in the better relative position.

49. a) 8, 6.5, 2, 10, 66   **b)** 18.5   **c)** $\approx$ 19.926   **d)** No

51. a)

Ruth	Mantle
.290	.300
.359	.365
.301	.304
.272	.275
.315	.321

b) Mantle's is greater in every case.
c) Ruth: .316; Mantle: .311; Ruth's is greater
e) Ruth: .307; Mantle: .313; Mantle's is greater

Section 13.6, Page 687

9. They would be the same since the spread of data about each mean is the same.

11. 9, $\sqrt{15} \approx 3.87$ **13.** 6, $\sqrt{4.67} \approx 2.16$

15. 5, $\sqrt{7} \approx 2.65$ **17.** 8, $\sqrt{11.5} \approx 3.39$

19. 24, $\sqrt{74} \approx 8.60$ **21.** 15, $\sqrt{21.56} \approx 4.64$

23. 7, $\sqrt{6.22} \approx 2.49$

25. e) Mean of first set is 9; mean of second set is 599. The standard deviation of both sets is 2.16.

27. a) The standard deviation increases. There is a greater spread from the mean as they get older.
b) $\approx$ 133 lb   **c)** $\approx$ 21 lb
d) Mean: $\approx$ 100 lb; normal range: $\approx$ 60 to 140 lb
e) Mean: $\approx$ 62 in.; normal range: $\approx$ 53 to 68 in.
f) 5%

29. c) $\approx$ 381.86 ft   **e)** $\sqrt{15.478} \approx 3.93$

31. c) Platform, 83.625°; outside, 77.375°

    d) Platform, $\sqrt{47.98} \approx 6.93$; outside, $\sqrt{85.70} \approx 9.26$

Section 13.7, Page 699

1. A rectangular distribution is one where all the values have the same frequency.

3. A bimodal distribution is one where two non-adjacent values occur more frequently than any other values in a set of data.

5. A distribution skewed to the left is one that has "a tail" on its left.

11. The mode is the lowest value, the median is greater than the mode, and the mean is greater than the median.

15. They all have the same value.

17. A z score will be negative when the data is less than the mean.

19. 0   **21.** 0.50   **23.** 0.818   **25.** 0.071   **27.** 0.037

29. 0.019   **31.** 0.113   **33.** 33.2%   **35.** 89.8%

37. 97.1%   **39.** 97.5%   **41.** 13.9%   **43.** 34.1%

45. 68.2%   **47.** 81.8%   **49.** 30.8%   **51.** 36.1%

53. 24.8%   **55.** 95.1%   **57.** 50.0%   **59.** 69.2%

61. ≈ 24 **63.** 29.1%   **65.** 59.9%   **67.** 69.2%

69. 42.1%   **71.** 54,560   **73.** 11.1%

75. 3.6%—A; 10.0%—B; 74.9%—C; 8.6%—D; 2.9%—F

77. 0.944

Section 13.8, Page 710

1. A correlation coefficient measures the strength of the relationship between the quantities.

3. 1   **5.** 0

7. A negative correlation indicates that as one quantity increases the other quantity decreases

9. No   **11.** Yes   **13.** No   **15.** No

The answers in the remainder of this section may differ slightly from your answers, depending upon how your answers are rounded and which calculator you used. The answers given here were obtained from a Texas Instruments TI-36x solar calculator.

17. a) 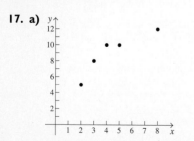   **b)** 0.903

    c) Yes   **d)** No

19. a) 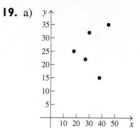   **b)** 0.220

    c) No   **d)** No

21. a)   **b)** 0.999

    c) Yes   **d)** Yes

23. a)   **b)** -0.968

    c) Yes   **d)** Yes

25. $y = 1.0x + 4.4$ **27.** $y = 0.2x + 20.5$

29. $y = 0.8x + 5.8$ **31.** $y = -0.1x + 9.5$

33. a) 0.998   **b)** Yes   **c)** $y = 4.4x - 267.8$

35. a) -0.946 **b)** Yes   **c)** $y = -1.9x + 16.5$

37. a) -0.782 **b)** Yes   **c)** $y = -0.7x + 22.3$ **d)** 12.5

39. a) -0.901 **b)** Yes   **c)** $y = -0.9x + 105.0$

    d) 24 min

41. a) 0.977   **b)** Yes   **c)** $y = -12.9x + 99.6$

    d) 41.6%

43. c)

Dry pavement

Wet pavement

    d) 0.999   **e)** 0.990   **g)** $y = 5.4x - 183.4$

    h) $y = 16.2x - 669.8$

    i) Dry, 232.4 ft; wet, 577.6 ft

47. a) 0.998   **b)** Should be the same   **c)** 0.998

Review Exercises, Page 715

1. a) A population is the entire group of items being studied.
   b) A sample is a subset of the population.

2. A random sample is one where every item in the population has the same chance of being selected.

3. The candy bars may have lots of calories, or fat, or salt. Therefore it may not be healthy to eat them.

4. Sales may not necessarily be a good indicator of profit. Expenses must also be considered.

5. a)

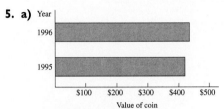

b)

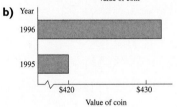

6. a)

Class	Frequency
35	1
36	3
37	6
38	2
39	3
40	0
41	4
42	1
43	3
44	1
45	1

b), c)

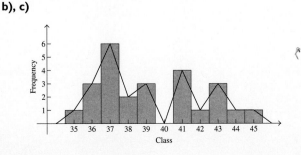

7. a)

Class	Frequency
30– 38	4
39– 47	4
48– 56	6
57– 65	7
66– 74	7
75– 83	7
84– 92	4
93–101	1

b) and **c)**

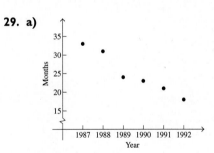

8. 77   **9.** 77.5   **10.** None   **11.** 77.5   **12.** 39

13. $\sqrt{193.2} \approx 13.90$ **14.** 13   **15.** 13   **16.** None

17. 13.5   **18.** 19   **19.** $\sqrt{40} = 6.32$ **20.** 68.2%

21. 95.4%   **22.** 94.5%   **23.** 5.5%   **24.** 72.6%

25. 34.1%   **26.** 34.5%   **27.** 33.7%   **28.** 27.4%

29. a)

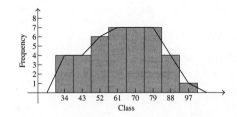

c) -0.972 **d)** Yes

30. $y = -3.0x + 296.1$

31. a) 0.922   **b)** Yes   **c)** $y = 0.03x + 5.6$ **d)** 5.81%

32. a)

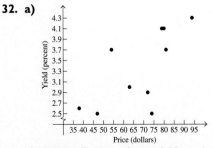

c) 0.686   **d)** Yes   **e)** $y = 0.03x + 1.42$

33. a)

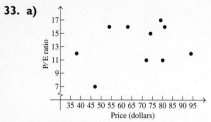

c) 0.270   **d)** No   **e)** $y = 0.05x + 9.91$

34. 175 lb   **35.** 180 lb   **36.** 25%   **37.** 25%   **38.** 14%

39. 18,700 lb   **40.** 233 lb   **41.** 145.6 lb   **42.** 3.610

43. 2   **44.** 3   **45.** 7   **46.** 14   **47.** $\sqrt{8.244} \approx 2.87$

48.

Class	Frequency
0– 1	8
2– 3	14
4– 5	10
6– 7	6
8– 9	1
10–11	1
12–13	0
14–15	1

49. and 50.

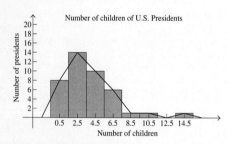

Number of children of U.S. Presidents

51. No   **52.** No   **53.** No

Chapter Test, Page 718

1. 20   **2.** 21   **3.** 25   **4.** 17.5   **5.** 15

6. $\sqrt{38} \approx 6.16$

7.

Class	Frequency
25–30	7
31–36	5
37–42	1
43–48	7
49–54	5
55–60	3
61–66	2

8.

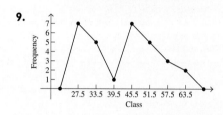

9.

10. $495   **11.** $470   **12.** 75%   **13.** 79%   **14.** $50,000

15. $540   **16.** $440   **17.** 31.8%   **18.** 89.4%   **19.** 10.6%

20. $\approx$ 69 cars

21. a)

c) 0.820   **d)** Yes   **e)** $y = 0.35x - 0.01$ **f)** 11.95%

Credits

INDEX